ADVANCES IN TURBULENCE VI

FLUID MECHANICS AND ITS APPLICATIONS
Volume 36

Series Editor: **R. MOREAU**
MADYLAM
Ecole Nationale Supérieure d'Hydraulique de Grenoble
Boîte Postale 95
38402 Saint Martin d'Hères Cedex, France

Aims and Scope of the Series

The purpose of this series is to focus on subjects in which fluid mechanics plays a fundamental role.

As well as the more traditional applications of aeronautics, hydraulics, heat and mass transfer etc., books will be published dealing with topics which are currently in a state of rapid development, such as turbulence, suspensions and multiphase fluids, super and hypersonic flows and numerical modelling techniques.

It is a widely held view that it is the interdisciplinary subjects that will receive intense scientific attention, bringing them to the forefront of technological advancement. Fluids have the ability to transport matter and its properties as well as transmit force, therefore fluid mechanics is a subject that is particulary open to cross fertilisation with other sciences and disciplines of engineering. The subject of fluid mechanics will be highly relevant in domains such as chemical, metallurgical, biological and ecological engineering. This series is particularly open to such new multidisciplinary domains.

The median level of presentation is the first year graduate student. Some texts are monographs defining the current state of a field; others are accessible to final year undergraduates; but essentially the emphasis is on readability and clarity.

For a list of related mechanics titles, see final pages.

Advances in Turbulence VI

Proceedings of the Sixth European Turbulence Conference,
held in Lausanne, Switzerland, 2–5 July 1996

Edited by

S. GAVRILAKIS

L. MACHIELS

P. A. MONKEWITZ

Fluid Mechanics Laboratory,
Swiss Federal Institute of Technology (EPFL),
Lausanne, Switzerland

KLUWER ACADEMIC PUBLISHERS

DORDRECHT / BOSTON / LONDON

A C.I.P. Catalogue record for this book is available from the Library of Congress

ISBN-13: 978-94-010-6618-1   e-ISBN-13: 978-94-009-0297-8
DOI: 10.1007/978-94-009-0297-8

Published by Kluwer Academic Publishers,
P.O. Box 17, 3300 AA Dordrecht, The Netherlands.

Kluwer Academic Publishers incorporates
the publishing programmes of
D. Reidel, Martinus Nijhoff, Dr W. Junk and MTP Press.

Sold and distributed in the U.S.A. and Canada
by Kluwer Academic Publishers,
101 Philip Drive, Norwell, MA 02061, U.S.A.

In all other countries, sold and distributed
by Kluwer Academic Publishers Group,
P.O. Box 322, 3300 AH Dordrecht, The Netherlands.

Printed on acid-free paper

TABLE OF CONTENTS

II. COHERENT STRUCTURES AND VORTICITY

Invited Lecture

Contributed Lectures

Posters

III. INDUSTRIAL AND ENVIRONMENTAL APPLICATIONS

Invited Lecture

Contributed Lectures

Posters

IV. HIGH REYNOLDS NUMBER TURBULENCE AND INTERMITTENCY

Invited Lecture

Contributed Lectures

V. COMPRESSIBLE TURBULENCE

Contributed Lectures

Posters

VI. TRANSITION AND DYNAMICAL SYSTEMS

Contributed Lectures

VII. EXPERIMENTS AND NOVEL EXPERIMENTAL TECHNIQUES

Invited Lecture

Contributed Lectures

VIII. TURBULENCE IN MULTI-PHASE FLOWS

Invited Lecture

Contributed Lectures

Posters

IX. TURBULENT MIXING

Contributed Lectures

Posters

PREFACE

This sixth volume in the series *Advances in Turbulence* presents an update on the state of turbulence research with some bias towards research in Europe, as it represents an almost complete collection of the paper presentations at the *Sixth European Turbulence Conference*, sponsored by EUROMECH, ERCOFTAC and COST, which will be held at the Swiss Federal Institute of Technology in Lausanne, July 2-5, 1996. The present volume is being produced before the conference in order to assure timely dissemination of the latest research results and replaces the usual collection of abstracts which is generally only accessible to participants. The volume is thus geared towards specialists in the area of flow turbulence who could not attend the conference as well as anybody who would like to quickly assess the most active current research areas and the groups associated with them.

Judging from the volume of contributions under each section it is obvious that, since the *Fifth European Turbulence Conference* in 1994, the problem of transition, together with the structural description of turbulence, and the scaling laws of fully developed turbulence have continued to receive most attention by the research community. It is equally clear that much remains to be done in these areas despite significant progress. In particular, the presumed duality of structural and statistical description of turbulence has not yet been established at a level even remotely comparable to the duality of particle and wave description in atomic physics.

Inevitably, we are also asked about the benefits of research to those who have to solve "real problems," i.e. to industry. The needs most often cited are for reliable predictions of transition and robust turbulence models for numerical codes. Transition prediction has recently made and is still making great strides due to fundamental research in the areas of "receptivity," "transient growth" and efficient parabolized schemes for stability computations. As the industrial "customer" slowly moves towards Large Eddy Simulations, the current efforts towards a better understanding of small-scale turbulence are equally necessary for the improvement of subgrid scale models. This is not to say that turbulence research is perfectly in tune with industry. The research community should in fact be well ahead, as the need for more sophisticated models of turbulence in multi-phase and non-Newtonian flows, for instance, will certainly become more pressing soon and fundamental investigations cannot be carried out on an industrial timescale. We look forward to a continuing increase of the number of contributions in these "new" areas.

After these few thoughts which did not profit from the expected lively presentations and discussions at the conference, we would like to thank all the contributors for keeping to the tight schedule for the submission of manuscripts and the Kluwer Academic Publishers for agreeing to produce this volume in the very short time of two months. We further thank Mrs. Eva Gasser for keeping in touch with all the authors as well as for receiving and filing the flood of papers with composure. Finally, thanks goes to all the financial sponsors who made the conference and this volume possible.

April 1996,   Spyros Gavrilakis, Luc Machiels and Peter A. Monkewitz

I

Numerical Simulation and Modeling of Turbulence

DIRECT NUMERICAL SIMULATION OF TRANSITIONAL BACKWARD-FACING STEP FLOW MANIPULATED BY OSCILLATING BLOWING/SUCTION

G. BÄRWOLFF

Hermann-Föttinger-Institut für Strömungsmechanik,
Technische Universität Berlin, D-10623 Berlin, Germany

H. WENGLE [1]

Institut für Strömungsmechanik u. Aerodynamik, LRT/WE 7,
Universität der Bundeswehr München, D-85577 Neubiberg, Germany,

and

H. JEGGLE

Fachbereich 3, Mathematik, Funktionalanalysis u. Numerische Mathematik ,
Technische Universität Berlin, D-10623 Berlin, Germany

1. Problem specification

Introduction: The backward-facing step flow with laminar inflow is of particular interest because of the fact that the shear layer separating from the edge of the step undergoes transition to turbulence (at appropriate Reynolds number). There are experimental investigations carried out on the controlled manipulation of this flow case (Re=3000) by the research group of Prof. H. Fernholz (TU Berlin, Germany). Related experimental work has been accomplished by Hasan (1992). For turbulent boundary layer inflow, and Re=5100, there are DNS results available from Le, Moin and Kim (1993), and LES results from Akselvoll and Moin (1993).

Computational grid and boundary conditions: The flow configuration selected is a backward facing step with an expansion ratio of 1.2. If all length scales are normalized by the step height, h, the dimensions of the computational domain are $(L_x, L_y, L_z) = (17.6, 6.0, 6.0)$ with an entry length of 5 reference lengths. Here, x is the main flow direction, y is the lateral direction, z is the vertical direction, and x/h=0 corresponds to the position of the step. Starting from an uniform velocity profile at the inlet section (at x/h= -5.0), a Blasius profile is developed with a boundary layer thickness of about $\delta/h = 0.2$ at the edge of the step (at x/h=0). This condition for laminar inflow, and the Reynolds number of $Re_h = U_o h/\nu = 3000$ (based on step height h and the inlet free stream velocity U_o) have been

---

[1] To whom correspondence should be sent

S. Gavrilakis et al. (eds.), Advances in Turbulence VI, 3-6.
© 1996 *Kluwer Academic Publishers.*

selected in accordance with the current experimental investigations carried out at Berlin. There are no slip boundary conditions along the lower wall, slip conditions along the upper boundary of the computational domain, periodic boundary conditions in the lateral direction and, for the results presented here, normal gradients of flow variables have been set to zero at the outflow cross section.

The flow is manipulated by periodic blowing/suction through a cross-wind slot at the edge of the step. The oscillating jet is inclined under 45° and its exit velocity can be described by $V_{jet} = A \cdot sin(2\pi f t)$, with a frequency f and an amplitude A as the two main parameters of the disturbation. The manipulation is implemented via time dependent boundary conditions for the horizontal (u) and vertical (w) velocity components on two neighbouring crosswind rows of grid points close to the edge of the step.

Numerical solution method: The Navier-Stokes equations for an incompressible fluid are solved numerically on a non-equidistant staggered grid with second-order finite-differencing in time and space (explicit leapfrog for time discretization, central differencing for convection and time-lagged diffusion terms). The problem of pressure-velocity coupling is solved iteratively by a pressure (and velocity) correction method. To resolve all the relevant scales of the flow we used $(N_x, N_y, N_z) = (512, 128, 160)$ grid points. Measured in units of the viscous length (using a wall shear velocity of $u^+ = 0.055$ from X=15.0) the distance of the first grid point to a rigid wall is about $\Delta Z^+ = 1.1$, the lateral resolution is about $\Delta Y^+ = 7.0$, and in the longitudinal direction the resolution is about $\Delta X^+ = 6.0$ (using an equidistant grid after the step).

2. Discussion of results and conclusions

Instantaneous flow field: From the instantaneous flow field it becomes evident that the most interesting and the most active flow regimes are the free shear layer (undergoing transition to turbulence) and the reattachment zone. Strong horizontal velocity fluctuations are created close to the wall by the splashing down of the shear layer onto the bottom plate. The instantaneous spanwise vorticity contours, ω_y, show the large scale roll-up of the shear layer which is strongly influenced by the reattachment of the flow, causing smaller scale structures of the vorticity (figure 1).

Manipulation of the flow: The flow has been disturbed by an oscillating (blowing/suction) wall jet close to the edge of the step. Always using a frequency of f=50 Hz (the most successful frequency found in the experimental investigations), we carried out three different simulations with amplitudes $A = 10^{-3}, A = 10^{-2}$, and $A = 10^{-1}$ of the time-periodic forcing, causing a significant reduction of the length of the mean recirculation zone. This becomes evident from a comparison of one of the manipulated cases (A=0.01) with the reference case (A=0.0), see figure 2. To determine more accurately the mean recirculation length, profiles of the mean streamwise velocity component along a wall-parallel line through the first grid point next to the wall have been drawn in figure 3. The zero-crossings of these longitudinal profiles are taken to be the positions of the mean recirculation lengths. With a disturbation amplitude $A = 10^{-3}$ a reduction to 84%, with $A = 10^{-2}$ a reduction to 66%, and with $A = 10^{-1}$ a reduction to 57% of the value of the reference case without excitation (A=0.0) has been achieved.

Comparison with experiment: The experimental investigations of Huppertz (1994) show a 30 % reduction of the mean recirculation length applying a similar time-periodic forcing in the frequency range between 40 and 60 Hz. This corresponds to the numerically simulated case with f=50 Hz and $A = 10^{-2}$. The measured mean recirculation length of the undisturbed reference case, $x_r/h = 6.4$, differs from the corresponding value, $x_r/h = 7.4$, in the numerical simulation. In the meantime, after having excluded possible effects of a slightly different geometry of the edge of the step, there is some experimental and numerical evidence available that the difference of about 15 % in the mean recirculation length may be caused by the different expansion ratio, ER, used in the DNS (ER=1.2) and the experiments (Er=1.09), respectively.

Conclusions: A first series of direct numerical simulations of manipulated transitional backward-facing step flow showed that a significant reduction of the mean recirculation zone can be obtained by a low-amplitude time-periodic blowing/suction excitation through a narrow crosswind slot at the edge of the step. The major effect of the forcing is an acceleration of the transition process. In the current and future work the evaluation of the flow fields will be continued to extract more information about the statistical properties and the dynamics of the dominant flow structures.

References

Akselvoll, K. and Moin, P. (1993): Large eddy simulation of a backward facing step flow, Engineering Turbulence Modelling and Experiments 2, W. Rodi and F. Martelli (Editors), Elsevier Science Publishers

Hasan, M. A. Z. (1992): The flow over a backward-facing step under controlled disturbation: laminar separation, J. Fluid Mech. 238, pp. 73-96

Huppertz, A. (1994): Beeinflussung der Strömung hinter einer rückwärtsgewandten Stufe durch dreidimensionale Anregung, Diploma-Thesis (in german), Hermann-Föttinger-Institut für Thermo- und Fluiddynamik, TU Berlin

Le, H., Moin, P. and Kim, J. (1993): Direct numerical simulation of turbulent flow over a backward-facing step, 9th Symp.Turb.Shear Flows, Aug. 16-18, 1993, Kyoto, Japan

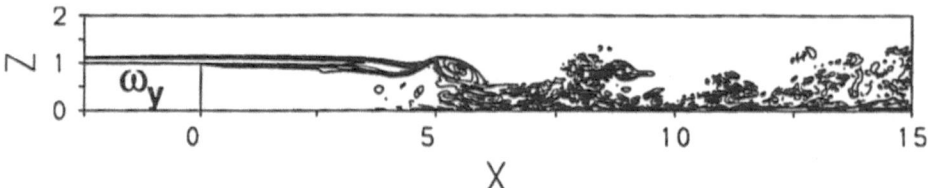

Figure 1. Isolines of instantaneous vorticity field :
lateral vorticity component $\omega_y(X, Z)$; min=-5.0, max=+5.0, increment=1.0,
in midplane, dashed lines: negative values, full lines: positive values

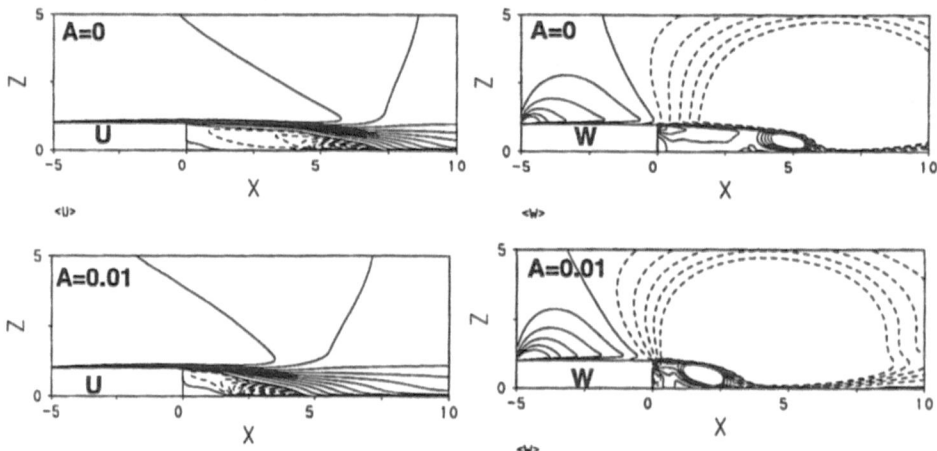

Figure 2. Isolines of mean streamwise and mean vertical velocity component:
above: reference case without manipulation (A=0.0)
below: manipulated case with A=0.01
left: mean streamwise velocity $< U(X, Z) >$; min=-0.2, max=+1.0
right: mean vertical velocity $< W(X, Z) >$;min=-0.02,max=+0.02,incr.=0.004
full lines: positive values, dashed lines: negative values

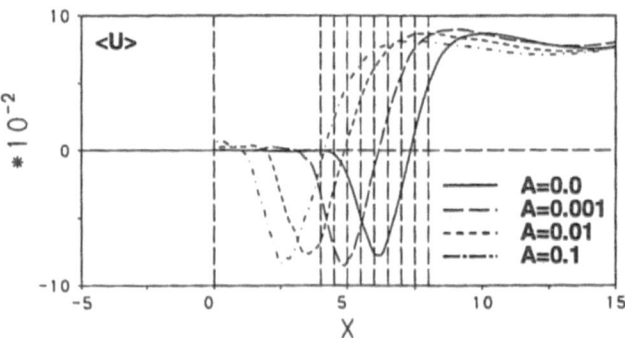

Figure 3. Evaluation of mean recirculation length x_r/h:
from mean streamwise velocity distribution along the bottom wall
at height $Z^+ = 1.1$ (first grid point next to wall)
for the reference case (without manipulation, A=0.0) and three cases
with different manipulation amplitudes: A=0.001, 0.01, and 0.1

MODELLING TURBULENT SHEARED CONVECTION

A.R.Brown and P.J.Mason
U.K. Meteorological Office
Bracknell, Berkshire, RG12 2SZ, United Kingdom

In recent years, the use of large-eddy simulation (LES) models to produce turbulence datasets describing the atmospheric boundary layer under a wide range of conditions has become increasingly popular. These datasets can then be used, in conjunction with data from other sources (e.g. observations and physical models) to evaluate turbulence closure schemes.

The large-eddy model used in the present study has a subgrid parametrization which incorporates the stochastic backscatter model of Mason and Thomson (1992), extended to include buoyancy effects by Brown et al. (1994). The flows are driven by an imposed surface buoyancy flux ($\langle wb \rangle_0$), and a horizontal pressure gradient which can be related to a geostrophic wind (surface speed, G, with shear $\partial G/\partial z$ at angle γ to the surface geostrophic wind). Here results are presented from a new series of simulations, spanning the entire range of stabilities between neutral and free convective conditions. Additionally, results obtained for the baroclinic boundary layer in which the driving pressure gradient is a function of height are also discussed.

The variation with stability of various scaled turbulence statistics (e.g. variances, non-dimensional gradients, entrainment flux) is discussed. The confidence in the results is increased by the good agreement found between them and available observational datasets. It is also noted that the results appear to be relatively insensitive to model resolution. This is encouraging as it provides at least limited evidence for convergence of results with increasing resolution. The use of the backscatter model is shown to have significant beneficial effects in neutral conditions, leading to mean field and variance profiles in better agreement with observations. For example, Figure 1 shows mean windspeed profiles (non-dimensionalized using the friction velocity, u_*) from simulations of the neutral boundary layer, capped by stress-free rigid lids at $z = 1000$ m. The simulation using backscatter can be seen to show a much more realistic logarithmic profile in the surface layer ($z < 100$ m). In contrast, the backscatter model has minimal impact in highly convective simulations. This is consistent with earlier model results (e.g. Nieuwstadt et al., 1991) which show that large-eddy simulation results for the convective boundary layer are relatively insensitive to the subgrid model, due to the dominance of large eddies which are easily resolved.

Accurate prediction of the surface stress is vital in weather forecasting and climate prediction models, as it directly affects near surface profiles (which are to be

7

S. Gavrilakis et al. (eds.), Advances in Turbulence VI, 7-10.
© 1996 Kluwer Academic Publishers.

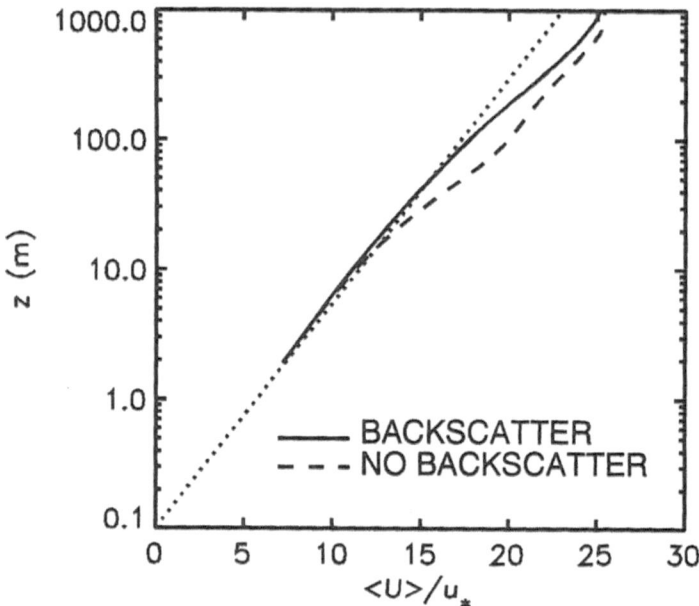

Figure 1: Mean windspeed profiles from simulations of the neutral boundary layer, with and without the backscatter model. The dotted line shows the expected logarithmic profile.

forecast) and also because it controls the frictional convergence in the boundary layer, which in turn affects large-scale development. In the atmospheric boundary layer, surface stress is commonly related to the geostrophic wind through Rossby similarity theory, but the two coefficients, conventionally denoted as A and B, must be determined empirically. The scatter in published estimates for A and B is testimony to the difficulties in making such measurements, and is, in part, due to the sensitivity of the coefficients to scale-height ratio, stability, and baroclinicity (Arya, 1977). Large-eddy simulations have the advantage of being better able to isolate these separate effects, and can also be run to long time to remove uncertainties due to non-stationarity.

In neutral conditions, without shear in the geostrophic wind, the stress predicted by the LES is within the range of values consistent with the observational results for A and B tabulated in Grant and Whiteford (1987). In convective conditions, the results are found to be consistent with the predictions of the simple mixed layer model of Garratt et al. (1982). This increases confidence in the value of the results, and attempts have been made to reproduce them using some simple closure models which are often used in boundary layer parametrization schemes in large-scale models.

As an example, the performance of a simple stability dependent mixing length

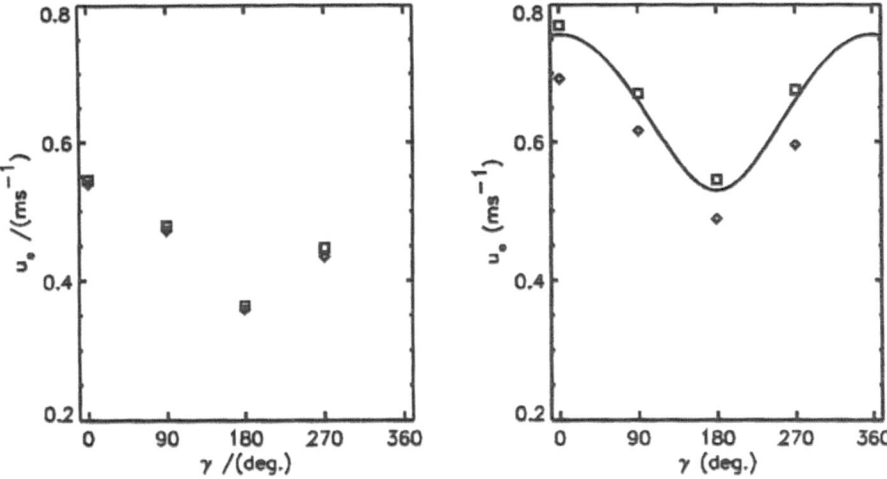

Figure 2: Friction velocity, u_*, versus γ, in the baroclinic boundary layer. Both plots are for $G = 10$ ms^{-1}, $\partial G/\partial z = 0.005$ s^{-1}, $z_i = 1000$ m, $z_0 = 0.1$ m. The plot on the left is for zero surface flux, while that on the right is for $\langle wb \rangle_0 = 10^{-2}$ m^{2}s^{-3}. Squares : LES; diamonds : mixing length model; solid line : Garratt et al. (1982).

closure is examined in baroclinic conditions in which the geostrophic wind varies with height. This is of particular interest as there is evidence (Hollingsworth, 1994) for systematic errors in numerical weather prediction model forecasts in conditions when the wind backs (turns anticlockwise) with height – a situation which is most likely to occur when the geostrophic wind backs with height. Figure 2 shows results for the variation of u_* with γ. All simulations have $G = 10$ ms^{-1}, $\partial G/\partial z = 0.005$ s^{-1}, boundary layer depth $z_i = 1000$ m, and roughness length $z_0 = 0.1$ m, giving values of non-dimensional baroclinicity ($M = (z_i/u_*)(\partial G/\partial z)$) of O(10). The plot on the left shows results with zero surface buoyancy flux. The squares are the LES results, and the diamonds are the predictions of a mixing length model which relates local stresses to local shear using a length scale which is proportional to height close to the surface, but is limited to $0.15z_i$ aloft. The agreement is excellent and suggests that inclusion of shear in the geostrophic wind does not, on its own, lead to failure of simple local closure models. The situation in convective conditions is rather different. The second plot in Figure 2 shows results with a fixed surface buoyancy flux of 10^{-2} m^{2}s^{-3}, which leads to values of $-z_i/L$ of O(10). The LES results are in very good agreement with the predictions of the mixed layer model of Garratt et al. (1982), but the mixing length closure systematically gives lower values. Note however, that a similar discrepancy is found without shear in the imposed geostrophic wind, suggesting that it is largely caused by the well-known failings of mixing length models in convective conditions, rather than by any intrinsic problem with using such models to represent a boundary layer which has shear in the geostrophic wind.

Other tests indicate that models which specify profiles of eddy coefficients as functions of boundary layer stability (e.g. Holtslag and Boville, 1993) perform at least as well as the local mixing length model in reproducing the present equilibrium large-eddy model results. The results of these closures may also be expected to be more robust as they are much less sensitive to small changes in mean gradients, and they also allow the introduction of counter-gradient flux correction terms which have been shown to be beneficial in the heat flux parametrization. In contrast, the carrying of an additional prognostic equation for the turbulent kinetic energy (TKE) has been found to have very little impact in the present equilibrium tests, although clearly it may be beneficial in models with sufficient resolution in space and time for the effects of non-homogeneity and non-stationarity to be significant.

References

Arya, S.P.S.: 1977, 'Suggested revisions to certain boundary layer parametrization schemes used in atmospheric circulation models', *Mon. Weather Rev.*, **105**, 215–227.

Brown, A.R., Derbyshire, S.H. and Mason, P.J.: 1994, 'Large-eddy simulation of stable atmospheric boundary layers with a revised stochastic subgrid model', *Quart. J. Roy. Meteor. Soc.*, **120**, 1485–1512.

Garratt, J.R. Wyngaard, J.C. and Francey, R.J.: 1982, 'Winds in the atmospheric boundary layer – prediction and observation', *J. Atmos. Sci.*, **39**, 1307–1316.

Grant, A.L.M. and Whiteford, R.: 1987, 'Aircraft estimates of the geostrophic drag coefficient and the Rossby similarity functions A and B over the sea', *Bound. Layer Meteor.*, **39**, 219–231.

Hollingsworth, A.: 1994, 'Validation and diagnosis of atmospheric models', *Dynamics of Atmospheres and Oceans*, **20**, 227-246.

Holtslag, A.A.M. and Boville, B.A.: 1993, 'Local versus non-local diffusion in a global climate model', *J. Climate*, **6**, 1825–1842.

Mason, P.J. and Thomson, D.J.: 1992, 'Stochastic backscatter in large-eddy simulations of boundary layers', *J. Fluid Mech.*, **242**, 51–78.

Nieuwstadt, F.T.M., Mason, P.J., Moeng, C.-H., and Schumann, U.: 1991, 'Large-eddy simulation of the convective boundary layer: a comparison of four computer codes', *Turbulent Shear Flows Vol. 8*, Springer Verlag, Berlin.

SPACE STRUCTURE OF THE FREE, UNSTEADY, ROUND, HOMOGENEOUS JET AT LOW REYNOLDS NUMBERS

I. DANAILA
I.N.E.R.I.S., Parc Tech. Alata, 60550 Verneuil en Halatte
I.R.P.H.E., 12 Av. Général Leclerc, 13003 Marseille

J. DUSEK
I.M.F., 2 rue Boussingault, 67000 Strasbourg

AND

F. ANSELMET
I.R.P.H.E., 12 Av. Général Leclerc, 13003 Marseille

Abstract. Three–dimensional direct numerical simulations of unforced, incompressible, free, spatially evolving round jets are used to investigate the onset of the instability at low diametral Reynolds numbers ($10 \leq Re \leq 500$). Compact, coherent structures are identified by means of iso–surfaces of the vorticity field. At the upper limit of the investigated range of Reynolds numbers, the present simulations are consistent the widely accepted scenario of the primary and secondary instability of the round jet. The appearance of pairs of axially counter–rotating vortex filaments is found (for the first time, to our knowledge, in unforced, spatial numerical simulations) to characterize the secondary instability. For lower Reynolds numbers, a superposition of symmetry–breaking (helical) modes is shown to prevail downstream of three nozzle diameters.

1. Numerical implementation

We present direct numerical simulations of three–dimensional, spatially evolving jets at low Reynolds numbers ($10 \leq Re \leq 500$), using the NEKTON code based on a spectral spatial discretization. An important computational effort has been made to optimize the discretization of the shear layer; the absence of spurious reflections at the boundaries, which can trig-

11

S. Gavrilakis et al. (eds.), Advances in Turbulence VI, 11-14.
© *1996 Kluwer Academic Publishers.*

ger global instabilities (Buelle & Huerre, 1988) was also verified. The results presented here use an axisymmetric domain ($r_{max} = 5.33$, $l_z = 15$) of 52 spectral elements and 17 836 nodes. The initial and inflow velocity profile is a 'top–hat' one ($U(r) = 1, r < 1/2$ and $U(r) = 0, r > 1/2$); non-reflecting boundary conditions are imposed elsewhere. The instantaneous vorticity field provides an effective mean of flow visualization and coherent structures identification.

2. Low Reynolds numbers turbulence

The widely accepted scenario of the early stages of the evolution of the round jet is the following: the shear layer originating from the nozzle lip is inviscidly unstable via the Kelvin–Helmholtz primary instability mechanism (Ho & Huerre, 1984); the instability waves grow downstream and roll up into coherent vortex rings (Yule, 1978); the structures merge as they are convected downstream and determine the shear layer spread (Ho & Huang 1982); stream-wise vortex structures develop through a secondary three–dimensional instability of the thin vorticity layer between two neighboring vortex rings (Liepmann & Gharib, 1992).

For a Reynolds number of 500, our numerical simulations provide a qualitative assessment of the formation and dynamics of large scale vortical structures described above. The azimuthal vorticity concentrates into ring vortices while the braid regions become depleted; the distribution of stream-wise vorticity shows the formation of pairs of counter–rotating vortex filaments (fig. 1). The form of the stream-wise structures agrees with that found in direct simulations of round jets evolving in time (Abid *et al.*, 1993), (Brancher *et al.*, 1994).

3. Primary instability

The presence of the symmetry–breaking modes in the jet flow is reported by both linear stability analysis and experimental studies. The inviscid stability analysis of Batchelor & Gill (1962) demonstrated the unstable character of the 'top–hat' velocity profile with respect to axisymmetric as well as helical modes. Mattingly & Chang (1974) considered the inviscid instability of the experimental velocity profiles for a Reynolds number of 300 and showed that the axisymmetric mode dominated close to the jet exit, whereas downstream of three nozzle diameters the helical mode achieved the maximum amplification.

For a Reynolds number of 240 the early stages of the primary instability in the jet have been captured. The presence of helical–like modes is identified three diameters and farther downstream of the nozzle. The antisymmetric structure displayed in this region (fig. 2) suggests that both, the clockwise and counterclockwise rotating, helical modes predicted by the stability theory coexist in the flow. A superposition of two counter-rotating helical modes leads to a selection of a fixed antisymmetry plane in the linear regime as visible in fig. 2. It has to be confirmed whether this selection is a result of initial conditions. Furthermore, this arbitrarily fixed symmetry has to be expected to disappear in the average on the final attractor. The non-linear regime is therefore presently under investigation with the objective to determine the nature of this primary attractor, which is not likely to be a simple limit cycle.

The convectively unstable nature of the incompressible, spatially evolving mixing layer (Huerre & Monkewitz, 1985), can explain the existence of few and less conclusive experimental studies at low Reynolds numbers. The experiments of Crow & Champagne (1971) showed a continuous evolution of the jet shape from a sinusoid to a helix, and finally to a train of axisymmetric waves, as the Reynolds number increases from 100 to 1000. Although the shift of the jet shape from a helix to a train of axisymmetric coherent structures was shown in our simulations, we have not seen any evidence of the sinus mode reported by Crow & Champagne.

4. Conclusions

Our simulations yield a very similar behavior to that reported from experiments and other simulations at significantly supercritical Reynolds numbers. At low Reynolds numbers (at $Re = 240$ and below) we observe characteristics of the instability which are in many respects contradictory to widely expected theories. Reliable experimental and numerical data being unavailable for these Reynolds number values, numerical testings and a theoretical analysis of the results is under way.

This work is funded by l'Institut National de l'Environnement Industriel et des Risques (I.N.E.R.I.S.). We are grateful for their support. Nekton is a registered trademark of Nektronics, Inc. and the Massachusetts Institute of Technology. ©1991 by creare.x, Inc., Hanover, New Hampshire, USA.

References

Abid, M., Brachet, M. (1993) Numerical characterization of the dynamics of vortex filaments in round jets, *Phys. Fluids* **Vol. A5 no. 11**, pp. 2582–2584

Batchelor, G. K., Gill, A. E. (1962) Analysis of the stability of axisymmetric jets, *J. Fluid Mech.* **Vol. 14**, pp. 529–551

Buelle, J. C., Huerre, P. (1988) Inflow/outflow boundary conditions and global dynamics of spatial mixing layers, *Summer Programm* **Rep. No. CTR–S88**, pp. 19–27

Brancher, P., Chomaz, J. M., Huerre, P. (1994) Direct numerical simulations of round jets: Vortex induction and side jets, *Phys. Fluids* **Vol. 6 no. 5**, pp. 1768–1774

14

Crow, S. C., Champagne, F. H. (1971) Orderly structure in jet turbulence, *J. Fluid Mech.* **Vol. 48**, pp. 547–591

Ho, C. M., Huerre, P. (1984) Perturbed free shear layers, *Ann. Rev. Fluid Mech.* **Vol. 16**, pp. 365–424

Ho, C. M., Huang, L. S. (1982) Subharmonics and vortex merging in mixing layers, *J. of Fluid Mech.* **Vol. 118**, pp. 443–473

Huerre, P., Monkewitz, P. A. (1985) Absolute and convective instabilities in free shear flows, *J. Fluid Mech.* **Vol. 159**, pp. 151–168

Liepmann, D., Gharib, M. (1992) The role of streamwise vorticity in the near–field entrainment of round jets, *J. Fluid Mech.* **Vol. 245**, pp. 643–668

Mattingly, G. E., Chang, C. C. (1974) Unstable waves on a axisymmetric jet column, *J. Fluid Mech.* **Vol. 65**, pp. 541–560

Plaschko, P. (1979) Helical instabilities of slowly divergent jets, *J. Fluid Mech.* **Vol. 92**, pp. 209–215

Yule, A. J. (1978) Large structures in the mixing layer of a round jet, *J. Fluid Mech.*

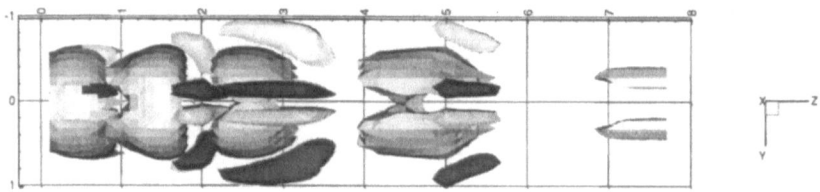

Figure 1. Re=500. Instantaneous surfaces of constant azimuthal vorticity (grey), streamwise vorticity (white for positive and black for negative) at a threshold corresponding to 40% of the respective maxima and minima.

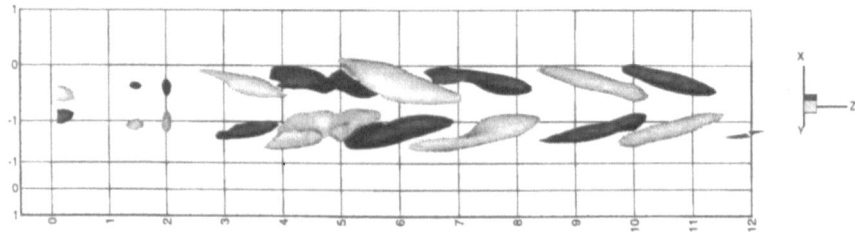

Figure 2. Re=240. Instantaneous surfaces of constant streamwise vorticity at a threshold corresponding to 50% of the respective maximum and minimum (white for positive and black for negative).

LARGE–EDDY SIMULATION OF AIRCRAFT EXHAUST PLUMES IN THE FREE ATMOSPHERE

T. DÜRBECK AND T. GERZ

DLR, Institut für Physik der Atmosphäre,
D-82230 Oberpfaffenhofen, Germany

1. Introduction

Aircraft emissions at cruising altitudes influence the physical and chemical state of the atmosphere and may therefore contribute to a possible global or regional climate change *(Schumann, 1994; WMO, 1995)*. It is crucial for the quantitative understanding of the influence of aircraft emissions on the state of the atmosphere to determine how they get dispersed in space and time. This is essential because chemical and microphysical processes in the atmosphere often evolve nonlinearly and obey similar timescales as the mixing motions.

Here, we investigate the effective diffusion of aircraft emissions at cruising levels, i.e. in the free, stably stratified atmosphere near the tropopause, by means of large–eddy simulation. At these altitudes fully developed turbulence is the exception rather than the rule. Although measured variance spectra exhibit a slope close to -5/3, vanishing turbulent dissipation rates suggest that the spectral turbulent energy transfer from large to small scales is almost entirely suppressed here *(Schumann et al., 1995)*. Hence, the atmosphere is best represented by a weakly turbulent flow under stratified conditions, in a domain of size 4.3×1.1^2 km^3. We start the simulations at the beginning of the diffusion regime, when the aircraft induced wake vortex and the turbulence have ceased and the mixing process is caused by atmospheric motions only. The exhaust plume of an aircraft is represented by a concentration of a passive tracer and initialized as a line source with Gaussian cross-sections.

2. Investigation of situations without mean–flow shear

Firstly, situations without mean–flow shear are studied. Simulations are carried out for seven different levels of stratification characterized by a

15

S. Gavrilakis et al. (eds.), Advances in Turbulence VI, 15-18.

16

Brunt–Väisälä–frequency N between 0.006 s^{-1} and 0.03 s^{-1}. At the time of plume initialization t_{init}, the turbulence velocity is $q = 0.39$ ms^{-1} with a ratio of horizontal to vertical variances as a measure of anisotropy ranging from 2:1 to 7:1 with increasing N. (Figure 1 shows the evolution of the rms velocities for $N = 0.019$ s^{-1}.) Likewise the integral length scale l ranges between 76.9 m and 94.4 m and the root mean square of the temperature fluctuations θ' between 0.04 K and 0.18 K. As required the flows are only weakly turbulent: The inverse Froude number $Fi = Nl/q$ at t_{init} is greater than 1 for all simulations and increases with time. This means, according to *Kaltenbach et al. (1994)*, that the flow has lost most of its stirring properties. The correlation coefficient $|\overline{w\theta}/w'\theta'|$ is smaller than 0.1 for all simulations. Correspondingly the vertical turbulent diffusivity for heat is less than 0.02 m^{2}s^{-1}. The ratio of the Ozmidov length scale to the Kolmogorov length scale is far below the range for active turbulence, implying that all scales are influenced by buoyancy.

At each given point of evaluation in time the concentration field of a given plume is averaged in the airplane's flight direction and the variance matrix of the resulting two–dimensional field is calculated. Thereupon the horizontal and vertical diffusion coefficients D_h and D_v are obtained using

$$\sigma_h^2 = 2D_h t + \sigma_{h,o}^2, \qquad \sigma_v^2 = 2D_v t + \sigma_{v,o}^2, \tag{1}$$

where σ_h^2 and σ_v^2 are the horizontal and vertical variances and the index o denotes the corresponding values at the time of initialization. D_h and D_v are assumed as constant *(Monin and Yaglom, 1971)*.

For each flow the diffusion of 30 different plumes with different positions within the flow–field are studied. Thus it is taken into account that the statistical variability of the mixing process is rather high because the plume size is of the same order of magnitude as the length scale of the decaying

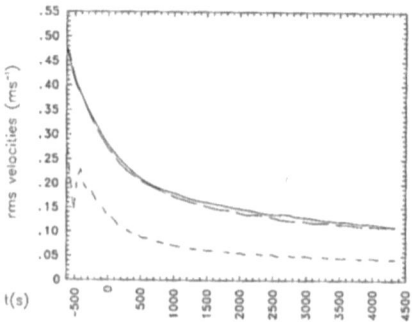

Figure 1. Timeseries of the rms velocities for $N = 0.019$ s^{-1}: u' (——), v' (- -), w' (- - -); $t = 0$ indicates the initialization time of the plumes.

turbulent motion. Mean values and standard deviations of the diffusion coefficients are calculated for this statistical ensemble.

It is found that σ_h^2 depends almost linearly on time, resulting in a single value for D_h. On the other hand, the evolution of σ_v^2 can be described linearly in an initial ($t \lesssim 600$ s) and a late time span ($t \gtrsim 1200$ s). resulting in two values for D_v. Table 1 shows the values of the diffusion coefficients and their standard deviation for all considered values of N obtained by least–square fits. D_h increases with increasing N. In the initial flow phase D_v decreases sharply with N. For $t \gtrsim 1200$ s, D_v is becoming very small and its value is almost independent of N.

The effective plume cross–section given by $A = \pi\sigma_h\sigma_v$ increases with time and stratification.

TABLE 1. Values of the effective diffusion coefficients and their standard deviation for various values of N.

N $(10^{-2}$ s$^{-1})$	D_h (m^{2}s^{-1})	D_v (m^{2}s^{-1}) $t \lesssim 600$ s	D_v (m^{2}s^{-1}) $t \gtrsim 1200$ s
0.6	10.9 ± 2.2	2.28 ± 0.60	0.14 ± 0.06
1.1	14.3 ± 2.6	1.11 ± 0.23	0.15 ± 0.04
1.5	16.7 ± 3.5	0.70 ± 0.18	0.15 ± 0.02
1.9	19.5 ± 4.4	0.52 ± 0.12	0.15 ± 0.01
2.3	21.2 ± 4.3	0.44 ± 0.10	0.15 ± 0.01
2.7	20.3 ± 3.6	0.39 ± 0.06	0.15 ± 0.01
3.0	20.0 ± 3.3	0.37 ± 0.04	0.14 ± 0.01

3. Investigation of situations with mean–flow shear

Secondly, the influence of mean wind shear is investigated. We perform simulations for typical atmospheric conditions in the lower stratosphere: $N = 0.019$ s^{-1} and a wind shear S between 0.001 and 0.007 s^{-1}. Furthermore, we carry out a simulation with $N = 0.006$ s^{-1} and $S = 0.005$ s^{-1} to verify whether our findings for the cases above remain valid for the case of an atmosphere that is only weakly stable.

The effective diffusion coefficients are now derived using

$$\sigma_h^2 = 2/3\, S^2 D_v t^3 + (2SD_s + S^2\sigma_{v,o}^2)t^2 + 2(S\sigma_{s,o}^2 + D_h)t + \sigma_{h,o}^2, \quad (2)$$

$$\sigma_s^2 = SD_v t^2 + (S\sigma_{v,o}^2 + 2D_s)t + \sigma_{s,o}^2, \quad (3)$$

$$\sigma_v^2 = 2D_v t + \sigma_{v,o}^2, \quad (4)$$

extending the approach of (1) to the shear case *(Konopka, 1995)*. (σ_s^2 denotes the skewed variance.)

It is found that under these typical conditions the diffusion coefficients are hardly influenced by shear. This is not surprising because the bulk Richardson numbers of the flows are high. Therefore wind shear has very little impact on the turbulence of the flows.

Nevertheless, shear is very important, because a significant shear controls the horizontal dispersion of the aircraft exhausts via height dependent advection. Furthermore, shear in combination with diffusion results in enlarged plume cross-sections, as depicted in Figure 2. This can be explained by inserting (2) to (4) into

$$A = \pi(\sigma_h^2 \sigma_v^2 - \sigma_s^4)^{1/2}. \tag{5}$$

leading to a formula of the type $A = \pi(a_4 t^4 + a_3 t^3 + a_2 t^2 + a_1 t + a_0)^{1/2}$, where a_0 and a_1 are independent of shear, but a_2 to a_4 depend heavily on shear.

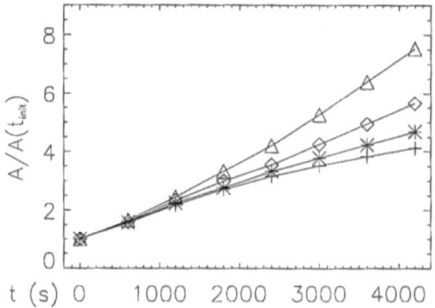

Figure 2. Evolution of the plume cross–sections, for various cases of S: 0.001 $(+)$, 0.003 $(*)$, 0.005 $(\diamond)$, 0.007 $(\triangle)$ s^{-1}.

References

Dürbeck, T. and T. Gerz: Large–eddy simulation of aircraft exhaust plumes in the free atmosphere: effective diffusivities and cross–sections, *Geophys. Res. Lett.* **22**, 3203–3206, 1995.

Kaltenbach, H.–J., T. Gerz and U. Schumann: Large eddy simulation of homogeneous turbulence and diffusion in stably stratified shear flow, *J. Fluid Mech.*, **280**, 1–40, 1994.

Konopka, P.: Analytical Gaussian solutions for anisotropic diffusion in a linear shear flow, *J. Non–Equilib. Thermodyn.*, **20**, 78–91, 1995.

Monin, A. S. and A. M. Yaglom: *Statistical fluid mechanics: Mechanics of turbulence*, MIT Press, Cambridge, Massachusetts, and London, 1971

Schumann, U.: On the effect of emissions from aircraft engines on the state of the atmosphere, *Ann. Geophys.*, **12**, 365–384, 1994.

Schumann, U., P. Konopka, R. Baumann, R. Busen, T. Gerz, H. Schlager, P. Schulte and H. Volkert: Estimation of diffusion parameters of aircraft exhaust plumes near the tropopause from nitric oxide and turbulence measurements, *J. Geophys. Res.*, **100**, D7, 14147–14162, 1995.

WMO: *Scientific Assessment of Ozone Depletion: 1994*, Global Ozone Research and Monitoring Project – Report **37**, WMO, Geneva, Switzerland, 1995.

LARGE-EDDY SIMULATION OF TURBULENT FLOWS IN BAFFLED STIRRED TANK REACTORS

J.G.M. EGGELS
Shell International Oil Products b.v.
Shell Research and Technology Centre, Amsterdam
P.O. Box 38000
1030 BN Amsterdam
The Netherlands

1. Motivation

Baffled stirred tank reactors are frequently used in the (petro)chemical industry for mixing purposes and detailed information on the (time-dependent) flow patterns and turbulence statistics in such reactors is important for an optimal design. The flow in these systems is highly unsteady and in particular near the turbine, quasi-periodic fluid motion can be observed. Since large-eddy simulation (LES) is inherently a time-dependent technique, this unsteady and quasi-periodic behavior of the flow is accounted for in a natural way. This is part of the motivation to use LES to this kind of flow configurations instead of the usual approach based on Reynolds-averaged Navier-Stokes (RANS) methods. In addition, the use of LES is supported by the complexity of the flow. In the reactor, and in particular near the turbine, the swirling motion of the flow is strong. Furthermore, a 'jet-like' behavior of the flow occurs near the turbine with impingement of the jet at the reactor wall. At the baffles, flow separation takes place which results in recirculation zones behind the baffles. All these phenomena occur *simultaneously* in the reactor. From the past, we have learned that conventional turbulence models have a hard job to do to deal properly with such flow phenomena. Specific and optimized models have been developed to tackle the individual phenomena more accurate than general models, but no 'general-purpose' turbulence model is yet available that can handle all these phenomena appropriately with a given set of coefficients. Since in LES the major fraction of the turbulent fluctuations is resolved explicitly in space and time, the details of the subgrid-scale (SGS) model do not play

19

S. Gavrilakis et al. (eds.), Advances in Turbulence VI, 19-22.
© *1996 Kluwer Academic Publishers.*

that much of an important role as in conventional turbulence models. It is also for this reason that we expect LES to do a better job in simulations of turbulent flows in baffled stirred tank reactors than can be achieved with RANS methods based on turbulence modeling.

2. Simulation tool

The large-eddy simulation technique is incorporated in our solver that is based on the lattice-Boltzmann scheme. This scheme, and the closely related lattice-gas automata, are a particular class of numerical techniques to solve the equations of motion for a time-dependent fluid flow (Somers and Rem, 1991), (Somers, 1993), (Eggels and Somers, 1995). The motivation for using the lattice-Boltzmann scheme are multiple: First, the lattice-Boltzmann scheme can easily handle complex geometries which are almost always encountered in industrial applications. Secondly, the method is efficient from a computational point of view in the sense that the number of operations per gridpoint is usually smaller than for other methods, such as finite difference or finite element methods. Finally, due to the simplicity of the scheme, it can easily be implemented in a code that runs with great efficiency on parallel computer platforms. Extended with the Smagorinsky SGS model (Smagorinsky, 1963), this solver enables us to look at high Reynolds number turbulent flows of industrial relevance.

3. Applications

Recently, a facility has been implemented in our solver to handle the motion of a stirring element of an arbitrary shape through the computational domain. The effect of the motion of this element on the fluid, i.e. the exchange of momentum, is accounted for via a moving (dynamic) body force model. The principle idea of this model is that *locally*, on a grid point basis, a body force is applied to the momentum equations of the fluid such that the velocity of the fluid is forced *locally* to the velocity of the turbine as if the turbine was present at that position at that time. To start off with a simple model first, this body force is assumed to be linearly dependent on the local velocity difference between fluid and turbine. For validation, the flows in two reference configurations, one with a standard disc turbine and one with a pitch blade turbine, have been computed. The geometry of the standard disc turbine configuration is sketched in Fig. 1. The Reynolds number, defined as $Re = ND^2/\nu$ with $N = 400$ [rpm] the rotational speed and $D = 100$ [mm] the diameter of the turbine, is equal to 10^5.

The results are compared with experimental data from the literature and very good agreement is obtained, not only for the mean flow but also for the root-mean-square (rms) velocities (Eggels, 1996). As an example,

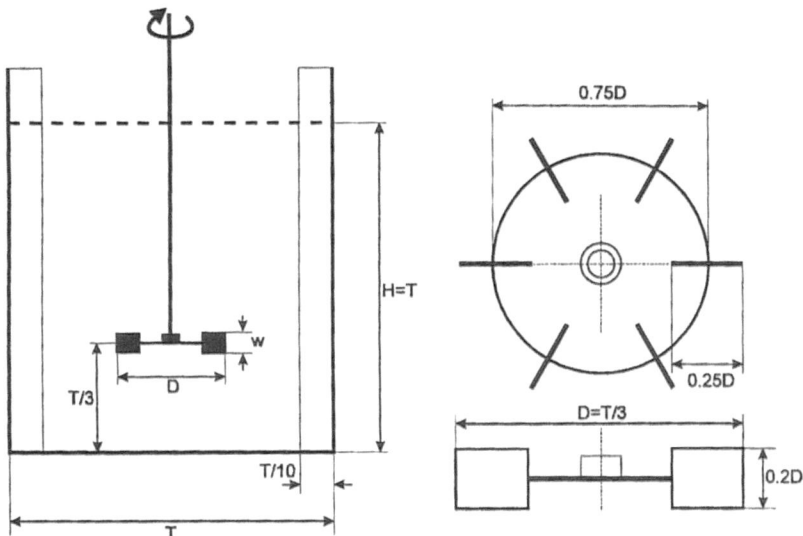

Figure 1. Reactor configuration and shape of the applied 6 blade disc turbine.

the mean and rms values of the radial and axial velocity are shown in Fig. 2 along a vertical traverse just outside the turbine swept volume, i.e. at $r/D = 0.5$. Each of the two lines denote the simulation data taken at opposite sides from the symmetry axis of the configuration and indicate the convergence of the time-averaged statistics. The markers denote the corresponding experimental data by Bakker (1995).

4. Conclusions

For the unsteady and quasi-periodic flow in the baffled stirred tank reactor considered here, the LES results are much better than those obtained so far using the RANS technique, especially those for the turbulence intensities. A few reasons for this have already been mentioned in Section 1. This clearly indicates the applicability (and potential) of LES for industrially relevant flow problems. Furthermore, the availability of accurate, dynamic, flow information will become more and more important in our design studies of equipment, such that there is a clear need for time-dependent simulation techniques such as LES.

References

Bakker, R.A. (1995) Laser-Doppler measurements in a baffled stirred tank reactor with a disc turbine. Experimental data to appear in Ph.D. thesis, Delft University of Technology, the Netherlands.

22

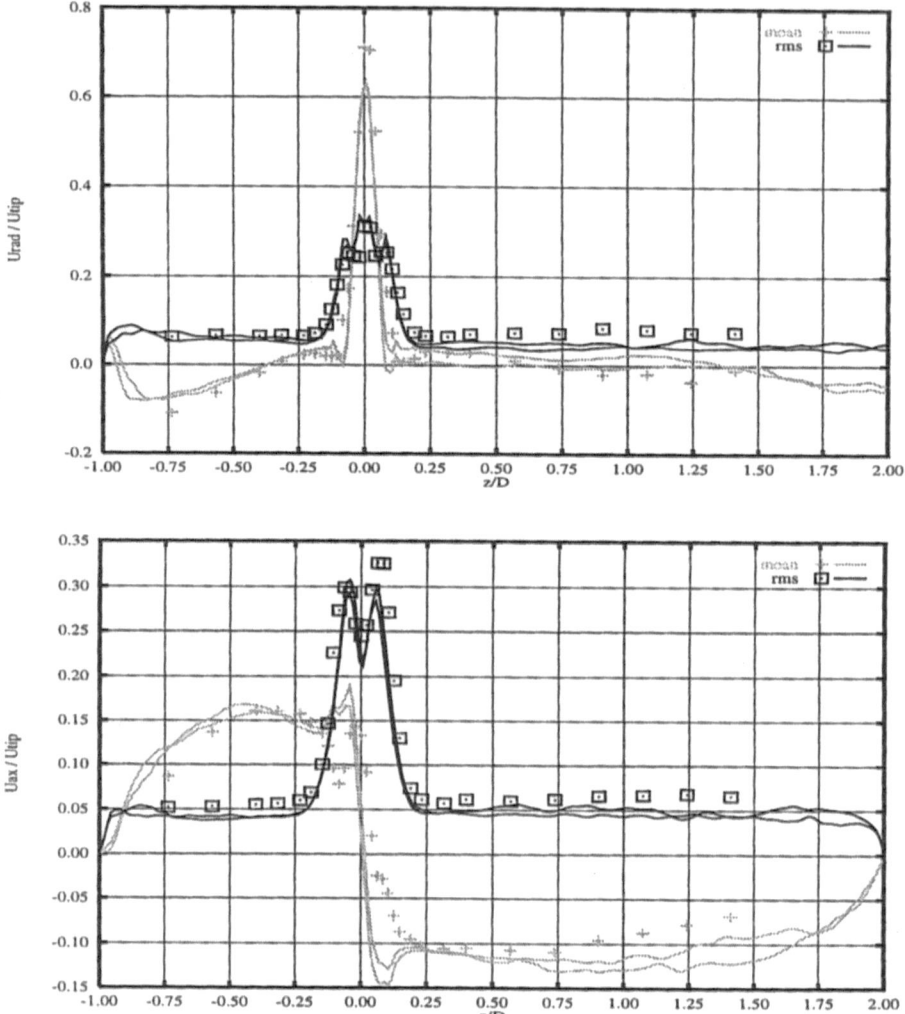

Figure 2. Mean and rms profiles of the radial (upper plot) and axial velocity (lower plot) along a vertical traverse at $r/D = 0.5$. $z/D = -1$ corresponds with the solid wall at the bottom of the reactor, $z/D = 2$ with the free surface at the top. All velocities are scaled with the tip velocity of the turbine.

Eggels, J.G.M. and Somers, J.A. (1995) Numerical simulation of free convective flow using the lattice-Boltzmann scheme *J. Heat and Fluid Flow* **16**, pp. 357–364.

Eggels, J.G.M. (1996) Direct and large-eddy simulation of turbulent fluid flow using the lattice-Boltzmann scheme. To appear in *J. Heat and Fluid Flow*.

Smagorinsky, J. (1963) General circulation experiments with the primitive equations: 1. The basic experiment, *Mon. Weather Rev.* **91**, pp. 99–164.

Somers, J.A. and Rem, P.C. (1991) Flow Computation with Lattice Gases, *Appl. Sci. Res.* **48**, pp. 391–435.

Somers, J.A. (1993) Direct Simulation of Fluid Flow with Cellular Automata and the Lattice Boltzmann Equation, *Appl. Sci. Res.* **51**, pp. 127–133.

AN EXACT SGS-MODEL FOR LES

L. FUCHS

Division of Fluid Mechanics,
Department of Heat- and Power-Engineering,
Lund Institute of Technology,
S-221 00 Lund, Sweden

Abstract. A new class of Sub-Grid-Scale (SGS) formulations is presented. This class of SGS formulations enables one to derive closed forms of the SGS terms, in contrast to other methods that *model* these terms. The main difficulty with the exact SGS formulation is the additional data required on the boundaries. A particular form of the SGS terms, belonging to the class, has been implemented for simulating different transitional and turbulent flows. Here, we present some results for wall-free jet flows. For this type of flows, one avoids the difficulties at the boundaries, and therefore may study directly the effects of the SGS formulation. Simulated results have been compared with experimental data and simulations using dynamic SGS model. The new SGS formulation seems to reproduce very well the experimental data.

1. The Filtering Process

The spatial filtering process that we consider here is such that it will "smooth" out high frequency components of a given function. This smoothing process is not necessarily strict in the sense that the the filter has a sharp "cut-off" frequency. Rather, the filter has to be positive in the real-space, in order to be realizable, Verman *et al.* (1994).

Consider a spatial filter kernel $G(\mathbf{x} - \mathbf{x}')$ applied to a function $\mathbf{q}(\mathbf{x}; t)$, resulting in a "smooth" function $\overline{\mathbf{q}}(\mathbf{x}, t; \Delta)$:

$$\overline{\mathbf{q}}(\mathbf{x}, t; \Delta) = \int G(\mathbf{x} - \mathbf{x}'; \Delta) \, \mathbf{q}(\mathbf{x}', t) \, d\mathbf{x}' \qquad (1)$$

23

S. Gavrilakis et al. (eds.), Advances in Turbulence VI, 23-26.
© 1996 *Kluwer Academic Publishers.*

The choice of the filter may be more or less arbitrary, unless one wants to utilise the properties of the chosen filter. The essence of LES is finding an expression for the SGS effects in terms of the filtered variables. Here, we propose a way of determining an exact form for the SGS expressions.

Suppose that one could define the inverse operation of the filter in (1). The inverse filter is required to be a differential operator with certain properties. To define such a filter assume that the inverse operator of the filtering is a polynomial of derivatives (D). We denote the inverse as $P(D)$. More precisely, we define the difference between the unfiltered quantity $\mathbf{q}$ and its filtered counterpart $\bar{\mathbf{q}}$ to be proportional to a differential operator $P(D)$ operating on the filtered field:

$$\mathbf{q} - \bar{\mathbf{q}} = -\mathbf{P(D)}\,\bar{\mathbf{q}} \tag{2}$$

Taking the Fourier transforms, denoted by $(\hat{*})$, of (1) and (2), respectively, one obtains:

$$\hat{\bar{\mathbf{q}}} = \hat{G}\hat{\mathbf{q}} \tag{3}$$

and

$$\hat{\mathbf{q}} - \hat{\bar{\mathbf{q}}} = -\widehat{P(D)}\mathbf{q} = -P(ik)\hat{\bar{\mathbf{q}}} \tag{4}$$

Using these relations one may express the filter kernel G in terms of the polynomial P:

$$\hat{G} = \frac{1}{1 - P(ik)} \tag{5}$$

Depending on the particular choice for the polynomial P, one may define different forms for the filtering kernel G. The simplest example is $P(D) = D^2$, which in the one dimensional case yields $P = d^2/dx^2$. For this case the form of G in the real space is $G(x - x') = \frac{1}{2\Delta}exp(-|x - x'|/\Delta)$. The corresponding three-dimensional form can be easily found. This is identical to the differential filter as proposed by Germano (1986). The polynomial P in this case is of order 2 and the difference between the filtered and the unfiltered quantities is:

$$q - \bar{q} = -\Delta^2\nabla^2\bar{q} \tag{6}$$

One may show that the filter of the above type can be related to a "smoothed" (molified) "top-hat" filter in the Fourier-space, such that in the real space it maintains its positivity. The SGS terms for the momentum equations $\tau_{ij,j}$, can be identified to be

$$\tau_{ij,j} = \Delta^2\Big[\frac{\partial \tilde{u}_i}{\partial t} + \frac{\partial \tilde{u}_i\,\overline{u}_j}{\partial x_j} + \frac{\partial \overline{u}_i\,\widetilde{u}_j}{\partial x_j} + \Delta^2\frac{\partial \tilde{u}_i\,\widetilde{u}_j}{\partial x_j} + \frac{\partial \tilde{p}}{\partial x_i} - \frac{1}{Re}\frac{\partial}{\partial x_j}\frac{\partial \tilde{u}_i}{\partial x_j}\Big] \tag{7}$$

where $\tilde{q} = \nabla^2 \bar{q}$. In contrast to the classical LES formulation we have even a SGS term in the continuity equation. There are several consequences of these SGS terms. One is related to the *number* of required boundary conditions, which is larger than the number required for the unfiltered equations. Secondly, the mean velocity vector does not vanish at solid walls. It can be shown to be $O(\Delta^2)$. Thirdly, the form of the SGS given in (7) is computationally less favorable due to the presence of the Laplacian of the pressure. This term has been replaced in the calculations by a mathematically equivalent term, containing the invariants of the deformation and rotation tensors. This latter approach has also the advantage that the required number of boundary conditions is reduced by one.

2. Large Eddy Simulations

Here, we consider a boundary free flow; i.e. circular jets. For these flows one can avoid the difficulties arising from the additional boundary conditions and the need for computing the space averaged boundary values. For the circular jet, we use cartesian grids with a sequence of local grid refinements, Fuchs (1986), so that adequate spatial resolution can be attained. The filtered equations, including the SGS terms are discretised using fourth order accurate finite-differences, with the exception of convective terms which are approximated by third order accurate upwind (not strictly positive) finite-differences. The SGS terms in (7) are approximated by second order finite-differences. However, since the terms are alredy of $O(\Delta^2)$, this lower order accuracy does not affect the total accuracy of the numerical results. The integration in time is carried out by an implicit integration. The implicit step is computed using a Multi-grid scheme. More details of the numerical scheme are given in Olsson and Fuchs (1996). To avoid interference between the numerical dissipation and the SGS terms, we require that Δ is larger than the grid size by a factor of at least 2.

Consider a circular jet with $Re = 10^4$. We compare the results of the new SGS formulation with the results obtained by using a "dynamic" model Olsson and Fuchs (1996) and experimental data Crow and Champagne (1971). The calculations "cover" both the transitional and the proximal turbulent regimes of the jet. Some results are shown in the enclosed figures. One figure depicts the radial variation of the fluctuation of the axial velocity component, at four distances from the jet nozzle. The axial development of these fluctuations is also shown in the other figure. We have also considered the effects of the mesh spacing and inlet perturbations on the transition and the proximal turbulence (detail not shown here). The overall picture is that the new SGS formulation yields results that is in very good agreement with experiments. A more detailed comparison with the dynamic SGS model

26

shows, if the same grids are used, that the new formulation yields better agreement with experiments

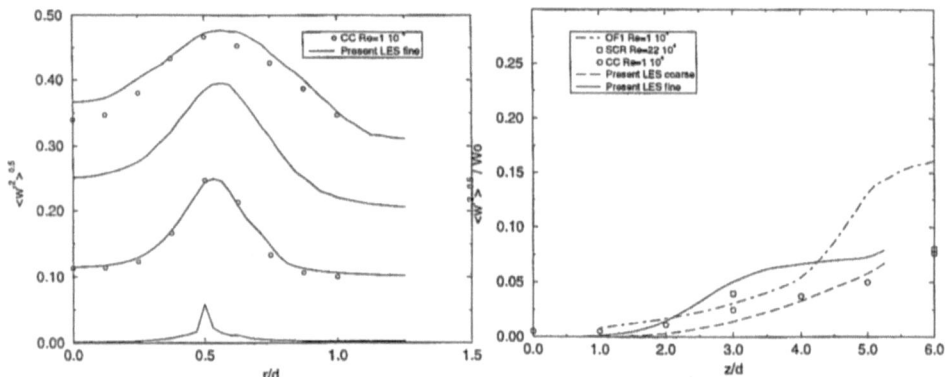

Figure 1. The fluctuating axial velocity component vs. radial distance at four axial locations; z/D=1,2,3 and 4 (left) and its variation along the jet axis (right). CC = Crow and Champagne (1971); OF1 = Olsson and Fuchs (1996); SCR = Sami *et al.* (1967)

3. Concluding Remarks

A new formulation of the SGS model, based on filters with explicit differential inverse, has been proposed and evaluated for simulating spatially developing turbulent jets. The results show very good agreement with experiments both in the transitional and the proximal turbulent regions. The new model has also been used for simulating swirling jet flows and certain wall bounded flows. These results are to be reported elsewhere.

Ackkowledgement: This work was partially supported by TFR, the Swedish Research Council for Engineering Sciences.

4. References

Crow, S.C. and Champagne, F. H., (1971) Orderly structure in jet turbulence. *J. Fluid Mech.* **48**, 547.

Fuchs, L. (1986) A local mesh refinement technique for incompressible flows. *Computers and Fluids*, **14** 69.

Germano, M. (1986) Differential filters for the large eddy numerical simulation of turbulent flows. *Phys. Fluids.* **29**, 1755.

Olsson, M. and Fuchs, L. (1996) Large Eddy Simulation of the proximal region of a spatially developing circular jet. Submitted for publication.

Sami, S. and Carmody, T. and Rouse, H. (1967) Jet diffusion in the region of flow establishment. *J. Fluid Mech.* **27**, 231.

Verman,B., Geurts, B. and Kuerten, H. (1994) Realizability conditions for the turbulent stress in LES. *J. Fluid Mech.* **278**, 351.

ALTERNATIVE APPROACH TO MODELLING
THE DISSIPATION EQUATION

K. HANJALIĆ
Faculty of Applied Physics, Delft University of Technology,
Lorentzweg 1, 2628 CJ Delft, The Netherlands

S. JAKIRLIĆ
Lehrstuhl für Strömungsmechanik, University of Erlangen,
Cauerstr. 4, 91058 Erlangen, Germany

AND

J. R. RISTORCELLI
Institute for Computer Applications in Science and Engineering,
NASA Langley Research Center, Hampton, VA 23681, USA

1. INTRODUCTION

Direct numerical simulations show that even for simple equilibrium flows the models of major interactions in the transport equation for turbulence energy dissipation, $\varepsilon = \nu \overline{(\partial u_i/\partial x_l)^2}$ do not adequately reflect the corresponding terms in the exact ε equation. This is particularly pronounced in low-Re-number flow regions where the dynamics of dissipative correlation is strongly affected by viscosity and vicinity of a solid wall. The success in reproducing ε reasonably well, even at low-Re-numbers and near walls is achieved by *ad hoc* tuning of the equation as a whole, concealing the compensating errors in term-by-term modelling. Hence, the application of the models in non-equilibrium situations, different from ones in which the models were calibrated, leads often to unsatisfactory performances.

Various extension of the ε equation have been proposed to account for extra strain effects, streamline curvatures, differentiating the effects of rotational and irrotational strains, etc. While such extensions yielded improvements of some classes of flows, they failed in others.

2. ANALYSIS

In this paper we revisit the dissipation equation and present some results from a different approach which focusses on the low-Re-number near wall flow. The novelty is in using the dissipation equation derived from the transport equation for the two-point correlation [1], and in a *term-by-term* modelling of the equation. The new equation for "homogeneous" dissipation rate $\varepsilon^h = \varepsilon - 1/2\nu(\partial^2 k/\partial x_l \partial x_l)$ differs from the standard one only in the factor $1/2$ in the viscous diffusion term, which is important only very close to a solid wall [2], i.e.

$$\frac{D\varepsilon^h}{Dt} = \frac{D\varepsilon}{Dt} - \frac{1}{2}\frac{\partial}{\partial x_l}\left(\nu\frac{\partial \varepsilon^h}{\partial x_l}\right) \tag{1}$$

where $D\varepsilon/Dt$ should be replaced by the right-hand-side of the conventional low-Re-number ε-equation. The reproduced ε in a plane channel is very similar to, and in some respect

27

S. Gavrilakis et al. (eds.), Advances in Turbulence VI, 27-30.

better than that obtained by the standard formulation. However, in addition to better physical foundation of the derivation of the ε^h equation, a major advantage is achieved in evaluating the individual dissipation rate components ε_{ij}, which becomes important for second-moment closures. First the same popular expression used in the conventional low-Re-number model is applied now to compute the homogeneous components, $\varepsilon_{ij}^h = 2/3\varepsilon^h\delta_{ij}(1 - f_\varepsilon) + f_\varepsilon(\overline{u_iu_j}/k)\varepsilon^h$, where f_ε is a damping function. The components of the full dissipation rate tensor are then obtained from $\varepsilon_{ij} = \varepsilon_{ij}^h + 1/2\mathcal{D}_{ij}^\nu$, where $\mathcal{D}_{ij}^\nu = \nu(\partial^2\overline{u_iu_j}/\partial x_l\partial x_l)$ is the viscous diffusion of the corresponding stress component. The obtained ε_{ij} satisfies exactly the wall-limits of each normalized dissipation component $\varepsilon_{ij}/\overline{u_iu_j}$, except for a slight discrepancy for $i = j = 2$, for which $\varepsilon_{22}/\overline{u_2^2} = 3.5$ instead of exact 4. Hence, the first benefit is that the special correction for wall values of ε_{ij} proposed by Launder and Reynolds (1983) and used in most low-Re-number second-moment closures, can be abandoned.

Reformulations of some of the terms in ε equation are now described, which apply irrespective of whether the conventional or new ε-equation is considered. First we consider the "mixed" production term

$$P_\varepsilon^1 + P_\varepsilon^2 = -2\nu\left(\overline{\frac{\partial u_i}{\partial x_l}\frac{\partial u_j}{\partial x_l}} + \overline{\frac{\partial u_l}{\partial x_i}\frac{\partial u_l}{\partial x_j}}\right)\frac{\partial U_i}{\partial x_j} \tag{2}$$

This term is regarded as negligible in high-Re-number flows. The major production of ε, associated with the self-stretching of the fluctuating vortices is usually modelled in terms of the energy production P_k, scaled with the characteristic turbulence time scale k/ε.

It should be recalled that the first term in the bracket is ε_{ij}. The second term is closely related to ε_{ij} - they contain common terms. DNS data for near-wall flows show that ε_{ij} remains anisotropic beyond the viscosity affected wall region, even at higher Re numbers. Now ε_{ij} can be modelled satisfactory [3] and this term can be retained in its exact form. Of course, away from the wall and at high Re numbers, ε_{ij} becomes isotropic irrespective of the stress anisotropy and this term is not sufficient to account for total production of ε so that the standard production term should be retained, although with a smaller coefficient. Fig. 1 presents a new production model in a plane channel. The new model consists of the sum of the new and standard term, with $C_{\varepsilon 1} = 1$,

$$P_\varepsilon^1 + P_\varepsilon^2 = -\varepsilon_{ij}\frac{\partial U_i}{\partial x_j} - 1.0\,\overline{u_iu_j}\frac{\partial U_i}{\partial x_j}\frac{\varepsilon}{k} \tag{3}$$

In addition to a better reproduction of the exact terms, than achieved by the standard term alone with $C_{\varepsilon 1} = 1.44$, the new model offers additional flexibility, which is particularly useful in non-equilibrium flows subjected to strong linear straining. This becomes obvious if P_ε^1 is expanded into components. For 2-D flows:

$$P_\varepsilon^1 = -\varepsilon_{12}\frac{\partial U_1}{\partial x_2} - (\varepsilon_{11} - \varepsilon_{22})\frac{\partial U_1}{\partial x_1} \tag{4}$$

While $\overline{u_1u_2}/(\overline{u_1^2} - \overline{u_2^2})$ is of the order of magnitude of 1, making the production by both the rotational and irrotational strain of equal importance, $\varepsilon_{12}/(\varepsilon_{11} - \varepsilon_{22})$ is much smaller than 1, except very close to the wall [3]. Hence, the term will itself distinguish the effect of the irrotational from the rotational strain in the production of ε.

The gradient production of ε is

$$P_\varepsilon^3 = -2\nu\overline{u_k\frac{\partial u_i}{\partial x_l}}\frac{\partial^2 U_i}{\partial x_k\partial x_l}. \tag{5}$$

Current practice assumes a simple gradient model $\overline{u_k\frac{\partial u_i}{\partial x_l}} \propto \tau\overline{u_ju_k}\frac{\partial^2 U_i}{\partial x_j\partial x_l}$, where $\tau = k/\varepsilon$, yielding the term with the squared second velocity derivative. The expression follows from the Taylor vorticity transport approach and is a rigid constraint because it does not allow for a proper sign of the curvature of the mean velocity profile. Bernards [4] vorticity transport theory update provides a more rational method. The turbulent velocity gradient flux is expanded into

$$\overline{u_k \frac{\partial u_i}{\partial x_l}} = \left(\frac{\partial \overline{u_k u_i}}{\partial x_l} - \overline{u_i \frac{\partial u_k}{\partial x_l}} \right) = \left(\frac{\partial \overline{u_k u_i}}{\partial x_l} - \overline{u_i s_{kl}} - \overline{u_i \omega_{kl}} \right) \tag{6}$$

where s_{kl} and ω_{kl} are the fluctuating strain rate and vorticity respectively. The first term is now exact. The second term needs modeling. The third term is omitted since it is antisymmetric in its indices while the velocity curvature term is symmetric. For the 2D near-equilibrium wall layer $\frac{\partial^2 U_i}{\partial x_k \partial x_l} = \frac{\partial^2 U_1}{\partial x_2 \partial x_2} \delta_{i1} \delta_{k2} \delta_{l2}$. The term $\overline{u_1 s_{22}}$ can be expanded, using the continuity, to produce

$$\overline{u_1 s_{22}} = -\frac{1}{2} \frac{\partial \overline{(u_1^2 + u_3^2)}}{\partial x_1} - \frac{\partial \overline{u_1 u_2}}{\partial x_3} - \overline{u_3 \omega_2} \tag{7}$$

where $\omega_i = \epsilon_{ijk} \frac{\partial u_j}{\partial x_k}$ The first two terms can be neglected because of spanwise and streamwise homogeneity. The results of Bernard [4] is used to close $\overline{u_3 \omega_2}$:

$$\overline{u_3 \omega_2} = \frac{1}{2} \frac{Q_4}{1 + Q_3 Q_4 (\partial U_1/\partial x_2)^2} \frac{\partial \overline{u_3^2}}{\partial x_2} \frac{\partial U_1}{\partial x_2} \tag{8}$$

where Q_3 and Q_4 are the components of the Lagrangian integral scales Q_{ikl} of the two-point third-order tensors, $\overline{u_i (\partial u_k/\partial x_l)}$.

The expression (8) reduces to the term proposed by Rodi and Mansour [5], designed in the framework of $k - \varepsilon$ model, if $\partial \overline{u_3^2}/\partial x_2$ is replaced by $\partial k/\partial x_2$ and the function in front of the term is expressed as proportional to k/ε. In the context of a $k - \varepsilon$ model in which $\overline{u_3^2}$ is not calculated we adopted Rodi and Mansour formulation of this term in combination with the above first term. The final form of the gradient production for near-equilibrium 2-D flows is

$$P_\varepsilon^3 = -2\nu \frac{\partial \overline{u_1 u_2}}{\partial x_2} \frac{\partial^2 U_1}{\partial x_2 \partial x_2} + 0.02\nu \frac{k}{\varepsilon} \frac{\partial k}{\partial x_2} \frac{\partial U_1}{\partial x_2} \frac{\partial^2 U_1}{\partial x_2 \partial x_2} \tag{9}$$

The plot of each term in the expression (8) and of their sum (the complete model) of P_ε^3 computed from DNS data for a plane channel is shown in Fig. 2.

For the two remaining terms, representing the difference between the turbulent production and viscous destruction of ε, which represents the major source of dissipation at high Re numbers, Hanjalić and Launder [6] proposed a joint model

$$P_\varepsilon^4 - Y = -2\nu \overline{\frac{\partial u_i}{\partial x_k} \frac{\partial u_i}{\partial x_l} \frac{\partial u_k}{\partial x_l}} - 2 \overline{\left(\nu \frac{\partial^2 u_i}{\partial x_k \partial x_k} \right)^2} = -C_{\varepsilon 2} f_\varepsilon \frac{\varepsilon \tilde{\varepsilon}}{k} \tag{10}$$

where $\tilde{\varepsilon} = \varepsilon - 2\nu (\partial k^{1/2}/\partial x_l)^2$. The plot of the model, with original function f_ε and modified one proposed by Coleman and Mansour (1993), shows poor agreement close to the wall, Fig 3a. For illustration, the proposal of Durbin to replace in the model (10) the time scale $\tau = k/\varepsilon$ by Kolmogorov scale $\tau_K = \sqrt{\nu/\varepsilon}$, when τ_K becomes larger than $\tau/6$, is presented, showing also poor agreement. In contrast, the application of the same model using the homogeneous dissipation rate, i.e. $-C_{\varepsilon 2} f_\varepsilon \varepsilon^h \tilde{\varepsilon}^h/k$, where $\tilde{\varepsilon}^h = \varepsilon^h - \nu (\partial k^{1/2}/\partial x_l)^2$, yields much better agreement with the DNS data, as shown in Fig. 3b. The final outcome is shown in Fig. 4 where the reproduction of ε^h and of total ε by the new equation is presented for a plane channel flow, using the DNS data for all input variables.

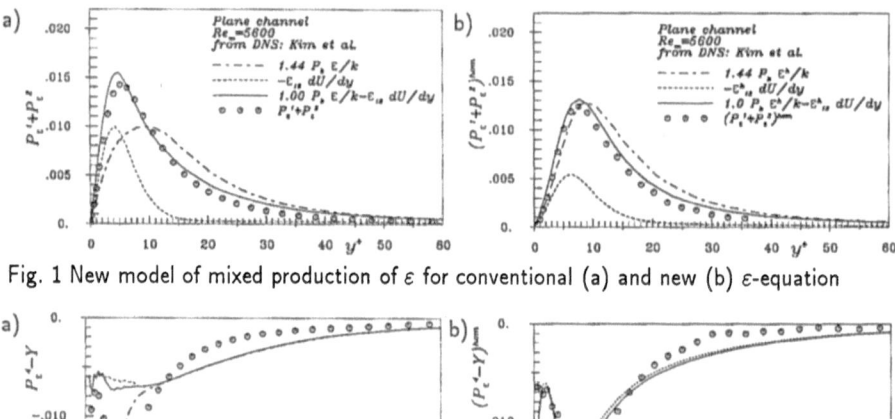

Fig. 1 New model of mixed production of ε for conventional (a) and new (b) ε-equation

Fig. 3 Model of $P_\varepsilon^4 - Y$ for conventional (a) and new (b) ε-equation

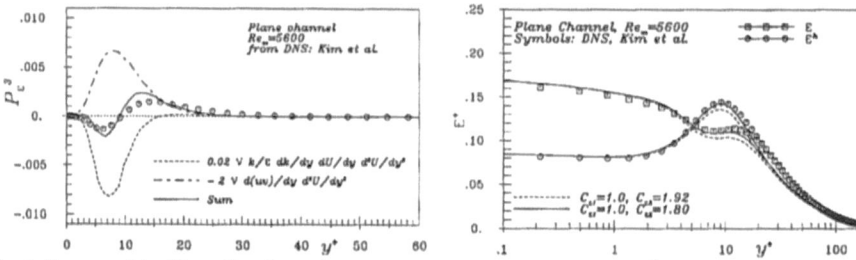

Fig. 2 New model of "gradient" production of ε Fig. 4 Computed ε^h and ε by new model.

3. CONCLUSION

The derivation of the dissipation equation from the two-point correlation and reformulating the models of each term using some novel arguments and the notions of the vorticity transport theory shows that it is possible to perform term-by-term modelling for near-equilibrium low-Re-number shear flows in good agreement with DNS data.

Acknowledgement: The authors acknowledge fruitful discussions with Dr. J. Jovanović of the University of Erlangen.

REFERENCES:

1. Jovanović, J., Ye, Q.-Y., Durst, F., J. Fluid Mech. **293**, 321-347, 1995.
2. Hanjalić, K., Jakirlić, S., *submitted for publication*
3. Hanjalić, K., Jakirlić, S., Appl. Scientific Research **51**, 513-518, 1993
4. Bernard, P. S., Theor. Comp. Fluid Dynamics, **2**, 165-183, 1990
5. Rodi, W., Mansour, N.N., J. Fluid Mech. **250**, 509-529, 1993.
6. Hanjalić, K., Launder, B.E., J. Fluid Mech. **74**, 593-610, 1976.

A NEW EXPLICIT ALGEBRAIC
REYNOLDS STRESS MODEL

ARNE V. JOHANSSON
Department of mechanics, KTH, S-100 44 Stockholm, Sweden

AND

STEFAN WALLIN
FFA, Box 11021, S-161 11 Bromma, Sweden

Abstract. A new explicit algebraic Reynolds stress model (EARSM) is presented, which represents an exact description of the implicit algebraic Reynolds stress model with linear models for the terms involved. In contrast to earlier published models there is here no need for ad-hoc assumptions about the production to dissipation ratio.

1. Introduction

The general platform for explicit algebraic Reynolds stress models (EARSM) is a two-equation turbulence model, but with the Boussinesq eddy-viscosity hypothesis replaced by a more general relation between the stress anisotropy ($a_{ij} \equiv \langle u_i u_j \rangle / K - 2\delta_{ij}/3$) and the mean flow quantities (only incompressible turburbulent flows are treated here).

$$a_{ij} = a_{ij} \left(S_{ij}, \Omega_{ij}, \text{scalars}, \ldots \right) \tag{1}$$

where $\mathbf{S}, \Omega$ are the symmetric and antisymmetric parts of the mean velocity gradient tensor, *i.e.* mean strain and rotation rate tensors, both here normalized by $\tau = K/\varepsilon$.

A general relation of the form (1) can be shown to be expressible in terms of ten tensorially independent groups, where the coefficients may be functions of the five independent invariants of (S_{ij}, Ω_{ij}).

In this short paper we will restrict ourselves to twodimensional mean flows where a_{ij} can be completely described by three groups and two invariants (*e.g.* $II_S \equiv S_{ik}S_{ki}, II_\Omega \equiv \Omega_{ik}\Omega_{ki}$). In order to keep a concise notation

31

S. Gavrilakis et al. (eds.), Advances in Turbulence VI, 31-34.

we will here use boldface to denote a second rank tensor (or matrix). Two consecutive boldface letters denote inner product ($\mathbf{AB} \equiv A_{ik}B_{kj}$). With this notation we may write the complete expression for the anisotropy tensor as ($\mathbf{I}$ is the identity matrix)

$$\mathbf{a} = \beta_1\mathbf{S} + \beta_2\left(\mathbf{S}^2 - \frac{1}{3}II_S\mathbf{I}\right) + +\beta_4\left(\mathbf{S\Omega} - \mathbf{\Omega S}\right) \tag{2}$$

We see that rotational effects enter in higher order terms, but may also enter through the invariants that contain the rotation rate tensor. The first coefficient, β_1, is equivalent to $-2C\mu$ in the linear ν_T-hypothesis.

The traditional ARSM idea (Rodi 1972,1976) to construct an algebraic relation for a_{ij} is equivalent to neglecting advection and diffusion terms in the exact transport equation for a_{ij}.

2. The new model

We here choose the two-equation model platform to be the K-ε model with the following transport equations

$$\frac{DK}{Dt} = \mathcal{P} - \varepsilon - \frac{\partial J_k}{\partial x_k} \qquad \frac{D\varepsilon}{Dt} = \frac{\varepsilon}{K}\left(C_{\varepsilon 1}\mathcal{P} - C_{\varepsilon 2}\varepsilon\right) - \frac{\partial J_k^\varepsilon}{\partial x_k} \tag{3}$$

The production, $\mathcal{P}$, does not need to be modelled since, in this context, we have an explicit algebraic expression for a_{ij} based on its modelled transport equation.

As models for the terms involved in this equation we choose the general linear form of $\Pi_{ij}^{(r)}$ (see Launder, Reece & Rodi 1975), the Rotta model (Rotta 1951) for $\Pi_{ij}^{(s)}$ (with $c_1 = 1.8$) and an isotropic dissipation rate tensor. This results in the following implicit equation for a_{ij}

$$\mathbf{a}\left(\frac{\mathcal{P}}{\varepsilon} + c_1 - 1\right) = -\frac{8}{15}\mathbf{S} - \frac{5 - 9c_2}{11}\left(\mathbf{aS} + \mathbf{Sa} - \frac{2}{3}\mathrm{tr}\{\mathbf{aS}\}\mathbf{I}\right)$$
$$+ \frac{7c_2 + 1}{11}\left(\mathbf{a\Omega} - \mathbf{\Omega a}\right) \tag{4}$$

where we should keep in mind that $\mathbf{S}, \mathbf{\Omega}$ are normalized by τ, so that $\mathcal{P}/\varepsilon = -a_{mn}S_{nm}$.

The value of c_2 in the $\Pi_{ij}^{(r)}$ model was originally suggested to be 0.4, but more recent studies of homogeneous shear flow indicate a higher value, close to 5/9. This suggests that the second group in equation (4) is of quite small influence (also noted by Taulbee 1992).

The interesting feature of the above system with $c_2 = 5/9$ is that insertion of the full a_{ij} expression (valid for 3D mean flows) leads to a nonlinear system of equations for the ten β_i:s, where the solutions for five of the β_i:s are identically zero. For 2D mean flows the system is reduced to three equations.

The nonlinearity of the system forms a major obstacle for this approach and the studies published so far have circumvented the problem by taking $\mathcal{P}/\varepsilon$ (or equivalently $\mathrm{tr}\{\mathbf{aS}\}$) as a constant during the solution of the system of equations, and thereafter making some *ad hoc* assumption for this quantity.

Here we first rerite eq. 4 (with $c_2 = 5/9$) in the following way

$$N\mathbf{a} = -\frac{6}{5}\mathbf{S} + \mathbf{a}\mathbf{\Omega} - \mathbf{\Omega}\mathbf{a} \tag{5}$$

where $N = c_1' - (9/4)\mathrm{tr}\{\mathbf{aS}\}$ ($c_1' = \frac{9}{4}(c_1 - 1)$). The solution is obtained with N assumed as known, which for 2D mean flows gives

$$\mathbf{a} = \frac{-6}{5}\frac{1}{N^2 - 2II_\Omega}(N\mathbf{S} + \mathbf{S}\mathbf{\Omega} - \mathbf{\Omega}\mathbf{S}) \tag{6}$$

(*i.e.* $\beta_2 = 0$) where the denominator always remains positive since $II_{\Omega^*} \leq 0$, and N is the real positive root of a cubic equation. The physical solution may be written in closed form as

$$N = \begin{cases} \frac{c_1'}{3} + (P_1 + \sqrt{P_2})^{1/3} + \mathrm{sign}\,(P_1 - \sqrt{P_2})\,|\,(P_1 - \sqrt{P_2})\,|^{1/3} & P_2 \geq 0 \\ \frac{c_1'}{3} + 2(P_1^2 - P_2)^{1/6}\cos\left(\frac{1}{3}\arccos\left(P_1/\sqrt{P_1^2 - P2}\right)\right) & P_2 < 0 \end{cases}$$

$$P_1 = c_1'\left(\frac{c_1'^2}{27} + \frac{9}{20}II_S - \frac{2}{3}II_\Omega\right) \qquad P_2 = P_1^2 - \left(\frac{c_1'^2}{9} + \frac{9}{10}II_S + \frac{2}{3}II_\Omega\right)^3$$

It is readily verified that N remains real and positive for all possible values of II_S and II_Ω. For 3D mean flows we have a sixth order equation for which an approximation of N easily is found by linearization around the 2D state. The removal of the need for *ad-hoc* relations for $\mathcal{P}/\varepsilon$ represents a substantial improvement for this type of modelling.

Different EARSM formulations can be compared by studying the behaviour of the β coefficients. In the figures below β_1 ($\equiv -2C_\mu$) is shown for 2-D mean flows in the $\sigma \equiv \sqrt{II_S/2}, \omega \equiv \sqrt{-II_\Omega/2}$ plane for the Gatski & Speziale, Taulbee and Shih *et al.* models. These fail to closely mimick the behaviour corresponding to the ARSM equations, whereas the present model gives an exact ARSM representation for 2-D mean flows.

For homogeneous shear flow the asymptotic value of σ (and ω) is close to 3.0 (see Tavoularis & Corrsin 1981, TC hereafter), for which value the present model gives a perfect agreement with TC for $\mathcal{P}/\varepsilon$ and also a good agreement for a_{12}, whereas the rest of the anisotropy state is less well predicted (see table 1).

34

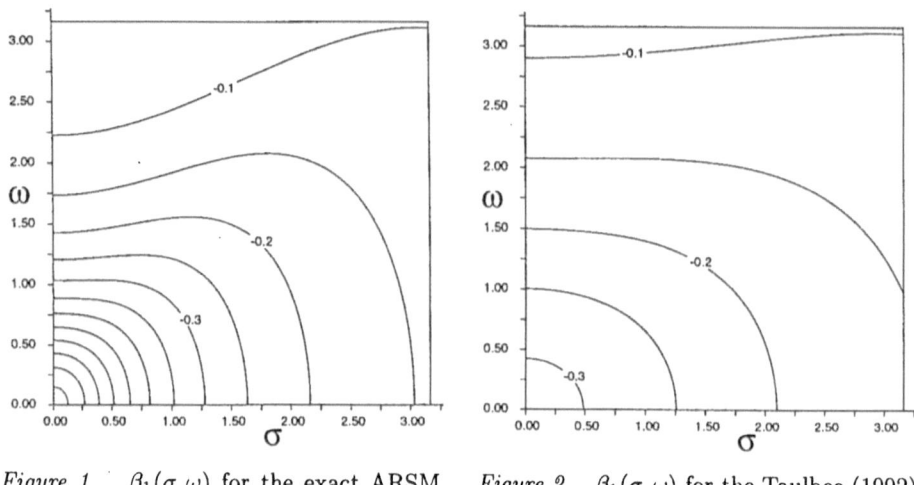

Figure 1. $\beta_1(\sigma, \omega)$ for the exact ARSM and present model

Figure 2. $\beta_1(\sigma, \omega)$ for the Taulbee (1992) model

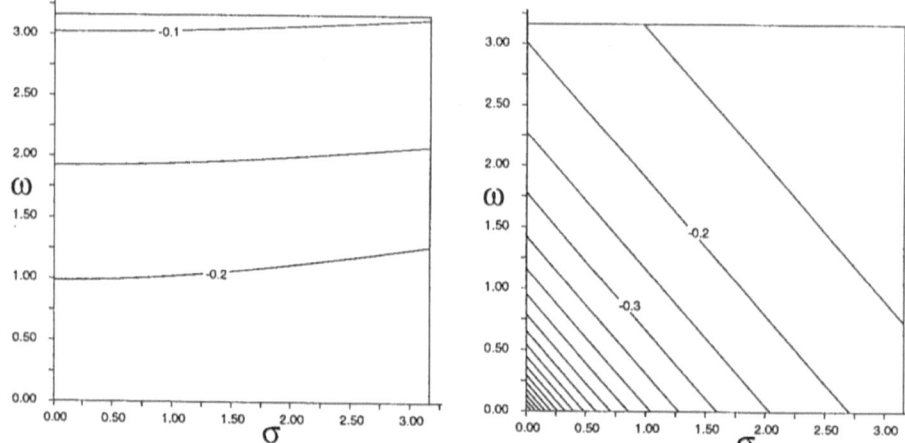

Figure 3. $\beta_1(\sigma, \omega)$ for the Gatski & Speziale (1993) model

Figure 4. $\beta_1(\sigma, \omega)$ for the Shih et al.(1994) model

TABLE 1.   Comparison between new model predictions and exp. results (TC) for hom. shear flow

	$\mathcal{P}/\varepsilon$	a_{12}	a_{11}	a_{22}	a_{33}
TC	1.8	-0.28	0.40	-0.29	-0.11
new model	1.8	-0.30	0.17	-0.17	0
$K - \varepsilon$ model	2.1	-0.43	0	0	0

EFFECTS OF STABLE STRATIFICATION UPON COHERENT VORTICES BEHIND A BACKWARD-FACING STEP

M. LESIEUR, B. FALLON and F. DELCAYRE

L.E.G.I.-Institut de Mécanique de Grenoble

BP 53, 38041 Grenoble-Cedex 9-France

1. Introduction

The stratified backward-facing step flow is a useful prototype configuration both in geophysical fluid dynamics (flow behind mountains for instance) and in nuclear engineering (heat exchangers in fast breeder reactors). The purpose of this paper is to use large-eddy simulations in order to characterize the dynamics of coherent structures shed behind the step. It will be done first in the non-stratified case. Then we will look at the effect of a stable stratification brought into the incoming flow.

2. Isothermal case

In the constant-density case, the nature of coherent vortices shed behind the step is still an open issue. Let $R = W/H$ be the geometric ratio, where W and H stand respectively for the outlet channel and step heights. In order to simulate an experiment developed by Jovic and Driver (1994), Le and Moin (1994) carried out a DNS in the case of a low step ($R = 6$, with an upper free-slip boundary and grid refinement at the lower wall). They could obtain a very good statistical agreement with experiments, provided the flow upstream of the step corresponds to a deterministic field obtained from LES of a boundary layer. The same was recovered by Akselvoll and Moin (1995) using various types of LES. None of them addressed topological features of the flow. LES of the backstep flow using the structure-function model (Métais and Lesieur 1992) had been previously carried out by Silveira et al. (1993), using a regular grid. For the high-step case ($R = 1.25$) and at a Reynolds $R_e = 6000$ (based on H), a flow regime was found where quasi two-dimensional Kelvin-Helmholtz vortices are shed downstream with longitudinal hairpin vortices stretched inbetween. In fact, the same calculations pushed for longer times show that the flow evolves into a dislocated pattern resembling the helical-pairing structure found in the tempo-

S. Gavrilakis et al. (eds.), Advances in Turbulence VI, 35-38.

ral mixing-layer simulations of Comte et al. (1992). More precisely, Figure 1-a represents low-pressure and streamwise vorticity in this simulation at $t = 70U_0/H$ (where U_0 is the incoming velocity). Large billows undergoing helical pairing are shed up to the reattachment, and then transform into weaker longitudinal Λ-shaped vortices. Since in this simulation the statistical agreement with the experiment of Eaton and Johnston (1980) is not very good for second-order quantities, we have redone with our code the simulation of Jovic and Driver's (1980) experiment, at $R_e = 5100$. We take the same grid-refinement at the wall as Le and Moin (1994). We use now the selective-structure model in its four-point formulation in planes parallel to the wall. No wall law is used. The upstream velocity profile is the same as the mean experimental velocity profile, perturbed by a weak 3D random white noise. Here, the mean quantities (velocity, rms velocity fluctuations, Reynolds stresses) are in good experimental agreement. Figure 1-b shows the low-pressure field, which indicates (as in Figure 1-a) the helical pairing of vortices before the reattachment. Vorticity maps downstream are similar to Figure 1-a.

Figure 1: LES of a non-stratified backstep; a) high step, regular grid, low-pressure and longitudinal vorticity; b) low step, grid refinement, low pressure.

3. Stratified case

We take the same code as used for the simulation of Figure 1-a (Boussinesq approximation, regular grid, $R = 1.25$. Reynolds number is 48.000, and Prandtl number 1. At the inlet, a constant velocity U_0 is imposed, with a stable temperature profile step (80% of hot fluid above, $\Delta\rho$ is the density difference). A wall law is used in the boundary layer. We define an upstream Richardson number $R_i = |\Delta\rho|gH/\rho_0 U_0^2$.

We have first carried out four LES using the structure-function model (six points), in order to study the effects of growing stratification on coherent vortices development. For $R_i = 0.25$, stratification does not affect significantly the dynamics. At $R_i = 0.5$, we observe a tendency towards two-dimensionalization upstream. Kelvin-Helmholtz vortices oscillate first in phase, then a strongly three-dimensional pairing occurs at $x \approx 7H$. At $R_i = 0.7$, the flow is now quasi two-dimensional. The pairing has been suppressed, and the reattachment length has become infinite. For larger Richardson numbers ($R_i = 1$), the flow loses any kind of vortices, and becomes laminar.

We present now LES using the selective structure-function model (SSF, six points) at $R_i = 0.7$, still with a regular grid. Visualization of the instantaneous vorticity field at t= 80 H/U_0 is provided on Figure 2. Compared with the structure-function (SF) case mentioned above at the same R_i, Kelvin-Helmholtz instability starts immediately behind the step, with quasi two-dimensional pairings. Further downstream, the mixing-layer growth is blocked. It is clear that the SSF model is here much more reliable than its SF counterpart, which dissipates too much the quasi two-dimensional fields induced by stratification.

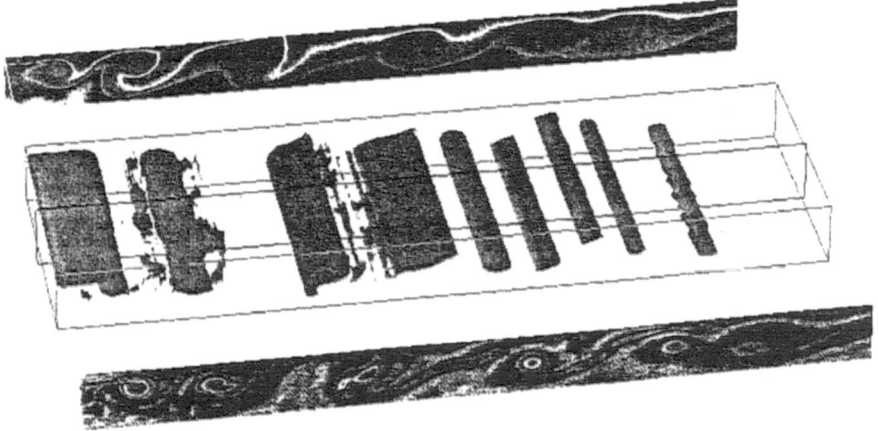

Figure 2: Stratified backstep (SSF model), high step, regular grid, vorticity modulus, together with temperature and vorticity cuts at mid planes.

We have also showed in this case evidence for two-dimensional baroclinic effects developing on density interfaces between two primary billows. This is particularly neat at $x = 7H$ on the vorticity cut of Figure 2. As is well known, Boussinesq approximation allows production of spanwise vorticity in the braids: the vorticity equation shows that a density interface tilted with respect to the gravity will create a baroclinic torque responsible for spanwise vorticity production. These baroclinic layers remain stable up to

$x \approx 7H$, and from there roll-up into secondary 2D vortices.

4. Conclusion

We have shown using LES that the coherent vortices behind a backward-facing step are, in the case of an isothermal flow, subject to helical pairing up the reattachment. Further downstream, they transform into weaker longitudinal Λ-shaped eddies. This result has been obtained for various configurations, subgrid models and computational grids.

Afterwards, a stable temperature step has been imposed upstream. By increasing the stratification, we show that it has a two-dimensionalizing effect on the vortices. However, their topology depends upon the subgrid-model chosen. The best results seem to be obtained with the selective-structure function model. For an intermediate stratification regime (upstream Richardson number of 0.7), one shows the occurrence of secondary baroclinic motions within the braids reconnecting the vortices.

This work was sponsored by CEA, CNRS and DRET.

5. References.

Akselvoll, K. and Moin, P., 1995, "LES of turbulent confined coannular jets and turbulent flow over a backward-facing step", Report TF-63, Stanford University.

Comte, P., Lesieur, M. and Lamballais, E., 1992, "Large and small-scale stirring of vorticity and a passive scalar in a 3D temporal mixing layer", Phys. Fluids, **A4**, pp 2761-2778.

Eaton, J.K. and Johnston, J.P., 1980, "Turbulent flow re-attachment: an experimental study of the flow and structure behind a backward-facing step", Stanford University, Rep. MD-39.

Jovic, S. and Driver, D.M., 1994, "Backward-facing step measurement at low Reynolds number $Re_h = 5000$", Nasa technical memorandum 108 807.

Le, H. and Moin, P., 1994, "D.N.S. of turbulent flow over a backward-facing step", Report TF-58, Stanford University.

Métais, O. and Lesieur, M., 1992, "Spectral large-eddy simulation of isotropic and stably stratified turbulence", J. Fluid Mech., **239**, pp 157-194.

Silveira-Neto, A., Grand, D., Métais, O. and Lesieur, M., 1993, "A numerical investigation of the coherent vortices in turbulence behind a backward-facing step", J. Fluid Mech., **256**, pp 1-25.

LARGE EDDY SIMULATION OF TURBULENT MIXING

F.MATHEY , J.P.CHOLLET

Laboratoire des Ecoulements Géophysiques et Industriels

BP. 53 X 38041 Grenoble

1. Introduction

The understanding of turbulent mixing is crucial in the context of chemical reactions. For example the knowledge of quantities like mixing rates are determinant in chemical engineering. Large eddy simulations are very efficient for the calculation of mixing at large scales since such scales are very dependent on initial and boundary conditions. For instance, in the mixing layers which are considered herein, the mixing between the two streams is quite well reproduced by computing explicitly the large scales. Because of the limited size of the computers, an explicit calculation of these large scales can not be run simultaneously with the calculation at smallest scales. Concerning the velocity field, these small scales can be modeled by an eddy viscosity concept, but flows with chemical reactions are typical of problems where mixing must be described at all scales . We consider here the linear model of Kerstein [1] used as a large eddy model. Using this model, we investigate the entrainment and mixing processes in a spatial three-dimensional turbulent mixing shear layer. The results from the simulation are discussed and compared with experimental results.

2. Large Eddy simulation

The large eddy fields, explicitly simulated in the LES, can be obtained by filtering the true fields. By applying this filter to the compressible Navier-Stokes equations, we obtain a set of equations describing the evolution of the resolved part of the flow. The unresolved quantities which appear in these equations are the subgrid stress tensor and the unresolved enthalpy flux term. These terms are modeled by an eddy-viscosity model. In the present simulation, we used the dynamical model of Germano[2], extended for compressible flows by Moin et al. [3]. In this model, the constants are dynamically estimated by introducing a test filter, with a larger filter width than the resolved grid filter. The main difficulty is then to evaluate the filtered reaction source term. This can be done with the Large Eddy Probability

S. Gavrilakis et al. (eds.), Advances in Turbulence VI, 39-42.

Density Function (LEPDF) approach [4] which consists in solving the large eddy PDF transport equation, or presuming this PDF [5]. The disadvantages of these methods is that there both requires the modelization of either molecular mixing terms or subgrid scalar fluctuations. The Linear Eddy Model (LEM) used here , developed by Kerstein , and extended for large Eddy Simulation by Menon et al.[6], will be compared in the last section with the presumed PDF approach. Because it solves all relevant length scales of the flow, from the integral lengthscale to the Kolmogorov scale, the LEM is able to separate turbulent stirring and molecular diffusion mechanisms which govern the evolution of a scalar field. This is realized through the consideration of a one dimensional domain, the linear domain, on which we solve a 1D diffusion equation. The turbulent stirring mechanism is modelized by a rearrangement process applied to the 1D-scalar field of the domain. This process consist of events, called triplet map, which rearranges the distribution of the scalar. In the large eddy simulation, the linear domain is the statistical representation of the sub-grid scalar fluctuations. The "splicing process" [6] is used to convect the mean and fluctuating scalar field across each cells according to the resolved velocity field. We have slightly improved this process by using the PPM advection scheme in order to ensure mass an species conservation.

3. Turbulent shear layer

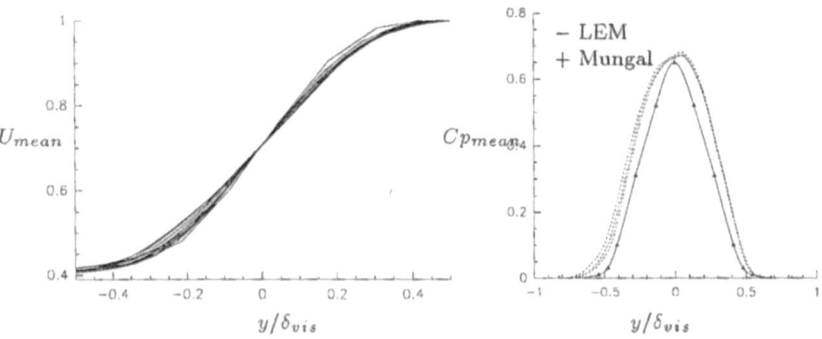

Figure 1: a) mean velocity profile b) mean product concentration profile

The turbulent shear layer investigated in this paper corresponds to the experimental investigations of Mungal et al. [7], and the recent numerical large eddy 2D-simulation of Calhoon et al. [8]. The reaction chemistry in the layer is assumed to be governed by the irreversible reaction $F+O-> P$, with the assumption of infinite rate chemistry. Equal diffusivities is assumed for the species and the passive scalar transported with this hypothesis is the Shvab-Zeldovich mixture fraction. The numerical resolution is 200x64x48 with a sub-grid domain containing 150 points. Normalized by the visual thickness of the layer (δ_{vis}), the mean velocity profile obtained from the simulation are found to be self-similar (figure 1.a). The mean concentration profiles across the layer are given figure (1.b), for different positions along the stream-wise axis, and are in good agreement with the experimental results. The figure (2.) displays the product thickness δ_p, defined from the relation $\delta_p = \int C_p(y)dy$ and normalized by δ_{vis}. On the same figure, we have plotted the experimental results of Mungal, The results are similar and show that the total amount of product formed in a shear layer is a decreasing function of the Reynolds number, as suggested by the Broadwell and Breidenthal (BBM) model of mixing [9].

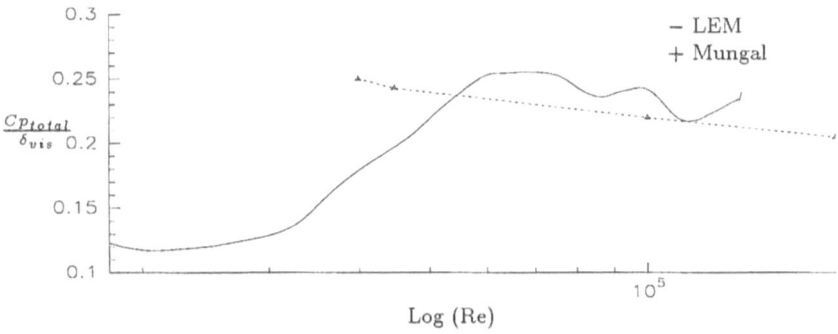

Figure 2: Product thickness versus the Reynolds number

4. Dynamical model

The three-dimensional shear layer calculation presented leads to a computational effort that can be very expensive. In order to provide a simple closure model for the sub-grid fluctuation, Cook and Riley [5] have suggested to evaluate the subgrid scalar fluctuation $\overline{\xi'^2}$, which is needed in the presumed LEPDF approach, according to a scale similarity model. We use the results obtained in the previous section as a data-base to test this formulation, and suggest an improvement, based on the the assumption that the subgrid scalar fluctuation scales like the resolved gradient. With this formulation the constants can be evaluated by the same kind of dynamical procedure used in the dynamical model of Germano. The results obtained in the previous section with the LEM are used as 'a priori' test in order to compare the two models, and the correlation are shown figure (3). The results are similar, but the dynamical formulation avoids the determination of any constants.

5. Conclusion

The application of the LEM model in this study demonstrate the ability of such a model to capture the trends observed in High-Reynolds number flows. The results

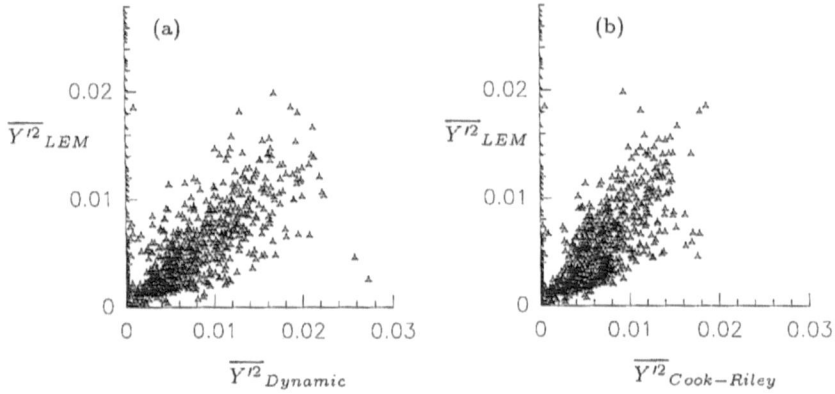

Figure 3: a) LEM versus the present model ; b) LEM versus Cook

obtained can play the role of a data base in order to test more simple closure assumption for the subgrid concentration fluctuations.

Acknowledgment

This work was supported in part by the SNECMA. The numerical calculations were carried out on the CRAY C98 of the I.D.R.I.S. (CNRS) and on the CRAY C94 of the CEA computing centers.

References

[1] Kerstein, A.R., 1988, "Linear-Eddy Model of Turbulent Scalar Transport and Mixing", Comb.Sci Tech.,60,pp.391-421.

[2] Germano, M., Piomelli, U., Moin, P., Cabot, W.H., 1991, "A dynamic subgrid-scale eddy viscosity model ", Phys.Fluids, A3, pp 1760

[3] Moin, P., Squires, K., Cabot, 1991, "A dynamic subgrid-scale model for compressible turbulence and scalar transport", Phys.Fluids, A3 (11), pp. 2746-2757.

[4] Gao, F., O'Brien, E., 1993, "A large-eddy simulation scheme for turbulent reacting flows" , Phys.Fluids, A 5 (6), pp 1282

[5] Cook, W., Riley, J.J., 1994, "A subgrid model for equilibrium chemistry in turbulent flows" , Phys.Fluids, Vol.6, No.8, pp 2868

[6] Menon, S., McMurtry,P.A., Kerstein,A.R., (1993). "A Linear Subgrid Model for Turbulent Combustion : Application to Premixed Combustion", AIAA paper 93-0107.

[7] Mungal, M.G., Hermanson, J.C., Dimotakis, P.E., 1985, "Reynolds Number effects on Mixing and Combustion in a Reacting Shear Layer", AIAA journal,Vol. 23, No. 9

[8] Calhoon, W.H., Menon, S., 1996, "Subgrid Modeling for Reacting Large Eddy Simulations", AIAA paper 96-0516.

[9] Broadwell, J.E., Breidenthal, R.E., 1982, "A simple model of mixing and chemical reaction in a turbulent shear layer", J.Fluid Mech. vol.125, pp. 397-410

STROUHAL NUMBER EVALUATION
BY 3D LATTICE GAS SIMULATIONS

J.P. RIVET

CNRS, Dept. G.D. CASSINI,
Observatoire de Nice, B.P. 229, 06304 Nice Cedex 04, France

1. Introduction

The lattice gas method is an alternative to classical "floating point based" simulation methods for fluid dynamics. It is particularly well suited to address fluid dynamical problems with complex physics (multiphase, multispecies, with chemical reactions, diffusion...), and/or complex geometry, such as open flows around obstacles of arbitrary shapes.

The basic idea of the method is to introduce a fictitious (fully discrete and boolean) microscopic structure for the fluid. Under some constrains, and after suitable limits are taken, the macroscopic behaviors of such a fictitious gas are identical to those of a real fluid (Frisch *et al.*, 1987).

The aim of our work is both to study numerically a realistic physical problem and to provide non-trivial quantitative validations for the lattice gas method by comparison with "real" laboratory experiments.

The physical problem is the unstationary incompressible three-dimensional viscous flow around a long cylinder with circular basis, at Reynolds numbers ranging between 40 and 140. In such regimes, the wake is known to display a succession of alternated vortices (Bénard-von Kàrmàn vortex streets) which detach from the obstacle with a Strouhal number (dimensionless frequency) that depends on the Reynolds number. Moreover, the wake can undergo spontaneous symmetry breaking leading to various fully three-dimensional vortex patterns, each one having a different Strouhal-Reynolds relationship.

S. Gavrilakis et al. (eds.), Advances in Turbulence VI, 43-46.

2. The numerical algorithm

The simulations were performed on a CRAY-C98, with the three-dimensional "FCHC" lattice gas model (d'Humières *et al.*, 1986). The collisions were done according to the so-called "FCHC-8" optimized rule due to M. Henon (Dubrulle *et al.*, 1990). The simulation domains had typically $512 \times 512 \times 128$ lattice nodes, and the space-averages were computed over small cells of $8 \times 8 \times 8$ nodes. The simulations have been running for 20000 to 40000 time steps. Figure 1 shows the typical geometry of such simulations.

3. Strouhal number evaluation

The numerical method to extract Strouhal numbers from lattice gas simulations is based on the same principle as the experimental one: at each time step, the transverse component of the local fluid velocity at some point in the wake is computed by space-averaging over a small cubic cell of $8 \times 8 \times 8$ lattice nodes. The time series so obtained is post-processed by Fourier analysis to get a first estimate of the fundamental wake frequency. Since the length of such time series is limited by the numerical resources needed, the estimate of f_0 by simple Fourier analysis is not sufficiently accurate. We thus use the numerical equivalent of an heterodyne detection to get a better estimate of the wake frequency.

To get the Strouhal number, we also need to know the $g(\rho)$ factor which is a lattice gas specific quantity that depends on the model used and on ρ, the average density of the fluid. This factor is roughly the ratio between the physical time and the lattice gas time step. In the context of lattice gases, the expression of the Strouhal number thus reads:

$$S_t = \frac{1}{g(\rho)} \cdot \frac{f_0 d}{U},$$

where d, u and f_0 are expressed in natural lattice units.

The accuracy of the result crucially depends on the care taken to evaluate the $g(\rho)$ factor, the upstream velocity U and the frequency f_0. The statistical noise due to the microscopic aspect of the lattice gas method can be evaluated to less than 5 % (typically 3 %).

Figure 2 shows a comparison between experimental results (Williamson, 1989) and lattice gas predictions. Empirical Strouhal-Reynolds relations (König *et al.*,) are also plotted on the same graph. The agreement is correct, except for a set of data that fall above the experimental values. These data correspond to a simulation where the obstruction effect due to the obstacle was too high.

4. Drag coefficient evaluation

The numerical evaluation of drag and lift coefficients requires to compute the momentum $\Delta \mathbf{P}$ exchanged per time step at the surface of the obstacle by fluid particles bouncing on it. In the lattice gas context, the expression of the drag coefficient is:

$$C_D = \frac{1}{a\,g(\rho)} \cdot \frac{\Delta P_z}{\frac{1}{2}\rho U^2 S},$$

where a is the volume of the primitive cell of the underlying lattice, ρ is the mass density per lattice node, U is the upstream fluid velocity, and S is the transverse section of the obstacle in lattice units.

Figure 3 shows a comparison between experimental results (Tritton, 1959) and lattice gas predictions. The agreement is correct, especially if one takes into account the natural dispersion of the experimental values.

5. Conclusion

The lattice gas method for three-dimensional incompressible fluid dynamics simulations is known to produce physically relevant qualitative results such as, for example, the spontaneous symmetry breaking of the translational invariance in the wake of a long cylinder (Rivet, 1991).

The numerical studies described above show that the lattice gas method also leads to quantitative predictions that are in correct agreement with experimental data. However, the precision presently obtained (a few percents) is still too low for a physically relevant exploration of the fine structure of the Strouhal-Reynolds relationship in a 3D wake (König *et al.*,).

References

Dubrulle, B., Frisch, U., Hénon, M. and Rivet, J.P. (1990) Low viscosity lattice gases, *J. Stat. Phys.* **59**, pp 1187–1226.

Frisch, U., d'Humières, D., Hasslacher, B., Lallemand, P., Pomeau, Y. and Rivet, J.P. (1987) Lattice gas hydrodynamics in two and three dimensions, *complex Systems* **1**, pp 649–707.

d'Humières, D., Lallemand, P. and Frisch, U. (1986) Lattice gas models for 3-D hydrodynamics, *Europhys. Lett.* **2**, pp. 291–297.

König, M., Eisenlohr, H. and Eckelmann, H. (1990) The fine structure of the Strouhal-Reynolds number relationship of the laminar wake of a cylinder, *Phys. Fluids A* **2**, pp. 1607–1614.

Rivet, J.P. (1991) Brisure spontanée de symétrie dans le sillage tri-dimensionnel d'un cylindre allongé, simulé par la méthode des gaz sur réseaux, *C.R. Acad. Sci. Paris II* **313**, pp. 151–157.

Tritton, D.J. (1959) Experiments on the flow past a circular cylinder at low Reynolds numbers, *J. Fluid Mech.* **6**, pp. 547–567.

Williamson, C.H.K. (1989) Oblique and parallel modes of vortex shedding in the wake of a circular cylinder at low Reynolds numbers, *J. Fluid Mech.* **206**, pp. 579–627.

46

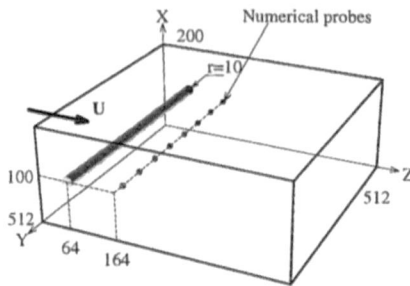

Figure 1. Typical geometry of lattice gas simulations.

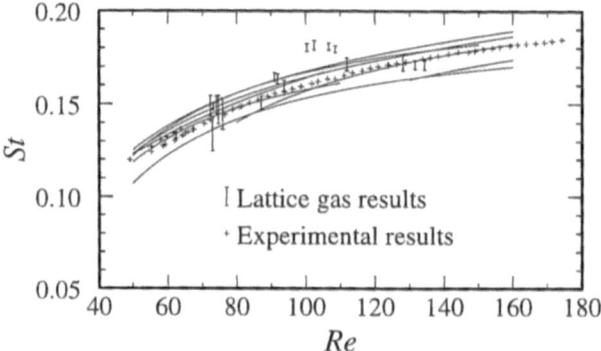

Figure 2. Strouhal number for a long cylinder *vs* Reynolds number. Continuous curves: empirical relations (König et al., 1990), + : experimental data (Williamson, 1989), error bars : lattice gas results (J.P. Rivet, 1995).

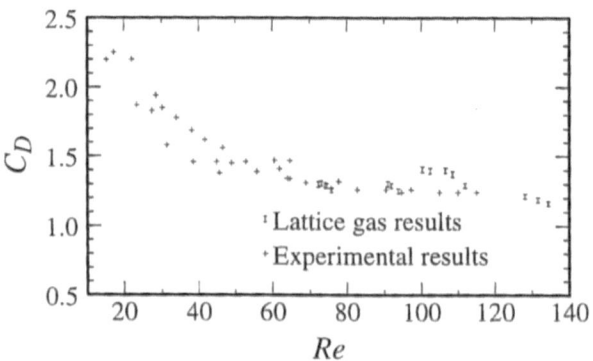

Figure 3. Drag coefficient for a long cylinder *vs* Reynolds number. + : experimental data (Tritton, 1959), error bars : lattice gas results (J.P. Rivet, 1995).

INSTABILITY CONSIDERATIONS FOR VELOCITY STREAKS IN NEAR-WALL TURBULENCE

N. D. SANDHAM
Department of Engineering
Queen Mary and Westfield College
Mile End Road
London E1 4NS, UK

1. Introduction

The dominant feature of turbulence close to a wall ($y^+ < 30$) is the streaky structure of the streamwise velocity field. Denoting the resulting flow pattern as $U(y, z)$ it is noted that the spanwise profiles $U(z)$ at a fixed distance y from the wall are inflectional, similar to periodic wake flow, while profiles $U(y)$ at fixed z may also have instantaneous inflection points. Instabilities resulting from such inflection points have been invoked by several authors (see the review by Robinson, 1991) as a fundamental part of the mechanism of near-wall turbulence. In particular Hamilton *et al.*(1995) use such instabilities to close a regeneration cycle for near-wall turbulence.

In the current paper the streak structure $U(y, z)$ is obtained from direct numerical simulation data. Stability calculations are made for profiles of $U(y)$ using the Orr-Sommerfeld equation, and for profiles of $U(y, z)$ using direct simulation with superimposed random disturbances.

2. Orr-Sommerfeld calculations

Figure 1 shows probability density functions (p.d.f.'s) of velocity at various distances from the wall, obtained from direct numerical simulation of turbulent channel flow at $Re_\tau = 180$ using a fully spectral method on a Cray T3D parallel computer. The p.d.f.'s have a wide, flat form in the sublayer, especially noticeable in the profiles at $y^+ = 8.8$ and $y^+ = 13.7$, reflecting the underlying streaky structure. An equation that approximately fits this presumed structure is

$$U^+(y, z) = \overline{U}^+(y) + cy^+ e^{-y^+/10} f(z^+), \tag{1}$$

S. Gavrilakis et al. (eds.), Advances in Turbulence VI, 47–50.
© 1996 *Kluwer Academic Publishers.*

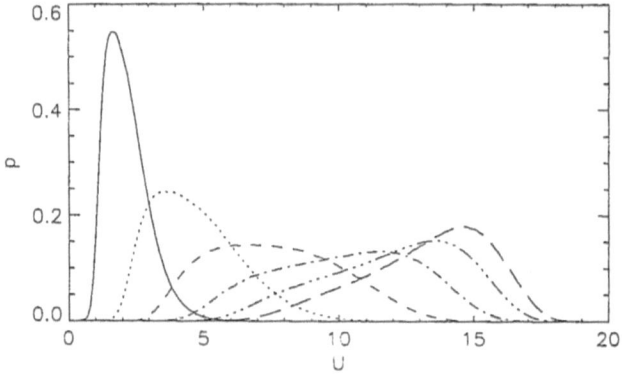

Figure 1. Probability density functions of streamwise velocity at various distances away from the wall. Reading from the left the curves are at $y^+ = 2.2$, 5.0, 8.8, 13.7, 19.6 and 26.5.

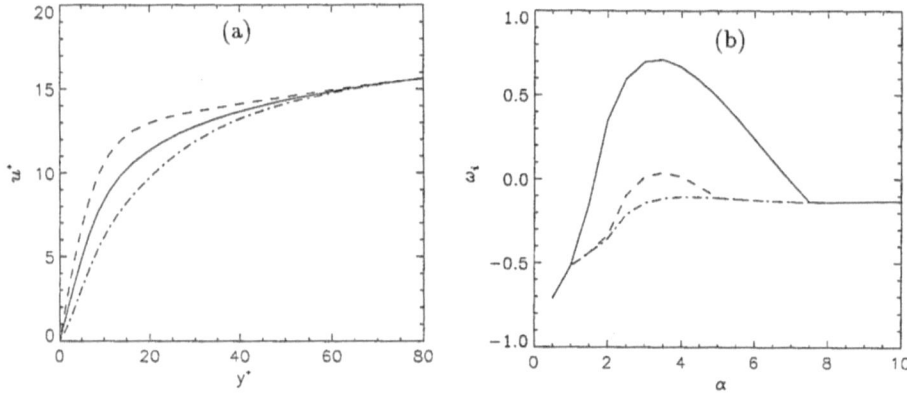

Figure 2. Streak properties from equation (1): (a) mean flow, low and high-speed streaks (b) linear amplification rate against wavenumber for values of $c = 0.6$ (chain dotted), $c = 1.95$ (dashed) and $c = 2.5$ (solid).

where $f(z^+)$ is a periodic function of z^+ varying between ± 1. Profiles of the mean, low-speed and high-speed streaks are shown on figure 2(a) for $c = 0.6$, which best fits the p.d.f. data. Temporal stability results of amplification rate ω_i against wavenumber α are shown on figure 2(b) for the low-speed streak profiles with various values of c. For $c = 0.6$ the profiles are stable. Much higher values of c are required for marginal instability ($c = 1.95$) and for strong instability ($c = 2.5$), but these values give unrealistic velocity perturbations that are not seen in the direct simulations.

An alternative approach of conditional sampling has been used to extract streak profiles directly from direct simulations (in this case small-domain calculations were used with turbulence statistics within 1% of the large domain calculations). The wall value of dU/dy was used for the con-

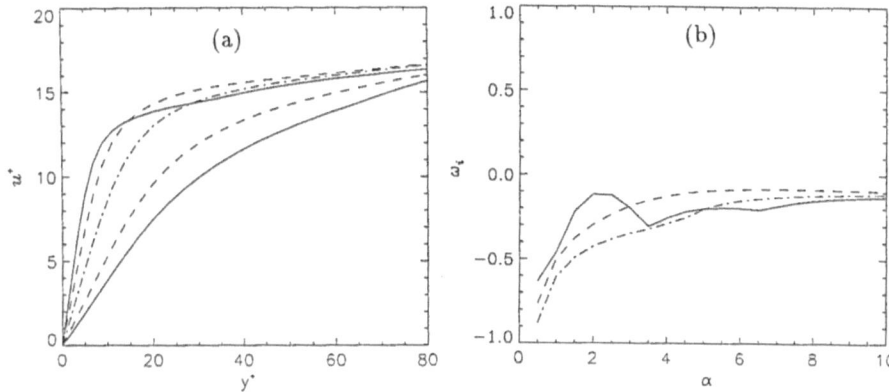

Figure 3.   Streaks deduced from conditional sampling: (a) profiles corresponding to fraction of area under the $(dU/dy)_w$ p.d.f. of 1% and 99% (solid), 10% and 90% (dashed) and 50% (chain dotted); (b) linear amplification rate ω_i against wavenumber α for 1% (dashed), 10% (chain dotted) and 99% (solid) profiles.

ditioning. Velocity profiles extracted in this way are shown on figure 3(a), with stability calculations of selected profiles on figure 3(b). The streak structure is well captured by this approach. In fact the 'organised' motion contains about 50% of the turbulence production for $y^+ < 40$. However none of the profiles has a strong inflection point and the stability calculations show that all the profiles are stable.

3.   Direct simulations with random noise forcing

The above work has shown that profiles $U(y)$ are stable. To consider the inflection points in the $U(z)$ profiles, direct numerical simulations with superimposed random noise have been carried out. These simulations were carried out with a forcing term added to the Navier-Stokes equations to preserve the streak structure given by equation (1). The streak shape was given by $f(z) = \cos(4\pi z^+/L_z^+)$ with $L_z^+ = 200$, which gives two streaks in the box to allow for possible subharmonic effects. The streamwise box length was $L_x^+ = 900$. Random noise was added to all velocity components and the growth of various Fourier modes of the kinetic energy $E(k_x, k_z)$ was monitored at $y^+ = 33$. Figure 4 shows results for $c = 0.6$ and $c = 2.5$. The former case is stable and the latter unstable, with a stability boundary at around $c = 2.4$. Other simulations without a mean $U(y)$ have been carried out. These have shown that the presence of $U(y)$ has a stabilising effect on inflection points in the $U(z)$ profiles. The general conculsion is that streamwise velocity streaks by themselves are linearly stable entities at the typical amplitudes found in near-wall turbulent flow.

50

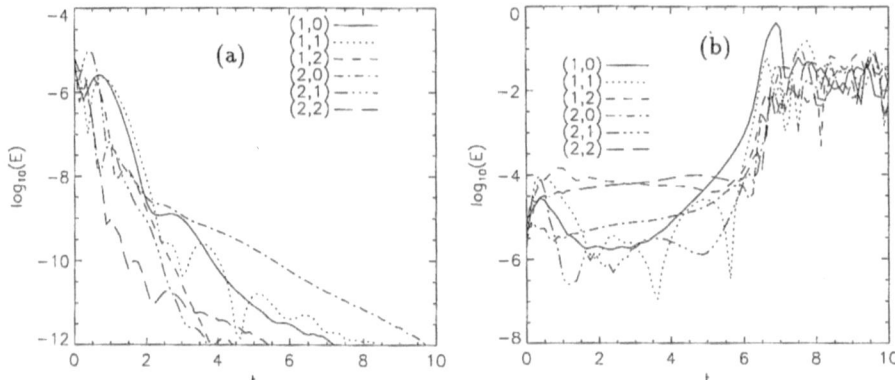

Figure 4. Kinetic energy in various Fourier modes at $y^+ = 33$ for simulations of streaks with superimposed random noise. Streak amplitude (a) $c = 0.6$ and (b) $c = 2.5$.

4. Discussion

Both methods used to investigate the instability of streaks in near-wall turbulence have shown that typical profiles $U(y, z)$ are stable. Besides the stabilising effect of $U(y)$, profiles with instantaneous inflection points may be stable or only very weakly unstable due to the low Reynolds number. As an example, working in non-dimensional wall variables, a shear layer with a thickness of 10 and a velocity difference of 2 has a local Reynolds number of only 20. Thus, one needs to take care in applying linear instability ideas to profiles in near-wall flows just because they may be inflectional.

An alternative feature that closes a regeneration cycle for near-wall turbulence is the nonlinear mechanism of vorticity stretching due to the presence of significant V-velocity in some portions of the low-speed streaks. This generates intense local shear layers above and at the edges of low-speed streaks which roll up into vortices deterministically (*i.e.* without the presence of background forcing). This regeneration mechanism was discussed in detail by Sandham and Kleiser (1993) with reference to end-stage transition as well as fully-developed turbulence near a wall.

Acknowledgement. Supercomputer time for this work was provided under EPSRC Grant GR/K43957.

References

Hamilton, J.M., Kim, J. and Waleffe, F. (1995) Regeneration mechanisms of near-wall turbulence structures, *J. Fluid Mech.*, **287**, 317–348

Robinson, S.K. (1991) Coherent motions in the turbulent boundary layer, *Ann. Rev. Fluid Mech.*, bf 23,601–639

Sandham, N.D. and Kleiser, L. (1993) The late stages of transition to turbulence in channel flow, *J. Fluid Mech.*, **245**, 319–348

Reynolds stress budgets of low Reynolds number pipe expansion flow

C. Wagner and R. Friedrich

Lehrstuhl für Fluidmechanik
Munich University of Technology
Arcisstr. 21, 80333 München, Germany

1. Introduction

The turbulent flow in a sudden pipe expansion is of theoretical and practical importance. Although the flow domain is quite simple and the separation point is fixed at the step, the flow itself is complex and its accurate prediction remains a challenge for turbulence modellers still today. Its complexity is due to the appearence of a mixing layer, a recirculation and a reattachment zone. Flow of this type occurs in engines, heat exchangers, sudden-expansion dump combustors and numerous other applications so that the understanding of its physics is a prerequisite of efficient engineering design and theoretical prediction.

Only few detailed experimental studies of turbulent sudden pipe expansion flow have been performed. Devenport and Sutton [1] e.g. used pulsed hot wires to investigate the near wall separated flow, while Papadopoulos et al. [2] performed precise LDA measurements by fully matching the refractive index of the working fluid with that of the glass test section.

Aim of the present computational data is to complement the experimental ones by forming a data base on the budgets of the Reynolds stress tensor which should be of considerable value for improvement of turbulence models. The results are obtained from direct numerical simulation of sudden pipe expansion flow for an expansion ration of $ER = 1.2$ and a Reynolds number of the incoming fully developed flow of 6950 (based on diameter and mean centerline velocity).

2. Numerics

DNS was performed with a finite volume technique that integrates the incompressible Navier-Stokes equations on a cylindrical (z, φ, r) grid system. The scheme is essentially second order accurate and energy conserving. Staggered grids and a direct Poisson solver for the pressure (FFT in φ-direction and the influence matrix technique) allow for an accurate integration of the mass balance. The reliability of the numerical method was demonstrated for fully developed pipe flow by comparison with experimental and other DNS data [4]. The computational domain in the present simulation comprises an upstream pipe section of diameter D and length 1.48D and a downstream section of diameter 1.2D and length 2.03D. An

S. Gavrilakis et al. (eds.), Advances in Turbulence VI, 51–54.

equidistant grid consisting of $180 \times 128 \times 115$ cells in (z, φ, r)-direction guarantees a proper resolution of all turbulent scales.

The boundary conditions for the inflow plane are obtained from an extra DNS of fully developed non-swirling pipe flow at a Reynolds number of 6950 or 360 when based on the friction velocity and pipe diameter. For statistical averaging 1050 independent realizations with a time lag of 1/20th of an eddy-turnover time D/u_τ have been taken, requiring 1500 CPU hours on a CRAY-YMP 8/864 and 30MW of memory. For more details, cf. [3].

3. The Reynolds stress components and their budgets

With the exception of the mean axial velocity in fig.1, which is made dimensionless with the centerline velocity of the incoming flow, all other velocity components are scaled to the inlet friction velocity. Position $z/D = 0.1$ represents fully developed pipe flow, while positions $z/D = 1.8$ and $z/D = 2.2$ are within the recirculation zone. (Positions of sudden expansion: $z/D = 1.48$ and mean reattachment: $z/D = 2.49$). The mean axial velocity is in good agreement with PIV data of Brouillette [5] for exactly the same flow. Profiles of the axial turbulence intensity and the Reynolds shear stress are presented in fig. 2 for the positions $z/D = 0.1$ and $z/D = 2.2$. They show that the highest turbulence activity occurs in the mixing layer. The maxima of the other two stress components, at any streamwise location, occur in the mixing layer as well (not shown). The absolute maxima of all components are found just before the reattachment, however at different positions ($(z/D)_{u_{z\,max}} \approx 2.25$). The budgets for each Reynolds stress component are plotted in figures 3-6 for the two locations, $z/D = 0.1$ and 2.2. No least-square fit has been applied to the budget terms which are normalized by $u_\tau^3(z = 0)/D$. Close to the entrance plane ($z/D = 0.1$) all terms in figs 3a-6a are very similar to those of turbulent channel flow (cf. [4]). Roughly 3 step heights prior to reattachment the longitudinal stress budget (fig. 3b) is dominated by production (PR) and redistribution effects in the mixing layer region (VPG is the sum of pressure strain PS and pressure diffusion PD terms). While PR gets negative near the wall because of the reverse flow, VPG becomes positive. In contrast to simple shear flow (fig. 3a), the VPG term

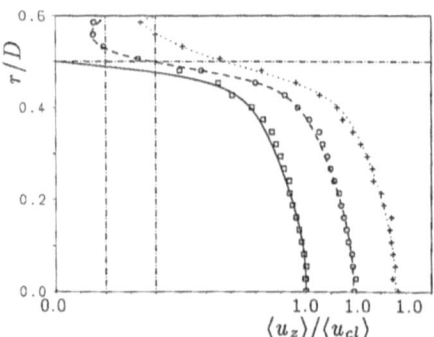

Fig.1 Mean axial velocity. Lines as in fig.2. Symbols: PIV-data of Brouillette [5].

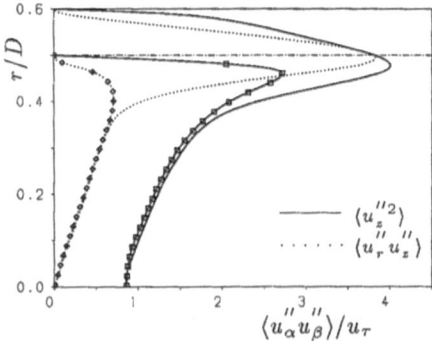

Fig.2 Profiles of axial turbulence intensity and Reynolds shear stress. Lines with/without symbols represent position $z/D = 0.1/2.2$

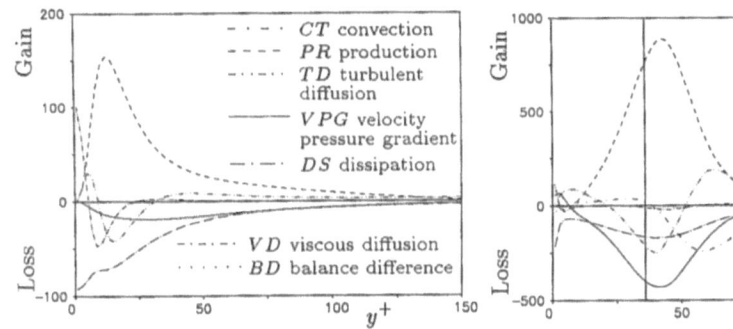

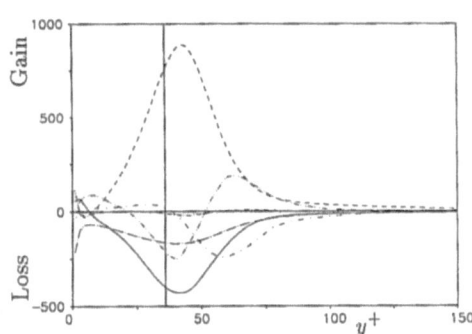

Fig.3a Budget of the Reynolds stress component $\langle u_z''^2 \rangle$ at $z/D = 0.1$.

Fig.3b Budget of the Reynolds stress compo-
nent $\langle u_z''^2 \rangle$ at $z/D = 2.2$. Lines as in fig.4a

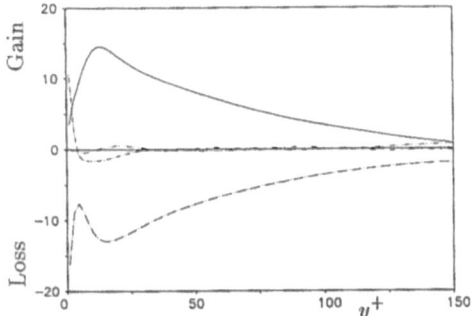

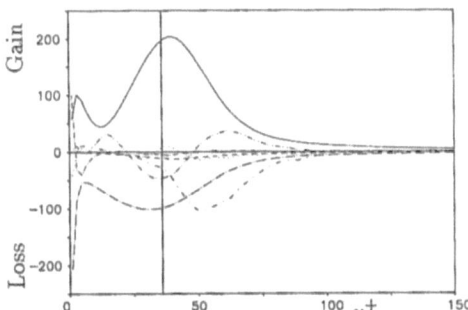

Fig.4a Budget of the Reynolds stress compo-
nent $\langle u_\varphi''^2 \rangle$ at $z/D = 0.1$.Lines as in fig. 4a

Fig.4b Budget of the Reynolds stress compo-
nent $\langle u_\varphi''^2 \rangle$ at $z/D = 2.2$. Lines as in fig.4a

is about two times as large in the free shear layer as the dissipation (DS) term
which peaks at the same location as PR. Turbulent diffusion (TD) and convection
(CT) are equally important. The u_φ''-budget in fig. 4b reflects the significance of
redistribution (PD is everywhere small) and dissipation of fluctuating energy.

In the high-speed region of the free shear layer CT and TD largely contribute to
the balance. In contrast to simple shear flow (fig. 4a) the near-wall peak in VPG
is six times as large and due to energy received from the radial component. The
radial stress budget (fig. 5b) must be interpreted with some caution, because the
balance difference BD is not zero everywhere. The amplitudes are, however, small
enough to allow for the following comments: VPG, turbulent diffusion, dissipation
and convection effects are dominant in the mixing layer zone. Near the wall,
dissipation and VPG effects essentially balance turbulent diffusion. Compared
to position $z/D = 0.1$ all terms (except CT, PR) are by an order of magnitude
larger. The budget of the Reynolds shear stress (fig. 6b) although affected by some
non-zero BD within the shear layer is, like the u_z''-budget, dominated by PR and
VPG terms. Viscous dissipation is small, but TD and CT are significant, at least
on the high-speed side.

An answer to the question why the longitudinal energy production (and likewise
PR of $\langle u_z'' u_r'' \rangle$) increases by nearly an order of magnitude within the mixing layer
compared to its maximum in the buffer layer upstream is quickly found by inspec-

54

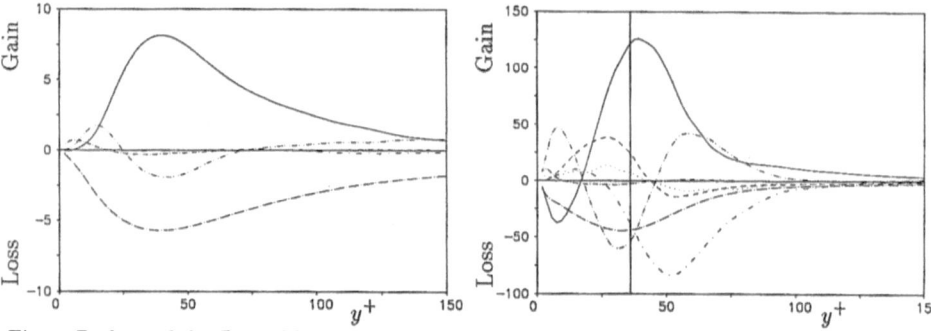

Fig.5a Budget of the Reynolds stress component $\langle u_r''^2 \rangle$ at $z/D = 0.1$. Lines as in fig.4a

Fig.5b Budget of the Reynolds stress component $\langle u_r''^2 \rangle$ at $z/D = 2.2$. Lines as in fig.4a

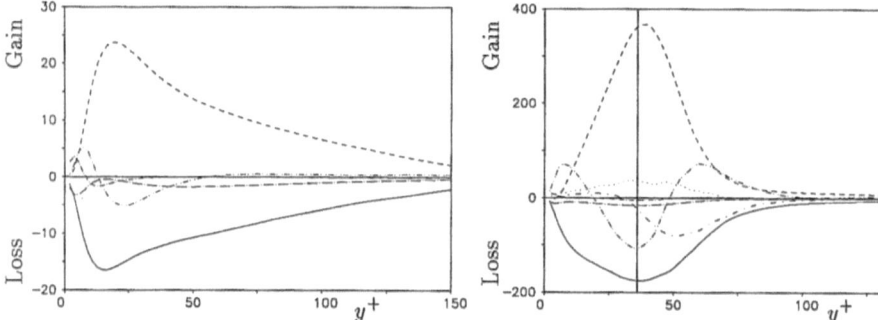

Fig.6a Budget of the Reynolds stress component $\langle u_r'' u_z'' \rangle$ at $z/D = 0.1$. Lines as in fig.4a

Fig.6b Budget of the Reynolds stress component $\langle u_r'' u_z'' \rangle$ at $z/D = 2.2$. Lines as in fig.4a

tion of the term itself:

$$PR_{zz} = -2\frac{\partial \langle u_z \rangle}{\partial r}\langle u_r'' u_z'' \rangle - 2\frac{\partial \langle u_z \rangle}{\partial z}\langle u_z'' u_z'' \rangle. \tag{1}$$

The second term on the r.h.s need not be discussed because it diminishes the magnitude of PR_{zz}. Since the velocity gradient is reduced in the mixing layer, the increase in PR_{zz} must be due to the rapid growth of $\langle u_r'' u_z'' \rangle$. This in turn can be explained by the fact that the radial velocity fluctuations relax from the impermeability constraint of the wall upstream. A DNS of the same flow with perfectly permeable walls is under way to demonstrate this.

References

[1] Devenport, W.J., Sutton, E.P. (1993): An experimental study of two flows through an axisymmetric sudden expansion. Exp. in Fluids, 14, pp. 423-432.

[2] Papadopoulos, G., Lekakis, I., Durst, F. (1996): Wall pressure and velocity characteristics of axisymmetric sudden-expansion flow. To appear in: Exp. in Fluids.

[3] Wagner, C. (1996): Direkte numerische Simulation turbulenter Strömungen in einer Rohrerweiterung. VDI Fortschrittsbericht Nr. 283, Reihe 7.

[4] Eggels, J.G.M., Unger, F., Weiss, M.H., Westerweel, J., Adrian, R.J., Friedrich, R., Nieuwstadt, F.T.M. (1994): Fully developed turbulent pipe flow. A comparison between direct numerical simulation and experiment. J. Fluid Mech. 268, pp. 175-209.

[5] Brouillette, J.N. (1994): Incompressible turbulent flow through a sudden axisymmetric expansion using particle image velocimetry, Master thesis, University of Illinois, Urbana-Champaign, Illinois, USA.

NUMERICAL STUDY OF MECHANISMS OF BOUNDARY-LAYER TRANSITION AFTER A SEPARATION BUBBLE

ZHIYIN YANG AND PETER R VOKE

Department of Mechanical Engineering, University of Surrey, Guildford GU2 5XH, U.K.

Abstract. Large-eddy simulation has been employed to study boundary layer transition on a flat plate immediately following a separation bubble induced by a change of curvature of the surface. The geometry is a flat plate with a semi-circular leading edge. The simulated results are compared with the available experimental data. The separation bubble at the leading edge is unstable and plays an important role in triggering transition for low or zero free stream turbulence. Data from the simulation has been gathered which allows us to extract all terms in the derived equations for the Reynolds stresses. Aspects of these balances are presented which may allow new insights into the physical mechanisms at work.

1. Introduction

It is well established that there are two kinds of transition on a flat plate without separation, i.e. natural and bypass transition. Large-Eddy Simulation (LES) and Direct Numerical Simulation (DNS) have become powerful tools for investigating boundary layer transition due to rapid advances of supercomputer power and numerical techniques. We have applied LES successfully to bypass transition of the boundary layer on a flat plate with zero pressure gradient [1], and DNS have been performed of natural boundary layer transition. However, little simulation work has been done on boundary layer transition with leading edge shapes which cause a separation bubble. It is suspected that the transition mechanism is different in this case and the present study helps to elucidate the mechanisms involved in such a case.

S. Gavrilakis et al. (eds.), Advances in Turbulence VI, 55-58.

2. Methods

The numerical methods employed for the present LES are direct descendants of well-known finite-volume techniques successfully used for many high-Reynolds-number LES studies, and recently for flat plate bypass transition [1]. The 2D curvilinear geometry requires the use of general coordinates in the x and y dimensions, though the simplicity of the geometry in the present case means that the coordinate system in the (x, y) plane can be orthogonal. The fully covariant general Navier-Stokes equations are discretised on a staggered mesh, using forms of the equations that employ the Jacobian of the transformation so that the advection term is relatively simple and involves no connection coefficients.

The computational box is nominally $16d \times 7.5d \times 2d$, curving around the bottom, front and top of the plate; d is the leading edge diameter, equivalent to 10 mm in an experiment. The meshing used in the simulation was $264 \times 72 \times 64$ stretched in the y direction away from the wall and also in the x direction to ensure sufficient resolution in the separation bubble; Δx^+ varies from 5.6 to 86, Δy^+ varies from 0.7 to 137 and $\Delta z^+ = 13.3$. (The wall units are based on the friction velocity at $x = 7d$ from the blend point where the cylindrical leading edge meets the flat plate.) The Reynolds number is 3333 based on the uniform inflow velocity of 5 m/s and the leading edge diameter of 10 mm.

The subgrid-scale model used is based on a Smagorinsky model modified to allow for low-Reynolds-number viscous effects. We find that using a single value for the Smagorinsky constant C_s in the whole computational domain does not give satisfactory results since there are two different flow regions. Several tests have been conducted for this geometry: a) Using a single value of $C_s = 0.1$; b) Using two values, $C_s = 0.23$ up to reattachment and $C_s = 0.1$ beyond; c) Using a variable C_s changing from 0.27 at separation to 0.1 beyond reattachment; d) Using a dynamic subgrid-scale model with the Smagorisky model as a base.

We find that the third of these methods gives the most satisfactory behaviour of the separation bubble, reattachment and transition, and the results presented in this report are extracted from this simulation. Further work using the dynamic approach is required to verify its utility for this flow.

3. Results

The simulation is started from an undisturbed mean flow with a very small level of pseudo-random disturbance at the inflow which is removed from the simulation before the first signs of hydrodynamic instability start to appear.

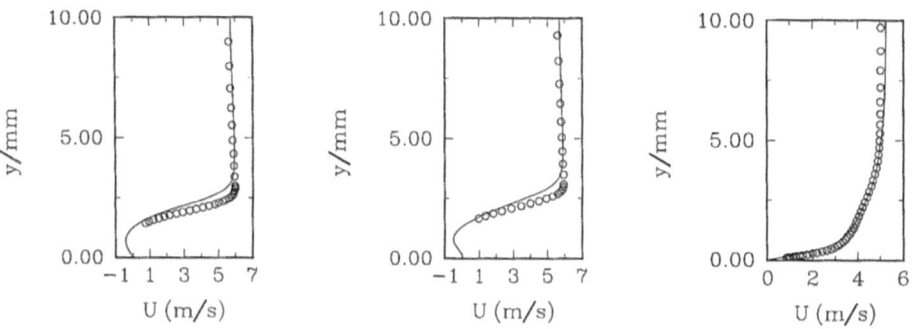

Figure 1. Mean U profiles at $x/x_r = 0.44$ (left), 0.67 (middle) and 1.64 (right). Solid line, LES; symbols, experimental data.

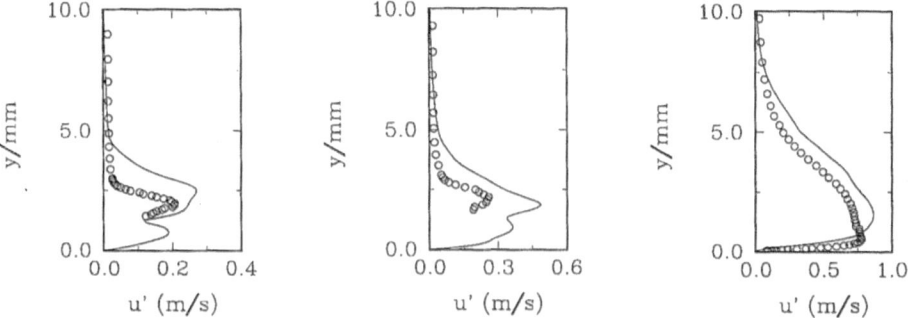

Figure 2. R.m.s fluctuation u' profiles. As Figure 1.

Figure 1 shows the comparison between the simulated mean streamwise velocity profiles and experimental data [2] at three streamwise stations, two in the bubble and one after the separation. Separation occurs immediately at the blend point and the streamwise distance is normalised by the bubble length. As can be seen from the figure good agreement has been obtained at all three stations and by $x/x_r = 1.64$ the numerical results and the experimental data both show a developed turbulent boundary layer velocity profile.

Figure 2 shows the simulated streamwise rms intensity and the corresponding experimental data at the same stations. Numerical results compare well with the experimental data at all stations apart from the near wall region where the numerical results indicate another peak. The double peaks predicted by the simulation within the separation bubble are genuine, not due to lack of statistical samples, but arise from two different sources, one in the free shear layer and another in the wall layer. This can be confirmed from the Reynolds stress balances.

58

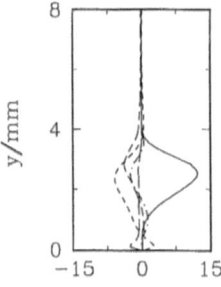 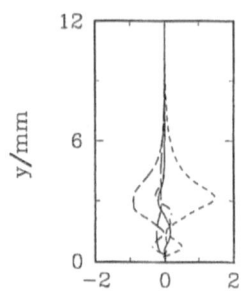 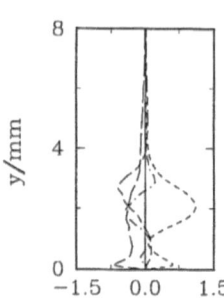

Figure 3. Terms in the balance of u'^2 (left), v'^2 (middle) and w'^2 (right) at $x/x_r = 0.76$. Solid line, production; alt dash, convection; dot dash, turbulent transport; short dash, pressure; long dash, viscous.

Figure 3 shows the Reynolds stress balance terms for u'^2, v'^2 and w'^2 at $x/x_r = 0.76$. Production of u'^2 is balanced mainly by the pressure term which redistributes u'^2 into v'^2 and w'^2. The initial v' comes from the unstable free shear layer, which provides a rich source of stimulus for the production of $u'v'$ and u'^2 fluctuations.

4. Conclusions

It is evident that the transition process initiates with 2D instability of large-scale unstable free shear layer in the separation bubble. Three-dimensional motions follow rapidly, in which pressure redistribution of streamwise disturbances into wall-normal and spanwise disturbances plays a key role. Full transition occurs around the reattachment point in a way similar in certain respects to bypass transition under free-stream turbulence though in this case the external disturbances originate in the large-scale instability of the separated shear layer. The boundary layer develops into a fully turbulent state a short way downstream.

5. Acknowledgements

The authors gratefully acknowledge the support of Rolls-Royce plc and U.K. EPSRC for this research under grant number GR/J 07334. The supercomputer time was provided on the Y-MP8 by DRAL.

References

1. Voke, P.R. and Yang, Z.Y.: Numerical study of bypass transition, *Phys. Fluids* **7** (1995), 2256-2264.
2. Coupland, J.: Private Communication (1994).

INVESTIGATING FLOW TOPOLOGIES IN BOUNDARY LAYER TRANSITION BY NUMERICAL SIMULATION

J. BULBECK AND M.S. CHONG
Department of Mechanical and Manufacturing Engineering
University of Melbourne, Parkville, Vic 3052, Australia

AND

J. SORIA
Department of Mechanical Engineering
Monash University, Clayton, Vic 3168, Australia

1. Introduction

In this work, the flow topologies in the early stages of transition in a boundary layer are determined and compared with flow topologies from other simulations of turbulent and transitioning flows (e.g. Soria *et. al* (1994)). Determination of the flow topology at a given point in the flow field requires determination of the invariants P, Q, and R of the velocity gradient tensor, $a_{ij} = \partial u_i / \partial x_j$ at that point.

2. Numerical Method

In order to calculate a_{ij} for boundary layer transition, a numerical method has been developed to simulate both spatially and temporally-evolving incompressible boundary layer flows. The equations to be advanced in time are derived by taking the curl of the incompressible Navier-Stokes equations twice, and retaining the wall-normal components, giving evolution equations for the wall-normal vorticity and the Laplacian of wall-normal velocity. At the end of each time-step, the remaining components of velocity and vorticity are recovered using the continuity equation and definitions of vorticity.

For temporal simulations, the flow is assumed periodic in the streamwise direction. In the spatial case high order compact differences are used in the streamwise direction. In the wall-normal direction, the semi-infinite domain is truncated at six displacement thicknesses from the boundary, and compact differences are used on an equispaced grid. The flow is assumed periodic in the spanwise direction, and a Fourier spectral method is used in this direction.

Time-stepping is implicit (Crank-Nicholson) for the wall-normal viscous terms, and explicit (third-order Runge-Kutta) for all other terms.

The numerical method has been tested by comparing growth rates of small two- and three-dimensional disturbances with growth rates calculated from linear stability theory. They are found to agree to within fractions of a percent.

S. Gavrilakis et al. (eds.), Advances in Turbulence VI, 59-60.
© *1996 Kluwer Academic Publishers.*

60

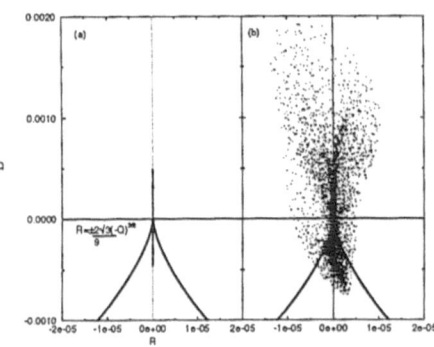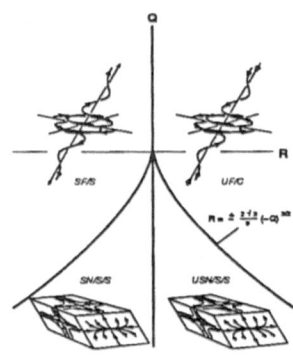

Figure 1. Distribution of QR *(a)* initially and *(b)* shortly before the first spike stage of transition. Also shown are flow topologies to which QR combinations correspond.

3. Results and Discussion

Results are presented for a temporal simulation with $Re_{\delta*} = 700$, with a streamwise wavelength $\lambda_x = 23.27$ and a spanwise wavelength $\lambda_z = 15.71$. A 2D Tollmien-Schlichting wave of amplitude 0.015 and a pair of oblique waves of amplitude 0.001 are the initial conditions. The mode shapes of the waves are calculated from linear stability theory. These parameters are the same as used by Laurien & Kleiser (1987). The grid comprises 32x48x16 points in the streamwise, wall-normal and spanwise directions respectively. The simulation is run until the first spike stage of transition at $t = 515$.

Figure 1(a) shows the distribution of the invariants Q and R at the initial condition. The points virtually lie on the Q-axis indicating that the flow is essentially two-dimensional. At a time just before the first spike stage of transition (Figure 1(b)), the flow has become three-dimensional. Concentrations of points are seen in the UNSS region, which is also observed in other turbulent simulations (Soria *et. al* (1994)). However, the tendency, observed in other simulations, of points to cluster in the SFS region is not observed with the current dataset. Instead, there is a weak tendency for points to cluster into the UFC region, indicating that many points in the domain are experiencing compressed vortical flow at this stage of the simulation.

4. Conclusion

A numerical method has been developed and used to simulate transitioning boundary layer flow. The velocity fields produced by the simulation have been analysed to determine the invariants of the velocity gradients tensor and to classify the topology at every point in the flowfield.

Acknowledgements

The leading author (J. Bulbeck) is very grateful to the Air Operations Division, Aeronautical and Maritime Research Laboratory for financial support of this research.

References

Laurien E., Kleiser, L (1989) Numerical Simulation of boundary-layer transition and transition control, *J. Fluid Mech.* **199**, 403-440.

Soria J., Sondergaard R., Cantwell B.J., Chong M.S., Perry A.E. (1994) A study of the fine-scale motions of incompressible time-developing mixing layers, *Phys. Fluids* **6(2)**, 871-884.

ASSESSMENT OF $K-\epsilon$ AND REYNOLDS-STRESS
TURBULENCE MODELS IN A NINETY-DEGREE PIPE BEND

D. COKLJAT
*CCLRC, Daresbury Laboratory, TCS Division
Warrington WA4 4AD, UK.*

AND

C. KRALJ
*UMIST, Mechanical Engineering Department
P.O. Box 88, Manchester M60 1QD, UK.*

1. Introduction and Methodology

In this paper predictions by the standard $k-\epsilon$ turbulence model and a full Reynolds-stress model (RSM) in a 90° pipe bend are compared with the experimental data of Enayet et al.(1982).

The solution procedure is based on a finite volume approach with co-located variable arrangement and segregated solution strategy. Pressure and velocity are coupled via the SIMPLE algorithm and the special Rhie and Chow (1983) interpolation practice is employed to avoid their decoupling. A similar practice is used to couple the mean velocity and turbulence fields in the RSM, Basara and Cokljat(1995). The pressure-strain term in the RSM is modelled according to Speziale et al.(1991), which allows for the wall-induced effects without the use of ad-hoc wall-damping terms.

2. Results

The computational domain, Figure 1, is split into 80000 hexahedral cells with 40x40 cells in the cross sections. Figure 2 depicts the pressure driven secondary flow (RSM results) at one pipe diameter downstream of the bend exit plane. The contours of the mean velocity, in cross section at 75° downstream from the bend inlet plane, are shown in Figure 3. The results obtained by RSM model show better agreement with the measurements than those of the $k-\epsilon$ model. This is particularly so for the extent of the mean-velocity deformation imposed by the pressure-driven secondary flow. The

S. Gavrilakis et al. (eds.), Advances in Turbulence VI, 61-62.

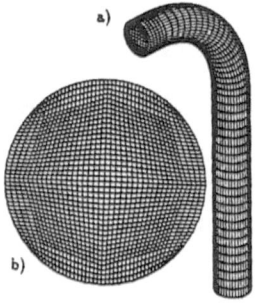

Figure 1 Computational grid.

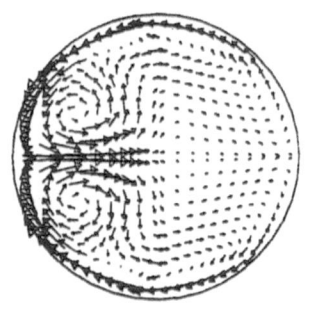

Figure 2 Secondary flow at L=d.

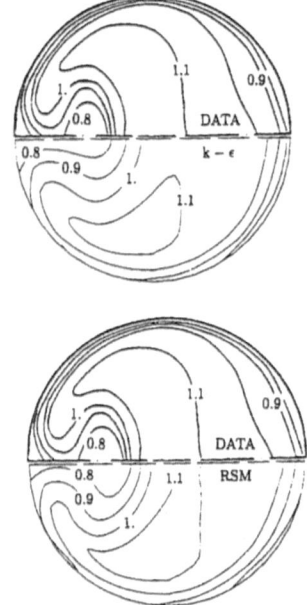

Figure 3 Mean-velocity contours.

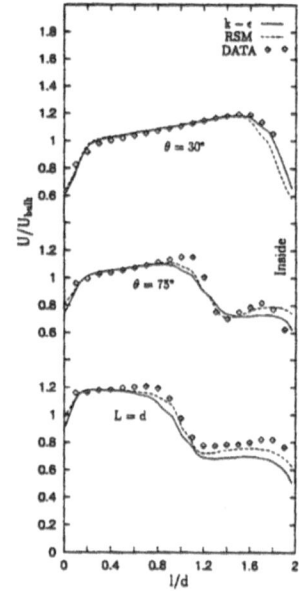

Figure 4 Mean-velocity profiles.

same conclusion can be made for the mean-velocity profiles shown in Figure 4.

References

Enayet,M.M., Gibson,M.M., Taylor,A.M.K.P. and Yienneskis,M. (1982), NASA, CR 3551.

Rhie,C.M. and Chow,W.L. (1983), *AIAA Journal*, Vol. **21**, pp. 1525-1532.

Speziale,C.G., Sarkar,S. and Gatski,T.B. (1991), *Journal of Fluid Mechanics*, Vol. **227**, pp. 245-272.

Basara, B. and Cokljat, D.(1995), *FED Conference* ASME, Vol. **221**, pp. 27-32.

ENHANCEMENT OF FINITE REYNOLDS NUMBER EFFECTS IN THE TURBULENT BOUNDARY LAYER DUE TO INNER-OUTER SUBLAYER INTERACTION

H. DEKKER
TNO Physics and Electronics Laboratory
P.O. Box 96864, 2509 JG Den Haag, and:
Institute for Theoretical Physics
University of Amsterdam, The Netherlands

A.M.J. van EIJK
TNO Physics and Electronics Laboratory

Abstract. We have re-examined the established data for the mean velocity profile for turbulent flow over a smooth surface and found its Reynolds number dependence in the inner region to be remarkably large. By combining two eddy-viscosity models (with overlapping validity in the inertial sublayer), this feature is shown (both analytically and numerically) to arise from inner-outer sublayer interaction, involving a nonzero viscous correlation length.

1. Introduction

Significant residual Reynolds number effects in turbulent boundary layers (and in pipe or channel flows) have been found in the inner region, i.e. close to the wall (e.g. [1], [2]). Indeed, near the wall the data are typically obtained for relatively small Reynolds numbers of the order of $Re \approx 10^3 - 10^4$ (based on bulk velocity and layer or channel width (δ)) and the revival in interest in these features is well justified.

In recent work (e.g. [3], [4]) we *inter alia* re-examined the established data for the mean velocity profile (e.g. [5], [6]) for turbulent flow over a smooth surface and found that the Reynolds number dependence can be described by the formula

$$\Delta \bar{U}+ \approx c_0 Re^{-1} + \mathcal{O}(Re^{-2}) \quad , \tag{1}$$

with $\Delta \bar{U}+ = \bar{U}+(Re) - \bar{U}+(Re=\infty)$, valid for $y+ > 5$ (and well into the logarithmic region). E.g., for channel flow $c_0 \approx (4 \pm 2) \times 10^3$. For pipes it is even larger. See [4], [7]. We clarify (1) by deriving it from eddy-viscosity theory, showing that the remarkably large value of c_0 (pertaining to the inertial sublayer) arises from the interplay between the outer part of the boundary layer (the wake) and its inner part (the viscous sublayer).

63

S. Gavrilakis et al. (eds.), Advances in Turbulence VI, 63-64.
© 1996 *Kluwer Academic Publishers.*

2. Analytical modeling

We combine two (rigorously solvable) eddy-viscosity models: one valid in the viscous sublayer (from [3],[4]), the other pertaining to the wake (from [8]), and both asymptotically containing the inertial sublayer ($y \to \infty$, respectively $y/\delta \to 0$). The wake model covers both channel (or pipe) flow and boundary layers *per se*. The combined model (not solvable analytically) admits a systematic perturbative calculation in powers of Re^{-1} (*viz.* $\mathcal{R}^{-1} = \nu/\kappa u*\delta$). The result reads

$$\Delta \bar{U}+ \approx (a^2/\kappa)\mathcal{R}^{-1} + O(\mathcal{R}^{-2}) \quad , \qquad (2)$$

where $a \approx 6$ is the viscous sublayer correlation length found in [3],[4] (the coefficient is obtained for $a \gg 1$). Relating $\mathcal{R}$ to Re (using $Re \approx (2/\kappa^2)\ln(e^{\gamma}\mathcal{R})$ with $\gamma \approx (2\pi/3^{3/2})a^{2/3}$; see [4]) for $Re \approx 10^3$-10^4 one finds $\mathcal{R}/Re \approx \kappa^2/8$. Hence, $c_0 \approx 8a^2/\kappa^3$ in (1). With $\kappa \approx 0.4$ this indeed yields $c_0 \approx 5 \times 10^3$.

3. Numerical results

A user-friendly Turbo Pascal 7.0 program was written to verify (2). It plots the eddy-vicosity and mean velocity (versus $\psi = \kappa y+$ or $\psi/\mathcal{R}$, and for fixed y+ versus $1/\mathcal{R}$). In nondimensional variables the only input parameters are $\mathcal{R} > 30$, $a \approx 6$-7, and a parameter ϕ. For boundary layers *per se* $\phi \approx 0$. For pipe or channel flow $\phi \approx 0.4$ [9], which indeed gives perfect agreement with (2).

References

1. Wei,T. and Willmarth,W.W.: J.Fluid Mech.**204**(1989)57-95.
2. Antonia,R.A., Teitel,M., Kim,J., and Browne,L.W.B.: *J.Fluid Mech.***236**(1992)579-605.
3. Dekker,H., de Leeuw,G., and Maassen van den Brink,A., in: R.Benzi (ed.), *Advances in Turbulence V*, p.p.100-104, Kluwer, Dordrecht, 1995.
4. Ibid.: *Physica* **A218**(1995)335-374.
5. Nikuradse,J.: *Gesetzmassigkeiten der turbulenten Stromung in glatten Rohren*, VDI-Forshungsheft **356** (1932).
6. Laufer,J.: Natl.Adv.Com.Aeron.Rep.**1033**(1951); **1174**(1954).
7. Patel,V.C. and Head,M.R.: J.Fluid Mech.38(1969)181-201.
8. Dekker,H.: *Diffusive K-theory: finite geometry effects*, Rep.**0310A**(1992).
9. Longwell,P.A.: *Mechanics of Fluid Flow*, McGraw-Hill, New York, 1966.

DYNAMICS OF COHERENT STRUCTURES NEAR A ROTATING BODY

M. FERLAUTO

Politecnico di Torino
Dipartimento di Ingegneria Aeronautica e Spaziale
Corso Duca degli Abruzzi, 24 - 10129 Torino, Italy

The aim of this work is to study the control performed by the wall shape on the vortical structures generated by a rotating cylindrical body. In the case of the flow generated from rest outside a rotating circular cilinder we know that the vorticity created at the solid surface all diffuses to infinity. A suction through the surface of the body can retain the vorticity near the wall [1]. We can suppose that for a complex geometry of the body (without suction) the vorticity is partly retained and partly diffused.

We assume the flow as bidimensional and inviscid. In fact vorticity structures develop essentially in an inviscid way for sufficiently high Reynolds numbers and for time intervals comparable to the convective time scale.

The vorticity is concentrated in a finite number of singular points (*'point vortices'*) or small regions with a compact support (*'blobs'*). On the body contour there are some cusps, periodically spaced, which are responsible for the injection of vorticity inside the fluid.

Briefly, in a pure inviscid solution the only boundary condition required at solid walls is impermeability, whereas in real flows the no-slip condition is responsible for the vorticity generation at the walls. Afterwards,this vorticity is diffused and convected into the main flow.

We use a simplified inviscid model to simulate such a complicated behaviour: in a discrete number of points of the body contour (generally sharp edges) we enforce a *Kutta condition*, by placing, close to them and at given time intervals, a point vortex (or a small vorticity blob) whose strength is determined by the no-slip condition. The Lagrangian time evolution of the vortices is described numerically.

According to this vortex method [2], we investigate numerically the impulsive start in the rotation of a cilyndrical body from rest.

By way of this model, clouds of vortices form structures that resemble the physical vortical structures inside an incompressible fluid.

Analitically we study the existence of solutions for the simple case of a point vortex steadily trapped between two contiguous cusps, testing the capability of such a rotor to retain vorticity in his neighbourhood.

S. Gavrilakis et al. (eds.), Advances in Turbulence VI, 65-66.
© *1996 Kluwer Academic Publishers.*

We are using a conformal transformation to map the exterior of the unit circle into a periodical sector of the rotating body [3].

The computation (see figure 1) shows the vortical structures generated by the separations occurring at the geometrical discontinuities. The transient is tracked from the impulsive start to the final flow characterized by vorticity bubbles steadily trapped between two contiguous edges.

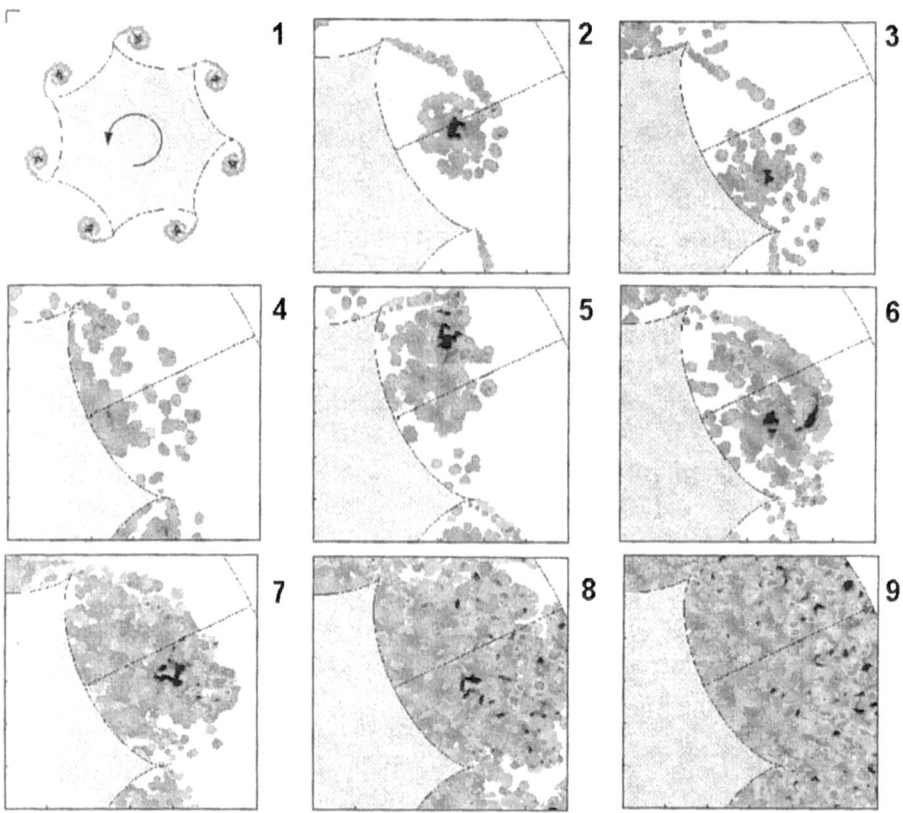

Figure 1: A sequence of vorticity maps of the flow generated from rest by a rotating cilyndrical solid body.

References

[1] G. K. Batchelor, *An introduction to Fluid Dynamics*, Cambridge Univ. press, 1967.

[2] K. Gustafson, J. Sethian, *Vortex Mathods and Vortex Motion*, SIAM, 1991.

[3] L. M. Milne-Thomson, *Theoretical Hydrodynamics*, McMillan, 1968.

A POINTWISE LOW-RE $K - \epsilon$ TURBULENCE MODEL

U.C. GOLDBERG, D.D. APSLEY AND F.-S. LIEN
UMIST, Mechanical Engineering Department
P.O.B. 88, Manchester M60 1QD, U.K.

1. Introduction

The standard high Reynolds number $k - \epsilon$ model is a good performer in free shear flows but its extension to near-wall flow regions proved over the years a rather difficult task with less than universal success. One approach is that of Jones-Launder[1] and Launder-Sharma [2], involving 1st and 2nd normal-to-wall derivatives. This imposes numerical "stiffness" and also prevents the model from being pointwise (local) because of the need for wall-normal unit vectors. The current work proposes a low Reynolds number $k - \epsilon$ model which retains the pointwise attribute without numerical stiffness by maintaining time scale realisability and utilising a simple wall boundary condition for ϵ. This results in a topology-free model, unlike most existing approaches which resort to geometry-dependent wall-distance functions.

2. Model Formulation

The proposed model provides the eddy viscosity $\nu_t = C_\mu f_\mu k^2 / \epsilon$ based on the following transport equations for k and ϵ:

$$\frac{D\rho k}{Dt} = \frac{\partial}{\partial x_j}\left[\left(\mu + \frac{\mu_t}{\sigma_k}\right)\frac{\partial k}{\partial x_j}\right] + P_k - \rho\epsilon \ . \tag{1}$$

$$\frac{D\rho\epsilon}{Dt} = \frac{\partial}{\partial x_j}\left[\left(\mu + \frac{\mu_t}{\sigma_\epsilon}\right)\frac{\partial\epsilon}{\partial x_j}\right] + (C_{\epsilon 1}P_k - C_{\epsilon 2}\rho\epsilon + E)T_t^{-1} \ . \tag{2}$$

Here P_k is the turbulence production, $-\overline{u_i'u_j'}U_{i,j}$, modelled in terms of the Boussinesq concept $-\overline{u_i'u_j'} = \nu_t(U_{i,j}+U_{j,i})-(2/3)k\delta_{ij}$. Time scale realisability is ensured by using $T_t - (k/\epsilon)\max\{1, 1/\xi\}$ where $\xi = \sqrt{R_t}/C_\tau$ and $R_t = k^2/(\nu\epsilon)$. This imposes the Kolmogorov scale $\sqrt{\nu/\epsilon}$ on Eq. (2) for $R_t \ll 1$.

S. Gavrilakis et al. (eds.), Advances in Turbulence VI, 67-68.

The effect of the additional source term $E = (2A_k C_{\epsilon 2})/(C_\tau A_\mu)\mu f_\mu \mathcal{S}^2 \exp\{-A_f\sqrt{\mathcal{E}_k/(\nu \mathcal{S})}\}$ is to cancel the destruction term $C_{\epsilon 2}\rho\epsilon$ at the immediate vicinity of walls. Here $\mathcal{S} = \sqrt{2S_{ij}S_{ij}}$, $S_{ij} = (U_{i,j} + U_{j,i})/2$ being the mean strain invariant. $\mathcal{E}_k = Q^2/2 + k$, the total kinetic energy, where $Q = \sqrt{U_i U_i}$. The near-wall damping function forces the relation $\nu_t \sim \overline{v'^2}T_t$ through the formula $f_\mu = (1 - \exp\{-A_\mu R_t\})/(1 - \exp\{-\sqrt{R_t}\})\max\{1, 1/\xi\}$. The model constants are $C_\mu = 0.09$, $A_\mu = 0.01$, $C_{\epsilon 1} = 1.44$, $C_{\epsilon 2} = 1.92$, $\sigma_k = 1.0$, $\sigma_\epsilon = 1.3$, $C_\tau = \sqrt{2}$, $A_k = 0.05$, $A_f = 0.214$. Wall boundary conditions: $k_w = 0$, $\epsilon_w = 2[\nu k/y^2]_1$ where "1" denotes the centroid of the 1st cell off walls. This boundary condition implies $(\partial k/\partial y)_w = 0$, satisfying the second boundary condition for k implicitly.

3. Results

Flow over a hill at Re=60,000 (Almeida et al.[3]) is shown here, using the STREAM code (Lien and Leschziner[4]). Fig. 1a is a streamline plot, showing reattachment at $x/H \approx 4.8$, in agreement with the data. Fig. 1b compares predictions using several turbulence models with velocity data at the reattachment point.

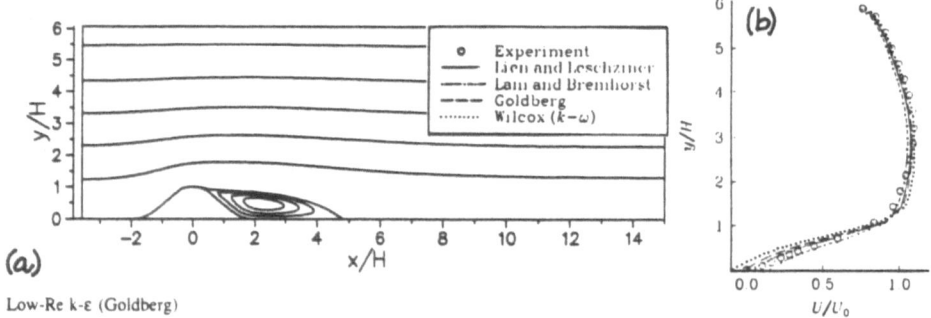

Low-Re k-ε (Goldberg)

Figure 1. Flow over Hill: streamlines and profiles

References

1. Jones, W.P. and Launder, B.E.: The prediction of laminarization with a two-equation model of turbulence, *Int. J. Heat Mass Transfer* **15** (1972), 301.
2. Launder, B.E. and Sharma, B.I.: Application of the energy-dissipation model of turbulence to the calculation of flow near a spinning disc, *Letters Heat Mass Transfer* **1** (1974), 131.
3. Almeida, G.P., Durao, D.F.G. and Heitor, M.V.: Wake flows behind two-dimensional model hills, *Experim. Thermal Fluid Sci.* **7** (1993), 87.
4. Lien, F.-S. and Leschziner, M.A.: A general non-orthogonal collocated finite-volume algorithm for turbulent flow at all speeds, incorporating second-moment turbulence-transport closure. Part I: computational implementation, *Computer Methods Appl. Mech. Engrg.* **114** (1994), 123.

LARGE EDDY SIMULATION OF A PLANAR CO-FLOWING JET

GUY HOFFMANN AND CARLO BENOCCI
von Karman Institute for Fluid Dynamics
B-1640 Rhode-St-Genèse, Belgium

1. Introduction

Turbulent jet flows are of considerable interest both for fundamental research and practical applications. However, in spite of extensive previous research, major deficiencies in the knowledge and understanding of this flow field still exist. Time-accurate computations, through direct or large eddy simulation (LES), have become increasingly viable as a means of investigating turbulent flow problems. As a preliminary step towards the application of these techniques to jet flows, the LES of a co-flowing jet is proposed at $Re = 30000$ (based on the uniform jet velocity and the nozzle width h) and a ratio of free-stream velocity U_1 versus jet velocity U_J of 0.16, reproducing the conditions of the experimental study by Bradbury 1965.

2. Large Eddy Model

The LES approach is based upon the application of a filtering operation to the three-dimensional unsteady Navier-Stokes equations. The large scale velocity field is then obtained by the direct numerical solution of the filtered Navier-Stokes equations. The contribution of the scales smaller than the filter size appear as an unknown term which has to be modelled. In a finite difference formulation, the filtering operation takes the form of a top-hat filter in physical space, which is applied implicitly by the finite difference operators. The incompressible Navier-Stokes equations are discretized over a staggered grid using 2nd order finite differences, apart from the advection term, where a 4th order central discretization with the addition of a 4th order dissipation term in order to control *aliasing errors* is used (see Hoffmann and Benocci 1994). The momentum equation is advanced in time using a three-step low-storage Runge-Kutta scheme in conjunction with a fractional step method. Continuity is satisfied by solving a Poisson equation with a direct Poisson solver.

The contribution of the unresolved small scales is modelled using the concept of an eddy viscosity. The presence of mixed laminar, turbulent and transitional flow regions requires the use of a turbulence model which is entirely based on local quantities and predicts vanishing turbulent viscosities in laminar flow regions. These requirements can be fulfilled by the recently proposed filtered structure function model (Ducros and Comte 1995); details of the implementation of this model and its validation can be found in Hoffmann 1996.

The main problem of the free jet simulation is related to the choice of boundary conditions at the open portions of the domain, which are crucial for the stability of the compu-

S. Gavrilakis et al. (eds.), Advances in Turbulence VI, 69-70.
© 1996 *Kluwer Academic Publishers.*

tation and the quality of results. The open boundary conditions used in the present study (Hoffmann 1996) allow fluid to enter the computational domain and impose the mean symmetry of the flow. It was found that in order to ensure global stability of the flow field, a small co-flowing stream had to be introduced.

3. Results

The computational domain of dimensions $30h \times 5h \times 50h$ is subdivided by $127 \times 24 \times 128$ grid points in the streamwise x-direction, the periodic y-direction and the normal z-direction. The results obtained for the time-averaged flow quantities agree well qualitatively with the experimental reference data ($cf.$ Figure 1 and Figure 2). Quantitatively, the turbulence

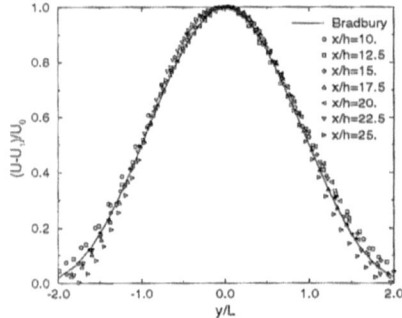
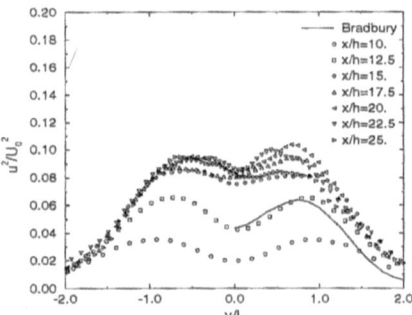

Figure 1. Time-averaged mean velocity profiles ($U_0 = U_{centreline} - U_1$).

Figure 2. Turbulence intensities in the streamwise x-direction ($u = $ rms turbulence intensity).

intensities and the Reynolds shear-stress are overpredicted by a significant amount. Analysis of the instantaneous flow features reveals that the numerical simulation captures correctly the main primary instability mechanisms. The most amplified frequency f gives a Strouhal number $St = f h/(U_J - U_1) = 0.25$, which lies well within the experimentally obtained range of 0.15 to 0.42 ($cf.$ Namer and Ötügen 1988). The flow field is characterized by the growth of large two-dimensional vortices which are located symmetrically in the upper and lower shear layers of the jet. At $x/h \approx 10$ the 2-d vortices break up into 3-d structures. The location of the 3-d break-up coincides approximately with the transition from the symmetrical to the asymmetrical mode of vortex formation in the upper and lower jet shear layers. A possible cause for the shortcomings of the numerical simulation is that the break-up of the primary vortices is delayed and the flow excessively retains a 2-d character.

References

L.J.S. Bradbury. The structure of a self-preserving turbulent plane jet. *J. Fluid Mech.*, 23:31–64, 1965.

G. Hoffmann and C. Benocci. Numerical simulation of spatially-developing planar jets. In *74th AGARD FDP Symposium*, Chania, Greece, April 18-21 1994.

F. Ducros and P. Comte. Large-eddy simulation of spatially-developing boundary layer. In *Advances in Turbulence V*, Roberto Benzi ed., Kluwer, 1995.

G. Hoffmann. Engineering application of large eddy simulation to turbulent free and wall-bounded shear layers. PhD thesis, TU München, 1996.

I. Namer and M.V. Ötügen. Velocity measurements in a plane turbulent air jet at moderate Reynolds numbers. *Exp. in Fluids*, 6:387–399, 1988.

PREDICTIONS OF HIGHER MOMENTS BY A REALIZABLE SECOND-MOMENT TURBULENCE MODEL

D. Lentini
Dipartimento di Meccanica e Aeronautica
Università degli Studi di Roma "La Sapienza"
Via Eudossiana 18, I-00184 Roma RM, Italy

For practical users, the most interesting output of prediction codes for turbulent flows resides in first-, or at most second-moments. However, closure hypotheses in turbulence models are effected on second- or third-moments (depending on the closure level), on the dissipation and the pressure–strain terms. Consideration of higher–moment predictions is therefore crucial to assess turbulence models.

Figure 1 shows comparisons of measured (Spencer and Jones, 1971) and predicted third–moments in a mixing layer involving two airstreams with velocity ratio 0.6. The prediction code adopts a realizable second–moment closure model, Launder's 'new model' (Launder, 1989). Thanks to the nonlinear (up to cubic) relationships adopted for unclosed terms, it effectively allows preventing the occurrence of unphysical situations, such as negative normal stresses, or shear stresses violating Schwarz's inequality. The generalized gradient transport hypothesis (Daly and Harlow, 1970) is adopted to model third–moments. Given the fact that such terms are wholly modeled, the reported agreement can be considered as quite reasonable.

The modelled conservation equation for the viscous dissipation is the one which implies the most drastic assumptions. It is therefore interesting to compare, as in Fig. 2, the predicted dissipation to that estimated by Spencer and Jones as $\bar{\epsilon} = 30\,\nu\,\overline{u'^2}\,/\,\lambda^2$, where $\overline{u'^2}$ and the Taylor scale λ are recovered from measurements. However, from a turbulent kinetic energy budget, they find that this expression underestimate $\bar{\epsilon}$ by a factor about 3.6. Predictions by Launder's 'new model' result in an excellent agreement with the corrected estimates; in particular, a marked improvement is observed with respect to the standard k–ϵ model.

REFERENCES

Daly, B.J. and Harlow, F.H. (1970) *Phys. Fluids* **13**, 2634–2649.

Launder, B.E. (1989) *Int. J. Heat and Fluid Flow* **10**, 282–300.

Spencer, B.W. and Jones, B.G. (1971) AIAA paper 71–613.

S. Gavrilakis et al. (eds.), Advances in Turbulence VI, 71-72.

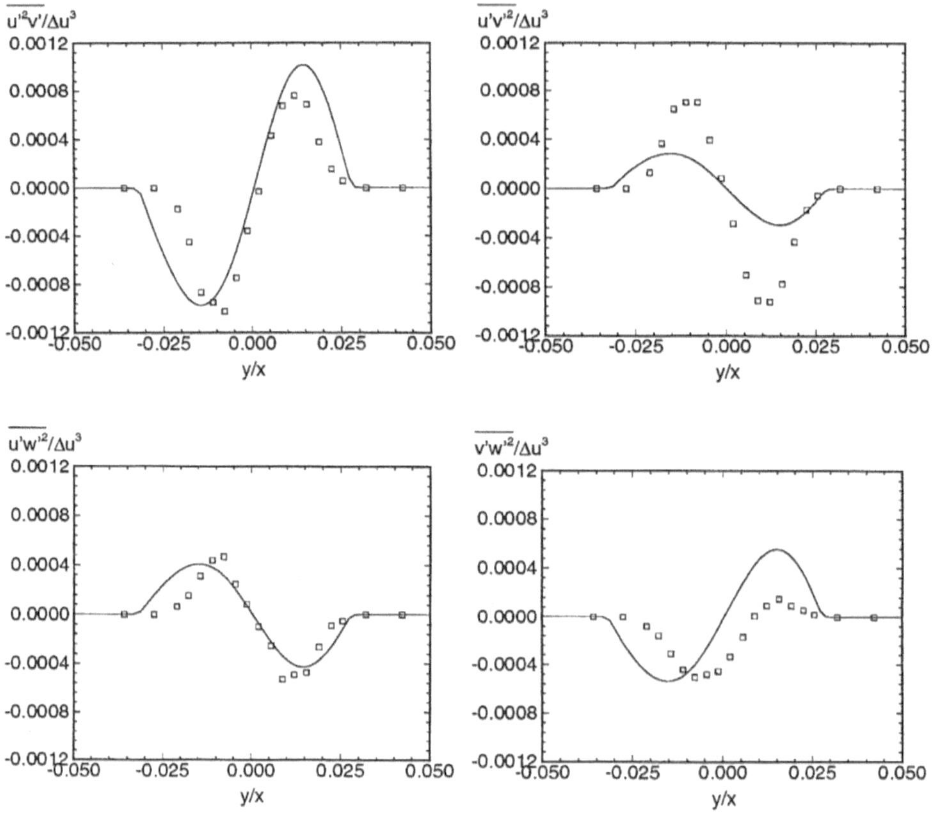

$\overline{u'^2v'}/\Delta u^3$

$\overline{u'v'^2}/\Delta u^3$

$\overline{u'w'^2}/\Delta u^3$

$\overline{v'w'^2}/\Delta u^3$

Figure 1: Transverse profiles of dimensionless third–moments $vs.$ similarity variable: □ measurements, ——— predictions by realizable second–moment closure.

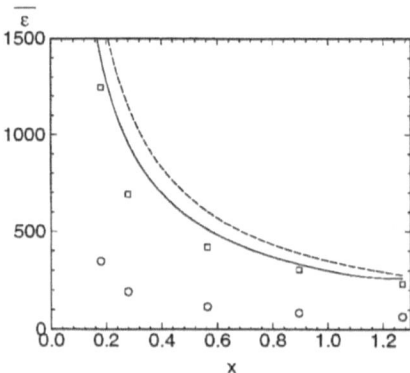

Figure 2: Axial profile of viscous dissipation rate: ○ estimates on the basis of measurements, □ corrected estimates, ——— predictions by realizable second–moment closure, — — — predictions by k–ϵ model. x in m, $\overline{\epsilon}$ in m^2/s^3.

NUMERICAL SIMULATIONS OF INHOMOGENEOUS TURBULENCE GENERATED BY AN OSCILLATING GRID AND SUBMITTED TO SOLID-BODY ROTATION

L. LOLLINI AND C. CAMBON

L.M.F.A. UMR 5509 CNRS - Ecole Centrale de Lyon
BP 163 - 69131 Ecully Cedex France

1. General background

An experiment made by Hopfinger *et al.*, 1982 (hereafter referred to as HBG), shows how a turbulence, generated by an oscillating grid in a rotating tank, is affected by rotation for various values of the Rossby number. Near the grid, the Rossby number is set to a large value so that the flow is unaffected by rotation. An abrupt transition is located at a distance from the grid corresponding to a local Rossby number of about 0.20. Then, the flow, which becomes dominated by rotation, consists of concentrated vortices having axes of rotation approximately parallel to the rotation axis but a pure two-dimensional state is not reached. This kind of experiment is of great interest in geophysical applications because the transition between a three-dimensional small scale turbulence to a quasi two-dimensional one with larger characteristic scale is observed in atmospheric or oceanic flows, but not completely understood. Furthermore, it should lead us to a better understanding of cyclones genesis.

Yang, 1992, performed a D.N.S. calculation to simulate the HBG's experiment. Although his calculation grid was very coarse in the vertical direction and his forcing was far from the physical one, he obtained a good qualitative agreement with the experimental results. This has led us to undertake DNS-LES pseudo-spectral computations which can reach a greater resolution in the inhomogeneous direction and where the grid parameters appear explicitly in the forcing. Furthermore, higher Reynolds calculations will be performed.

2. Forcing method

We solve the forced Navier-Stokes equations for an incompressible fluid of viscosity ν between two parallel planes on which no-slip conditions are imposed. The axis of rotation is chosen normal to the planes (vertical). The forcing process, which is localized in a plane corresponding to the mean position of the oscillating grid in the HBG experiment ($x = x_G$), consists of the superposition of a deterministic signal exhibiting explicitly the grid parameters (frequency n, mesh size M, amplitude S) and of a stochastic component which intensity is to be adjusted :

$$F_i(x_G, y, z, t) = \delta_{i1} \cos(\frac{2\pi}{M}y) \cos(\frac{2\pi}{M}z) \sin(t) + \text{White Noise}$$

73

S. Gavrilakis et al. (eds.), Advances in Turbulence VI, 73-74.
© 1996 *Kluwer Academic Publishers.*

74

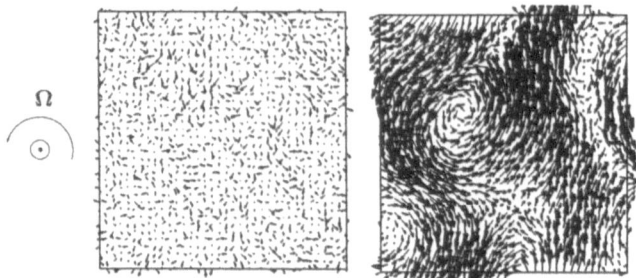

Figure 1. Horizontal flow field in a homogeneous plane parallel to the forcing plane for $\Omega = 0$ (left) and $\Omega = 0.151$ (right). The energetic vortex generated in the second case is cyclonic.

The equations are non-dimensionalized by a velocity and a length scale given by the forcing parameters. Hence, the two independent relevant dimensional parameters that appear are the Rossby number $Ro = n/(2\Omega)$ and the Reynolds number $Re = nS^2/\nu$.

3. Preliminary results and perspectives

A pseudo spectral method is used in the LES-DNS calculations (Fourier modes in two homogeneous directions and Chebyshev polynomials in the inhomogeneous one). Without rotation, that is to say in the case of pure 'diffusive turbulence', LES allow us to tune the forcing term and compare the spatial variations of statistical quantities with those observed in Hopfinger & Toly, 1976. Some DNS calculations are undertaken with several grid Rossby numbers as in Hopfinger *et al.*, 1982. With respect to the free decay of rotating turbulence, extensively studied by related experiments, theoretical models and numerical simulations (see Cambon *et al.*, 1995, for a review), the vertical confinement and the forcing yield modified characteristics (sharper transition, stronger quasi-2D vortices, Ekman layer). These DNS calculations should confirm or not the critical value of the local Rossby number for which the flow pattern changes dramatically. We will be able to investigate the effect of the increase of the Reynolds number which is directly linked to the grid frequency, by using a subgrid-scale model (optional in the present code) of the Smagorinski kind corrected for walls. The capture by the numerical code of inertial waves possibly emanating from the forcing zone, and embedded in the turbulent field, presents a particular interest, as well as a clarification of their dynamics.

Some information about more complex rotating systems of industrial or geophysical interest could be gained from this study. Furthermore, in a near future, the code should take into account thermal effects which involve internal gravity waves.

This work is supported under DRET grant 95-2550A.

References

Cambon, C., Mansour, N. N., & Godeferd, F.S. 1995. Energy transfer in rotating turbulence. *Submitted to J. Fluid Mech.*

Hopfinger, E.J., Browand, F.K., & Gagne, Y. 1982. Turbulence and waves in a rotating tank.

Hopfinger, E.J. & Toly, J.-A. 1976. Spatially decaying turbulence and its relation to mixing across density interfaces. *J. Fluid Mech.*, 78:155–175.

Yang, G. 1992. *DNS of boundary forced turbulent flow in a non-rotating and a rotating system.* Cornell University. Ph. D. Thesis dissertation.

A SPECTRAL CLOSURE FOR INHOMOGENEOUS TURBULENCE APPLIED TO TURBULENT CONFINED FLOW

S. PARPAIS AND J.P. BERTOGLIO
Laboratoire de Mécanique des Fluides de d'Acoustique UMR 5509 CNRS
Ecole Centrale de Lyon 69130 Ecully - FRANCE

1. Introduction

The aim of the paper is to present a spectral model for inhomogeneous turbulence, which provides information on the turbulent kinetic energy spectrum and is used to close the averaged Navier-Stokes equations. This model, called S.C.I.T. for Spectral Closure for Inhomogeneous Turbulence, was first introduced by Bertoglio & Jeandel, 1987, in an earlier version. New developments of the production term and of the transport term now allow an Eulerian formulation of this model. This new model has been implemented in a two-dimensional averaged Navier-Stokes equation solver, using a finite element method. Computations are performed on several confined flows.

2. Spectral Model for Inhomogeneous Turbulence

The basic equation for the present turbulence model is the transport equation for the turbulent kinetic energy spectrum $E(K, \vec{X})$ (Bertoglio & Jeandel, 1987). This equation reads:

$$\left((\partial_t + U_i \partial_{X_i}) - 2\nu \left(\frac{1}{4} \partial_{X_i} \partial_{X_i} - K^2 \right) \right) E(K, \vec{X}) = P(K, \vec{X}) + T(K, \vec{X}) + D(K, \vec{X})$$

In this equation $P(K, \vec{X})$ is the **production term**, $T(K, \vec{X})$ is the **non-linear transfer term** (accounting for the energy cascade in spectral space) and $D(K, \vec{X})$ is the **transport term** (transport by triple velocity correlations, lumped together with transport by pressure-velocity correlations).

The closed form of the **non-linear transfer term** is identical to the classical expression obtained when the E.D.Q.N.M. theory (Orszag, 1970) is applied to the case of homogeneous isotropic turbulence.

The **production term** also requires closure. This is done using a spectral extension of the work of Shih *et al.*, 1994 on the non-linear $k - \varepsilon$ model. One obtains:

$$P(K, \vec{X}) = -\varphi_{ij}(K, \vec{X}) \partial_{X_j} U_i$$

where, $\varphi_{ij}(K, \vec{X}) = 2 \left(E(K, \vec{X}) \delta_{ij}/3 - \nu_{ts}(K, \vec{X}) S_{ij} \right)$ and $\nu_{ts}(K, \vec{X})$ is found, by dimensional analysis, to be of the form:

$$\nu_{ts}(K, \vec{X}) = \frac{E(K, \vec{X}) \tau_s(K, \vec{X})}{A_o + A_r \sqrt{S_{ij} S_{ij} + \Omega_{ij} \Omega_{ij}} \tau_s(K, \vec{X})}$$

75

S. Gavrilakis et al. (eds.), Advances in Turbulence VI, 75-76.
© 1996 *Kluwer Academic Publishers.*

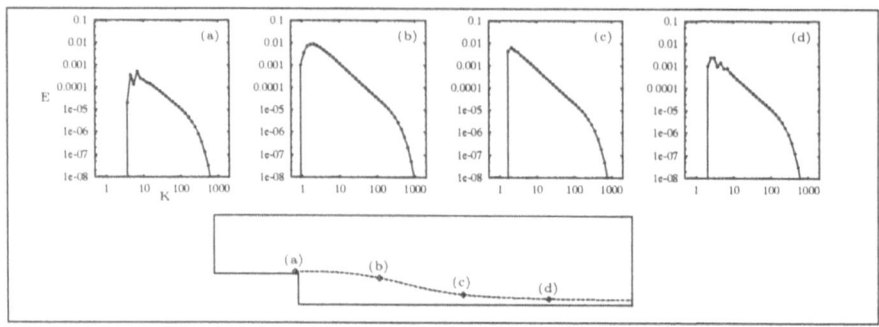

Figure 1. Backward facing step - Energy spectra for various points of the same stream line

with $\tau_s(K, \vec{X}) = (K^3 E(K, \vec{X}))^{-1/2}$, $S_{ij} = (\partial_{X_j} U_i + \partial_{X_i} U_j)/2$ and $\Omega_{ij} = (\partial_{X_j} U_i - \partial_{X_i} U_j)/2$. The constant A_o is adjusted to give correct results for homogeneous shear flows and A_r is found to be equal to $\sqrt{3}/2$ in incompressible, two dimensional (with respect to the mean velocity) flows, for realisability reasons (Reynolds, 1987).

The **inhomogeneous transport term** is obtained using the local diffusive expression originally proposed by Besnard *et al.*, 1992, modified by introducing the variable $\nu_{ts}(K, \vec{X})$. It reads:

$$D(K, \vec{X}) = \partial_{X_i} \left(\left(\int_0^\infty \nu_{ts}(R, \vec{X}) \, dR \right) \partial_{X_i} E(K, \vec{X}) \right)$$

Finally, to take into account the wall effect, a low wave-number cut-off $K_c(\vec{X})$ (geometrically defined) is introduced as in Bertoglio & Jeandel, 1987 - i.e. the energy $E(K, \vec{X})$ of wave numbers K lower than $K_c(\vec{X})$ is taken equal to zero.

Last but not least, the averaged Navier-Stokes equations is closed using the relation:

$$\overline{u_i u_j}(\vec{X}) = \int_0^\infty \varphi_{ij}(K, \vec{X}) \, dK$$

Computations using a finite element method have been performed on a fully-developed turbulent flow in a channel and give good results. The case of a backward facing step has also been treated and gives satisfactory results. The model also provides energy spectra for every point of the domain. In figure 1 we exhibit four spectra along the same stream line.

References

Bertoglio J.-P. and Jeandel D. (1987) A Simplified Spectral Closure for Inhomogeneous Turbulence: Application to the Boundary Layer, *Fifth Turb. Shear Flows, Springer-Verlag*

Besnard D.C., Harlow F.H., Rauenzahn R.M. and Zemach C. (1992) Spectral Transport Model for Turbulence *Los Alamos National Laboratory Report LA-UR-92-1666*

Orszag S. A. (1970) Analytical Theories of Turbulence *J. Fluid Mech.* **Vol. 41 part 2**, *pp. 363-386*

Reynolds W. C. (1987) Fundamentals of turbulence modeling and simulation *Lecture Notes for Von Karman Institute, Agard Report # 755*

Shih T.-H., Liou W.W., Shabbir A., Yang Z., Zhu J. (1994) A New $k - \varepsilon$ Eddy Viscosity Model for High Reynolds Number Turbulent Flows - Model Development and Validation *NASA Technical Memorandum 106721, ICOMP-94-21, CMOTT-94-6*

REYNOLDS STRESS MODELING OF SEPARATED FLOWS OVER CURVED SURFACES

S. PERZON AND L. DAVIDSON
Chalmers University of Technology
S-412 96 Göteborg, Sweden

AND

M. RAMNEFORS
Volvo Data AB
S-405 08 Göteborg, Sweden

1. Introduction

Reynolds Stress Models (RSMs), are sensitive to streamline curvature and should mimic the turbulent behaviour more accurately than eddy viscosity models, EVMs, when the flow is heavily curved.

This work is focused on flows including separation. Two testcases are regarded and these are the flow over a 2D hill and the flow over a backward facing step. Comparisons are made between simulations and experiments or direct numerical simulations, DNS, respectively.

2. Turbulence models

All RSMs used in this work models the turbulent diffusion by the generalized gradient hypothesis by Daly & Harlow. Furthermore High-Re models assumes $\varepsilon_{ij} = \frac{2}{3}\varepsilon\delta_{ij}$ and low-Re models assumes an anisotropic dissipation rate model according to Hanjalić & Launder[1]. Thus the models differ mainly in the modeling of the pressure strain interaction term ϕ_{ij}. The simple isotropization of production model by Naot *et al.* 1970, (IP), is used and the model of Gibson and Launder, (GL), has been added on to IP. In addition to this model a nonlinear quadratic model by Speziale, Sarkar and

[1]HANJALIĆ K. and LAUNDER B.E., Contribution towards a Reynolds-stress closure for low-Reynolds-number turbulence., *J. Fluid Mech.*, 74:593–610, 1976.

S. Gavrilakis et al. (eds.), Advances in Turbulence VI, 77-78.
© 1996 *Kluwer Academic Publishers.*

Gatski[2] (SSG), and a cubic proposal by Launder and Li[3] (LLI) has been tested. Wall-functions are used for high-Re models. The low-Re RSM used was proposed by Hanjalić and Launder, (HL). This model was also tested when replacing its pressure strain model for the more elaborate cubic realizable model by Launder & Li, (LLIHL).

3. Results and concluding remarks

2D hill flows includes separation at a curved boundary. The results for the U velocity compared with experiments by Almeida[4] et al. can be seen in fig 1. High Reynolds number models completely fails in the prediction of the separation point which occurs too late.

In the flow over a backward facing step, comparisons are made between simulations and DNS, by Le & Moin[5] . All high-Re models reattach too fast and the k-ε-model reattached faster than any RSM. All these models had an erroneous flow angle over the step corner. The LLIHL reattached somewhat too late which is because of an underprediction of the shear stress, $\overline{uv}$, in the shear region, see fig 2.

It is shown that it is hazardous to use wall functions in any flow where separation occurs. Low-Re models is necessary in order to capture the features of the separation region properly.

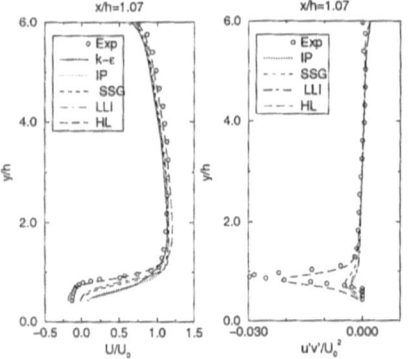

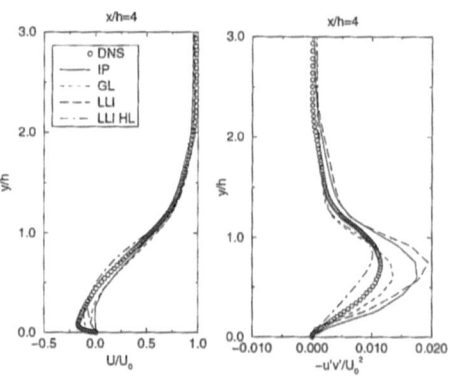

Figure 1 *U-velocity and shear stress profiles for the 2D hill flow at $x/h = 1.07$.*

Figure 2 *U-velocity and shear stress profiles for the backward facing step flow at $x/h = 4$.*

[2]SPEZIALE C.G., SARKAR S. and GATSKI T.B. , Modeling the pressure-strain correlation of turbulence. , *J. Fluid Mech.*, 227:245–272, 1991.

[3]LAUNDER B.E. and LI S.-P. , On the elimination of wall-topography parameters from second-moment closure. , *Phys. Fluids*, 6(2):999–1006, 1994.

[4]ALMEIDA M.V., DURAU D.F.G. and HEITOR M.V., Wake flows behind two-dimensional hills., *Exp. Thermal and Fluid Science*, 7:87–101, 1993.

[5]LE H. and MOIN P. , Direct numerical simulation of turbulent flow over a backward facing step. , Report no. tf-58, Stanford University, Dept. Mech. Eng., 1994.

PREDICTING TRANSITION WITH TURBULENCE MODELS

The Effect of Variable Free-Sream Turbulence Length-Scale

A.M. SAVILL
Senior Research Associate   &   Visiting Research Fellow
Department of Engineering,   Mech. Eng. Department
University of Cambridge   UMIST
Cambridge   CB2 1PZ   Manchester M60 1QD
England UK   UK North ERCOFTAC Pilot Centre

1. Introduction

This paper presents new results obtained by the COST-ERCOFTAC Transition Special Interest Group as part of a project to evaluate and refine turbulence models for predicting by-pass transition under a wide range of free-stream conditions. Co-operative research carried out by the SIG over the last 5 years [1] has identified optimum model approaches to be employed at each level of closure from simple integral/correlation methods up to Large Eddy Simulations. In particular a low-Re Reynolds Stress Transport (RST) model developed at Cambridge and UMIST [2], which represents a fairly simple extension of the 'best' (Launder-Sharma) k-ε scheme, has been shown to predict the onset and the length of transition surprisingly accurately for sharp leading edge test cases with 1-10% intensity isotropic free-stream turbulence.

The same model can also capture the correct influence of free-stream anisotropy, and effects of pressure-gradient, convex stream-line curvature & Reynolds number, but, like most other low-Re turbulence models, it is sensitive to initial and free-stream boundary conditions for dissipation.

Until recently it had not been possible to properly examine the ability of any proposed prediction methods to capture the effect of variations in free-stream length scale on transition, although these are known to have a strong influence in the case of turbulent boundary layers [3]. However, COSTEC-funded SIG participants working at the Prague Institute of Thermomechanics have now conducted a series of transition experiments in which the wind-tunnel and turbulence-generating grid configurations were carefully controlled to allow a range of imposed length-scales to be investigated, without simultaneously varying the free-stream turbulence intensity [4].

2. Results with Discussion

The first experimental results have only just been reported and further data-taking is still underway at Prague. However SIG Test Cases are always done 'blind' in the first instance, and a test case specification was released to participants at the time the experiments were started, so it is already possible to present a first evaluation of the six sets of 'blind' predictions submitted to the SIG co-ordinator prior to receipt of the initial data.

79

S. Gavrilakis et al. (eds.), Advances in Turbulence VI, 79-80.
© 1996 *Kluwer Academic Publishers.*

The computations all assumed a free-stream turbulence intensity of 3% at the sharp leading edge of the nominally zero-pressure gradient test-plate, with the standard -5/7 power law decay, and an initial free-stream turbulence dissipation length scale of 8.2mm or 26mm (generated by experimental grids 1 & 5 [4]), with a 1/2 power law growth.

The comparisons presented in Figure 1 shows that the SLY model correctly predicts onset of transition (minimum Cf location) and a reasonable variation of H through transition for both cases. In particular the extent of the shift in the Cf & H distributions with varying free-stream length-scale is predicted almost exactly correct. In these respects the SLY results are better than five other sets of model predictions.

The Integral Method of Johnson and the νt-92 one-equation eddy-viscosity transport model of Secundov & Vasiliev at CIAM appear to be far too sensitive to the change in length scale; whereas the Launder-Sharma k-ε model (as tested by Henkes & Westin at Delft) and the k-ε-γ model of Steelant & Dick from Ghent appear to be too insensitive to the length-scale variation. Only the two-layer k-ε/surface-renewal model developed by the Prague group itself, displays the correct sensitivity, but their computations were not strictly blind since they had access to their own data.

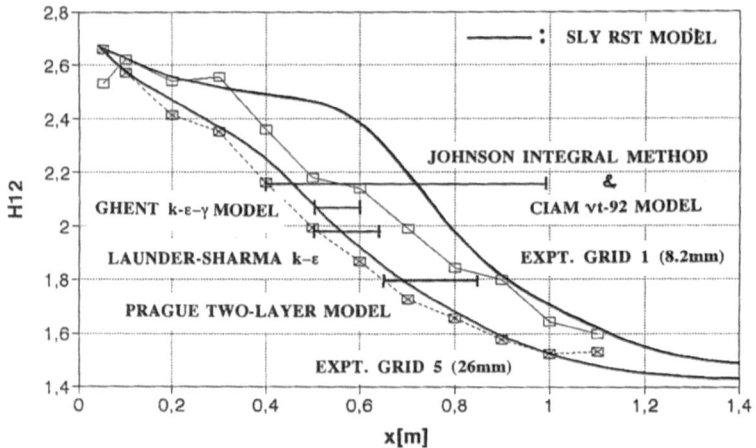

Figure 1: Predicted versus experimental Shape Factor variation with fst length-scale

3. References

1. Savill, A.M. (1995) A Summary Report on COST-ERCOFTAC Transition SIG Project Evaluating Turbulence Models for Predicting Transition. *ERCOFTAC Bulletin* 2 4, 57-59.
2. Savill, A.M. (1995) The SLY RST Intermittency Model for Predicting Transition. *ERCOFTAC Bulletin* 2 4, 37-39.
3. Savill, A.M. (1987) Algebraic and Reyolds Stress Modelling of Manipulated Boundary Layers Including Effect of Free-stream Turbulence. Proceedings International Conference on *Turbulent Drag Reduction by Passive Means*, RAeS, 1 pp. 89-143.
4. Pittaluga, F. (1995) Experimental Activities Within COST-F1, Sub-Group A-7, on Laminar-Turbulent Transition: A Summary Report. Proceedings IMACS-COST Conference on *Computational Fluid Dynamics Three-Dimensional Flows*, EPF Lausanne

II

Coherent Structures and Vorticity

COHERENT STRUCTURES:
PAST, PRESENT AND FUTURE.

Invited paper

J.-P. BONNET
CEAT/LEA, CNRS 191
Université de Poitiers

J. LEWALLE
Syracuse University
on leave at CEAT/LEA

AND

M.N. GLAUSER
Clarkson University

1. Background

It is now commonplace to recognize that coherent structures (CS's), beautifully illustrated in the art of many cultures, but long held to be marginal (transitional, low-Reynolds-number) in the science and engineering of fluid turbulence, were brought back to the core of the field by the work of Townsend [31] and the explosion of literature following the paper of Brown & Roshko [10] and e.g Hussain [20].

Here, we define a coherent structure as a fluctuation, identified by a concentration of large-scale vorticity and significant energy levels, organized by a mean velocity gradient, that maintains its identity for several turnover times. Several variants of the concept have been proposed, including the definition of a mean and a fluctuation, the identification of a vortex, the energy content, the spatial and temporal extent. These variants reflect partly differences between physical models, partly a pragmatic acknowledgement of experimental, numerical or analytical potentialities and limitations. While CS's occur in large Reynolds number flows, stability studies [23] and LES [22] show that a subgrid or effective viscosity captures the effect of small scales on the CS's.

S. Gavrilakis et al. (eds.), Advances in Turbulence VI, 83-90.
© *1996 Kluwer Academic Publishers.*

Furthermore, it seems generally accepted that the CS's are distinctive by flow type [33, 34, 29, 6], reflecting of the distinctive organizing velocity gradient (RDT [30], stability [25], linearized NS); that CS's are non-linear phenomena; that CS's are Lagrangian flow markers [27], characterized by low diffusion, associated with their lifespan (relation to the vorticity theorems [21, 32]); that they constitute a significant part of the large eddy population, which makes them important for turbulent transport and engineering applications; that they can be triggered, phase-locked, and manipulated; and consequently that their modeling and control is of technological value.

We will use an incompressible mixing layer as the example of application on which we are pursuing a collaborative approach combining different techniques.

2. Coherent structures in a mixing layer

This collaborative effort has been devoted to processing of data obtained with rakes of hot wires (at most 24 X wires) in a subsonic, plane mixing layer originated from fully turbulent boundary layers. Several groups from Europe, Japan and USA are engaged in this collaborative project and covers the essential aspects of CS detection and analysis. Some methods are not described here, such as Pseudo-Flow-Visualization or Pattern Recognition Technique [14]. The methods of CS eduction are based on complementary ideas. We will briefly describe the methods, the major results and attempt an evaluation of their effectiveness in providing physical insight of CS's, their modeling and control.

2.1. CONDITIONAL SAMPLING

Conditional sampling techniques have been traditionally related to the 'conventional' definitions of CS's. These techniques give the instantaneous location of a given CS [17], and also yield the average (most 'dominant' or most 'likely') structure. The Houston group [7] has developed an approach based on the Vorticity Based Conditional Technique. A version of this approach, based on weighted average gradient (WAG), has been applied by the Newcastle group [4]. A different approach from the Tokyo group [26] uses velocity based conditional sampling to obtain CS. Another conditional sampling scheme (Poitiers) [6] uses the multipoint data made available by the use of hot-wire rakes: this allows for the defintion of 'delocalized' conditions. The Berlin group uses a correlation technique specifically designed for the mixing layer, and this is used in association with the POD (below). This is equivalent to a Pattern Recognition Analysis. Topology theory was used by the Berlin group to determine the vortex regions in the mixing layer.

The criterion for the dividing line between vortex and saddle points [19, 28] was applied by the Berlin group [7]. All these methods have allowed to give quite comparable multiple decomposition analysis of the mixing layer.

2.2. THE WAVELET TRANSFORM

The combination of temporal and spectral discrimination has been made possible by the Wavelet Transform [15] (WT). The technique was used in this context by Kevlahan [7], who identified a wavelet 'signature' of coherent structures and calculated conditional power spectra exhibiting different slopes inside and outside the structures. Lewalle et al. [24] devised and intermittent filtering algorithm that yields a decomposition of the velocity traces into a coherent and a random part (Fig. 1). The associated power spectra (Fig. 2) show clearly that the random field exhibits a 'normal' turbulent spectrum, while the coherent field accounts for the broad peak at the dominant frequency. This accomplishes a decomposition similar to the one advocated by Brereton & Kodal [9]. In addition the application of topological techniques [28] to the coherent components permitted the identification of vortices and saddle points with single- or multiple cores, and conditional statistics have been calculated.

2.3. PROPER ORTHOGONAL DECOMPOSITION (POD)

As introduced by Lumley, the POD [3] consists in the identification of the coherent structure as that which has the largest mean-square projection on the flow field. Several approaches can be retained for the implementation of the method: snapshot POD was applied by the Berlin group [18], while the Clarkson group [7] and the Poitiers group [13] apply the conventional POD based on scalar and vector approaches. Further developments, combining conditional methods or Linear Stochastic Estimation are also developed by these groups. It appears that about 45 % of the turbulent energy is captured by the first mode. Temporal reconstruction of the flow field, made possible thanks to the instantaneous knowledge of the individual signals from hot wire rake, shows that the CS are well captured by the first mode, making realistic the use of low dimensional models that will be described later on.

2.4. LINEAR STOCHASTIC ESTIMATION (LSE)

Linear Stochastic Estimation (LSE), as introduced by Adrian [1], uses the conditional information specified about the flow at one or more locations in conjunction with its statistical properties (the two-point correlation tensor) to estimate the information at the remaining locations. Here the thrust will be to utilize the instantaneous velocity at selected y locations in the mixing layer to estimate the instantaneous velocity for all y locations. It has been shown that two well-located reference time histories are sufficent to provide an excellent estimation of the large scale behaviour of the flow. As a result of the above, the estimated time series can be used in conjunction with the POD to implement the so-called complementary technique [6] resulting in a very efficient tool and was used by the Clarkson and Poitiers groups.

2.5. OVERVIEW OF COMMON RESULTS

The applications of several conditional methods and/or concepts on the same database has proven that they are in close agreement as well as for qualitative or quantitative description of CS. As an example, Fig. 3 shows that the detection instants are equivalent. Fig. 4 shows that the average structures are quite comparable, with less than ±6%. All the methods have been proved to be robust, in the sense that the average structures are not very sensitive to threshold level[1]. For further understanding of the physics (topology, dynamics etc.) of the CS, 3D data have to be processed. These data can come from simulations (DNS, LES) or experiments using some special probe arrangement and making use of LSE [13]. The combination of several methods (POD/LSE, WAG/vorticity, wavelets/conditional, etc.) is now possible and can allow to perform detailed validations of various prediction methods and can open some new roads of prediction and control.

3. Prospects for modeling and control

In the end, the knowledge acquired about turbulent flows and coherent structures will be used for engineering predictions of flow properties, and for the management or control of specific flow features to improve system performance.

3.1. MODELING

Turbulence computing ranges (in sophistication, physical realism and required hardware) from algebraic models to DNS. Most engineering models of turbulent flows do not yet include specific structural information. As new models emerge, we need to assess what parameters of the stuctures are essential in affecting flow statistics. Our ability to simulate (DNS, LES) coherent structures and to manipulate them numerically, will complement experiments in the identification of key parameters.

---

[1]However, the number of structures identified (or efficiency) can vary by a ratio of the order of 4, depending on threshold level; this issue is not central to some published results

From that perspective, we anticipate that wavelet methods will be efficient in the extraction of conditional statistics and spectra, i.e. the information necessary for model development and testing. Already, we have the ability to extract coherent structures from high-Reynolds number flows, to categorize them according to various topological criteria, and to calculate the pdf's, turbulent fluxes, spectral content and intermittency associated with each category. The algorithms should apply to numerical as well as to experimental data.

This information will hopefully feed directly into hybrids of proven models, such as the so-called semi-deterministic model [16], and possibly lead to the emergence of new modeling concepts. Also, the spectral interactions between the coherent and random contributions to the velocity field will be compared to spectral interactions in LES.

3.2. CONTROL

Turbulence manipulations predate the acceptance of coherent structures. The list includes polymer additives or suspensions for drag reduction, temperature effects or Lorentz forces for transition control, acoustic excitation, LEBU and MEM devices for CS modification. The control of turbulent flows can be local or global, with feedback or 'blind'. Recent trends favor local controls with feedback. To this end, better knowledge of the dynamics of the CS is essential.

The POD is particularly well suited to this goal, because of its ability to yield low-dimensional dynamical systems that capture the dominant features of non-homogenous flows. The approach of Lumley [8] has been developed by several authors, in boundary layer flows [2], in jet flows [12] or in mixing layers [11]. As an example, the dynamical systems that can be used for control are based on the first mode of the POD. This projection results in a set of ODE's for the random coefficients. The motions of CS (1st mode) can be examined through bifurcation theory. It has to be outlined that this approach requires closure of a new kind and then specific models have to be conceived and tested. Fig. 5 [12] shows the typical behaviour of the time dependent coefficients. From these results new spectral energy transfers can be observed and can lead to control strategies that were not devisable from conventional (Fourier as an exampe) approaches.

3.3. CONCLUSIONS AND EXPECTATIONS

In recent years, we have witnessed and participated in the emergence of new measurement techniques (not all of which could be included in this paper, e.g PIV), new processing algorithms and more detailed and precise simulated (DNS) and modeled (LES) flow data, including non-stationary

88

conditions. The interplay of these techniques, and their mutual validation, has greatly contributed to our growing knowledge of CS's. However, we should not overlook unanswered questions, fundamental and practical, or ignore new questions generated by the new data.

As the syndrome of CS's is better documented, we believe that a fundamental understanding of their role in fluid turbulence will develop. Although we 'see' CS's coming out of the equations of motion, a clear connection between Navier-Stokes dynamics and CS dynamics is an important missing link.

Modeling trade-offs are still hard to evaluate, mainly because no *CS-based* model suitable for engineering applications has yet been proposed. The identification of key parameters, the evaluation of the qualitative and quantitative importance of CS's for a given application, the use of modeling as a form of experimentation, are thus delayed. However, we believe that the information necessary to develop and test such models — either by adaptation of existing models, or by the creation of entirely new modeling tools, or the combination of both[2] — is now becoming available. While wavelets are not a panacea, they will probably be a dominant tool of data reduction in this context.

When it come to CS control, CS dynamics (parameters, stability, etc) and implementation are the central issues. POD appears to be the most promising method in terms of the necessary predictive capabilities. Thus, a clear line of study connects the eigenfunctions, the dynamical system and its properties. The actual manipulation of CS's, based on such predictions, will require much ingenuity. Truly 'smart' (active) control requires the interplay of dynamical understanding of the role of governing parameters, and of a matching action on the flow itself. Among the specific new questions that arise in this context are the problem of closure in "POD" space, and the management and manipulation of very large data sets (time-dependent 3-D velocity fields).

Acknowledgements

This work is based on data collected by Dr. J. Delville, who contributed many useful suggestions and remarks. JL gratefully acknowledges the support of the Université de Poitiers and CNRS.

---

[2]The question remains open as to when CS must be explicitly part of a model or when they can be ignored for a given application

References

1. R.J. Adrian & P. Moin, 1988, J. Fluid Mech. 190, 531-559.
2. N. Aubry, P. Holmes, J.L. Lumley and E. Stone, 1988, J. Fluid Mech. 192, 115-173.
3. G. Berkooz, P. Holmes & J.L. Lumley, 1993, Ann. Rev. Fluid Mech. 25, 539-575.
4. D.K. Bisset, R.A. Antonia & L.W.B. Browne, 1990a, J. Fluid Mech. 212,349-361.
5. J.P. Bonnet & M.N. Glauser Eds., 1992, Eddy structure identification in free turbulent shear flows, Proceedings of the IUTAM Symposium – Poitiers October 1992, Kluwer Academic Publishers
6. J.P. Bonnet, D.R. Cole, J. Delville, M.N. Glauser & L.S. Ukeiley, 1994, Exp. Fluids 17, 307-314.
7. J.P. Bonnet, M.N. Glauser, R.A. Antonia, S. Bellin, D.K. Bisset, D.R. Cole, J. Delville, H.E. Fiedler, J.H. Garem, D. Hilberg, J. Jeong, N.K.R. Kevlahan, M. Maekawa, H. Takahashi, L. Ukeiley & E. Vincendeau, 1996, submitted, Exp. in Fluids.
8. J.P. Bonnet Ed., 1996, Eddy structure identification techniques for turbulent shear flows, CISM Course Series 353, Springer Verlag.
9. G.J. Brereton & A. Kodal, 1994, Phys. Fluids 6, 1775-1786.
10. G.L.Brown & A. Roshko, 1974, J. Fluid Mech. 64, 775-816.
11. L. Cordier, R. Manceau, J. Delville & J.P Bonnet, 1996, this meeting.
12. T. Corke, M.N. Glauser & G. Berkooz, 1994, Appl. Mech. Rev. 6, S132-S138.
13. J. Delville, 1993, in [5], 225-238.
14. J.A. Ferre & F. Giralt, 1993, in [5],181-194.
15. A. Grossman & J. Morlet, 1984, S.I.A.M. J. Math. Anal. 15, 723-736.
16. Ha Minh H., 1994, Order and disorder in turbulent flows: their impact on turbulence modelling, Osborne Reynolds Centenary Symposium, UMIST – Manchester.
17. M. Hayakawa, 1993, in [5], 45-64.
18. D. Hilberg, W. Lazik & H.E. Fiedler, 1993, in [5], 251-260.
19. J.C.R. Hunt, C.J. Abell, J.A. Peterka & H. Woo, 1978, J. Fluid Mech. 212, 497-532.
20. A.K.M.F. Hussain, 1983, Phys. Fluids 26, 2816-2850.
21. A.Leonard, 1985, Ann. Rev. Fluid Mech. 17:523-559.
22. M. Lesieur, 1990, Turbulence in Fluids, Kluwer Ac. Pub.
23. M. Lessen, 1978, J. Fluid Mech. 88:535-540.
24. J. Lewalle, J.-P. Bonnet & J. Delville, 1996, in preparation.
25. J.T.C. Liu, 1989, Ann. Rev. Fluid Mech. 21:285-35.
26. H. Maekawa, T. Nozaki & S. Ohsako, 1988, 11th Symposium on Turbulence, Missouri-Rolla.
27. H.K. Moffatt & A. Tsinober, 1992, Ann. Rev. Fluid Mech. 24:281-312.
28. A.E. Perry & M.S. Chong, 1987, Ann. Rev. Fluid Mech. 19:125-155.
29. S.K. Robinson, 1991, Ann. Rev. Fluid Mech. 23:601-639.
30. A.M. Savill, 1987, Ann. Rev. Fluid Mech. 19:531-575.
31. A.A. Townsend, 1956, The structure of turbulent shear flows, Cambridge University Press.
32. J.M. Wallace & J.F. Foss, 1995, Ann. Rev. Fluid Mech. 27:469-514.
33. C.D. Winant & F.K. Browand, 1974, J. Fluid Mech. 63, 237-255.
34. K.B.M.Q. Zaman & A.K.M.F. Hussain, 1984, J. Fluid Mech. 138, 325-351.

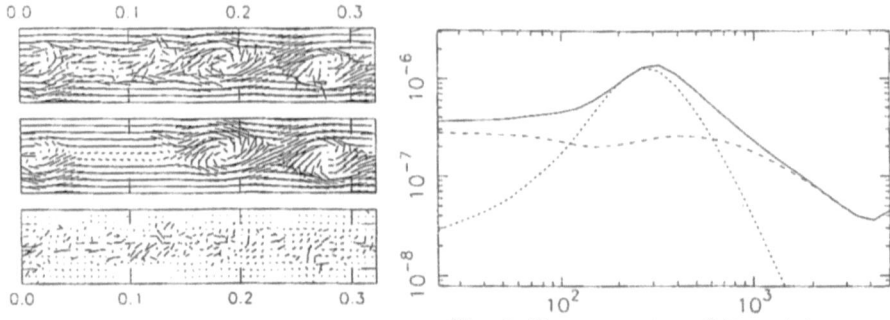

Fig. 1: Vector plots of original velocity field, wavelet-filtered (coherent) field and random field.

Fig. 2: Power spectra of the original (solid), coherent (dotted) and random (dashed) traces of the v component at the mixing layer center line.

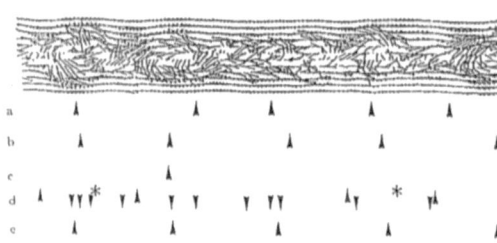

Fig. 3: Comparison of detection stamps from various conditional sampling methods (after [7]).
a) Wavelet; b) Correlation;
c) Vorticity; d) WAG; e) Delocalized.

Fig. 4: Average conditional structure.
a) From 1st POD mode (correlation sampling, Berlin group); b) Vorticity detection (Houston group); c) Vorticity detection (Newcastle group); d) Delocalized (Poitiers group) (after [7]).

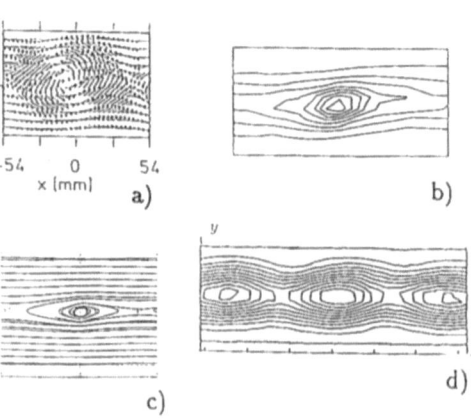

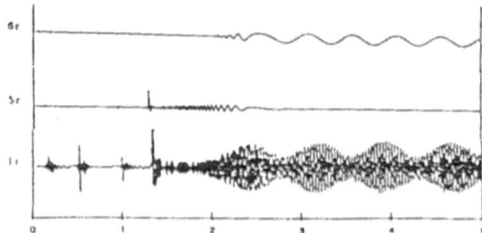

Fig.5: Time evolution of azimuthal modes in a jet, showing energy exchanges (after [12]).

GROWTH AND DECAY OF LONGITUDINAL ROLL CELLS IN ROTATING TURBULENT PLANE COUETTE FLOW

K.H. BECH & H.I. ANDERSSON
Division of Applied Mechanics
Norwegian University of Science and Technology
N-7034 Trondheim, Norway

1. Background and Motivation

The plane Couette flow, i.e. the shear-driven fluid motion between two parallel planes in relative motion, is one of the fundamental prototype flows in classical fluid mechanics. The presence of the roll-cell instability in *laminar* plane Couette flow subject to system rotation has been explored by Lezius & Johnston (1976) and Speziale & Wilson (1989), while the existence of persistent pairs of counterrotating longitudinal vortices or roll cells in a weakly rotating *turbulent* Couette flow was first observed by Bech & Andersson (1995). In the latter numerical study, roll cells were found to be present for weak anticyclonic rotation, i.e. the imposed background or system vorticity is antiparallel to the mean flow vorticity. More surprisingly, however, was that in spite of the destabilizing influence of the Coriolis force with respect to roll cell formation, the turbulence level was actually reduced.

The objective of the research to be reported herein was to explore the effect of stronger rotation on the behaviour of the roll cells and the turbulence structure. In contrast with rotating Poiseuille flow, which due to the shape of the mean velocity profile exhibits both a cyclonic and an anticyclonic side, the rotating Couette flow enables us to study the influence of cyclonic and anticyclonic rotation separately (simply by changing the sense of rotation) since the Coriolis force has the same effect on both sides of the Couette channel. This attractive feature is not shared by rotating turbulent Poiseuille flow, in which roll cells have been observed experimentally by Johnston *et al.* (1972) and numerically by Kristoffersen & Andersson (1993).

2. Problem Formulation

The problem considered is that of statistically steady turbulent flow bounded by two infinite parallel planes separated a distance 2h, the fluid motion being induced solely by the velocity difference $2U_w$ between the planes. The incompressible flow is assumed to be fully developed in the streamwise x-direction and the Couette channel rotates with constant angular velocity Ω about the spanwise z-axis of a Cartesian coordinate system, i.e. in orthogonal mode rotation. The two independent dimensionless parameters inherent in the flow problem are thus the Reynolds number $Re = U_w h/\nu$ and the rotation number $Ro = 2\Omega h/U_w$.

S. Gavrilakis et al. (eds.), Advances in Turbulence VI, 91-94.
© *1996 Kluwer Academic Publishers.*

3. Mathematical Modelling and Computational Approach

Fully developed flows are in general difficult to realize in a laboratory since a certain length is needed in order to develop the flow. This would be further complicated by the need for a rotating apparatus. We have therefore adopted the direct numerical simulation (DNS) approach, and integrated the complete time-dependent three-dimensional Navier-Stokes equations numerically on a discrete $256 \times 70 \times 256$ grid system sufficiently fine to resolve all essential scales of the turbulence. The computational domain was $10\pi h$ long and the aspect ratio of its cross-section was $4\pi h/2h = 2\pi$, thereby leaving room for three pairs of counter-rotating vortical cells. The simulation code was an adapted version of the finite-difference code ECCLES developed by Gavrilakis *et al.* (1986) with second-order central-difference approximations in space and a second-order explicit Adams-Bashforth scheme in time. System rotation was readily accounted for by implementation of extra body force terms (Coriolis) in the governing momentum equations, after first having demonstrated the physical realism of a numerically simulated Couette flow in a fixed coordinate system, cf. Bech *et al.* (1995).

The Reynolds number Re was kept constant and equal to 1300 throughout the investigation and statistically steady flow fields were generated for positive rotation numbers equal to 0.10, 0.20 and 0.50, whereas results for weak cyclonic (Ro = -0.01) and anticyclonic (Ro = +0.01) rotation have been reported elsewhere (Bech & Andersson, 1995). To facilitate the analysis of the simulated flow fields the instantaneous velocity u_i' was decomposed into three parts

$$u_i' = U_i(y) + \tilde{u}_i(y,z) + u_i(x,y,z,t)$$

where U_i is the one-component mean or background flow, $\tilde{u}_i$ represents the three-component secondary flow field associated with the roll cells, and u_i denotes turbulent fluctuations. This decomposition has previously been applied by Moser & Moin (1987) to explore the roll cells in numerically simulated Dean flow, i.e. curved channel flow. The secondary flow field $\tilde{u}_i$, which is of particular interest in the present study, can readily be obtained as $\tilde{u}_i = \overline{u_i'} - U_i$ where the overbar denotes an average in the homogeneous x-direction and in time.

4. Results and Discussion

The secondary flow field $\tilde{u}_i(y,z)$, which may result from the imposed system rotation, consists of a streamwise component $\tilde{u}$ and the cross-flow components $\tilde{v}$ and $\tilde{w}$. Since $\partial \tilde{u}/\partial x = 0$, the cross-flow obeys the continuity constraint $\partial \tilde{v}/\partial y + \partial \tilde{w}/\partial z = 0$ and may therefore be associated with a stream function. Contour lines for the latter, scaled with $U_w h$, are shown in Figure 1, from which it can be observed that the roll cell patterns for the intermediate rotation numbers 0.10 and 0.20 are far more regular than for weak rotation Ro = 0.01. When the rotation number was further increased to 0.50, a disordering of the predominant roll cell pattern occurred during the transient stage until the flow field settled at a new statistically steady state substantially less affected by roll cells, cf. Figure 1.

The fluid motion associated with the roll cells as well as the velocity fluctuations associated the turbulence represent kinetic energy. The partition of the overall energy among the two kinds of kinetic energy can be seen in Table 1. Although a substantial fraction of the kinetic energy was associated with the roll cells at low rotation (Ro = 0.01), the roll cells at the moderate rotation numbers 0.10 and 0.20 were far more energetic than at low Ro and, moreover, contained more than 4 times the energy of the turbulence itself. It is also noteworthy that the peculiar tendency for the turbulence to be reduced for Ro = 0.01 as compared to the non-rotating case is growing further as the rotation rate is increased. The turbulence level was, in fact, reduced by some 50 per cent at Ro = 0.10 as compared to the case Ro = 0. As far as the secondary flow $\tilde{u}_i$ is concerned all three velocity components contribute to the kinetic energy associated with the roll cells. At Ro = 0.01 only 10 per cent of the energy is contributed by the cross-flow components $\tilde{v}$ and $\tilde{w}$ which have been visualized in Figure 1 and the remaining 90 per cent stems from the streamwise component $\tilde{u}$. At the higher rotation rate Ro = 0.10 the contribution from $\tilde{u}$ to the kinetic energy of the roll cells is reduced to 40 per cent.

When the rotation number was raised to 0.50 and the roll cell breakdown set in, the overall energy content decreased but the turbulence level was substantially enhanced (by about 200 per cent) and even exceeded that for Ro = 0. An unusually high and uniform

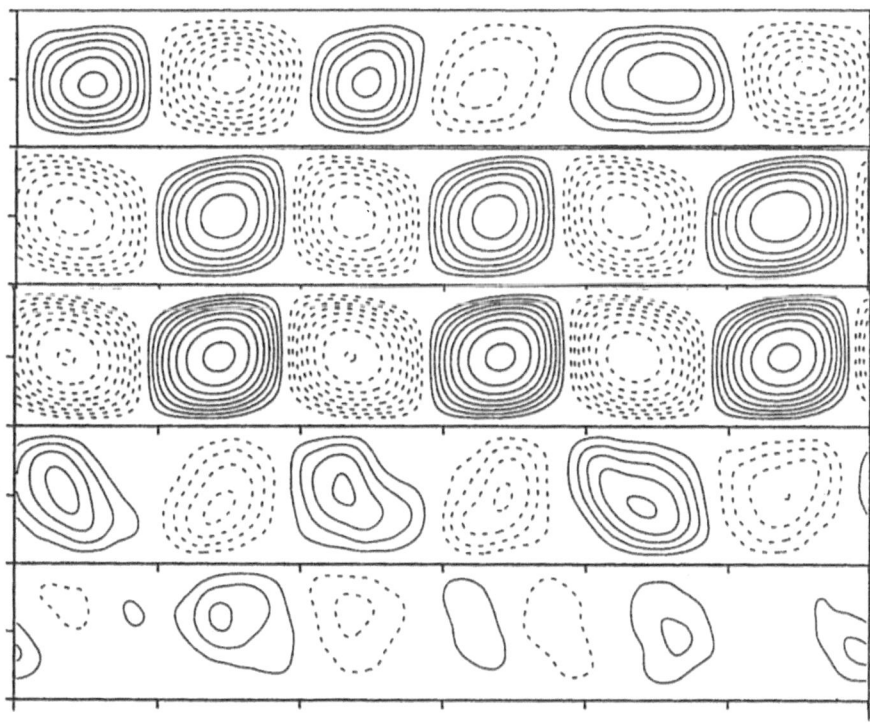

Figure 1. Contour lines of the cross-flow stream function. From top to bottom: Ro = 0.01; Ro = 0.10; Ro = 0.20; transient roll cell breakdown; Ro = 0.50. The increment between the contour lines for Ro = 0.01 is 1/4 of that for the other cases.

TABLE 1. Kinetic energy of the roll cells and the turbulence.
Volume-averaged data scaled with u_τ^2 for Ro = 0.

Ro	Re_τ	Kinetic energy		
		Roll cells	Turbulence	Sum
0	82.2	-	3.28	3.28
0.01	84.7	1.28	2.59	3.87
0.10	106.7	5.98	1.41	7.39
0.20	107.2	7.34	1.57	8.91
0.50	91.0	0.54	4.84	5.38

enstrophy level could then be observed in a wide central core region. This behaviour was ascribed to a substantial amplification of the streamwise vorticity and was accompanied by an anomalous variation of the mean turbulent kinetic energy which peaked in the middle of the channel rather than near the walls.

Finally, it should be recalled that the Reynolds number Re was equal to 1300 for all simulations. The variation of $Re_\tau = u_\tau h/\nu$ with Ro in Table 1 therefore reflects the variation of the wall friction velocity $u_\tau = (\tau_w/\rho)^{1/2}$ with rotation. The increasing tendency up to Ro = 0.20 should obviously be ascribed to the presence of the quite energetic roll cells.

The computer simulations were supported by The Research Council of Norway (Programme for Supercomputing) through a grant of computing time.

References

Bech, K.H. and Andersson, H.I. (1995) Secondary flow in weakly rotating turbulent plane Couette flow, in *Proc. 10th Symposium on Turbulent Shear Flows*, Pennsylvania, pp. 8.1-8.6. To appear in *Journal of Fluid Mechanics*.

Bech, K.H., Tillmark, N., Alfredsson, P.H., and Andersson, H.I. (1995) An investigation of turbulent plane Couette flow at low Reynolds numbers, *Journal of Fluid Mechanics* **286**, 291-325.

Gavrilakis, S., Tsai, H.M., Voke, P.R., and Leslie, D.C. (1986) Large-eddy simulation of low Reynolds number channel flow by spectral and finite difference methods, in U. Schumann and R. Friedrich (eds.), *Direct and Large Eddy Simulation of Turbulence*, Notes on Numerical Fluid Mechanics Vol. 15, Vieweg Verlag, Braunschweig, pp. 105-118.

Johnston, J.P., Halleen, R.M., and Lezius, D.K. (1972) Effects of spanwise rotation on the structure of two-dimensional fully developed turbulent channel flow, *Journal of Fluid Mechanics* **56**, 533-557.

Kristoffersen, R. and Andersson, H.I. (1993) Direct simulations of low-Reynolds-number turbulent flow in a rotating channel, *Journal of Fluid Mechanics* **256**, 163-197.

Lezius, D.K. and Johnston, J.P. (1976) Roll-cell instabilities in rotating laminar and turbulent channel flows, *Journal of Fluid Mechanics* **77**, 153-175.

Moser, R.D. and Moin, P. (1987) The effect of curvature in wall-bounded turbulent flows, *Journal of Fluid Mechanics* **175**, 479-510.

Speziale, C.G. and Wilson, M.B. (1989) Numerical study of plane Couette flow in a rotating framework, *Acta Mechanica* **77**, 261-280.

VORTEX PERSISTENCE

A New Parameter for Vortices near Interfaces

A.J. COTEL and R.E. BREIDENTHAL
University of Washington
Box 352400, Seattle, WA 98195-2400, USA

1. Introduction

When a turbulent vortex is near a thin, stratified interface, the entrainment rate across the interface depends on three conventional parameters: the Richardson, Reynolds, and Schmidt numbers (Turner 1973). However, recent experiments have revealed that these parameters by themselves are not sufficient to determine the entrainment rate. For example, a horizontal jet parallel to and below the interface entrains at a rate proportional to $Ri^{-3/2}$ (Schneider 1980), while the same jet, operating at identical values of these parameters but oriented to impinge vertically on the interface, entrains at a rate proportional to $Ri^{-1/2}$ (Cotel 1995). Therefore, an additional parameter must be necessary to determine the entrainment rate. This paper presents the arguments for one particular candidate.

2. A new parameter

Cotel found that the vertical jet creates an impingement dome, generating persistent lateral vortices which are responsible for the entrainment. On the other hand, the entraining eddies in the horizontal jet move rapidly with respect to the interface. This suggests that motion of the vortices *with respect to the interface* may be important in selecting the entrainment regime.

A new parameter has been proposed in a model to describe entrainment when a turbulent vortex is near an interface (Breidenthal 1993 & 1994, Cotel & Breidenthal 1994, Cotel 1995). The persistence parameter T is defined to be the number of times a vortex rotates while it moves a distance with respect to the interface equal to its own diameter.

95

S. Gavrilakis et al. (eds.), Advances in Turbulence VI, 95-98.
© *1996 Kluwer Academic Publishers.*

This is proportional to the eddy velocity ratio, i.e., the ratio of the rotational to the translational speed of the eddy.

To any inertial observer, the rotational speed is unambiguous, but the translational speed must be defined in the frame of reference of the interface. In principle, the interface could be any definable surface, such as the stratified interface discussed above, or even the surface of a nearby vortex. It could also be a solid wall.

3. A model for stratified entrainment

Figure 1 is a theoretical entrainment diagram of the different entrainment regimes in the model for a thin, stratified interface at Sc>1. The entrainment rate is labeled in each regime. For example, Regime I (Ri<1) is the unstratified limit, where the entrainment rate is a constant, independent of Ri and the other parameters. For simplicity, all coefficients here are presumed to be one. In Regime II, the entraining eddies are persistent, and the rate is proportional to $Ri^{-1/2}$ as in the case of the stationary vertical jet discussed in the Introduction.

At Ri=T, there is a discontinuous transition, indicated by the double line, to nonpersistent Regime III. The entrainment rate goes as $Ri^{-3/2}$, corresponding to the case of the aforementioned horizontal jet. Following Linden (1973), the model assumes that nonpersistent entrainment occurs when impinging vortices rebound from the interface, entraining tongues of fluid across it as they rebound away.

With increasing stratification, eventually the entrainment rate from this rebound process is exceeded by Kolmogorov microscale nibbling of the Taylor layers. Since the latter are diffusive, the Schmidt or Prandtl number now becomes important. The transition to Regime IV occurs at Ri=Sc.

If $Ri > Re^{1/4}$, even the smallest eddies at the Kolmogorov microscale do not have enough kinetic energy to engulf a tongue of buoyant fluid across the interface. Then the interface is flat for all eddy scales, and the Richardson number drops out of the problem. Entrainment occurs via molecular diffusion across the flat interface, just as in the case of heat transfer at a solid wall. According to surface renewal theory, the entrainment velocity is

$$w_e = \sqrt{\mathcal{D}/\tau_\lambda},\qquad(1)$$

where $\mathcal{D}$ is the molecular diffusivity and τ_λ is the time scale of the rate-limiting eddy. In the nonpersistent limit (Regime V), τ_λ is the smallest eddy time, and in the persistent limit (Regime VI), it is the largest eddy time.

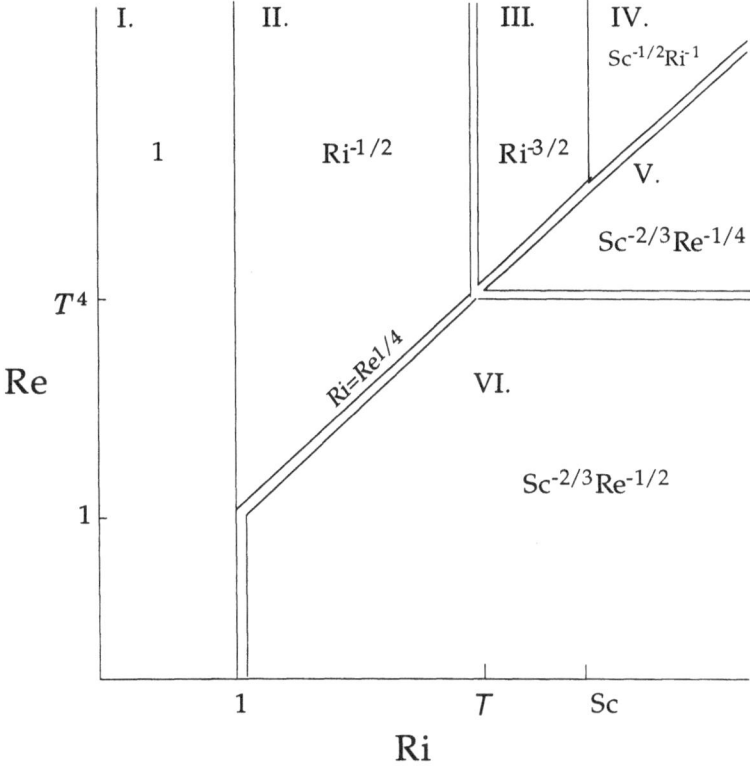

Figure 1. Entrainment diagram for a thin, stratified interface

4. Comparison with experiment

Approximately 85% of the known laboratory experiments seem to be in accord with the model, as is discussed in detail by Cotel (1995). To mention a few examples here, Turner (1968) observed that an oscillating grid in water entrains fluid across a heat-stratified interface at a rate that changes at about Ri=7, which is the Prandtl number for water. This is consistent with Regimes III and IV and the transition between them.

Edwards and Furber (1956) found that heat transfer in a forced, laminar boundary layer is completely independent of the amplitude of the fine scale, freestream turbulence. The mean flow is persistent, so the small scale turbulence has no effect on the flux (Regime VI).

Gharib *et al.* (1994) concluded that the trajectories of starting jets in laboratory simulations of human heart flows are governed by what they call a 'formation number'. This is precisely proportional to the persistence parameter.

5. Conclusions

Experiments of stratified entrainment, wall heat transfer, and starting jets in confined regions all support the notion that the entrainment rate near *any* type of interface is strongly influenced by a new parameter, the vortex persistence T. Most of the experimental observations are in accord with an entrainment model based on this concept.

A measure of the stationarity of the largest eddies, the persistence parameter is independent of Richardson, Reynolds, and Schmidt numbers. When a vortex is near an interface, the vortex persistence plays a major role, equal to that of any of the conventional parameters.

6. References

Breidenthal, R.E. (1993) Simple models of stratified and compressible entrainment, *Workshop on Compressible Turbulent Mixing*, Cambridge University.

Breidenthal, R.E. (1994) Entrainment and mixing in stratified flows, *Appl. Mech. Rev.* **47**(6), part 2, S108-S112.

Cotel, A.J. (1995) Entrainment and detrainment of a jet impinging on a stratified interface, Ph.D. thesis, University of Washington.

Cotel, A.J. and Breidenthal, R.E. (1994) Persistence theory of stratified entrainment, *Preprints of the Fourth International Symposium on Stratified Flows*, **4**, Grenoble and submitted to *Journal of Fluids Engineering*.

Edwards, A. and Furber, B.N. (1956) The influence of free stream turbulence on heat transfer by convection from an isolated region of a plane surface in parallel air flow, *Proc. Inst. Mech. Eng.* **170**, 941.

Gharib, M., Rambod, E., Dabiri, D., Shiota, T., and Sahn, D. (1994) Pulsatile heart flow: A universal time scale, *International Conference for Experimental Fluid Mechanics*, Torino.

Linden, P.F. (1973) The interaction of a vortex ring with a sharp density interface: A model for turbulent entrainment, *J. Fluid Mech* **60**, 467-480.

Schneider, H-H. (1980) Laboratory experiments to simulate the jet-induced erosion of pycnoclines in lakes, *Proc. Int. Sympos. Stratif. Flows*, Nor. Inst. Technol., Trondheim.

Turner, J.S. (1968) The influence of molecular diffusivity on turbulent entrainment across a density interface, *J. Fluid Mech.* **33**, 639-656.

Turner, J.S. (1973) *Buoyancy Effects in Fluids*, Cambridge University Press.

THE RELATION BETWEEN THE INTERNAL STRUCTURE OF THE COHERENT VORTEX AND HEAT TRANSFER PROCESSES

STANISŁAW DROBNIAK[*], ELSNER J.W.[*],
EL-SAYED ABOU-EL-KASSEM[**]
[*]Institute of Thermal Machinery
Technical University of Częstochowa
Al.Armii Krajowej 21, 42-200 Częstochowa, Poland
[**]Cairo University, Faculty of Engineering, Giza, Egypt

The paper presents an experimental analysis concerning the mutual relation between the structure of coherent vortices existing in the round, free jet and their involvement into the transport processes realized in this type of flow. The heat transfer processes are usually treated in a manner analogous to momentum transport phenomena, even in the flows containing the organized vorticity, where much larger scales are encountered than in the classical, random turbulence. The above processes become a complex matter when the existence of coherent structures is taken into account, because in this case the turbulent diffusion is being replaced by large scale convection, as the major transport mechanism. The axisymmetric free jet has been chosen as the flow case studied, because of the simultaneous presence of several different modes of organized motion characterized by substantially different scales. The advanced visualization experiments used contemporarily [1] allow to trace the dynamics of coherent vortices but they can not give the quantitative information concerning the transport properties of the phenomena studied. That is why it has been decided to apply the proper computer-aided data processing that would enable to reconstruct the flowfields related to oganized and random components of turbulent motion, respectively. Furthermore the small overheat of the flow has also been applied, which enabled to trace the mutual interactions of velocity and temperature fields.

According to the above idea it was decided to explore the initial region of the round, free jet issuing from the contoured nozzle D = 0.04 m, at the $Re_D = 4 \cdot 10^4$, that corresponded to the jet exit velocity U_o = 19.4 m/s. The overheat of the flowing medium (i.e. the difference between the exit θ_o and ambient θ_a temperatures) was kept at the constant level $\Delta \theta = 40° C$. Following the previous works of Hussain [2] and Favre-Marinet [3] the column-mode structures, characterized by Strouhal number $St_D = 0.42$, have been chosen as an object of the present study.

S. Gavrilakis et al. (eds.), Advances in Turbulence VI, 99-102.

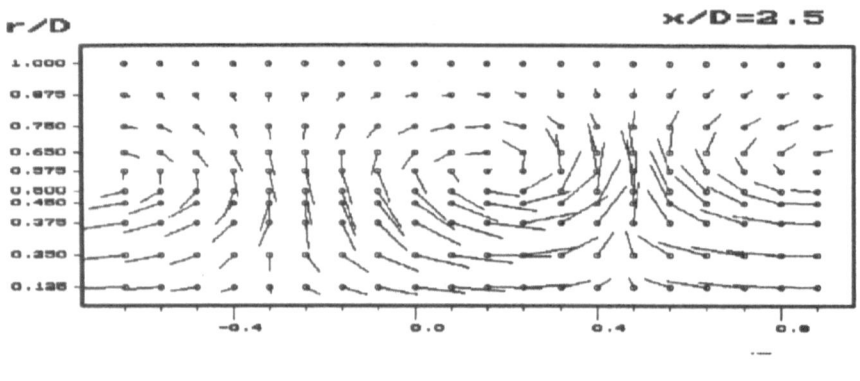

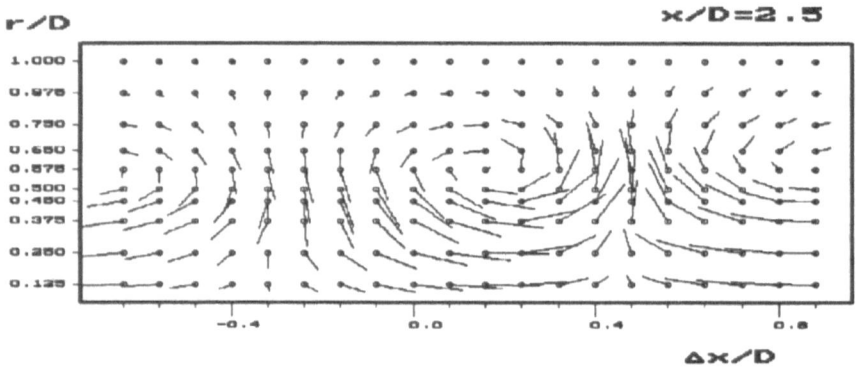

Fig.1. Phase-averaged total (a) and periodical (b) flow-field patterns at the cross-section x/D=2.5

The above structures were stabilized by pure - tone acoustic field, which cancelled the random dispersion of their size, orientation and shedding frequency and furthermore enabled one to use acoustic pressure as a reference signal in phase-averaging procedure. Velocity U_i and tempera-ture θ fields were treated as a superposition of mean-time $(\bar{\ })$, periodic (~) and random (') components, i.e.:

$$U_i = \overline{U}_i + \tilde{u}_i + u_i' \quad ; \quad \theta = \overline{\theta} + \tilde{v} + v'$$

The simultaneous measurements of three velocity components and instan-taneous temperature have been performed with the use of four-wire combined HWA probe; more details concerning the experimental facility and data processing techniques may be found in [4].

The phase averaged pictures presenting the vortex flowfield have been presented in Fig.1 for the sample cross-section x = 2.5D, where the maximum spatial coherence of the structure could be observed. One may

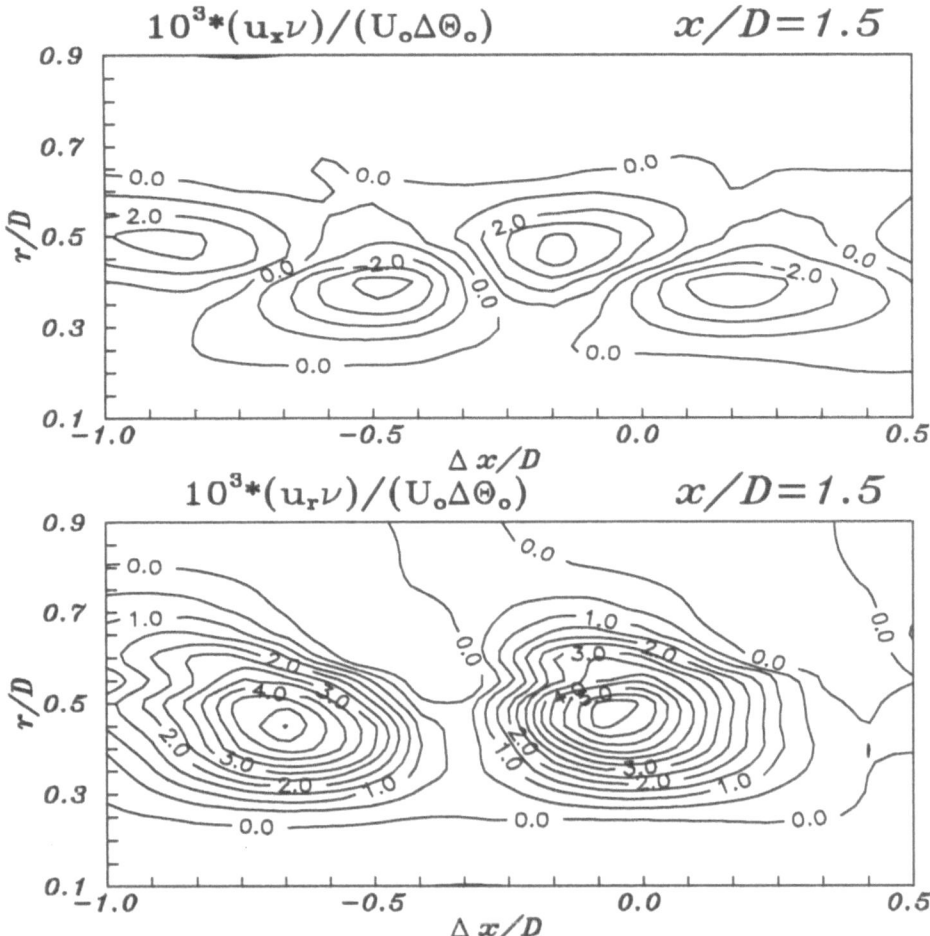

Fig.2. Isoline contours of phase averaged longitudinal (a) and radial (b) oscillatory heat fluxes at x/D=1.5 cross-section (centre of vortex in the middle of the picture).

easily notice the regions of intense radial transport at the front and rear sides of the vortex (centre of the structure in the middle of the picture), as well as its overall flow-field pattern. The triple decomposition applied for all the components of turbulent stress tensor as well as for instantaneous heat fluxes allowed to trace the role of coherent vortex in the energy transfer processes realized in the flow considered.

As the most valuable quantities in the analysis of heat transfer processes, the periodical $\langle \tilde{u}_i \tilde{v} \rangle$ and random $\langle u_i' v' \rangle$ phase-averaged heat fluxes have been selected. As can be seen in Fig. 2a, the $\pi/2$ phase-shift between the $\tilde{u}_x$ and $\tilde{v}$ results in doubling the frequency of longitudinal oscillatory heat flux $\langle \tilde{u}_x \tilde{v} \rangle$ in comparison with the basic shedding frequency of column-mode vortices (length of the picture frame corresponds to one wavelength of the structure). One may easily notice here the four regions of alternately positive (located at the high-speed side of the mixing layer) and negative (in outer region i.e. at low speed side) isolines of $\langle \tilde{u}_x \tilde{v} \rangle$, which create the regular mirror reflections with equal absolute peak values. Quite a different pattern may be seen in Fig. 2b, where one may observe the two regions of positive radial oscillatory heat fluxes $\langle \tilde{u}_r \tilde{v} \rangle$, located at the front and at the back of the vortex, respectively. In consequence, the two positive regions of $\langle \tilde{u}_r \tilde{v} \rangle$ result in the "unidirectional" action of the organized structure, giving in effect the amplification of the radial heat transport processes. On the contrary, the simultaneous existence within one structure wavelength of the positive and negative areas of $\langle \tilde{u}_x \tilde{v} \rangle$ means, that the particular oscillatory streams of heat cancel themselves and that is why no visible increase of longitudinal heat transport may be observed in the presence of coherent structures.

During the next stages of the research programme the phase-averaging was applied to random heat fluxes that allowed to extract the part of $\langle u_i' v' \rangle$ which phase-related to the coherent structure. As a result it was found possible to confirm the double role of organized vorticity, which on one hand directly stimulated the heat transfer processes and on the other "indirectly" triggered the increased production of random vorticity, being phase-related to coherent structure.

The additional evidence concerning the role of organized vorticity in heat transfer processes was obtained from the numerical analysis of kinetic and internal energy transport realized in the flow considered that gave the quantitative description of the shares of particular energy fluxes, transferred in the flow analyzed.

As a result, a detailed description of mechanism, governing the transport processes realized by coherent structures could be obtained.

The research has been founded by State Committee for Scientific Research under the grants KBN 310199101 and 7TO7C 050 08.

References

Paschereit C.O., et al (1992) Flow visualization of interaction among large coherent structures in an axisymmetric jets, *Experiments in Fluids*, vol.12.

Hussain A., (1986) Coherent structures and turbulence, *J.F.M.*, vol. 173.

Favre-Marinet M. (1989) Coherent structures in a round jet, *Proc. Von Karman Inst.*, Lecture Ser. 1989-03.

El-Sayed Abou El-Kassem (1995) *An analysis of heat transfer processes in the coherence region of a round, free jet*, PhD Thesis, Technical Univ. of Częstochowa.

DISSIPATIVE FLOW STRUCTURES WITHIN NEAR-WALL TURBULENCE-PRODUCING EVENTS

S. GAVRILAKIS

IMHEF-DGM, Swiss Federal Institute of Technology
1015 Lausanne, Switzerland

1. Introduction

The high turbulent energy production rates in near-wall shear-driven turbulence are attributed to two types of transient flow events commonly termed ejections and sweeps. These motions possess length and time scales that make them accessible to direct measurements at low and moderate Reynolds numbers. Although the energy-producing aspects of the ejection/sweep flow patterns have been emphasized, it will be shown that these are also associated with intense and localized viscous dissipation which often result in the net removal of turbulent energy from the fluctuating field. The dissipative aspects of the ejections and sweeps contribute to the intensification of turbulent vorticity field in the vicinity of the wall.

2. Turbulent database

The present analysis is based on the near-wall flow field derived form the direct simulation of a low Reynolds number incompressible turbulence in a straight duct of rectangular cross section. The aspect ratio of the duct is 2:1. Only data close to the two longer sides of the duct is used - excluding segments within 100 wall units from each of the shorter sides. The numerical method used for the simulation is based on spectral elements and Fourier expansions. Time advancement is by fractional step method where the non-linear and viscous terms are advanced by an explicit fourth order Runge-Kutta scheme. The flow field is driven by imposing a constant flow rate at every time step. Physical and numerical parameters of the simulation are shown in table 1.

3. Statistics of sweeps and ejections

The presence of sweep and ejection events was sought using the instantaneous values of the turbulent energy production rate $P = u_i u_j (\partial U_i / \partial x_j)$ as

S. Gavrilakis et al. (eds.), Advances in Turbulence VI, 103-106.
© *1996 Kluwer Academic Publishers.*

TABLE 1. Physical and numerical parameters of the simulation : U_b=Bulk velocity; D=Hydraulic diameter of the duct; u_τ=Average wall friction velocity

$Re = U_b D/\nu$	7200
$Re^+ = u_\tau D/\nu$	450
Domain size (x,y,z)	$4\pi h(streamwise) \times 2h \times 4h$
Number of collocation points	$96(x) \times 92(y) \times 176(z)$
Time span of database	$5\ h/u_\tau$

a marker. (Capital letters are used to denote mean quantities; lower case are used for the fluctuating part.) Detection was determined when the value of P at a point exceeded a preselected threshold. In order to avoid counting multiple ejection/sweep events as separate, the allowed points had the highest value for P (which exceeded the threshold about 10 times the long-time average value) within a volume of $200 \times 60 \times 130$ (streamwise,normal to the wall,spanwise) wall units and centered in the x and z directions (but not in y) about the detection point.

The instantaneous dissipation rate at a point was calculated from $\epsilon = \nu \partial_j u_i (\partial_j u_i + \partial_i u_j)$. Figure 1 shows the distribution the integral of turbulent energy production and dissipation rates over the above-stated volume surrounding ejections (with detection point at $y^+ = 14$) and sweeps (at $y^+ = 9$) near one of the duct boundaries. Sweep events appear to be more numerous than ejections. This is due to that the majority of the latter type events are associated with sweeps whereas sweeps can occur without any strong ejections in their vicinity. The diagonal $P = \epsilon$ separates the $P - \epsilon$ plane into regions of net energy production (below the diagonal) and net energy dissipation (above the diagonal.) It appears that for a substantial part of the time when P exceeds the threshold value, turbulent energy is on balance dissipated. The balance over all the data points shown is slightly negative – though this small imbalance has not statistical significance for this sample. The data in figure 1 therefore suggests that the energy required to maintaining turbulent motion away from the boundaries is mainly produced at the early or late stages of the eject/sweep cycle (i.e. when P is below some threshold value). The correlation coefficient for these integral values are greater than 85%.

The contributions of high production events to the overall maintainance of the turbulent field in the duct does not seem substantial, but their contribution in another respect is significant. In figure 2 scatter plots of volume integrals of P and ϵ against the integral of the enstrophy production rate, given by $S = \omega_i \omega_j s_{ij}$, is shown. s_{ij} is the fluctuating strain rate. This repre-

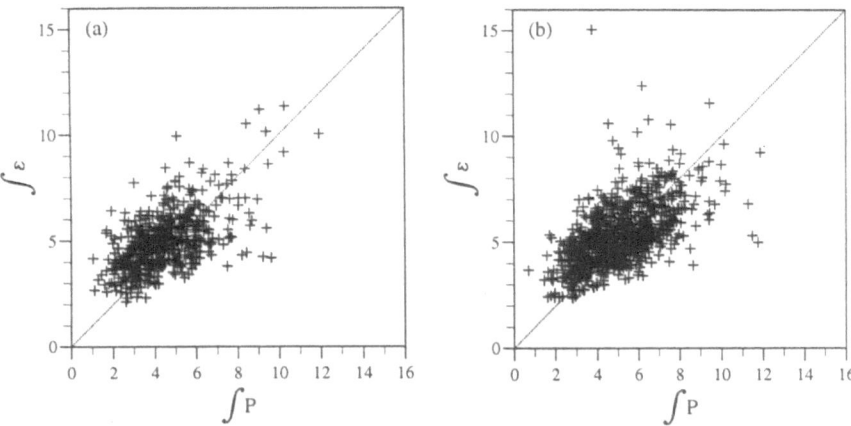

Figure 1. Scatter plots of the integrals of production and dissipation rates near the smooth boundaries; (a) Ejections; (b) Sweeps

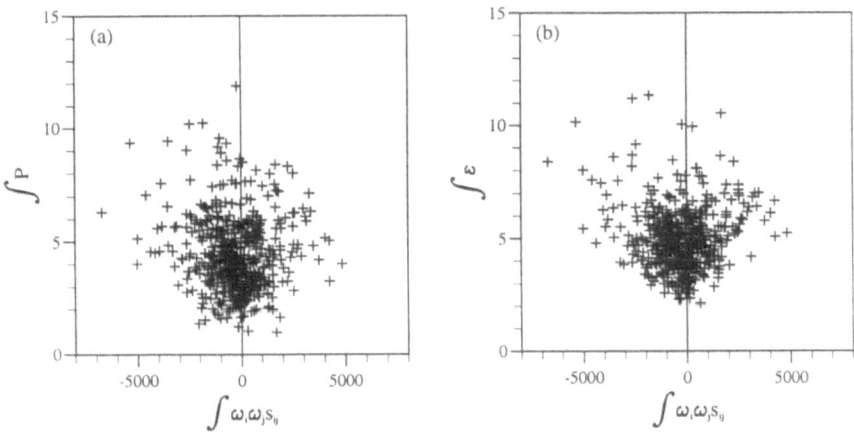

Figure 2. Scatter plots of integrals of P, ϵ against $\omega_i \omega_j s_{ij}$; Ejection events only.

sents only part of the turbulent enstrophy production rate and concerns the intensification of turbulent vorticity by the fluctuating strain field; there is a mean field contribution which has not been yet explored.

A degree of correlation between P and S is observable in figure 2a, but the distribution of the points in ϵ versus S is more telling. Even though the points have been implicitly selected for having high dissipation, at the lowest values of ϵ in figure 2b there virtually no enstrophy production. Increasing enstrophy production rates (both positive and negative) are associated with higher rates of viscous dissipation. It appears that the high

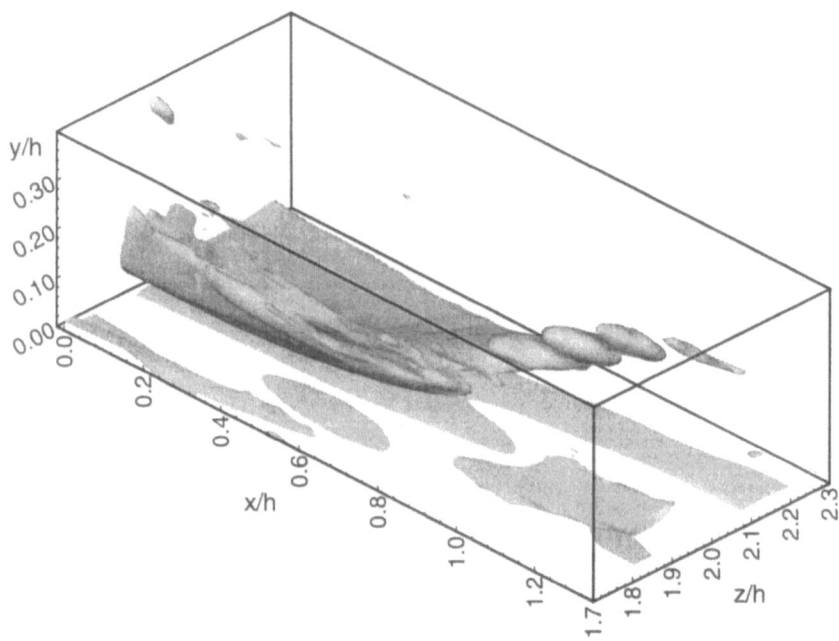

Figure 3. Surfaces of constant dissipation in the instantaneous near-wall flow field. Only the x-z plane at y=0 is on the wall. Mean flow is in the x-direction.

values of ϵ reflect the cost of generating enstrophy although the process seems quite inefficient – high dissipation is also associated with negative enstrophy production which is mainly associated with the interaction of the turbulent field with the wall.

High instantaneous value of the dissipation is mainly due to the streamwise and spanwise fluctuating velocity components. At the walls where the highest instantaneous values of ϵ are to be found, the normal gradients of the streamwise and spanwise velocities are (by far) the largest and second largest components respectively. Outside the viscous sublayer the crossstream gradients of the streamwise velocity are dominant. Figure 3 shows the spatial distribution of the instantaneous dissipation rate near one of the duct walls. Thin sheets of high dissipation at the wall and with the main flow may be seen. The break up of the dissipation layer separating from the wall is associated with the shear layers detected by the VITA technique.

Acknowledgments: M. Deville, L. Machiels and E. Leriche have helped with the development of the spectral element DNS code.

EJECTIONS, BURSTS AND SWEEPS
IN STOCHASTIC SIGNALS

M. HARTMANN AND D. RONNEBERGER
Drittes Physikalisches Institut der Universität Göttingen
Bürgerstraße 42-44, D-37073 Göttingen

1. Introduction

It is well established now that large scale coherent structures are essential features of turbulent shear flows. These structures can be observed and have been detected by means of visual studies. However, extended quantitative investigations rely on measurements at isolated locations in most cases. Then the problem arises that the information pertaining to the investigated structures has to be extracted from time records of the velocity or of other quantities like wall shear stress or pressure. For this purpose detection schemes have been invented in order to fix the instances when structures pass the probe. Sometimes conditional averaging is used to reconstruct the flow field associated with the structure.

Another common way to characterize the turbulent flow is to evaluate power spectra and probability density distributions of the measured time records. So the question arises whether independent pieces of information can be obtained by the two evaluation schemes, i.e., by the detection of isolated events and, on the other hand, by long–time averaging the statistical properties of the measured signals. It has been presumed that the statistical characteristics of the turbulent quantity under consideration depend on whether or not a structure is passing the probe.

2. Surrogate signals

In order to investigate this question surrogate signals have been generated that have some long time averaged statistic properties in common with measured signals but more or less differ from the measured signals in all other respects. In particular no isolated events have been built in the surrogate signals. Actually, the streamwise component of the flow velocity $u(t)$ has

S. Gavrilakis et al. (eds.), Advances in Turbulence VI, 107-110.
© 1996 *Kluwer Academic Publishers.*

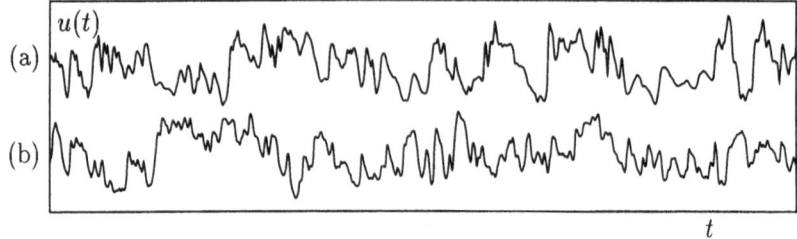

Figure 1. Short portion of (a) the measured signal and (b) the surrogate signal

been measured in a 2D turbulent channel flow at various distances from the wall, and the power spectral density $S_u(f)$ as well as the probability density functions $p(u)$ and $p(\partial u/\partial t)$ have been evaluated. An iterative algorithm has been developed in order to impose these three statistical characteristics on a given random signal (sequence of statistically independent numbers) without scrambling the given signal more than necessary. Short samples of the measured and the surrogate signals are shown on Figure 1.

Both the measured and the surrogate signals have been examined for the occurrence of special events by means of various frequently used detection schemes including the u–level method (Lu & Willmarth, 1973) and the well–known VITA technique (Blackwelder & Kaplan, 1976). It has been speculated that these techniques might respond to different structural features like ejections or sweeps.

3. Results

No salient differences have been found between the measured and the surrogate signals in the light of the investigated detection schemes. On the contrary, the differences are unexpectedly small. This will be demonstrated for at least a few examples pertaining to the VITA and the u–level technique:

The VITA technique depends on two parameters namely the integration time T and the threshold $k \cdot u_{\mathrm{rms}}^2$. In order to reasonably fix the integration time, Alfredson & Johansson (1984) have maximized the frequency of the detected events which is a function of T at a given threshold parameter k. Figure 2a shows such dependencies for two different values of k. The differences between the measured and the surrogate signals at the higher threshold (lower pair of curves) are comparatively large, nevertheless these differences are far too small to enable a significant differentiation between measured and surrogate signals.

This is even more so in the next example (Figure 2b). With regard to the grouping of the events into bursts Barlow & Johnston (1985) and

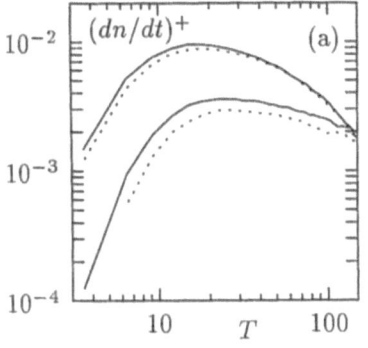 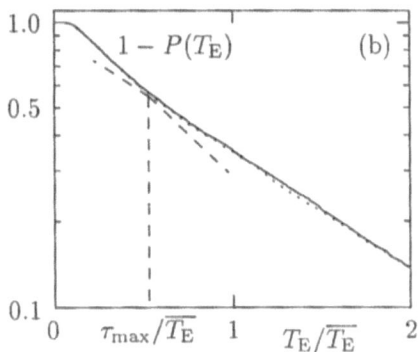

Figure 2. (a) Frequency of VITA events in viscous time units as a function of the integration time ($k = 0.5$ upper pair, $k = 1.0$ lower pair of curves), (b) Distribution of time between u–level events ($L = 1.2$);— measured signal, - - - surrogate signal

Luchik & Tiederman (1987) have investigated the distribution of the time between successive u–level events. The probability distribution $1 - P(T_{\mathrm{E}})$ of finding an interval that is greater than the time T_{E} has been found to be particularly meaningful in this respect. While $1 - P(T_{\mathrm{E}})$ decays exponentially for a Poison process, two such processes with different mean frequencies seem to be superimposed in the case of the u–level events. This is illustrated by the two straight dashed lines in Figure 2b. Hence the question whether or not two events belong to the same burst has been reduced to the question whether or not the time between the events is smaller than a time $\tau_{\max}$ that has been defined by the intersection of the two asymptotes in Figure 2b. Again, practically no differences have been found between this kind of evaluation of the measured and the surrogate signals.

A striking agreement has also been found for conditionally averaged time histories of the events. Two examples are shown in Figure 3.

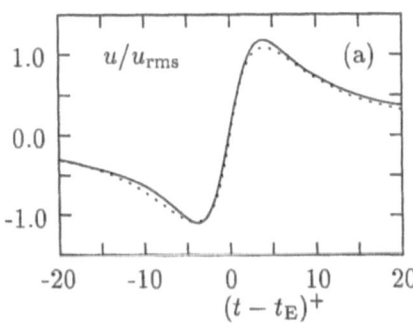 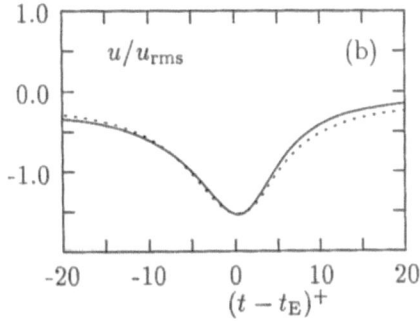

Figure 3. Conditionally averaged velocity signals using (a) the VITA technique and (b) the u–level technique; — measured signal, - - - surrogate signal

In an extended investigation surrogate signals have also been constructed from stochastical raw signals that are more orderly than the output of a random number generator used for the presented examples. Though the differences between the measured and the surrogate signals are mostly larger for the orderly raw signals than for the completely stochastic raw signal the similarity is still the salient feature of the results (Hartmann & Ronneberger, 1996).

4. Concluding remarks

It turns out that the VITA technique and all the other investigated detection schemes do not provide information on the signal other than that that is more or less contained in a few long–time averaged simple statistical characteristics of the examined signals. In a slightly different version of the present study only two characteristics namely the power spectral density $S_u(f)$ and the probability density function $p(\partial u/\partial t)$ had been imposed on the surrogate signal. Even then the measured and the surrogate signals do not differ significantly in the light of investigated detection techniques.

On the other hand the question arises whether it is possible at all to have a signal that consists of stochastically occurring events embedded in another stochastic signal, and, in particular, to eliminate this kind of events from all surrogate signals that have a few long–time statistical properties in common with the original signal. For the u-level technique, e.g., the mean frequency of the events can be unambiguously predicted from the joined probability density function $p(u, \partial u/\partial t)$, and we suspect that similar relations between statistical properties of the detected events and the long–time averaged statistical properties of the signal exist also for other detection schemes.

This does not exclude that events can be detected by means of suited schemes provided these events really exist, however, the detection of events is no proof of the existence of the events.

References

ALFREDSON, P. H., JOHANSSON, A. V. 1984 On the detection of turbulence-generating events *J. Fluid Mech.* **139**, 325.

BARLOW, R. S., JOHNSTON, J. P. 1985 Structure of turbulent boundary layers on a concave surface *Rept. M D-47*, Thermosciences Div., Dept. of Mech. Engng. Standford University.

BLACKWELDER, R. F., KAPLAN, R. E. 1976 On the wall structure of the turbulent boundary layer *J. Fluid Mech.* **76**, 89.

LUCHIK, T. S., TIEDERMAN, W. G. 1986 Timescale and structure of ejections and bursts in turbulent channel flows *J. Fluid Mech.* **174**, 529.

HARTMANN, M., RONNEBERGER, D. 1996 The detection of the bursting phenomenon in stochastic signals *(to be published elsewhere)*

LU, S., WILLMARTH, W. W. 1973 Measurements of the structure of Reynolds stress in a turbulent boundary layer *J. Fluid Mech.* **60**, 481.

APPLICATION OF THE VORTEX PATCH THEORY TO FORCED TWO-DIMENSIONAL TURBULENCE

B. JÜTTNER†‡, A. THESS†

†Institut für Strömungsmechanik
Technische Universität Dresden
01062 Dresden, Germany
‡e-mail: juettner@fz-rossendorf.de
i-net: http://www.tu-dresden.de/mwism/juettner/home.html

Abstract

Employing both direct numerical simulations (DNS) and a statistical approach we study large scale structures evolving in a forced two-dimensional (2D) turbulent flow in a box. Although the vorticity field is driven only in a small part of the lower left quarter of the box, the DNS predicts for the long term behaviour an annular vorticity distribution in the domain. Applying the statistical theory to this case we find the same annular structure for the coarse grained vorticity. Moreover the statistics predict a stream function-vorticity relation which is in agreement with the observations of the DNS.

1. Introduction

Large scale structures in quasi-2D flows can be observed in planetary atmospheres and in oceans [1] as well as in laboratory experiments on rotating flows [2] or on magnetohydrodynamic flows [3] [4]. The dynamics of such quasi-2D flows can be modeled by a generalized 2D Navier-Stokes-equation including the Rayleigh friction term [5]. In the case of a stationary chess-board like acceleration of a quasi-2D magnetohydrodynamic flow, this model has been successfully tested for a large range of the bottom friction parameter [6]. Because DNS and laboratory experiments require high numerical or "experimental" resolution, the existence of an efficient tool for the prediction of large scale structures in 2D turbulence has practical importance. A promising candidate for such a tool is the vortex patch theory (VPT) of the 2D Euler equation, independently proposed by Kuzmin, Miller, and Robert & Sommeria, which can be understood as a generalization of the statistical mechanics of point vortices of Onsager, and Montgomery & Joyce (for references see [7]). There are several successful quantitative tests of the predictions of the VPT in the case of *free* turbulence (for references see [8]). However, for the case of *forced* 2D turbulence the material from the statistics is rare. Although there is a recently proposed relaxation equation method for the forced case which has

111

S. Gavrilakis et al. (eds.), Advances in Turbulence VI, 111-114.
© *1996 Kluwer Academic Publishers.*

been successfully tested on a geophysical application [9], there are open questions. In the following we will focus on the problem whether the spatial structure of the forcing in 2D turbulence is reflected by the evolving large scale structures or not. Therefore in the next section we present the results of a long term DNS which we compare in the third section with predictions of the VPT.

2. DNS of a locally driven turbulent 2D flow

We have performed a DNS of the generalized 2D Navier-Stokes-equation [5] [6] [9]

$$\frac{\partial \omega}{\partial t} = \frac{\partial \psi}{\partial x}\frac{\partial \omega}{\partial y} - \frac{\partial \psi}{\partial y}\frac{\partial \omega}{\partial x} - \mu\omega - \nu_2\Delta^2\omega + f \tag{1}$$

with $\omega = -\Delta\psi$ in a box $(x, y) \in D = [-0.5, 0.5]^2$. On the boundaries we assume $\psi(\partial D) = \omega(\partial D) = 0$. We include Rayleigh friction ($\mu = 5 \times 10^{-4}$) and hyperviscouse dissipation ($\nu_2 = 10^{-8}$). The flow is driven with $f = 1$, for $(x, y) \in [-0.3, -0.2]^2$ and $f = 0$ elsewhere. We chose this type of forcing because (i) it produces only positive vorticity, (ii) it is localized in a small spatial range, and (iii) it has a lower symmetry then the domain. This forcing seems to be suitable for a test whether the evolving large scale structure has the same spatial properties as $f(x, y)$. The DNS is realized with a modification of the pseudo-spectral code [8]. Because the vorticity field changes its graining during the evolution, we start with a spatial resolution of 64^2 and change to the resolution of 128^2 when small scales appear which cannot be resolved by the 64^2-grid. The time steps are in the range $4 \times 10^{-4} < dt < 10^{-3}$. The simulation is started with the flow at rest. During the first, very short period the vorticity field is accelerated in the domain $(x, y) \in [-0.3, -0.2]^2$. The total circulation, which is always positive, increases and produces an annular velocity field which transports the vorticity away from its source. As result, a ring of vorticity is formed, which is located at the same radius $R_1 \approx 0.35$ as the range of forcing. In the center of the box the vorticity has a local minimum with $\omega_{min} = 0$. This structure does not change significantely up to a time $t \approx 6000$ while the maximum of vorticity, the enstrophy and the energy are increasing. Then, when the maximum of vorticity and the enstrophy are saturated, only the energy increases further. The growth of energy is accompanied by a change of the vortical structure: the local minimum of vorticity increases, begins to turn around the center of the box, and small scale fluctuations develop inside a circle with radius R_1 which is shown in Fig. 1a. In the balanced forcing case the counterpart of the coarse grained vorticity of the VPT is the time averaged vorticity of the DNS. We calculate the time averaged vorticity at the end of observation ($t \approx 10000$) when the increment of energy per time becomes small. The intervall of averaging has a length $T = 10$. Finally we get an annular time averaged vorticity distribution which is shown in Fig. 1b. Inside the circle of radius R_1 the vorticity distribution is flat, outside the vorticity decreases rapidly towards the boundaries. We find that this vorticity distribution does not reflect the spatial distribution of the forcing which in fact is concentrated in a very small spatial range. Therefore we conclude that the distribution of coarse grained vorticity calculated from a valid statistical theory should not locally depend on the function $f(x, y)$.

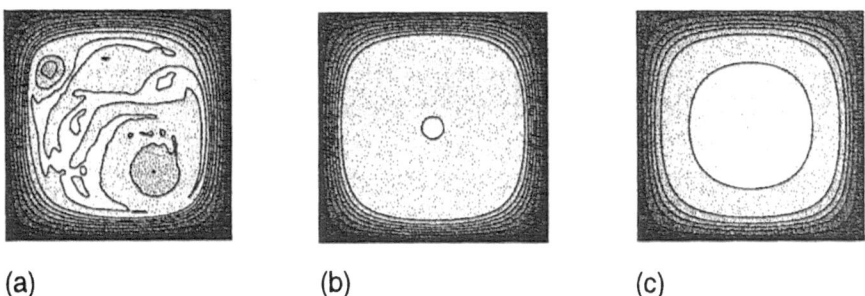

| (a) | (b) | (c) |

Fig. 1   Contour plots of a vorticity snapshot from DNS at $t \approx 10000$ (a), of the time averaged vorticity from DNS at $t \approx 10000$ (b), and of the predicted coarse grained vorticity from VPT (c).

3. Application of the VPT

The VPT was originally proposed for the 2D Euler equations under the assumtion that the flow consists of vorticity patches of distinct vorticity levels q_i which are strongly mixed. It predicts spatial probability distributions p_i for the levels q_i under the constraints that the areas A_i of the vorticity patches and the energy E are conserved. The probabilities p_i are used to define a locally averaged vorticity Ω and a related mean field stream function Ψ which fulfil a relation $\Omega = -\Delta\Psi = F(\Psi)$. In the DNS the counterparts of Ω and Ψ are the time averaged vorticity and stream function, respectively. The shape of F depends strongly on the values of A_i and E. However, while in the Euler case A_i and E are given a priori, in forced 2D turbulence the values of the integral quantities are the result of the dynamical equilibrium between forcing and dissipation. Because an efficient solver of the variational problem including the equilibrium conditions for A_i and E is not available yet, we use an indirect way [10]. We take measured values of A_i and E from the last time of the DNS and put them in the iterative solver [11] of the variational problem of the Euler case. If the probability distributions in the forced 2D turbulence are determined in the same way as in the case of free turbulence then the predicted coarse grained fields Ω and Ψ should agree with time averages from the DNS. Doing this procedure we get an annular distribution of the coarse grained vorticity which is flat in the center of the box and decreases rapidly near the boundaries (Fig. 1c). The good agreement of the statistics with the observations of the DNS is visible when looking at the stream function-vorticity plot in Fig. 2. We find that both the shape of the function F and the absolute values of Ω and Ψ are correctly reflected. The small differences at $\Psi = 0$ result from the boundary condition on the vorticity in the DNS, which does not exist in the Euler case. The differences at $\Psi \approx 1$ seems to result from aborting the DNS at $t \approx 10000$ when the energy is still increasing with small amounts per time.

114

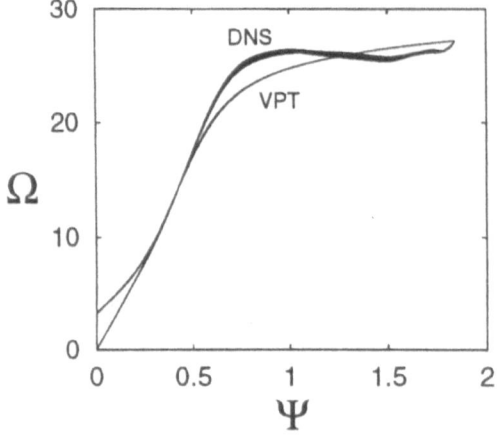

Finally we can conclude for balanced forcing 2D turbulence (i) that the statistical theory of vortex patches can be extended to this case and (ii) that the applied mean field equations should not depend *locally* on the forcing function.

Fig. 2   Stream function-vorticity scatter plot: results of DNS versus prediction of VPT.

Acknowledgement

We are grateful to the Deutsche Forschungsgemeinschaft for financial support under Grant Th497/4-2.

References

1.   Pedlosky J.: *Geophysical Fluid Dynamics* , Springer, New York, 1979.

2.   Sommeria J., Meyers S. D., Swinney H. L.: Laboratory simulation of Jupiter's Great Red Spot, *Nature* **331** , 1986, 689-693.

3.   Sommeria J.: Experimental study of the two-dimensional inverse energy cascade in a square box, *J. Fluid Mech.* **170** , 1986, 139-168.

4.   Kljukin A., Uspenski W., Moreau R., Bolcato R.: An experimental study of a quasi-2D turbulence in MHD free shear layers, *The Eight Beer-Sheva Intern. Seminar on MHD-Flows and Turbulence 25.-29. 2. 1996 Jerusalem* .

5.   Danilov S., Dolzhanskii F., Krymov V.: Quasi-two-dimensional hydrodynamics and problems of two-dimensional turbulence, *Chaos* **4** , 1994, 299-304.

6.   Verron J., Sommeria J.: Numerical simulations of a two-dimensional turbulence experiment in magnetohydrodynamics, *Phys. Fluids* **30** , 1987, 732-739.

7.   Frisch U.: *Turbulence* , University Press, Cambridge, 1995.

8.   Jüttner B., Thess A., Sommeria J.: On the symmetry of self-organized structures in two-dimensional turbulence, *Phys. Fluids* **7** , 1995, 2108-2110.

9.   Kazantsev E., Sommeria J., Verron J.: Subgrid Eddy Parametrisation by Statistical Mechanics in a Barotropic Ocean Model, submitted *J. Phys. Ocean.* 1995.

10.   Jüttner B., Thess A.: Inertial organization in Forced Two-dimensional Turbulence, preprint for *Progress in Astronautics and Aeronautics* 1996.

11.   Whitaker N., Turkington B.: Maximum entropy states for rotating vortex patches, *Phys. Fluids* **6** , 1994, 3963-3973.

DYNAMICS OF A STRONG VORTEX TUBE
IN A UNIFORM SHEAR FLOW

S. KIDA
Theory and Simulation Center, National Institute for Fusion Science, Nagoya 464-01, Japan

G. KAWAHARA
Department of Mechanical Engineering, Ehime University, Matsuyama 790-77, Japan

M. TANAKA
Department of Mechanical and System Engineering, Kyoto Institute of Technology, Kyoto 606, Japan

AND

S. YANASE
Department of Engineering Sciences, Okayama University, Okayama 700, Japan

1. Introduction

The time-evolution of a strong straight vortex tube in a uniform shear flow is studied analytically and numerically. A vortex filament of circulation Γ is set at an initial instant with it being inclined from the streamwise (X_1) direction both to the spanwise (X_3) and vertical (X_2) directions. The uniform vorticity of the shear flow is pointed to the negative X_3-direction. Hereafter, a vortex tube, whose spanwise vorticity component is negative (positive), is called a cyclonic (anti-cyclonic) vortex.

An asymptotic analysis at large Reynolds number $\Gamma/\nu \gg 1$, ν being the kinematic viscosity of fluid, and at the initial time of evolution $St \ll 1$, S being the shear rate, is performed by extending Moore's method (1985). This leads us to the following findings. (1) The uniform shear vorticity normal to the vortex tube is expelled from the core, in which only the tangential component survives. (2) The remaining tangential vorticity is stretched by the uniform shear, so that a cyclonic vortex tube is intensified

S. Gavrilakis et al. (eds.), Advances in Turbulence VI, 115-118.
© 1996 Kluwer Academic Publishers.

while an anti-cyclonic one is weakened. (3) At the periphery of the vortex tube, vorticity of opposite sign to the tube itself is generated. (A similar phenomenon was observed in the regeneration process of longitudinal vortices in a turbulent channel flow (Sendstad & Moin 1992)).

2. Formulation

Let the coordinate system $OX_1X_2X_3$ be at rest. The uniform shear velocity U is taken as $U = SX_2\hat{X}_1$, where $S\,(>0)$ denotes the shear rate, which is constant in time, and $\hat{X}_i$ is the unit vector in the X_i-direction ($i = 1, 2, 3$).

We shall formulate the problem in the rotating system $Ox_1x_2x_3$, the x_1 axis is in the direction of a straight vortex tube. Rotating the coordinate system $OX_1X_2X_3$ by an angle of β around the X_1-axis, we set the x_3-axis as the new X_3-direction. Further rotating $OX_1X_2x_3$ by an angle of α around the x_3-axis, we have the system $Ox_1x_2x_3$.

Let us decompose the velocity field into the uniform and fluctuating parts as

$$u = \overline{u} + u', \quad \overline{u} = U - \Omega \times x, \tag{1}$$

where Ω is the angular velocity of the rotating coordinate system which is expressed in terms of α and β.

In general, $\overline{u}_2$ and $\overline{u}_3$ include the coordinate x_1 explicitly. In order that u' and $\omega' = \nabla \times u'$ are uniform in the x_1-direction, however, it is necessary for $\overline{u}_2$ and $\overline{u}_3$ to be independent of x_1. This requires

$$\Omega_1 = \Omega_2 = 0, \quad \Omega_3 = -S\sin^2\alpha\cos\beta, \tag{2}$$

where $\alpha = \arctan\left(1/(\cot\alpha_0 + St\cos\beta)\right)$ and $\beta = const.$, with α_0 denoting the value of α at $t = 0$.

Supposing that the fluctuating field be independent of x_1, we obtain closed equations for ω_1' and u_1'. Now we introduce plane polar coordinates (r, θ) with $(x_2 = r\cos\theta, x_3 = r\sin\theta)$ and adopt Lundgren's (1982) transformation for radial coordinate and time as

$$R = A(t)^{1/2}r, \quad T = \int_0^t A(t')\,dt', \tag{3}$$

where

$$A(t) = \exp\left(S\int_0^t \gamma(s)\,ds\right), \quad \gamma(t) = \cos\alpha\sin\alpha\cos\beta. \tag{4}$$

Then, we normalize the variables using $A(t)$ as

$$\omega(R, \theta, T) = \omega_1'(r, \theta, t)/A(t)(= -\nabla^2\psi), \tag{5}$$

$$u(R, \theta, T) = A(t)(u_1' + S\cos\beta\, x_2 - S\cos\alpha\sin\beta\, x_3), \tag{6}$$

where ψ is the streamfunction and $\nabla^2 = \partial_2^2 + \partial_3^2$.

The early-time evolution of a strong straight vortex tube can be investigated analytically in the high-Reynolds-number limit. A vortex filament with strength Γ is put in a uniform shear flow at an initial instant $T = 0$. The Reynolds number Re is then defined by $Re = \Gamma/(2\pi\nu)$. We suppose that $\epsilon = (2\pi Re)^{-1} = \nu/\Gamma \ll 1$. An asymptotic analysis will be performed at a high Reynolds number ($\epsilon \ll 1$) and at an early time of evolution ($ST \ll 1$).

After all the variables are nondimensionalized by the use of S, ν and Γ, the basic equations are written as

$$\frac{1}{R}\frac{\partial(\psi,\omega)}{\partial(R,\theta)} - \epsilon(\partial_T - \nabla^2)\omega = \epsilon L_1\omega + \epsilon^2 L_2 u - \epsilon^2 \frac{2\gamma(t)\lambda(t)}{A(t)^2}, \qquad (7)$$

$$\frac{1}{R}\frac{\partial(\psi,u)}{\partial(R,\theta)} - \epsilon(\partial_T - \nabla^2)u = \epsilon L_1 u \qquad (8)$$

where $\lambda(t) = -\sin\alpha\sin\beta$, $\xi(t) = 2\sin^2\alpha\cos\beta$,

$$L_1 = \frac{1}{2A(t)}[\gamma(t)(\sin 2\theta\partial_\theta - R\cos 2\theta\partial_R) - \lambda(t)(\cos 2\theta\partial_\theta + R\sin 2\theta\partial_R - \partial_\theta)],$$
$$(9)$$

$$L_2 = \frac{\xi(t)}{RA(t)^{5/2}}(\cos\theta\partial_\theta + R\sin\theta\partial_R). \qquad (10)$$

Considering the early period of the time-evolution of a strong vortex tube, we seek for analytical solutions of (7) and (8) in an expansion form by taking the LHS as the leading order and the RHS as higher orders. We also performed direct numerical simulations of a straight vortex tube in a uniform shear flow by solving the Navier-Stokes equation. The early stage of its evolution agrees with the above asymptotic results.

3. Results

The spatial distribution of the absolute value of vorticity component perpendicular to the vortex tube is drawn in figure 1, where the vortex tube is viewed from the upstream and it rotates clockwise. Vorticity is stronger in the black regions and is weaker in the white. The vortex tube wraps and stretches the uniform shear vorticity around itself to form two spiral layers of high vorticity normal to its axis. However, in the inner region of the tube an excessive stretching enhances the viscous diffusion, which leads to expelling of the normal vorticity from the tube core, where only the tangential component remains to be stretched by the uniform shear. As a result, a cyclonic (anti-cyclonic) vortex is strengthened (weakened) (see figure 2).

118

Cyclonic   Anti-cyclonic

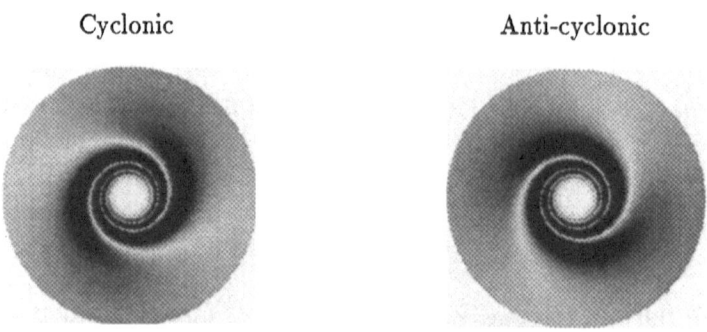

Figure. 1   Absolute value of vorticity component perpendicular to the vortex tube

Cyclonic   Anti-cyclonic

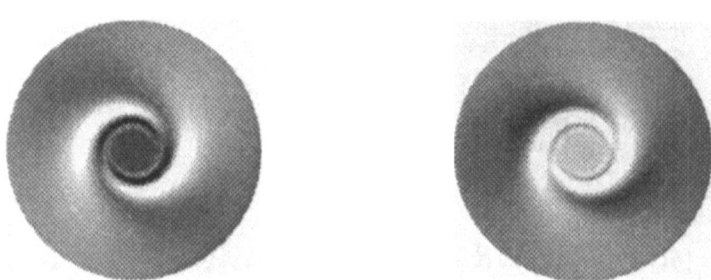

Figure. 2   The tangential vorticity component

Figure 2 shows the spatial distribution of the tangential vorticity component (subtracted the Gaussian core). The black and white regions represent
positive and negative vorticity, respectively. It can be seen that there appears negative vorticity around the core where vorticity is positive. Sendstad & Moin (1992) observed a similar phenomenon around longitudinal
vortices in a turbulent channel flow, and they found that the streamwise
vorticity of opposite sign develops to form new streamwise oriented vortices.
The spirals of high normal vorticity induce flows twisting the uniform shear
vorticity toward the tangential direction, so that the tangential vorticity
component of opposite sign is generated around the vortex tube.

References

Lundgren, T. S. (1982) Strained spiral vortex model for turbulent fine structure, *Phys.
Fluids*, **25**, pp. 2193–2203

Moffatt, H. K., Kida, S. & Ohkitani, K. (1994) Stretched vortices - the sinews of turbulence; large-Reynolds-number asymptotics, *J. Fluid Mech.*, **259**, pp. 241–264

Moore, D. W. (1985) The interaction of a diffusing line vortex and an aligned shear flow,
Proc. R. Soc. Lond. A, **399**, pp. 367–375

Pearson, C. F. & Abernathy, F. H. (1984) Evolution of the flow field associated with a
streamwise diffusing vortex, *J. Fluid Mech.*, **146**, pp. 271–283

Sendstad, O. & Moin, P. (1992) The near wall mechanics of three-dimensional turbulent
boundary layers, Department of Mechanical Engineering Report No. TF-57, Stanford
University, Stanford, CA.

LONG-RANGE ORDER AND DEFORMATION RADIUS EFFECTS IN 2-D TURBULENCE

N.N. KUKHARKIN, S.A. ORSZAG, and V. YAKHOT
Fluid Dynamics Research Center
Princeton University, Princeton, NJ 08544, U.S.A.

The most prominent feature of a 2-D flow excited at a small scale l_f is an inverse cascade, i.e. creation of large-scale motions from small-scale eddies. While in an infinite system with no external fields this process lasts forever, in a finite system of size $L \gg l_f$, energy eventually starts accumulating at L ("condensation") and the energy spectrum becomes steeper, thus slowing down the inverse cascade. For a flow in an external field (including rotation, stratification, and magnetic fields), where a substantial part of the energy goes into waves, the inverse cascade often slows down after waves of the scale of the deformation radius L_R, determined by the external field, are created. In these systems the energy spectrum first steepens at scale $O(L_R)$, and then slowly larger motions are generated due to energy leakage through the "shield" at L_R. The basic questions are: What are the dominant structures in this phenomenon? If structures exist, how are they distributed in physical space? Can the flow achieve a stable or a quasistationary configuration? We attempt to give at least partial answers to these questions.

We report novel features in vortex formation and dynamics in the systems where a characteristic length scale interferes with the inverse cascade [1]. In this context, we study the driven-damped Charney-Hasegawa-Mima model for Rossby-wave turbulence in the atmosphere and drift-wave turbulence in plasmas [2]

$$\partial_t(\nabla^2 \phi - \lambda^2 \phi) + [\phi, \nabla^2 \phi] + \beta \partial_x \phi = Diss + F \tag{1}$$

and show how a vortex "quasi-crystal" is formed in this system. For this purpose we consider the space-time averaged even-order moments of potential vorticity increments (structure functions) $S_{2n}(r) = \langle [\xi(\mathbf{x} + \mathbf{r}) - \xi(\mathbf{x})]^{2n} \rangle$, where $\xi = \nabla^2 \phi - \lambda^2 \phi$, $L_R \sim \lambda^{-1}$ is the Rossby or Larmor radius. The results for $S_{2n}(r)$ demonstrate the presence of long-range order, characteristic of the quasi-crystalline phase (Fig.1). The first peak of $S_{2n}(r)$ with maxima at r slightly larger than L_R denotes a shell of nearest vortex neighbors, and there are oscillations representing more distant neighbors.

As was discussed for the first time in [3], the governing equation for vortices larger than the deformation radius L_R should include an additional nonlinear term $\beta \phi \partial_x \phi$, physically analogous to the scalar nonlinearity of the Korteweg-de Vries equation which describes solitons. The scalar nonlinearity is known to introduce a cyclone/anticyclone asymmetry into the Eq.(1). We study the question

S. Gavrilakis et al. (eds.), Advances in Turbulence VI, 119-120.
© 1996 *Kluwer Academic Publishers.*

120

determining the combined influence of the β-effect (both Rossby waves and KdV nonlinearity) and deformation radius on the coherent structures [4]. It is shown that for a certain set of parameters (close to those of giant planets) large-scale zonal flows created by small-scale forcing tend to preferentially form anticyclonic vortices (Fig. 2). Their emergence can be explained by the modified Rayleigh-Kuo instability criterion [5] which takes into account the deformation of free surface. This work was partially supported by ARPA/ONR N00014-92-J-1796 and IBM.

References

1. Kukharkin, N., Orszag, S.A., and Yakhot, V. (1995) Quasicrystallization of vortices in drift-wave turbulence, *Phys. Rev. Lett.* **75**, 2486-2489.
2. Hasegawa, A., Maclennan, C.G., Kodama, Y. (1979) Nonlinear behavior and turbulence spectra of drift waves and Rossby waves, *Phys. Fluids* **22**, 2122-2129.
3. Petviashvili, V.I. (1980) Red spot of Jupiter and the drift soliton in plasma, *Sov. Phys. JETP Lett.* **32**, 619-622.
4. Kukharkin, N. (1996), in preparation.
5. Nezlin, M.V., Snezhkin, E. N. (1993) *Rossby Vortices, Spiral Structures, Solitons,* Springer-Verlag, Heidelberg.

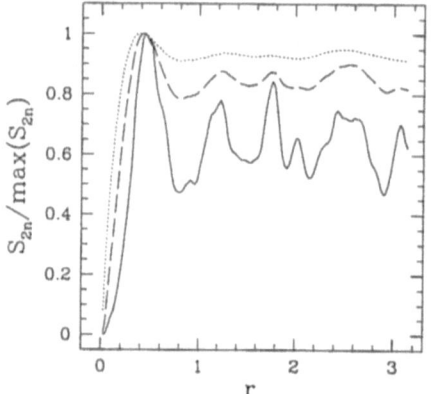
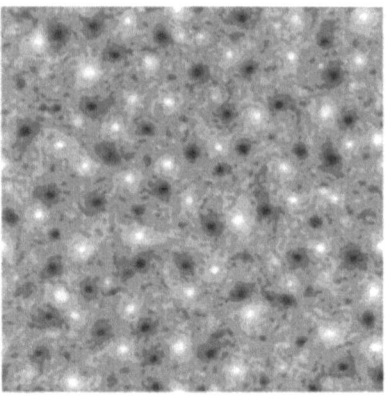

Figure 1. Structure functions S_2 (dotted), S_4 (dashed), and S_8 (solid) normalized to their maximum values and the potential vorticity field: $\lambda = 20$, $L_R = 2\pi/\lambda \simeq 0.31$.

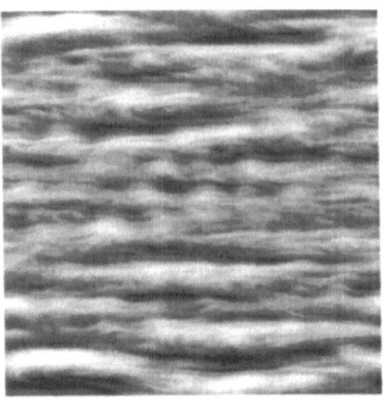
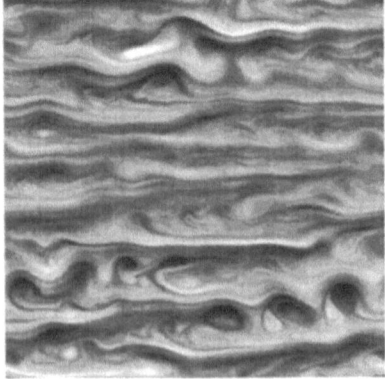

Figure 2. Presence of the deformation radius makes shear flows unstable with respect to generation of anticyclones (darker colors – right picture), box size is $L = 10L_R$.

A SUBLAYER BURSTING MODEL FOR TURBULENT BOUNDARY LAYERS

M. T. LANDAHL
Massachusetts Institute of Technology
Cambridge MA 02139, USA

1.Introduction

Modelling of turbulence near a wall requires the accurate representation of the statistics of the fluctuating velocities and pressures with as few empirical parameters as possible.Commonly used boundary layer turbulence models such as Prandtl's mixing-length model, large-eddy simulation models or the k- ε model attempt to do this essentially without taking into account the basic structure of the turbulent fluctuating field. Landahl (1975, 1990) suggested that the stress-carrying eddies could be modelled as localized disturbances created by small-scale breakdowns due to a local instability or other strong nonlinearity. He showed (1993) that the arising initial-value problem could be solved analytically through a time sequential procedure, in which the linear shear interaction stage will govern the flow during the early times of the order of 1/U'(y),U(y) being the mean velocity distribution, (treated as approximately parallel), with their long-time asymptotes serving as the initial values for the subsequent viscous and nonlinear stages. Calculations presented in Landahl (1990) demonstrated that initial disturbances due to oblique instability waves without spanwise symmetry gave as a result streaky structures with good qualitative agreement with those observed in the near-wall region of a turbulent boundary layer. Here we present a simple model along such lines showing how Prandtl's (1925) mixing-length theory comes out from averaging a random distribution of initially localized three-dimensional disturbances.

2. Space-time evolution of a 3-D initial disturbance

We consider the evolution of an unsteady pressure $p(x_j,t)$ and velocity field

$$U_i(x_j,t) = U(y) \ \delta_{i1} + u_i(x_j,t) \tag{1}$$

due to a weak three-dimensional disturbance introduced at t=0 into a steady parallel and incompressible shear flow above a solid wall.. The effect of nonparallelism of the mean flow is ignored. The initial disturbance is assumed to be localized in space so and has a characteristic spatial scale L. The perturbed flow is governed by the equations for conservation of volume, $\partial u_i/\partial x_i=0$, and momentum

$$\frac{Du_i}{Dt}+U'(y)u_i = \frac{1}{\rho}\{-\frac{\partial p}{\partial x_i}+\frac{\partial}{\partial x_j}[\mu\frac{\partial u_i}{\partial x_j}+ \tau_{ij}]\} \tag{2}$$

where $D/Dt=\partial/\partial t+U(y)\partial/\partial x$ and $\tau_{ij}=u_iu_j-<u_iu_j>$. With a rigid wall located at y=0, the boundary conditions are that the velocity perturbations must vanish at y=0, y=∞, and at x =±∞. In order to properly account for the role of the continuous spectrum, we will avoid the traditional spatial Fourier decomposition of the problem together with a Tollmien-Schlichting eigenfunction expansion , and will instead follow Landahl's (1993) procedure and analyze the evolution of the weak disturbance directly in space-time, assuming it to occur in three different temporal stages; I) a linear shear interaction stage, $t=O(t_s)$, $t_s =S^{-1}$; II) a viscous linear stage, $t=O(t_v)$, $t_v= (\Lambda^2/vS^2)^{1/3}$; and III) an inviscid nonlinear stage $t=O(t_n)$, $t_n=S\Lambda/v_0$, where S denotes a characteristic shear rate, for example the wall shear rate, U'(0). The nonlinear viscous stage, for $t=O(t_n)=O(t_v)$, will not be considered here. On the assumption that $t_s \ll t_v \ll t_n$, the solutions for the three different stages I--III may be obtained analytically.

121

S. Gavrilakis et al. (eds.), Advances in Turbulence VI, 121-124.
© 1996 *Kluwer Academic Publishers.*

3. Shear interaction stage

In this stage of disturbance development viscous and nonlinear terms in (1), (2) may be neglected. Elimination of the pressure between the component equations, and the use of continuity, gives that

$$\frac{DL}{Dt} = U'' \frac{\partial v}{\partial x}, \quad L(x,y,z,t) = \nabla^2 v \tag{3}$$

It is convenient to formulate this problem in terms of then fluid element liftup, l, and the pressure impulse, p, defined as follows:

$$l = \int_0^t v Dt', \quad p = \int_0^t p Dt' \tag{4}$$

Here, $\int_0^t .Dt'$ denotes time integration in a mean linear Lagrangian frame, i.e., with $\xi = x - tU(y)$ held constant,

$$l(\xi,y,z,t) = \int_0^t v[x-(t-t')U(y), y,z,t'] Dt' = \int_0^t v[\xi - t'U(y), y,z,t'] Dt' \tag{5}$$

In terms of l and p one has

$$u = -p_x - U'(y) \, l + u_0(\xi,y,z) \tag{6}$$

$$v = \frac{Dl}{Dt} = l_t (\xi,y,z,t) \tag{7}$$

$$w = -p_z + w_0(\xi,y,z) \tag{8}$$

Integration of (3) over time, following the fluid element (ξ =const.), gives

$$L = L_0(\xi,y,z) + U''(y)\frac{\partial l}{\partial x} \tag{9}$$

where $L_0 = \nabla^2 v_0$. The formal solution for v satisfying the boundary condition that v should vanish at infinity and at $y = 0$ reads

$$v = \frac{Dl}{Dt} = -\frac{1}{4\pi} \int_{-\infty}^{\infty} dx' \int_0^{\infty} dy' \int_{-\infty}^{\infty} dz'[L_0(\xi',y'z')+U''(y')l_x(x',y',z',t)]G(X,Y,Y_*,Z) \tag{10}$$

where $\xi' = x'-tU(y'), X = x-x', Y = y-y', Y_* = y+y', Z = z-z'$, and

$$G = \frac{1}{\sqrt{X^2+Y^2+Z^2}} - \frac{1}{\sqrt{X^2+Y_*^2+Z^2}} \tag{11}$$

or, expressed in terms of Lagrangian coordinates

$$l_t (\xi,y,z,t) = -\frac{1}{4\pi} \int_{-\infty}^{\infty} dx' \int_0^{\infty} dy' \int_{-\infty}^{\infty} dz'[L_0(\xi',y'z')+U''(y')l_\xi(\xi',y',z',t)]G(\Xi,\eta,Y,Y_*,Z) \tag{12}$$

where $\eta = t[U(y)-U(y')], \xi' = x'-tU(y'), \Xi = \xi - \xi'$, and

$$G(\Xi,\eta,Y,Y_*,Z) = \frac{1}{\sqrt{(\Xi+\eta)^2+Y^2+Z^2}} - \frac{1}{\sqrt{(\Xi+\eta)^2+Y_*^2+Z^2}} \tag{13}$$

Near the wall, a Taylor series expansion around y = 0 gives for the liftup

$$l \approx y(\frac{\partial l}{\partial y})_{y=0} = -\frac{y}{2\pi}\int_0^t dt' \int_{-\infty}^{\infty} d\Xi' \int_0^{\infty} dy' \int_{-\infty}^{\infty} dz''$$

$$[L_0(\xi',y',z') + U''(y') l_\xi(\xi',y',z',t')] \frac{y'}{[(\Xi-t'U(y'))^2+y'^2+Z^2]^{3/2}} \tag{14}$$

We need to follow the development of the disturbance over a long time. As $T = tU'(y) \to \infty$ the integrand in (14) will tend to zero as T^{-3}, except near y' = 0. Therefore, the major contributions to the integral (14) will come from regions near the wall, for which the term $\partial l / \partial \xi$ will become unimportant, since $l \to 0$ for y = 0. After neglect of this term and integration with respect to time one finds that

$$l \approx -\frac{y}{2\pi}\int_{-\infty}^{\infty} d\Xi' \int_0^{\infty}\frac{y'dy'}{U(y')(y'^2+Z^2)} \int_{-\infty}^{\infty} dz' L_0(\xi',y',z') [\frac{\Xi-tU(y')}{\sqrt{(\Xi-tU(y'))^2+y'^2+Z^2}}) - \frac{\Xi}{\sqrt{\Xi^2+y'^2+Z^2}}] \tag{15}$$

In the limit $T \to \infty$, $\sqrt{(\Xi-tU(y'))^2+y'^2+Z^2} \to |\Xi-tU(y')|$, so that

$$l \to -\frac{y}{2\pi}\int_0^{\infty}\frac{y'dy'}{U(y')(y'^2+Z^2)} \int_{-\infty}^{\infty} dz' \int_0^{\infty} d\xi L_0(\xi',y',z')[\text{sgn}(\Xi-tU(y')) - \frac{\Xi}{\sqrt{\Xi^2+y'^2+Z^2}}] \tag{16}$$

From (16) one finds that the liftup varies near the wall linearly with y and will vanish for $|\xi| \to \infty$ with ξ as

$$l \to -\frac{y}{2\pi}\int_0^{\infty}\frac{y'dy'}{U(y')(y'^2+Z^2)} \int_{-\infty}^{\infty} dz' F_\infty(y',z')[1 - \frac{\xi}{\sqrt{\xi^2+y'^2+Z^2}}]\text{sgn}(\xi) \tag{17}$$

where

$$F_\infty(y',z') = \int_{-\infty}^{\infty} L_0(\xi',y,z)d\xi' \tag{18}$$

so that the neglect of of the term $U''(y') l_\xi$ compared to L_0 near the wall is justified. The streamwise length of the disturbance will grow like $tU(\Lambda)$, i.e., a streaky structure will result. Inside the streak, away from its ends, the liftup will become independent of ξ and given by

$$l(y,z,t) \to yf_\infty(z,t) = \frac{y}{2\pi}\int_{-\infty}^{\infty} dz' \int_{y_0(t)}^{\infty}\frac{y'dy'}{U(y')(y'^2+Z^2)}F_\infty(y',z') \tag{19}$$

where $y_0(t)$ is given by $U(y_0) = |\xi|/t$. Thus the liftup inside the streak will become independent of ξ for $y > y_0$. In the inside region the pressure, as well as the streamwise and spanwise vorticity components will become zero, only vertical shear layers on the spanwise sides of the streak will remain. In the absense of a streamwise pressure gradient inside the streak the streamwise perturbation component u will become simply

$$u \approx -l U'(y) = -yU'(y) f_\infty(z,t) \tag{20}$$

with its length growing like tU(Λ). However, in the streak end regions the spanwise and streamwise vorticity layers will intensify like tU'(y), leading to eventual breakdown due to secondary instability or other strong nonlinear mechanism..

3. Reynolds shear stress

For the Reynolds shear stress due to a random distribution of streaks we need to evaluate

$$\tau_{12}(y) = -\rho <uv> \tag{21}$$

with the ensemble average <> to be taken here over both time and space.

. From (20) and the definition of the liftup we have

$$\tau_{12}(y) = \rho U'(y) < \frac{Dl}{Dt}> = \frac{1}{2}U'(y) < \frac{D}{Dt}l^2> \tag{22}$$

Because of inflectional instability of shear layers that will intensify like tU'(y) near the end

regions, the streak will have a finite lifetime, which on the average will scale as $t_m = \bar{t}_m/U'_{,}(y)$. The average number of streaks being created per unit time will hence be proportional to $1/t_m$.   At the end of the liftup phase, the liftup will be given by (19), so that the contribution to the Reynolds stress from the streak will be of the form

$$\tau_{12}(y) = \rho y^2 U'(y)^2 Q(z) \tag{23}$$

where

$$Q(z) = \frac{1}{2}<\alpha(z)^2>/\bar{t}_m \tag{24}$$

For the constant-stress near-wall region the total shear stress is

$$\tau = \mu U'(y) + \rho y^2 U'(y)^2 Q(z) = \mu U'_w \tag{25}$$

For large y , where the Reynolds stress dominates over the viscous stress, one thus obtains that $U' \sim y^{-1}$, which leads directly to Prandtl's (1925) mixing-length theory.

4. Conclusions

The analysis presented here suggests that the primary mechanism for producing the Reynolds shear stress near a wall is primarily inviscid and can be analyzed with the aid of linear theory to determine the part played by interaction of the fluctuations with the mean shear. From the linear shear interaction results, the viscous near-wall region may be determined simply as an unsteady Stokes' flow driven by the inviscid liftup and the pressure. The effects of nonlinearity in limiting the life-time of the stredaks needs further study. The treaks are also possibly the most important factor in transition , as has been demonstrated through direct numerical simulations by Lundbladh and Johansson (1993).

5 References

Landahl, M.T (1975) Wave breakdown and turbulence, *SIAM J. Applied Math* **28**, 775-756.
Landahl, M.T (1980) A note on algebraic instability of inviscid parallel shear flows. *J. Fluid Mech.* **98**, 243-251
Landahl, M.T. (1990) On sublayer streaks. *J. Fluid Mech.*212, 593-614.
Landahl, M.T. (1993) Model for the wall-layer structure of a turbulent shear flow. *Eur. J. Mech.B/Fluids 12*, 85-96
Lundbladh, A. annd Johansson, A. (1993) Direct simulation of turbulent spots in plane Couette flow. *J. Fluid Mech. 229*, 499-516.
Prandtl, L. (1925) Bericht uber Untersuchungen zur ausgebildeten Turbulenz. *Z. Angew. Math. Mech5*, 136-137.

COHERENT CONTRIBUTION TO THE TURBULENT MIXING
OF A BUOYANT JET IN CROSS FLOW

S.G. NYCHAS, J.N.E. PAPASPYROS,
P.N. PAPANICOLAOU, E.G. KASTRINAKIS
Department of Chemical Engineering, Aristotle University of Thessaloniki
Univ. Box 453, 54006 Thessaloniki, GREECE

1. Introduction

A jet issuing in cross flow is encountered in several engineering applications. The description of the jet structure and the identification of the sources of the jet wake vorticity, however, are still open questions. This work is part of a series of experimental studies on turbulent buoyant jets in cross flow [1, 2]; here, a coherent contribution to the turbulent entrainment and mixing of a buoyant air jet issuing in a cross wind is reported. Further, this work offers support to a theory on the generation of the jet wake vorticity.

2. Experimental

In this work, an air jet with initially passive buoyancy was studied. The jet issued from a nozzle of diameter D, placed in front of a flat plate with a sharp leading edge; the nozzle exit was at the level of the flat plate. The ratio R of jet exit velocity to cross flow velocity had a rather high value of R=7. The measurements presented here were anemometric, and covered a long distance downstream from the jet origin, down to the position where the buoyancy starts dominating the flow (the role of buoyancy is enhanced downstream). Temperature was used as a convenient scalar. Instantaneous temperature and two velocity components were simultaneously measured by a triple-wire probe at several positions across the jet cross section (150-300 point measurements per cross section). In total, seven jet cross sections were scanned at positions from x=18 D to x=100 D downstream from the nozzle exit. The experimental set-up and the probe used are sketched in Fig.1.

3. Results and discussion

The measurements provided contour maps of several quantities, including the Reynolds stress ρuw and the turbulent scalar transport terms $\rho c_p uT'$ and $\rho c_p wT'$ (u and w are streamwise and vertical velocity fluctuation, respectively, T' is temperature fluctuation, ρ is density and c_p is specific heat at constant pressure); these quantities are necessary for realistic numerical modelling of the flow field. Typical contour maps of the scalar rms and the vertical turbulent scalar transport term are shown in Fig.2. The analysis of the data

125

S. Gavrilakis et al. (eds.), Advances in Turbulence VI, 125-128.

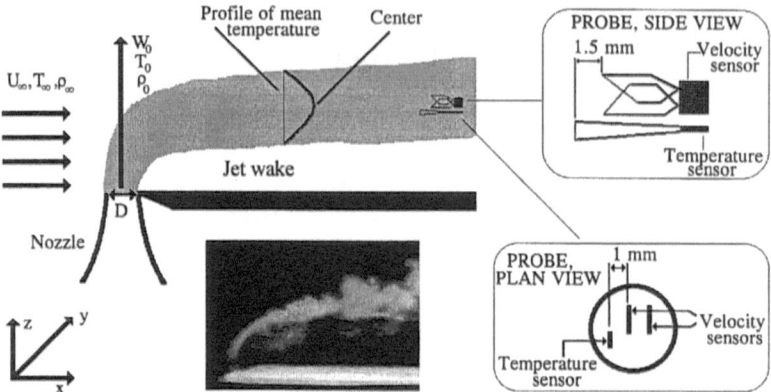

Figure 1. The flow studied and probe used (not to scale). The origin of the axis is located at the center of the nozzle exit. In this work, D=2.65 mm, U_∞=3.2 m/s, W_o=22.4 m/s and T_o=80 °C. The sampling frequency was 4000 Hz, and the sampling time was 25 s. The embedded picture was recorded during a flow visualization study; it depicts the important role of large-scale structures in the studied flow field.

provided a picture of strong entrainment of ambient air at the bottom of the jet. This entrainment is related to a complex interaction between the jet and the jet wake.

An open question is whether Karman-like vortices (KLV hereafter) are shed in the jet wake and, if so, whether they originate from jet vorticity. According to Fric and Roshko [3], KLV are shed in the wake of a wall jet in cross flow only if the jet issues in a boundary layer; the vorticity of KLV originates in the boundary layer, as it separates at the nozzle exit. In this work, the jet issued in a *homogeneous* cross flow, and KLV were not detected. This supports the theory of Fric and Roshko; if the jet was behaving like a solid cylinder, or if KLV were formed by jet vorticity shed in the wake, as other theories

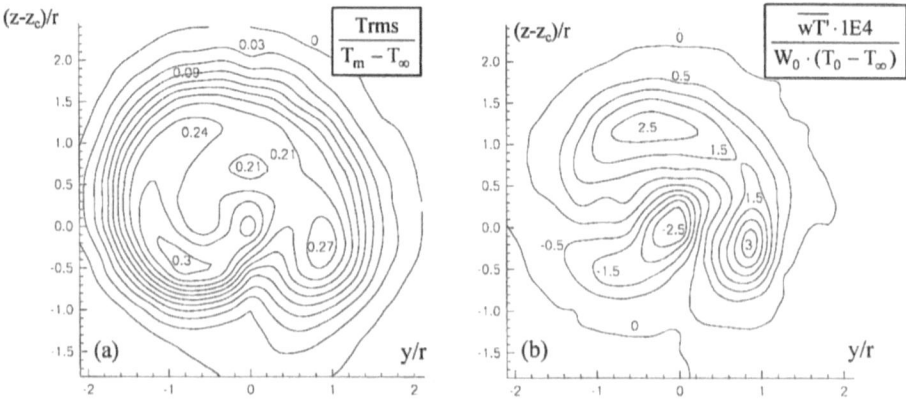

Figure 2. Typical contour maps of (a) the temperature rms and (b) the fluctuating part $\overline{wT'}$ of the vertical turbulent scalar transport term, at x=18 D. The coordinates are expressed as distances from the jet center, i.e., the point along y=0 where the time-averaged temperature $\overline{T}$ is maximum; they are normalized by the jet half-width r, defined as the distance above the center, along y=0, where $\overline{T} - T_\infty = 1/2 \left(\overline{T}_c - T_\infty \right)$. Subscript 'c' denotes values at the jet center, while 'm' implies maximum time-averaged value across the jet cross section.

postulate, KLV should be detected in any case, be the oncoming cross flow homogeneous or of boundary layer type. Further, Fric and Roshko stated that the jet does not shed any of its vorticity in the jet wake. In this work, it is considered probable that the jet *does* shed part of its vorticity in the wake (notice the jet fluid that seems to be "pulled" from the jet in Fig.1); this agrees with most of the references in the literature [5].

Although KLV were not noticed in the present work, the power spectra displayed a peak at f=26 Hz (KLV would give a peak around f=180 Hz). This peak was attributed to a coherent structure, and the acquired time signals were analyzed by a triple decomposition [4]: $Q = \overline{Q} + <q> + q_r$, where Q is an instantaneous quantity, $\overline{Q}$ is its time-average, $<q>$ is the coherent fluctuation and q_r is the random fluctuation. The coherent parts were considered to be the outputs of digital band-pass filtering of the signals around f=26±5 Hz. Vertical profiles of the total and coherent parts of several quantities are shown in Fig.3. Clearly, the coherent structure plays an important role in the energetics of the flow field. The structure is more evident along y=0; the horizontal profiles (not shown) display weaker contribution of the coherent parts (with the notable exception of $<w><T'>$). The structure is active even at x=100 D, where buoyancy starts dominating the flow.

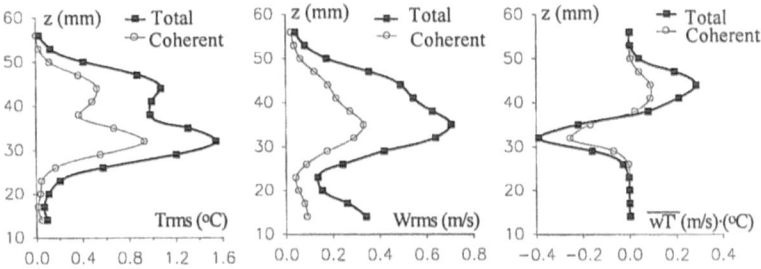

Figure 3. Total and coherent parts of measured quantities along the jet plane of symmetry (y=0), at x=18 D (the coherent product terms are defined as the products of the coherent parts of the corresponding signals, i.e.,
$$(wT')_{coherent} \equiv <w><T'> \text{, etc.}).$$

Next, the coherent signals were phase-averaged as described in Matsumura and Antonia [4], the $<w>$ signal being used as a reference signal. Selected contour maps of phase-averaged quantities are shown in Fig.4. It is confirmed that there is a coherent structure in the studied flow field. This structure could be related to entrainment of ambient air at the bottom of the jet as it interacts with the jet wake; notice the occurence of "cold" (negative $<T'>$), "slow" (negative $<u>$) instances at a phase value of 90 deg., where the reference signal $<w>$ displays its positive peak.

According to a recent study of Eiff *et al.* [5], downstream of a turbulent round jet issuing in cross flow from a *stack* there are two wake regions, the stack wake and the jet wake. These authors report that there are KLV in both wakes; the two sets of KLV are locked-in, and have different source of vorticity, the jet wake vorticity being shed from the jet. Eiff *et al.* provided contour maps of phase-averaged coherent streamwise velocity that resemble Fig.4b. Accordingly, the coherent structure identified in the present work could be associated with vorticity either shed by the jet, or produced in the boundary layer that is developing in the experiments of the present work.

128

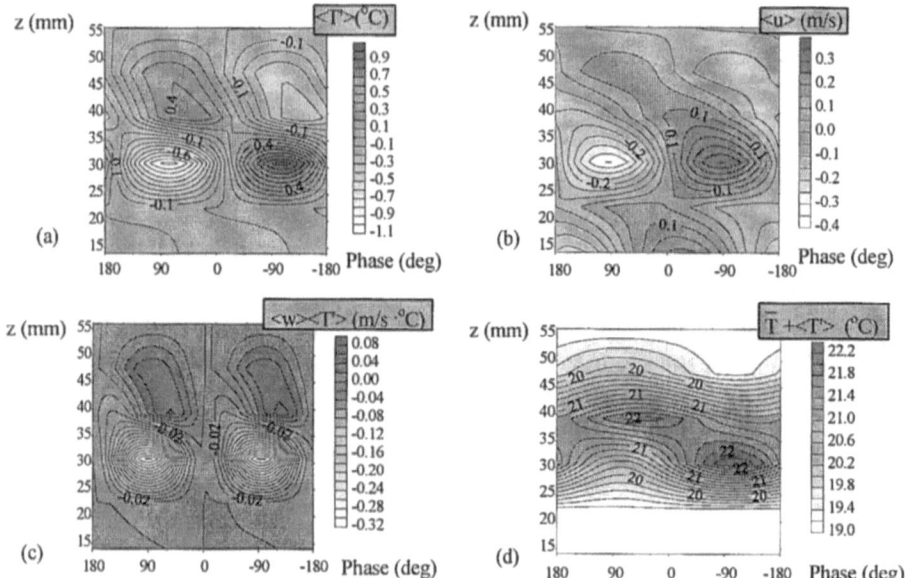

Figure 4. Typical contour maps at x=18 D (flow is from left to right) of phase-averaged coherent parts of:
(a) temperature fluctuations, (b) streamwise velocity fluctuations, (c) vertical turbulent scalar transport term
and: (d) sum of time-averaged temperature and coherent part of temperature fluctuations.

4. Conclusions

The main conclusions of the present work are the following:

- A coherent structure was identified in buoyant jets issuing in cross flow. This structure plays an important role in the energetics of the flow field; it is related to entrainment of ambient air at the bottom of the jet, as it interacts with the jet wake.
- A jet in cross flow sheds part of its vorticity in the jet wake. This vorticity is not organized in Karman-like vortices, except perhaps if it is locked-in with other vorticity.
- The coherent structure identified here could be associated with vorticity either shed in the jet wake, or produced in the developing boundary layer of this study.

5. References

1. J.N.E. Papaspyros, P.N. Papanicolaou, E.G. Kastrinakis and S.G. Nychas: Mixing of a turbulent round buoyant jet in cross flow, in R. Benzi (Ed.), *Advances in Turbulence V.*, Kluwer Academic Publishers, Dordrecht, 1995, pp. 403-407.
2. J.N.E. Papaspyros, P.N. Papanicolaou, E.G. Kastrinakis and S.G. Nychas: Buoyant jet in cross flow: study of turbulent transport terms. Accepted for presentation in the "3rd International Symposium on Engineering Turbulence Modelling and Measurements", May 27-29 1996, Heraklio, Crete, Greece.
3. Fric, T.F. and Roshko, A.: Structure in the Near Field of the Transverse Jet, in *Turbulent Shear Flows 7*, Springer-Verlag, Berlin, Heidelberg, 1991, pp. 225-237.
4. Matsumura, M. and Antonia, R.A.: Momentum and heat transport in the turbulent intermediate wake of a circular cylinder, *J.Fluid Mech* **250** (1993), 651-668.
5. Eiff, O.S., Kawall, J.G. and Keffer, J.F.: Lock-in of vortices in the wake of an elevated round turbulent jet in a crossflow, *Exp. Fluids* **19** (1995), 203-213.

This work has been supported by the EEC STEP program under contract number #STEP-PL-900597

EFFECTS OF SHEAR-LAYER ROLL-UP ON AXISYMMETRIC DNS VELOCITY SIGNALS IN A COAXIAL JET CONFIGURATION

M. ONORATO Jr., P. PETAGNA, M. V. SALVETTI, G. BURESTI
Dipartimento di Ingegneria Aerospaziale, Università di Pisa
Via Diotisalvi 2, 56126 Pisa, Italy

1. Introduction

For the characterization of developing turbulent flows, which are dominated by the dynamics of vorticity structures of different scale, it would be very useful to connect the features of the velocity fluctuations detected at a certain point with the evolution of the structures that are present in the nearby regions. To this end, outputs of DNS calculations, which can provide simultaneously both the evolution of the vorticity and the "velocity signals" corresponding to the different regions of the flow field, may be used. Obviously, only low Reynolds number conditions can be analysed, but the influence of basic mechanisms, as roll-up, passage and pairing of vortical structures, can be singled out, both as regards their "signatures" on the velocity time histories, and their effect on different statistical quantities. In particular, indications may be obtained on the processes that contribute to the production of Reynolds stresses, and thus to the average mass transport between two streams at different velocities. This type of analysis might then be considered as a very first step towards the identification of possible procedures for the recognition of the presence and evolution of the dominating vortical structures from velocity signals obtained experimentally at higher Reynolds number.

In the present paper, we use the results of an axisymmetrical DNS of a coaxial jet configuration, with outer to inner velocity and diameter ratios $U_o/U_i = 0.71$ and $D_o/D_i = 1.41$, and $Re = (U_i D_i)/v = 6000$, which was studied experimentally in [1]; the numerical procedure is described in [2]. Velocity signals with a sampling frequency of $400\ Hz$ are available from the DNS at 10 radial positions in 15 axial sections. Attention is focused on the analysis of these signals in the region between the roll-up and the first pairing, so that the assumption of axisymmetry should not inficiate the description of the main flow features (see e.g. [1] and [3]). A wavelet co-spectral analysis is used in order to associate the roll-up of the shear layer, the passage of formed vortices, and their pairing, with average, spectral and instantaneous contributions to the Reynolds stresses.

2. Wavelet co-spectral analysis

Let $W_u(a,\tau)$ and $W_v(a,\tau)$ be the wavelet transforms of the axial and radial velocities $u(t)$ and $v(t)$; a is the scale parameter. We define the wavelet cross-scalogram as:

S. Gavrilakis et al. (eds.), Advances in Turbulence VI, 129-132.

$$W_{uv}(a,\tau)=W_u{}^*(a,\tau)W_v(a,\tau), \tag{1}$$

where $*$ means complex conjugate. If the analysing wavelet is complex, then the cross-scalogram is also complex, and can be written in terms of its real and imaginary parts:

$$W_{uv}(a,\tau)=CoW_{uv}(a,\tau)-iQuadW_{uv}(a,\tau). \tag{2}$$

The angle of phase between the $u(t)$ and $v(t)$ velocity signals is then given by:

$$\Theta_{uv}(a,\tau)=\tan^{-1}\big(QuadW_{uv}(a,\tau)/CoW_{uv}(a,\tau)\big). \tag{3}$$

From (3), it is clear that if $u(t)$ and $v(t)$ have the same phase, $QuadW_{uv}(a,\tau)$ is zero and $CoW_{uv}(a,\tau)$ is large; the situation is reversed when the two signals are in quadrature. Starting from the definition of wavelet transform, it can be shown that [4]:

$$\overline{uv}=(1/T)\int_0^T u(t)v(t)dt=(1/C_g)\int_0^\infty\int_0^T CoW_{uv}(a,\tau)d\tau da\,/\,a^2\,, \tag{4}$$

where C_g is the characteristic constant of the wavelet. Thus the integrand function (the co-scalogram) on the right hand side gives the contribution to $\overline{uv}$ as a function of time and scale. Therefore, the wavelet transform allows the instantaneous contribution of the different scales to the Reynolds stresses to be obtained. Using the relation between scales and frequencies of the chosen wavelet, by integration of $CoW_{uv}(a,\tau)$ in τ we may also obtain a wavelet co-spectrum, which basically gives the same information as the Fourier co-spectrum, but is smoother and clearer, particularly when the time series are too short to obtain adequately averaged Fourier spectra or co-spectra, [5].

3. Results

In figure 1 a contour map of the correlation $\overline{uv}$ is shown for an intermediate region of the coaxial jet field; x and y are the axial and radial coordinates. As can be seen a maximum is present at $y/D_i\cong0.7$, $x/D_i\cong3$; moving downstream along the shear layer of the outer jet, $\overline{uv}$ decreases, reaching a negative minimum at $x/D_i\cong6$. Other negative and positive peaks are also present.

These results may be interpreted in terms of the behaviour of the vorticity field. Indeed, the typical instantaneous vorticity map of figure 2 shows that three regions can be qualitatively recognized: in the first one $(x/D_i\cong3)$ the roll-up of the outer shear layer takes place, in the second one $(x/D_i\cong4.5)$ there is the passage of formed vortices, and in the third one $(x/D_i\cong6)$ the vortices tend to interact and pair. The inner vorticity layer remains almost unperturbed up to $x/D_i\cong4.5$, when it starts rolling-up and interacting with the outer one. However, the time and position of occurrence of all these events, as well as the shape and size of the vortices, are only approximately repetitive.

As pointed out in [6], the value and sign of the Reynolds stresses at a certain point may be connected with the orientation of the passing vorticity structures, and with the consequent phase differences between the u and v fluctuations.

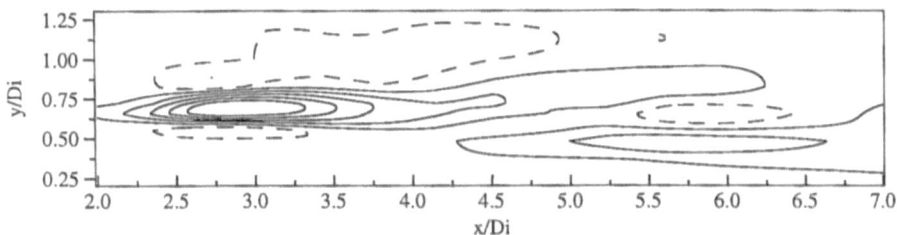

Figure 1. Contour map of $\overline{uv}$ (full line: positive, dashed line: negative)

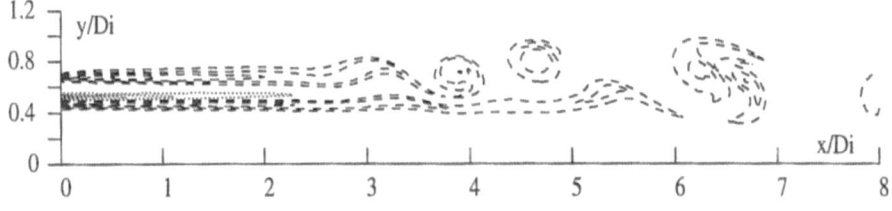

Figure 2. Instantaneous vorticity map at $t = 51.5\ s$

This connection between phase lag of the vorticity-induced fluctuations and Reynolds stresses may better be highlighted through the wavelet analysis. We use a particular Morlet wavelet (see e.g. [7]), for which the following relation exists between scales and frequencies: $f = 2\pi / (6a)$. Since most of the energy of the signals is contained at low frequencies, to achieve a good spectral resolution we restrict our wavelet analysis to the interval between $0\ Hz$ and $2\ Hz$. In figure 3 the wavelet co-spectrum and quad-spectrum are presented for two signals, respectively taken at $x/D_i = 3$, $y/D_i = 0.66$ and at $x/D_i = 4.3$, $y/D_i = 0.82$. In the first position (figure 3a) the maximum of the co-spectrum is near $0.7\ Hz$, which corresponds to the roll-up frequency of the shear layer of the outer jet. It can also be noted that the co-spectrum is higher than the quad-spectrum, which implies that the average phase difference between the two signals is not large. This is found to be typical of the roll-up region, where $u(t)$ and $v(t)$ have usually a small phase lag, giving a strong contribution to the Reynolds stresses. In figure 3b the situation is reversed: the quad-spectrum is much larger than the co-spectrum, so that $u(t)$ and $v(t)$ are almost orthogonal. This condition corresponds to a region where vortices pass or pair, and are not inclined in the flow direction; thus, even though the fluctuations (which have now a dominating frequency close to 0.4 Hz) are of the same order as those in the roll-up region, their contribution to the Reynolds stresses is small.

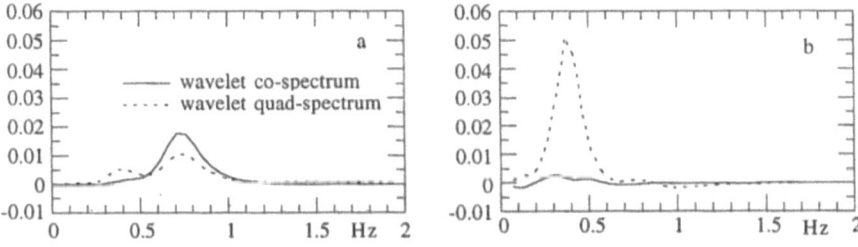

Figure 3. Wavelet co-spectra and quad-spectra. a) x/Di=3, y/Di=0.66, b) x/Di=4.3, y/Di=0.82

132

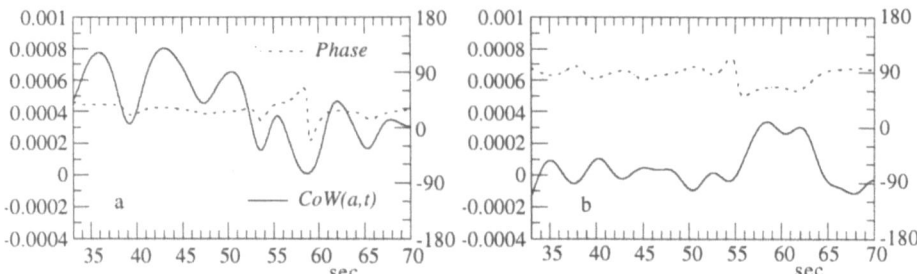

Figure 4. Cross-sections of wavelet co-scalograms and phase maps of $u(t)$ and $v(t)$.
a) x/Di=3, y/Di=0.66, f=0.66Hz; b) x/Di=4.3, y/Di=0.82, f=0.40Hz

A more detailed analysis of the instantaneous contribution of the different scales to the Reynolds stresses may be obtained from the wavelet co-scalograms. In figure 4 we show, for the two cases of figure 3, the cross-sections of the co-scalograms and phase maps at scales nearly corresponding to the relevant dominating frequencies. The differences between the two positions as regards both the value of $CoW_{uv}(a,\tau)$ and the phase are clearly apparent; in particular it is confirmed that the low contribution to $\overline{uv}$ in figure 3b is due to a phase lag that is prevailingly near 90°. However, it is also clear that in both cases significant time variations occur, which can be shown to be related to irregular behaviour of the vortical structures (e.g. delayed or anticipated roll-ups and pairings, variations in shape, etc.). The cross-sections at constant scale of the wavelet scalograms may then be treated as time-signals, and their statistical moments used to characterize different types of flow, and in particular the amount of intermittency in the occurrence of certain flow features.

A preliminary analysis of a second DNS calculation, in which a random perturbation is superposed to the inlet velocity profile, confirms the significant production of $\overline{uv}$ connected with the roll-up process, which however occurs more upstream. The wavelet co-spectral analysis allows again a clear description to be obtained of the average and instantaneous contribution of the different vorticity structures to the Reynolds stresses.

Acknowledgements
The present work was financially supported by the Italian M.U.R.S.T. We also wish to thank P. Orlandi and R. Verzicco for making the DNS data base available.

References
1. Dahm, W.J.A., Frieler, C.E., Tryggvason G.: Vortex structure and dynamics in the near field of a coaxial jet, *J. of Fluid Mech.* **241** (1992), 371-402.
2. Salvetti, M.V., Orlandi, P., Verzicco, R.: Direct simulations of transitional axisymmetric coaxial jets, to be published in *AIAA J.*, **34**, n.12 (1996).
3. Grinstein, F.F., Hussain, F., Oran, E.S.: Vortex-ring dynamics in a transitional subsonic free jet. A numerical study, *Eur. J. Mech., B/Fluids* **9**, n. 6 (1990), 499-525.
4. Daubechies, I.: *Ten Lectures on Wavelets*, SIAM, 1992.
5. Katul, G.K., Parlange, M.B., Chu, C.R.: Intermittency, local isotropy, and non Gaussian statistics in atmospheric surface layer turbulence, *Phys. Fluids* **6** n. 7 (1994), 2480-2492
6. Ho, C.M., Huerre, P.: Perturbed free shear layers, *Ann. Rev. Fluid Mech.* **16** (1984), 365-424
7. Farge, M.: Wavelet transforms and their applications to turbulence, *Ann. Rev. Fluid. Mech.* **24** (1992), 395-457

VORTEX STRETCHING AND FILAMENTS

P. PETITJEANS and J. E. WESFREID

Laboratoire de Physique et de Mécanique des Milieux Hétérogènes
URA CNRS 857
Ecole Supérieure de Physique et de Chimie Industrielles
10, rue Vauquelin 75005 Paris, France

1. Introduction

Understanding the formation and of dynamics of filamentary vortices with intense vorticity in turbulent flows is of great fundamental interest. Most of the energy transfer towards dissipation occurs through these structures and their breakdown. It has been shown numerically by Siggia [1] that the regions of largest vorticity concentration in a turbulent flow look like filaments. Experimentally filaments have been observed by Couder et al. [2] in a turbulent flow in a cylinder between two counter-rotating stirrers. They observed filaments corresponding to low pressure regions. They are generated by local stretching of the vorticity sheets [3].

In this communication we present a new set-up to experimentally study the effects of stretching on a controlled vorticity sheet coming from a laminar boundary layer flow on a flat plate. This flow is not turbulent. Therefore, the characteristic time and length are large enough so as to enable us to study the instability. At the same time it provides a fairly good model of vorticity instabilities occuring in fully turbulent flows. Thus, this experiment opens the possibility to accomplish quantitative measurements of the dispersion relation of vortex instabilities.

2. Experiments and results

The experimental set-up consists of a low-velocity water channel with very well controlled flow (Fig. 1). The flow develops laminar boundary layer with a velocity of order 1 cm/s far away from the wall. The Reynolds number based on the boundary layer thickness is $Re_\delta \approx 100$. The boundary layers on the lower and upper walls form the initial vorticity: $\vec{\omega} = \pm \omega \vec{z}$. Then the vorticity sheet is stretched parally to the vorticity vector by means of the flow deviation on each lateral wall. We control the total flow rate through the channel as well as the flow rate in these deviations; consequently, we can control the initial vorticity (via the thickness of the boundary layer) as well as the stretching. When the stretching is large enough, we obtain the formation of strong periodic vortices by roll-up of fluid sheets. The formation, development and

133

S. Gavrilakis et al. (eds.), Advances in Turbulence VI, 133-136.

destabilization of these vortices is visualized by injection of a sheet of fluorescent dye (fluoresceine) on the wall, 2 cm upstream from the stretching position, as shown Fig. 1.

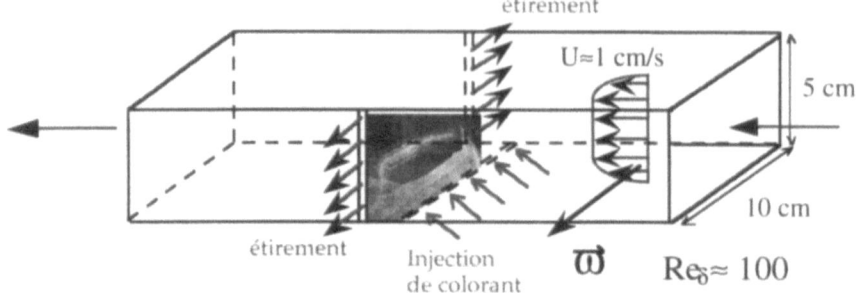

Fig. 1: Experimental set-up.

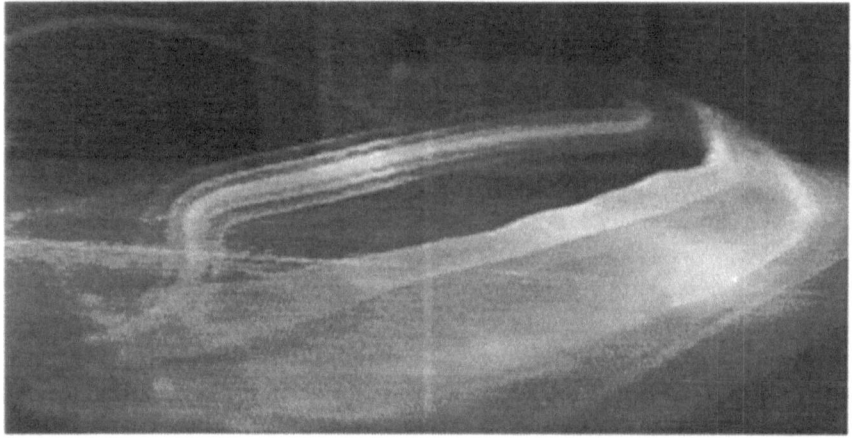

Fig. 2: Visualization of a vortex created by stretching.

Figure 2 shows such a visualization. The dye sheet separates from the injector, and penetrates into the vortex by winding around the axis. Then a very thin dye filament that materializes the core of the vortex can be observed. In the core, the dye is evacuated on the two sides towards the deviations on the lateral walls of the channel. Another vortex is then generated, and so on, with an excellent regularity due to the permanent longitudinal flow.

We can deduce from the fact that the advection velocity $U_a = U_a(y)$ that vortices have an elliptical form and are tilted. This is because of the advection velocity that still has a boundary layer profile due to the wall. By the way, this mechanism is also responsible for the vortex pairing instability [4] that is observed very clearly in this experiment. A perturbation of the fastest moving part of a vortex (furthest from the wall) catches up with the slowest moving part of the preceding vortex. Then the two vortices first rotate around each other, and then coalesce into a stronger vortex. We can control this instability by means of the advection velocity, i. e. the flow rate of the channel.

Another observed instability concerns the vortex line that can be deformed sinuously (sequence Fig. 3).

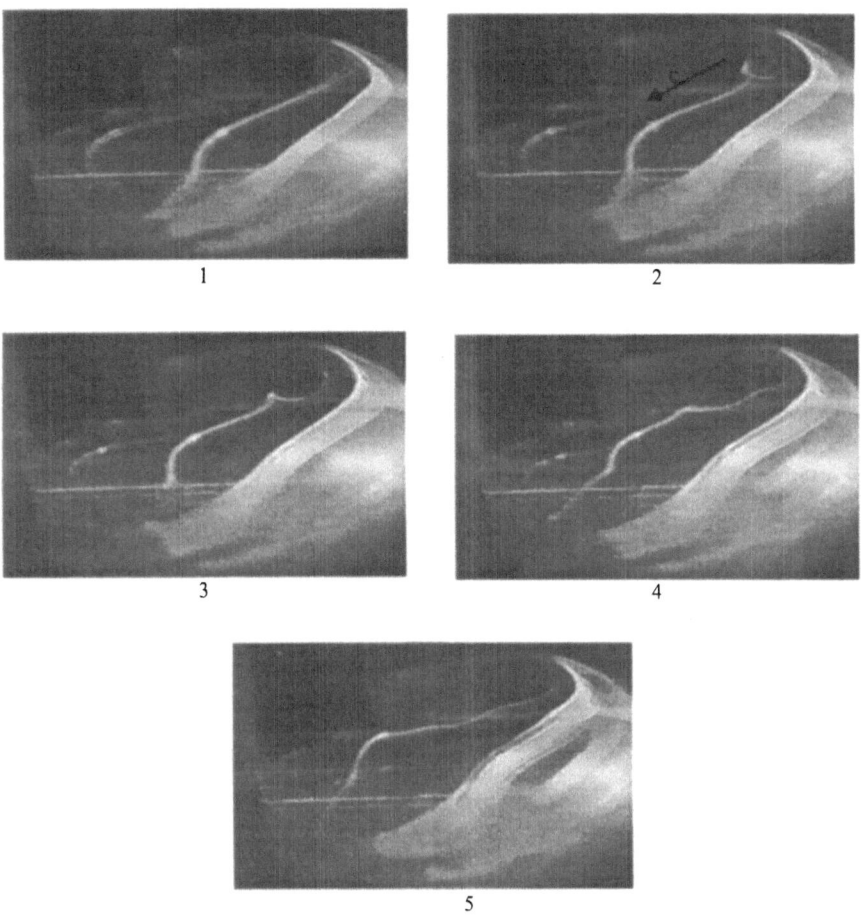

Fig. 3: Sequence of 5 views during 3 secondes showing the propagation of a wave along the filament.

The transversal perturbation is amplified and the difference in vertical velocities produces an enhancement of one of the lobes that is then highly developed and shifted to a localized longitudinal line (hairpin). The effect is magnified by the attachment of the vortex to the openings on the walls which are responsible for stretching.

The last instability we can observe is what is generally called "vortex breakdown" [5]. It happens when a vortex loses its axial symetry, deforms helicoïdally and collapses into a "coil shape" (Fig. 4). This transition is clearly observed in this experiment. This gives a picture of some intermittent features of turbulent flows.

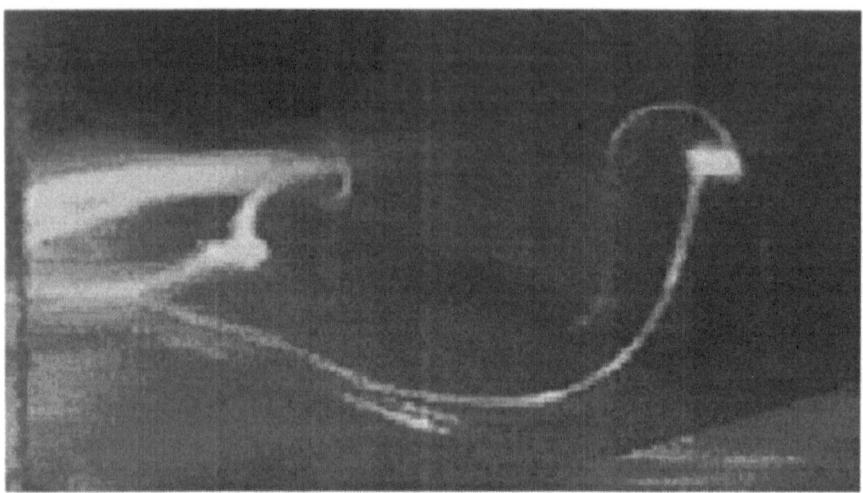

Fig. 4: View of a "coil shape" during the breakdown of a vortex.

3. Conclusion

The main point of this preliminary work is the design of an experimental set-up which enables direct observation of the dynamics of vortices under stretching. This allows to understand and measure the typical mechanisms of vortex instability and gives information on the evolution of stretched vortices. We clearly observe the formation of vortices by rolling-up.

Later, we observe several forms of destabilization by pairing, by sinuous deformation, or by breakdown. We can observe in detail the intermittent "coil shape" transition as the ultimate evolution of vortex breakdown. We have noticed that shear and velocity profile U(y) deform the vortex which assumes an elliptical form. As it is known from the inviscid theory of stability, circular vortices are stable, and as in the experiment, the deformed elliptical profile produces three-dimensional instability.

References:

[1] E. D. Siggia, "Numerical study of small-scale intermittency in three-dimensional turbulence", J. Fluids Mech. **107**, 37 (1981)
[2] O. Cadot, S. Douady, and Y. Couder, "Characterization of the low pressure filaments in three-dimensional turbulent shear flow", Phys. Fluids. **7**, (3), 630-646 (1995)
[3] J. C. Neu, "The dynamics of stretched vortices", J. Fluid Mech. **143**, 253 (1984)
[4] P. G. Saffman and G. R. Baker, "Vortex interactions", Ann. Rev. Fluid Mech., **11**, 95 (1979)
[5] M. G. Hall, "Vortex breakdown", Ann. Rev. Fluid Mech., **4**, 195 (1972)

FINAL STATES OF INCOMPRESSIBLE TWO DIMENSIONAL DECAYING VORTICITY FIELDS

ENRICO SEGRE

DIAS, Politecnico di Torino, Italy

AND

SHIGEO KIDA

National Institute for Fusion Sciences, Nagoya, Japan

1. Introduction

We consider the free decay of the vorticity $\omega(\vec{x})$ of a two dimensional incompressible fluid, described by the Navier-Stokes equation. Stable large scale structures form and survive for a long time, before being ultimately damped by the dissipation. Most of the numerical simulations in the literature start from random initial conditions, which lead to isolated vortex cores and their progressive merging, until to the formation of a final dipole. Less attention has been devoted to the latest state of the field, for which a few theories, which we recall, are found in literature.

Point vortices systems   The mean field equilibrium configuration of an ensemble of point vortices satisfies the differential equation [4]:

$$\omega_0(\vec{x}) = -\nabla^2 \psi_0(\vec{x}) = c_1 e^{-\beta\psi_0(\vec{x})} - c_2 e^{\beta\psi_0(\vec{x})}. \tag{1}$$

With some connection to this, [5] propose a decomposition of the vorticity in four non-physical subfields, which implies an $\omega_0(\vec{x})$ which differs from (1) by an additive constant. The functional dependence implies the stationarity of the motion, for null dissipation.

Minimum enstrophy principle   The "selective decay hypothesis" [1] argues that the final state should minimize the enstrophy Q with constrained energy E. For doubly periodic boundary conditions the solution $\omega_0(\vec{x}) = \pi k \sqrt{E/2} \sin \vec{k} \cdot \vec{x}$, with integer vector $\vec{k}$, is immediately found.

S. Gavrilakis et al. (eds.), Advances in Turbulence VI, 137-140.
© *1996 Kluwer Academic Publishers.*

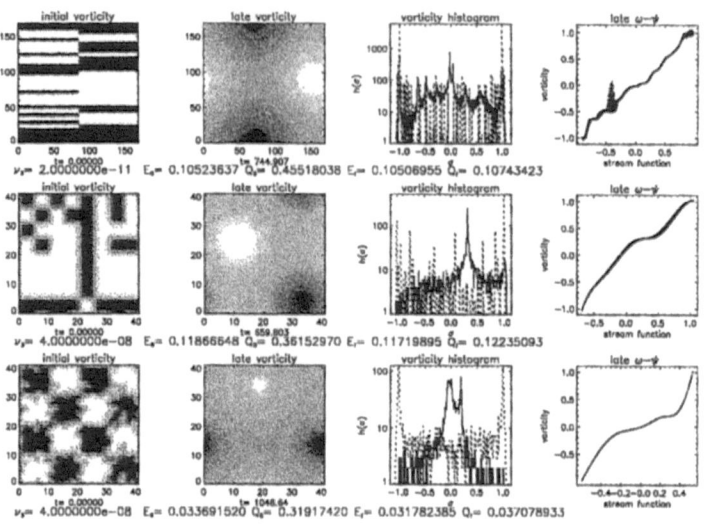

Figure 1. arbitrary 2-level initial conditions which decay in final dipoles.

Maximum entropy theory   The theory introduced by [3] and [6] is intended for an underresolved Euler equation. With the assumption of ergodicity of the mixing, the fluid is expected to relax to in its most probable way. An implicit system is obtained. Particular forms of $\psi_0(\omega_0)$ are found with additional hypoteses. Specifically, if the vorticity takes only two values $\Omega_1 = -\Omega_2$,

$$\omega_0(\vec{x}) = -\tanh\beta(\Omega_1\psi_0(\vec{x}) - \frac{\mu(\Omega_1) - \mu(-\Omega_1)}{2}),\qquad(2)$$

where $\mu(\Omega_1)$ and $\mu(\Omega_1)$ are implicitly related to Q. Furthermore, the 'dressed vorticity' corollary implies that ω_0 has maximal energy among all the fields with the same distribution of vorticity.

2. Numerical experiments

We integrated in time several vorticity fields, using a pseudospectral code on the periodical square $(2\pi)^2$ with a small hyperviscosity. The choice of the periodic domain, as well as our initial conditions, which consist of arbitrarily arranged patches of opposite vorticity, are legitimate as test cases, and allow a direct connection with the theoretical formulas. The patches intertangle in a complicate way; filamentary features are blurred because of the finite resolution and of the dissipation, until a simpler state stabilizes. The late configuration undergoes no further evolution, apart of a slow viscous diffusion, which can in principle be made very small; this state can be therefore called 'final'.

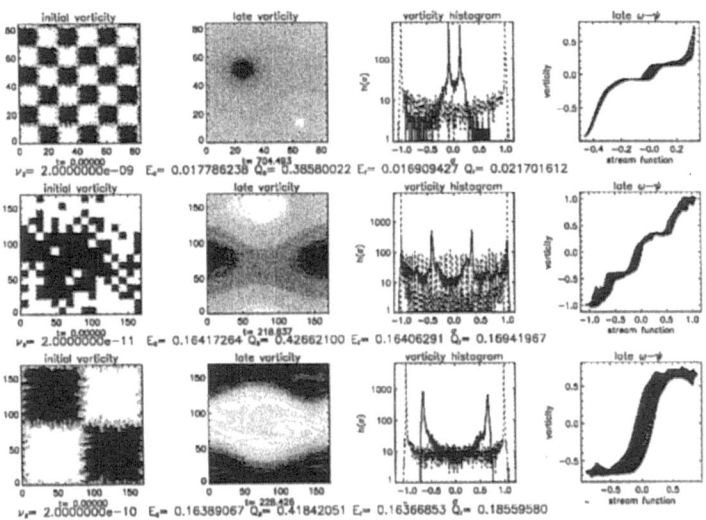

Figure 2. arbitrary 2-level initial conditions which decay in nonstationary configurations.

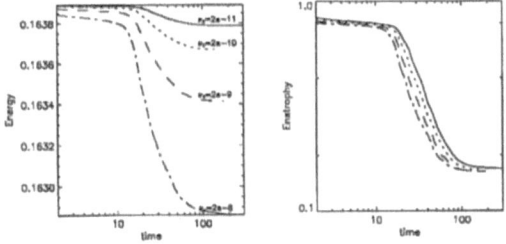

Figure 3. Energy and enstrophy decay for different values of the hyperviscosity

Sometimes the final state is a stationary dipole, as in the accepted point of view (figure 1). But the profiles of cores are particular of each decay. The scatterplot of $\omega - \psi$ tends to a line, but does not match any of the expected functional relations.

Other cases decay in nonstationary final configurations (figure 2), with a variety of forms. A configuration that prevents further mixing may develop. In those cases the scatterplot of $\omega - \psi$ never thins out.

The final states appear to be quite independent from the resolution and the (small) values of the viscosity. We plot in figure 3 $E(t)$, $Q(t)$ for the last case of figure 2 and different values of hyperviscosity. The loss in E is indeed negligible; Q instead decreases significantly, but stabilizes on a final value independent on the hyperviscosity. Apart of the values and relaxation times, the same behavior is seen for the other decays, quite independently of the viscous or hyperviscous term used. In contrast, the theories take into account a lowering of Q, but fix it in some unique way. We do not see an universal final state emerging from the relaxations. The dynamical path underwent by the system during its relaxation cannot be neglected: initial states which have equal energy, enstropy and values of vorticity decay to different final states.

3. Energy maximization and stability

We can add some remarks on the implications of energy maximization. First, by variational differentiation with respect to a generic area preserving infinitesimal deformation it can be proven that, for vanishing dissipation, the "local" extremal energy states, and only them, are stationary.

On the other side, requiring that E is absolutely maximal with respect to discrete pointwise exchanges of values of vorticity, it follows that a relation $\psi(\omega)$ has to exist, to be single valued and monotonically increasing.

Moreover, the Rayleigh-Arnold criterion [2] states that the maximal energy arrangement of a vorticity field is unique and stationary. If $d\psi_0(\omega_0)/d\omega$ is strictly positive and limited, then the field is also nonlinearly stable.

These arguments do not guarantee that *any* initial field decays to such energy maximizing configuration. It is possible compute numerically the energy of the rearrangements. The highest energy configuration of equal area tiles of vorticity $\pm\Omega_1$ on the periodic square, is a subdivision in two stripes parallel to the sides. If we rearrange the final states obtained in section 2, we see that for the stationary final states the highest energy is achieved for a 'square dipole' rearrangement, with minimal increase of energy. For the nonstationary ones, we often see that the stripe rearrangement is preferred, with a significant increase in E. The choice among the two maximal profiles is basically a function of the available vorticity.

References

1. F. P. Bretherton and D. B. Haidvogel. Two-dimensional turbulence above topography. *Journal of Fluid Mechanics*, 78(1):129–154, 1976.
2. D. D. Holm, J. E. Marsden, T. Ratiu, and A. Weinstein. Nonlinear stability of fluid and plasma equilibria. *Physics Reports*, 123(1-2):1–116, 1985.
3. J. Miller, P. B. Weichman, and M. C. Cross. Statistical mechanics, Euler's equation and Jupiter's red spot. *Physical Review A*, 45(4):2328–2359, 1992.
4. D. Montgomery and G. Joyce. Statistical mechanics of negative temperature states. *Physics of Fluids*, 17:1139–1145, 1974.
5. D. Montgomery, X. Shan, and W. H. Matthaeus. Navier-Stokes relaxation to sinh-Poisson states at finite Reynolds numbers. *Physics of Fluids A*, 5(9):2207–2216, 1993.
6. R. Robert and J. Sommeria. Statistical equilibrium states for two-dimensional flows. *Journal of Fluid Mechanics*, 229:291–310, 1991.

THE INFLUENCE OF SHALLOWNESS ON LARGE-SCALE COHERENT STRUCTURES

J. TUKKER and R. BOOIJ
Faculty of Civil Engineering
Delft University of Technology
P.O. Box 5048, 2600 GA Delft
The Netherlands

1. Introduction

In turbulent shallow-water flow, transverse gradients in the horizontal velocity generate large, quasi-two-dimensional, horizontal coherent structures. They give rise to some problems in modelling the turbulent flow with standard turbulence models. The essence of these problems arises from the shallowness of the flows considered. On the one hand, the relatively small depth restricts the large coherent structures to basically two-dimensional horizontal motions, (presumably) characterized by suppressed vortex stretching, and an inverse energy cascade. On the other hand, the bottom friction gives rise to small-scale three-dimensional turbulence. This system of coexisting different turbulent scales (Babarutsi and Chu, 1991) cannot be modelled with conventional turbulence models which use a single, scalar eddy viscosity based on a single length scale. Large, quasi-two-dimensional structures can be found for example in wakes behind islands in shallow seas, and in mixing layers between two shallow-water flows.

The aims of the present research are to gain insight into the influence of the shallowness on the evolution of large, horizontal coherent structures in a transverse mixing layer in a free-surface flow, and to detect the presumed gap between large horizontal scales and small three-dimensional scales in the turbulent energy spectrum.

2. Experiments

Experiments have been executed in a glass-bottom shallow water channel with a length of 20 m, a width of 3 m and a height of 0.20 m. in the Laboratory of Fluid Mechanics of the Delft University of Technology. Experimental research on transverse mixing layers in shallow water is executed in the Laboratory of Fluid Mechanics of the Delft University of Technology. The presence of the glass bottom allows Laser-Doppler measurements from below. This was chosen because the width and the shallowness of the flow preclude measurements from the sides. The flow chosen in this investigation

S. Gavrilakis et al. (eds.), Advances in Turbulence VI, 141-144.

is a transverse mixing layer developing behind a splitter plate between two adjacent turbulent streams with an initial flow velocity of 0.10 m/s and 0.30 m/s respectively, and a water depth h of 0.06 meter. Turbulence data were obtained with a fibre-backscatter Laser-Doppler-Anemometry system and three Burst Spectrum Analyzer processors (Dantec). The combination of a one-dimensional probe and a two-dimensional probe allowed the measurements of spatial correlations between horizontal velocity components.

3. Mixing layer development

Behind the splitter plate instabilities in the vertical shear layer result in the generation of horizontal coherent structures and a developing mixing layer. The dimensions of these coherent structures are proportional to the local width of the mixing layer δ, taken equal to the maximum-slope thickness. The production P of turbulent energy per unit area of the horizontal coherent structures is proportional to the product of the transverse turbulent shear stress τ_t and the transverse velocity gradient

$$P = \tau_t \frac{\Delta U}{\delta} h \qquad (1)$$

with ΔU the depth-averaged velocity difference across the mixing layer. The transverse shear stress τ_t is proportional to the velocity difference squared,

$$\tau_t = \chi \rho (\Delta U)^2 \qquad (2)$$

in which χ is a proportionality constant, and ρ the fluid density.

As the mixing layer spreads, its width and the scale of the horizontal coherent structures become much larger than the water depth. The shallowness of the flow restricts the scales of the vertical turbulence movements and therewith interrupts the normal energy cascade. The presence of the large structures leads to a local change of the bottom shear stress, which results in a dissipation of energy of these large-scale structures. The time-averaged work (per unit area) done by the bottom friction consists of two contributions. The first contribution $c_w \rho U^3$ is the work done against the mean flow, with c_w the bottom friction coefficient and U the time- and depth-averaged horizontal velocity. The second contribution F,

$$F = c_w \rho U \left(3\overline{U'^2} + \frac{3}{2}\overline{V'^2} \right) \qquad (3)$$

is the work done against the large-scale coherent structures. In this relation the contribution $\tfrac{1}{2}\rho\overline{U'^2}$ is the longitudinal component and $\tfrac{1}{2}\rho\overline{V'^2}$ the transverse component of the depth-averaged turbulent energy density of the large-scale coherent structures.

Because of the dissipation F the growth of the quasi-two-dimensional structures is suppressed and therewith the spreading rate of the mixing layer decreases downstream.

This decrease is estimated by (see Chu and Babarutsi, 1988)

$$\frac{d\delta}{dx} = \left(\frac{d\delta}{dx}\right)_{x=0} \left(1 - \frac{F}{P}\right) \qquad \text{for} \quad F \leq P \qquad (4)$$

4. Separation of scales

A method to investigate the presence of two different turbulent length scales is the measuring of spatial correlations. Transverse spatial correlations between longitudinal turbulent velocities were measured far from the splitter plate ($x = 15.4$ m) at two levels: $z/h = 0.5$ and 0.9 in shallow water (see Figure 1). During these measurements the reference point was fixed in the centre of the mixing layer and the second measuring point was at various positions in both transverse directions. In Figure 1 the peak in the correlation for small distances (smaller than the water depth) represents the small-scale turbulence, which is superposed on the large-scale turbulence apparent over all distances. The sharpness of the small-scale peak in the correlation makes it possible to extrapolate the large-scale part of the correlation to zero-distance (Townsend, 1976), and so to separate the turbulent energy and the spatial correlation in a large-scale and in a small-scale contribution. The measured spatial correlations yield about the same integral length scale of the large-scale structures of 0.6δ for both levels.

The turbulence intensity of the large-scale structures, scaled with the slightly depth-dependent velocity difference across the mixing layer is plotted in Figure 2. This scaled turbulence intensity is nearly constant in the mixing layer for $z/h > 0.3$, in accordance with the notion of quasi-two-dimensional structures. The scaling with the velocity difference corrects for the influence of the bottom friction. Below $z/h = 0.3$ the dominating small-scale turbulence precludes the accurate determination of the intensity of the large-scale turbulence. The contribution of the large-scale structures to the work done by the bottom friction and hence to the production of small-scale turbulence is small in shallow water. Hence, the intensity of small-scale turbulence depends on the local depth-averaged velocity U. The measured distributions of the small-scale turbulence intensity over depth agree well with the expressions presented by Nezu and Nakagawa (1993) for uniform, free surface flows.

5. Conclusions

- In the considered shallow mixing layer flow the large-scale, horizontal coherent structures have length scales much larger than small-scale, bed-generated turbulence. These large-scale turbulence structures have a quasi-two-dimensional structure.
- The large-scale structures lose turbulent energy because of the presence of the bottom friction. In this process, energy is directly transferred from the large-scale eddies to the small ones.
- This loss of energy of the large-scale structures limits the transverse development of the mixing layer.

144

References

Babarutsi, S. and V.H. Chu (1991) A two-length-scale model for quasi-two-dimensional turbulent shear flows, *Proceedings 24th IAHR Congress, Madrid*, C-53-C-60.

Chu, V.H. and S. Babarutsi (1988) Confinement and Bed-Friction Effects in Shallow Turbulent Mixing Layers, *Journal of Hydraulic Engineering* **114-10**, 1257-1274.

Nezu, I and H. Nakagawa (1993), *Turbulence in Open-Channel Flows*, IAHR Monograph series.

Townsend, A.A. (1976) *The structure of turbulent shear flow*. Cambridge University Press.

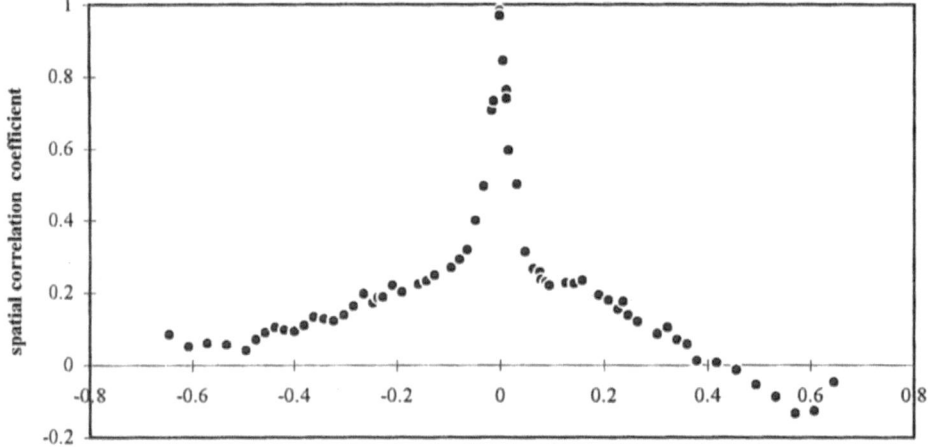

transverse distance scaled with layer width

Figure 1.   Transverse spatial correlation function between longitudinal turbulent velocities, measured at $z/h = 0.9$ at a distance of 15.4 m behind the splitter plate where $\delta/h = 13$. The reference probe was fixed in the centre of the mixing layer. The transverse distance is scaled with the layer width δ (= 0.79 m).

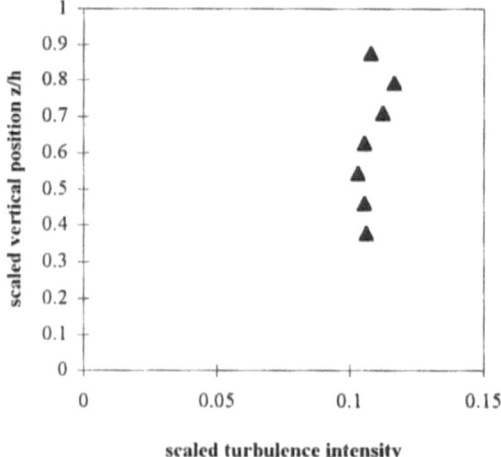

scaled turbulence intensity

Figure 2.   Vertical distribution of turbulence intensity of the large-scale coherent structures, measured in the centre of the mixing layer at a distance of 5.8 m behind the splitter plate, in shallow water. The turbulence intensity is scaled with the local velocity difference.

KOLMOGOROV CAPACITIES OF STREAMLINES AROUND TURBULENT VORTEX TUBES

J.C. VASSILICOS
DAMTP University of Cambridge
Silver Street, Cambridge CB3 9EW, UK

AND

JAMES G. BRASSEUR
Department of Mechanical Engineering
The Pennsylvania State university
University Park, PA 16802, USA

1. Introduction

Two issues relevant to small-scale turbulence are addressed together in this paper. Firstly, recent laboratory and numerical experiments have revealed the existence of vortex tubes in the small scales of the turbulence (see Villermeaux *et al.* (1995) and references therein). Secondly, Hunt & Vassilicos (1991) have shown that a self-similar high-wavenumber energy spectrum $E(k) \sim k^{-2p}$ with *non-integral* exponent p implies the existence of near-singularities that cannot be near-discontinuities in the turbulent velocities or their derivatives. Kolmogorov's self-similar spectrum $E(k) \sim k^{-5/3}$ corresponds to $p = 5/6$.

Two questions arise. Firstly, is it possible to define vortex tube flows topologically in such a way as to detect vortex tubes irrespective of their enstrophy level and independently of their enstrophy profile? If an entity such as a vortex tube exists in a turbulent flow, it is not necessarily characterised by high enstrophy throughout its extent. However, vortex tubes in real flows are generally characterized by spiral streamlines surrounding the core. Vassilicos & Brasseur (1996) have developed a systematic algorithm for the detection of spiral-like streamlines in Direct Numerical Simulations (DNS) of small-scale turbulence. This algorithm successfully identifies many high-enstrophy vortex tubes in the small-scale turbulence irrespective of the

S. Gavrilakis et al. (eds.), Advances in Turbulence VI, 145-148.
© 1996 *Kluwer Academic Publishers.*

enstrophy profile across and modulations along the vortex tube and gives direct access to the topology of the flow in and around these vortex tubes.

The second question that arises is to know whether some of the near-singularities of the turbulence reside in these vortex tubes. An attempt to address this second question can be made by measuring the Kolmogorov capacity of streamlines around turbulent vortex tubes on the basis of the following theorem.

2. The theorem of self-similar streamlines

In an axisymmetric incompressible flow with bounded vorticity at infinity, when the Kolmogorov capacity D_K of a spiral-helical streamline is strictly larger than 1, there must exist a velocity singularity at the axis of symmetry of the flow.

To prove this theorem we prove the contrapositive statement: *in an axisymmetric, incompressible flow with bounded vorticity at infinity, if the velocity field is regular (no singularities) at the axis of symmetry, then the Kolmogorov capacity D_K of spiral-helical streamlines is necessarily equal to 1.*

A function $u(r)$ is regular around $r = 0$ if it can be Taylor expanded around $r = 0$, i.e. $u(r) = u(0) + ru'(0) + \frac{r^2}{2}u''(0) + ...$, where u' and u'' are, respectively, the first and second derivatives of u. In an axisymmetric flow, the velocity field is conveniently decomposed in azimuthal, radial and axial components u_ϕ, u_r and u_z and the spatial coordinates are r, the distance from the axis of symmetry, ϕ, the angle around the axis of symmetry, and z, the distance along the axis of symmetry. Our basic premises of axisymmetry and incompressibility imply the existence of a Stokes stream function $\psi(r, z)$, where

$$u_r(r, z) = -\frac{1}{r}\frac{\partial}{\partial z}\psi(r, z), \tag{1}$$

$$u_z(r, z) = \frac{1}{r}\frac{\partial}{\partial r}\psi(r, z). \tag{2}$$

The velocity field's regularity at the axis of symmetry means that, as $r \to 0$,

$$u_\phi(r, z) \approx r^m G(z), \tag{3}$$

$$\psi(r, z) \approx r^l F(z), \tag{4}$$

where $G(z)$ and $F(z)$ are regular functions of z for all values of z, m is an integer greater or equal to 1 for regularity and incompressibility, and l is an integer which must be greater or equal to 2 in order for u_z to be regular at the axis.

It turns out that regularity is too strong a requirement for the spiral-helical streamline's D_K to be equal to 1; *we prove that $D_K = 1$ provided that m and l are such that $m > -1$ and $l > 0$ and not necessarily integers.* Hence, there can exist singular flows where $D_K = 1$. But there can be no regular flow where $D_K > 1$. (Note that the existence of a spiral-helical streamline is assumed.)

The boundedness of vorticity at infinity and the regularity of $G(z)$ imply that $G(z)$ is a bounded function of z. By examining the spiral-helical streamline in the (r, z) plane we find that $F(z)$ must be a monotonically increasing function of z. Indeed, as $r \to 0$ *along* the spiral-helical streamline,

$$\frac{dr}{dz} = \frac{u_r}{u_z} \approx -\frac{r}{lF(z)} \frac{dF}{dz}(z) \tag{5}$$

which can be integrated to yield

$$r^l F(z) \approx Const \tag{6}$$

as $r \to 0$ where r and z are streamline coordinates and one is therefore a function of the other (a more detailed discussion is given in Vassilicos & Brasseur 1996). It follows that $F(z) \to \infty$ as $r \to 0$ on the spiral-helical streamline, and since $F(z)$ is regular, $F(z)$ must be a monotonically increasing function of z.

Examining the spiral-helical streamline in the azimuthal plane (r, ϕ), we see that

$$\frac{d\phi}{dr} = \frac{u_\phi}{ru_r} \approx -r^{m-1} \frac{G(z)}{\frac{dF}{dz}(z)} \tag{7}$$

and

$$2\pi = \int_{n2\pi}^{(n+1)2\pi} d\phi \approx -\int_{r_n}^{r_{n+1}} dr \, r^{m-1} \frac{G[z(r)]}{\frac{dF}{dz}[z(r)]} \sim \int_{z_n}^{z_{n+1}} dz \frac{G(z)}{F^\beta(z)}, \tag{8}$$

where r_n and z_n are radial and axial coordinates of the spiral-helical streamline on the nth turn of the spiral, and $\beta = \frac{m+1}{l}$. Since β is strictly positive, $\frac{G(z)}{F^\beta(z)}$ is monotonically decreasing as $z \to \infty$, and therefore $\Delta z_n \equiv z_{n+1} - z_n \to \infty$ as $n \to \infty$.

The distance Δ_n between successive turns after the nth turn of the spiral streamline in 3-D space is given by $\Delta_n^2 = \Delta z_n^2 + \Delta r_n^2$ where $\Delta r_n \equiv r_n - r_{n+1}$. Because $\Delta z_n \to \infty$ as $n \to \infty$, $\Delta_n \to \infty$ as $n \to \infty$ and therefore, *no accumulation of spiral turns exists on the spiral streamline, which implies that $D_K = 1$* (see Vassilicos & Hunt 1991). Indeed the Kolmogorov capacity D_K of streamlines is measured by covering the 3-D space with boxes, and even though the projection of the spiral-helical streamline on the azimuthal

148

plane does accumulate ($\Delta r_n \to 0$ as $n \to \infty$), the spiral-helical streamline in 3-D space does not. It converges towards the central axis of symmetry, but as in does, the distance Δ_n between successive turns increases indefinitely. Hence, there is no accumulation that can be reflected in a non-integral D_K.

We have proved that *in an axisymmetric incompressible flow with bounded vorticity at infinity, if there is no singularity on the axis of symmetry of the flow or if there is a singularity of the type* (3) *and* (4) *where m and/or l are non-integral and* $m > -1$, $l > 0$, *then the Kolmogorov capacity* D_K *of a spiral-helical streamline is equal to* 1. The theorem of self-similar streamlines follows:

In an axisymmetric incompressible flow with bounded vorticity at infinity, when the Kolmogorov capacity D_K *of a spiral-helical streamline is strictly larger than* 1, *there must be a singularity on the axis of symmetry of the flow which is not a singularity of the type* (3) *and* (4) *where m and/or l are non-integral and such that* $m > -1$ *and* $l > 0$.

3. Application to highly resolved DNS of low Reynolds number turbulence

Evidence has been obtained from a series of direct simulations of decaying isotropic turbulence with exceptionally fine small-scale resolution at low Reynolds number ($k_{max}\eta \approx 6$ in 512^3 DNS at $Re_\lambda \approx 21$) demonstrating that a significant number of streamlines around small-scale vortex tubes have a well-defined $D_K > 1$ in a range of length-scales between the Taylor microscale λ and the Kolmogorov length-scale η. In these simulations there is one decade of length-scales between λ and η. Furthermore, evidence based on the spatial correlation of enstrophy with viscous force indicates that the spatial vorticity profile across the vortex tubes is not a well-resolved gaussian even when the resolution of the DNS is half η. The flow is therefore not totally smoothed out by viscosity around scales of order η in some of the vortex tubes. Details can be found in Vassilicos & Brasseur (1996).

This study suggests that in low Reynolds number isotropic turbulence there exist near-singular vortex tubes at the small scales. The study leaves open the question of whether these near-singular vortical events persist in high Reynolds number turbulence.

References

Hunt J.C.R. & Vassilicos J.C. (1991) *Proc. R. Soc. Lond.* A **434**, 183-210.
Vassilicos J.C. & Brasseur J.G. (1996) *Phys. Rev.* E (to appear).
Vassilicos J.C. & Hunt J.C.R. (1991) *Proc. R. Soc. Lond.* A **435**, 505-534.
Villermeaux E., Sixou B. & Gagne Y. (1995) *Phys. Fluids* **7** (8), 2008-2013.

VORTEX QUADRUPOLES AND PROPAGATION OF GRID TURBULENCE

S.I. VOROPAYEV* and H.J.S. FERNANDO**
*Institute of Oceanology, Russian Academy of Sciences,
Moscow, Krasikova st.23, 117851, Russia
**Arizona State University,
Tempe, AZ 85287-562106 U.S.A.

1. Introduction

The purpose of this communication is to present the results of experiments dealing with the propagation of turbulent fronts induced by oscillating grids in homogeneous fluids and to explane the experimental results theoretically. In most of the previous experiments grids made with large square bars were used to study a variety of problems ranging from velocity decay law to mixing across density interfaces (e.g., see Fernando, 1991). Barenblatt (1977) modeled the grid forcing by a source of turbulent kinetic energy distributed homogeneously in the grid's plane and analyzed the propagation of a turbulent front . Voropayev et al. (1980) demonstrated experimentally that this modeling leads to the conclusion that the turbulent kinetic energy flux from the grid, oscillating with constant frequency and amplitude, rapidly decreases with time, which seems unrealistic (also see Barenblatt and Voropayev, 1983; Benilov et al. ,1983). Dickinson and Long (1978) studied the propagation of a turbulent layer by employing a fine grid with small mesh and bar diameter. Considering such a grid, it is possible to simplify the problem and develop an idealized model to describe the grid forcing on the fluid by using some singularities distributed in the grid's plane. Such an attempt was made by Long (1978), who modeled the flow near the grid by a system of point source-sink doublets. An essential shortcoming of this model is the absence of vorticity in the flow, and in a recent paper Long (1995) attempted to modify this model because the source-sink doublets produce no vorticity -- "the *sine qua non* of turbulence".

2. Quadrupolar flow

The main idea of the present study is to consider first in detail the flows induced by a single grid element with the purpose to understand what kind of singularities distributed in the grid's plane may be used to model correctly the oscillating grid forcing on the fluid. The simplest grid element is a small cylinder oscillating in the direction normal to its axis. A cylinder oscillating with a high frequency and a small amplitude in a fluid of kinematic viscosity ν exerts a force on the fluid in alternating, opposite, directions during each half-period. Hence, the action of this oscillating force may be thought of as being equivalent to the action of a line force doublet of intensity Q. Using the analytical results obtained by Stokes in 1850 (Stokes, 1966), Q can be estimated. Based on these ideas, the experimental data on the evolution of vortex quadrupoles induced by an

149

S. Gavrilakis et al. (eds.), Advances in Turbulence VI, 149-152.

oscillating cylinder (Fig.1a) had been interpretted in terms of the Reynolds number of the flow, $Re = Q / 4\pi v^2$, (Voropayev *et al.*,1995).

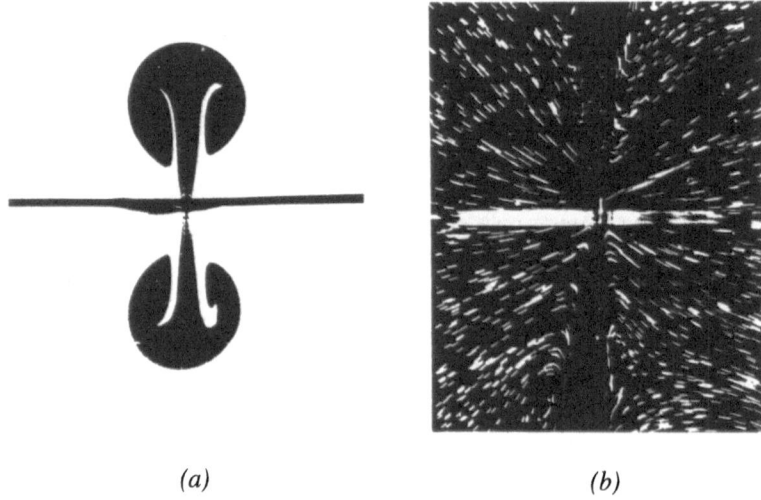

Figure1. Starting (a) and steady (b) quadrupolar flow induced by a vertical cylinder oscillating horizontally in the direction normal to the supporter (horizontal line). Top view.

In the mathematical model, a line force doublet is used as the source of motion. In an initially quiescent unbounded fluid this source produces zero net momentum and generates symmetrical quadrupolar flow similar to that shown in Fig.1a. For starting flow a second-order unsteady solution is obtained in terms of a power series of Re. This solution demonstrates that, as time $t \rightarrow \infty$, the flow becomes steady and radial. To describe this steady asymptote, a particular nonlinear solution is derived. It is shown that the problem permits a similarity solution for all values of Re when a mass sink of prescribed intensity, $q = q(Re)$, is added to the flow. This steady asymptote is reproduced experimentally (Fig.1b), using a vertical porous cylinder that oscillates horizontally in the shallow upper layer of a two-layer fluid and sucks fluid through its porous walls (Voropayev *et al.*, 1996).

3. Grid turbulence

Then the results of experiments on the propagation of turbulent fronts induced by oscillating grids in homogeneous fluids are presented. The symmetrical quadrupolar mode of the flow, induced by the grid elements, was studied in the range $KC = 4 - 45$, $S = 1 - 20$ (KC is the Keulegan-Carpenter number, S is the Stokes number). Two cases are considered: two-dimensional turbulence induced in a thin horizontal layer of fluid by a grid of vertical bars oscillating horizontally (Fig.2a) and three-dimensional turbulence induced in a deep container by a horizontal grid oscillating in a vertical direction (Fig.2b). As a first approximation, the effect of the grid on an initially still fluid can be modeled by a system of line force doublets (with point cross-section) of intensity Q located in the grid's plane with a typical spacing equal to the mesh of the grid. They generate an array of quadrupolar vortices and the interest is to find the dependence of the location of the vorticity front on the external parameters.

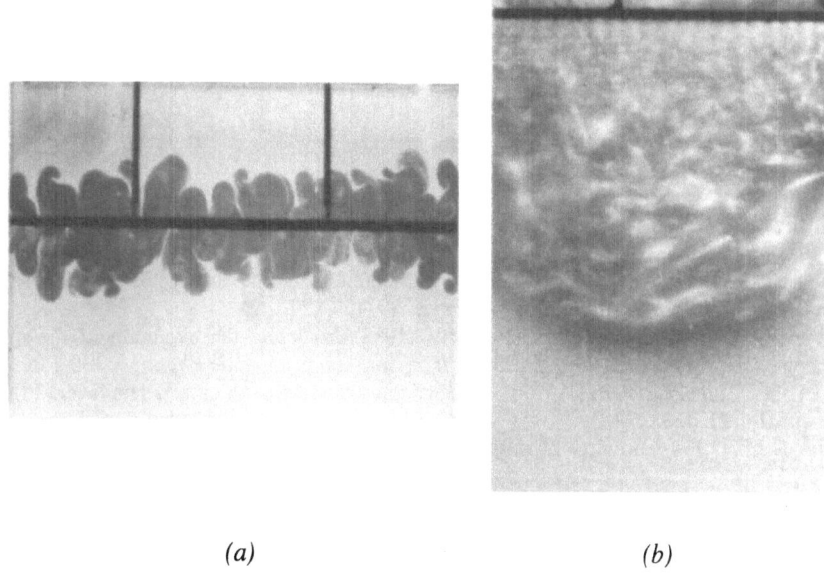

(a)   (b)

Figure 2. Propagation of turbulent front in two-dimensional (a) (top view) and three-dimensional (b) (side view) geometry.

For both cases theoretical models are given (Voropayev *et al.*,1995; Voropayev and Fernando,1996). It is shown that the thickness of the turbulent layer increases with time as $H = \text{const } t^{1/2}$. The results are presented in nondimensional form, in terms of the nondimensional thickness of the turbulent layer and time. Some of the external parameters, such as the diameter of grid rods, the amplitude and frequency of oscillations, do not directly come into play but are included in the nondimensional intensity of the forcing, Re, while the spacing between grid elements is included into the nondimensional time. Theoretical arguments based on the properties of quadrupolar flows are given to explain the observed turbulent-layer growth. A comparison of results with available previous experimental data is also made.

4. Conclusions

Finally, note that we have considered only the simplest symmetrical mode of the flow induced by the grid elements. Preliminary experiments demonstrate that, under certain conditions, more complicated asymmetrical modes can be induced causing the front to become unstable, sometimes sliding down (e.g., regime E of Tatsumo and

152

Bearman, 1990) or rolling up along the walls, forming a large coherent vortex propagating from the grid (e.g., regime F). Considering that at least eight different flow regimes may be realized behind an oscillating cylinder, it is hardly possible to expect that a "universal" law for the propagation of grid turbulence may exist i a broad range of external governing parameters -- and this may explain why strong unwarranted secondary circulations appear in some oscillating grid experiments.

The authors are indebted to Prof. R.R. Long who stimulated this study and made many useful comments on this work.

References

Barenblatt, G.I. (1977) Strong interaction of gravity waves and t urbulence, *Izvestiya, Atmos. Oceanic Phys.,* **8**, 581-583.

Barenblatt, G.I. and Voropayev, S.I. (1983) A contribution to the theory of a steady-state turbulent layer, *Izvestiya, Atmos. Oceanic Phys.,* **19**, 126-129.

Benilov, A.Yu., Voropayev, S.I. and Zhmur, V.V. (1983) Modeling the evolution of the upper turbulent layer of the ocean during heating, *Izvestiya, Atmos. Oceanic Phys.,* **19**, 130-136.

Dickinson, S.C. and Long, R.R. (1978) Laboratory study of the growth of turbulent layer of fluid, *Phys. Fluids*, **21**, 1698-1701.

Fernando, H.J.S. (1991) Turbulent mixing in a stratified fluid, *Ann. Rev. Fluid Mech.,* **23**, 455-493.

Long, R.R. (1978) Theory of turbulence in a homogeneous fluid induced by an oscillating grid, *Phys. Fluids*, **21**, 1887-1888.

Long, R.R. (1995) A theory of grid turbulence in a homogeneous fluid, to be submitted.

Stokes, G.G. (1966) On the effect of the internal friction of fluids on the motion of pendulums, *Mathematical and Physical Papers*, **3(33)**, 1-141.

Tatsumo, M. and Bearman, P.W. (1990) A visual study of the flow around an oscillating circular cylinder at low Keulegan-Carpenter numbers and low Stokes numbers, *J. Fluid Mech.*, **211**, 157-173.

Voropayev, S.I., Afanasyev, Ya. D. and van Heijst, G.J.F. (1995) Two-dimensional flows with zero net momentum: evolution of vortex quadrupoles and oscillating grid turbulence, *J. Fluid. Mech.,* **282**, 21-44.

Voropayev, S.I., Gavrilin, B.L., Zatsepin, A.G. and Fedorov, K.N. (1980) A laboratory study of the deepening of a mixed layer in a homogeneous liquid, *Izvestiya, Atmos. Oceanic Phys.,* **16** 126-128.

Voropayev, S.I. and Fernando, H.J.S. (1996) Propagation of grid turbulence in homogeneous fluids, *Phys. Fluids,* in press.

Voropayev, S.I., Fernando H.J.S. and Wu, P.C. (1996) Starting and steady quadrupolar flow, *Phys. Fluids*, **8**, 384-396.

MOMENTUM AND HEAT TRANSPORT IN A QUASI 2-D FLOW.

Turbulent Wake of a Cylinder Interacting with a Turbulent Boundary Layer.

S.G. NYCHAS, G.A. SIDERIDIS and E.G. KASTRINAKIS

Department of Chemical Engineering, Aristotle University of Thessaloniki
Univ. Box 453, 54006 Thessaloniki, Greece.

Summary

An experimental investigation of the transport characteristics of momentum and heat in a turbulent wake of a cylinder interacting with a turbulent boundary layer has been carried out (Figure 1). The cylinder was placed just outside the plate's boundary layer. The von Karman street of vortices produced behind the cylinder interacted with the boundary layer downstream of x/D=10. Heat was supplied to the boundary layer flow by means of the line heat source shown. Simultaneous recording of the instantaneous values of temperature and two velocity components were performed using hot wire anemometry employing a triple-wire probe. Time variations of one Reynolds stress component and two turbulent scalar flux components were obtained directly from the experimental data, by digital post-processing.

The flow within the intermediate wake (15<x/D<35), exhibited a distinct quasi two-dimensional character at the Reynolds number of 1800 selected [1]. This was demonstrated by: (i) suppression of one fluctuating velocity component and (ii) deviation of the velocity spectrum scaling region from the -5/3 power law. These features were enhanced using a cylinder with rough surface and a spectral slope of -7/3 was achieved. This type of flow can simulate the flow encountered at high atmospheric altitudes, where coherent eddy structures are also present, as in the laboratory flow.

In order to obtain a physical insight into the mechanisms of transport processes, the present data were analyzed by the quadrant splitting technique. The momentum and heat flux signals were conditionally averaged with respect to the four elementary fluid motions assumed by that method. It was found that hot masses of fluid move in a direction away from the plate with lower streamwise velocity than the local average. Cold fluid moves towards the plate, primarily with higher streamwise velocity than the local average. Further investigation of the transport processes followed, with a phase-averaging method similar to that used by Matsumura and Antonia [2]. It involved triple-decomposition of the recorded signals resulting in a time-mean value, a coherent part and a random part. The coherent and random parts were then phase-averaged relative to the vortex shedding period. Typical results from the present data are given in Figure 2.

S. Gavrilakis et al. (eds.), Advances in Turbulence VI, 153-154.

154

As shown, coherent heat flux is present only at the up-going alleyway between adjacent vortices and it is approximately aligned with the coherent motion. Random heat flux has constant orientation relative to the coherent motion.

This work was supported by the European Communities Commission, under contract AVI-CT92-0017.

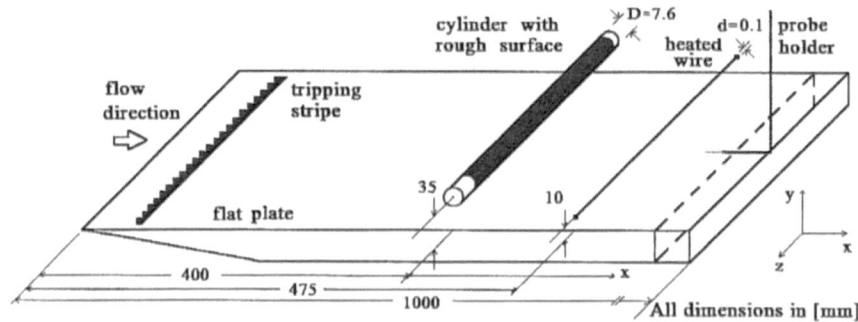

Figure 1: The experimental set-up for the present work.

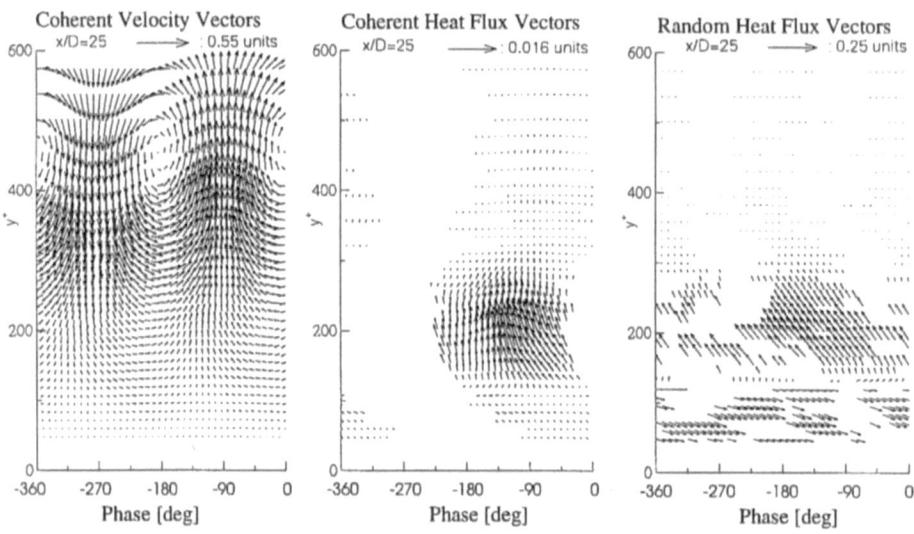

Figure 2: Vector plots of phase-averaged coherent velocity, coherent heat flux and random heat flux (reference frame is moving with the vortex street; flow direction is left to right).

References

1. Sideridis, G. A., Kastrinakis, E. G., and Nychas, S. G.: Turbulent heat flux measurements in a quasi two-dimensional flow and in presence of large-scale structures. Paper presented at the *8th Beer-Sheva International Seminar on MHD-flows and turbulence*, Jerusalem, Israel, 1996.
2. Matsumura, M., and Antonia, R. A.: Momentum and heat transport in the turbulent intermediate wake of a cylinder, *J. Fluid Mech.*, 250 (1993), 651-668.

III

Industrial and Environmental Applications

SIMULATION OF TURBULENT FLOWS FOR INDUSTRIAL APPLICATIONS: A FLUID ENGINEER'S VIEW

M.V. CASEY

Fluid Dynamics Laboratory, Sulzer Innotec Ltd
CH-8401 Winterthur, Switzerland

1. Turbulence models in industrial applications

Industrial fluid dynamics has undergone a major revolution in recent years. Ten years ago most optimisation of fluid dynamic components in industry required expensive and time-consuming experiments in models, prototypes or test rigs. Today the design of many varied fluid engineering components is undertaken with the help of CFD codes solving the 3D Reynolds-averaged Navier-Stokes equations, with substantially less need for prototype testing.

The progress in industrial flow simulations has not been brought about by any massive improvement in our ability to model turbulent flows. Improvements in turbulence modelling over the last twenty years have been pretty small compared to the growth of computer speed and memory capacity and the increase in complexity attainable by industrial CFD simulations.

Industrial CFD simulations rely on relatively primitive turbulence models, either one-equation models or two-equation models. These were developed for simple low-speed shear flows and their limitations for complex flows is clear. For example, the standard k-ε model (Launder *et al.*, 1974) is known to exhibit the following weaknesses:

- overprediction of turbulent kinetic energy in impinging regions
- underestimation of any regions of recirculation in a swirling flow
- underestimation of turbulent mixing in buoyant flows
- poor estimation of separation in adverse pressure gradients (curvature)
- poor predictions of flows in small gaps (non-isotropic turbulence)

Despite these inadequacies, the k-ε model has become the accepted standard for industrial CFD simulations. This lecture examines the reasons for this by considering some recent published applications of CFD on fluid engineering products within the Sulzer technology corporation.

S. Gavrilakis et al. (eds.), Advances in Turbulence VI, 157-162.
© *1996 Kluwer Academic Publishers.*

2. Calibration and Validation

Because of the weaknesses of CFD codes, no self-respecting fluids engineer normally applies CFD to solve a problem without first validating the method (Casey, Borth *et al.*, 1995). When accurate answers are needed the code must first be assessed on a suitable test case involving similar flow structures to those of the component being designed. Test cases with extensive measurement data giving details of flow velocities and pressure fields need to be compared with predictions of the code, see, for example, the work of ERCOFTAC (Bonnin *et al.*, 1996).

In an industrial environment, it is often necessary to design new experiments on purpose built models using non-intrusive laser flow anemometry to provide accurate test data for the validation of codes. The engineers involved in the validation process benefit greatly from the combination of CFD modelling and experiments, as the CFD often sheds light onto flow features that the limited test instrumentation cannot possibly reveal.

3. Some recent examples of CFD applications within Sulzer

3.1. DIFFUSER FLOW IN CENTRIFUGAL PUMPS

Sulzer has recently carried out a substantial CFD validation program to examine the flow in impellers and vaned diffusers of radial pumps (Casey, Eisele *et al.*, 1995). As part of this project the flow in one of the simplest possible fluid dynamic devices was examined: the two dimensional flow in a planar diffuser. The objective was to examine how well the standard k-ε model was able to predict the classical diffuser design charts (Kline, 1967).

The static pressure rise in the diffusers was well predicted at low diffuser loadings (small opening angles). At high loading levels typically found in radial pump diffusers, the pressure rise was too optimistic; as the k-ε model tended not to allow the flow to separate early enough. Given that highly efficient pumps have been designed for over a century on the basis of relatively inaccurate empirical models then any inaccuracy of the turbulence model poses no new problem for the practical pump designer, he simply adjusts the prediction on the basis of his experience.

3.2. TRANSONIC FLOW IN AN AXIAL COMPRESSOR ROTOR

Transonic axial compressors achieve their pressure rise by flow deceleration in a rotating curved blade row and across shocks. To test standard methods on this problem Sulzer took part in a recent ASME code validation exercise (Dalbert and Wiss, 1995). In this blind test case, the participants calculated the flow and performance of a transonic compressor rotor (NASA rotor 37) at various operating points. The measurement data was not made available until after the calculations were submitted. The contributions submitted by Sulzer used the commercial software packages BTOB3D with the Baldwin-Lomax turbulence model (Dawes, 1988) and TASCflow with the standard k-ε turbulence model (ASC, 1995).

The good agreement between the test data and the Sulzer simulations is somewhat surprising given the simplicity of the turbulence models (no account of rotation, curvature and periodic unsteadiness from upstream blade wakes). It would appear that the turbulence model is not the most critical feature of turbocompressor flow calculations. A good engineering prediction of the performance of a compressor requires some fundamental engineering features to be well predicted, for example:

- the turning of the flow by the blade rows (essentially inviscid)
- the blockage of the boundary layers (displacement thickness)
- the formation of tip clearance vortices (essentially inviscid)

and for this a relatively crude turbulence model suffices. Improvements in the turbulence model would help, but they are probably less important than the attention given to other effects such as:

- the careful design of blading to avoid flow separation (Casey, 1994)
- the laminar sublayer, where most losses are generated (Dawes, 1990)
- the role of transition on performance (Denton, 1993)
- the unsteady nature of the flow due to the passage of wakes from upstream blade rows, which gives rise to additional stress terms in the mean flow, very like Reynolds stresses (Adamczyk, 1985)
- the steady-state interaction of adjacent blade rows (Sick, 1996)

3.3. HYDROABRASIVE EROSION

Another exciting recent application of CFD within Sulzer is to predict the hydroabrasive sand erosion due to the passage of sand particles through water turbines. This is particularly relevant to hydroelectric power generation in Switzerland, as the streams in the Alps often carry a high load of highly abrasive sand particles.

Three separate empirical models are used to predict the rate of material erosion due to sand in the water flowing through hydraulic machines. Firstly the mean velocity flowfield in the component is calculated using the k-ε turbulence model. Secondly, the particle paths of sand through the component are calculated by Langrangian tracking models under the assumption that the sand particles are sufficiently sparsly distributed that they do not interact with each other and they have no effect on the flowfield. In order for this calculation to be statistically significant, the calculation is repeated for a wide range of positions and sizes of sand particles in the inlet flow, typically 200,000 particles. Each separate calculation of the path of a sand particle takes into account in a statistical way the small-scale random nature of the turbulence, such that similar particles starting from the same point in the inlet do not necessarily follow the same path through the machine. Finally, an empirical model is used to relate the material erosion rate due to collisions between the sand particles and the turbine components from the kinetic energy and incidence angle of the impinging sand particles.

Good quantitative agrement between the predictions and measurements of erosion can be obtained by this method (Drtina *et al.*, 1995). It would appear that the most important flow processes that determine the erosion (i.e. the velocity levels, particle paths and their statistical variations) are all predicted well enough and are not currently limited by the turbulence models. Accuracy can be improved by better particle tracking and calibrated empirical erosion models. This example is typical of engineering applications of CFD in two-phase flows, where it is often our weak modelling capability for the two-phase nature of the flow that limits the ultimate accuracy of the calculation.

3.4. MIXING IN INDUSTRIAL STATIC MIXERS

Static mixers are used in various processes which include mixing, dispersion, polymerisation and chemical reaction. They mix the flow by guiding part of the flow across the flow channel to stretch and fold the interface between two components (macromixing). This is followed by micromixing, involving mixing from small-scale turbulent eddies down to mixing through molecular diffusion. The energy for mixing is obtained from the pressure drop over the static mixer element. The flow and geometry in static mixers is so complex that until recently the development of such devices relied heavily on experimental testing, but it is now quicker to design them with the help of CFD, see (Lang *et al.*, 1995).

CFD works remarkably well at predicting the mixing in these devices, even with the k-ε turbulence model, because a good static mixer relies on macromixing by means of large-scale flow stuctures with low pressure loss, such as longitudinal vortices. A simple mixer that relied purely on small-scale turbulent mixing, such as a pipe, would be roughly one hundred times less effective, so the engineering design of the static mixer automatically makes it somewhat insensitive to the detail of the turbulent mixing. Improvements to the turbulence models would only be of additional use in CFD calculations of mixers if they were able to analyse some of the micromixing features related to details of the flow in the turbulent eddies, (Wehrli *et al.*, 1996).

3.5. FLOWS IN LARGE ENCLOSURES

CFD calculations are used to examine the ventilation flows in rooms and flows in large enclosures. The velocity levels and air movements in buildings can be used to estimate comfort levels and the age of air.

Predicting the fine detail of the turbulence in rooms is difficult. First, many of the flow fluctuations within a room are large scale low frequency eddies and should be modelled by an unsteady flow calculation rather than a turbulence model. Second, the advection air flows in the boundary layers on the walls cannot be modelled with the logarithmic law of the wall and no other general law is available. The use of more sophisticated turbulence models (low Re k-ε model or two layer model) is limited by the grid needed to capture the very thin boundary layer. The heat exchange to the wall is determined by the wall boundary layers and is crucial to a correct prediction of the room temperature distribution so that empirical relationships for the heat transfer are used (Borth *et al.*, 1994). Flows caused by multiple air inlets to the room can also be a problem and may need special modelling.

In general the level of accuracy required from such calculations allow the k-ε model to be used with some reliance. The air-conditioning engineer is generally not interested in the detail of the turbulent eddy motion within the room, but needs an engineering estimate of the peak summer temperature or maximum air velocity.

4. Conclusions

CFD with the standard k-ε turbulence models is now widely used in industry to examine flow phenomena in engineering flows. Despite the limitations of the k-ε model both qualitative and quantitative details of complex flows can be predicted. The inaccuracy of the turbulence model is not always a serious problem because:

- In fluid engineering design, detailed knowledge of turbulence is less important than prediction of global performance parameters.
- The inadequacies of the turbulence models in predicting global flow structures are compensated by calibration of the CFD codes against suitable test data in advance of their use for engineering design.
- Other aspects of the physical modelling often play a more significant role than the turbulence model, especially in turbomachinery and two-phase flows.

Improved turbulence models are needed, even if only to reduce the large industrial effort required for validation of the codes on new applications. Progress on validation is slow, so that LES may reach industrial maturity before the next generation of eddy-viscosity models is fully validated.

5. Acknowledgements

The author would like to thank the many colleagues referred to below for their discussions and stimulation during the preparation of this talk.

References

Adamczyk, J.J. (1985) "Model equation for simulating flows in multistage turbomachinery", ASME paper 85-GT-226.

Bonnin, J.Ch., Buchal, T. and Rodi, W. (1996) "Data bases and testing of calculation methods for turbulent flows", ERCOFTAC Bulletin, Vol. 28, March, pages 48-54.

Borth, J. and Suter, P. (1994) "Influence of Mesh Refinement on the Numerical Prediction of Turbulent Air Flow in Rooms", Roomvent '94, Fourth International Conference on Air Distribution in Rooms, Krakow, Poland June 15-17

Casey, M. V. (1994) " The industrial use of CFD in the design of turbomachinery", in AGARD lecture series "Turbomachinery Design using CFD", AGARD LS-195

Casey, M.V., Eisele, K., Muggli, F.A., Guelich, J. and Schachenmann, A. (1995) Flow Analysis in a Pump Diffuser, Part 2: Validation of a CFD Code for Steady Flow, ASME FED-vol. 227, Numerical Simulations in Turbomachinery, ASME

Casey, M.V., Borth, J., Drtina, P., Hirt, F., Lang, E., Metzen, G. and Wiss, D. (1995) "The application of computational modelling to the simulation of environmental, Medical and Engineering Flows", SPEEDUP Journal, Volume 9 number 2, pages 62-69

Dalbert, P. and Wiss, D. (1995) Numerical transonic flowfield predictions for NASA compressor rotor 37, ASME GT Congress, Houston, ASME paper 95-GT-326

Dawes, W.N. (1988) "Development of a 3D Navier-Stokes solver for application to all types of turbomachinery", ASME GT Congress, Amsterdam, ASME Paper 88-GT-70

Dawes, W.N. (1990) "A comparison of zero and one equation turbulence modelling for turbomachinery calculations", ASME GT Congress, Brussels, ASME Paper 90-GT-303

Denton, J.D. (1993) "Loss Mechanisms in Turbomachines", ASME GT Congress, Ohio, ASME Paper 93-GT-435

Drtina, P. and Krause, M. (1995) "Numerical Prediction of Abrasion for Hydraulic Turbine Guide Vanes", IMACS-COST Conference on CFD, Lausanne 13.-15.Sept.

Lang, E., Drtina, P., Streiff, A. and Fleischli, M. (1995) "Numerical simulation of the fluid flow and the mixing process in static mixers", Int.J.Heat Mass Transfer, Vol.38, No.12, pp.2239-2250.

Reneau, L.R., Johnston, J.P., and Kline, S.J. (1967) "Performance and design of straight two-dimensional diffusers", ASME Jnl. of Basic Enginnering, pp141-150

Launder, B.E. and Spalding, D.B.(1974) "The numerical computation of turbulent flows", Comput. Methods Applied Mech. and Engrg., vol 3, pages 269-289

Sick, M., Casey, M. and Galpin, P. (1996) "The validation of a stage calculation in a Francis Turbine", IAHR Symposium, Valencia, September

TASCflow Theory Documentation, Version 2.4. (1995) Advanced Scientific Computing Ltd., Waterloo, Ontario, Canada

Wehrli, M., Borth,J., Drtina, P., Lang, E. and Mack, R. (1996) "Industrial Application of CFD for Mass Transfer Processes, SPEEDUP Journal, Vol. 10, No. 1

AIRFOIL STALL PREDICTION USING A TWO-EQUATION AND AN EXPLICIT ALGEBRAIC STRESS MODEL

THOMAS B. GATSKI
NASA Langley Research Center
Hampton, VA 23681-0001 USA

I INTRODUCTION

The purpose of this study is to assess the predictive capabilities of two types of turbulence models on flow over an airfoil near stall, and at different Reynolds numbers. An isotropic, eddy-viscosity $K - \varepsilon$ model and an explicit algebraic stress model (EASM) are used in the analysis.

Other studies have examined the performance of $K - \varepsilon$ and algebraic stress models in predicting the aerodynamic characteristics of airfoil flows. Although the exact form of the two-equation models and ASM's differs between these studies (as well as this one), the intent is to examine whether the increased physics introduced into the ASM have a significant impact on the results. Davidson and Rizzi[1] applied a two-equation model and an ASM to the ONERA-A airfoil flowfield at a single chord Reynolds number Re of 2.1×10^6. Their results showed that the ASM predicted the variation of C_l and the stall location better than the $K - \varepsilon$ model (and the Baldwin-Lomax model, which was also tested). Recently, Davidson[2] performed a follow-on study with a $K - \varepsilon$ and a Reynolds stress model at the same conditions as the Davidson and Rizzi work. Once again, the higher order model outperformed the isotropic eddy-viscosity $K - \varepsilon$ model. Stall location was accurately predicted, as were the velocity profiles on the airfoil and in the wake.

Lien and Leschziner[3] compared the performance of a $K - \varepsilon$, a nonlinear $K - \varepsilon$ (ASM), and a Reynolds stress model on the ONERA-A airfoil as well, at the same chord Reynolds number used in the Davidson studies. Their study included the development of an improved ASM and Reynolds stress model to better predict the flow field. Their results suggested that the Reynolds stress model outperforms both the $K - \varepsilon$ and nonlinear $K - \varepsilon$ models in predicting the aerodynamic characteristics of the airfoil.

S. Gavrilakis et al. (eds.), Advances in Turbulence VI, 163-166.
© *1996 Kluwer Academic Publishers.*

II RESULTS AND DISCUSSION

An isotropic, eddy-viscosity $K - \varepsilon$ model and a nonlinear $K - \varepsilon$ or algebraic stress model are evaluated based on their performance in predicting the turbulent flow over a 13 percent-thick general-aviation airfoil (GA(W)-2). The $K - \varepsilon$ model of Speziale, Abid and Anderson[4](SAA) is one of the models; the other is the explicit algebraic stress model (EASM) developed by Gatski and Speziale[5, 6]. The incompressible ($M_\infty = 0.15$) experimental data of McGhee et al.[7] is used to assess the models in predicting the aerodynamic characteristics of the airfoil over a range of chord Reynolds numbers.

Figure 1(a) shows the computed lift coefficient for various angles of attack ($\alpha = 0°$, $8°$, $14°$, $16°$, and $18°$) at $Re = 2.1 \times 10^6$. At this chord Reynolds number, both models correctly predict the stall location; however, the EASM performs better than the $K - \varepsilon$ model in predicting the magnitude of C_l; the largest departure from the experimental values occurs at the higher angles of attack.

Figure 1(b) shows the C_l for the $Re = 4.3 \times 10^6$ case. At the lower angles of attack ($\alpha = 0°$ and $8°$), the EASM and to a slightly lesser extent the $K - \varepsilon$ model do a good job of predicting the C_l. At higher angles of attack ($\alpha = 16°$, $18°$, $20°$, and $22°$), both models do a poor job of predicting C_l. At $\alpha = 16°$ and $18°$, both models yield the same C_l, but at $\alpha = 20°$ and $22°$, the EASM prediction for C_l is less than that for the $K - \varepsilon$ model and is closer to experimental results.

At $Re = 5.3 \times 10^6$, Fig. 1(c) shows that at $\alpha = 0°$ and $8°$, both models correctly predict the C_l; at the larger angles of attack ($\alpha = 15°$, $17°$, $19°$, and $21°$), neither model accurately predicts the C_l levels. The EASM yields results that are closer to the experimental levels than those of the $K - \varepsilon$ model; however, these predictions are still not acceptable. In addition, the stall angle predicted with the EASM is shifted slightly from both the $K - \varepsilon$ model predictions and the experimental value.

A clearer picture of this Reynolds-number scaling effect is shown in Fig. 1(d) in which the observed experimental behavior can be seen to be linear over the range of Reynolds-numbers studied. The $K - \varepsilon$ model displays neither the correct slope nor linear behavior in this region, which suggests that a key physical element is not being correctly incorporated into either the calculation or the turbulence model. The EASM also predicts the incorrect slope; however, the variation across the Reynolds-number range is much closer to linear than for the $K - \varepsilon$ model. Nevertheless, in both computations the qualitative features are predicted inaccurately.

Figure 2 shows the separation zone predicted by each model along the trailing edge of the upper surface near the stall angle of the EASM. At $Re = 2.1 \times 10^6$, the isotropic model does not predict a separation zone along the surface which is consistent with the results of Lien and Leschziner[3]; however, the EASM is found to predict separation along the airfoil. At the intermediate Reynolds number $Re = 4.3 \times 10^6$, both models predict separation zones along the surface. Fig. 1(b) shows in this case that the $K - \varepsilon$ model predicts the stall angle more accurately; the EASM overpredicts the separation zone at this Reynolds number. This correlation of separation-zone size and lift coefficient suggests that at higher angles of attack both models grossly overpredict the size of the separation zone at the trailing edge. The results at $Re = 6.3 \times 10^6$ show that the EASM model more closely approximates the experimental C_l distribution near stall [Fig. 1(c)], although the stall angle is underpredicted by both models. Because both models overpredict the C_l distribution at angles of attack greater than the stall angle, it can be assumed that at these higher angles of attack the separation-zone size would be underpredicted by both models. The reduction in separation-zone size near the trailing edge, between the $Re = 4.3 \times 10^6$ and 6.3×10^6 cases, correlates with the decrease in $C_{l\max}$.

These results have shown that neither model can accurately predict the lift coefficient near stall over a range of Reynolds numbers, however, areas where the models do perform well relative to one another and in comparison with experiment have been identified.

REFERENCES

[1]Davidson, L. and Rizzi, A. (1992) Navier-Stokes Computation of Airfoil in Stall Using Algebraic Reynolds-Stress Model, *AIAA 30th Aerospace Sciences Meeting* Paper No. 92-0195, January 6–9, Reno, Nevada.

[2]Davidson, L. (1995) Prediction of the Flow Around an Airfoil Using a Reynolds Stress Transport Model, *Transactions of the ASME – Journal of Fluids Engineering* **117**, 50–57.

[3]Lien, F. S. and Leschziner, M. A. (1995) Modelling 2D Separation from a High Lift Aerofoil with a Non-Linear Eddy-Viscosity Model and Second-Moment Closure, *Aeronautical Journal* **99**(984), 125–144.

[4]Speziale, C. G., Abid, R., and Anderson, E. C. (1992) Critical Evaluation of Two-Equation Models for Near-Wall Turbulence, *AIAA J.* **23**(9), 1308–1319.

[5]Gatski, T. B. and Speziale, C. G. (1993) On Explicit Algebraic Stress Models for Complex Turbulent Flows, *J. Fluid Mech.* **254**, 59–78.

[6]Gatski, T. B. (1996) Prediction of Airfoil Characteristics Using Higher-Order Turbulent Models, *NASA Technical Memorandum 110246*.

[7]McGhee, R. J., Beasley, W. D., and Somers, D. M. (1977) Low-Speed Aerodynamic Characteristics of 13-Percent-Thick Airfoil Section Designed for General Aviation Aviation Applications, *NASA Technical Memorandum X-72697*.

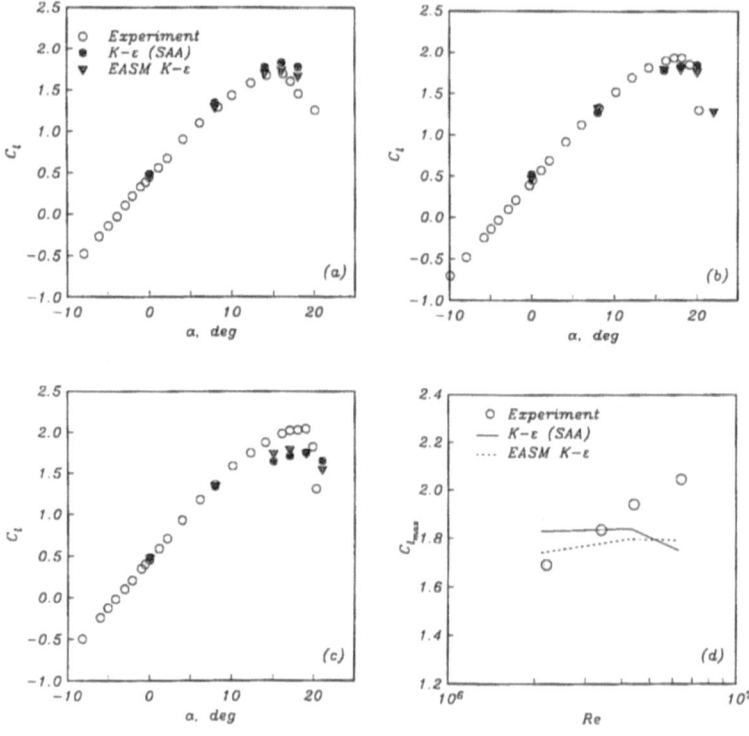

Figure 1. Lift coefficient comparisons: a) $Re = 2.1 \times 10^6$; b) $Re = 4.3 \times 10^6$; c) $Re = 6.3 \times 10^6$; d) Maximum lift coefficients

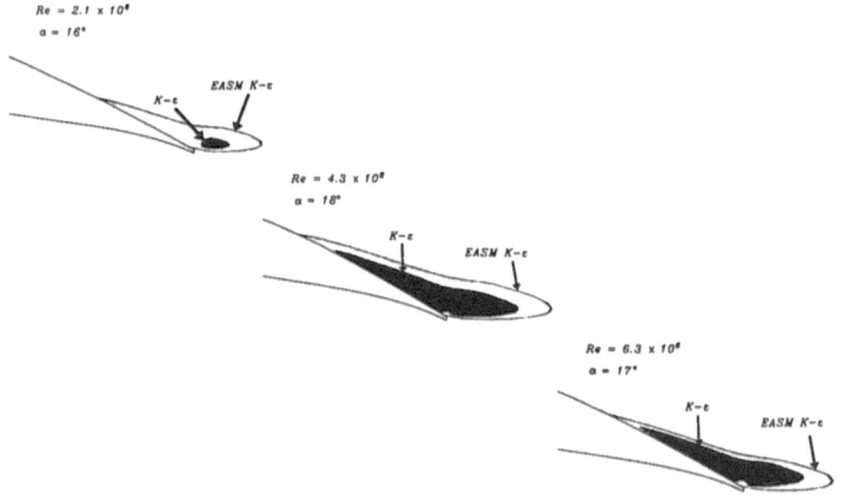

Figure 2. Comparison of stall separation zone size at airfoil trailing edge

TURBULENCE AND FINANCIAL MARKETS

A Transaction Cascade in Foreign Exchange Markets

S. GHASHGHAIE
Fürstensteinerstrasse 4, 4053 Basel, Switzerland

W. BREYMANN
Institut für Physik der Universität Basel,
4056 Basel, Switzerland

J. PEINKE
Experimentalphysik II, Universität Bayreuth,
95440 Bayreuth, Germany

AND

P. TALKNER
Paul Scherrer Institut, 5232 Villigen, Switzerland

Price dynamics of speculative markets is one of the most complex phenomena in economics. Already the statistical description turns out to be difficult. The most prominent characteristic of the distribution of logarithmic price differences (returns) Δy for a given time delay Δt is its leptokurtosis, i.e., the pronounced frequencies with which both small and large returns occur. Proper modelling of this effect is of practical relevance for risk management. The kurtosis of the return distribution is largest for Δt of the order of minutes and decreases monotonically with increasing Δt, accompanied by an according change in the form of the distribution [1, 2]. Simultaneously, the variance of the distribution increases: it depends on the time delay according to a power law $\langle (\Delta y)^2 \rangle \sim \Delta t^{\xi_2}$.

An adequate model of the return distribution must reproduce both the scaling behavior and the change in form of the distribution in the relevant range of time delay. In this paper we show that these features can be accounted for by a mixture of distributions. Our main idea is that the similarity between the return distribution and the distribution of velocity differences in hydrodynamic turbulence gives a hint to the existence of a hierarchical structure in foreign exchange (FX) markets.

The main characteristic of fully-developed hydrodynamic turbulence is the energy cascade, which manifests itself in a scaling of the moments of

S. Gavrilakis et al. (eds.), Advances in Turbulence VI, 167-170.

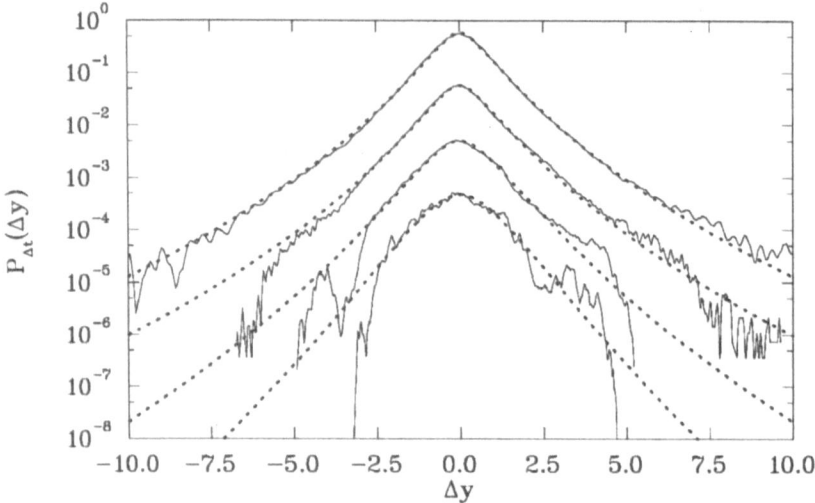

Figure 1. Data points: Standardized probability density $P_{\Delta t}(\Delta y)$ of logarithmic price changes $\Delta y \equiv \log x(t) - \log x(t + \Delta t)$ for time delays $\Delta t = 640s, 5120s, 40960s, 172800s$ (from top to bottom). Full lines: observed data. Dotted lines: results of (least squares) fits carried out according to [6]; $\lambda^2 = 0.27, 0.25, 0.160, 0.10$ (from top to bottom). The data are provided by *Olsen & Associates*, Zürich.

differences Δv of velocities at points separated by a distance Δr in the flow as $\langle (\Delta v)^n \rangle \sim (\Delta r)^{\zeta_n}$ [3]. If the energy-dissipation rate ϵ resulting from this downward energy flow was homogeneously distributed, $\zeta_n = n/3$ [4], and the probability densities $P_{\Delta r}(\Delta v)$ would be scale invariant (i.e., the *standardized* probability densities would not depend on Δr). However, there is experimental evidence for intermittency, leading to values of ζ_n which follow a concave curve definitely below the straight line $\zeta_n = n/3$ [5]. For different experimental situations the change in shape of $P_{\Delta r}(\Delta v)$ has been successfully parameterized using a superposition of Gaussian densities with log-normally distributed variances [6, 7, 8]. The variance λ^2 of this log-normal distribution is a measurable form parameter and in turbulence corresponds to $\langle (\log \epsilon)^2 \rangle$ [5]. Furthermore, λ^2 contains information on the depth of the energy cascade [9].

The probability density function $P_{\Delta t}(y)$ of the FX returns (Fig. 1) exhibits the same qualitative characteristics as the turbulent analogue $P_r(\Delta v)$ of velocity differences. The higher order moments of the DEM–USD returns scale as $\langle (\Delta y)^n \rangle \propto (\Delta t)^{\xi_n}$, for Δt varying from about 5 minutes up to several hours. As in turbulence, the scaling exponents ξ_n of the nth order moments $\langle (\Delta y)^n \rangle$ depend on n in a nonlinear way. Furthermore the scaling exponents ξ_n of the FX data are close to the scaling exponents ζ_n of the turbulent data [10]. This is also in agreement with the observation that the shapes of $P_{\Delta r}(\Delta v)$ and $P_{\Delta t}(\Delta y)$ depend on their respective scale parame-

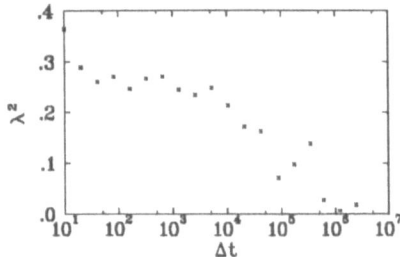

Figure 2. Dependence of the form parameter λ^2 on Δt (in seconds).

ters in a similar way: both show a decrease of kurtosis with increasing scale parameter [11, 10]. However, note that there is no reason for the 4/5 law of Kolmogorov predicting the linear dependence of the third moment to hold for the FX data.

For time delays Δt ranging from about 2 minutes up to more than 1 day the distribution of returns for DEM–USD FX rates can well be described by the same superposition of Gaussian densities as the distribution of velocity differences in turbulence (Fig. 1). Although the data in the center dominate the fit, the agreement is reasonable also as regards the tails of the distributions. This is in contrast to fits with Levy distributions which significantly deviate in the tails [12]. It follows in particular that the distribution of FX returns cannot be described by stable laws. The parameter λ^2 which measures the spread of the variances of the superimposed Gaussians, decreases linearly with $\log \Delta t$, see Fig. 2. This corresponds to intermittency in turbulence and confirms the similarity of the statistical behavior of FX markets with the classical picture of turbulence as given by Kolmogorov [5, 6]. In FX markets, intermittency corresponds to clusters of high and of low volatilities.

The similarity in the statistical behavior of FX markets and hydrodynamic turbulence could be caused by a cascade in FX markets generating hierarchical features analogously to the energy cascade in turbulence. A corresponding hint in the description of intra-day dynamics of FX markets is provided by an asymmetry in information flow between market components with different time horizons: the statistical volatility defined on a coarse time grid significantly predicts the volatility defined on a fine time grid [13]. We suppose that a cascade in the FX markets similar to the energy cascade in turbulence is generated by the inter-dealer transactions, which make up about 90% of the whole trading volume. A dealer having received an order of large size will devide it into orders of smaller sizes and sell most of them to other dealers so as to provide cover against market risks. In most cases the subsequent dealers may do the same, thus dividing an initially large order into still smaller ones. Driven by the risk tolerance of dealers, this process of dividing will typically be repeated several times, thus generating a hierarchically structured equilibrating process to a new fundamental price.

For turbulence, the picture of the energy cascade as a relevant driving mechanism is widely accepted. It therefore seems natural to search for a similar mechanism which produces the apparently related distributions of returns with the same scaling behavior as in turbulence. However it is unlikely that there is a set of a few partial differential equations like the Navier Stokes equations in hydrodynamics which might serve as a model of FX markets dynamics. In this paper we suggest a "transaction cascade" generated by inter-dealer trading as a possible mechanism, which plays the same role in FX markets as does the energy cascade in turbulence.

References

1. Müller, U. A., Dacorogna, M. M., Olsen, R. B., Pictet, O. V., Schwarz, M., and Morgenegg, C.: Statistical study of foreign exchange rates, empirical evidence of a price change scaling law, and intraday analysis, *Journal of Banking and Finance* 14 (1990), 1189–1208.
2. Baillie, R. T. and Bollerslev, T.: The message in daily exchange rates: a conditional variance tale, *Journal of Business and Economic Statistics* 7 (1989) 297–305; Intra-day and inter-market volatility in foreign exchange rates, *Review of Economic Studies* 58 (1990), 565–585.
3. Monin, A.S. and Yaglom, A.M.: *Statistical Fluid Mechanics Vol. 1 & 2*, MIT Press, Cambridge (MA), 1971 & 1975.
4. Kolmogorov, A. N.: The local structure of turbulence in a viscous incompressible fluid for very large Reynolds numbers, *Dokl. Akad. Nauk. SSSR* 30 (1941), 301–305.
5. Kolmogorov, A. N.: Refinement of previous hypotheses concerning the local structure of turbulence in a viscous incompressible fluid at high Reynolds number, *J. Fluid Mech.* 13 (1962) 82–85. ·Obukhov, A. M.: Some specific features of atmospheric turbulence, *J. Fluid Mech.* 13 (1962), 77–81.
6. Castaing, B., Gagne, Y., and Hopfinger, E.: Velocity probability density functions of high Reynolds number turbulence, *Physica D* 46 (1990), 177–200.
7. Chabaud, B., Naert, A., Peinke, J., Chillà, F., Castaing, B., and Hébral, B.: Transition toward developed turbulence, *Phys. Rev. Lett.* 73 (1994) 3227–3230.
8. Peinke, J., Castaing, B., Chabaud, B., Chillà, F., Hébral, B., and Naert, A.: On a fractal and an experimental approach to turbulence, *Fractals in the Natural and Applied Sciences* A41 (1994), 295–304.
9. Naert, A., Puech, L., Chabaud, B., Peinke, J., Castaing, B., and Hébral, B.: Velocity intermittency in turbulence: how to objectively characterize it?, *J. Phys. II France* 4 (1984), 215–224.
10. Ghashghaie, S., Breymann, W., Peinke, J., Talkner, P., and Dodge, Y.: Turbulent cascades in foreign exchange markets, submitted to Nature (1995).
11. Vassilicos, J. C.: Turbulence and intermittency, *Nature* 374 (1995), 408–409.
12. Mantegna, R. N. and Stanley, H. E.: Scaling behaviour in the dynamics of an economic index, *Nature* 376 (1995), 46–49.
13. Müller, U. A., Dacorogna, M. M., Davé, R. D., Olsen, R. B., Pictet, O. V., and von Weizsäcker, J. E.: Volatilities of different time resolutions — analyzing the dynamics of market components, *Journal of Empirical Finance*, in press.

A SIMPLE MODEL OF TURBULENCE INTENSITY AND TURBULENCE SCALE DISTRIBUTION IN GRAVEL BED RIVERS

V. I. NIKORA and G. M. SMART
National Institute of Water and Atmospheric Research,
PO Box 8602, Christchurch, New Zealand

1. Introduction

Turbulence intensity is a simple and widely investigated river turbulence characteristic. It is often used in both basic research and practical applications. Many researchers have tried to find some universal functions for describing its vertical distribution in open channel flows. In most cases the attempts were empirical (Grinvald, 1974; Iwasa and Asano, 1980, among others). At the same time some semi-empirical models were developed by Nakagawa, Nezu, and Ueda (1975), Li, Schall, and Simons (1980), and Nezu and Nakagawa (1993). The latter derived their relationships on the basis of the K-ε turbulence model and an assumption that the turbulent energy is in local equilibrium (turbulence generation G = turbulence dissipation ε). Also, Nezu and Nakagawa (1993) recognize that the eddy viscosity vertical distribution in their model differs somewhat from the parabolic distribution that follows from the logarithmic velocity law.

While Nezu and Nakagawa (1993) recommend use of their relationships for both smooth and rough beds irrespective of Reynolds and Froude numbers, Grinvald and Nikora (1988) and Nikora (1992) showed great scattering of experimental points on the graph $\sigma_{i=u} / U_* = f(Z/H)$ for real river conditions and explained it by the influence of bottom configuration, roughness, suspended sediments, bed-load intensity and three-dimensionality of the flow structure (σ_i is the standard deviation of the i velocity vector component, u is the longitudinal velocity vector component, U_* is the shear velocity, Z is the distance from the bottom, and H is the flow depth).

So we need to account for some additional factors which influence the turbulence regime in real river flows. An attempt to derive a relationship for σ_i / U_* which takes into account the parabolic distribution of the eddy viscosity, characteristic turbulence scale and bottom roughness is presented below.

171

S. Gavrilakis et al. (eds.), Advances in Turbulence VI, 171-174.
© 1996 *Kluwer Academic Publishers.*

2. Model

The main assumptions of our model are: *(1)* river flow is two-dimensional (change of turbulence properties vertically is much stronger than cross-sectionally); *(2)* in the intermediate flow region the turbulence energy generation G is equal to the turbulence energy dissipation ε ; *(3)* velocity distribution with depth is logarithmic; *(4)* applicability of Taylor's 'frozen' turbulence hypothesis to mean flow quantities (spectra, structure functions etc.); *(5)* structure function $D_i(l) = \overline{[u_i(x+l) - u_i(x)]^2}$ for the i velocity vector component consists of two main regions (the smallest dissipative scales are not important for this problem and can be neglected): *(i)* inertial subrange, where $D_i(l) = c_i \varepsilon^{2/3} l_i^{2/3}$ (Monin and Yaglom, 1975), and *(ii)* saturation region with $D_i(l) = 2\sigma_i^2$ (Figure 1).

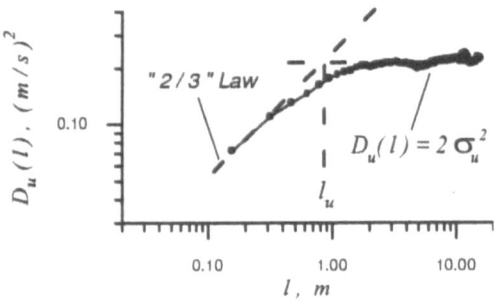

Figure 1. Explanation of assumption (5) for the case of longitudinal velocity.

From assumptions *(1)*, *(2)*, and *(3)* it follows:

$$\varepsilon = G = -\overline{u'v'} \frac{d\overline{U}}{dZ} = \frac{U_*^3}{kH}(\frac{1-\eta}{\eta}) \tag{1}$$

where k is the von Karman constant, v is a vertical component of the velocity vector, $\overline{U}$ is the local mean velocity, and $\eta = Z/H$. According to assumption *(5)* the velocity structure function at $l = l_i$ can be presented as:

$$D_i(l_i) = 2\sigma_i^2 = c_i \varepsilon^{2/3} l_i^{2/3} \tag{2}$$

where c_i is a Kolmogoroff constant, l_i is the scale which divides the inertial and saturation regions of the function $D_i(l)$ (Figure 1). It should be noted that the scale l_i has much clearer physical meaning for geophysical turbulent flows (as an external scale of the inertial subrange for the i velocity component) than other widely used integral scales. Together with (1) equation (2) gives:

$$\frac{\sigma_i}{U_*} = (\frac{c_i}{2k^{2/3}})^{1/2} (\frac{l_i}{H})^{1/3} (\frac{1-\eta}{\eta})^{1/3} \tag{3}$$

Equation (3) shows that the turbulence intensity σ_i / U_* depends not only on the distance from the bottom η (as, for instance, in Nezu and Nakagawa (1993) formulas) but also on the turbulence scale, l_i.

Another useful relationship which follows from equation (2) is relationship (4):

$$\varepsilon = (\frac{2}{c_i})^{3/2} \frac{\sigma_i^3}{l_i} \tag{4}$$

which relates the turbulence dissipation with the turbulent energy and characteristic scale.

To close the problem we need to parametrize the relative turbulence scale l_i / H in relationship (3). From physical and similarity considerations and some laboratory experiments we can assume for the intermediate flow region:

$$l_i / H = f(Z / Z_0) = b_i (Z / Z_0)^{\alpha_i} \tag{5}$$

where Z_o is the roughness parameter in the logarithmic law $\overline{U} / U_* = 1 / k \ln(Z / Z_o)$, and an exponent α_i and factor b_i should be determined from experiments.

3. Model test and conclusions

To test our model for the case of longitudinal velocity we used field measurements in three New Zealand gravel bed rivers. Instantaneous longitudinal velocities were measured by electronic Pitot tubes developed and manufactured by the National Institute of Water and Atmospheric Research, NZ (Smart, 1991, 1994). The ranges of the main hydraulic and morphometric characteristics of the investigated river reaches were: width 9-95 m, average depth 0.2-1.2 m, water discharge 5.5-250 m^3/s, water surface slope 0.003- 0.012, and bottom particle size d_{90}=130-250 mm. No bedload was observed during the measurements. The test of the initial model assumptions against field data showed that they are in good agreement with measurements. On the

174

basis of the same dataset we also obtained the exponent $\alpha_u \approx 1.0$ and factor $b_u = 0.037$ in relationship (5) (Figure 2). The exponent α_u proved to be close to 1.0, which is in good agreement with the old Prandtl hypothesis of a linear increase in mixing length with distance from the bottom.

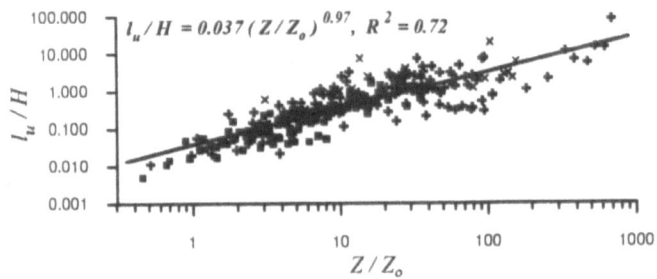

Figure 2. Graph $l_u / H = f (Z / Z_o)$ for New Zealand gravel bed rivers,
+ Waiho River, ■ North Ashburton River, ✕ Hurunui River

Good agreement between measurements, model assumptions and theoretical considerations allow us to conclude that relationships (3), (4) and (5) can be used in sediment transport and water quality models for gravel bed rivers.

Acknowledgments. The research was conducted under contract CO1512 from the Foundation for Research, Science and Technology (New Zealand). The authors are grateful to D. G. Goring and G. Carter for useful comments.

4. References

Grinvald, D. I. (1974) *Turbulence in Open Channel Flows* (in Russian). Hydrometeoizdat, Leningrad (former USSR).

Grinvald, D.I. and Nikora, V. I. (1988) *River Turbulence* (in Russian). Hydrometeoizdat, Leningrad (former USSR).

Iwasa, Y. and Asano, T. (1980) Characteristics of turbulence in rivers and conveyance channels. *Proc. of the 3rd Intern. Symposium on Stochastic Hydraulics*, Tokyo, D1-1-D1-12.

Li, R., Schall, J. D., and Simons, D.B. (1980) Turbulence prediction in open channel flow. *J. Hydraulics Div., ASCE* **106**, HY-4, 575-587.

Monin, A.S. and Yaglom, A.M. (1975) *Statistical Fluid Mechanics: Mechanics of Turbulence*, vol. 2, MIT Press, Boston, Mass.

Nakagawa, H., Nezu, I., and Ueda, H. (1975) Turbulence of open channel flow over smooth and rough beds. *Proc. of Japan Soc. Civil Engrg.* **241**, 155-168.

Nezu, I. and Nakagawa, H. (1993) *Turbulence in Open-Channel Flows*. A. A. Balkema, Rotterdam.

Nikora, V. I. (1992) *Channel Processes and Hydraulics of Small Rivers* (in Russian). Shtiintsa, Kishinev (former USSR).

Smart, G.M. (1991) A P.O.E.M. on the Waiho (electronic gauging of rivers). *Journal of Hydrology* (NZ), **30**, 1, 37-44.

Smart, G.M. (1994) Turbulent velocities in a mountain river. *Hydraulic Engrg. '94*, Proc. ASCE Nat. Conf. on Hydraulic Engrg., Buffalo (NY), 844-848.

THE SYMMETRY PROPERTIES OF THE FLOW
IN A NUCLEAR REACTOR VESSEL

A round U-turn jet

PIERRE ALBARÈDE

Commissariat à l'Energie Atomique
Cadarache-13108 St Paul les Durance-France
pierre.albarede@cea.fr

The turbulent flow in a Pressurized Water Reactor vessel is modeled in a small scale experiment. Careful observations and flow control experiments, driven by considerations of symmetry, show that this flow of industrial importance has quite weird properties.

A schematic description is given by fig. 1. The four inlet ducts, called "branches" (1), have nominally identical flow rates. Water is fed from top, then flows down into an annular duct or"downcomer". undergoes a U-turn in a nearly hemispheric "plenum". to eventually flow up through a grid (3) and into a cylindrical duct.

The nominal symmetry group is the four order group, generated by the two mirror symmetries shown on fig. 1 and fig. 2 (dashed lines). The flow exhibits spontaneous symmetry breaking, as shown on fig. 2: one transverse (horizontal) vortex and two longitudinal (vertical) vortices appear. Four configurations are observable, any pair of them being reciprocally symmetrical. The installed configuration depends on initial conditions, and flow history: a commutation technique has been devised, involving a modification of inlet flow rates.

As the vortex pattern is quite indifferent to some large external perturbations, the actual flow would be similar to an easily conceived ideal flow, i. e. a "round U-turn jet", with nominal axial and mirror symmetries. The ideal flow would have only one mirror symmetry (dotted line on fig. 2), and its neutral azimuthal position could be taken as a phase parameter.

The observed phenomena would result from a spontaneous symmetry breaking of the round U-turn jet, under restricted symmetry and special phase dynamics, due to perturbating technical details.

This work results from a cooperative research with FRAMATOME.

S. Gavrilakis et al. (eds.), Advances in Turbulence VI, 175-176.
© 1996 *Kluwer Academic Publishers.*

fig. 1

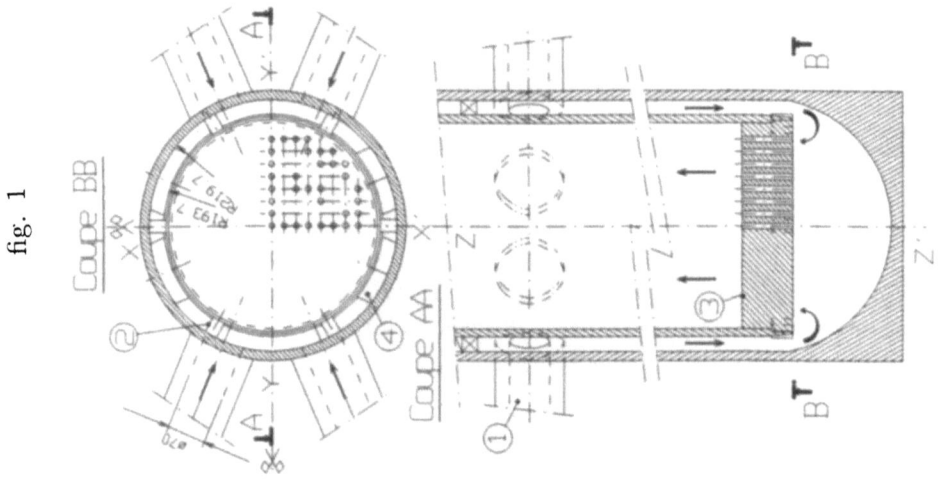

fig. 2

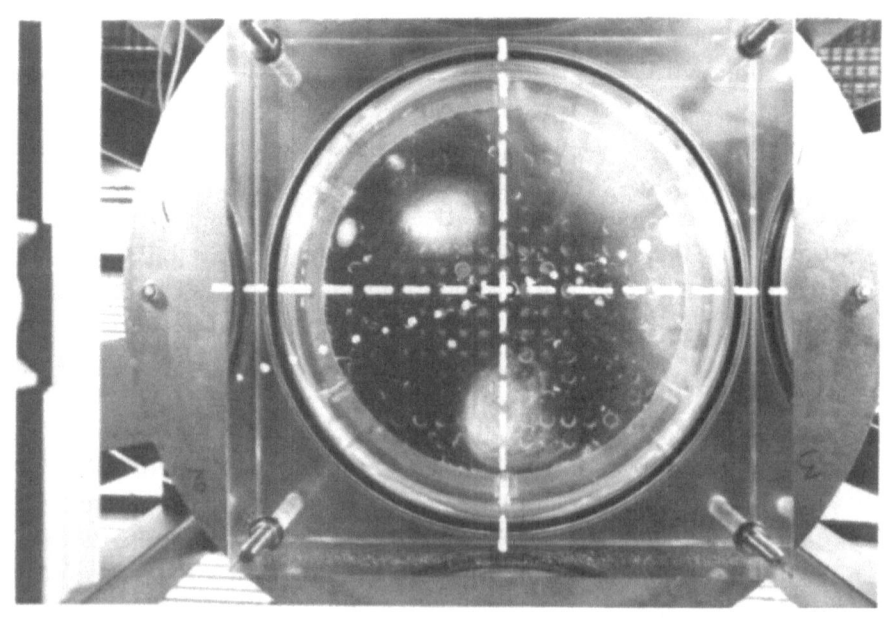

INVESTIGATIONS IN THE NEAR FIELD
OF A NOMINALLY TWO-DIMENSIONAL
STRONG ACOUSTIC PERTURBATION SOURCE

P. ERK, F. BÉRAUD AND K. GRAICHEN

Hermann-Föttinger-Institut für Strömungsmechanik
Technische Universität Berlin
Müller-Breslau Straße 8, 10 623 Berlin, FRG

Flow control, especially separation control, has been accomplished by the introduction of acoustically generated perturbations into the flow. Most of these experiments follow the same design principle: One or more loud-speakers generate the perturbations and are connected by a wave-duct to a slot in the surface from which the flow separates. In this experimental investigation we look more closely at the near field of the flow at the slot with and without cross-flow.

The experimental device consisted of a plane surface with a 900 mm long slot where the width could be adjusted. Underneath the slot, a circular wave-duct was connected to two horn drivers. Perturbation frequencies ranged from 800 Hz to 2000 Hz at sound pressure levels up to 155 dB.

Phase-averaged flow velocities were measured with single hot-wire and two-component laser-Doppler anemometry. Flow visualizations using dye and aluminum particles were performed in a water tunnel with a model preserving Strouhal and Reynolds similarity to the experiments in air.

The oscillating flow through the slot without cross-flow is dominated by two symmetrical, counter-rotating vortices (Fig. 1a). They form as the flow separates from the slot edges during outflow. Large-scale, circulatory streaming takes place where fluid along the wall is moving towards the slot and fluid along the slot axis is moving away from the slot.

The peak outflow velocity exceeds the peak inflow velocity (Fig. 1b). With increased forcing, the difference between the inflow and the outflow velocity and the phase difference between the outflow maximum and the pressure minimum becomes smaller. The phase shift between pressure and velocity is around $\pi/2$.

S. Gavrilakis et al. (eds.), Advances in Turbulence VI, 177-178.
© 1996 *Kluwer Academic Publishers.*

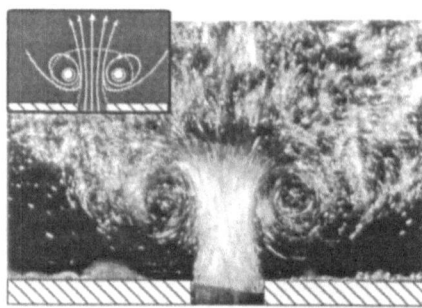
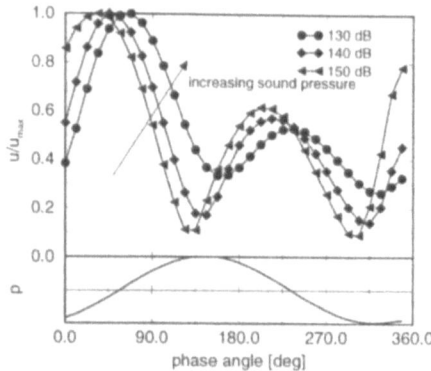

Figure 1. Flow without cross-flow: (a) Vortex formation during outflow phase. (b) Phase-averaged velocity at center of slot (hot-wire, frequency 800 Hz, slot width 0.25 mm), lower trace: pressure.

The presence of the cross-flow changes the flow topology: During inflow the flow separates at the upstream edge of the slot and forms a vortex within the slot which is ejected out of the slot (Fig. 2a). At high forcing amplitudes, two additional vortices are generated at the edges of the slot during outflow. All three vortices interact strongly with the cross-flow and induce flow towards the wall.

Laser-Doppler measurements show that above the slot the cross-flow is alternately accelerated and retarded (Fig. 2b). The vertical velocity oscillations extend several slot widths into the cross-flow boundary layer.

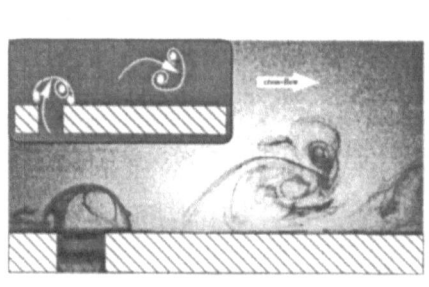
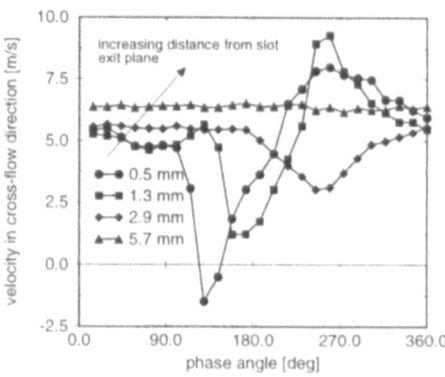

Figure 2. Flow with cross-flow: (a) Vortex formation and convection during outflow phase. (b) Phase-averaged velocity at center of slot (laser-Doppler, frequency 1000 Hz, slot width 1.6 mm, cross-flow velocity 6.5 m/s).

INTERACTION BETWEEN THE TURBULENT BOUNDARY LAYER AND A THREE-DIMENSIONAL HILL

S. FRANCHIN
Department of Civil Engineering
via S. Marta, 3 - 50139- Firenze - Italy

A wind tunnel study aimed at the description of the flow around a three-dimensional isolated hill-shaped obstacle in a fully developed turbulent boundary layer is presented. The presence of a hill modifies the mean velocity and turbulence fields; these effects have not so far been studied in detail and the boundary layer response to this strong distortion in the flow is particularly complex and not yet fully understood. A description of flow patterns and velocity distribution around three-dimensional hills is relevant for boundary layer parametrization in large-scale numerical models, predictions of dispersion of pollutants in complex terrain and wind energy utilization.

Experiments were performed in the C.R.I.A.C.I.V. wind tunnel (Department of Civil Engineering, University of Firenze) under neutral stability condition. It is a low-speed open circuit wind tunnel with a rectangular test section 2.4 m wide, 1.6 m high. The distance from the inlet to the test section is 10 m. The variable pitch axial flow fan allows a continuous variation of velocity from 2 to 35 m/s. The free-stream turbulence intensities at the speeds of these experiments were about 1%. The model is a circular axisymmetric gaussian surface of equation $z=h \cdot \exp(-r^2/\sigma^2)$ where h=0.3 m is the model height and $\sigma^2=0.055$ m^2; the hill diameter is 1.0 m.

Flow around hills of different shapes has been studied experimentally by Hunt [2], a numerical approach at different Reynolds numbers can be found in [3] and [4]. Mean flow separation over three-dimensional hills is not generally accompanied by recirculating flow with closed streamlines [1] and presents a complex topology which can be identified only with a complete 3D picture of the flow.

The separated region is here quantitatively described by means of surface pressure measurements supported by flow visualizations. Pressure measurements were made through a pressure scanning and data acquisition system realized by Pressure Systems. It is supported by 4 multi-channel miniature EPS Pressure Scanners, each one has 16 channels, allowing 64 contemporary measurement points.

The Reynolds number of the flow based on the velocity outside the boundary layer and on the hill height varies in the narrow range between 2×10^5 and 5×10^5, the flow is nearly independent of this parameter.

In *Figure 1* the pressure coefficients field in the higher part of the hill together with the measurement points is plotted. The Reynolds number of the flow is 2.1×10^5. The

179

S. Gavrilakis et al. (eds.), Advances in Turbulence VI, 179-180.
© *1996 Kluwer Academic Publishers.*

180

levels of pressure coefficients contours are equally spaced and projected vertically onto a horizontal plane. The flow moves from right to left. The separated region, detected using flow visualization, is characterized by a nearly constant pressure value which can be seen clearly on the left side of the picture. As it can be deduced from the visualization, separation occurs in proximity of the maximum of the adverse pressure gradient, immediately downward the top of the hill; this is the same location where the standard deviation of the pressure has its maximum. From Reynolds equations, written on a rigid boundary [3], it can be shown that the tangential pressure gradient is proportional to the normal gradient of vorticity in the other tangential direction; as a consequence the main generation of vorticity normal to the surface takes place where the maximum values of the pressure gradient realize.

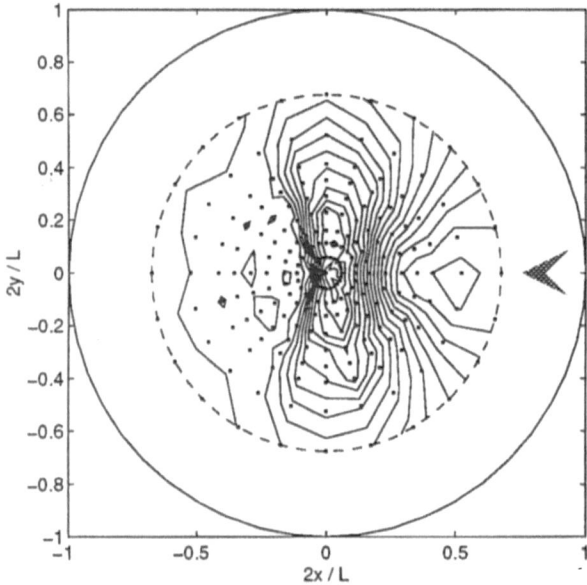

Figure 1 - Pressure coefficients field

1. Hunt, J. C. R., Abell, C. J., Peterka, J. A. & Woo, H. (1978) Kinematical studies of the flows around free or surface-mounted obstacles; applying topology to flow visualization, Journal of Fluid Mechanics, vol. 86, part 1, pp. 179-200.
2. Hunt, J. C. R. & Snyder, W. H. (1980) Experiments on stably and neutrally stratified flow over a model three-dimensional hill, Journal of Fluid Mechanics, vol. 96, part 4, pp. 671-704.
3. Mason, P. J. & Morton, B.R., (1987) Trailing vortices in the wake of surface-mounted obstacles, Journal of Fluid Mechanics, vol. 175, pp. 247-293.
4. Suzuki, M. & Kuwahara, K. (1992) Stratified flow past a bell-shaped hill, Fluid Dynamics Research, 9, pp. 1-18.

EXPERIMENTAL AND NUMERICAL INVESTIGATIONS OF ROTATING
AND SWIRL EFFECTS IN A RAPID-COMPRESSION-MACHINE (RCM)

K. HANJALIĆ

Faculty of Applied Physics, Delft University of Technology,
Lorentzweg 1, 2628 CJ Delft, The Netherlands

AND

S. JAKIRLIĆ, S. PARKS, C. TROPEA, J. VOLKERT

Lehrstuhl für Strömungsmechanik, University of Erlangen,
Cauerstr. 4, 91058 Erlangen, Germany

1. Introduction

A rapid compression machine (RCM) has been built in order to investigate the effects of a one-dimensional compression upon swirling flow in an environment typical for Diesel engines. In compliance with the experimental program, the numerical investigations consist of both a statistically stationary and a transient spin down behaviour of the induced swirling flow. Additionally, the instationary decaying flow is compressed by a 1D piston movement, which results in an extremely complex nature of the flow field under investigation. An integral part of such a flow is the swirl, which is used to enhance turbulent mixing and to stabilize the combustion process in IC engines. In the RCM, the swirl is generated by the rotation of the cylinder wall, providing a well defined initial state of the flow prior to the piston movement, i.e. compression.

2. Experimental Work

The experimental program consists of two phases. The first phase is that related to the generation of swirl inside the cylinder. The cylinder consists of a rotating and a non-rotating part, where the rotating section is driven by an external motor. The second phase relates to the 1D compression of the flow. The hardware available to realize this phase consists of a programmable hydraulic/pneumatic driving unit, which allows constant strain rates over the entire compression stroke to be achieved. The experimental investigation of the flow is based on both Laser Doppler Anemometry (LDA) and Particle Image Velocimetry (PIV). High temporally and spatially resolved measurements of turbulence quantities by LDA are presented for the cases of stationary swirl and spin-down without compression.

3. Numerical Work

It is well known that swirl causes a strong deviation of the velocity profile from the logarithmic law, such that the widely used wall functions are no longer valid for the treatment of wall boundary condition. Therefore the application of a turbulence model accounting for the low Reynolds number and wall proximity effects is necessary, making possible the

S. Gavrilakis et al. (eds.), Advances in Turbulence VI, 181-182.
© *1996 Kluwer Academic Publishers.*

182

integration of governing equations up to the wall. This is illustrated in Fig. 1a, in which the computed axial, circumferential and resultant velocity profiles are compared with the logarithmic law and with the LDA measurements for the case of steady rotation.

It is also well known that swirl causes a very high anisotropy of both the stress and dissipation tensors, causing at the same time a highly anisotropic eddy viscosity. This leads to the failure of all models based upon the eddy viscosity concept. The turbulence model used in the present study is therefore a second-moment (Reynolds stress) closure which satisfies all important invariance requirements, with a specific emphasis on: the asymptotic and limiting states of turbulence (vanishing and very high turbulence Re numbers, two-component limit, ...), reproducing the laminar-to-turbulent and reverse transition (by-pass and shear-generated transition with minimum background turbulence), appropriate reproduction of effects of extra strain rates (transverse shear, skew-induced three-dimensionality), high acceleration (including laminarization), high deceleration (approaching separation), swirl effects, flow separation, recirculation and reattachment [1, 2, 3, 4].

Selected results are shown in Figs. 1 and 2, with corresponding experimental data. Experiments and computations will be presented for three swirl rates. An evaluation of the employed turbulence models based on this and other experiments will be made.

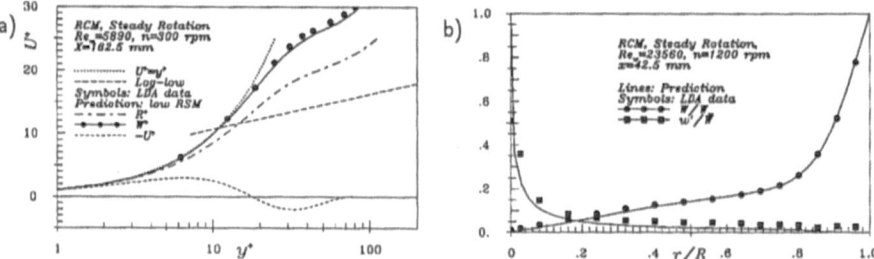

Fig. 1 Steady Rotation: a) Departure of the mean axial $(U^+ = U/U_\tau)$, the mean circumferential $(W^+ = (W_w - W)/W_\tau)$ and the 'resultant' $(R^+ = \sqrt{U^2 + (W_w - W)^2}/\sqrt{U_\tau^2 + W_\tau^2})$ velocity from the logarithmic law and b) Mean circumferential velocity and turbulence intensity

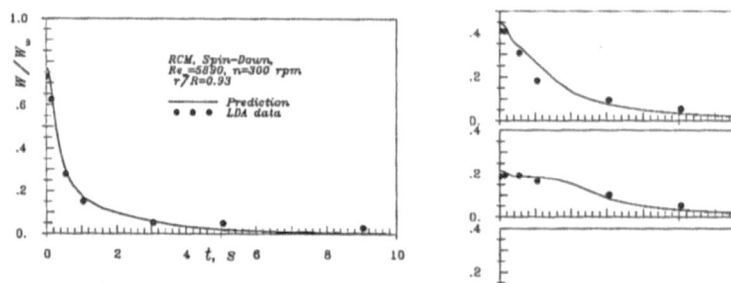

Fig. 2 Spin-Down: time decay of the mean circumferential velocity at different radial positions

4. References

1. Hanjalić K., Jakirlić S., Durst F., (1994.): Int. J. of Heat and Fluid Flow, Vol. 15, No.4, pp 269-282
2. Hanjalić K., Jakirlić S., Hadžić I., (1995.): Turbulent Shear Flows 9, Eds. F.Durst et al., Springer Berlin
3. Jakirlić S., Hanjalić K., (1994.): Proc. Int. Symp. on Turbulence, Heat and Mass Transfer, Lisbon,
4. Jakirlić S., Hanjalić K., (1995.): Proc. 10th Symp. on Turbulent Shear Flows, The Penn-State University.

TURBULENT FLOW AROUND CIRCULAR CYLINDER AFFEC-
TED BY INCIDENT STREAM OSCILLATIONS

ALICJA JARŻA
Institute of Thermal Machinery
Technical University of Częstochowa
Al.Armii Krajowej 21, 42-200 Częstochowa, Poland

The report deals with the experimental analysis of the flow around stationary circular cylinder in an oscillatory incident stream conditions.

The organized, periodic shedding of vortices is a predominant cause of the self-induced resonant oscillations of the flexible body immersed in the flow. The coincidence of the vortex and vibration frequencies is commonly termed lock-on [1]. Its occurence is accompanied by substantial increase in overall aerodynamical loading.

As results from [2] and the experimental study presented here, the phenomenon of vortex-shedding lock-on can also be observed in the case of stationary, rigid cylinder when the incident mean flow has a periodical component characterized by properly matched frequency f_o and amplitude ΔU.

In the present experiment the controlled oscillations of the incident flow were introduced by means of a set of two shutters, rotating in phase at the down stream part of the wind tunnel test section. The complex flow field around the cylinder was studing using Honeywell pressure transducers and multichannel DISA 55 CTA System. A method of signal processing based on the phase averaging technique was applied.

The effects of incident flow oscillations are visible in Fig.1. The lock-on phenomenon takes a form a plateau near $f_s/f_o=0,5$ and this value is kept constant over relatively wide range of reduced velocity $U_o/f_o \cdot D$. The zone of lock-on conditions is clearly associated with the higher levels of the vortex shedding peaks E_s in the power spectra of pressure fluctuations on the cylinder surface. The influence of external flow disturbances is particularly strong near the separation point ($\theta=90°$). The phase averaging technique made it also possible to consider the instantaneous values of velocity U_i as superposition of the mean $\bar{U}$ as well as periodic $\tilde{u}_i$ and random u_i' turbulent components. The obtained results indicate that incident flow osillations considerably change the the turbulent structure of the cylinder-wake flow (Fig.3). The substantial influ-

S. Gavrilakis et al. (eds.), Advances in Turbulence VI, 183-184.
© *1996 Kluwer Academic Publishers.*

184

ence is observed with respect to the contribution of periodic motion to the overall fluctuation energy (Fig.2).

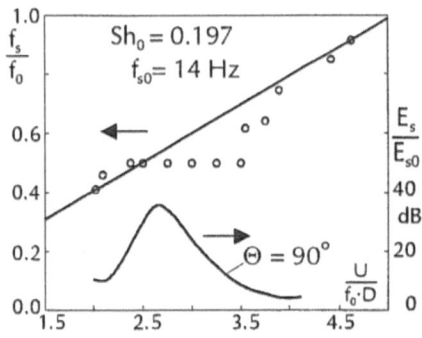

Fig.1. Frequency ratio f_s/f_0 and normalized pressure spectra peaks E_s/E_{s0} versus reduced velocity $U/f_0 \cdot D$ (Θ - angular position with respect to the mean flow direction; index "so" referes to uniform inlet conditions).

Fig.2. Share of periodic motion energy in the total turbulent kinetic energy. (ε_1 and ε_2 referes to longitudinal and cross component respectively).

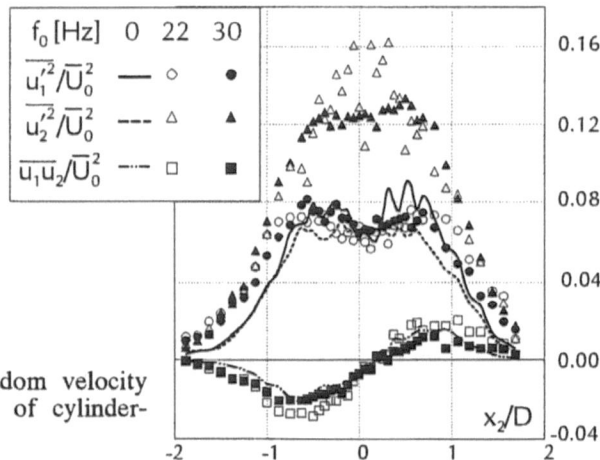

Fig.3. Distributions of random velocity variances in cross section of cylinder-wake.

References

Griffin O.M., Hall M.S. (1991) Review-Vortex Shedding Lock-on and Flow Control in Bluff Body Wake, Transm.ASME, *J.of Fluids Eng.*, vol.113, pp.526-537.

Jarża A., Słomczyński K. (1996) Vortex Shedding from Circular Cylinder in Oscillatory Incident Flow. *J.Teoret. and Applied Mechanics.*, vol. 2, N.34.

Heat transfer in a stably stratified grid-generated flow

T. Kanzaki and Y. Ichikawa
Central Research Institute of Electric Power Industry
Komae, Tokyo 201 Japan

1. Introduction

The evolution of grid-generated turbulence is affected by buoyancy force in a stratified flow, where heat transfer is controlled by turbulent motion and stratification. It is known that the counter-gradient scalar transfer occurs in a turbulent flow with strongly stable stratification. In previous works, the effects of the degree of stratification and the Prandtl (or Schmidt) number on the heat transfer mechanism have been investigated (Lienhard and Van Atta 1990; Yoon and Warhaft 1991; Komori et al. 1994). The results show that the counter gradient transfer is initiated at small scales in high Prandtl number water flows (Komori et al. 1994). However, it remains to be determined experimentally and numerically how small-scale turbulent motion affects the counter-gradient heat transfer in air flow of low Prandtl number. The purpose of the present paper is to investigate the counter-gradient heat transfer in stably stratified air flow.

2. Experiments

The experiments were conducted in a wind tunnel. Figure 1 shows the experimental setup and the measuring system. Stable stratification was achieved with a temperature step profile by heating the upper half-layer while turbulence was produced by a square mesh placed behind the heater .

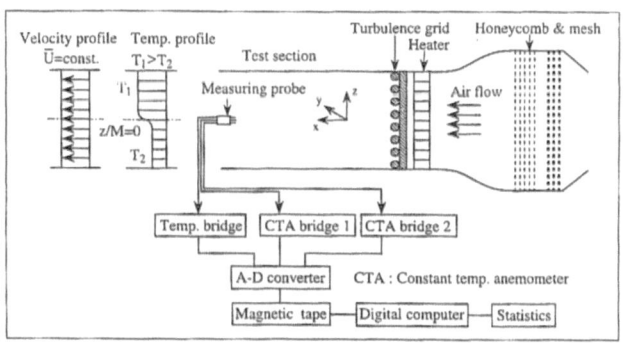

Figure 1 The experimental setup and the measuring system.

Instantaneous streamwise and vertical velocity fluctuations and temperature fluctuations were simultaneously measured by the combined use of a hot-wire anemometer and a cold-wire thermometer. To investigate the effects of small-scale turbulent motion on the counter-gradient heat transfer, measurements were conducted in three stratified flows with different turbulent Reynolds numbers, Re_λ, in the range of 41 to 96. Here, the Brunt-Väisälä frequency $N=[(g/T) (dT/dz)]^{1/2}$ was 1.7 to 1.8 s^{-1}.

S. Gavrilakis et al. (eds.), Advances in Turbulence VI, 185-186.
© 1996 *Kluwer Academic Publishers.*

186

3. Experimental Results and Discussion

Figure 2 shows the profiles of the vertical heat flux correlation coefficient for three different turbulent Reynolds numbers as a function of the buoyancy time Nt. The heat flux correlation coefficient is a positive value at Nt>1.6. This means that the counter-gradient heat transfer occurs at Nt>1.6. There is a discernible dependence of the correlation coefficient on the turbulent Reynolds number. The higher the value of Re λ of the flow, the smaller is the value of Nt at which the counter-gradient heat transfer actively occurs. Figure 3 shows the profiles of the ratio of vertical kinetic energy to potential energy, (VKE/PE), for the three flows.

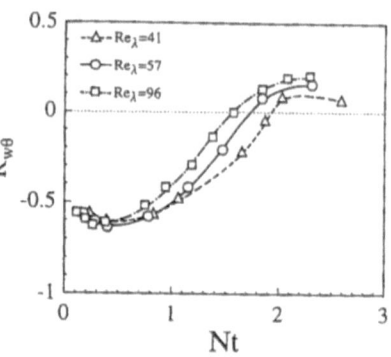

Figure 2 Profiles of the vertical turbulent heat flux correlation coefficient.

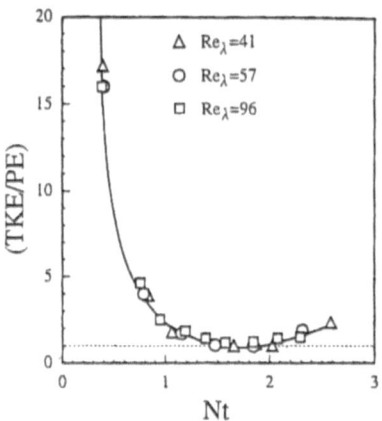

Figure 3 Ratio of vertical kinetic energy to potential energy.

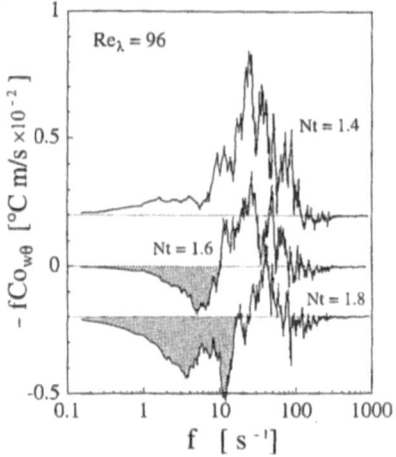

Figure 4 Cospectra of the vertical turbulent heat flux.

The profiles almost coincide. Figure 4 shows the cospectra of the vertical turbulent heat flux at Nt=1.4, 1.6, and 1.8 for Re λ =96. The positions of the cospectra of Nt=1.4 and 1.8 are shifted 0.2 and -0.2 relative to 0. The behavior of the cospectrum suggests that the counter-gradient heat transfer appears in the low-frequency region for air flows. The tendencies of the profiles for the other two flows are almost the same. The results show that the counter-gradient heat transfer occurs only on a large scale in air flow.

4. References

Komori, S., Nagata, K. and Murakami, Y. (1994) Heat and mass transfer in stably stratified flow. *In Proc. of the 4th Int. Symp. on Stratified Flows*, Grenoble, France.
Lienhard, V, J. H. and Van Atta, C. W. (1990) The decay of turbulence in thermally stratified flow. *J. Fluid Mech.*. **210**, 57-112.
Yoon, K. and Warhaft, Z. (1991) Turbulent mixing and transport in a thermally stratified interfacial layer in decaying grid turbulence. *Phys. Fluids A* **3**, 1143-1155.

TURBULENCE IN THE WAKE OF A DELTA WING

G.D. MILLER& C.H.K. WILLIAMSON
Sibley School of Mechanical & Aerospace Engineering
Cornell University, Ithaca, NY 14853 USA

Although the aerodynamics of delta wings has been studied extensively, remarkably little attention has been devoted to the downstream wake. In the present work we study the large- and small-scale turbulent structure in the trailing vortex wake, with a view to understanding the basic instabilities, vertical mixing and decay rate of the primary vortex pair. In our experimental work we use both a water towing tank and a wind tunnel. Our approach involves flow visualization, including PIV and also point-velocity measurements, as well as analytical studies. Initial experiments employed a novel technique of flying delta wings in free flight through water to completely eliminate any disturbances from struts or the towing mechanism.

We have made clear observations of the downstream development of the wake of a delta wing, for what appears to be the first time, in laboratory conditions. Our visualizations and velocity spectral measurements have led us to discover a direct link between structure in the "braid wake" and structure in the primary vortex pair. This result corresponds well with the recent turbulence measurements in a single vortex of Devenport *et al.* (1994). Figure 1 shows an example of the wake near the wing, which demonstrates both an extremely periodic structure in the braid wake, and also shows the direct link between it and the structure on the primary vortices.

We have discovered a surprisingly large vertical extent of the delta wing wake, due to the distribution of vortical structures within and outside the separatrix of the vortex pair. This "curtain" of fluid remnants left behind the descending vortex pair is evident in Figure 2. We propose that the mechanism which is responsible for the "curtain" involves three-dimensional dynamics of vorticity in the secondary vortices which are shed from the wing surface, underneath the primary vortices.

We have found discrete wavelengths of instability as the trailing wake travels downstream, whose lengthscale increases in size. In Figure 2 we see the development of the large-scale instability of the vortex pair far behind the wing. Ultimately, we find an interconnected vortex ring-like structure which is of a scale markedly smaller than the classical Crow (1970) instability wavelength, and over which we have distinct control.

Finally, we were interested in controlling the rate of decay and break-up of the coherent vortex pair wake by influencing the three-dimensional instabilities in these flows. A novel technique using image processing to measure the amplitude of the large-scale instability of the vortex pair has enabled us to measure the growth rate of this instability, which we found to compare very well with the analysis of Crow (1970). In Figure 3 we show these results and demonstrate the direct correspondence between the growth rate and the observed instability wavelength in the wake. Very slight forcing can strongly control the wake lifetime, which has significant practical application to aircraft wake control.

In conclusion, we have made what appear to be the first clear observations of the downstream wake of a delta wing in laboratory conditions. We have discovered a direct link between turbulent stucture between the primary vortices and the structure on the primary vortices. We have found a very periodic structure in the "braid wake". We have found a remarkably large vertical extent to the wake of the delta wing. We have observed the large-scale instablity of the vortex pair and vortex ring-like structures which result from it, and we have demonstrated control over the instability wavelength and lifetime of the vortex pair. In addition to our scientific interest, there is great practical importance to understanding the wake and, thus, we are also motivated by the need to understand the behavior of aircraft wakes and also the mixing as a result of these wakes in our atmosphere.

References

CROW, S. C. 1970 Stability theory for a pair of trailing vortices *AIAA J.* **8**, 2172
DEVENPORT, W. J., RIFE, M. C., LIAPAS, S. I. & MIRANDA, J. 1994 Turbulent Trailing Vortices, AIAA Paper 94-0404

S. Gavrilakis et al. (eds.), Advances in Turbulence VI, 187-188.
© *1996 Kluwer Academic Publishers.*

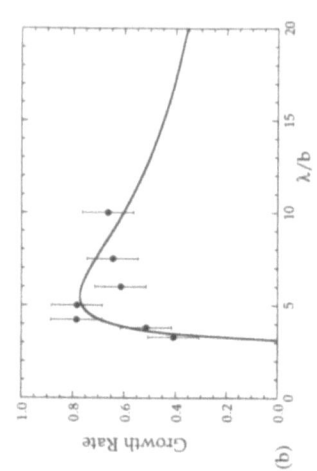

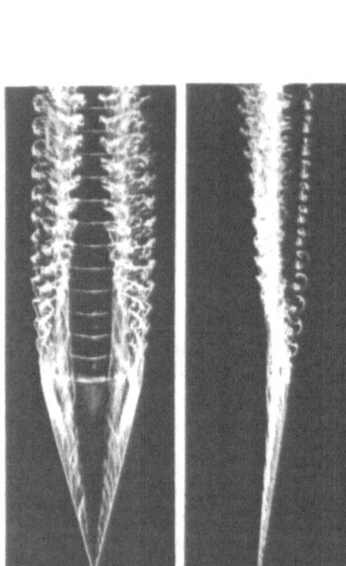

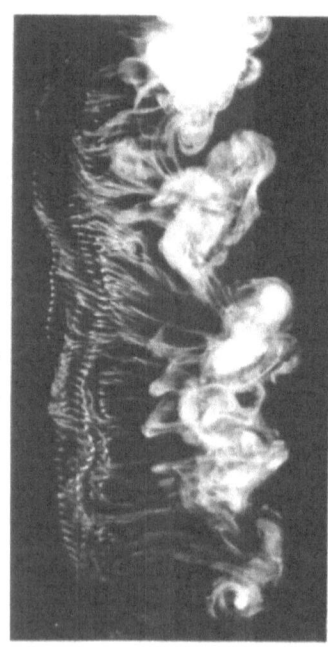

Figure 3. **Response of the wake to forcing.** (a) The shaded region is the envelope marking synchronization of the vortex pair instability with the applied forcing. Measurements made with no forcing show the natural instability wavelength to be $\lambda/b = 4.5$, which corresponds to the wavelength with the largest growth rate, shown in (b), both from our measurements (symbols) and from linear stability theory (solid line) such as Crow (1970). However, given sufficient initial amplitude, a more slowly growing wave can dominate the pattern. It is clear from these results that greater initial amplitude is required to produce a response further from the natural wavelength, and that even large amounts of forcing cannot produce a response at wavelengths shorter than the stability limit.

Figure 1. The near wake of a delta wing in free flight. The principal structures in the near wake comprise the primary vortex pair, and the small-scale periodic spanwise vortices within the "braid wake". These are clearly observed in the top and side views. There is a clear link between the scale of the small-scale instabilities with the "braid wake" and the primary cores.

Figure 2. The far wake of a delta wing in free flight. In the far wake the primary vortices have undergone a remarkable evolution to become a series of interconnected vortex ring-like structures under the action of an instability like the one described by Crow (1970). There is also evidence of fluid left behind the descending vortex pair in what is known as the "curtain". This is vortical fluid which either originated outside the separatrix of the primary vortex pair or crossed the separatrix during the descent of the pair.

IV

High Reynolds Number Turbulence and Intermittency

SCALING LAWS IN FULLY DEVELOPED TURBULENCE

S. CILIBERTO
Laboratoire de Physique
Ecole Normale Supérieure de Lyon (URA 1325 CNRS)
46, Allée d'Italie 69364 Lyon Cedex 07 France

R. BENZI, L. BIFERALE, M.V. STRUGLIA
Dipartimento di Fisica, Universitá "Tor Vergata"
Via della Ricerca Scientifica 1, I-00133 Roma, Italy

AND

L. TRIPICCIONE
INFN, Sezione di Pisa, S. Piero a Grado, 50100 Pisa, Italy

1. Introduction

In order to characterize the statistical properties of fully developed turbulence[1], one usually studies the scaling properties of moments of velocity differences at the scale r:

$$S_p(r) = < |v(x+r) - v(x)|^p > = < |\delta v(r)|^p > \tag{1}$$

where $< \cdots >$ stands for ensemble average and v is the velocity component parallel to r. At high Reynolds number $Re = U_0 L / \nu$ the $S_p(r)$ satisfies the relation

$$S_p(r) \propto r^{\zeta(p)} \tag{2}$$

for $L > r >> \eta_k$ where L is the integral scale, $\eta_k = (\nu^3/\epsilon)^{1/4}$ is the dissipative (Kolmogorov) scale, ϵ is the mean energy dissipation rate, ν the kinematic viscosity and U_0 the R.M.S. velocity of the flow. The range of length $L > r >> \eta_k$, where the scaling relation (2) is observed, is called the inertial range. The Kolmogorov (K41) theory predicts $\zeta(p) = p/3$, but experimental[2] and numerical [3] results show that $\zeta(p)$ deviates substantially from the linear law. This phenomenon is believed to be produced by the intermittent behaviour of the energy dissipation which can be taken

S. Gavrilakis et al. (eds.), Advances in Turbulence VI, 191-196.

into account by rewriting eq.(2) in the following way:

$$S_p(r) \propto < \epsilon_r^{p/3} > r^{p/3} \propto r^{\tau(p/3)+p/3} \tag{3}$$

where ϵ_r is the average of the local energy dissipation $\epsilon(x)$ on a volume of size r centered on a point x. A comparison of eq.(1) and eq.(3) leads to the conclusion that the scaling exponents $\tau(p/3)$ of the energy dissipation are related to those of S_p by $\zeta(p) = \tau(p/3) + p/3$.

Since the Kolmogorov (K62) theory many other models, (see ref.[1] and references therein), [5], have been suggested to describe the behaviour of the $\zeta(p)$. However, it turns out that the $\zeta(p)$ may be not universal in non homogeneous, anisotropic flow and may depend on the location where measurements are done. Specifically, they may have different values if one measures either far away from boundaries, where turbulence is almost homogeneous and isotropic, or in locations of the flow where a strong mean shear is present. The $\zeta(p)$ depend also on the way in which turbulence is produced, for example 3D homogeneous turbulence, boundary layer turbulence, thermal convection and MHD. Thus there is the fundamental question of understanding in which way all these parameters influence the scaling laws. Furthermore all the above mentioned models assume the existence of two well defined intervals of lengths that are the inertial range and a dissipation range. According to idea of multiscaling these two ranges may eventually be connected by an intermediate region where the viscosity begins to act [1]. However this idea of a well defined inertial range, where viscosity does not act at all, and the idea of multiscaling turns out to be incompatible with the recently introduced new form of scaling, which has been named Extended Self Similarity (ESS) [6] [7](see section 2 below).

ESS has been observed in 3D homogeneous and isotropic turbulence both at low and high Re and for a wide range of scales r with respect to scaling (2). In contrast ESS is not observed when a strong mean shear is present, [8],[9]. All these experimental observations show that also the mechanisms by which energy is actually dissipated in a flow are very poorly understood. Specifically one would like to understand how viscosity acts on different scales. This is clearly an important point in order to safely use large eddy simulations in real applications.

Starting from these observations we have proposed [10], [11] a generalized form of ESS which has been checked in many different flows. Our interpretation of ESS and of this generalized scaling suggest that there is no sharp viscous cut-off in the intermittent transfer of energy.

In this paper we summarize only the main results, all the details can be found in ref.[11].

2. Extended Self Similarity

Extended Self Similarity (ESS) is a property of velocity structure functions of homogeneous and isotropic turbulence [6, 7]. It has been shown using experimental and numerical data [12] that the structure functions present an extended scaling range when one plots one structure function against the other, namely:

$$S_n(r) \propto S_m(r)^{\beta(n,m)} \qquad (4)$$

where $\beta(n,m) = \zeta(n)/\zeta(m)$. The details of ESS have been reported elsewhere [7]. In the following we describe only the main features.

The ESS scaling has been checked both on numerical data and in experiments, in a range $30 < R_\lambda < 2000$ (R_λ is the Reynolds number based on the Taylor scale), finding that the scaling (4) is satisfied for $5\eta_k < r < L$ within 2%.

ESS has been also checked for the temperature and velocity fields in Rayleigh-Benard convection [13] and in the case of a passive scalar [14], [15]. It turns out that ESS is a very useful tool in order to distinguish between Kolmogorov and Bolgiano scaling [13], [16].

Another interesting observation concerns the behaviour of $\beta(p,3)$ with respect to R_λ. The values of $\beta(p,3)$ have been measured in the range $30 < R_\lambda < 5\ 10^6$, without noticing, within error bars, any change or trend of $\beta(p,3)$ (a similar result has been reported in ref. [18]). This means that far away from boundaries the $\beta(p,3)$ are constants which do not depend on Re and on the way in which turbulence has been generated. Specifically it has been found that the $\beta(p,3)$ have the same values in 3D homogeneous and isotropic turbulence, in MHD [17] and in thermal convection in the case of the Bolgiano scaling [11].

A final point regarding ESS, concerns the generalization of the Refined Kolmogorov Similarity Hypothesis (RKSH).
The RKSH states that $\epsilon_r \sim \delta v^3/r$, as far as concern the dependence on the scale r, and supports eq.(3). We can generalize the RKSH by introducing an effective scale $L(r) = S_3(r)/\epsilon$, as suggested by ESS, and we obtain the following relation: $\epsilon_r = \delta v^3\ \epsilon/S_3$.
Generalization of RKSH simply states that:

$$S_p(r) =< |\delta v(r)^p| >= \frac{< \epsilon_r^{p/3} >}{\epsilon^{p/3}}\ S_3(r)^{p/3}. \qquad (5)$$

Equation (5) has been proposed in [7] and carefully checked in ref.[19]. One can argue that eq.(5) is a trivial one because for $r < \eta_k$, ϵ_r is constant and $S_p \propto r^p$, thus the scaling $S_n \propto S_3^{p/3}$ is obviously satisfied. Furthermore for r in the inertial range eq.(5) is certainly verified because $(S_3/\epsilon) \propto r$. However in principle the proportionality constant of eq.(5) in the inertial and in the dissipative range could be different. The fact that experimentally they are found equal has several important consequences which have been discussed in details in ref. [11].

3. A generalized form of ESS

In the previous section we have seen that, when the boundary effects are negligeable, not only the ESS works but also the exponents $\beta(n,3)$ are universal because they do not depend on the systems and on Re. We want to stress that this kind of universality, observed in different flows, disappears if the system is influenced by the presence of a strong mean shear. In this case ESS does not work, because an extended range of scaling is not present when S_n is drawn as a function of S_3. Violation of ESS has been observed experimentally in boundary layer turbulence [9] [20] and in the shear behind a cylinder [8].

However one finds that generalized Kolmogorov similarity hypothesis eq.(5) is satisfied also for values of r where ESS is no longer satisfied. These results suggest that the concept of ESS could be generalized in such a way to take into account the scaling relation, eq.(5).

For this purpose we introduce the dimensionless structure function

$$G_p(r) = \frac{S_p(r)}{S_3(r)^{p/3}} \qquad (6)$$

According to Kolmogorov theory (6) should be a constant both in the inertial and in the dissipative range, although the two constant are not necessarily the same. Because of the presence of anomalous scaling $G_p(r)$ are no longer constants and by using (5) we have:

$$G_p(r) = < \epsilon_r^{p/3} > \qquad (7)$$

Equation (7) is valid for all scales even in cases where ESS is not verified. Therefore, it seems reasonable to study the self scaling properties of $G_p(r)$ or, equivalently, the self-scaling properties of the energy dissipation averaged on an interval of size r:

$$G_p(r) = G_q(r)^{\rho(p,q)} \qquad (8)$$

where we have by definition:

$$\rho(p,q) = \frac{\zeta(p) - p/3\ \zeta(3)}{\zeta(q) - q/3\ \zeta(3)} \qquad (9)$$

$\rho(p,q)$ is given by the ratio between deviation from the K41 scaling. It will play an essential role in our understanding of energy cascade. Indeed, it is easy to realise that it is the only quantity that can stay constant along all the cascade process: from the integral to the sub-viscous scales. It is reasonable to imagine that the velocity field becomes laminar in the sub-viscous range, $S_p(r) \propto r^p$, still preserving some intermittent degree parametrized by the ratio between corrections to K41 theory. The validity of eq.(8) have been checked for many different experimental set-up [8],[20], [13], done at different Reynolds numbers and for some direct numerical simulation with and without large scale shear. Within experimental errors (about 3%), no deviations from the scaling regime are detected.

It is important to stress that the funtions $G_p(r)$ satisfy the hierarchy of the energy dissipation moments recently proposed in ref. [5]. This hierarchy has been tested experimentally in ref.[22] and rewritten in terms of velocity structure functions in ref. [23]. This hierarchy determines also the properties of the probability distribution function of the velocity field [24], [25], [26]. This property has been used in the construction of a model which takes into account all the observations described in this paper [11].

4. Conclusions

In this paper we have reviewed several new results concerning the scaling behaviour of small scale statistical properties of turbulence. Specifically we have shown that in homogeneous and isotropic turbulent flows, Rayleigh Benard convection and solar wind magnetohydrodynamics, the ratio $\zeta(p)/\zeta(3)$ seems to have an universal behaviour. This is a rather striking and unexpected result which implies that anomalous violation of dimensional scaling may be explained in an universal way. We have shown that ESS is not observed when relatively strong shear flows are present. Instead the generalized refined Kolmogorov similarity is true, also in cases where ESS is not observed. Based on these results we have introduced a generalization of ESS. This generalization is supported both by experimental and numerical data and it seems not affected by viscous cutoff.

We have proposed a model [11] which unifies all the previous points. This model suggests that all the above mentioned results can be explained by assuming that, once the large scale properties of the flow are subtracted, the cascade process, with which the energy is transfered from large to small scales, is a universal property, which does not depend on the specific turbulent flow.

References

1. U. Frish, *Turbulence*, Cambridge University Press 1995.
2. F. Anselmet, Y. Gagne, E. J. Hopfinger, R.A. Antonia, " High order velocity structure functions in turbulent shear flow",*J. Fluid Mech.* 140, 63 (1984)

196

3. A. Vincent, M. Meneguzzi, " The spatial structure and statistical properties of homogeneous turbulence", *J. Fluid Mech.*,225, 1 (1991).

4. B. Castaing, Y. Gagne, E. Hopfinger, "Velocity probability density functions of high Reynold number turbulence",*Physica D* 46, 177 (1990).

5. Z. S. She, E. Leveque, "Universal scaling laws in fully developed turbulence", *Phys. Rev. Lett.* 72, 336 (1994).

6. R. Benzi, S. Ciliberto, R. Tripiccione, C. Baudet, . F. Massaioli, S. Succi," Extended Self Similarity in turbulent flows", *Phys. Rev. E* 48 (1993) R29.

7. R. Benzi, S. Ciliberto, C. Baudet, G. Ruiz Chavarria," On the scaling of three dimensional homogeneous and isotropic turbulence", *Physica D* 80, 385 (1995).

8. R. Benzi, S. Ciliberto, C. Baudet, G. Ruiz Chavarria, R. Tripiccione, " Extended Self Similarity in the dissipation range of fully developed turbulence", *Europhys. Lett.* , 24, 275 (1993)

9. G. Stolovitzky, K. R. Sreenivasan, *Phys. Rev. E* , 48, 32 (1993).

10. R. Benzi, L. Biferale, S. Ciliberto, A. Struglia, L. Tripiccione, "On the intermittent energy transfer at viscous scales in turbulent flows" *Europhysics Lett.* 32, 709 (1995).

11. R. Benzi, L. Biferale, S. Ciliberto, A. Struglia, L. Tripiccione, "Generalized scaling in fully developed turbulence ", to be published in *Physica D* .

12. M. Briscolini, P. Santangelo, S. Succi, R. Benzi, "Extended Self Similarity in the numerical simulation of three dimensional flows", *Phys. Rev. E* 50, 1745 (1994).

13. R. Benzi, R. Tripiccione, F. Massaioli, S. Succi, S. CIliberto," On the scaling of the velocity and temperature structure functions in Rayleigh-Benard convection", *Europhys. Lett.* 25, 331 (1994).

14. G. Ruiz Chavarria, C. Baudet, S. Ciliberto, "Extended Self Similarity of passive scalars in fully developed turbulence", *Europhys. Lett.* 32, 319 (1995).

15. G. Ruiz, C. Baudet, S. Ciliberto, "Scaling laws and dissipation scale of a passive scalar in fully developed turbulence" to be published in *Physica D* .

16. S. Cioni, S. Ciliberto, J. Sommeria, "Temperature structure functions in turbulent convection at low Prandtl number" *Europhysics Lett.* 32, 413 (1995).

17. R. Grauer, J. Krug, C. Marliani," SCling of high order structure functions in magnetohydrodynamic turbulence", *Physics Lett. A* 195, 335 (1994).

18. F. Belin, P.Tabeling, H. Willaime, " Exponents of structure functions in a low temperature helium experiment ", to be published *Physica D* .

19. G. Ruiz Chavarria, " Anomalous scaling of velocity structure functions in turbulence: a new approach", *J. Physique* , 4, 1083 (1994).

20. G. Ruiz Chavarria "Scaling law in boundary layer turbulence ", in preparation (1995).

21. R. Benzi, M. V. Struglia, R. Tripiccione " ESS in numerical simulations of 3D anisotropic turbulence " preprint 1995, submitted.

22. G. Ruiz Chavarria, C. Baudet, S. Ciliberto, "Hierarchy of the energy dissipation momentsin fully developed turbulence" *Phys. Rev. Lett.* 74, 1986 (1995).

23. G. Ruiz Chavarria, C. Baudet, R. Benzi, S. Ciliberto, " Hierarchy of the velocity structure functions in fully developed turbulence",*J. Physique* 5, 485 (1995).

24. B. Dubrulle," Intermittency in fully developed turbulence: Log-Poisson Statistics and generalized scale-covariance", *Phys. Rev. Lett.* 73, 959 (1994).

25. Z.-S. She and E.C. Waymire, "Quantized energy cascade and log-Poisson statistics in fully developed turbulence", *Phys. Rev. Lett.* 74 262 (1995)

26. B. Castaing, B. Dubrulle,"Fully developed turbulence: a unifying point of view", *J. Phys. II France* 5, 895 (1995).

HELICAL SHELL MODELS FOR THREE DIMENSIONAL TURBULENCE

L. BIFERALE AND R. BENZI
Dipartimento di Fisica, Università di Tor Vergata
Via della Ricerca Scientifica 1, I-00133 Roma, Italy

R. M. KERR
Geophysical Turbulence Program
NCAR, P.O. Box 3000, Boulder, CO 80307-3000

AND

E. TROVATORE
INFM-Dipartimento di Fisica, Università di Cagliari
Via Ospedale 72, I-09124, Cagliari, Italy.

1. Introduction

The celebrated Kolmogorov theory of fully developed turbulence predicts for the scaling properties of structure functions in the inertial range the following form:

$$S_p(l) \equiv \langle (\delta v(l))^p \rangle \sim \langle (\epsilon(l))^{p/3} \rangle l^{p/3} \qquad (1)$$

where $\epsilon(l)$ is the energy transfer trough a scale l. If $S_p(l) \sim l^{\zeta(p)}$ and $\langle \epsilon^p(l) \rangle \sim l^{\tau(p)}$ then $\zeta(p) = p/3 + \tau(p/3)$. In K41 the $\epsilon(l)$ statistic is assumed to be l-independent, implying $\zeta(p) = \frac{p}{3}, \forall p$, in particular the energy spectrum going as $k^{-5/3}$.

While from a qualitative point of view Kolmogorov's intuition was a true breakthrough in the understanding of turbulence, his theory lacks quantitative agreement with experimental measurements of intermittency in physical space, in particular there are non-trivial scaling corrections to the "p over 3" Kolmogorov prediction for the $\zeta(p)$ exponents. There are many experimental and numerical [1] results telling us that energy is transferred intermittently.

In order to understand energy transfer dynamics and related intermittent effects, besides analytical and direct numerical approaches, it has been proposed to study dynamical deterministic models (shell models). Shell

S. Gavrilakis et al. (eds.), Advances in Turbulence VI, 197-200.
© *1996 Kluwer Academic Publishers.*

models concentrate all the dynamical interactions into a few degrees of freedom at different scales, retaining the non-linear structure of Navier-Stokes equations but neglecting completely their spatial location and losing most of their three-dimensional vector properties.

The most popular shell model is the Gledzer-Ohkitani-Yamada (GOY) model ([2]-[4]). It has been shown to predict scaling properties for $\zeta(p)$ similar to what is found experimentally. Recently, it has been pointed out that the set of free parameters for GOY with scaling exponents closest to the experimental measurements conserves in the inviscid, unforced limit two quadratic quantities. The first quantity is *energy*, while the second is, roughly speaking, the equivalent of *helicity* in 3D turbulence [4]. It has been suggested that both the *GOY-helicity* [7] and the helicity in the N-S equations [5] play roles in triggering the intermittent nature of the energy cascade.

From the new emphasis on the role played by the helicity a new class shell models based upon a helical decomposition of interactions between modes in the Navier-Stokes equations [5] has been suggested [7]. In this way, it is possible to obtain a second non-positive defined invariant closer to the definition of helicity in the Navier-Stokes equations. The models, which have two complex variables per shell, are generalizations of the GOY model with helical structures that include all the possible helical interactions in Navier-Stokes.

2. The GOY-helical shell model.

The GOY model can be seen as a severe truncation of the Navier-Stokes equations: it retains only one complex mode u_n as a representative of all Fourier modes in the shell of wave numbers k between $k_n = k_0\lambda^n$ and k_{n+1}, λ being an arbitrary scale parameter ($\lambda > 1$), usually taken equal to 2.

The dynamics is governed by the following set of complex coupled ODEs where only couplings with the nearest and next nearest shells are kept:

$$\frac{d}{dt}u_n = i\,k_n\left(au^*_{n+1}u^*_{n+2} + bu^*_{n+1}u^*_{n-1} + cu^*_{n-1}u^*_{n-2}\right) - \nu k_n^2 u_n + \delta_{n,n_0}f, \quad (2)$$

where ν is the viscosity, f the external forcing acting on a large scale n_0 and a,b,c are three free parameters. By adjusting the time scale we can always fix $a = 1$ while the other two parameters fix the form of the inviscid conserved quantities. It turned out that when the two quadratic inviscid quantities are similar to the energy and helicity Navier-Stokes expression the model has the same **quantitative** intermittency of a true turbulent flow. The importance of a "helicity-like" invariant in determining intermittent properties of the energy transfer inspired us to introduce helical-shell models such as to stay closest as possible to the original N-S eqs.

Starting from the Helical-fourier decomposition of N-S eqs. [5]. we have defined GOY-helical shell models by using two dynamical variables per shell, u_n^+ and u_n^-, transporting positive and negative helicity respectively. Being the nonlinear term in NS quadratic, we have only for different non-equivalent classes of triad interactions distinguished form their helical structure. One of these four classes leads to a dynamical model which is the equivalent of the old GOY rewritten in terms of helical variables, other two would transfer energy mainly backward while the last one has all the interesting properties: net forward and intermittent energy transfer and an helical structure which is different from the old GOY model. This last model has been, therefore, our candidate for a detailed numerical investigation.

The dynamical equations are the following:

$$\dot{u}_n^+ = ik_n(a u_{n+2}^- u_{n+1}^+ + b u_{n+1}^- u_{n-1}^+ + c u_{n-1}^- u_{n-2}^-)^* - \nu k_n^2 u_n^+ + \delta_{n,n_0} f^+. \quad (3)$$

and the same but with all helicity reversed for $\dot{u}_n^-$. The coefficients a, b, c are determined imposing conservation of energy anf of a generalized Helicity:

$$\frac{d}{dt}E = \frac{d}{dt}(\sum_n (|u_n^+|^2 + |u_n^-|^2)) = 0; \quad \frac{d}{dt}H_\alpha = \frac{d}{dt}\sum_n k_n^\alpha (|u_n^+|^2 - |u_n^-|^2) = 0$$

$$(4)$$

Let us notice that the choice $\alpha = 1$ would correspond to the Helicity for NS. We integrated eqs.(4) using the standard parameters $a = 1$. $\lambda = 2$. $k_0 = 2^{-4}$. $f^\pm = 5(1+i)10^{-3}$, $n_0 = 1$, $\nu = 10^{-7}$ and a total number of shells equal to $N = 22$ and $N = 26$. In the numerical integration we used a fourth-order Runge-Kutta method, with a time step varying between $dt = 10^{-5}$. Most of the results presented here are for the case $N = 22$. with a number of iterations of the order of hundred millions, which correspond roughly to several thousands of eddy turnover times at the integral scale. Stationarity is checked by monitoring the total energy evolution.

The quantities we have looked at are the structure functions: $S_p(n) = \langle|\tilde{u}_n|^p\rangle$. where $\tilde{u}_n = \sqrt{|u_n^+|^2 + |u_n^-|^2}$.

Fig. (1) shows the $\zeta(p)$s compared with those of the GOY model with the same parameters $\alpha = 1$ and $\lambda = 2$ (corresponding to the classical choice which conserves the analog of the 3D helicity). The scaling is nearly the same. What occurs is a strong similarity for $\alpha = 1$; nevertheless we expect a different behavior when this parameter is allowed to vary. For the α dependence of the models we have explored two other different values: $\alpha = 0.5$ and $\alpha = 2$ (keeping fix $\lambda = 2$). It is well known [4, 7] that the GOY model shows a strong dependence of its statistical properties on the α value. For example, if $\alpha < 2/3$ the dynamics is attracted toward a fixed point with Kolmogorov scaling, $\zeta(p) = p/3$. For $\alpha > 1$ intermittency become more important than what is usually measured in turbulent flows [4]. On

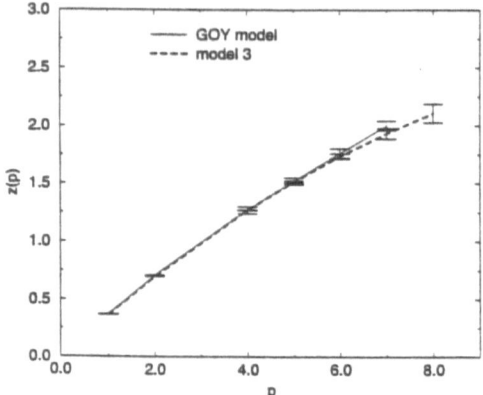

Figure 1. The $\zeta(p)$s of the old GOY model (from [4]) and of the new model (called model 3 here). The parameters values are $\alpha = 1$ and $\lambda = 2$. Error bars for data concerning model 3 take into account both statistical and power-law fit errors.

the other hand, the statistical properties of model 3 turn out to be robust under changes of the α parameter.

The robustness of this model , with respect to variation in α, gives to this model an important role among the possible shell models of turbulence.

The main result presented here is the strong universality shown by this new helical-shell model at varying the physical dimension of the second invariant. This can be partially understood by invoking the tendency of each single shell to transfer helicity backward and forward simultaneously, leading to a net vanishing of helicity flux. There is nothing from these studies to suggest any preference between the old GOY model and this new model inside the NS decomposition. The remarkable fact is that both models come from a NS decomposition in helical modes. In shell model one can play with the physical dimension of the second invariant, while for NS this is obviously impossible. The amazing fact is that for the correct choice of the invariants dimension all this models have an intermittent energy transfer very similar to what one can find in the original NS equations.

References

1. R. Benzi, S. Ciliberto, R. Tripiccione, C. Baudet, C. Massaioli & S. Succi, Phys. Rev. E **48**, R29 (1993).
2. E.B. Gledzer, Sov. Phys. Dokl. **18**, 216 (1973).
3. M. Yamada & K. Ohkitani, Prog. Theor. Phys. **81**,
4. L. Kadanoff, D. Lohse, J. Wang and R. Benzi; Phys. Fluids **7** 617 (1995).
5. F. Waleffe, Phys. Fluids A **4**, 350-363 (1992).
6. R. Benzi, L. Biferale, R. M. Kerr & E. Trovatore, Phys. Rev. E. (1996).
7. L. Biferale & R. Kerr, Phys. Rev. E **52** 6113 (1995).

SUBGRID-RESOLVED SCALE DYNAMICS IN ISOTROPIC TURBULENCE[†]

JAMES G. BRASSEUR
Department of Mechanical Engineering
The Pennsylvania State University
University Park, PA 16802, USA

HUNGRUI GONG[‡]
Theoretical Division and Center for Nonlinear Studies
Los Alamos National Laboratory
Los Alamos, NM 87545, USA

SHIYI CHEN
IBM Research Division
T.J. Watson Research Center, P.O. Box 218
Yorktown Heights, NY 10598, USA

1. Introduction

In large eddy simulation (LES) the nonlinearities in the evolution of the low-pass filtered velocity $\mathbf{u}^r(\mathbf{x})$ can be separated into dynamic interactions within the resolved velocity field $[N_R(\mathbf{x})]$ and those between resolved-scale (RS) and subgrid-scale (SGS) motions $[N_S(\mathbf{x})]$:

$$\frac{\partial}{\partial t} \mathbf{u}^r(\mathbf{x}) = N_R(\mathbf{x}) + N_S(\mathbf{x}) + \nu \nabla^2 \mathbf{u}^r(\mathbf{x}). \tag{1}$$

Whereas the influence of SGS motions is typically parameterized through a SGS flux tensor, we focus in this study on "SGS-RS dynamics" given by $N_S(\mathbf{x})$.

In high Reynolds number turbulence with the filter cutoff deep within the inertial range, SGS-RS dynamics does not directly influence integral-scale evolution (§4), so the primary constraint on the SGS closure is correct prediction of SGS-RS energy flux (i.e., the dissipation-rate). However, the existence of a boundary reduces the typical scale of energy-containing motions, leading to under-resolved large-scale motions near the boundary. In these regions the filter cutoff is well into the integral scales (Khanna & Brasseur 1996), the RS motions are directly influenced by SGS-RS dynamics (§4), and the detail to which the SGS model represents true SGS-RS dynamics is of much greater concern.

With an eye towards the identification of the more essential attributes required of SGS parameterizations in under-resolved regions of high Reynolds number turbulent flows, we analyze RS-SGS dynamics and energy transfer in direct numerical simulations (DNS) of forced and decaying isotropic turbulence in a 256^3 computational domain. Combining the forced DNS with isotropic decay we approximate, as best as possible on

---

[†]Supported by DOE at Los Alamos National Laboratory (LANL) and ARO/URI grant DAAL03-92-G-0117. Computations were performed at the Advanced Computing Laboratory of LANL.
[‡]currently Depart. of Mechanical Engrg, Texas A & M Univ., College Station, TX 77843.

S. Gavrilakis et al. (eds.), Advances in Turbulence VI, 201-204.
© 1996 *Kluwer Academic Publishers.*

a computer, the effect of the filter cutoff moving from the inertial range to the integral scales. These comparisons are contrasted with the cutoff within the dissipation-dominated scales, as analyzed, for example, by Domaradzki et al. (1993) and Kerr et al. (1996).

We distinguish, in this study, RS-SGS dynamics from RS-SGS energy transfer. To examine the dynamics underlying "backscatter" we separate spectral RS-SGS triadic interactions into "forward" and "inverse" triads (§2) and consider separately their relative contributions to both interscale energy transfer and $N_S(x)$. We consider the potential value in parameterizing separately "forward" and "inverse" dynamics.

The current analysis fits within a series of related studies which have focused, by in large, on energy transfer between resolved and subgrid scale motions, specifically Domaradzki, Liu & Brachet (1993), Domaradzki et al. (1994), Kerr, Domaradzki & Barbier (1996). These studies draw from analyses of interscale dynamics in fully developed turbulence within the Fourier framework by Domaradzki & Rogallo (1990), Yeung & Brasseur (1991), Domaradzki (1992), Zhou (1993), Brasseur & Wei (1994) and Yeung, Brasseur & Wang (1995) and Zhou, Yeung & Brasseur (1996).

2. Dynamics vs. energetics: equations and decompositions

In (1) the nonlinear interactions between resolved and SGS motions are given by

$$N_S(x) = -\nabla \cdot (u^r u^s + u^s u^r + u^s u^s)^r - \frac{1}{\rho} \nabla p_S , \qquad (2)$$

where p_S is the RS-SGS contribution to the pressure. In the Fourier description the evolution of an elemental resolved scale is given by

$$\frac{\partial}{\partial t} \hat{u}(k_r) = \hat{N}_R(k_r) + \hat{N}_S(k_r) - \nu\, k_r^2\, \hat{u}(k_r), \qquad (3)$$

where corresponding terms in (1) and (3) are FTF pairs. In the Fourier-spectral view, RS-SGS dynamics is given by triadic sums between resolved (k_r) and SGS (k_s) modes:

$$\hat{N}_S(k_r) = 2 \sum_{k_r=p_r+q_s} \hat{N}(k_r|p_r,q_s) + \sum_{k_r=p_s+q_s} \hat{N}(k_r|p_s,q_s) \equiv \hat{N}_{sr}(k_r) + \hat{N}_{ss}(k_r). \qquad (4)$$

$\hat{N}(k_r|p,q)$ represents a single interaction between a resolved mode and two other modes within a triad (see Brasseur & Wei 1994), one or both of which must be SGS. $\hat{N}_{sr}(k_r)$ and $\hat{N}_{ss}(k_r)$ each form FTF pairs with corresponding parts of $N_S(x)$. Including pressure in (2) implies that $N_S(x) \Leftrightarrow \hat{N}_S(k_r)$ each satisfy incompressibility. Energy transfer to RS motions from RS-SGS interactions is given by $T_S(k_r) = \hat{u}^*(k_r) \cdot \hat{N}_S(k_r) + cc$, where

$$T_S(k_r) = 2 \sum_{k_r=p_r+q_s} T(k_r|p_r,q_s) + \sum_{k_r=p_s+q_s} T(k_r|p_s,q_s) \equiv T_{sr}(k_r) + T_{ss}(k_r), \qquad (5)$$

The triadic geometry of "ss" and "sr" interactions in (4) and (5) implies that whereas the ss term in (1) can couple RS motions with all subgrid scales, the sr term involves only SGS motions between k_c and $2k_c$. We ask here whether ss and sr dynamics are sufficiently different to suggest separate modeling strategies for each term in (4).

Whereas "backscatter" in physical space has been described as those spatial regions where RS-SGS flux is into the resolved scales (Piomelli et al. 1991), in the spectral view inverse energy transfer is given by $T_S(k_r) > 0$. Here we further decompose the triadic sums in (5) into "forward" interactions $T_S^+(k_r)$ and "inverse" interactions $T_S^-(k_r)$ given by the partial sums where $T(k_r|p,q) < 0$ and $T(k_r|p,q) > 0$, respectively. In a similar manner, RS-SGS *dynamics* $\hat{N}_S(k_r)$ is correspondingly decomposed by conditioning individual triadic terms $\hat{N}(k_r|p,q)$ in (4) on the sign of $T(k_r|p,q)$ for that triad. Thus we write:

$$T_S(k_r) = T_S^+(k_r) + T_S^-(k_r) , \quad \hat{N}_S(k_r) = \hat{N}_S^+(k_r) + \hat{N}_S^-(k_r). \tag{6}$$

$T_S(k_r)$, $T_S^+(k_r)$ and $T_S^-(k_r)$ are calculated as normalized eddy viscosities (Kraichnan 1976).

The physical-space structure of "forward" and "inverse" dynamics is given by the inverse transform of $\hat{N}_S^+(k_r) \Rightarrow N_S^+(x)$ and $\hat{N}_S^-(k_r) \Rightarrow N_S^-(x)$. We analyze the time rates of change in *direction* and *magnitude* of $N_S(x)$, $N_S^+(x)$ and $N_S^-(x)$ separately:

$$N_S(x) \equiv \left[\frac{\partial}{\partial t}(U_r e_r)\right]_{SGS} = e_r \left(\frac{\partial}{\partial t} U_r\right)_{SGS} + U_r \left(\frac{\partial}{\partial t} e_r\right)_{SGS} , \tag{7}$$

where $e_r(x)$ is a unit vector in the direction of u^r at x and $U_r(x) \equiv |u^r(x)|$.

3. The direct numerical simulations and filter cutoffs k_c

The 3D turbulent velocity fields were obtained from 256^3 pseudo-spectral simulations (a) of isotropic decaying turbulence at $R_\lambda = 32$ and (b) stationary isotropic turbulence forced in shells $k=1,2$ with a $k^{-5/3}$ slope ($R_\lambda = 151$). The latter dataset is described in Wang *et al.* (1996). Whereas the decaying turbulence simulates the full range of integral scales, the forced simulation is an attempt to model inertial-range dynamics by replacing the integral scales with low wavenumber forcing. To simulate RS-SGS dynamics when k_c is in the integral scales, the filter cutoff is placed at $k_c=5$ relative to the spectral energy peak $k_E=4$ in the decay simulation. To simulate inertial-range dynamics, k_c is placed at 5 and 10 in the forced simulation. To simulate dissipation-range dynamics the cutoff is placed at $k_c=20$ relative to dissipation peaks $k_D \approx 14$ in both simulations.

4. Summary of results

We find that the SGS terms interact directly only with those resolved-scale motions at most a factor of 2-3 larger than the filter scale. Thus, the integral scales are only indirectly affected by the SGS model when the inertial range is well resolved and the primary demand on the SGS closure is accurate prediction of net RS-SGS energy flux. However, when the filter cutoff is in the energy-containing range we find that the *entire* range of resolved scales is dominated by RS-SGS dynamics. Thus, when the integral scales are under-resolved (e.g., near a boundary), the SGS model determines *directly* resolved-scale evolution, and a higher level of accuracy in the SGS closure is needed.

We find that when the filter cutoff is in the dissipation range, the RS-SGS dynamics is dominated by $T_{sr}(k_r)$ and $\hat{N}_{sr}(k_r)$, indicating that SGS motions at most a factor of two smaller than k_c are involved. When the integral scales are under-resolved, however, $T_{ss}(k_r)$ and $\hat{N}_{ss}(k_r)$ dominate RS-SGS dynamics. When the filter cutoff is in the inertial range, both the *ss* and *rs* dynamics contribute significantly. We conclude that there may be value in parameterizing the *ss* and *rs* contributions separately.

We find that the physical space structure of the time change in magnitude vs. direction of $u^r(x)$ due to RS-SGS dynamics (see §2) is very different. Changes in direction of $\mathbf{u}^r(\mathbf{x})$, for example, are found to be strongly correlated with low resolved-scale velocity magnitude, whereas time changes in magnitude (related to resolved-subgrid energy flux) tends to be more strongly correlated with regions of high $U^r(\mathbf{x})$.

Whereas a great deal of both "forward" and "inverse" activity (as defined in §2) takes place locally, most of this activity cancels, leaving a relatively small residual of net forward dynamics and energy transfer. Furthermore, $\hat{N}_S^+(\mathbf{x})$ and $\hat{N}_S^-(\mathbf{x})$ have very similar physical space structure, indicating that cancellations in forward and inverse dynamics tend to occur within local overlaps of high activity. We conclude that whereas approximating some details of "forward-inverse" dynamics may be important in SGS parameterization where the energy-containing scales are under resolved, the large cancellations makes it unwise to model separately "forward" and "inverse" dynamics.

References

Brasseur, J.G. and Wei, C. 1994 Interscale dynamics and local isotropy in high Reynolds number turbulence within triadic interactions. *Phys. Fluids* 6: 842-870.

Domaradzki, J.A. and Rogallo, S.S. 1990 Local energy transfer and nonlocal interactions in homogeneous, isotropic turbulence. *Phys. Fluids A* 2: 413-426.

Domaradzki, J.A. 1992 Nonlocal triad interactions and the dissipation range of isotropic turbulence. *Phys. Fluids A* 4: 2037.

Domaradzki, J.A., Liu, W. and Brachet, M.D. 1993 An analysis of subgrid-scale interactions in numerically simulated isotropic turbulence. *Phys. Fluids A* 5: 1747-1759.

Domaradzki, J.A., Liu,W., Härtel, C., Kleiser, L. 1994 Energy transfer in numerically simulated wall-bounded turbulent flows. *Phys. Fluids* 6: 1583-1599.

Kerr, R.M., Domaradzki, J.A., Barbier 1996 Small-scale properties of nonlinear interactions and subgrid-scale energy transfer in isotropic turbulence. *Phys. Fluids* 8: 197-208.

Khanna, S. and Brasseur, J.G. 1996 Analysis of Monin-Obukhov similarity from large-eddy simulation. manuscript under review.

Kraichnan, R.H. 1976 Eddy viscosity in two and three dimensions. *J. Atmos. Sci.* 38: 2747.

Piomelli, U., Cabot, W.H., Moin, P., Lee, S. 1991 Subgrid-scale backscatter in turbulent and transitional flows. *Phys. Fluids A* 3: 1766-1771.

Wang, L.-P., Chen, S., Brasseur, J.G., Wyngaard, J.G. 1996 Detailed analysis of the Kolmogorov refined turbulence theory through high-resolution simulations. Part 1. Velocity field *J. Fluid Mech.* 309: 113-156.

Yeung, P.K., Brasseur, J.G. 1991 The response of isotropic turbulence to isotropic and anisotropic forcing at the large scales. *Phys. Fluids A* 3: 884-897.

Yeung, P.K., Brasseur, J.G. and Wang, Q. 1995 Dynamics of direct large-small scale couplings in coherently forced turbulence: concurrent physical and Fourier-space views. *J. Fluid Mech.* 283: 43-95.

Zhou, Y. 1993 Interacting scales and energy transfer in isotropic turbulence. *Phys. Fluids A* 5: 2511-2524.

Zhou, Y., Yeung, P.K., Brasseur, J.G. 1996 Scale disparity and spectral transfer in anisotropic numerical turbulence. *Physical Rev. E* 53: 1261-1264.

EVOLUTION OF A PERTURBATION
IN A TURBULENT JET FLOW

R. CAMUSSI*, S. CILIBERTO† & C. BAUDET†
*On leave from Terza Università di Roma,
Dipartimento di Meccanica ed Automatica,
via Segre 60, 00146 Roma, Italy.
†Ecole Normale Superieure, Physique Recherche
46 Alleé d'Italie, 69364 Lyon Cedex 06, France.

Abstract

Experimental results concerning the interaction of a fully developed turbulent flow with a perturbation properly generated in the physical domain, are presented. By means of a conditional averaging technique, it has been possible to extract informations on the response function of the turbulent flow. Satisfactory agreement with the DIA Kraichnan's theory prediction, is also observed.

1. Introduction

The study of the dynamics of turbulent velocity fields can be conducted by the analysis of the response of the velocity field to an appropriate perturbation. This approach permits the knowledge of response functions, interscale dynamics and interactions between different scales which are fundamental aspects for turbulent models, theories and flow control problems.

Among various theoretical works on this subject (e.g. [1] and [2]), we focus attention on the Direct Interaction Approximation (DIA) theory [3], that is based on the determination of the impulse-response tensor and, in particular, on its linear approximation, $g(k, \tau)$, that can be written as:

$$g(k, \tau) = e^{[-c \ k^2 \ \tau^2]} \ . \tag{1}$$

k represents the wavenumber and τ the time. c is a constant parameter proportional to the velocity variance. Few experimental analyses of the evolution of a perturbation in turbulent flows, have been conducted so far and only in the wavenumber domain, [4] and [5], or concerning convective flows in Rayleigh-Bénard configurations [6].

The experimental study of the evolution of a perturbation (in the present case a velocity front) generated in the physical domain and moving through

S. Gavrilakis et al. (eds.), Advances in Turbulence VI, 205-208.

a fully developed turbulent velocity field and the experimental validation of eq. (1), are the main purposes of the present work.

Measurements are conducted with a single probe hot wire anemometer in 4 axial positions x_i downstream of a jet of diameter $D = 120mm$. Specifically the measurements points x_i/D, with $i = 1, ..., 4$, were equal to $12, 16, 19$ and 22 respectively. The jet was mounted inside a low-speed wind tunnel. Two different outflow velocities are considered, 15 and $30m/s$ corresponding, at the measuring locations, to a velocity V_o of the order of $4m/s$ and $2.5m/s$ respectively. The perturbation is generated by a pressure valve commanded by an HP workstation and connected to a small jet, coaxial with the larger one, with diameter d (two sizes were considered: 5 and $10mm$). The small jet is driven by compressed air which is switched on and off by an electrically controlled valve. The opening of the valve was controlled by the data acquisition system. Hereafter we consider the large jet turbulence as the $back - ground$ turbulence.

2. Principal results

The analysis in the wavenumber domain has been conducted by the use of a local Taylor hypothesis to convert time in space. Preliminary checks have shown that, in absence of back-ground turbulence, the perturbation is only convected by the mean flow and no interactions are observed. The expected k^{-2} behavior (the Fourier spectrum of a step function) is in fact always observed even when very far from the ejection orifice. When back-ground turbulence is $active$ the time and space evolutions of the perturbation are detected by means of conditional averages of the velocity signals. The hot wire acquisitions and the perturbation ejection are activated simultaneously since they are commanded by the same trigger pulse generated by the data acquisition system. In Fig. 1a we show an example of the turbulent signals acquired at $x/D = 12$ whereas in Fig. 1b the result of the ensemble averaging is shown. Accounting for the local Taylor hypothesis, it is found that, in all cases, the energy spectra do not follow anymore the k^{-2} power law, indicating that turbulence and velocity front are interacting. By fixing a certain value of the wavevector k it is possible to analyze the evolution of a mode as a function of the distance from the jet. It is found that the spatial evolution of modes amplitudes follows, in terms of space variable x, a square exponential law as $\sim exp[-A(k) \ x^2]$ where A is a quadratic function of k. This behavior is observed in all flow configurations considered. In Fig. 2a we give an example of the evolution of different modes. In this figure also the least square approximation is reported (each curve corresponds to different k). The agreement with the experimental point is satisfactory in all cases confirming the validity of eq. (1). The observed results indicate that the evolution law of the modes amplitude is the effect of the interaction between the back-ground turbulence and the perturbation. In fact, a pure diffusive behavior should lead to an exponential decay law of the form $\sim exp(-xK^2\chi/V)$ where χ is the thermal diffusivity. The question which arises is whether the evolution law, that in terms of time (or space) dependence is correctly represented by eq. (1), can be assumed to be the response function of the turbulent system. This point can be addressed by verifying its predictability properties. Once the mode evolution law is known, it is possible in fact to reconstruct the spectra at each position starting with a k^{-2} spectrum at $x = 0$, using eq. (1) with the proper time delay and coefficients, and comparing the reconstructed spectra with the measured ones. In Fig. 2b, the reconstructed spectrum, at $x/D = 22$ and $V_o = 4m/s$ is compared with the measured one. The agreement between the two curves is quite good and it is confirmed the capability of a correct prediction of the actual turbulent spectra by the knowledge of the response function and of the perturbing spectrum.

As a conclusion, preliminary results obtained from an experimental analysis of the interaction between a step-function perturbation and a turbulent velocity field have been reported. A conditional averaging technique has been used in order the perturbation to be observed in spite of the background turbulence. The response function of the turbulent system has been evaluated by the estimation of each mode evolution law. It has been found that the functional form predicted by the linear approximation of Kraichnan's DIA response function, is in good agreement with present results both in the time and the wavenumbers domain. These results have been checked by analyzing two different mean velocities and two different sizes of the small jet which generates the perturbation. In any case the predictability properties of the linear response function are confirmed.

References

1 Lin C.C., "A critical discussion of similarity concepts in isotropic turbulence", Proc. Symp. on Appl. Math. (Amer. Math. Soc.) 4 (1951).

2 Townsend A.A., *The structure of turbulent shear flows*. Cambridge University Press (1956).

3 Kraichnan R., "The structure of isotropic turbulence at very high Reynolds numbers", J. Fluid Mech. 5, 497-543 (1959).

4 Kellog R.M. and Corrsin S., "Evolution of a spectrally local disturbance in grid-generated turbulence", J. Fluid Mech. 96, 641-669 (1980).

5 Itsweire E.C. and Van Atta C.W., "An experimental study of the response of nearly isotropic turbulence to a spectrally local disturbance". J. Fluid Mech. 145, 423-445 (1984).

6 M. Caponeri and S. Ciliberto, "Thermodynamic aspects of the transition to spatiotemporal chaos " Physica D, 58, 365 (1992).

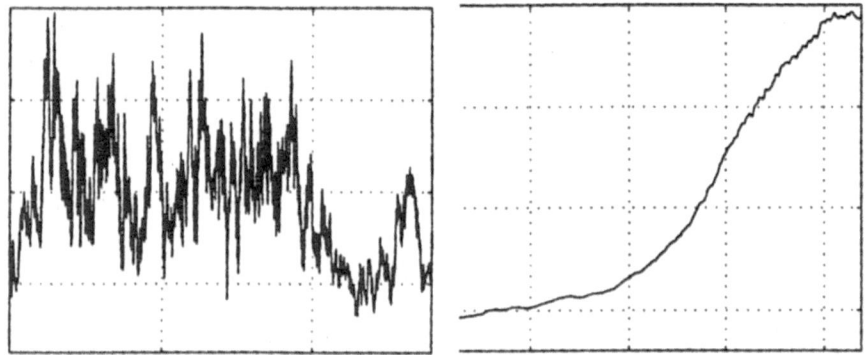

Figure 1. Velocity signal representing the perturbation before (**a**) and after (**b**) the synchronized averages.

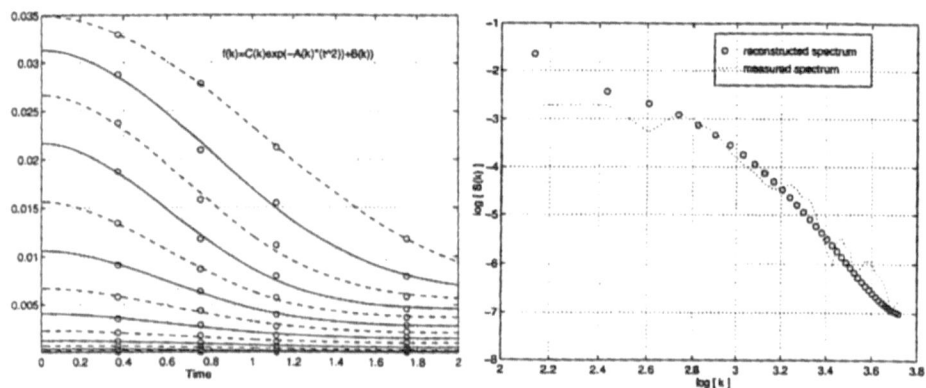

Figure 2. (**a**): evolution of modes of different amplitudes and exponential fit. (**b**): comparison between the reconstructed and measured spectrum at $x/D = 22$.

EXPERIMENTAL ANALYSIS OF INTERMITTENCY IN THE DISSIPATION RANGE OF TURBULENT FLOWS

R. CAMUSSI & G. GUJ

Terza Università di Roma, Dipartimento di Meccanica ed Automatica, via Segre 60, 00146 Roma, Italy.

Abstract

Experimental data obtained in various turbulent flows are analyzed by means of orthogonal wavelet transform in combination with the Extended Self Similarity form of scaling. It is shown that some statistical properties of fully developed turbulence can be extended to low Re_λ flows.

1. Introduction

In the last years many experimental and numerical investigations (e.g. [1]), have evidenced the limits of the Kolmogorov theory [2], which does not take into account the intermittent nature of the turbulent energy dissipation. Furthermore, some numerical (e.g. [3]) and experimental (e.g. [4]) analyses, have evidenced the connection between coherent structures and intermittency in fully developed turbulence. Filamentary structures have been observed but it is still far to be understood which is the role and the actual shape of such structures in real turbulent flows and, in particular, in the dissipation range. In the present paper we analyze intermittency and turbulent structures of velocity signals obtained by single and double probes hot wire anemometry in low Re_λ flows (Re_λ ranging from ~ 3 up to ~ 250). The study is performed by the application of the Extended Self Similarity (ESS) form of scaling, recently introduced in [5], in combination with the wavelets decomposition, in the form given in [6]. The first is adopted in order the intermittency exponents to be accurately measured, whereas the second seemed necessary since the analysis of turbulent structures can be facilitated by the time-frequency signal representation (or space-wavenumbers if Taylor hypothesis is adopted).

2. Main results

The principal indicators that we have used for the analysis of intermittency and turbulent structures are the following:

S. Gavrilakis et al. (eds.), Advances in Turbulence VI, 209-212.

Flatness factor, defined as: $F(r) = \frac{<[w^{(r)}(i)]^4>}{<[w^{(r)}(i)]^2>^2}$ where $w^{(r)}(i)$ represents the wavelet coefficient at the scale (or resolution) r in the position (time or space) i. This function indicates the level of intermittency at different scales and at each position. The average $< . >$ is performed over i that is the discretized version of the position x. In Fig. (1) the Flatness factor is reported for homogeneous and non-homogeneous grid turbulence. At the smallest scales it is observed the increase in magnitude that indicates the departure from Gaussian distribution. It has to be pointed out that for the homogeneous case $Re_\lambda = 12$ whereas in non-homogeneous turbulence $Re_\lambda \sim 36$. This indicates that the effect of intermittency is significant at small scales even for very low Re_λ. Furthermore it is observed a decrease of intermittency for increasing non-homogeneity that is for decreasing x/M. This behavior indicates the effect of the characteristic scales of the turbulence *generator* which influence the scaling behavior.

Local Intermittency Measure (LIM), defined as: $LIM(r, i) = \frac{[w^{(r)}(i)]^2}{<[w^{(r)}(i)]^2>}$. This function enhances non-uniform distributions of energy in space. In Fig. (2) LIMs calculated in homogeneous and non-homogeneous grid turbulence are reported. The energy intermittent nature is well evidenced by the non uniform distribution of the peaks. Therefore LIM can be used for the selection, by choosing a proper threshold level t, of those *events* which yield the strongest energy bursts and which are supposed to be generated by the passage of turbulent structures. If, for $x = x_0$, $|LIM(x_0, \bar{r})| > t$, it can be assumed that a flow singularity has been detected in the position x_0 at the scale $\bar{r}$. By varying the trigger level, one can select singularities of different *strength* that lead to different energy levels. The time signature of such flow singularities can be obtained by the ensemble averaging of the velocity signal centered in the positions x_0. If $j = i_0^{(t)}$ is the set of samples where singularities are detected for a certain trigger level t, the averaging procedure can be written as: $V^{(t)}(i) = < V(j, i) >_j$ where i extends over an appropriate interval (to be chosen in correspondence of the width of the singularity). In Fig. (3), results obtained in homogeneous grid turbulence at $Re_\lambda \sim 10$ is presented for different trigger levels. The passage of a turbulent structure has been clearly identified in all cases and it looks like a vortex tube passing close to the measurement position. Similar results are obtained at higher Re_λ grid flows and confirm that also coherent structures conserve in low turbulence the same properties of high turbulent flows. In Fig. (4) the structures identified for the longitudinal and transverse velocities in jet turbulence are reported. In these cases the averaged velocity signals, always in a qualitative point of view, seems induced by the passage of a vortex ring around the probe. The longitudinal component shows the expected behavior since the velocity maximum is in correspondence of the passage of the center of the vortex through the probe position whereas, as expected, the transverse component is always close to zero. This indicates that turbulent structures can be strongly affected by the turbulence *generator*.

ESS applied to wavelet coefficients: it is possible to verify whether the ESS applies for the wavelet coefficents and to check the validity of ESS when only the flow singularities are selected. In Fig. (5), it is given an example of the wavelet coefficients' scalings at $Re_\lambda \sim 10$. The scaling exponents are in agreement with previous results at higher Re_λ (see e.g. [1]). Since intermittency is related to coherent structures, that in the present case are interpreted as singularities in the LIM distribution, ESS should apply also when considering only those *events* if they are responsible of intermittency. In Fig. (6) the fourth versus third order moment is shown for the same case of Fig. (5) at different trigger levels. The collapse of the points is satisfactory in all cases and the linear fit gives the same exponent (1.28 for $p = 4$) of fully developed turbulence. Analogous results are obtained for different p and in the other flow conditions. In particular, at $Re_\lambda = 3$, the expected Kolmogorov scaling is observed as an effect of the scaling *regularization*.

As a conclusions, these results support the robustness of ESS as a correct way for observing scaling laws and confirms that only singularities are responsible of the observed intermittent behavior. Furthermore, various indicators considered have evidenced that intermittency is a property of flows with $Re_\lambda > 10$ and that turbulent structures are affected by the turbulence generator even when apparently in homogeneous and isotropic conditions.

References

1 Anselmet F., Gagne Y., Hopfinger E.J. and Antonia R.A., " High-order velocity structure functions in turbulent shear flows", J. Fluid Mech. **140**, 63-89, 1984.

2 Kolmogorov A., " The local structure of turbulence in incompressible viscous fluid for very large Reynolds numbers", C. R. Akad. Sci. SSSR **30**, 301-305, 1941.

3 Jimenez J., Wray A.A., Saffman P.G. and Rogallo R.S., "The structure of intense vorticity in isotropic turbulence", J. Fluid Mech., **255**, 65-90, 1993.

4 Cadot O., Douady S. and Couder Y., "Characterization of the low pressure filaments in a 3D turbulent shear flow" Phys. of Fl. **7** (3), 630, 1995.

5 Benzi R., Ciliberto S., Tripiccione R., Baudet C., Massaioli F. and Succi S., " Extended self-similarity in turbulent flows", submitted to Phys. Rev. E, **48**, 29, 1993.

6 Meneveau C, "Analysis of turbulence in the orthonormal wavelet representation", J. Fluid Mech. **232**, 469-520, 1991 and also "Appendices A,B and C to CTR manuscript 120" CTR Summer Prog., 1991.

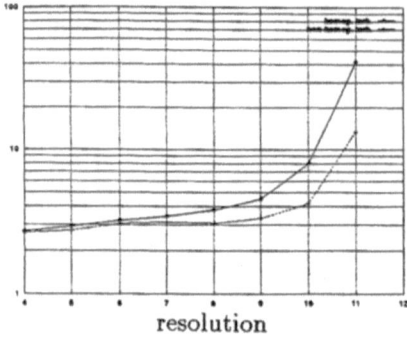

resolution

Figure 1. Flatness factor calculated for homogeneous and non homogeneous turbulence.

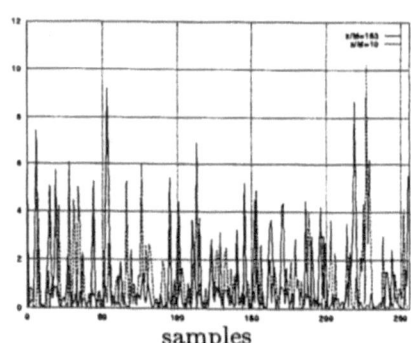

samples

Figure 2. LIM calculated for homogeneous and non homogeneous grid turbulence.

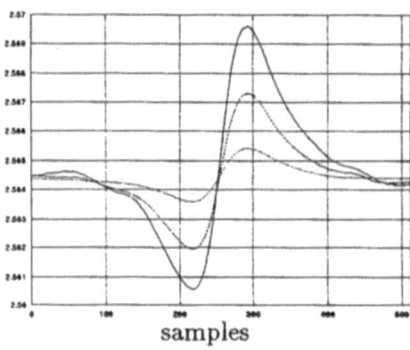

samples

Figure 3. Turbulent structure in homogeneous turbulence at $Re_\lambda = 12$.

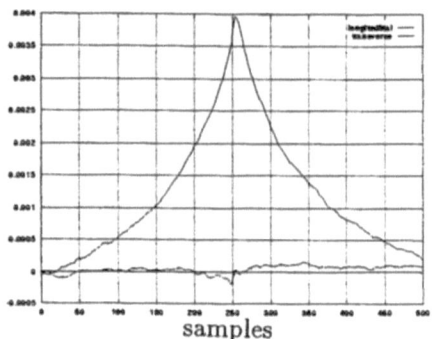

samples

Figure 4. Longitudinal and transverse velocity signature of turbulent structures in jet turbulence at $Re_\lambda = 250$.

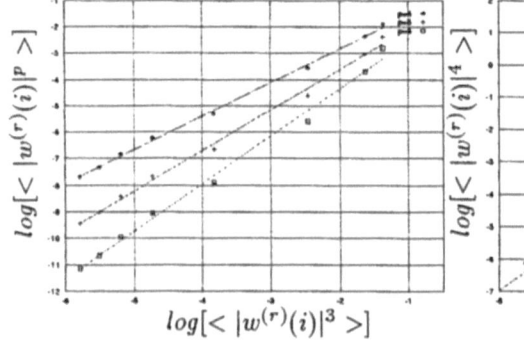

$log[< |w^{(r)}(i)|^3 >]$

Figure 5. Scaling ranges obtained by the ESS in homogeneous grid turbulence ($Re_\lambda = 20$).

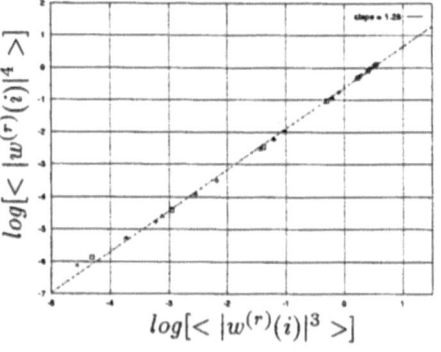

$log[< |w^{(r)}(i)|^3 >]$

Figure 6. Same as before but for $p = 4$ and different trigger levels.

HIGH RAYLEIGH NUMBER CONVECTION WITH GASEOUS HELIUM AT LOW TEMPERATURE

X. CHAVANNE, F. CHILLÀ*, B. CHABAUD, B. CASTAING,
J. CHAUSSY, B. HÉBRAL
CRTBT, laboratoire associé à l'UJF, CNRS, BP 166,
38042 Grenoble-Cedex 9, France
*Present address : Ecole Normale Supérieure de Lyon,
46 Allée d'Italie, 69364 Lyon-Cedex 7, France

We study Rayleigh-Benard convection in a cell which consists of a cylinder of 20 cm height and 10 cm diameter filled with gaseous helium between 4 and 5 K. The new feature compared to previous studies comes from a very accurate thermocouple measurement of the temperature difference across the cell ΔT (resolution of 20 μK). The Rayleigh number Ra ranges from the conductive regime to 5×10^{12} by changing the gas density from 3 g/m^3 to 30 kg/m^3 and by scanning ΔT from 1 mK to 800 mK for each density. On the figure the Nusselt number Nu versus Ra is presented. The high sensitivity measurement of low ΔT allows to evidence and measure the adiabatic gradient influence on the convective threshold (in our geometry $Ra = 30000 \pm 4000$ without this influence); the adiabatic gradient slows down the convection for points at low ΔT as seen on the figure.

The study is focused on the turbulent regime and particularly the hard turbulent one ($Ra \leq 10^7$). Up to $Ra \approx 3 \times 10^{10}$ the Nu variation scales as a Ra power law with a 0.295 ± 0.005 exponent which is close to the 2/7 law previously reported. Above this Ra value the local "exponent" $\delta \ln Nu / \delta \ln Ra$ rises continuously up to 0.4 at $Ra \approx 5 \times 10^{12}$. We do not notice any plateau at the 1/3 value often proposed. Rather, we suggest that this growth agrees with the transition to an ultimate regime where Nu should scale as $Ra^{0.5}$, as proposed by Kraichnan. The corresponding evolution of temperature fluctuations in the cell will be discussed.

S. Gavrilakis et al. (eds.), Advances in Turbulence VI, 213-214.
© 1996 Kluwer Academic Publishers.

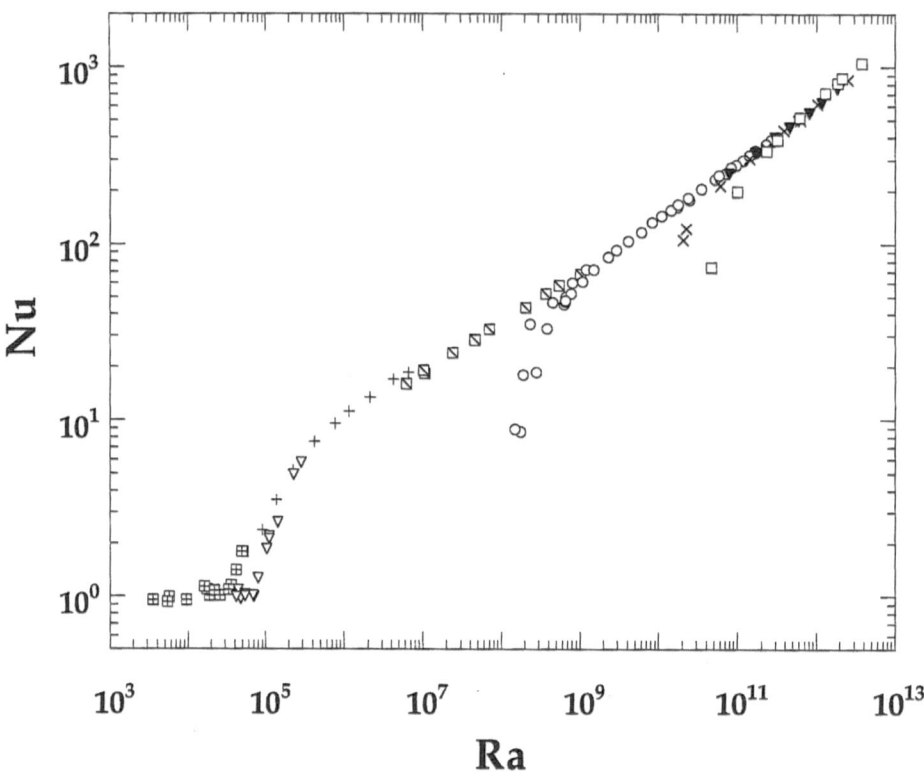

Nu vs Ra for different gas densities:

⊞	3.44 g/m³	▽	88 g/m³	○	5.82 kg/m³	×	18.9 kg/m³
+	30.9 g/m³	◧	384 g/m³	▼	14.5 kg/m³	□	29.1 kg/m³

INFLUENCE OF A LARGE SCALE VORTEX ON TURBULENCE SMALL SCALE PROPERTIES

F. CHILLÀ, J.-F. PINTON
École Normale Supérieure de Lyon, France

AND

R. LABBÉ
Universidad de Santiago de Chile

Abstract. We investigate the flow between coaxial co-rotating disks in the situation where a strong axial vortex is present over a turbulent background. It seems that the cascade process is preserved although modified by the large scale coherent structure [1].

1. Introduction

One of the long standing goals of the study of *homogeneous turbulence* has been to derive tools that can be applied to solve the problems of *inhomogeneous turbulent flows* which have a wider practical importance. In the analysis of the flow field behavior, it is customary to represent the velocity of a high Reynolds number flows as the sum of two contributions: $\vec{v} = \vec{U} + \vec{u}$, where $\vec{U}$ is the mean flow field and $\vec{u}$ are the turbulent fluctuations. In inhomogeneous flows, $\vec{U}$ is not constant in space and time and the question arises about the interaction and equilibrium between the large scale flow and the turbulent fluctuations. We show that the tools recently introduced for the analysis of the turbulent energy cascade in homogeneous isotropic flows apply in this case and indicate how the presence of the coherent vortex influences the turbulent background flow.

2. Flow and Experimental set-up

The flow is produced in the gap between two coaxial corotating disks driven at regulated constant speeds. Air is the working fluid, in a free geometry in

S. Gavrilakis et al. (eds.), Advances in Turbulence VI, 215-218.

order to eliminate side-wall effects. Local velocity measurements are performed using hot-wire anemometry. The detailed experimental set-up is described in [2] where it is also shown that the structure of the flow is made up of a large scale strong vortex superimposed to a turbulent background. Figure 1(a) shows the coherent vortex as reconstructed from 3D velocity measurements (the core size is about 2cm). The vortex, being located at a

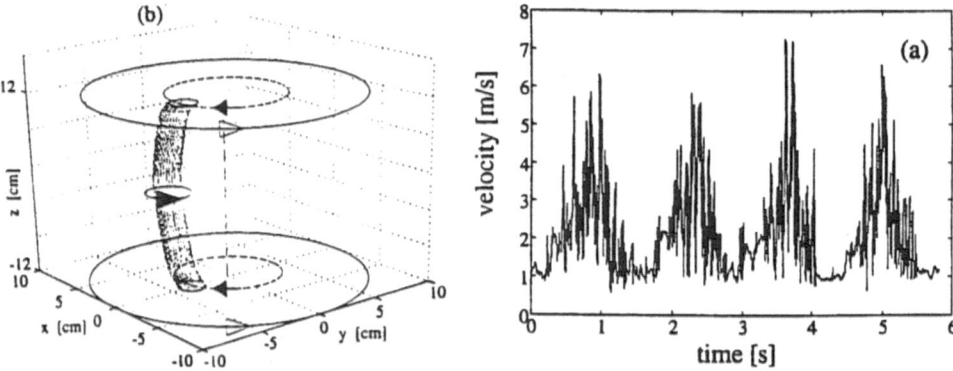

Figure 1. (a): 3D plot of the velocity modulus. (b): time recording of the velocity for a fixed location of the probe

distance from the axis, has a slow precession motion and sweeps almost periodically the hot-wire probe. Indeed, in the instantaneous velocity recording of figure 1(b), one observes the periodic sweeping of the vortex. The corresponding frequency spectrum shows the precession frequency as well as a clear Kolmogorov 5/3 scaling region. The associated period is much longer than the all turbulent time scales and it is thus possible to perform a local Reynolds-like decomposition first to extract the time-periodic mean flow due to the precession of the axial vortex, and second to characterize the statistical properties of the turbulent fluctuations at different phases of the mean-flow cycle. We have thus divided one period of the vortex motion into 50 intervals to condition the turbulent fluctuations statistical properties to the distance d to the vortex core.

3. Results

We have analyzed the characteristics of the velocity fluctuations, conditioned to the distance d. Figure 2(a) shows a plot of the third order structure function $- <|v(x+r) - v(x)|^3 >_x$; an inertial range plateau may be observed at all d. We have also calculated the local dissipation rate and the local turbulent Reynolds number R_t based on the Taylor microscale – shown in figure 2(b). We see that traditional turbulence analysis may be applied. when conditioned to the proximity of the large scale structure.

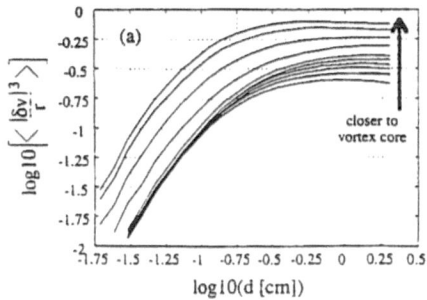
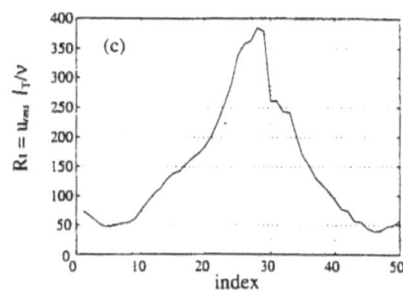

Figure 2. (a): Third order structure function at various d. (b): Evolution of the turbulent Reynolds number $R_t = u_{rms}\ell_T/\nu$ (ℓ_T is Taylor's microscale).

We thus proceed to the detailed analysis of the turbulent cascade using the framework recently proposed by B. Castaing for homogeneous isotropic flows. He and co-workers [3] have shown that the PDF P_r of the velocity differences δu_r at scale r may be reconstructed as a superposition of functions having the shape of the integral scale PDF and a range of width σ. Each P_r is characterized by $\lambda_r^2 = <(\Delta\ln\sigma)^2>$. It has been shown recently [5] that λ_r^2 may be easily computed from the kurtosis K of $P_r(\delta u)$: $\lambda_r^2 = D\ln\frac{K}{K_L}$, where K_L is the kurtosis of the large scale distribution and D is a proportionality constant whose expression depends on the intermittency model ($D = 3$ in the lognormal model). Figure 3(a) we show the $\lambda_r^2(r)$ curves

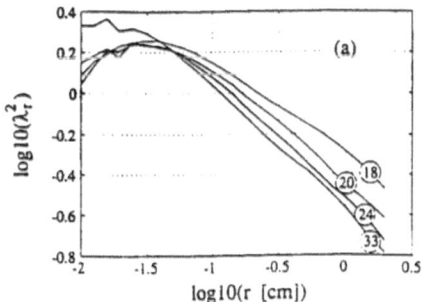
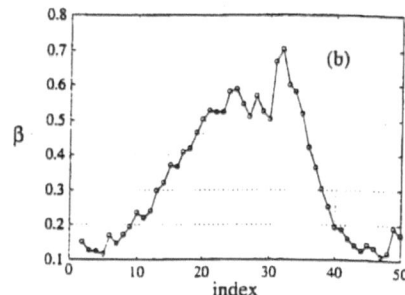

Figure 3. (a): λ_r curves for $d = 10.3; 9.2; 4.4; 4$cm. (b):variation of the magnitude of the slope β at different location from the vortex core

computed at distances $d = 4.4, 9.2$ and 10.3cm before the vortex and 4cm after. An almost linear region in $\ln(\lambda_r^2)$ *vs* $\ln(r)$ coordinates is observed at each distance to the vortex, indicating a power law behaviour ($\lambda_r^2 \propto r^{-\beta}$) as in homogeneous turbulent flows [4] but with a varying slope. As may be expected, the intermittency at small scales is little affected by the vortex (as much as the variations of η, the Kolmogorov dissipation length), whereas next to it the intermittency at large scale is much reduced due to the 'organizing' effect of the coherent vortex structure. Accordingly, β increases as one gets closer to the vortex core – see figure 3(b). This behavior is different from what is observed in homogeneous turbulent flows, where β decreases with increasing Reynolds numbers [6].

These findings are not an effect of the above model of the energey cascade. Indeed, we have computed the structure functions exponents ζ_p using the ESS ansatz [7], conditioned to the distance to the vortex – see figure 4(a). Again, the overall intermittency is reduced as one gets closer

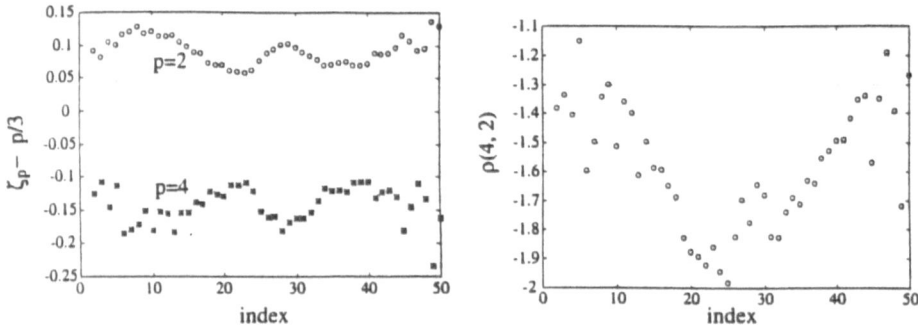

Figure 4. (a): Structure function exponents. (b): generalized exponent.

to the vortex (the exponents get closer to their K41 value $\zeta_p = p/3$). In addition the average (not conditioned) generalized self-similarity exponent $\rho(4, 2) = \frac{\zeta_4 - 4\zeta_3/3}{\zeta_2 - 2\zeta_3/3}$ has the "universal" (-1.58) value proposed recently by R. Benzi *et al.* [8]. However, when conditioned to the distance to the vortex the value of $\rho(4, 2)$ is not constant showing again that the coherent vortex proximity modifies the local energy cascade – see figure 4(b).

Acknowledgment
We are indebted to Bernard Castaing for enlightening comments.

References
1. Chillà F., Pinton J.-F., Labbé R., submitted to *Europhys. Lett.*, january 1996.
2. Labbé R., Pinton J.-F., Fauve S., *Phys. Fluids*, 8(4), 914-922 (1996).
3. Castaing B., Gagne Y. and Hopfinger E.J., *Physica D*, **46**, 177-200, (1990).
4. Castaing B., Gagne Y., Marchand M., *Physica D*, **68**, 387-400, (1993).
5. Chillà F., Peinke J., Castaing B., *J. Phys. II France*, **6**, 455-460, (1996).
6. Chabaud B. *et al.*, *Phys. Rev. Lett.*, **73**, (1994).
7. Benzi R. *et al.*, *Europhys. Lett.*, 24(4), 275-279, (1993).
8. Benzi R. *et al.*, *Europhys. Lett.*, **32**, 709-713, (1995).

INFLUENCE OF PRANDTL NUMBER ON STRONGLY TURBULENT RAYLEIGH-BÉNARD CONVECTION

Mercury convection

S. CIONI, S. CILIBERTO AND J. SOMMERIA

Laboratoire de Physique
Ecole Normale Supérieure de Lyon (URA 1325 CNRS)
46, Allée d'Italie 69364 Lyon Cedex 07 France

Experimental results of Rayleigh-Bénard convection in the strongly turbulent regime are presented. The liquids tested are water ($Pr \sim 7$) and mercury ($Pr = 0.025$), covering a range of Rayleigh numbers $4.6 \times 10^6 < Ra < 6.3 \times 10^9$. The convective chamber consists of a cylindrical cell of aspect ratio $\Gamma = 1$ ($\Gamma = \frac{diameter}{depth}$). The heat flux, the temperature fluctuation spectra [1] and the structure of the mean flow have been measured [2]. Here, we discuss the heat flux results as well as the properties of the structure of the large scale flow inside the cell such as the frequency recirculation ω_p (see Fig.2) as function of Ryleigh.

S. Gavrilakis et al. (eds.), Advances in Turbulence VI, 219-222.
© *1996 Kluwer Academic Publishers.*

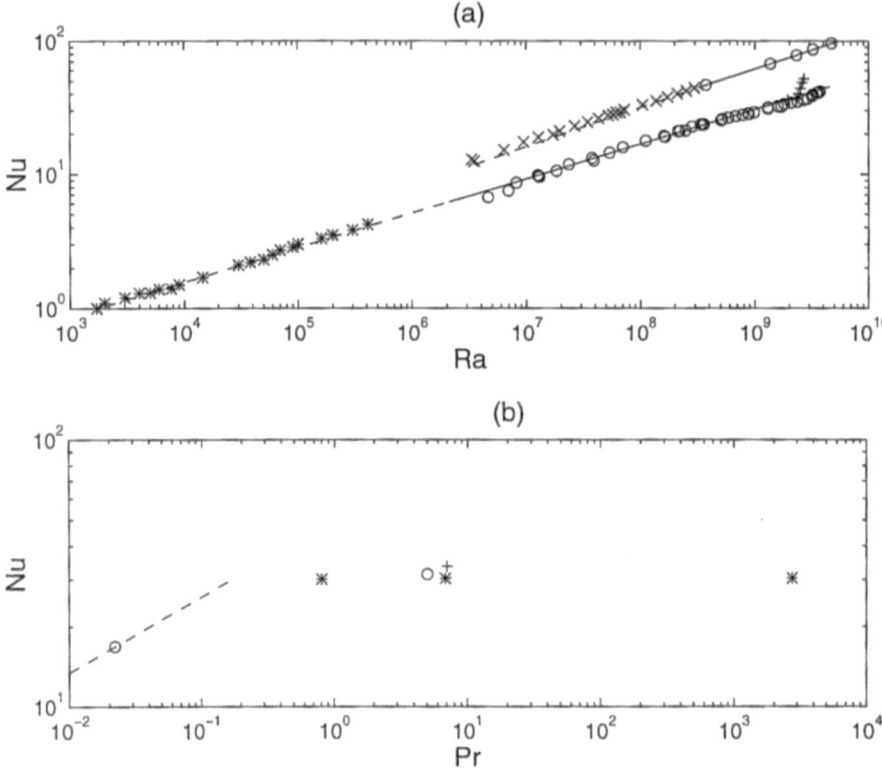

Figure 1. (a) Nusselt *versus* Rayleigh. *Mercury:* previous experiments: (*) Rossby (1969), (o) Present Experiment; *Water:* (x) Chillá *et al.*, (o) Present Experiment. The two straight lines are best fits by power laws (see below). Notice the excellent agreement with previous experiments. (b) Experimental data for the Nusselt number at $Ra = 1 \times 10^8$ *versus* Prandtl. (*) data compiled by Goldstein *et al.* [3], (o) data from this work, (+) data from Chillá *et al.* [4]

The results of this experiment can be summerize as follow:

Heat transfer measurements:

- $Nu_{H_g} = 0.14 Ra^{0.26}$ $4.6 \times 10^6 < Ra < 4.5 \times 10^8$
  $Nu_{H_g} = 0.44 Ra^{0.20}$ $4.5 \times 10^8 < Ra < 2.1 \times 10^9$
- For $Ra > 2 \times 10^9$ a transition to a new turbulent regime has been observed with considerable increase of the Nusselt number (see Fig.1a).
- $Nu_{H_2o} = 0.145 Ra^{0.292}$ $3.7 \times 10^8 < Ra < 7 \times 10^9$
- A decrease of Nusselt with decreasing Prandtl number is observed at given Ra (see Fig.1b) in contrast with some recent models [5],[6]. A simple model based on *mixing zone model* proposed by Castaing et al. proposed; giving the following scaling:

$$Nu \sim (RaPr)^{2/7}$$

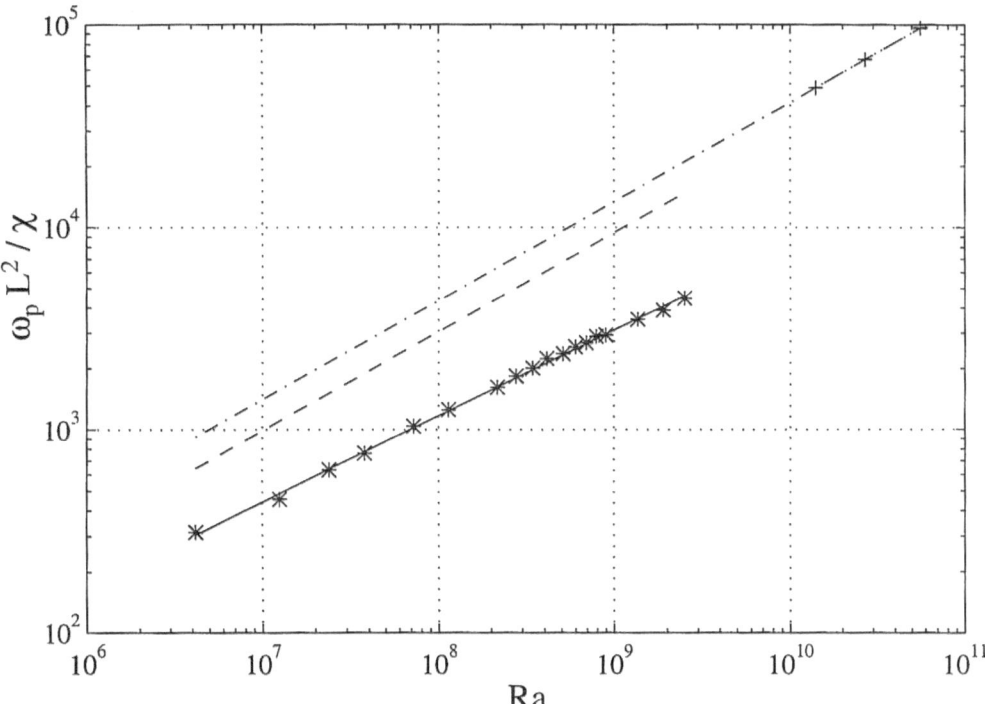

Figure 2. Log-log plot of the characteristic frequency ω_p, normalized by κ/L^2 *versus Ra* showing a clear scaling to the 3/7 power law continuos line. The dashed lines indicates the scaling of the normalized frequency for the Chicago experiment (+) [7] and (*) for Ciliberto *et* al. experiment [8].

References

1. Cioni, S., Ciliberto, S. & Sommeria, J. 1995 Temperature structure functions in turbulent convection at low Prandtl number. *Europhys. Lett.* **32**, 413-418.

2. Cioni, S., Ciliberto, S. & Sommeria, J. 1996 Experimental study of High-Rayleigh-Bénard convection in mercury and water. *Dynamics of Atmospheres and Oceans, Special issue.* **24**, 117-127.

3. Goldstein, J.R., Chiang, H.D. & See, D.L. 1990 High-Rayleigh-number convection in a horizontal enclosure. *J. Fluid. Mech.* **123**, 111-126.

4. Chillá, F., Ciliberto, S., Innocenti, C. & Pampaloni, E. 1993 Boundary layer and scaling properties in turbulent thermal convection. *Il Nuovo Cimento* **15D**, 1229-1249.

5. Castaing, B., Gunaratne, G., Heslot, F., Kadanoff, L., Libchaber, A., Thomae, S., Wu, X., Zaleszi, S. & Zanetti G. 1989 Scaling of hard thermal turbulence in Rayleigh-Bénard convection. *J. Fluid Mech.* **204**, 1-29.

6. Shraiman, B.I. & Siggia, E.D. 1990 Heat transport in high-Rayleigh-number convection. *Phys. Rev. A* **42**, 3650.

7. Sano, M., Wu, X.Z. & Libchaber, A. 1989 Turbulence in helium-gas free convection. *Phys. Rev. A* **40**, 6421.

8. Ciliberto, S., Cioni, S. & Laroche, C. 1996 On the properties of large scale flow in turbulent thermal convection. Submitted to *Phys. Rev. Lett.*

A SIMILARITY HYPOTHESIS FOR THE TWO-POINT VELOC-ITY CORRELATIONS IN A TEMPORALLY EVOLVING WAKE

D. EWING[1], W. K. GEORGE[2], R. D. MOSER [3], & M. M. ROGERS[4]

1 Introduction

It has long been recognized (*e.g.* Towsend, 1956) that the governing equations for the single-point moments in several free-shear flows admit similarity solutions if the flows evolve from virtual sources. These similarity solutions, however, do not provide information about many important features of the flow, such as how the turbulent kinetic energy is distributed amongst the different scales of motion. In order to gain this information it is necessary to consider more complex statistical measures of the flow, such as the two-point correlations.

Most of the previous analyses of the two-point equations have considered decaying isotropic turbulence (*e.g.*, Batchelor, 1948 or George, 1992) or homogeneous shear turbulence (George and Gibson, 1992). Ewing (1995) later demonstrated that the equations for the two-point velocity correlation in the far fields of the axisymmetric and planar jets admit similarity solution. However, no attempt was made to test the hypotheses using data from these non-homogeneous flows. Here, data from two Direct Numerical Simulations of the temporally evolving wake (Moser and Rogers,1994) are used to test the similar hypothesis for the two-point correlations in this flow.

2 Analysis of the Governing Equations

The temporally evolving wake is a non-homogeneous shear flow with a momentum deficit that spreads in non-homogeneous x_2-direction as the flow evolves. The other two directions, including the mean flow direction x_1, are statistically homogeneous. Thus, for this flow, the governing equation for the two-point velocity correlation reduces to

$$\frac{\partial \overline{u_i u_j'}}{\partial t} + (U_1 - U_1') \frac{\partial \overline{u_i u_j'}}{\partial \varsigma} =$$

---

[1] *Department of Mechanical Engineering, Queen's University, Kingston, Ont., Canada. K7L 3N6*

[2] *Department of Mechanical and Aerospace Engineering, State University of New York at Buffalo, Amherst, NY, 14260*

[3] *Department of Theoretical and Applied Mechanics, University of Illinois at Urbana-Champaign, Urbana, IL 61801*

[4] NASA-Ames, Moffett Field, CA.

S. Gavrilakis et al. (eds.), Advances in Turbulence VI, 223-226.
© *1996 Kluwer Academic Publishers.*

$$-\frac{1}{\rho}\left[\frac{\partial}{\partial\varsigma}\left(\overline{pu'_j}\delta_{i1} - \overline{p'u_i}\delta_{j1}\right) + \frac{\partial\overline{pu'_j}}{\partial x_2}\delta_{i2} + \frac{\partial\overline{p'u_i}}{\partial x'_2}\delta_{j2} + \frac{\partial}{\partial\gamma}\left(\overline{pu'_j}\delta_{i3} - \overline{p'u_i}\delta_{j3}\right)\right]$$

$$-\frac{\partial}{\partial\varsigma}\left(\overline{u_1 u_i u'_j} - \overline{u'_1 u_i u'_j}\right) - \frac{\partial\overline{u_2 u_i u'_j}}{\partial x_2} - \frac{\partial\overline{u'_2 u_i u'_j}}{\partial x'_2} - \frac{\partial}{\partial\gamma}\left(\overline{u_3 u_i u'_j} - \overline{u'_3 u_i u'_j}\right)$$

$$-\overline{u'_j u_2}\frac{\partial U_1}{\partial x_2}\delta_{i1} - \overline{u_i u'_2}\frac{\partial U'_1}{\partial x'_2}\delta_{j1} + \nu\left(2\frac{\partial^2}{\partial\varsigma^2} + \frac{\partial^2}{\partial x_2^2} + \frac{\partial^2}{\partial x_2'^2} + 2\frac{\partial^2}{\partial\gamma^2}\right)\overline{u_i u'_j}, \quad (1)$$

where $\varsigma = x_1 - x'_1$, $\gamma = x_3 - x'_3$, U_1 is the mean velocity in the x_1-direction, u_i are the fluctuating components in the x_i-directions, and the primed and unprimed variables are evaluated at two different arbitrary points in the flow.

Following the approach outlined by George (1989), it is hypothesized that these equations have solutions of the form

$$U_{1\infty} - U_1(x_2, t) = U_s(t)f(\eta)$$

$$\overline{u_i u'_j} = Q^{i,j}(t)q_{i,j}(\xi, \eta, \eta', \varsigma, *),$$

$$\overline{pu'_j} = P_1^{\cdot j}(t)p_{\cdot j}^1, \qquad \overline{p'u_i} = P_2^{i\cdot}(t)p_{i\cdot}^2,$$

$$\overline{u_k u_i u'_j} = T_1^{ki,j}(t)tt_{ki,j}^1, \qquad \overline{u_i u'_k u'_j} = T_2^{i,kj}(t)tt_{i,kj}^2, \quad (2)$$

where the solutions are functions of similarity variables given by

$$\xi = \frac{\varsigma}{\delta_1(t)} = \frac{x_1 - x'_1}{\delta_1(t)}, \quad \eta = \frac{x_2}{\delta(t)}, \quad \eta' = \frac{x'_2}{\delta(t)}, \quad \zeta = \frac{\gamma}{\delta_3(t)} = \frac{x_3 - x'_3}{\delta_3(t)}.$$

In order to ensure that these solutions are consistent with the similarity solution for the single-point moments it follows that $\delta \propto (t - t_o)^{1/2}$ (Moser et al., 1996).

Ewing (1995) demonstrated that these solutions are consistent with the governing equations if $\delta_1 \propto \delta \propto \delta_3$ and the ratio of three characteristic time scales in the flow are constant; i.e.,

$$\beta = \frac{1}{U_s}\frac{d\delta}{dt} \propto const \quad (3)$$

and

$$Re_\delta = \frac{U_s\delta}{\nu} \propto const \quad (4)$$

(all of which are satisfied for this flow Ewing, 1995). The choice for $Q^{i,j}$ must be consistent with the solutions for the single-point Reynolds stresses (Moser et al., 1996) so it follows that $Q^{i,j} \propto U_s^2$. The scales for the other terms are given by $P_1^{i\cdot} \propto U_s Q^{i,k}$, $P_2^{\cdot j} \propto U_s Q^{k,j}$, and $T_1^{ki,j} \propto T_2^{i,kj} \propto U_s Q^{i,j}$. The resulting governing equations for the similarity solutions include both β and Re_δ so these solutions may depend on both these constants.

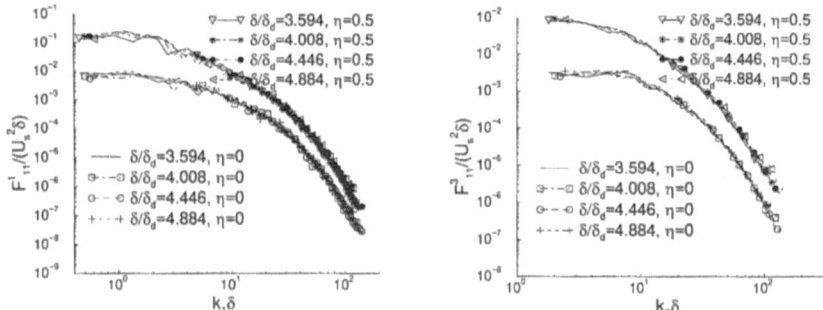

Figure 1: One-dimensional spectra from the unforced wake simulation.

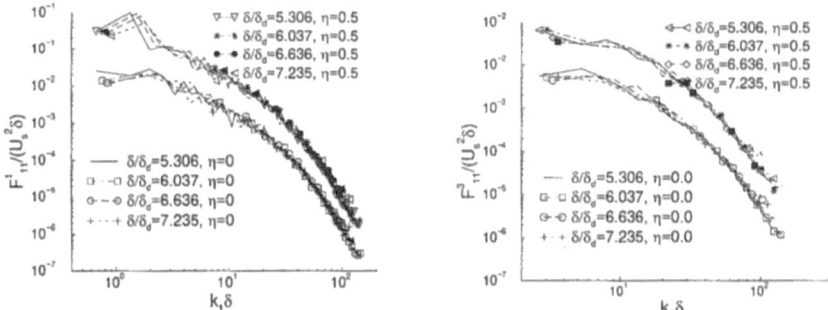

Figure 2: One-dimensional spectra from the forced wake simulation.

3 Comparison with Experimental Data

The predictions of the similarity analysis were tested using data from two Direct Numerical Simulations of the temporally evolving wake (Moser and Rogers, 1994). In the first simulation the initial conditions were generated using two realizations from a turbulent boundary layer simulation to yield a wake with $Re_{\delta_d} = [\int_{-\infty}^{\infty}(U_1 - U_{1\infty})dx_2]/\nu = 2000$. In the second, forced wake, simulation the u_1 and u_2 components in two dimensional modes (with $k_3 = 0$) were amplified by a factor of five. Moser $et\ al.$ (1996) have previously demonstrated that the data from these simulations agree with the predictions of the similarity hypothesis for the single-point moments.

These simulations were computed with periodic boundaries in the two homogeneous directions so the length scales in these directions can not grow continuously as predicted in the similarity hypothesis. However, it is generally argued that the motions 'much' smaller than the box should not be significantly effected by these finite boundaries. Thus, the comparison of the predictions from the hypothesis and the data test both this argument and the theory.

The value of the one-dimensional spectra F_{11}^1 and F_{11}^3 at the centerline and the half-deficit point, $\eta = 0.5$, in the unforced wake, (scaled using the appropriate similarity variables Ewing, 1995), are illustrated in figure 1, while these

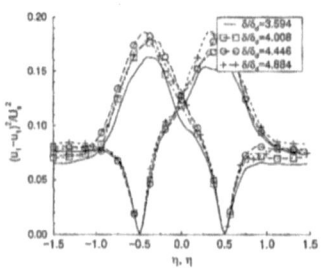 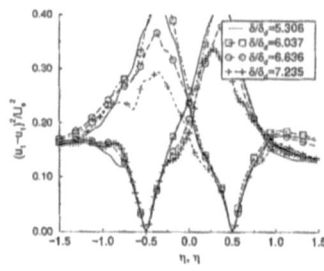

Figure 3: Second-order structure function from: left - the unforced wake simulation and right - the forced wake simulation

spectra at the same positions in the forced wake are illustrated in figure 2. All of these one-dimensional spectra collapse for all but the lowest wavenumbers. In addition, the second-order structure function for the u_1 components in the x_2 direction; $(i.e., \overline{[u_1 - u_1']^2}|_{\xi=0,\zeta=0})$ centered at the half-deficit points, $\eta = \pm 0.5$, from both simulations are illustrated in figure 3. The data from both simulations collapse for small and intermediate separation distances.

4 Discussion and Summary

The comparison of the predictions from the theory and the data indicate that the similarity solutions provide a good description of all but the largest scales of motion in the turbulent field. The poor collapse for the measures of these motions are due in part to the finite boundaries in the simulation and in part due to a lack of statistical convergence. In general, the predictions of the similarity hypothesis can be used to examine the effect that finite boundary conditions have on the evolution of different scales of motion if sufficient statistical convergence can be achieved.

5 References

Ewing, D. (1995) On multi-point similarity solutions in turbulent free-shear flows, *Ph. D. Dissertation*, SUNY at Buffalo, Amherst, New York.

George, W. K. (1992) Decay of homogeneous isotropic turbulence, *Phys. Fluids A*, **4**, 1492-1508.

George, W. K. (1989) The self-similarity of turbulent flows and its relation to initial conditions and coherent structures. In *Advances in Turbulence*. edit. W. K. George and R. E. Arndt, Hemisphere Publishing, 39-73.

George, W. K. & Gibson, M. M. (1992) The self-preservation of homogeneous turbulence, *Expt. in Fluids*, **13**, 229.

Moser, R. D. & Rogers M. M. (1994) Direct simulation of a self-similar plane wake, *NASA Tech. Memo.* TM 108815.

Moser, R. D., Rogers, M. M., & Ewing, D. (1996) *J. Fluid Mech.*, in preparation.

Townsend, A. A. (1956) *The Structure of Turbulent Shear Flow*, Cambridge University Press.

DISSIPATION PROPERTIES OF NEAR-SINGULARITIES IN SMALL-SCALE TURBULENCE

P. FLOHR AND J.C. VASSILICOS

DAMTP, Silver Street, Cambridge CB3 9EW, U.K.

1. Introduction

The inertial range of scales in the turbulent velocity and scalar power spectra is well-described by $k^{-5/3}$ laws. Hunt & Vassilicos (1991) have shown that a self-similar power spectrum, such as $E(k) \propto k^{-2p}$ with non-integral scaling exponent p, implies the existence in the turbulence of near-singularities[1] that cannot be mere isolated discontinuities in the velocities or their derivatives.

In the present study we investigate, both theoretically and numerically, the effects of singular flow structures on the small-scale dissipation properties of a scalar field $\theta(\mathbf{x}, t)$. As an example of a near-singularity that is caused by a well-defined fluid dynamical process we consider the spiral structure adopted by a scalar field under the winding action of a steady vortex. This spiral geometry interacts with the molecular diffusion of the scalar in such a way as to dramatically increase the scalar dissipation rate. Spiral fields that are more space-filling are more super-diffusive. The degree to which the spiral geometry is space-filling can be conveniently and directly measured by the Kolmogorov capacity D of the spiral ($0 \le D \le 1$) and the decay of the scalar variance (or "average energy") $\overline{\theta^2}(t) \sim \int |\theta(\mathbf{x}, t)|^2 \, d\mathbf{x}$ is quantitatively determined by this spiral geometry as follows:

$$\overline{\theta^2}(0) - \overline{\theta^2}(t) \sim t^{3(1-D)} \tag{1}$$

in a range of times t that is defined in the sequel.

---

[1] "Near"-singularities are singular only up to a small scale η where molecular diffusion smears out the singular structure.

S. Gavrilakis et al. (eds.), Advances in Turbulence VI, 227-230.
© *1996 Kluwer Academic Publishers.*

2. Scalar diffusion in a steady vortex

We study the interaction between advection and diffusion of a passive scalar θ in a vortex. In an oversimplified picture of a turbulent eddy we consider a steady velocity field with azimuthal component $u(r) = r\Omega(r)$ and vanishing radial and axial components. Such velocity fields may arise in the vicinity of strong vortices in 2-D turbulence, and in locally 2-D flows around vortex-tubes in 3-D turbulence. We therefore seek a solution of

$$\partial_t\theta + \Omega(r)\partial_\phi\theta = \kappa\nabla^2\theta \tag{2}$$

in the azimuthal plane (r, ϕ), with angular velocity $\Omega(r) = \Omega_0(r/R_0)^{-s}$, and molecular diffusivity κ. The length scale R_0 is related to the initial condition of θ as schematically specified in Figure 1a. Note that we consider a family of velocity fields that are incompressible solutions of the Euler equation.

Searching for a solution of the form $\theta(r, \phi, t) = \sum_n f_n(r, t) \exp[in(\phi - \Omega(r)t)]$, it follows from (2) that for large enough times, the amplitudes f_n decay according to (see for example Lundgren 1982)

$$f_n(r, t) = f_n(r, 0) \exp[-(ns(R_0/r)^{s+1})^2(t\Omega_0)^3/(3Pe)], \quad n \neq 0, \tag{3}$$

for arbitrary initial conditions $f_n(r, 0)$, and $Pe = \Omega_0 R_0^2/\kappa$ is a Peclet number. The scalar variance $\overline{\theta^2}(t) \equiv \frac{1}{R_0^2}\int|\theta(r, \phi, t)|^2 r dr d\phi$ can be calculated from the amplitudes $f_n(r, t)$ by

$$\overline{\theta^2}(t) = \frac{1}{R_0^2}\sum_n\int_{r_0}^{R_0}|f_n(r, t)|^2 r\, dr, \tag{4}$$

and the initial condition in Figure 1a is given exactly by

$$f_n(r, 0) = i/(2\pi n)\, (\exp(-in\pi) - 1)H(R_0 - r)H(r - r_0), \quad n \neq 0 \tag{5}$$

where H denotes the Heaviside function.

Using (3), (5) in (4) we calculate $\Delta E(t) \equiv \overline{\theta^2}(0) - \overline{\theta^2}(t)$. Omitting the details of the analysis, the main result is the following power-law[2]:

$$\Delta E(t) \propto t^{3(1-D)} \tag{6}$$

valid in the range of times t

$$(r_0/R_0)^{2(s+1)} Pe^{1/3} \ll t\Omega_0 \ll Pe^{1/3}. \tag{7}$$

---

[2]The power-law is defined as $\Delta E(t) \propto t^\beta$. We distinguish: $\beta = 1/2$ (classical diffusion), $\beta > 1/2$ (super-diffusion), $\beta < 1/2$ (sub-diffusion).

D denotes a Kolmogorov capacity of the spiral structure and is directly related to the accumulation rate s by $D = \frac{s}{s+1}$. The lower time limit of this law depends on the initial condition in θ and approaches zero when r_0 tends to zero, and the upper time limit determines the diffusive time scale of the entire structure. Note also that during the anomalous diffusion given by (6) the scalar spiral structure can be shown to have a power spectrum $E(k, t) \propto k^{-3+2D}$ by using arguments similar to Gilbert (1988).

We now want to illustrate the regime in (6) by looking at the evolution of a scalar spiral structure in a numerical experiment. (1) has been solved on a cylindrical coordinate frame using a second-order accurate finite-volume discretisation based on a limited TVD scheme. Figures 1 and 2 illustrate the following findings for times (i)-(iii) as specified in Figure 2:

(i) No spiral structure has yet formed and the scalar variance decays according to the classical $t^{1/2}$.

(ii) At the core of the spiral fine structure has formed and super-diffusive behaviour is observed. The asymptotic law for these early times $t\,\Omega_0 < (r_0/R_0)^{2(s+1)} Pe^{1/3}$ can be shown to be $\Delta E \propto t^{3/2}$ and is clearly observed in a set of initial conditions different from those of Figure 1a.

(iii) The spiral structure expands throughout the scalar field and diffusion acts according to (6). In this range the anomalous diffusion is directly linked to the dimensionality of the spiral and the spiral structure is rapidly wiped out by the interplay of molecular diffusion with the spiral accumulation.

3. Anomalous diffusion of spiral or fractal structures

In the previous section the fractal property D of the scalar field was intimately linked to a particular flow structure. We now briefly address the question of the diffusive decay of more general fractal or spiral fields. Consider, for example, a spiral scalar singularity generated by the continuous action of a steady vortex, or fractal scalar singularities generated by chaotic advection. Both these singular structures may be characterised by a Kolmogorov capacity $D \in [0, 1]$ as before. If the flow is somehow suddenly removed and diffusive attrition proceeds to destroy the structure, the scalar variance decays at early times $\frac{\kappa t}{L^2} \ll 1$ as follows (Vassilicos, 1995):

$$\Delta E(t) = E_0\, \Gamma(\frac{D+1}{2})\, \left(\frac{\kappa t}{L^2}\right)^{\frac{1-D}{2}} \tag{8}$$

where L is an integral initial length scale like R_0, and Γ is Euler's gamma function. This result is valid for any homogeneous fractal or homogeneous collection of spirals. (8) has been verified numerically and exemplary results are shown in Figure 3.

This anomalous dissipation can be understood in terms of a diffusive length-scale $\delta(t)$ which is a measure of the distance over which the effects of molecular diffusion are appreciable on the structure at time t. It is found that $\delta(t) \propto \sqrt{\kappa t}\, N(\sqrt{\kappa t})$ where $N(\sqrt{\kappa t})$ is the minimum number of segments of length $\sqrt{\kappa t}$ needed to cover the points of high scalar gradients (Vassilicos 1995). The fractal geometry of the structure implies $N(\sqrt{\kappa t}) = N(L)\,(\sqrt{\kappa t}/L)^{-D}$. The closer D is to 1, the more space-filling the covering by segments of size $\sqrt{\kappa t}$, the larger the total diffused length δ, and therefore the faster the early decay. This dependence of the diffusive length scale $\delta(t)$ on D is related to the observation of Frisch & Vergassola (1991) that singularities with different Hölder exponents have different inner dissipation length-scales in a multifractal picture of turbulence.

(a)   (b)   (c)   (d)

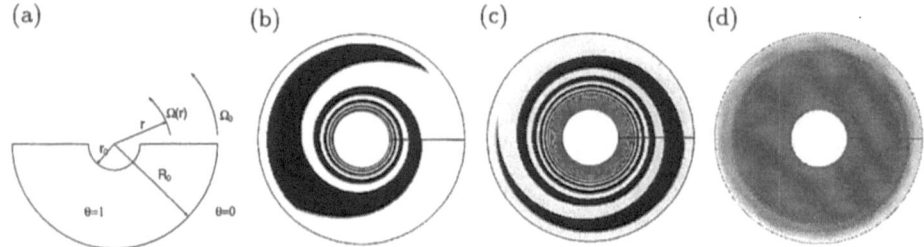

Figure 1. Evolution of a scalar spiral structure with accumulation rate $s = 4$. Different periods of time are depicted. (a)-(d), refer to Figure 2.

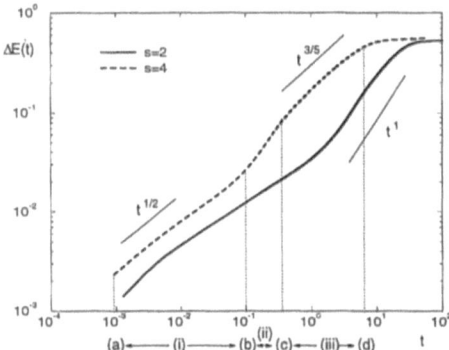

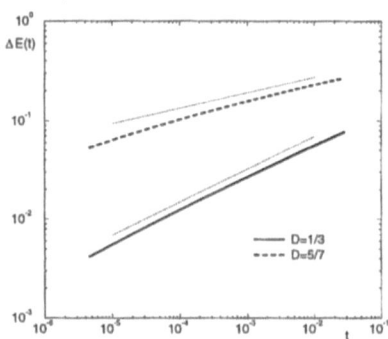

Figure 2. Energy decay of the scalar field from Figure 1a. Results for s=2,4 are shown here. (a)-(d), refer to Figure 1; (i)-(iii), explanations in section 2.

Figure 3. Energy decay of a scalar field, subject to initial conditions with $D_1 = 1/3, D_2 = 5/7$. Good agreement with (8) can be observed (dotted curves correspond to theoretical finding).

References

1. Frisch, U. & Vergassola, M. (1991), *Europhys. Lett.*, **14**(5), 439-444.
2. Gilbert, A.D. (1988), *J. Fluid Mech.*, **193**, 475-497.
3. Hunt, J.C.R. & Vassilicos, J.C. (1991), *Proc. R. Soc. London Ser. A*, **434**, 183-210
4. Lundgren, T.S. (1982), *Phys. Fluids*, **25**(12), 2193-2203
5. Vassilicos, J.C. (1995), *Phys. Rev. E*, **52**(6), R1-R4
6. Vassilicos, J.C. & Hunt, J.C.R. (1991), *Proc. R. Soc. London Ser. A*, **435**, 505-534

PROBABILITY DISTRIBUTION FUNCTIONS
FOR NAVIER-STOKES TURBULENCE

M J GILES
Department of Mathematics and Statistics
The University of Northumbria
Newcastle upon Tyne, UK.

1. Introduction

The probability distribution function (*pdf*) of a two point velocity difference in a turbulent fluid is known from experimental measurements and computer simulations to possess strongly non-Gaussian features. These features have attracted considerable interest, because they are associated with the phenomenon of intermittency, to which the failure of the classical Kolmogorov K41 theory is attributed. However, as existing theoretical work is based on phenomenological models, the problem of how to derive *pdf*s directly from the Navier-Stokes (NS) equations remains unsolved, and it is this question which the present paper addresses.

2. The Probability Distribution

Consider a velocity field $\mathbf{u}$, which is driven by a random stirring force $\mathbf{f}$, having a known functional probability distribution $P_0(\mathbf{f})$ (assumed Gaussian). From the NS equations, the corresponding functional probability distribution of $\mathbf{u}$ is $P(\mathbf{u}) = P_0(\mathbf{f}(\mathbf{u})) \, \delta\mathbf{f} / \delta\mathbf{u}$. Now suppose that one wants to derive the *pdf*, $P(\Delta u_r)$, of a longitudinal velocity increment Δu_r, in, say, the x-direction. With $P(\mathbf{u})$ known, this can be calculated from the functional integral

$$P(\Delta u_r) = \int \delta(\Delta u_r - \Delta u_1(r)) \, P(\mathbf{u}) \, \delta\mathbf{u} \, , \tag{1}$$

where $\Delta u_1(r) = u_1(x+r, y, z, t) - u_1(x, y, z, t)$. But, can one derive any useful results from this formalism? In particular, is it possible to account for the strongly non-Gaussian features of $P(\Delta u_r)$? We shall attack this question in two ways to show that some progress can be made.

231

S. Gavrilakis et al. (eds.), Advances in Turbulence VI, 231-234.
© 1996 *Kluwer Academic Publishers.*

3. The Saddle-Point Method

Given that $\mathcal{P}(\mathbf{u})$ will be of exponential form, an obvious way of tackling the evaluation of (1) is to apply the field-theoretic version of the well-known saddle-point technique. This entails locating the field configuration at which the exponent of $\mathcal{P}(\mathbf{u})$ attains a minimum, and then expanding the integrand about this field configuration. This leads to a tractable Gaussian functional integral from which one can demonstrate [1] that the tail of the distribution has the characteristic exponential form

$$P(\Delta u_r) \sim \frac{1}{\sqrt{2\pi a \, \Delta u_r}} \exp\left(-\frac{\Delta u_r}{a}\right) . \tag{2}$$

The velocity scale a can be expressed in terms of the self-energy function associated with $\mathcal{P}(\mathbf{u})$, but it seems difficult to progress usefully beyond this point, which, nevertheless, is of interst in demonstrating from the NS equations how exponential tails can arise.

4. The Renormalisation Group Approximation

To proceed further with an analysis of the non-Gaussian features of $P(\Delta u_r)$, on the basis of (1), one must recognise that, in the limit of high Reynolds numbers, the statistics of fluctuations must be treated differently according as they are dominated by low, medium, or high wavenumbers, because the fine detail cannot be captured if all ranges are dealt with simultaneously. To accomplish this we attempt to achieve an approximate factorisation of $\mathcal{P}(\mathbf{u})$, each factor containing the dependence of the process on phenomena that happen on one particular scale. This approach has the following elements.

First, as shown in [2], the fluctuations of Δu_r at small scales are intrinsically Gaussian, in as much as the non-Gaussian features are attributable to the intermittency of the dissipation ε_r, averaged over a sphere of radius r, with the consequence that, when conditioned to a given value of ε_r, $P(\Delta u_r)$ collapses to a Gaussian *pdf*. Second, as the effect of short wavelength fluctuations is contained in ε_r, we can derive an approximate expression for this underlying conditional *pdf*, $P(\Delta u_r|\varepsilon_r)$, by using renormalisation group (RG) averaging to eliminate all scales below the scale at which the flow is being observed. Third, to derive the required approximation for $P(\Delta u_r)$, we must average $P(\Delta u_r|\varepsilon_r)$ with respect to the *pdf* of ε_r, $F(\varepsilon_r)$, for which purpose a complementary approximation is needed.

To implement the foregoing mathematically, we recognise that $P(\Delta u_r)$ and $F(\varepsilon_r)$ are the marginal *pdfs* of the bi-variate *pdf* of Δu_r and ε_r, which yields the exact result

$$P(\Delta u_r) = \int P(\Delta u_r|\varepsilon_r) F(\varepsilon_r) d\varepsilon_r . \tag{3}$$

We then apply to this representation a course-graining procedure based on the RG transformation ℓ as defined in [1]. In particular, we approximate $P(\Delta u_r | \varepsilon_r)$ in terms of the fixed point distribution of ℓ, $P_*(\mathbf{u} | \varepsilon_r)$, applying to a dissipation rate ε_r, by using ℓ to successively eliminate all modes having wavenumbers above r^{-1}. This yields an approximately Gaussian *pdf* for $P(\Delta u_r | \varepsilon_r)$, having a variance of the form appearing in Kolmogorov's K62 refined similarity hypothesis [1], viz. $\langle \Delta u_r^2 \rangle \propto (r \varepsilon_r)^{2/3}$.

5. The Poisson Approximation

The result of coarse-graining $P(\Delta u_r)$ is to concentrate the effect of short wavelengths into the corresponding approximate expression for the *pdf* of the dissipation field fluctuations.

$$F(\varepsilon_r) = \int \delta(\varepsilon_r - \mathcal{E}_r(\mathbf{u})) P_*(\mathbf{u} | \bar{\varepsilon}) \delta\mathbf{u} \quad . \tag{4}$$

This again is a coarse-grained approximation, obtained by eliminating all modes down to the wavelength r^{-1}, in which $\mathcal{E}_r(\mathbf{u})$ is the dissipation rate of a particular realisation of the flow averaged over a sphere of radius r. Notice here that the fixed point distribution is that applying to the mean dissipation rate $\bar{\varepsilon}$. The calculation is completed by evaluating the above integral in the limit of zero viscosity. This gives $F(\varepsilon_r)$ in terms of a continuous Poisson distribution for a dissipation field growth defect index s defined by $\varepsilon_r = \varepsilon_\infty(r) \beta^s$, [1], where $\varepsilon_\infty(r)$ represents the maximum amplitude attainable by ε_r in intense fluctuation structures, such as filamentary vortices, while β is an intermittency parameter.

6. Discussion

What we have obtained from the RG coarse-grained approximation is the refined similarity hypothesis form for $P(\Delta u_r | \varepsilon_r)$ coupled with the log-Poisson model of intermittency for $F(\varepsilon_r)$, essentially as discussed in [3,4]. Indeed, a simple calculation leads to the general formula for the anomalous scaling exponents obtained for this model. With the divergent power law scaling $\varepsilon_\infty(r) \sim r^{-\Delta}$, we are left with the two parameters Δ and β to determine. In fact, one can advance at least three different arguments to show that these parameters must be connected by the relation $\beta = 1 - \Delta/2$, [1]. To date a value for Δ has been inferred from physical and dimensional arguments [3]. Its calculation from the NS equations via $P(\mathbf{u})$ entails determining the scaling of the two-point correlation function of the the local dissipation rate $\varepsilon(\mathbf{x})$. The value $\Delta = 2/3$ would imply the scaling $r^{-\mu}$ with $\mu = 2/9$. This is a problem that one might hope to resolve using an RG approach. However, particular difficulties attach to the treatment of correlation functions involving two fields at the same point, as their combination, in this case $\varepsilon(\mathbf{x})$, acts as an object in its

234

own right (composite operator). We are currently investigating this problem from the present point of view.

Numerical evaluation of (3) using the Gaussian approximation for $P(\Delta u_r | \varepsilon_r)$ and the Poisson approximation for $F(\varepsilon_r)$ produces results which are in good agreement with profiles derived from experimental and computational work, as illustrated in Fig. 1. This shows $P(\Delta u_r)$ as a function of Δu_r (in units of $\sigma = \langle \Delta u_r^2 \rangle^{1/2}$) for $\beta = 2/3$ and $\varepsilon_\infty(r)/\overline{\varepsilon} = 10$ (dashed) and 250 (dotted). The full line shows the Gaussian distribution corresponding to $\beta = 1$.

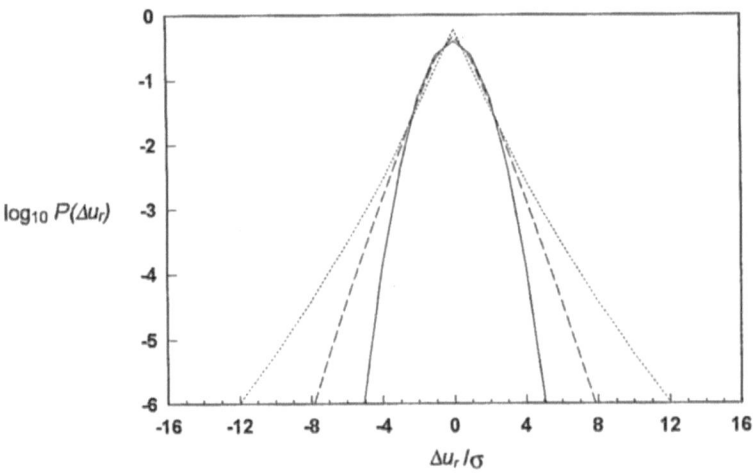

Figure 1. The Probability Distribution $P(\Delta u_r)$

Undoubtedly, the early optimism that field theoretic methods might provide the tools needed to deal with turbulent flows has been tempered in recent years by a more cautious assesment in the light of their failure to explain small scale phenomena. It is felt, however, that an approach based on approximate factorisation of $\mathcal{P}(\mathbf{u})$ in conjunction with RG methods might enable some of these obstacles to be overcome.

7. References

1. M.J. Giles, "Probability distribution functions for Navier-Stokes Turbulence," Phys. Fluids 7, 2785 (1995).
2. Y. Gagne, M. Marchand, and B. Castaing, "Conditional velocity pdf in 3-D turbulence," J. Phys. II France 4, 1 (1994).
3. Z.S. She and E. Lévêque, "Universal Scaling Laws in Fully Developed Turbulence," Phys. Rev. Lett. 72, 336 (1994).
4. B. Dubrulle, "Intermittency in Fully Developed Turbulence: Log-Poisson statistics and Generalised scale Covariance," Phys. Rev. Lett. 73, 959 (1994).

TRANSVERSE VELOCITY STRUCTURE FUNCTIONS IN DEVELOPED TURBULENCE

H.KAHALERRAS, Y.MALECOT AND Y.GAGNE
LEGI/IMG-CNRS, PB 53X, 38041 Grenoble Cedex, France.

1. Introduction

Velocity structure functions based on the longitudinal increment $\delta u_\parallel(r) = u(x+r) - u(x)$ (in which the component u is aligned with the direction of the separation r) have been extensively measured to study the scaling laws of small scale intermittency of fully developed turbulence.

Concerning the transverse velocity increments $\delta u_\perp(r)$ (with u normal to the direction of r) there are few experimental data. Experimentally, two kinds of transverse velocity increments can be obtained, one based on the streamwise component u which requires two single wire probes without any assumptions [1] and another based on the spanwise component v measured with an X-wires probe but using the Taylor hypothesis [2].

Experimental data indicate that the velocity structure functions of order p have scaling properties for the separation r lying in the inertial range, namely:

$$< (\delta u_\parallel(r))^p >\sim r^{\zeta_{p\parallel}} \tag{1}$$

and

$$< (\delta u_\perp(r))^p) \sim r^{\zeta_{p\perp}} \tag{2}$$

Today there is a controversy about these scaling exponents, some of the experimental results suggest that the $\zeta_{p\perp}$ are different from the $\zeta_{p\parallel}$ ones [1] and others show the opposite [2].

In section 2, we present experimental data on scaling exponents ζ_p of longitudinal and transversal velocity structure functions (only based on the streamwise component), measured in two different turbulent flows, by using the Extended Self Similarity technique [3]. In section 3, we compare the probability density functions of the longitudinal and transverse velocity

S. Gavrilakis et al. (eds.), Advances in Turbulence VI, 235-238.
© 1996 *Kluwer Academic Publishers.*

increments. Finally, in section 4, we analyze them in the context of the variational approach [4].

2. Scaling exponents ζ_p

Velocity fluctuations have been measured in a pseudo-grid turbulence ($R_\lambda \simeq$ 280) and in an axisymmetric jet ($R_\lambda \simeq 850$). Longitudinal velocity increments have been obtained by using the Taylor approximation; but instead of converting the time series $u(t)$ to the spatial file $u(x)$ with the mean velocity $< U >$ as usual, we have used the true instantaneous component velocity $u(t)$ [5]. Transverse increments of the streamwise component velocity have been measured by using two single parallel hot wires and by varying the separation between them (for more details, see [6]).

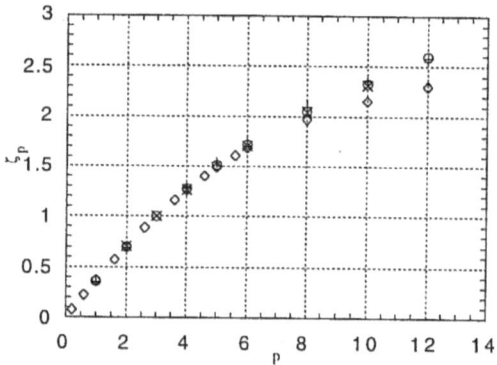

Figure 1. Variation of the exponent ζ_p as a function of p for both longitudinal ($\times$: jet, $R_\lambda \simeq 850$ - $\circ$: grid turbulence, $R_\lambda \simeq 280$) and transverse ($+$: jet - $\diamond$: grid turbulence) increments.

In order to extend the power law behaviour range of structure functions (in particular in the grid turbulent flow), we have used the Extended Self Similarity technique which permits us to get reliable values of the scaling exponents ζ_p even for the large values of the order p ($p \leq 12$).

Fig. 1 gives the $\zeta_{p\parallel}$ and $\zeta_{p\perp}$ obtained in the jet and in the grid turbulence. Our results clearly show no differences between the transverse and the longitudinal exponent values (for $p \geq 10$, the differences between $\zeta_{p\parallel}$ and $\zeta_{p\perp}$ for grid turbulence are of the same order as the experimental errors). With rather the same experimental conditions, Herweijer et al.[1] obtained an opposite result; but the measurements of Camussi et al. [2] are, on the contrary, in agreement with ours, even though they correspond to the transverse increment of the spanwise velocity component. Today, we still don't have clear explanation of this discrepancy and the only theoretical predictions concern the second order, $\zeta_{2\parallel} = \zeta_{2\perp}$ [7].

3. Velocity pdf

Fig. 2 shows, in the case of the grid turbulent flow, the pdfs for two different separations, one corresponding to the dissipative range ($\frac{r}{\eta} \simeq 3$) and the other to the inertial one ($\frac{r}{\eta} \simeq 30$). We clearly observe that the transverse velocity distribution is symmetric and that the skewness of the longitudinal velocity pdf is mainly due to the fact that the negative tail is more spread and stretched. It has been noticed that Herweijer et al. [1] obtained exactly the opposite; the negative tails of the transverse and longitudinal pdf collapse and the positive tail of the transverse distribution is more spread than the longitudinal one. Fig. 2, also shows how the skewness of $\delta u_{\parallel}(r)$ is decreasing down to zero when the inertial separation r tends to the integral scale.

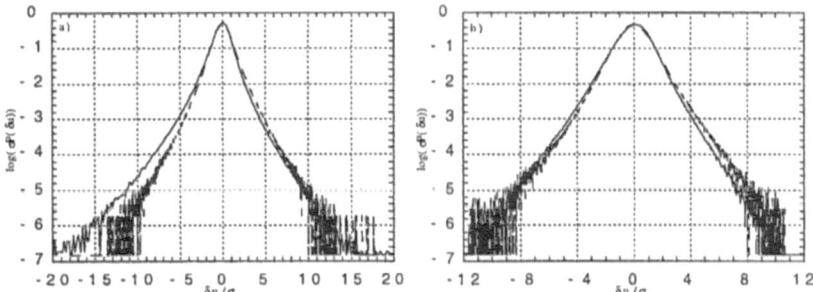

Figure 2. Comparison between longitudinal (full line) and transversal (dashed line) pdf's in the case of grid turbulence: a):$\frac{r}{\eta} \simeq 3$ - b):$\frac{r}{\eta} \simeq 30$.

4. "Depth" of intermittency cascade

In the context of the variational approach [4], it is shown that any velocity pdf, at a given scale r, can be fitted by a superposition of gaussian pdf's with a quasi-lognormal (or log Poisson) distribution of their standard deviations. Consequently, at a given scale r, the shape of the global velocity pdf can

238

be characterized by the variance of the latter distribution, called $\lambda^2(r)$. From the transverse and longitudinal velocity pdf's, we have calculated $\lambda_\parallel^2$ and $\lambda_\perp^2$ using an improvement of such a fitting. Fig. 3 shows the dependence of these variances $\lambda^2(r)$ on the scale, in the case of the grid turbulence. Obviously, there is no difference between the longitudinal and the transversal increment velocity field.

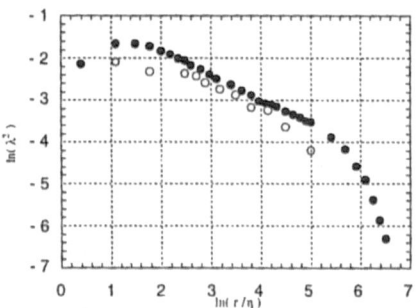

Figure 3. Behavior of λ^2 with the scale r plotted in natural logarithmic scales for both longitudinal (•) and transverse (o) increments (grid turbulence experiment).

5. Conclusion

This study brings new experimental data which suggest that, the transverse and longitudinal velocity increments have the same global statistical properties. This result can be somewhat surprising, considering that, from a dynamical point of view, the transverse increments rather corresponds to the vorticity, whereas the longitudinal ones play a major role in the energy dissipation.

References

1. Herweijer, J. & van de Water, W. (1995) *Advances in turbulence V*, ed.R.Benzi, Kluwer Academic Publishers , pp. 210–216.
2. Camussi, R., Barbagallo, D., Guy, G. & Stella, F. (1995), submitted to *Physics of Fluids*.
3. Benzi, R., Ciliberto, S., Tripiccione, R., Baudet, C., Massaioli, F. & Succi, S. (1993), *Phys.Rev.E*, **48**, pp.29.
4. Castaing, B., Gagne, Y. & Hoppfinger, E.J (1990), *Physica D*, **46**, pp. 177–200.
5. Gledzer, E. (1995), submitted to *Physica D*.
6. Kahalerras, H. (1994) D.E.A INPG.
7. Frish, U. (1995) *Turbulence the legacy of A.N. Kolmogorov*, Cambridge University Press.

NONLINEAR INTERACTIONS IN TURBULENCE WITH STRONG IRROTATIONAL STRAINING

N. K-R. Kevlahan
LMD-CNRS, Ecole Normale Supérieure
24, rue Lhomond, 75231 Paris cedex 05, France

J.C.R. Hunt
DAMTP, University of Cambridge
Silver St., Cambridge CB3 9EW, United Kingdom

Abstract

The rate of growth of the nonlinear terms in the vorticity equation are analysed for a turbulent flow with r.m.s. velocity u_0 and integral length scale L subjected to a strong uniform irrotational plane strain S, where $(u_0/L)/S = \epsilon \ll 1$. The Rapid Distortion Theory (RDT) solution is the zeroth order term of the perturbation series solution in terms of ϵ. We use the asymptotic form of the convolution integrals for the zeroth order nonlinear terms when $\exp(St) \gg 1$ to determine when (in wavenumber k and time t) the perturbation series in ϵ fails, and hence estimate precisely the domain of validity of inviscid and viscous RDT.

1 Introduction

In most turbulent flows the large scale velocity field is a straining motion, either irrotational or rotational, that changes slowly on the time scale of the turbulent eddies — for example flow over waves, turbulence entering engines etc. The main practical and fundamental question concerns how the statistical and eddy structure of the turbulence is distorted by the strain and how its other properties are changed, such as mixing caused by the separation of fluid elements, or dissipation caused by transfer of kinetic energy to small scales. A key problem of turbulence research is to study how the nonlinear interaction between eddies or between Fourier components are affected by distortion.

In a review by Hunt & Carruthers (1990) it was shown that many aspects of distorted turbulent flows are determined by the *linear* interaction between the turbulence and the large scale mean straining flow and for which the nonlinear interaction can be neglected. Rapid Distortion Theory (RDT) is the term used to

S. Gavrilakis et al. (eds.), Advances in Turbulence VI, 239-242.

describe these methods and detailed assumptions of this simplified, though where appropriate quite powerful, approach.

The only theoretical calculations of the nonlinear interaction for turbulent flows have been based on statistical physics concepts, such as Direct Interaction Approximation and EDQNM (reviewed recently by McComb 1990; Lesieur 1990). They have not shown how and when the nonlinear interaction dominates linear distortion effects, and therefore have not provided insight into the limitations of RDT.

In this paper we explore a different approach based on a general asymptotic analysis of the nonlinear terms (expressed in terms of convolution of Fourier transforms) for the fluctuating vorticity field; no assumptions are made about its initial form provided its amplitude is small compared with the mean strain. This allows us to calculate the next term in the expansion of which RDT is the zeroth order term, and hence to estimate accurately the period of validity of RDT.

The nonlinear terms in the equation for the fluctuating vorticity when the turbulence $\boldsymbol{\omega}$ is undergoing a large scale plane strain $\boldsymbol{U}(\boldsymbol{x})$ are those representing the vortex lines being randomly rotated and stretched (by the terms $(\boldsymbol{\omega} \cdot \nabla)\boldsymbol{u}$) and advected (by the term $(\boldsymbol{u} \cdot \nabla)\boldsymbol{\omega}$) by the small scale turbulence $\boldsymbol{u}$. The approach hitherto for estimating the nonlinear terms relative to the linear terms $(\boldsymbol{\omega} \cdot \nabla)\boldsymbol{U}$, has been to assume either that $(\boldsymbol{\omega} \cdot \nabla)\boldsymbol{u}$ has the same value as in undistorted turbulence $(\sim (u(l)^2/l^2)_{t=0})$, in which case the nonlinear terms are always greater than any exponentially decreasing linear term, or that the magnitude of $(\boldsymbol{\omega} \cdot \nabla)\boldsymbol{u}$ is determined by the maximum value of $\boldsymbol{\omega}$ and $\boldsymbol{u}$ in the distorted flow in which case $(\boldsymbol{\omega} \cdot \nabla)\boldsymbol{u}$ could grow even *faster* than the fastest growing linear term. Neither of these is likely because, as seen in the numerical simulation, the eddy structure changes in such a way as to reduce the nonlinear vorticity distortion terms! In the most frequently used Reynolds stress models, these nonlinear terms are based on estimates of the local distorted value of the turbulence covariance components and not on the distorted eddy structure. However the latter point has been demonstrated as crucial in modelling the nonlinear rotating and stretching effect, through an idealised analysis of large scale and small scale turbulence undergoing distortion (Kida & Hunt 1989) and through a statistical analysis of Direct Numerical Simulations by Mansour, Shih & Reynolds (1991) using a similar tensorial representation of the distorted spectra.

This paper summarises the main results of a detailed analysis of the distortion of turbulence caused by a large scale irrotational plane straining flow; the full details of this investigation will appear in a later article. This type of distortion was chosen because of its importance in a number of engineering problems and because of its fundamental importance in developing the basic theory of turbulence structure. We make the interesting discovery that the RDT assumptions remain valid for a relatively long time, usually until the final period of viscous decay, provided that the average in the compressed direction of the initial turbulent velocity component in this direction is zero. This condition is satisfied, for example, if the flow is bounded in the compressed direction. Estimates of the validity time of RDT are made and compared to the classical results.

2 Rapid Distortion Theory

The conditions for RDT are usually defined for eddies of length scale l and velocity scale $u(l)$ undergoing some kind of distortion over a time T_D, where the distortion may be an imposed strain of strength S, the sudden introduction of a boundary (in the frame of reference moving with the mean flow) or body forces etc. RDT is stated to be valid *either* if T_D is so rapid that the nonlinear terms in the vorticity equation have a negligible effect on the vorticity of the eddy ($\sim u(l)/l$), on the relevant time scale $\tau(l)$ of the eddy $l/u(l)$, *or* if the linear effects on $\boldsymbol{\omega}$ of the distortion (e.g. $\sim S(u/l)$ for a straining distortion) are much stronger than the nonlinear self-induced straining by the turbulence ($\sim (u/l)^2$).

Taking the integral scale $l = L$ and r.m.s. velocity u_0 this leads to two possible conditions for RDT

$$ t \quad \ll \quad \tau(L) \sim L/u_0(t), \tag{1} $$

$$ \text{or} \quad S \quad \gg \quad u_0(t)/L. \tag{2} $$

The latter condition for the strength of the strain rate may be satisfied even when the distortion is applied over a long period i.e. for slowly changing turbulence, and *may* be a valid condition for the accurate use of RDT. Thus we may combine the above conditions to give $u_0(t)/L \ll \max(S, 1/t)$. Hence in any given context one must take care to define precisely the term 'RDT'.

As explained in the introductory section, these conditions have been derived from the vorticity equation by inspection, and by assuming that the nonlinear terms are of the same order as in the undistorted turbulence. However, where the turbulence is strongly distorted the nonlinear stretching term $(\boldsymbol{\omega} \cdot \nabla)\boldsymbol{u}$ can be much less than the undistorted estimate of $(u(l)/l)^2$, a good example being where the distorted flow forms into strong straight vortex tubes where vorticity is greater than the applied strain rate S; in this case the nonlinear vortex stretching term vanishes identically.

The purpose of this paper is to refine these estimates for the validity of RDT by considering the RDT solution to be the zeroth order term in an asymptotic expansion for the distorted turbulence. The validity range of RDT (in time and wavenumber) may then be found by determining when this expansion breaks down. This method should lead to much more precise estimates for the validity of RDT and to an understanding of how the nonlinear terms are affected by irrotational straining.

3 Results

The magnitude of the nonlinear terms depends sensitively on the amplitude of the eddies with large length scales perpendicular to the direction x_1 of positive strain. If the average of the velocity component u_2 in the convergence direction

242

x_2 is initially zero

$$\int_{-\infty}^{+\infty} u_2(\boldsymbol{x})\,\mathrm{d}x_2 = 0, \qquad (3)$$

then the zeroth order nonlinear terms always remain smaller than the linear terms, even those that decrease exponentially. In this case RDT fails at a relatively long time $t \sim L/u_0 k^{-3}$ independent of ϵ (where k is the wavenumber), and the maximum amplification of vorticity under RDT is $\omega_1/S \sim \epsilon \exp(\epsilon^{-1}) \gg 1$. If (3) does not apply, the zeroth order nonlinear terms increase faster than the linear terms by a factor $O(\exp(St))$. RDT then fails at a relatively short time $t \sim 1/S\ln(\epsilon^{-1}k^{-3})$, and the maximum amplification of vorticity under RDT is $\omega_1/S \sim 1$. The analytical results on the growth of the nonlinear terms have been confirmed by numerical evaluation of the integrals for a particular form of eddy.

Viscous effects dominate when $t \gg 1/S\ln(k^{-1}(Re/\epsilon)^{1/2})$ (where Re is the Reynolds number), and RDT fails immediately in this range.

Thus we find that the usual order of magnitude estimate for the time period of the validity of RDT, namely that $u_0(t)/L \ll \max(S, 1/t)$, is an underestimate since $u_0(t)/L$ increases exponentially in time. Expressed in similar terms, we find instead that $u_0/L \ll 1/t$ if (3) is satisfied, and $u_0/L \ll S/\exp(St)$ otherwise, where $u_0 = u_0(t = 0)$. Interestingly, the 'crudest' order of magnitude estimate, $u_0/L \ll 1/t$, is also the most accurate if (3) is satisfied (e.g. bounded flows)!

Perhaps the most general point to emerge is that a weak random vorticity field can be amplified by a larger scale strain so that the strain rate can become of the same order as that of the applied strain. This is because the nonlinear processes, which might have inhibited this growth, are themselves inhibited by the straining. In other words strained turbulence adjusts itself so as to reduce to a consistent extent the 'scrambling' effects on its own amplified vorticity. This helps explain why weak turbulence can be so strongly amplified at the stagnation point of cylinders (Sadeh & Brauer 1980).

References

Hunt, J.C.R. and Carruthers, D.J. (1990) Rapid distortion theory and the 'problems' of turbulence. *J. Fluid Mech.* **212**, 497–532.

Kida, S. and Hunt, J.C.R. (1989) Interaction between different scales of turbulence over short times. *J. Fluid Mech.* **201**, 411–445.

Lesieur, M. (1990) *Turbulence in fluids: stochastic and numerical modelling.* Martinus Nijhoff.

McComb, W.D. (1990) *Physics of turbulence.* Clarendon Press.

Mansour, N., Shih, T.-H., Reynolds, W.C. (1991) The effects of rotation on initially anisotropic turbulent flows. *Phys Fluids* A **3**, 2421–2437.

Sadeh, W.Z., Brauer, H.J. (1980) A visual investigation of turbulence in the stagnation flow about a circular cylinder. *J. Fluid Mech.* **99**, 53–64.

SMALL SCALE INTERMITTENCY AND
THE RENORMALIZATION GROUP

G.A. KUZ'MIN
Institute of Thermophysics
630090 Novosibirsk, Russia

1. Introduction

The energy of the fully developed turbulence is excited at some large scale and is transferred through the inertial range to the viscous scales. While the transfer, the spatial intermittency is growing up and the small scale fluctuations occupy very small fraction of space [1]. Natural theoretical framework for the non intermittent pulsations is the original Kolmogorov (1941) scaling hypothesis [1] and the renormalization group technique [2], [3], [4], [5]. Intermittency induces corrections to simple scaling. Teodorovich [6] studied intermittency corrections by analyzing the infrared divergence within the framework of the field-theoretical renormalization group.

We consider another approach. The main difficulty is that the intermittent fluctuations are poorly described by correlation functions of low order. On the basis of the Kolmogorov [7] improved hypothesis, we express the strongly intermittent fields in terms of fields of small intermittency. As a result, a nonlinear Langevin model for small scale fluctuations arises. In the model, small scale velocity pulsations are excited by the non intermittent multiplicative noise. To close the model, the distribution of the noise have to be determined. Due to low intermittency, it may have simple scaling properties. We consider the scaling hypothesis for the noise and an application of the renormalization group technique to find its statistics.

2. Intermittency and the Scaling Theory.

As an alternative to the logarithmic normal theory, Kolmogorov [7] formulated a scaling theory in terms of the ratios of the velocity differences. This theory does not use any specific managing field such as the energy

S. Gavrilakis et al. (eds.), Advances in Turbulence VI, 243-246.
© 1996 *Kluwer Academic Publishers.*

dissipation. The ratios of the velocity differences are not tensors. This complicates applications of the modified similarity hypotheses. We reformulate the Kolmogorov modified theory in terms of the tensor dimensionless fields.

Let $u(k)$ be the Fourier transform of the velocity field. For a given scale l, we define the two components of the velocity

$$v^{(l)}(x,t) = \sum_{k>1/l} u(k,t)\exp(ik.x), \quad V^{(l)}(x,t) = \sum_{k<1/l} u(k,t)\exp(ik.x)$$

Note that $v^{(l)}(x)$ is similar to the Kolmogorov's velocity difference $v(x+r)-v(x)$. Another local field of scale l is the strain tensor $D_{ij}^{(l)}(x) = \partial V_i/\partial x_j + \partial V_j/\partial x_i$. The dimensionless vector $\mathcal{U}^{(l)}(x) = v^{(l)}(x)/\sqrt{l^2 D_{ij}^{(l)} D_{ij}^{(l)}}$ is similar to the Kolmogorov's ratios [7] of the velocity differences.

We reformulate the Kolmogorov hypotheses as follows.

- If l is much less than the main spatial scale L, the probability distributions for the dimensionless fields are supposed to depend only on the local Reynolds number $R = v^{(l)}(x)l/\nu$.
- The distributions given by the first hypothesis is independent on R if $R \gg 1$.
- Any two dimensionless fields are supposed to be statistically independent if their scales l, l' are of different order of magnitude.

For example, let $\Phi(r,l)$ be a correlation function of a dimensionless scalar field: $\Phi(r,l) = <\phi^{(l)}(x,t)\phi^{(l)}(x+r)>$. The above scaling hypothesis demands that $\Phi(r,l) = \Phi(\lambda r, \lambda l)$ for arbitrary λ. So $\Phi(r,l) = \Phi(r/l)$.

3. Turbulence Driven by a Multiplicative Force.

Let us consider the smoothed Navier - Stokes equation

$$\frac{\partial V_i^{(l)}}{\partial t} + \frac{\partial}{\partial x_j} V_i^{(l)} V_j^{(l)} = -\frac{\partial \Pi/\rho}{\partial x_i} - \frac{\partial}{\partial x_j} R_{ij} + \nu \Delta V_i^{(l)}. \tag{1}$$

In the right -hand side of (1) one has a divergence of the tensor $R_{ij} = \{v_i v_j\}^{(l)} - V_i^{(l)} V_j^{(l)}$, where $\{v_i v_j\}^{(l)}$ is the effective Reynolds tensor that is obtained after Fourier - smoothing of the Navier-Stokes equation. The tensor R_{ij} is the same in all Galilean systems of reference and is a small scale characteristic of the flow. Let us consider the ratio $h_{ij}(x) = R_{ij}(x)/l^2 D^2(x)$. The typical values of the nominator and denominator in h_{ij} depend on the position of the point x. If x is inside the region of intensive motion of scale l, they are large. Outside those regions they are both small but of an equal order.

We suppose that the ratio $h_{ij}(x)$ is less intermittent than the usual local characteristics such as R_{ij} or D_{ij}. In addition, the ratio $h_{ij}(x)$ is similar to the ratio of the velocity differences. According to the above scaling hypotheses, $h_{ij}(x)$ is expected to have an universal distribution that does not depend on l in the inertial range.

Let us substitute $R_{ij} = l^2 D^2 h_{ij}$ into (1).

$$\frac{\partial V_i^{(l)}}{\partial t} + \frac{\partial}{\partial x_j} V_i^{(l)} V_j^{(l)} = -\frac{\partial \Pi/\rho}{\partial x_i} - \frac{\partial}{\partial x_j}\left(l^2 D^2 h_{ij}\right) + \nu \Delta V_i^{(l)}. \tag{2}$$

It may be supposed that the main complication of the small scale statistics is owing to intermittency. It is to be hoped, that the field $h_{ij}(x)$ has no such the complications and is a field of a simple statistics. We suppose that the Gaussian $h_{ij}(x)$ is a good starting point for the renormalization group calculations.

4. Renormalization Group.

We consider the coarse grain transformations with $l_n > l_0$, $n = 1, 2, \ldots$ that progressively reduce the number of degrees of freedom in the inertial range. Main aim is to find the "fixed point" of the transformations. By a fixed point one mean an equation that is unchanged under the transformations. The equation is considered at scales that are large when compared to the cut off scale l. In the limit, $h_{ij}(x)$ is expected to have a distribution that does not depend on any details of the large scale component of the flow. Thus it may be treated as an external force. Various fixed points represent the different possible types of the scaling properties. Previous applications of the renormalization group to turbulence gave a fixed point that is the Navier - Stokes equation with additive forcing term.

The equation (2) contains the Fourier modes with the wave numbers $k < \Lambda = l^{-1}$. Let us shift the cutoff wave number Λ to the smaller value $\Lambda' = (l')^{-1}$. We obtain the equation for the smoothed velocity $V^{(l')}$ that contains the modes with wave numbers $k < \Lambda'$. The new equation may be written in the same form as (2) with l' substituted for l.

On the other hand, this equation may be obtained by repeated smoothing of (2). Equating the nonlinear terms, one obtains the relation

$$l'^2 D'^2 h'_{ij} + V_i^{(l')} V_j^{(l')} = \left(l^2 D^2 h_{ij}\right)^{(l')} + \left(V_i^{(l)} V_j^{(l)}\right)^{(l')}. \tag{3}$$

The above hypothesis demands that h'_{ij}, h_{ij} are statistically the same (after a scaling transformation) in the inertial range. This condition defines a new fixed point that differs from the previously considered ones.

The essential features of the new fixed point are the following ones. Ab initio, it is devised to describe self-similar intermittent cascades. So one needs not to seek for the "intermittency corrections" to simple scaling. One should equate the correlation functions for h'_{ij}, h_{ij}. It is to be hoped that h_{ij} will be approximately Gaussian, so one should compare only the first correlation functions. The linear viscous term and the additive subgrid forcing are not essential in the nonlinear Langevin equation at the fixed point. Note also that the scaling hypothesis gives an expression that is similar to the known Smagorinsky formula for subgrid modelling, but is more general.

For calculations, one may use some sort of the perturbation theory. The so called ϵ -expansion was discussed. The parameter ϵ depends on the spatial dimension d and the exponent of the additive force spectrum y : $\epsilon = y - d + 4$ [4]. Another possible parameter is the inverse dimension $1/d$. It was shown that the expansion parameter is determined by the field values v_l, h_l at the cut off scale l.

In the present model, the motion is excited by the parametric noise. In the zeroth approximation the viscous term is supposed to be balanced by the parametric one. Equating the viscous and the parametric term in (2) gives an estimation for velocity: $v \sim \nu/lh_l$. Thus, an expansion parameter in (2) is the Reynolds number $Re \sim 1/h_l$. We also argue that the parametric term is simplified at $d \to \infty$. In this case, the second small parameter $1/d$ arises.

This work was supported by the Russian Fund of Basic Recearch under Grant Number 94-01-00081.

References

1. Monin, A.S. and Yaglom, A.M. (1975) *Statistical Fluid Mechanics,* Vol. 2, MIT Press, Cambridge, Massachusetts.
2. Forster, D., Nelson, D.R., and Stephen, M.J. (1977) Large distance and long - time properties of a randomly stirred fluid, *Physical Review A,* **16**, 732–749.
3. Adjemyan, L.Ts., Vasilyev, A.N., and Pis'mak, Yu.M. (1983) Renormalization group approach to turbulence: The dimension of composite operators, *Teoreticheskaya i Matematicheskaya Fizika,* **57**, 268 –281. (In Russian)
4. Yakhot, V. and Orszag, S.A., (1986) Renormalization group analysis of turbulence. 1. Basic theory, *Journal of Scientific Computing* **1**, 3-51.
5. Teodorovich, E.V. (1988) The turbulent transfer phenomena and the renormalization group method, *Prikladnaya Matematika i Mekhanika* **52**, 218 –224. (In Russian).
6. Teodorovich, E.V. (1992) Use of the renormalization group theory for describing the intermittency and deriving corrections to the exponents in the Kolmogorov turbulence theory, *JETP* **102**, 863– 875. (In Russian).
7. Kolmogorov, A.N. (1962) A refinement of previous hypotheses concerning the local structure of turbulence in a viscous incompressible fluid at high Reynolds number, *J.Fluid Mech.* **13**, 82-85.

ON KOLMOGOROV'S THIRD ORDER STRUCTURE FUNCTION LAW, THE LOCAL ISOTROPY HYPOTHESIS AND THE PRESSURE-VELOCITY CORRELATION

ERIK LINDBORG

Department of mechanics, KTH, S-100 44 Stockholm, Sweden

Abstract. We show that Kolmogorov's (1941 b) inertial range law for the third order structure function can be derived from a dynamical equation including pressure terms and mean flow gradient terms. A new inertial range law, relating the two-point pressure-velocity correlation to the single-point pressure-strain tensor, is also derived. This law shows that the two-point pressure-velocity correlation, just as the third order structure function, grows linearly with the separation distance in the inertial range. The physical meaning of both this law and Kolmogorov's law is illustrated by a Fourier analysis.

1. Introduction

Kolmogorov (1941 a,b) developed the universal equilibrium theory for the small scales in turbulence by first making the assumption of 'local isotropy'. Local isotropy means that the statistical distribution of the velocity difference $\delta \mathbf{u} = \mathbf{u}' - \mathbf{u}$, of two points, is invariant under rotations and reflections, if the distance r between the points is small, that is if $r \ll L$, where L is the turbulence integral length scale. In his definition of local isotropy Kolmogorov also included steadiness in time of this distribution. Local isotropy implies that the n:th order statistical moment, or structure function,

$$B_{ij\ldots k}^{(n)} = \langle \delta u_i \delta u_j \ldots \delta u_k \rangle \tag{1}$$

is an isotropic tensor.

In his first paper (1941 a) Kolmogorov introduced two similarity hypotheses for the locally isotropic turbulence field; first that the $\mathbf{B}^{(n)}$ of

247

S. Gavrilakis et al. (eds.), Advances in Turbulence VI, 247-250.

different orders are determined by the kinematic viscosity ν, the average dissipation rate ϵ and the distance r; secondly that if there is a range where $r \gg \eta = \nu^{3/4}/\epsilon^{1/4}$ and still $r \ll L$ - that is an inertial range - then the $\mathbf{B}^{(n)}$ are determined only by ϵ and r in this range.

In his second paper (1941 b) Kolmogorov derived the inertial range law

$$B_{lll}^{(3)}(r) = -\frac{4}{5}\epsilon r \tag{2}$$

for the third-order longitudinal structure function. Here the index 'l' corresponds to the velocity component in the same direction as the separation vector $\mathbf{r}$ between the two points with velocities $\mathbf{u}$ and $\mathbf{u}'$.

Kolmogorov (1941 b) used the Kármán-Howarth (1938) equation for the two-point velocity correlation to derive (2). This equation presupposes global isotropy, or isotropy not only of the small scales of turbulence but also of the large scales. Therefore it contains no pressure terms, since these must be zero for globally isotropic turbulence. Monin & Yaglom (1975) have made an attempt to derive (2), using only an assumption of local isotropy, and not an assumption of global isotropy. However, the pressure terms which appear in their derivation are by an erroneous argument set to zero with reference to local isotropy, which makes it impossible to draw any well-founded conclusion about the behaviour of the two-point pressure correlation in the inertial range.

Here we will derive the third order structure function law for a homogeneous shear flow, and at the same time an inertial range law for the two-point pressure velocity correlation.

2. Derivation

The derivation starts by rewriting the equation for the two-point velocity correlation tensor $\langle u_i u_j' \rangle$, into structure function form. This equation can be found in for example Hinze (1975). By using homogeneity and incompressibility it is possible to rewrite this equation into an evolution equation for $\mathbf{B}^{(2)}$, as

$$2\Pi_{ij} - 2\epsilon_{ij} - \frac{\partial}{\partial t}B_{ij}^{(2)} = \frac{\partial}{\partial r_s}B_{sij}^{(3)} + \frac{\partial U_i}{\partial x_s}B_{sj}^{(2)} + \frac{\partial U_j}{\partial x_s}B_{si}^{(2)} + \frac{\partial U_m}{\partial x_s}r_s\frac{\partial}{\partial r_m}B_{ij}^{(2)}$$
$$- \frac{\partial}{\partial r_j}P_i - \frac{\partial}{\partial r_i}P_j - 2\nu\frac{\partial^2}{\partial r_s \partial r_s}B_{ij}^{(2)}, \tag{3}$$

where repeated indices are contracted and where we have introduced the notation

$$P_i = \frac{1}{\rho}\left(\langle u_i p' \rangle - \langle u_i' p \rangle\right) \tag{4}$$

for the two-point pressure-velocity correlation. $\mathbf{\Pi}$ is the (single-point) pressure-strain tensor and ϵ is the dissipation tensor. Half the trace of ϵ is equal to the average dissipation rate ϵ.

From the Reynolds stress equation it can be argued that $|\mathbf{\Pi}| \sim \epsilon$, for a homogeneous shear flow. Therefore, the pressure terms of (3) cannot be neglected. On the other hand, it can be shown that the viscous term, the time derivative and the mean flow gradient terms of (3) can be neglected for small separations. By neglecting these terms and by integrating (3) over the volume of a sphere with radius r in the inertial range, we find by virtue of the divergence theorem

$$\frac{8\pi r}{3}\left(\Pi_{ij} - \epsilon_{ij}\right) = \int \left(n_s B^{(3)}_{sij} - n_j P_i - n_i P_j\right) d\Omega, \tag{5}$$

where $\mathbf{n} = \mathbf{r}/r$, and $d\Omega$ is the element of solid angle.

By taking the trace of (14), the pressure terms of both the left hand side and the right hand side disappear, since they are traceless due to the condition of incompressibility. Thus we find

$$-\frac{16\pi r}{3}\epsilon = \int n_s B^{(3)}_{sii}\, d\Omega. \tag{6}$$

If we now assume that the vector $B^{(3)}_{sii}$ is isotropic, then the integrand of (20) is independent of angle and we immediately find that

$$n_s B^{(3)}_{sii} = -\frac{4}{3}\epsilon r \tag{7}$$

in the inertial range. If we further assume that the tensor $\mathbf{B}^{(3)}$ is isotropic, which of course is a much stronger assumption, then any component of $\mathbf{B}^{(3)}$ can be uniquely related to (21), since $\mathbf{B}^{(3)}$ in this case has only one independent component (see for example Landau & Lifshitz 1987). From well-known isotropic relations Kolmogorov's law (2) follows, and also the relation

$$B^{(3)}_{ltt}(r) = -\frac{4}{15}\epsilon r, \tag{8}$$

where the index 't' indicates a direction perpendicular to $\mathbf{r}$.

The pressure terms of both the left hand side and the right hand side of (14) are traceless and thus they contain no isotropic part. If local isotropy holds, the pressure terms must therefore balance each other. Since $|\mathbf{\Pi}| \sim \epsilon$, this yields

$$|\mathbf{P}| \sim \epsilon r \tag{9}$$

in the inertial range.

250

3. Fourier Analysis

By expressing $\mathbf{B}^{(3)}$ and $\mathbf{P}$ as Fourier-integrals and inserting these expression into (5), it is possible to derive the relation

$$r\left(\Pi_{ij} - \epsilon_{ij}\right) = \frac{3i}{2} \int \frac{j_1(kr)}{k} \left(k_s \widehat{B}_{sij} - k_j \widehat{P}_i - k_i \widehat{P}_j\right) \mathrm{d}^3 k, \qquad (10)$$

where '^' has meaning of the Fourier transform and j_1 is the first order spherical Bessel function. By taking the Fourier transform of (3), and by considering the fast oscillation of j_1 for high wave numbers, it can be shown that the integral of the right hand side is totally dominated by low wave number contributions. The local isotropy hypothesis implies that the pressure terms of (10) must balance each other. By expanding j_1 for low wave numbers, we find

$$\frac{1}{2}\epsilon_{ij} = -\frac{i}{4} \int_{k<k_0} k_s \widehat{B}_{sij} \, \mathrm{d}^3 k, \qquad (11)$$

$$\Pi_{ij} = -\frac{i}{2} \int_{k<k_0} \left(k_j \widehat{P}_i + k_i \widehat{P}_j\right) \mathrm{d}^3 k, \qquad (12)$$

where k_0 is a wavenumber in the lower end of the inertial range of Fourier space.

The relation (11) is basically equivalent to the familiar statement that 'the flow of energy into the inertial range is equal to the dissipation'. From our derivation it is clear that (11) is basically the same law as Kolmogorov's law (2), but formulated in Fourier space.

The relation (12) states that the pressure-strain tensor can be evaluated by integrating the pressure-velocity spectrum over the lower wave numbers. The pressure-strain tensor is the term in the Reynolds stress equation which is responsible for intercomponent energy transfer. Thus, according to (12) the intercomponent energy transfer takes place in the large scales of turbulence. This is, of course, in full agreement with the local isotropy hypothesis.

References

Kármán , T. von & Howarth, L. 1938 'On the Statistical Theory of Isotropic Turbulence' *Proc. Roy. Soc.*, A, **164**, 192-215

Kolmogorov, A.N. 1941 a 'The local structure of turbulence in incompressible viscous fluid for very large Reynolds number' and 1941 b 'Dissipation of energy in the locally isotropic turbulence' English translation in 'Turbulence and stochastic processes: Kolmogorov's ideas 50 years on' *Proc. Roy. Soc.*, A **434**, The Royal Society, London (1991).

Landau, L.D. & Lifshitz, E.M. 1987 'Fluid Mechanics' 2nd edition Pergamon press

Monin, A.S. & Yaglom, A.M. 1975 Statistical fluid mechanics II, The MIT Press

EXPERIMENTAL CHECK OF INFINITE DIVISIBILITY FOR THE VELOCITY CASCADE IN DEVELOPED TURBULENCE

A. NAERT, B. CHABAUD, B. HÉBRAL, B. CASTAING
CRTBT, laboratoire associé à l'UJF, CNRS, BP 166,
38042 Grenoble-Cedex 9, France

Interest raised recently about Infinite Divisibility [1, 2] of dissipation logarithm distributions (IDD) and Extended Self Similarity (ESS) [3] of velocity difference moments [4]. However the largest confusion exists about experimental tests. The relation between IDD and ESS lie on the Kolmogorov refined hypothesis [5] whose verification is not exempt of problems [6, 7]. On the other hand, the multifractal approach [8] and others focused on a distribution of local velocity amplitudes independently of the dissipation field. Some generalizations [9] have been shown [10] to yield both ESS and infinite divisibility for the distributions of logarithms of these velocity amplitudes (the last one hereafter refered as ID).

We thus are concerned with longitudinal velocity differences $\delta v = \vec{e} \cdot [\vec{v}(\vec{x} + \vec{e}r) - \vec{v}(\vec{x})]$ in homogeneous isotropic turbulent flows [11]. Most of the analysis of the δv statistics lie on the hypothesis that there exists a relation between the probability density function (pdf) P_r at the scale r and that at the integral scale L :

$$P_r(\delta v) = \int G_{rL}(\ln \alpha) \frac{1}{\alpha} P_L\left(\frac{\delta v}{\alpha}\right) d\ln \alpha. \tag{1}$$

The conjecture we check in this paper is that the G_{rL} distributions are infinitely divisible, i.e. for any integer n there exists H such that [12] :

$$G_{rL} = H \otimes H \otimes \ldots \otimes H = H^{\otimes n}, \tag{2}$$

where $\otimes$ stands for the convolution product. We have access to the shape of G through the moments of δv. Indeed, using Eq. (1) :

$$\langle |\delta v|^p \rangle_r = \langle \alpha^p \rangle_{rL} \langle |\delta v|^p \rangle_L.$$

With the absolute value [3], we focus on the symmetric part of the pdf. We put aside the interesting problem of the skewness of these pdf, which will

251

S. Gavrilakis et al. (eds.), Advances in Turbulence VI, 251-254.
© 1996 *Kluwer Academic Publishers.*

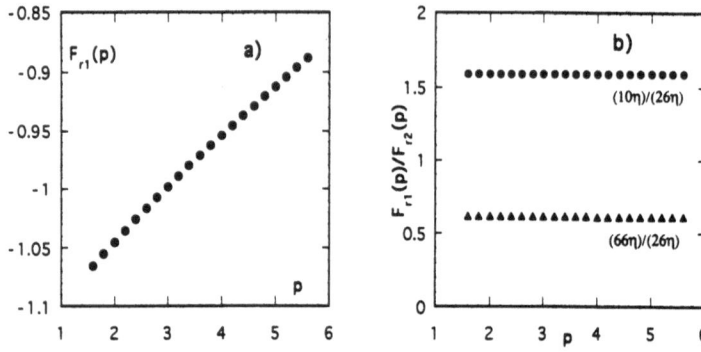

Figure 1. a) $F_{r_1 L}(p)$ versus p for $r_1 = 23\eta$ and $R_\lambda = 320$. The L index has been droped. b) $F_{r_2 L}(p)/F_{r_1 L}(p)$ for $r_2 = 66\eta$ (triangles) and $r_2 = 10\eta$ (circles).

not be discussed in this paper. As $\langle \alpha^p \rangle = 1$ for $p = 0$, we define the function $F_{rL}(p) = (\ln \langle \alpha^p \rangle)/p$. Convoluting the distributions G is equivalent to add the functions F, and the ID results in all the F_{rL} being proportional. This is the property we check. If F can be expanded in Taylor series :

$$F_{rL}(p) = \frac{1}{p} \ln \frac{\langle |\delta v|^p \rangle_r}{\langle |\delta v|^p \rangle_L} = C_1 + C_2 \frac{p}{2} + C_3 \frac{p^2}{6} + \ldots = \sum C_i \frac{p^{i-1}}{i!},$$

defines the cumulants C_i of the distribution G_{rL}. ID means that the ratio of two cumulants is independent of r. On the other hand $F_{rL}(p)$ can be singular in $p = 0$. This is the case if the G are Lévy distributions [12, 13] which gives a $p^{\gamma-1}$ term in F where $0 < \gamma < 2$.

Our results are from a low temperature gaseous ^{4}He jet experiment [14] for Taylor scale based Reynolds numbers R_λ from $R_\lambda = 90$ to $R_\lambda = 600$. For each R_λ, 10^7 velocity measurements were recorded. The hot wire probe size [15] was smaller than the Kolmogorov scale η for $R_\lambda \leq 400$. Figure 1a shows $F_{rL}(p)$ for $r = 23\eta$ and $R_\lambda = 320$. Note first that F is close to linear. This corresponds to G_{rL} being close to gaussian, that is a log-normal distribution [5, 9] for the "multipliers" α. However a downward curvature is visible. Second the extrapolated value for $p = 0$ is much larger than the slope. Thus $|C_1| \gg C_2$ (typically $\left|\frac{C_1}{C_2}\right| \simeq 10$), and F is dominated by its first term C_1. Figure 1b shows the ratio $F_{r_2 L}/F_{r_1 L} = Q(r_2, r_1)$ where r_1 is the preceding scale ($r_1 = 23\eta$) and two different values are chosen for r_2 ($r_2 = 66\eta$ and $r_2 = 10\eta$). These ratio are nicely independent of p, which can be shown equivalent to ESS [3]. As can be observed, $Q(r_2, r_1)$ is not linear in

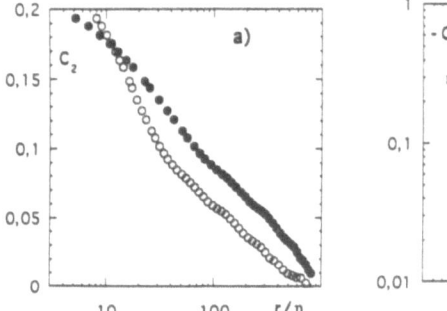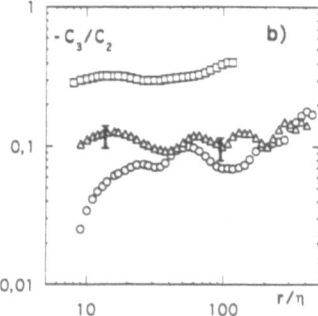

Figure 2. a) Behavior of $C_2 = \langle (\delta \ln \alpha)^2 \rangle$ versus $\ln r$ (open symbols) or $10^3 \ln \frac{\langle |\delta v|^3 \rangle}{\langle |\delta v|^3 \rangle_L}$ (dark symbols). The linearity of the last representation is a test of infinite divisibility. b) The ratio $-C_3/C_2$ versus $\ln r$ for $R_\lambda = 90$ (squares), $R_\lambda = 320$ (triangles) and $R_\lambda = 600$ (circles). In the last case, the drop at small scales corresponds to the lack of resolution of the hot wire. The error bars have a statistical meaning (see text).

$\ln \frac{r_2}{r_1}$ showing that $r_2 = 10\eta$ is already out of the traditional scaling range. The precision of the Figure 1b test is equivalent to that of plotting one moment versus the other in log coordinates. But this precision is very poor due to the large value of C_1. Each F function has a typical 20 % variation on the whole p scale. That their ratio is constant is not surprising !

A much finer test consists in looking for the proportionality of C_2 and $\ln \langle |\delta v|^3 \rangle$ as shown on Figure 2a. The linearity is clearly observed on a larger range than with a $\ln r$ abscissa, also shown in Figure 2a.

A further step in testing the ID is looking for the ratio C_3/C_2. This quantity represents the deviation from the log normality. If the distribution G is Poisson distribution [2, 4], $\frac{C_3}{C_2}$ is the quantum in $\ln \alpha$. In the "thermal" distribution [1, 10], the jumps x in $\ln \alpha$ have a Gibbs-like distribution $\exp \frac{x}{T}$. Then $\frac{C_3}{C_2} = -3T$, measuring the "temperature" T. A real discrimination between these possibilities (and other) would come from C_4 measurement, which is out of reach with the present statistics (10^7 samples). Figure 2b shows how C_3/C_2 behaves with $\ln r$ for various Reynolds numbers. The error bar shown on the $R_\lambda = 320$ results have been estimated by dividing our 10^7 record in ten files of 10^6 samples, and looking at the dispersion of the results [14]. Within these error bars the ratio C_3/C_2 can be considered as constant, confirming the conjecture of infinite divisibility. Strong deviations may be observed at large scale, when the measured cumulants are very small, or small scale when the dissipative length or the size of the hot wire are reached. The trend is clear for a variation of C_3/C_2 with the Reynolds number. However the uncertainties are too large for a law being clearly established. The tendency is to be closer and closer to a log-normal distribution when the Reynolds number is raised.

To conclude, the infinite divisibility hypothesis is experimentally confirmed up to an unprecedented accuracy. As already reported the parameter of this infinitely divisible family [14, 16] is not linear with $\ln r$ which explains the lack of scaling of moments $\langle|\delta v|^p\rangle$ with r and the "extended" aspect of their mutual scaling [3, 17]. This family is always close to the (stable) gaussian law, the closer for the higher Reynolds number. This Reynolds dependence is easily measurable. It is a new manifestation [14, 16] of the subtle effect of viscosity on the whole range of scales, up to rather high Reynolds numbers.

References

1. Saito, Y., *J. Phys. Soc. Jpn* **61** (1992) 403.
2. She, Z.-S. and Lévêque, E., *Phys. Rev. Lett.* **72** (1994) 336; She, Z.-S. and Waymire, E.C., *Phys. Rev. Lett.* **74** (1995) 262.
3. Benzi, R., Ciliberto, S., Baudet, C., Tripiccione, R., Massaioli, F., Succi, S., *Phys. Rev. E* **48** (1993) R29.
4. Novikov, E.A., *Phys. Rev. E* **50** (1994) R3303; Dubrulle, B., *Phys. Rev. Lett.* **73** (1994) 959; Chen, S. and Cao, N., *Phys. Rev. E* **52** (1995) R5757; Pedrizzetti, Novikov, E.A., Praskovsky, A.A., *Phys. Rev. E* **53** (1996) 475; Nelkin, M., *Advances in Physics* **43** (1994) 143.
5. Kolmogorov, A.N., *J.F.M.* **13** (1962) 82; Obukhov, A.M., *J.F.M.* **13** (1962) 77.
6. Stolovitzky, G., Sreenivasan, K.R., *Reviews Mod. Phys.* **66** (1994) 229; Borue, V., Orszag, S., *Phys. Rev. E* **53** (1996) R21.
7. Gagne, Y., Marchand, M., Castaing, B., *J. Phys. II (France)* **4** (1994) 1.
8. Parisi, G. and Frisch, U., in *"Turbulence and Predictability"*, Varenna, Summer School, M. Ghol, R. Benzi, G. Parisi Eds (North-Holland, Amsterdam 1985) p. 84.
9. Castaing, B., Gagne, Y., Hopfinger, E.J., *Physica D* **46** (1990) 177; Castaing, B., Gagne, Y., in *"Turbulence in Spatially Extended Systems"*, R. Benzi, C. Basdevant, S. Ciliberto Eds (Les Houches 1993).
10. Castaing, B., Dubrulle, B., *J. Phys. II (France)* **5** (1995) 895; Castaing, B., *J. Phys. II (France)* **6** (1996) 105.
11. Monin, A.S., Yaglom, A.M., in *"Statistical Fluid Mechanics"* (MIT Press, Cambridge, Massachussetts 1975).
12. Feller, W., in *"An Introduction to Probability Theory and its Applications"* (Wiley, New York 1971) Vol. 2.
13. Schertzer, D., Lovejoy, S., Schmitt, F., in *"Small Scale Structures in 3D and MHD Turbulence"*, P.L. Sulem, M. Meneguzzi, A. Pouquet eds (Springer 1995).
14. Chabaud, B., Naert, A., Peinke, J., Chillà, F., Castaing, B., Hébral, B., *Phys. Rev. Lett.* **73** (1994) 3227.
15. Castaing, B., Hébral, B., Chabaud, B., *Rev. Scient. Inst.* **63** (1992) 4167.
16. Castaing, B., Gagne, Y., Marchand, M., *Physica D* **68** (1993) 387.
17. Benzi, R., Bifferale, L., Ciliberto, S., Struglia, M.V., Tripiccione, L., *Physica D*, in press (1996).

REYNOLDS NUMBER DEPENDANCE OF THE VORTICITY ALIGNMENT WITH THE THREE PRINCIPAL RATES OF STRAIN

A. OOI AND M.S. CHONG
Department of Mechanical and Manufacturing Engineering
University of Melbourne, Parkville, Vic 3052, Australia

AND

J. SORIA
Department of Mechanical Engineering
Monash University, Clayton, Vic 3168, Australia

1. Introduction

Data obtained from direct numerical simulation of the Navier-Stokes equations have shown a tendency for the vorticity vector to align itself in the direction of the intermediate rate of strain. This result was first observed by Ashurst *et al.* (1987) and later by Jiminez *et al.* (1993) and Ooi *et al.* (1994). In this paper, the Reynolds number dependance of the alignment of vorticity with the three principal strain directions is discussed.

2. Data

Numerical data for this study was obtained by solving the Navier-Stokes equations with periodic boundary conditions in the three spatial directions. Two different data sets were considered. In the first instance, a pseudospectral method was used to simulate the Taylor-Green vortex in the domain $x_i \in [0, 2\pi]$. The Taylor-Green vortex is a flow that evolves from the initial conditions $u_1 = \sin(x_1)\cos(x_2)\cos(x_3)$, $u_2 = -\cos(x_1)\sin(x_2)\cos(x_3)$ and $u_3 = 0$, where u_1, u_2 and u_3 are the three orthogonal velocity components in the Cartesian coordinate system. The second set of data is from a direct numerical simulation of forced homogeneous isotropic turbulence. In these simulations, energy is injected into the large scale motions using the Uhlenbeck-Ornstein process (Eswaran & Pope (1988)).

S. Gavrilakis et al. (eds.), Advances in Turbulence VI, 255-258.
© 1996 *Kluwer Academic Publishers.*

3. Results and Discussion

Table 1 shows the percentage of the total enstrophy contained in regions of the flow which have $|\cos(\theta_2)|$ and $|\cos(\theta_3)|$ between 0.98 and 1.00 at $Re = 100$ and $Re = 800$ at the same stage in the evolution of the flow of the Taylor-Green vortex. θ_2 and θ_3 are the angle between the vorticity vector and the intermediate and the largest rate of strain respectively and Q_w is proportional to the enstrophy (Soria et al. (1994)).

TABLE 1. Table showing the dependance of the percentage of the total enstrophy in regions of the flow which has $|\cos(\theta_2)|$ and $|\cos(\theta_3)|$ between 0.98 and 1.00 for different Re.

Re	$Q_w\,\mathrm{P}(0.98 < |\cos(\theta_2)| < 1.00)$	$Q_w\mathrm{P}(0.98 < |\cos(\theta_3)| < 1.00)$
100	0.074	0.140
800	0.175	0.048

TABLE 2. Table showing the dependance of the percentage of the total enstrophy in regions of the flow with $|\cos(\theta_2)|$ between 0.98 and 1.00 for different Re_λ.

Re_λ	$Q_w\mathrm{P}(0.98 < |\cos(\theta_2)| < 1.00)$
41.5	0.11
48.1	0.12
70.9	0.13

For the Taylor-Green vortex simulations the Reynolds number is defined as $Re = 1/\nu$ with ν being the kinematic viscosity. This table shows that at this stage in the evolution and with $Re = 100$, 7.4% of the total enstrophy of the flow are in regions of the flow which has the vorticity vector aligning with the direction of the intermediate rate of strain and 14.0% of the total enstrophy are in regions where the vorticity vector aligns with the largest rate of strain. However, at $Re = 800$, regions of the flow which have the vorticity vector aligning with the direction of the intermediate rate of strain tensor have 17.5% of the total enstrophy while regions of the flow which have the vorticity vector aligning with the largest rate of strain contain 4.8% of the total enstrophy. Table 2 shows the percentage of the total Q_w that is contained in regions of the flow where $0.98 < |\cos(\theta_2)| < 1.00$ for various

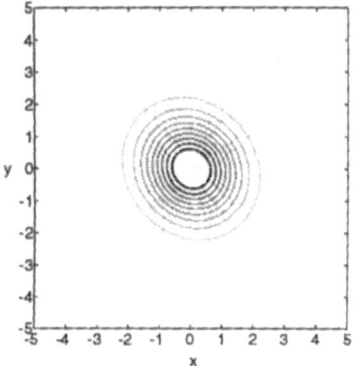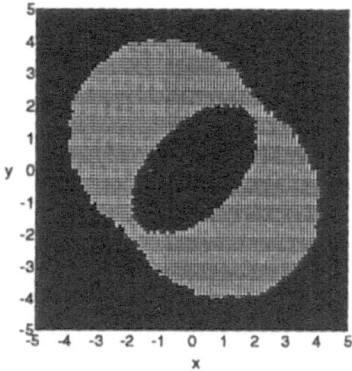

Figure 1. Vorticity countours (left) and the alignment of vorticity with the intermediate rate of strain tensor (right). $Re_\Gamma = 150$, $\lambda = 0.5$. White hatched regions are regions in the vortex where the vorticity vector aligns itself with the intermediate rate of strain tensor and black regions are regions where the vorticity vector align itself with the largest principal rate of strain.

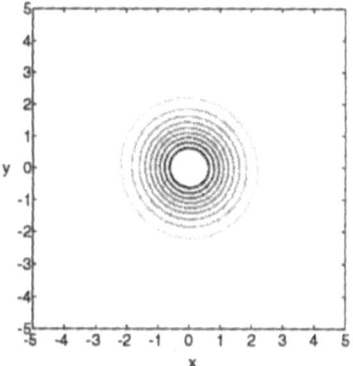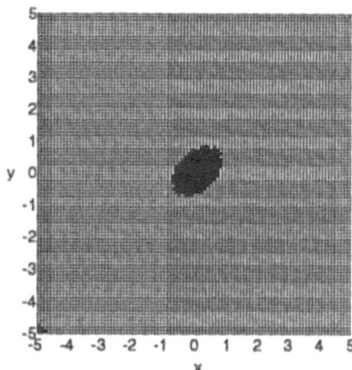

Figure 2. Vorticity countours (left) and the alignment of vorticity with the intermediate rate of strain tensor (right). $Re_\Gamma = 500$, $\lambda = 0.5$. White hatched regions are regions in the vortex where the vorticity vector aligns itself with the intermediate rate of strain tensor and black regions are regions where the vorticity vector align itself with the largest principal rate of strain.

Re_λ calculated using the data of forced homogeneous isotropic turbulence. Re_λ is the Reynolds number based on the Taylor-microscale and the root mean square velocity fluctuations. The data in this table shows that as Re_λ increases, more of the total enstrophy of the flow is found in regions where the vorticity vector is aligned with the intermediate rate of strain tensor. Thus in both flows there is a tendency that as Reynolds number increases, more of the total enstrophy of the flow is found in regions where the vorticity vector is aligned with the intermediate rate of strain.

Moffat (1994) suggested that the fine scale structures of turbulence can be modeled by two-dimensional vortices stretched by a uniform strain field.

These vortices have a Gaussian distribution of vorticity. The temporal and spatial vorticity distribution of such a vortex is governed by the following partial differential equation

$$\frac{\partial \omega}{\partial t} + (\alpha x + u_x)\frac{\partial \omega}{\partial x} + (\beta y + u_y)\frac{\partial \omega}{\partial y} = \gamma \omega + \nu \left(\frac{\partial^2 \omega}{\partial x^2} + \frac{\partial^2 \omega}{\partial y^2} \right) \qquad (1)$$

where α, β and γ are the applied strain rates in the x, y and z directions respectively and ν is the kinematic viscosity. Moffat (1994) provided an asymptotic solution to the steady-state problem of equation (1) for large Re_Γ and a non-axisymmetric strain field, where $Re_\Gamma = \Gamma/\nu$ is the Reynolds number based on the total circulation and the non-axisymmetry of the strain field is measured by the parameter $\lambda = (\alpha - \beta)/(\alpha + \beta)$. Figures 1 and 2 show the spatial distribution of vorticity and the alignment of the vorticity vector with the intermediate rate of strain using the asymptotic solution of Moffat (1994) at $Re_\Gamma = 150$ and $Re_\Gamma = 500$. This data shows that there are some regions close to the core of the vortex where the vorticity vector align itself in the direction of the intermediate rate of strain. This area is found to increase with increasing Re_Γ.

4. Conclusion

From the analysis of the data of forced homogeneous isotropic turbulence and the Taylor-Green vortex, it appears that for higher Reynolds number, more of the total enstrophy in the flow can be found in regions where the vorticity vector is aligned with the intermediate rate of strain. It is shown here that this result is consistent with the properties of stretched tube-like vortices at high Reynolds number.

References

Ashurst W.T., Kerstein A.R., Kerr R.M., and Gibson C.H. (1987) Alignment of vorticity and scalar gradient with strain rate in simulated Navier-Stokes turbulence, *Phys. Fluids* **25**, 2343-2353.

Eswaran V. & Pope S. B. (1988) An examination of forcing in direct numerical simulations of turbulence, *Computers & Fluids* **16**, 257-278.

Jiminez J., Wray A.A., Saffman P.G., Rogallo R.S. (1993) The structure of intense vorticity in isotropic turbulence, *J. Fluid Mech.* **255**, 65-90.

Moffat H.K., Kida S., Ohkitani K. (1994) Stretched vortices-the sinews of turbulence; large-Reynolds-number asymptotics, *J. Fluid Mech.* **259**, 241-264.

Ooi A., Chong M.S., Soria J. (1995) The study of vorticity-strain field alignment characteristics using direct numerical simulation of turbulent flows, *Proceedings of the twelfth Australasian fluid mechanics conference.* **2**, 747-750.

Soria J., Sondergaard R., Cantwell B.J., Chong M.S., Perry A.E. (1994) A study of the fine-scale motions of incompressible time-developing mixing layers, *Phys. Fluids* **6(2)**, 871-884.

SCALING PROPERTIES OF THE VELOCITY CIRCULATION IN A TURBULENT SHEAR FLOW

M. V. STRUGLIA
Gruppo di Modellistica Numerica
ENEA, C. R. Casaccia S.P. 110,
Via Anguillarese 301, S. Maria di Galeria
I-00060 Roma, Italy

R. BENZI AND L. BIFERALE
Dipartimento di Fisica, Università di Roma Tor Vergata,
Via della Ricerca Scientifica 1, I-00133 Roma, Italy

AND

R. TRIPICCIONE
INFN, Sezione di Pisa,
S. Piero a Grado, 50100 Pisa, Italy

Intermittency of the velocity field inertial range statistics has been widely investigated by means of the scaling laws of the velocity increments structure functions

$$F_n(r) \equiv \langle |\delta_r v|^n \rangle \sim r^{\zeta_n} \tag{1}$$

where the anomalous exponents ζ_n differ from the Kolmogorov[1] law $\zeta_n = n/3$.

Recently[2], it has been suggested that the velocity circulation could be a good variable respect to whom we can investigate the statistical properties of a turbulent flow.

The velocity circulation around a closed contour C is defined as

$$\Gamma(C) \equiv \oint_C \vec{v} \cdot \vec{dl} = \int_\Sigma \vec{\omega} \cdot d\vec{\sigma}, \tag{2}$$

where $\vec{\omega}$ is the vorticity field and Σ is any surface, lying on the contour C. The most natural ansatz, by following dimensional arguing, is that circulation structure functions, $G_n(r)$, scale as:

$$G_n(r) = \langle |\Gamma(r)|^n \rangle \sim< |\delta_r v|^n > r^n \tag{3}$$

S. Gavrilakis et al. (eds.), Advances in Turbulence VI, 259-262.
© 1996 *Kluwer Academic Publishers.*

where $\Gamma(r)$ means the circulation evaluated around a contour of radius r. The aim of this work is to study the scaling properties of the structure functions $G_n(r)$ in the case of a turbulent shear flow, and to check if the velocity structure functions and the circulation structure functions have the same intermittent behaviour, as suggested by equation (3).

Our data set comes from a simulation[3] of a 3-D turbulent shear flow, in a volume of $V = L^3$, with $L = 160$, (with our choice of parameters one lattice spacing is about 1 η_k scale and the local Reynolds number is $R_\lambda \sim 40$). The flow is forced such that the stationary solution of the N-S equations is:

$$U_x \sim sin(k_z z) \qquad U_y = 0 \qquad U_z = 0. \tag{4}$$

with $k_z = \frac{8\pi}{L}$ being the wave vector corresponding to the integral scales. In this way the shear has a spatial dependence $S(z) \sim cos(k_z z)$, and we can access both zones where the shear is maximum and locally homogeneous, and zones where the shear is minimum.

We evaluate $\Gamma(r)$ according to the definition (2) for all squared contours with fixed area $A = r^2$, with r extending from the dissipative range to the integral scales, chosen in the x-y plane at two different z-levels corresponding to the maximum and to the minimum shear level, respectively.

In the inertial range a scaling law for $G_n(r)$ is expected to hold

$$G_n(r) \sim r^{\chi_n} \tag{5}$$

Due to the moderate Reynolds number of our simulation we are unable to detect a scaling law of $G_n(r)$ respect to the scale r, therefore we use Extended Self Similarity (ESS) arguments [4, 5, 6] in order to improve the quality and the extension of the scaling regime. We consider the scaling of the structure functions $G_n(r)$ against $G_3(r)$:

$$G_n(r) \sim (G_3(r))^{\gamma_n} \tag{6}$$

At the maximum shear level equation (6) holds from 9 η_k to the integral scales, allowing us to estimate the corresponding scaling exponents, whereas at the minimum shear level the scaling region increases, extending from the dissipative range (4 η_k) to the integral scales.

In table 1 we show the scaling exponents γ_n for the minimum (first line) and maximum shear (second line).

Let us recall that, if the Kolmogorov scaling might be valid, we should obtain, for the scaling exponents defined in (6), the relation $\gamma_n = n/3$.

The values in table 1 give a first positive evidence for the anomalous scaling of the G_n, with an increasing degree of intermittency for the maximum shear case.

TABLE 1. Scaling exponents γ_n

γ_1	γ_2	γ_4	γ_5	γ_6
0.35	0.68	1.30	1.60	1.89
0.36	0.69	1.29	1.56	1.81

Moreover, we estimate the Probability Distribution Function $P(\Gamma)$ for various contours of fixed area $A = 256$ both at the maximum and minimum shear. All the points, within the error bars, collapse on a unique, not Gaussian, curve as it is expected because of the anomalous scaling laws already detected, regardless the presence of shear.

The results so far exposed give a clear indication of the fact that the velocity circulation shows some degree of intermittency.

Now we can ask ourselves how it is related to the velocity increments one. To this end, we turn our attention to relation (3). From (3) it follows that, in the inertial range,

$$\chi_n = \zeta_n + n \tag{7}$$

We cannot check this relation because we are not able to measure directly the exponents χ_n and ζ_n, but just the ratios χ_n/χ_3 and ζ_n/ζ_3, via the ESS. In order to establish if velocity circulations show any new kind of intermittent behaviour respect to the velocity increments one, we consider the following quantity:

$$\psi_n(r) \equiv \frac{G_n(r)}{F_n(r)} \tag{8}$$

If the velocity circulation and velocity increments structure functions have the same degree of intermittency their ratio will not exhibit an anomalous scaling law.

Indeed, we verify that the following scaling law holds,

$$\frac{G_n(r)}{F_n(r)} \sim \left(\frac{G_m(r)}{F_m(r)}\right)^{d(n,m)} \tag{9}$$

with the simple dimensional scaling $d(n, m) = n/m$ for any n, m. In Fig. 1 we show relation (9) for $n = 6$ and $m = 3$ for the maximum (crosses) and minimum shear (diamonds). The dashed lines are the best fits corresponding to the slopes $d(6, 3) = 1.98$ for the minimum shear and $d(6, 3) = 1.97$ for the maximum shear.

We have achieved two main results: we have shown, via the ESS, that the circulation structure functions have anomalous scaling laws, whereas

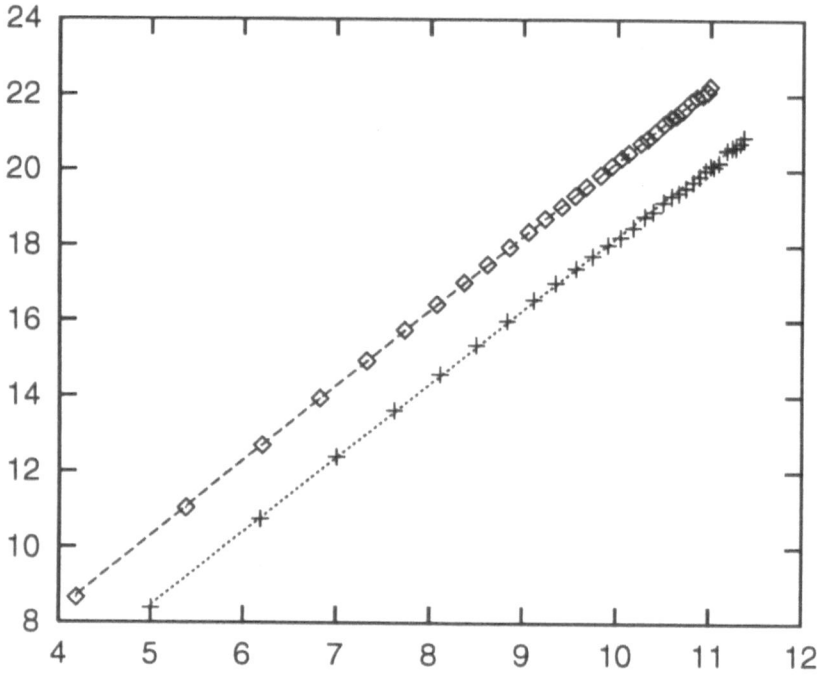

Figure 1. Log-Log plot of relation (9) for $n = 6$, $m = 3$ at the maximum (crosses) and minimum shear (diamonds)

the ψ_n are not anomalous quantities.

It turns out that the anomalous part of the scaling exponents ζ_n and χ_n is the same, confirming the existence of a link between the statistics of the velocity increments and velocity circulation structure functions.

References

1. A. N. Kolmogorov, C. R. Acad. Sci. (USSR) 30 (1941) 299
2. N. Cao, S. Chen, K. R. Sreenivasan *Phys. Rev. Lett.* **76** (4) 1996.
3. R. Benzi, M. V. Struglia, R. Tripiccione, "Extended Self Similarity in numerical simulations of 3D anisotropic turbulence", to appear in *Phys. Rev. E.*
4. R.Benzi, S.Ciliberto, R.Tripiccione, C.Baudet, F.Massaioli, S.Succi, *Phys.Rev E* **48** (1993) R29.
5. M.Briscolini, P.Santangelo, S.Succi, R.Benzi *Phys. Rev. E* **50** (1994) R1745.
6. R.Benzi, S.Ciliberto, C.Baudet, G.Ruiz Chavarria *Physica D* **80** (1995) 385.

GEOMETRICAL STATISTICS IN TURBULENCE

Invariant quantities and relations: alignments

A. TSINOBER
Faculty of Engineering, Tel-Aviv University,
Tel-Aviv 69978, Israel

1. What is geometrical statistics and why is it important?

Geometrical invariant quantities (such as enstropy generation, helicity, etc.) and relations (such as alignments between various vectors) - being independent of the frame of reference - are among the most appropriate for studying physical processes and characterization of the structure of turbulent flows. Moreover, just like phase relations these are *the* quantities and relations of utmost dynamical significance. An overview of a variety of alignments is given below along with applications to basic issues[1] with the emphasis on the physical aspects on the basis of both laboratory and numerical experiments.[2, 3]

1.1. REPRESENTATIVE EXAMPLES

• - The widely known example of the utmost importance of geometrical relations in turbulence is the *qualitative* difference between the dynamics of 3D and 2D turbulence. Since the essential dynamics of 3D-turbulence is contained in the interaction between vorticity $\boldsymbol{\omega}$ and the rate of strain s_{ij}, it depends strongly on the geometry of the field of velocity derivatives, e.g. on the mutual orientation of vorticity $\boldsymbol{\omega}$ and the eigenframe $\boldsymbol{\lambda}_i$ of s_{ij}.

• - It is rather common to use 'surrogates' of the type $(\partial u_1/\partial x_1)^n$ to represent the 'true' quantities such as dissipation, enstrophy ($n = 2$), enstrophy generation ($n = 3$), etc. However, this is true only of their means, whereas

---

[1] Such as: active versus passive, weak versus strong, Gaussian versus non-Gauassian, structured versus nonstructured and some others.

[2] The term **geometrical stastistics in turbulence** was introduced independently in Constantin 1994 and Tsinober *et al.* 1995a. We use this term here in much broader contexts.

[3] A more detailed, comprehensive review of a variety of issues of geometrical statistics in turbulence will be published elsewhere (Tsinober 1996).

S. Gavrilakis et al. (eds.), Advances in Turbulence VI, 263-266.
© *1996 Kluwer Academic Publishers.*

other properties of the 'surrogates' and of the 'true' quantities are generally different. Their PDFs' are essentially different [4] and so are their spectra and fractal properties.

● - It is also a common argument that the prevalence of vortex stretching is due to the predominance of stretching of material lines. However, this argument is, at best, true in part only, since there exists several *qualitative* differences between the two processes, which cannot be seen without using the 'real' quantities.[5]

2. Why Alignments?

Since the discovery of the preferential alignment between vorticity ω and the intremediate eigenvector λ_{int} of s_{ij} (Ashurst *et al.* 1987) a number of alignments in turbulent flows attracted considerable attention.[6] The importance of studying of the statistics of alignments has many aspects. For example, alignments between vectors composed of velocity derivatives are of particular interest, in the first place, since the very existence of such alignments points to the presence of internal organization at small scales. An important aspect is that alignments belong to the rare *objective quantitative statistical* manifestation of the existence of such an organization. Finally, alignments are suitable for events of *any* magnitude.

3. Alignments created by the dynamics of turbulence

The first example of this kind are the alignments between ω and the eigenframe λ_i of s_{ij}, in the sense that for a Gaussian velocity field the PDFs' of $cos(\omega, \lambda_i)$ are *precisely* flat. Another important example is the alignment of vorticity ω and the vortex stretching vector $\mathbf{W}(W_i \equiv \omega_j s_{ij})$ intimately related to the alignments between ω and λ_i and the predominance of enstrophy generation. In real turbulent flows the PDF of $cos(\omega, \mathbf{W})$ is strongly asymmetric in full conformity with the prevalence of vortex stretching, whereas it is symmetric for a random Gaussian field. To dynamically relevant belong alignments between vorticity ω and the eigenframe π_i of the pressure hessian $\Pi_{ij} \equiv \frac{\partial^2 p}{\partial x_i \partial x_j}$ (related to the rate of production of enstrophy generation)[7], between $\mathbf{u}$ and ω, and between ω and $curl\omega$ (both related to helicity), between $\mathbf{u}$ and the external force $\mathbf{f}$, and some others.

---

[4] Hence most their statistical properties are different too.

[5] More details on the two last aspects of geometrical statistics are given in Tsinober 1995, 1996.

[6] For references see Tsinober *et al.* 1995a,b, Tsinober 1996.

[7] Just like the PDF of $cos(\omega, \mathbf{W})$ the PDF of $cos(\omega, \mathbf{W}^\Pi)$, where $W_i^\Pi \equiv \omega_j \Pi_{ij}$, is strongly asymmetric in real turbulent flows, whereas it is symmetric for a random Gaussian field.

Using the dynamically relevant alignments it was possible to show that the structure of vorticity in quasi-homogeneous/isotropic turbulent flow is associated with strong alignments rather than with strong vorticity only. Consequently, much larger regions of turbulent flow than just those with intense vorticity are spatially structured. In particular, large regions with weak vorticity are not structureless, non-Gaussian and are dynamically not passive. [8]

4. Kinematically imposed alignments

Along with the dynamically induced alignments, there exists alignments mostly or completely of kinematical nature, e. g. between velocity $\mathbf{u}$ and $curl\boldsymbol{\omega}$, between the Lamb vector $\mathbf{u} \times \boldsymbol{\omega}$ and pressure gradient (potential part of Lamb vector), between velocity $\mathbf{u}$ and the eigenframe $\boldsymbol{\lambda}_i$ of the rate of strain tensor s_{ij}, and some others. Though imposed mostly by pure kinematic constraints these alignmens are also of dynamical significance, e. g. reduction of nonlinearity, vortex reconnection and direct interaction between large and small scales.

5. Alignments related to problems of stretching of material lines and sufaces, and turbulent diffusion and mixing

Typical examples are the angle between a material line vector $\boldsymbol{\ell}$ and the eigenframe $\boldsymbol{\lambda}_i$, the angle between the material area normal $\mathbf{n}$ and $\boldsymbol{\lambda}_i$, the angle between two material line elements that are initially orthogonal, between the normal to a material plane and a material line that is initially its normal (Girimaji and Pope 1990). In the context of relative diffusion an important angle is the one between the vector $\boldsymbol{\ell}$ and the relative velocity vector of the two points which are connected by $\boldsymbol{\ell}$ (Yeung 1994). MHD-dynamo is another important example (Brandenburg et $al.$ 1996).

6. Alignments as a diagnostic means

Alignments are useful also as an $objective$ diagnostic tool. For example, using the PDF of the cosine of the angle between $\boldsymbol{\omega}$ and $\nabla\omega^2$ (Tsinober et $al.$ 1995b) it was shown that the variation of enstrophy along the vorticity direction $\hat{\omega}$ is slow in regions with intense vorticity as should be in vortex tubes and/or sheets. This example is a part of a broader issue of

---

[8]Note that if ω is strictly parallel to one of the eigenvectors λ_i (i. e. $s_{ij}\omega_j = \lambda\omega_i$) then $D\omega/Dt = \lambda\omega$, and the direction of ω is constant (Okhitani 1993, Constantin 1994). This may be seen as an indication that the preferential alignment between vorticity and the intermediate eigenvector of the rate of strain tensor is closely related to the phenomenon of reduction of nonlinearity.

geometrical statistics of vortex lines and surfaces. In a similar way one can obtain statistical measures of the anisotropy of the flow structure, e.g. in cases when the flow acquires a quasi-two-dimensional structure (in MHD, turbulent convection, stably stratified or flows in a rotating frame). This is a particular case of a more general application (not only as a diagnostic means) in turbulent flows with imposed external constraints/influences. Kinematic alignment between velocity $\mathbf{u}$ and $curl\omega$ appeared useful for checking the performance of complicated multi-hot-wire probes (Tsinober and Vaisburd 1995).

7. Beyond Alignments

Alignments belong to the simplest local aspects of geometrical statistics of turbulent flows. More complicated local aspects include statistics of various properties (curvature, torsion, etc.) of lines and surfaces (material, vorticity). The most difficult are global, e.g. topological aspects, such as those related to helicity and related issues. Almost nothing is known about the statistics of these with the exception of statistics of some local properties of material lines and surfaces. It seems also that the question *Fractals: Where is the physics?* (Kadanoff 1986) cannot be answered without geometrical statistics.

References

Ashurst, Wm.T., Kerstein, A.R., Kerr, R.A. and Gibson, C.H. (1987) Alignment of vorticity and scalar gradient with strain rate in simulated Navier-Stokes turbulence, *Phys. Fluids*, **30**, 2343 -2353.

Brandenburg, A., Jennings, R.L., Nordlund, A., Rieutord, M., Stein, R.F., Tuominen, I. (1996) Magnetic structures in a dynamo simulation, *J. Fluid Mech*, **306**, 325 - 352.

P. Constantin (1994) Geometrical statistics in turbulence, *SIAM Rev.*, **36**, 73 - 98.

Girimaji, S.S. and Pope, S.B. (199) Material-element deformation in isotropic turbulence, *J. Fluid Mech.*, **220**, 427 - 458.

Kadanoff, L.P. (1986) Fractals: Where is the physics?, *Phys. Today*, **39**, 6-7.

Okhitani, K. (1993), Eigenvalue problems in three-dimensional Euler flows, *Phys. Fluids*, **5**, 2570 - 2572.

Tsinober, A. (1995) On geometrical invariants of the field of velocity derivatives in turbulent flows, *Actes du 12^2 Congrès Français de Mécanique*, **3**, 409-412.

Tsinober, A. (1996) Geometrical statistics in turbulence, to be published.

Tsinober, A., Eggels, J.G.M. and Nieuwstadt, F.T.M. (1995a) On alignments and small scale structure in turbulent pipe flow, *Fluid Dyn. Res.*, **16**, 297.

Tsinober, A., Shtilman, L., Sinyavskii, A. and Vaisburd, H. (1995b) Vortex stretching and enstrophy generation in numerical and laboratory tublulence, in *Small-scale structures in three-dimensional hydro- and magnetohydrodynamic turbulence, Lect Notes Phys.*, **462**, 17 - 24, eds. M. Meneguzzi, A. Pouquet and P.-L. Sulem, Springer.

Tsinober, A. and Vaisburd, H. (1995) On the asymptotic behaviour of intermittency exponents of turbulent palinstrophy, *Phys. Lett.*, **A197**, 293-296.

Yeung, P.K. (1994) Direct numerical simulation of two-particle relative diffusion in isotropic turbulence, *Phys. Fluids*, **6**, 3416 - 3428.

NUMERICAL SIMULATION OF TURBULENT LOW–PRANDTL CONVECTION IN A CYLINDRICAL CELL

R. VERZICCO*, R. CAMUSSI*, M. FATICA+ & G. LABONIA*

*Università di Roma "La Sapienza" Dipartimento di Meccanica
e Aeronautica, via Eudossiana 18, 00184 Roma, Italy
+ Center for Turbulence Research, Bldg. 500, Stanford University,
Stanford, CA 94305 USA.

1. Introduction

Low Prandtl number thermal convection is being studied, owing to the interest in geophysical applications and in many technical problems. Earth's liquid–core convection (at $Pr \sim 0.1$) and sodium heat exchangers are only two examples among many. As pointed out by Thual (1992), however, large attention is devoted to this problem also because its study contributes to a large amount of theoretical progress in nonlinear dynamics and turbulence. In addition, many numerical and experimental works are available in literature for Prandtl numbers in the range 0.7-7, while less is known when this parameter assumes lower values; this is the main motivation for the present study.

In this work, a numerical study of Rayleigh–Bénard convection in mercury ($Pr = 0.022$) is performed in a cylindrical cell with aspect ratio (diameter–to–height ratio) $\Gamma = 1$. The choice of this flow configuration has been suggested by the availability of experimental results by Cioni, Ciliberto & Sommeria (1995), obtained in highly controlled conditions, that can be used to validate the numerical simulations. A further point is that, the most of the theoretical predictions for the the convective turbulence (Chandrasekar, 1961) assume infinite aspect–ratio geometries, therefore an intriguing question concerning the present flow is whether or not the convection in small–aspect–ratio geometries satisfies the theoretical predictions.

Finally, the numerical results can provide a lot of information impossible to get in experimental studies since liquid metals do not allow a direct visualization of the flow structures. Contour plots of the average and rms velocity and temperature fields are an example and they will be discussed in the results.

S. Gavrilakis et al. (eds.), Advances in Turbulence VI, 267-270.

2. Numerical model and initial conditions

The numerical code solves the three–dimensional incompressible Navier–Stokes equations with the Boussinesq approximation, written in cylindrical coordinates (Verzicco & Orlandi, 1996). The equations are discretized on a staggered grid and the resulting system of equation is solved by a fractional–step method. The discretization of viscous and advective terms is performed by central second–order accurate finite–difference approximations. The time advancement of the solution is obtained by a third order Runge-Kutta scheme which calculates the non-linear terms explicitly and the viscous terms implicitly. The code has been validated by comparing the value of the critical Rayleigh number (Ra_c) for several aspect ratios Γ with the analytical results available in literature. In addition, for the aspect ratio $\Gamma = 1$, it has been checked that the maximum amplitude of the vertical velocity grows as $[(Ra - Ra_c)/Ra_c]^{0.5}$, for $Ra_c < Ra \leq 6Ra_c$.

Simulations with a fine grid ($97 \times 85 \times 193$ gridpoints in the azimuthal, radial and vertical directions respectively) have been run with the Rayleigh number in the range $10^5 \leq Ra \leq 6 \cdot 10^5$. The spatial and temporal resolution requirements given by Domaradski & Metcalfe (1988) have been satisfied in all runs. The numerical simulations were run in time until a statistical steady state was reached, then data were accumulated for 2 to 4 large–eddy–turnover times and the fields analysed.

3. Results and discussion

The large–scale motion, usually underlying every turbulent flow, becomes particularly important in Rayleigh–Bénard convection since the large eddies carry the heat from the bottom to the top of the cell thus contributing significantly to the heat transfer. Also, some heuristic explanations of the $Nu = N_0 Ra^\beta$ scaling consider the shearing winds produced by the large eddies and blowing across the boundary layers a fundamental mechanism in the determination of the β exponent. Castaing et al. (1989) refer that a careful study, including visualizations, to ascertain the real geometry of the flow was lacking in their paper, although experimental evidence suggests its importance. Indeed the details of the mean flow depend on the cell geometry and on the Rayleigh and Prandtl number, even if the first factor is the most effective. In large–aspect–ratio geometries the flow tends to develop convective rolls whose vertical extension spans the whole cell and with even larger horizontal dimension. This was recently confirmed by Kerr (1996) who observed in rectangular cells the appearance of large–scale patterns aligned along diagonals once the aspect ratio of the cell was ≥ 3. In contrast, when the cell becomes narrow large rolls are not observed, while smaller eddies with limited vertical extension are present. This is shown in figure 1 where velocity vectors and temperature fields are reported. This field has been obtained as an azimuthal average of three–dimensional flow fields and as time average over 2 large–eddy–turnover times. We wish to stress that occasionally snapshots of the temperature fields have shown hot plumes rising from one side of the cell and cold plumes going down along the other side, thus yielding a large recirculating eddy as big as the cell. However, if such a structure were steadily maintained in the flow its azimuthal average would give no motion, in contrast with the eddies shown in figure 1a.

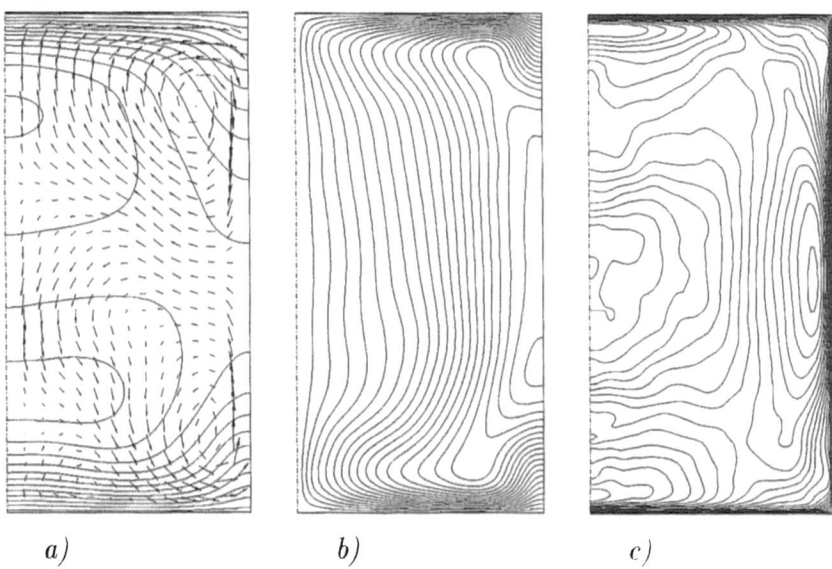

a)   *b)*   *c)*

Figure.1 Flow maps averaged in the azimuthal direction and in time for the Rayleigh–Bénard convection at $Ra = 6 \cdot 10^5$ and $Pr = 0.022$: *a)*, contour plots of temperature ($\Delta T = 0.1$) and velocity vectors; *b)*, contour plots of temperature rms ($\Delta T_{rms} = 0.0035$); *c)*, contour plots of velocity rms magnitude ($\Delta U_{rms} = 0.01$).

The persistence of two counter–rotating eddies in the mean flow, has been observed over the whole range of Rayleigh number investigated and this allows us to withdraw some considerations about their effect on the flow. When estimating the large–eddy–turnover time for the Rayleigh–Bénard convection, one usually assumes that the largest eddy has a size comparable with the distance between the horizontal walls H. Figure 1a shows that this is not true at least for this range of Rayleigh numbers and, since smaller eddies have faster revolution periods, the time needed to accumulate enough statistics might be considerably reduced (by a factor 2). This is of great help for the numerics, considering the high computational cost of three–dimensional numerical simulations.

Also, the convective motion of the two counter–rotating eddies is such that outside the boundary layer there is an inverted stable thermal stratification close to the axis of symmetry, while the stratification is unstable in the region close to the vertical wall. This is confirmed by the panels b and

270

c of figure 1 showing that the highest temperature fluctuations occur close
to the vertical walls where also strong velocity fluctuations are present.

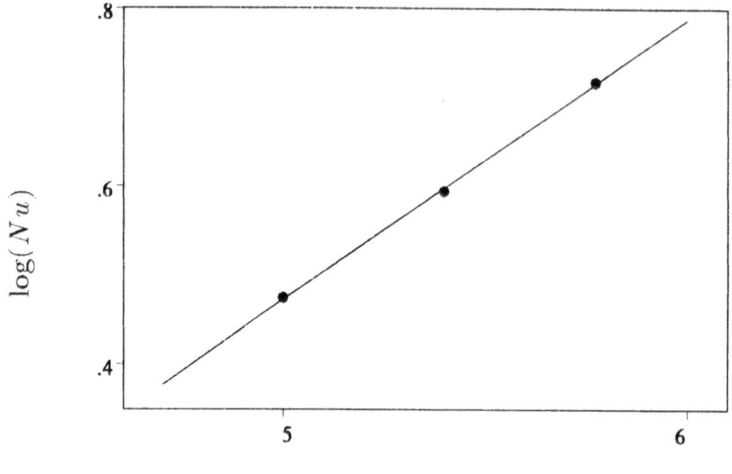

$$\log(Ra)$$

Figure.2 Nusselt versus Rayleigh plot; the linear fit law yields $Nu = 0.0786 Ra^{0.3154}$.

The final result that we wish to show is the power law of the Nu vs Ra.
A linear fit of the computed points yields an exponent $\beta = 0.3154$ which
is in between the classical value $\beta = 1/3$ and the value $\beta = 2/7$ found in
many experiments. This value, however, is consistent with the theory since,
Castaing *et al.* (1989) found that the value $\beta = 2/7$ pertains to the 'hard
turbulence' regime that, for the present geometry, holds only for $Ra \geq 10^7$.

The authors wish to acknowledge the Dr. M. Briscolini of IBM STSS Europe
Rome for useful advice during the preparation of the results. Dr. S. Cioni is also
acknowledged for the advice and the discussion of the results.

References
1. Castaing, B., Gunaratne, G., Heslot, F., Kadanov, L. Libchaber, S.T., Wu, X.Z.,
Zaleski, S. & Zanetti, G., 1989, "Scaling of hard turbulence in Rayleigh–Bénard
convection", *J.Fluid.Mech*, **204**, 1–30.
2. Cioni, S., Ciliberto, S. & Sommeria, J., 1995, "Temperature Structure Functions
in turbulent Convection at Low Prandtl Number", *Europhys. Lett.*, **32** (5), 413–
418.
3. Domaradski, J. A. & Metcalfe, R.W., 1988, "Direct numerical simulations of
the effects of shear on turbulent Rayleigh–Bénard convection", *J.Fluid.Mech*, **193**,
499–531.
4. Kerr, R. M., 1996, "Rayleigh number scaling in numerical convection", *J.Fluid.Mech*,
310, 139–179.
5. Thual, O., 1992, "Zero–Prandtl–number convection", *J.Fluid.Mech*, **240**, 229–
258.
6. Verzicco, R. & Orlandi, P., 1996, "A finite–difference scheme for three-dimensional
incompressible flow in cylindrical coordinates", *J. Comp. Phys.* , **123**. 402–414.

HIGH REYNOLDS NUMBER EXPERIMENT :
TRANSITION AND STRUCTURES

H. WILLAIME, F. BELIN, J. MAURER, P. TABELING

Laboratoire de Physique Statistique, ENS, CNRS,

24 rue Lhomond 75005 Paris, France

We present a high Reynolds number experiment. The flow takes place in a cylindrical vessel and is confined and forced by two counter-rotating disks. We use two different cells, the dimensions are for the first cell (the second resp) : a radius of 3.3 cm (10 cm resp) and a height of 5.5 cm (13.1 cm for the second one). The fluid used is helium gas at low temperature (4.2 to 8 Kelvin). At this temperature, by varying the pressure from 60 mbar to 6 bars, the kinematics viscosity varies over 3 orders of magnitude, and so the Reynolds number varies from 10^4 to 10^7 (in term of microscale Reynolds number R_λ, the system explores a range of value from 150 to 3200).

In this experiment, the local velocity is measured by a 'hot wire' anemometer: a 7 μm thick carbon fiber is stretched across a rigid frame, and is covered by a metal everywhere except on a spot at the centre, which thus defines the active lengths of the probe (from 7 to 50 μm). Depending on the experimental conditions, the spatial resolution of the probe varies from 0.2 (in the worth case) up to 3 times the Kolmogorov scale. The frequency response is good enough to study the dissipative range for R_λ up to 1500 and the inertial range for R_λ up to 3200. The velocity is measured at several points. The signal is digitized on a 16 bit converter. and the ratio signal over noise is typically 70 dB.

In this paper, we focus ourselves on the study of the probability density function (PDF) of the velocity derivatives. In particular, we analyse the evolution of the skewness S and the flatness F of these PDF with R_λ. These two quantities are

S. Gavrilakis et al. (eds.), Advances in Turbulence VI, 271-274.
© 1996 *Kluwer Academic Publishers.*

characteristic of intermittency in turbulent system by measuring differences between the PDF (which look like stretched exponential) and the gaussian distribution predicted by Kolmogorov.

One way to determine S and F, is to determinate the limit of :

$$s(r) = \frac{<\Delta V^3_r>}{<\Delta V^2_r>^{3/2}} \tag{1}$$

$$\text{and} \qquad f(r) = \frac{<\Delta V^4_r>}{<\Delta V^2_r>^2} \tag{2}$$

when r becomes small (in units of the Kolmogorov scale).

In figure 1, F is plotted in function of R_λ, -S has a similar behaviour. By analysing this plot, one can see that for R_λ above 700, F (and -S) increases with R_λ and the values obtained are in good agreement with previous experiments [1] and numerical simulations [2]. However the striking observation we make, is that above $R_\lambda = 700$, these 2 quantities decrease with R_λ which is in contradiction with the results of compilation [1] . We can so define 2 regimes separated by a transition. This result is somehow surprising because that means that the system becomes less and less intermittent; it is still to be understood. At higher R_λ, F and -S seem to increase again.

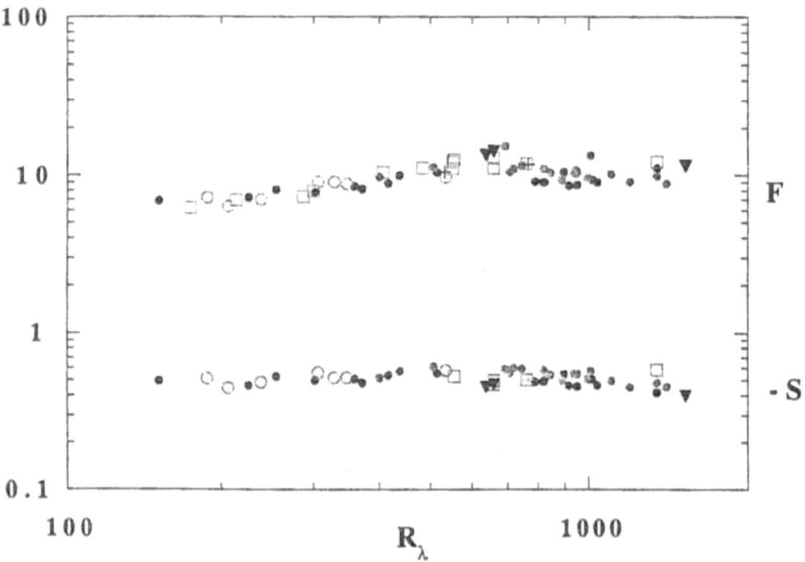

Figure 1 : -S and F versus R_λ in the 2 cells and at different fluctuation rate.

To complete this analysis of the intermittent part of the turbulent flow, we study intense events that appear in the signal, which correspond to the one that contribute to the tails of the PDF (i.e., in the dissipative range). For $R_\lambda < 700$, one can clearly identify structures that are intense filaments (around 40% of the events) such as the one observed numerically by Jimenez [2]. The distributions of the radius in term of Kolmogorov scale (the mean radius is around 3 Kolmogorov scales) of such filaments are independent of R_λ and their sizes are in good agreement with the one observed numerically (see figure 2). By suppressing these events of the signal, we show that these events control significantly the intermittency in the inertial range and in particular the exponents of structure functions are closer to the Kolmogorov value p/3. Above $R_\lambda = 700$ the structures described before are less well defined, the radius distributions are much wider and change with R_λ : filaments become unstable.

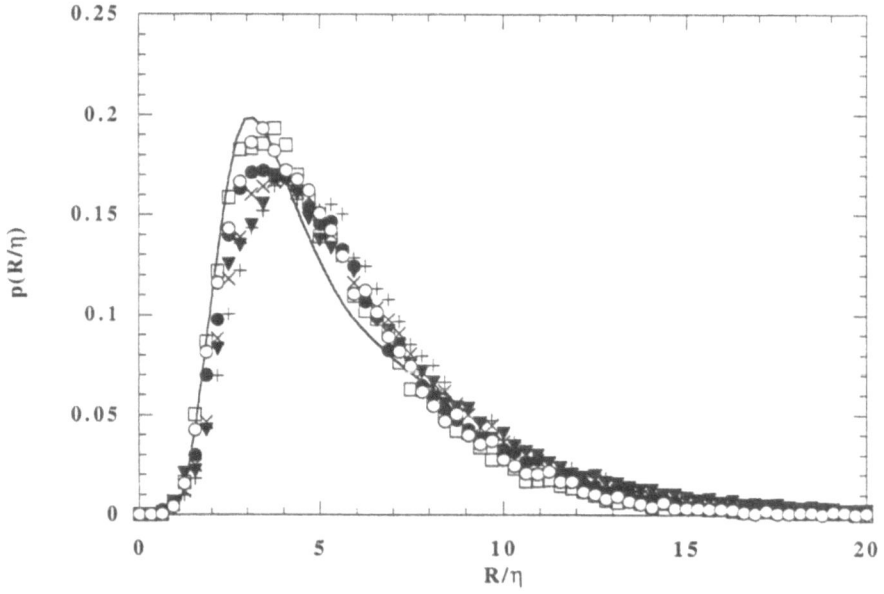

Figure 2 : Size of distributions of worms (a) for different R_λ : $+$ 151; $\bullet$ 225; $\sqsupset$ 255; $\bigcirc$ 359 ; $\blacktriangledown$ 507; $\times$ 718 ; full line is an averaged distribution, obtained from Ref. [2].

In conclusion, we have shown that for microscale Reynolds numbers below 700, the flatness and the skewness of the velocity derivatives increase with R_λ. But above $R_\lambda = 700$, the system shows a transitional behavior. This observation is somehow consistent with the worm study we have performed. Below $R_\lambda = 700$, the distributions of the radius (in unit of Kolmogrov scale) of the worms are independent of the Reynolds number. Above this threshold, the distributions change, and the worms seem to become larger and less well defined.

These structures, that are dissipative, play a non negligible role in the inertial range; this result may not survive at infinite Reynolds number. One may ask whether such structures exist in other systems (i.e. atmospheric boundary layers, jets or wind-tunnels).

References :

1. Van Atta, C.W. , and Antonia, R.A. : Reynolds number dependence of skewness and flatness factors of turbulent velocity derivatives, *Physic of Fluids* **23** (1980), 252-257

2. Jiménez, J., Wray, A.A., Saffman, P.G., and Rogallo, R.S. : The structure of intense vorticity in isotropic turbulence, *J. Fluid Mech* **153** (1985), 31-58

INVESTIGATION OF THE SPATIAL STRUCTURE OF TURBULENCE IN THE WAKE BEHIND BLUFF BODIES

E. Gaudin, S. Goujon-Durand
University Paris XII, Faculte de Sciences et Technologie
61, av. du General de Gaulle, 94010 CRETEIL cedex, FRANCE
B. Protas, J. Wojciechowski
Warsaw University of Technology, Inst. of Aeronautics & Appl.
Mechanics
24, Nowowiejska St, 00-665 Warsaw, POLAND
J.E. Wesfreid
PMMH ESPCI
10, rue Vauquelin, 75231 PARIS cedex, FRANCE

In the analysis of developed turbulence it is customary to make use of the universal scaling of the ensemble averaged moments of the structure function defined as

$$S_n(r) = \left\langle \left| V(x+r) - V(x) \right|^n \right\rangle \tag{1}$$

where r denotes the separation distance. Usually, the structure function is recovered by means of the Taylor Hypothesis from a time series recorded at a given point.

Concerning the scaling properties of structure functions, the celebrated K41 theory predicts that:

$$S_n(r) \propto r^{\frac{n}{3}} \tag{2}$$

With the recently developed concept of Extended Self Similarity (ESS) (Benzi *at al.*, 1995) it is possible to determine the scaling exponents when the inertial range is relatively short (i.e. the case of low Reynolds numbers flows):

$$S_n(r) \propto S_3(r)^{\zeta_n} \tag{3}$$

The deviation of the scaling exponent ζ_n from the K41 values is a plausible means of investigating the effects of intermittence (Anselmet *at al.*, 1984). Within the framework of isotropic and homogeneous turbulence, ζ_n's are assumed to be universal. Their specific numerical values are given by a number of theories, e.g. β-models, K62, multifractal models, random cascade models, etc. (Frisch, 1995). An interesting problem is how the presence of physically relevant boundary conditions affects the statistical properties of turbulence and thus the behavior of the scaling exponent ζ_n? Interaction with boundaries is responsible for the production of large scale shear and

275

S. Gavrilakis et al. (eds.), Advances in Turbulence VI, 275-276.
© 1996 *Kluwer Academic Publishers.*

276

coherent structures. In such cases, substantial degree of anisotropy is introduced at large scales, but locally the flow still remains isotropic. Therefore the turbulent wake behind bluff bodies was investigated both experimentally (in similar set-ups in two different laboratories) and numerically (2D direct numerical simulation with RANDOM VORTEX BLOB METHOD) (e.g. Chorin 1973).

Owing to relatively short time series obtained both from the numerical simulation and the experiments, only low order ($n=2,4$) statistical moments of the structure function could be obtained with required accuracy. The scaling exponents were ζ_n found to strongly vary in the near wake, where strong shear and coherent structures are ubiquitous. On the contrary, in the far wake, where the flow becomes more homogeneous, they approach some asymptotic values (Fig.1):

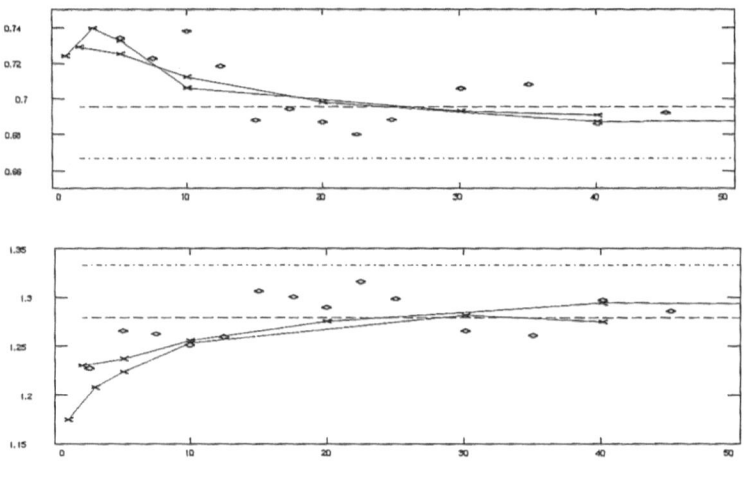

Fig.1

Dependence of the scaling exponents ζ_2 (upper) and ζ_4 (lower) on the non-dimensionalized distance from the obstacle (diamonds ( )- numerical simulation, solid line with "x's" - experiments, dash dot - K41 prediction, dashed line - She & Leveque model).

It is clearly visible that in the vicinity of boundaries the exponents ζ_n ($n=2,4$) cannot be regarded as universal, i.e. independent of the large flow structure. Work is going on on possible explanations of the discovered anomalies, as well as their relationship to other parameters of the flow.

References:
Anselmet, F., Gagne, Y., Hopfinger, E.J. and Antonia, R.A. (1984) High-order velocity structure function in turbulent shear flows, *J. Fluid Mech.*, **140**, 63-89
Benzi, R., Ciliberto, S., Baudet, C., Chavarria, G.R., (1995) On the scaling of three-dimensional homogeneous and isotropic turbulence, *Physica D 80*, 385-398
Chorin A.J. (1973), Numerical Study of Slightly Viscous Flow, *J. Fluid Mech.*, **53**, 785
Frisch, U., (1995) *Turbulence: The legacy of A.N. Kolmogorov*, Cambridge University Press, Cambridge

ON EXPONENT OF VORTEX NUMBER IN 2D TURBULENCE

X. HE
Mathematics Institute
University of Warwick
Coventry CV4 7AL, England

Coherent vortices play a dominant role in dynamics of incompressible fluid turbulence (see [C1], [M]). In two-dimensional (2D) unforced turbulence, numerical simulations [B],[D],[M] and laboratory experiments [F],[T] found that the total vortex number $N(t)$ decreases as $N \sim t^{-\xi}$, where different values of ξ were obtained. An explanation for ξ has remained a theoretical challenge.

In this note, we present an analytical study on the time evolution of the total vortex number N in a simple field model, using a probabilistic method recently developed [H1]. This new method is based on two elementary assumptions: (i) The vortex centre is subject to a Brownian motion in the presence of an external strain-rotation field, $F(\mathbf{x}, t)$; and (ii) The path of the vortex centre satisfies a stochastic variational principle similar to that in [N]. From these assumptions, an evolution equation for a 2D vortex was derived in [H1],

$$i\frac{\partial \psi}{\partial t} = [\, -\Gamma \triangle + F(\mathbf{x}, t) \,] \, \psi, \tag{1}$$

where ψ is the position probability amplitude, Γ is the initial circulation of the vortex, and F is the strain-rotation field. This evolution equation has been applied to various problems in vortex dynamics, and reasonable results have been obtained [H2–H4]. Here we shall quantize (1), to give a field model for a many-vortex system. Then by calculating the transition probability, we show that $N \sim t^{-\xi}$, where the decay exponent ξ is a function of initial conditions.

Let ψ be an operator, $\psi^\dagger$ be its adjoint. Assume the commutation relations

$$[a_l, a_m] = [a_l^\dagger, a_m^\dagger] = 0, \quad [a_l, a_m^\dagger] = \delta_{lm}, \tag{2}$$

where $a_l(t) = \int g_l^*(\mathbf{x})\psi(\mathbf{x}, t) \, d\mathbf{x}$, $a_l^\dagger(t) = \int g_l(\mathbf{x})\psi^\dagger(\mathbf{x}, t) \, d\mathbf{x}$. The commutation (2) was solved in [H3], so that in this representation the total vortex number operator is $\hat{N} = \sum_l \hat{N}_l = \sum_l a_l^\dagger a_l$.

We simplify coherent vortices as circular patches of uniform vorticity, which is relevant to fluid turbulence [C2],[D]. These patches form a finite set $\mathcal{A}$, and $\mathcal{A}$ is partitioned into disjoint subsets $\mathcal{A}_l$ ($l = 1, 2, \cdots k$) which contains elements of circulation Γ_l. Let $\tilde{f}$ be a perturbation to lth subset resulted from a straining flow

$$\tilde{f} = \hat{f}_0(\mathbf{x}) - 2\gamma(t)(x + y), \tag{3}$$

where $\gamma(t)$ is the strain rate divided by a length scale. Then in the subset, the probability of vortex number changing from time t_0 to t is [H4]

$$P_l(t) = \frac{\beta^{2m}}{(2^m m!)^2}(t - t_0)^2 \gamma^{2m}(t)e^{-\gamma^2(t)\beta^2/2}, \tag{4}$$

277

S. Gavrilakis et al. (eds.), Advances in Turbulence VI, 277-278.
© *1996 Kluwer Academic Publishers.*

where β is a constant. In [D] it shows the strain rate is slowly decreasing in time as $\gamma \sim (t - t_0)^{-1/2}$; thus using this form of $\gamma(t)$, for large time (4) approximates to $P_l(t) \sim t^{-m_l}$, where m_l denotes the initial number of vortices with circulation Γ_l in the subset $\mathcal{A}_l$. Now the total vortex number in the field can be written as

$$N(t) = \sum_l^k c_l P_l(t) \approx \sum_l^k c_l\, t^{-m_l}, \tag{5}$$

where c_l are normalization constants. Further one obtains from (5) that $N(t) \sim t^{-\frac{1}{k}\sum_{l=1}^{k} m_l} \equiv t^{-\xi}$, where $\frac{1}{k}\sum_{l=1}^{k} m_l = \xi$ is a function of initial conditions in terms of vorticity ω and vortex radius r ($\Gamma_l = \omega_l \pi r_l^2$). Suppose that initially the vortex number distribution is a function of vortex radius r only, then one can put $\xi = \frac{1}{k}\sum_{i=1}^{k} m_i \approx \int m(r)dr$, such that

$$N(t) \sim t^{-\int_{r_1}^{r_2} m(r)dr}, \quad r_1 < r_2. \tag{6}$$

Consider (a) a power law distribution $m(r) = r^{-\alpha}$, and (b) vortices of the same radii $m(r) = \delta(r_0)$, $r_0 \in (r_1, r_2)$. In case (a) $\xi \approx 0.5$ if $\alpha = 3$ and $r_2 = 10r_1$. In case (b), $\xi = 1$. These results are consistent with [B], [D] in case (a), and with [F],[M] and [T] in case (b). Hence the present work suggests that the decay exponent ξ depends on initial conditions.

References

[B] Benzi, R., Colella, M., Briscolini, M., Santangelo, P. (1992) A simple point vortex model for 2D decaying turbulence, Phys. Fluids **4**, 1036.

[C1] Chorin, A.J. (1994) Vorticity and Turbulence, Applied Mathematical Sciences **103**, Springer.

[C2] Constantin, P., Wu, J. (1995) Inviscid limit for vortex patches, Nonlinearity **8**, 735.

[D] Dritschel, D.G. (1993) Vortex properties of 2D turbulence, Phys. Fluids **5**, 984.

[F] Fine, K.S., Cass, A.C., Flynn, W.G., Driscoll, C.F. (1995) Relaxation of 2D turbulence to vortex crystals, Phys. Rev. Lett. **75**, in press.

[H1] He, X. (1994) On the evolution equation of an inviscid vortex in 2D turbulence, Chaos, Solitons and Fractals **4**, 693.

[H2] He, X. (1994) The most probable critical distance for two unequal vortices in 2D turbulence, Chaos, Solitons and Fractals **4**, 1183.

[H3] He, X. (1995) On vortex entities of 2D turbulence in wavenumber space, Chaos **5**, 687.

[H4] He, X. (1995) Dispersion of a vortex trajectory in 2D turbulence, Physica D (accepted for publication); —— and Valcke, S. (1996) Vortex merging in stratified fluids, submitted.

[M] McWilliams, J.C. (1990) The vortices of 2D turbulence, J. Fluid Mech. **219**, 361.

[N] Nelson, E. (1988) Stochastic mechanics and random fields, in Lecture Notes in Mathematics **1362**, 428.

[T] Tabeling, P., Burkhart, S., Cardoso, O., Willaime, H. (1991) Experimental study of freely decaying 2D turbulence, Phys. Rev. Lett. **67**, 3772.

WAVELET ANALYSIS OF HIGH-RESOLUTION TURBULENCE DATA

F. NICOLLEAU AND J.C. VASSILICOS
DAMTP University of Cambridge
Silver Street, Cambridge CB3 9EW, UK

1. The zero-crossings method

The zero-crossings of a Mexican-hat wavelet transform detect inflection points in the velocity signal $u(x)$. These zero-crossings can be used to define turbulent eddies that have a well-defined length-scale but are also *positioned* in x-space. The eddy capacity D_E is a box-dimension of these zero-crossings conditioned on wavelet-scale and characterises the distribution in x-space of turbulent eddies of all sizes (Kevlahan & Vassilicos 1994). We find that D_E is a direct geometrical measure of intermittency. Indeed, the flatness factor $F(r)$ of velocity differences $\Delta u(r)$ of self-similar signals is related to D_E by

$$F(r) \sim r^{D_E - 1}. \tag{1}$$

This relation is established empirically on a variety of different self-similar signals (spiral functions, random-phase signals, fractional Brownian motions, Weierstrass functions and other fractal constructions based on Cantor-like multiplicative processes in which case (1) can also be proved analytically). A self-similar signal is *not* intermittent if $D_E = 1$. We also find in all these cases that D_E is equal to the Kolmogorov capacity D_c' of the zero-crossings of the second derivative $\frac{d^2}{dx^2}u(x)$ of the signal which is important because D_c' is algorithmically easier to calculate than D_E. Furthermore, D_E and D_c' can be determined from data that are far less well resolved than is necessary for an accurate and converged determination of $F(r)$.

Finally we measure D_E and D_c' of a high resolution (1.2η, where η is the Kolmogorov length-scale) and high Reynolds number ($Re_\lambda = 3050$) turbulence 1-D velocity signal obtained by Y. Gagne at the wind-tunnel of Modane. We find that D_E and D_c' are between 0.92 and 0.97 in the

S. Gavrilakis et al. (eds.), Advances in Turbulence VI, 279-280.

upper length-scales of the inertial range. The r-dependence of $F(r)$ agrees with such values of D_E and D'_c in these scales. However, as noted also by Kevlahan & Vassilicos (1994), D_E and D'_c are very sensitive to noise and cannot be measured with confidence at the smaller length-scales of the inertial range.

2. The average wavelet method

These small inertial turbulence scales can be studied with a different wavelet technique which is based on successive averages of wavelet transforms of different realisations of the turbulence and which is fairly insensitive to noise (Kevlahan & Vassilicos 1994). The central idea of this technique is that the asymptotic convergence of these averages is directly linked to the distribution of the wavelet intensity in "wavelet space". This wavelet technique is developed here and leads to a distinction between three different types of such distributions: scale-dependent, scale-independent, and space-filling.

Scale-dependent distributions of wavelet intensity are such that the portion of space where the wavelet intensity lies decreases with wavelet scale. This behaviour reflects non-isolated near-singularities and in that sense is characteristic of an intermittent fractal topology. By contrast, scale-independent distributions of wavelet intensity are such that below a certain length scale the wavelet intensity fills the same amount of space at all wavelet scales without filling the entire space. This behaviour is characteristic of isolated near-singularities, in particular spiral-like structures. Finally, space-filling distributions of wavelet intensity fill the entire wavelet space at all wavelet scales, a behaviour which is characteristic of non-intermittent fractal structures. Thus the difference between non-isolated (fractal) and isolated near-singularities as well as the difference between intermittent and non-intermittent fractals are attainable by this method.

We apply the method to a variety of spiral and fractal signals, and to the high Reynolds number turbulence 1-D velocity signal obtained by Y. Gagne. The results of this wavelet analysis suggest that a length-scale exists above which the distribution of wavelet intensities of the the turbulence is very closely space-filling and below which it is scale-independent. This analysis having been conducted for only one Reynolds number does not allow a decisive identification of this length-scale. However, in the present Reynolds number-specific turbulence data, this length-scale is found to be close to the Taylor microscale.

References

Kevlahan N.K.-R. & Vassilicos J.C. (1994) *Proc. R. Soc. Lond.* A **447**, pp. 341-363.

THE CONTINUUM LIMIT
IN A SHELL MODEL OF TURBULENCE

E. TROVATORE

INFM-Dipartimento di Fisica, Università di Cagliari
Via Ospedale 72, I-09124, Cagliari, Italy

AND

R. BENZI AND L. BIFERALE

Dipartimento di Fisica, Università di Tor Vergata
Via della Ricerca Scientifica 1, I-00133 Roma, Italy

The aim of this contribute is to get some insight on the dependence of shell model intermittency on one of the free parameters entering the shell-modelization: the scale separation between neighbouring shells, λ. In particular we will address the problem of what happens in the so-called continuum limit, i.e. in the limit of infinitely close shells ($\lambda \to 1$).

The shell model we have considered (Benzi *et al.*, 1996a) belongs to a class of helical shell-models characterized by having two complex dynamical variables per shell, u_n^+ and u_n^-, transporting positive and negative helicity respectively. This helical model conserves two quadratic quantities in the inviscid and unforced limit, i.e. the energy $E = \sum_n(|u_n^+|^2 + |u_n^-|^2))$ and a generalized helicity $H_\alpha = \sum_n k_n^\alpha(|u_n^+|^2 - |u_n^-|^2)$ and was shown to have an intermittent forward energy cascade that, analyzed by means of the scaling properties of the structure functions $S_p(n) = \langle|u_n|^p\rangle \sim k_n^{-\zeta(p)}$, exhibits the same exponents $\zeta(p)$ of the true Navier-Stokes equations.

The main characteristic of the model is the strong universality of its exponents against variations of the helicity dimension α. On the other hand, the $\zeta(p)$s depend on the scale separation λ.

We have performed a detailed analysis of the self similarity properties of the structure functions $S_p(n)$ at varying λ, by using Extended Self Similarity (ESS), Generalized Extended Self Similarity (GESS) and the hierarchy relation (She and Leveque, 1994) based on the log-poisson hypothesis, as discussed in (Benzi *et al.*, 1996b).

S. Gavrilakis et al. (eds.), Advances in Turbulence VI, 281-282.
© *1996 Kluwer Academic Publishers.*

In particular, in the limit $\lambda \to 1$, the $\zeta(p)$ exponents tend toward the K41 values, indicating an intermittent behaviour of the model which depends on how much local the interactions between adjacent shells are.

This tendency toward a non intermittent behaviour letting the interactions being more and more local can be understood in the framework of a closure theory which was developed in (Benzi *et al.*, 1993).

References

Benzi, R., Biferale, L. and Parisi, G. (1993) On intermittency in a cascade model for turbulence, *Physica D*, **Vol. 65**, p. 163.

Benzi, R., Biferale, L., Kerr, R. and Trovatore, E. (1996a) Helical shell models for three dimensional turbulence, *Phys. Rev E*, to be published.

Benzi, R., Biferale, L., Ciliberto, S., Struglia, M.V. and Tripiccione, R. (1996b) Generalized scaling in fully developed turbulence, *Physica D*, submitted.

She, Z.S. and Leveque, E. (1994) Universal scaling laws in fully developed turbulence, *Phys. Rev. Lett.*, **Vol. 72**, p. 336.

A NUMERICAL STUDY OF UPPER BOUNDS ON THE HEAT TRANSPORT BY TURBULENT CONVECTION IN A FLUID LAYER HEATED FROM BELOW

N.K.VITANOV AND F.H. BUSSE

Physikalisches Institut, Universität Bayreuth, 95440 Bayreuth, Germany

Since it is not possible to derive asymptotic relationships for turbulent transports without arbitrary assumptions, it is of interest to derive upper bound for these quantities. In the case of no-slip boundaries Howard [1] and Busse [2] developed this idea into an asymptotic theory of upper bounds on the turbulent transport of heat and momentum based on the solutions with a countable set of wavenumber α_i. In the case of the stress-free boundaries this method seems to fail but one can obtain bounds on the heat transport by solving numerically the Euler equations for the following variational problem [3]:

For a given value of the convective heat transport $\mu > 0$, find the the minimum $R(\mu)$ (corresponding to the Rayleigh number) of the functional:

$$R(\mathbf{u}, \theta, \mu) = \frac{\langle |\nabla\theta|^2 \rangle \langle |\nabla\mathbf{u}|^2 \rangle + \mu \langle (\overline{w\theta} - \langle w\theta \rangle)^2 \rangle}{\langle w\theta \rangle^2} \tag{1}$$

among all solenoidal vector fields $\mathbf{u} = (u, v, w)$ and scalar fields θ that satisfy stress-free boundary conditions at the boundaries $z = \pm 1/2$.

The investigations show that the upper bound for the convective heat transport approaches closely an asymptotic power law of the form $\mu = qR^\sigma$ for the 1-,2- and 3-α solutions of the variational problem. These solutions are special cases of the N-α-solutions of the Euler equations for the variational problem: $\theta = \sum_{n=1}^{N} \Phi_n(x, y)\theta_n(z)$, $w = \sum_{n=1}^{N} \Phi_n(x, y)w_n(z)$ with $\Delta_2\Phi_n = -\alpha_n^2\Phi_n$ The exponents σ differ considerably from the exponents of the rigid-boundary case which indicates that the boundary layer scaling introduced from Howard and Busse for the case of rigid boundaries cannot be

S. Gavrilakis et al. (eds.), Advances in Turbulence VI, 283-284.
© *1996 Kluwer Academic Publishers.*

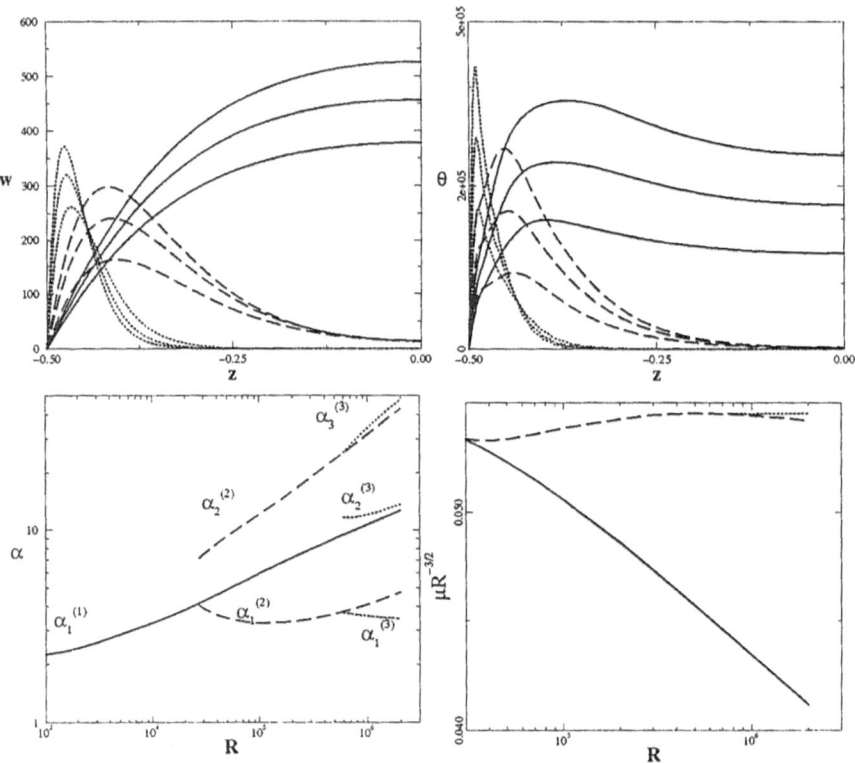

Figure 1. The function $w_1(z)$(solid lines), $w_2(z)$(dashed lines) and $w_3(z)$(dotted lines) of the 3-α-solution for the Rayleigh number $R=10^6, 1.5 \cdot 10^6, 2 \cdot 10^6$ (from bottom to top)

Figure 2. The function $\theta_1(z)$(solid lines), $\theta_2(z)$(dashed lines) and $\theta_3(z)$(dotted lines) of the 3-α-solution for the Rayleigh number $R=10^6, 1.5 \cdot 10^6, 2 \cdot 10^6$ (from bottom to top)

Figure 3. The wavenumbers $\alpha_i^{(N)}$, $i=1,..N, N=1,2,3$ of the extremalizing solutions as a function of the Rayleigh number

Figure 4. The upper bound of the convective heat transport as a function of the Rayleigh number R in the case of $1-\alpha$-(solid line), $2-\alpha$-(dashed line) and $3-\alpha$-solutions (dotted line)

used in the case of stress-free boundaries. Similar asymptotic relationships are obtained for the wave numbers $\alpha_i^{(N)}$, $i=1,...,N$.

References

1. Howard, L.N., Ann. Rev. Fluid. Mech. **4**, 473 (1972)
2. Busse, F.H., Adv. Appl. Mech. **18**, 77 (1978)
3. Vitanov, N.K., Busse, F.H., ZAMP (submitted)

V

Compressible Turbulence

LARGE-EDDY SIMULATIONS OF WEAKLY COMPRESSIBLE ISOTROPIC TURBULENCE

L. SHAO and J.P. BERTOGLIO

Laboratoire de Mécanique des Fluides et d'Acoustique, UMR CNRS 5509,
Ecole Centrale de Lyon, 36 Avenue Guy de Collongue, 69130 Ecully - France

1. Introduction

The aim of the present paper is to use Numerical Simulations to investigate in details the dynamics of compressible isotropic turbulence, and to provide understanding of the spectral mechanisms involved in these dynamics, with a view to help developing statistical spectral models for compressible turbulence as well as engineering models for high Mach number flows. It is well known that isotropic compressible turbulence is very sensitive to the choice of initial conditions. In order to overcome this difficulty, the present work is focussed on the asymptotical behaviour, at large time, of a turbulent field maintained statistically stationary by injecting energy in the large scales. A forcing technique is therefore introduced.

In the present paper, Large-Eddy Simulation (L.E.S.) is applied, not to the full Navier-Stokes and energy equations for a compressible fluid, but to a low Mach number approximation. A linearization of the density fluctuation is introduced and a barotropic assumption is retained. The reasons for this choice are the following :

(i) This approximation considerably simplifies the problem but preserves some of the important features of compressible turbulence, like the generation of acoustic energy by non linear interaction between solenoidal modes.

(ii) Analysing the results of existing DNS applied to the full set of equations, leads to the conjecture that the entropic mode, neglected in the present approximation, plays a limited role in the case of homogeneous turbulence at moderate Mach numbers, and that the coupling between entropic and acoustic modes is weak.

(iii) This approximation exactly corresponds to the one introduced as the starting point of the closure approach of Bataille et al [1]. The present results will provide data to assess the validity of closure models for compressible turbulence.

The present study is aimed at investigating the behaviour of compressible turbulence in the low Mach number limit, where the turbulent Mach number is defined as the ratio between the r.m.s. value u' of the fluctuation velocity and the sound speed : $M_t = u'/C_0$. The dilatational dissipation is analysed as a function of the Mach number in order to check whether the M_t^2 dependency that was deduced from DNS of sheared or decaying homogeneous turbulence, and used in one-point models, is valid. A particular attention is paid to the forcing technique.

2. Governing equations, Large-Eddy Simulation and forcing technique

Under the low Mach number approximation and splitting the velocity with Helmholtz decomposition into a solenoidal component u^s and a dilatational component u^c, one gets the following set of equations written in Fourier space :

$$\frac{\partial \hat{p}}{\partial t} = -i\,\rho_0 C_0^2 K_i \hat{u}_i^c \tag{1}$$

S. Gavrilakis et al. (eds.), Advances in Turbulence VI, 287-290.

$$\left(\frac{\partial}{\partial t}+v\ K^2\right)\hat{u}_i^s=-i\int P_{ij}\,(\vec{K})\,Q_l\hat{u}_j\,(\vec{P})\,\hat{u}_l(\vec{Q})\delta(\vec{K}-\vec{P}-\vec{Q})d\vec{P}d\vec{Q}+\hat{f}_i^s \tag{2}$$

$$\left(\frac{\partial}{\partial t}+v'\ K^2\right)\hat{u}_i^c=-i\int \Pi_{ij}\,(\vec{K})\,Q_l\hat{u}_j\,(\vec{P})\,\hat{u}_l(\vec{Q})\delta(\vec{K}-\vec{P}-\vec{Q})d\vec{P}d\vec{Q}-i\frac{K_i}{\rho_0}\hat{p}+\hat{f}_i^c \tag{3}$$

where P_{ij} and Π_{ij} respectively denote the projector onto the plane orthogonal to the wave-vector, and the projector along the wave-vector direction. The pressure fluctuation is split into a more detailed form : $p=p^{Pois}+p^{ac}$ where p^{Pois} is the "incompressible" pressure deduced from a Poisson equation applied to u^s and p^{ac} is the "acoustic" pressure. $\vec{f}$ represents the external force which supplies energy into the turbulent field at large scales.

For Large Eddy Simulations, in (2) and (3) the molecular viscosity v is replaced by $v+v_t$ and v' by $v'+v_{tc}$. The two subgrid viscosities v_t and v_{tc} are expressed as functions of the energy spectrum at the cut-off wave-number using an extension of the Chollet and Lesieur [2] formulation applied separately to the solenoidal and purely compressible modes. Comparisons between runs performed on a coarse and a fine grid show that this formulation leads to acceptable results (see Shao et al [3]).

In order to avoid excessive disturbance of the dynamics of the compressible mode, the forcing is applied only to the solenoidal part of the field. The external force is solenoidal : $f_i=f_i^s$ and $f_i^c=0$. It is obtained by a discretized Langevin equation :

$$\hat{f}_i^s(\vec{K},t+\Delta t)=\left(1-\frac{\Delta t}{T_{nl}}\right)\hat{f}_i^s(\vec{K},t)+\lambda\sqrt{\frac{F(K)}{4\pi k^2}}\sqrt{(2-\frac{\Delta t}{T_{nl}})\frac{\Delta t}{T_{nl}}}g_i^s(t) \tag{4}$$

where $g_i^s(t)$ is a constant amplitude random phase, solenoidal white noise process satisfying $\langle g_\alpha^s(t)g_\beta^s(t)\rangle=\delta_{\alpha\beta}$ in the local Craya frame; Tnl is the memory time of the external force $\vec{f}$. $F(K)$ characterises the specified spectrum of energy injection. As the forcing is applied only at small wave numbers, $F(K)$ is non zero only for K≤Kmax where Kmax denotes the wave number corresponding to the maximum of the initial spectrum of the solenoidal velocity. λ is a parameter determined to minimise the dependency of the energy effectively injected when Tnl varies. It can be noticed that in the limit where the memory time is taken equal to the integration step, Tnl --> Δt, the present force becomes a white noise process. Otherwise it is time correlated with a two-time correlation varying as exp(-(t-t')/Tnl). We also investigated the effect of a forcing with a correlation (1+(t-t')/Tnl)exp(-(t-t')/Tnl) having a zero derivative at t=t' (generated by using two Langevin equations). It can be noticed that in the case of compressible turbulence, Kida and Orszag [4] have used a white noise forcing (on a single Fourier mode, see also Miura and Kida [5]). Time correlated forces were only used in the literature in the case of incompressible turbulence (Eswaran and Pope [6]).

The numerical code uses a desaliased pseudo spectral method and a second order Runge-Kutta time integration scheme.

3. Results

In figure 1 the influence of the forcing memory time, Tnl, is shown. It appears that the compressible kinetic energy level is very sensitive to this parameter and is found to decrease when Tnl increases. However for large enough memory times, an asymptote is reached. All the results presented in the paper are given for Tnl large enough for the asymptote to be reached, so that they are considered as independent of the forcing memory time. In figure 2, the Mach number dependencies of the compressible kinetic energy E^c and the dilatational dissipation ε^c are analysed for different turbulent Reynolds numbers (Re_L) based on the integral length scales L. Due to the large integration times necessary before asymptotic energy levels were reached, most of the results concerning one point quantities given in the present paper correspond to computations on a coarse grid (16^3). It appears that the M_t^2 law is neither observed on the compressible kinetic energy nor on the dilatational dissipation. The slopes of the dilatational dissipation and the compressible kinetic energy as a function of M_t in a log-log representation are found to be constant at low Mach number : a M_t^4 dependence is observed for Mach number below 0.2. At larger Mach numbers, the behaviour of the dilatational dissipation changes : it increases more rapidly. Concerning the Reynolds number effects, the dilatational dissipation is found to be inversely proportional to the Reynolds number for small M_t ($M_t < 0.2$). In figure 3, the dilatational dissipation normalised by the solenoidal dissipation ε^s divided by the turbulent Reynolds number ($\varepsilon^c/(\varepsilon^s /Re_L)$) is given. For M_t less than 0.2, all curves collapse. The M_t^4 scaling for the dilatational dissipation was recently conjectured by Ristorcelli [7]. It is here confirmed by our results. In figure 4 the spectra corresponding to computations on a 64^3 grid are given. $E^{ss}(K)$ is the spectrum of the double velocity correlation of solenoidal velocity, $E^{PoisPois}(K)$ is the spectrum of the "incompressible" pressure, $E^{cc}(K)$ and $E^{PacPac}(K)$ respectively stand for the compressible velocity spectrum and the "acoustic" pressure spectrum or potential energy density spectrum. In this case, the Reynolds number is moderate (nearly equal to 100) and M_t is about 0.13. On can see that within the inertial range ($K^{-5/3}$ on $E^{ss}(K)$ spectrum), the spectrum of $E^{PoisPois}(K)$ exhibits a slope close to -7/3 which is consistent with classical results for incompressible turbulence. Some interesting comments should be made concerning the behaviour of $E^{cc}(K)$ and of $E^{PacPac}(K)$: the equipartition between spectral kinetic energy and potential energy is observed only at very small K; otherwise, $E^{cc}(K)$ is clearly one order higher than $E^{PacPac}(K)$, this suggests that the equipartition phenomena relate to the purely "acoustic" dynamics or the non-local interaction between different scales and that the local interaction is dominant. The slope of $E^{cc}(K)$ is found to be -3 while the slope of $E^{PacPac}(K)$ is found to be closer to -4. An analytical model predicting the K^{-3} behaviour of $E^{cc}(K)$ and the non-equipartition between $E^{cc}(K)$ and $E^{PacPac}(K)$, based on the response of the compressible modes to the excitation produced by "incompressible" pressure fluctuations and under the local interaction assumption, can be proposed [8].

4. Discussion and direction for future works

The present study shows the importance of the forcing scheme to study the asymptotical states of weakly compressible isotropic turbulence. The Mach number and the Reynolds number dependencies of the compressible fields were investigated in terms of compressible kinetic energy, dilatational dissipation and spectra. A M_t^4/Re_L scaling for dilatational dissipation is found at low M_t. Spectral analysis shows K^{-3} behaviour of the compressible velocity spectrum and non-equipartition between kinetic and potential energy.

From the recent work of Bertoglio et al. [8], it appears that the two time correlations of the incompressible field play an essential role in the dynamics involved in the compressible field, these two time correlations will also be evaluated. Comparisons between the presents results and those of Bataille [1] confirm the importance of the choice of the form of the two time correlations used in the derivation of the EDQNM model. A modified EDQNM model [9] is presently under investigation. Preliminary comparisons between this modified EDQNM theory and present results show promising results.

5. Acknowledgement

The authors would like to thank the CNRS IDRIS computer centre for facility and assistance provided on a Cray C98 computer.

290

6. References

[1]   Bataille F. and Bertoglio J.P., 1993, "The long-time behaviour of weakly compressible turbulence", 9th Turbulent Shear Flow Symp., Kyoto, Vol. 2, pp. 22.11-22.16.

[2]   Chollet J.P. and Lesieur M., 1981; Parameterization of Small Scales of Three-Dimensional Isotropic Turbulence utilising Spectral Closures, Journ. of Atm. Sci., Vol. 38, pp. 2747-2757.

[3]   Shao L., Perlat J.P. and Bertoglio J.P., 1996, « Large-Eddy Simulation of the longtime behaviour of weakly compressible isotropic turbulence », 9th ISTP, to be held in Singapore, June 96

[4]   Kida S. and Orszag S.A., 1990, « Energy and spectral dynamics in forced compressible turbulence », J. Sci. Comp., Vol. 5, N°2, pp. 85-125.

[5]   Miura H. and Kida S., « Acoustic energy exchange in compressible turbulence », Phys. Fluids A, Vol. 7 N°7, pp. 1732-1742.

[6]   Eswaran V. and Pope S.B., 1988, « An examination of forcing in direct numerical simulation of turbulence », Computer & Fluids, Vol. 16, N°3, pp 257-278.

[7]   Ristorcelli J.R., 1995 « A pseudo-sound constutive relationship for the dilatational covariances in compressible turbulence : an analytical theory », ICASE Report N° 95.

[8]   Bertoglio J.P. and Wunenburger R., 1996, In preparation.

[9]   Bertoglio J.P. and Fauchet G., 1996, In preparation.

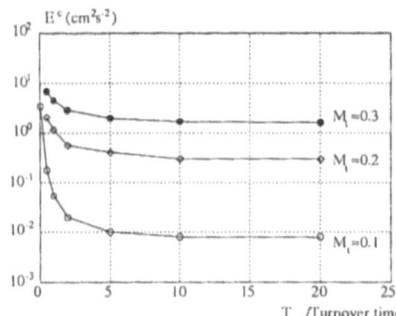

Fig. 1 Compressible kinetic energy as function of T_{nl}

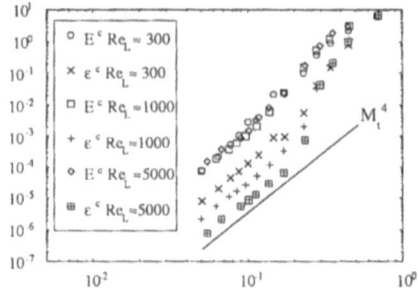

Fig. 2 Compressible kinetic energy, dilatational dissipation as functions of M_t for differents Reynolds numbers

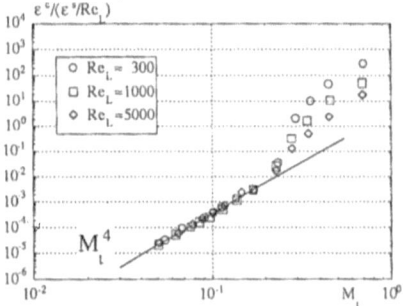

Fig. 3 Normalised dilatational dissipation as function of M_t

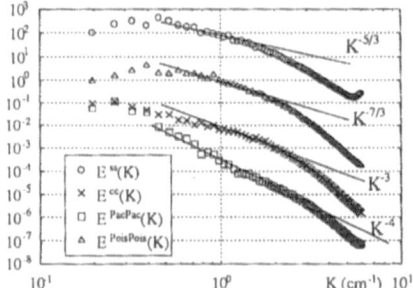

Fig. 4 Spectra of different modes

Multi-fractals in Compressible Turbulence

Bhimsen K. Shivamoggi

University of Central Florida, Orlando, FL 32816

Abstract:

Some aspects of multifractal scaling at the Kolmogorov microscale in fully developed compressible turbulence are considered.

1. Scaling Laws in Compressible Turbulence

The problem of spectral laws in compressible turbulence was tackled by Moiseev et al. (1981) and Shivamoggi (1992). By postulating the invariance of the Navier-Stokes equations under the scaling transformation, Shivamoggi (1992) deduced the following scaling behaviors for the increments in velocity v and the density ρ over a small distance l:

$$v(l) \sim l^{\alpha}, \rho(l) \sim l^{\frac{2\alpha}{\gamma-1}} \tag{1}$$

where γ is the ratio of a specific heats of the fluid.

The scaling exponent α was determined by postulating the invariance of mean rate of kinetic energy dissipation $\hat{\epsilon}$ under the scaling transformation, which gives

$$\alpha = \frac{\gamma-1}{3\gamma-1} \tag{2}$$

In the limit $\gamma \Rightarrow \infty$, (2) reduces to the well known result $\alpha = \frac{1}{3}$ for incompressible turbulence.

2. The Multifractal Scaling at the Kolmogorov Microscale:

Numerical simulations of compressible turbulence by Passot and Pouquet (1991) and Lee et al. (1991) have revealed the presence of spatial intermittency in compressible turbulence. One may follow Mandelbrot [12] and argue that the spatial intermittency effects in compressible turbulence are related to the fractal aspects of the geometry of turbulence. Shivamoggi (1995a) considered a multifractal model to describe the effects of spatial intermittency in fully developed compressible turbulence. The multifractal scaling in the inertial range was extrapolated down to the Kolmogorov microscale in compressible turbulence (Shivamoggi, 1995b).

S. Gavrilakis et al. (eds.), Advances in Turbulence VI, 291-294.
© 1996 Kluwer Academic Publishers.

The fluctuations in the kinetic energy dissipation rate $\hat{\epsilon}$ lead to fluctuations in the Kolmogorov microscale $\hat{\eta}$, so we have, following Paladin and Vulpiani (1987), and Meneveau and Nelkin (1989),

$$\hat{\eta} = \left[\frac{v_0^3 \rho_0}{\hat{\epsilon}(\hat{\eta})} \right]^{1/4}$$

(3)

where the subscript 0 denotes some reference values.

On using (2), $\hat{\eta}$ then exhibits the following scaling behavior:

$$\left(\frac{\hat{\eta}}{L} \right) \sim \langle \hat{R} \rangle^{-\frac{1}{1+\left(\frac{3\gamma-1}{3\gamma-3}\right)\alpha}}$$

(4)

where

$$\langle \hat{R} \rangle = \frac{\left(\langle \hat{\epsilon} \rangle L^4 / \rho_0 \right)^{1/3}}{v_0},$$

$\langle \hat{\epsilon} \rangle$ being the global average rate of kinetic energy dissipation.

Let us now assume that the kinetic energy dissipation is concentrated on a multifractal object for which one has a continuous spectrum of scaling exponents α, with each α distributed on a set $S(\alpha)$ of Hausdorff dimension $f(\alpha)$ embedded in the d-dimensional space (Frisch and Parisi, 1985).

3. Moments of the Velocity Gradient Distribution

Let us now consider moments of the velocity gradient distribution,

$$A_p \equiv \left\langle \left| \frac{\partial v}{\partial x} \right|^p \right\rangle$$

(5)

Assuming that multifractal scaling in the inertial range can be extrapolated down to the Kolmogorov microscale $\hat{\eta}$, we have for A_p,

$$A_p \sim \int \langle \hat{R} \rangle^{-\frac{1}{1+\left(\frac{3\gamma-1}{3\gamma-3}\right)\alpha}[p\alpha-p+d-f(\alpha)]} d\mu(\alpha)$$

(6)

where the factor

$$\langle \hat{R} \rangle^{-\frac{d-f(\alpha)}{1+\left(\frac{3\gamma-1}{3\gamma-\gamma}\right)\alpha}}$$

(or the factor $\langle \hat{\eta} \rangle^{d-f(\alpha)}$) is the probability of encountering the set $S(\alpha)$ within a d-dimensional sphere of radius $\langle \hat{\eta} \rangle$.

The integral in (6) may be evaluated in the limit of small $\langle \hat{\eta} \rangle / L$ (or large $\langle \hat{R} \rangle$) by the method of steepest descent. This corresponds to maximizing the exponent, or

$$\left[1 + \frac{3\gamma - 1}{3\gamma - 3}\alpha\right]\left[p - \frac{df}{d\alpha}\right] = \left(\frac{3\gamma - 1}{3\gamma - 3}\right)\left[p\alpha - p + d - f(\alpha)\right] \tag{7}$$

The singularity spectrum $f(\alpha)$ is related to the generalized fractal dimension D_q of the kinetic energy dissipation field (Halsey et al. 1986) by the following Legendre transformation:

$$\left[\left(\frac{3\gamma - 1}{\gamma - 1}\right)\alpha_* - 1 + d\right]q - f(\alpha_*) = (q - 1)D_q \tag{8}$$

$$\frac{df(\alpha_*)}{d\alpha} = \left(\frac{3\gamma - 1}{\gamma - 1}\right)q \tag{9}$$

Equation (7) - (9) imply that $q = Q$, where Q is now given by

$$Q = \frac{D_Q + p\left(1 + \dfrac{3\gamma - 3}{3\gamma - 1}\right) - 1}{D_Q + 4 - d}. \tag{10}$$

Using (7) - (9) and (10), (8) becomes

$$A_p \sim \langle \hat{R} \rangle^{\frac{D_Q\left[\left(\frac{3\gamma - 3}{3\gamma - 1}\right)p - 3\right] + \left[\frac{3\gamma - 3}{3\gamma - 1}(1 - d) - 3\right]p + 3d}{D_Q + 4 - d}} \tag{11}$$

For $p = 2$, (11) yields

$$A_2 \sim \langle \hat{R} \rangle^{\left(\frac{3\gamma + 3}{3\gamma - 1}\right)\left[\frac{D_Q + \frac{4\gamma}{\gamma + 1} - d}{D_Q + 4 - d}\right]}. \tag{12}$$

(12) implies that the mean kinetic energy dissipation, namely, $\rho_0 v_0 A_2$, is not independent of v_0 as $v_0 \Rightarrow 0$, unlike the case with incompressible turbulence $(\gamma \Rightarrow \infty)$. Therefore, the concept of inviscid dissipation of kinetic energy appears not to be valid in compressible turbulence.

In order to see the transition dimensionality of $d = 4$, let us write the exponent in (11) as follows:

$$-\frac{D_Q\left[\left(\frac{3\gamma - 3}{3\gamma - 1}\right)p - 3\right] + \left[\frac{3\gamma - 3}{3\gamma - 1}(1 - d) - 3\right]p + 3d}{D_Q + 4 - d} = -\frac{1}{D_Q}\left[D_Q\left\{\left(\frac{3\gamma - 3}{3\gamma - 1}\right)p - 3\right\}\right.$$

$$-3p\left(1+\frac{3\gamma-3}{3\gamma-1}\right)+12\right]+\frac{3\left[p\left(1+\frac{3\gamma-3}{3\gamma-1}\right)-4\right]}{D_\varrho\left(D_\varrho+4-d\right)}(d-4) \qquad (13)$$

(13) shows that $d=4$ is a transition dimensionality for higher

$$\left(p>p_* = \frac{4}{1+\frac{3\gamma-3}{3\gamma-1}} = \frac{2(3\gamma-1)}{3\gamma-2}\right)$$ moments of the velocity gradient distribution.

However, for $p=p_*, d=4$ is not a transition dimensionality. Observe that for incompressible turbulence $(\gamma\Rightarrow\infty)$, $p_*=2$ so that the inviscid energy dissipation property of the incompressible turbulence is a special event that is associated with the loss of transition dimensionality $d=4$ for the corresponding velocity gradient moment (namely, $p=2$).

4.   Discussion

The concept of inviscid dissipation of kinetic energy in incompressible turbulence is seen above not to survive the presence of compressibility effects in the fluid. On the other hand, the inviscid dissipation of kinetic energy is quite a special event even in incompressible turbulence because it depends on some remarkable properties of the turbulence dynamics at the Kolmogorov microscale in the hyperspace with $d>3$!

References

[1]   Frisch, U. and Parisi, G.: in Turbulence and Predictability in Geophysical Fluid Dynamics and Climate Dynamics, Ed. M. Ghil R. Benzi and G. Parisi, North-Holland (1985).

[2]   Halsey, T.C., Jensen, M.H., Kadanoff, L., Proccacia, I. and Shraiman, B.I.: Phys. Rev. A 33, 1141, (1986).

[3]   Lee, S., Lele, S.K. and Moin, P.: Phys. Fluids A 3 657 (1991).

[4]   Mandelbrot, B.: in Turbulence and Navier-Stokes Equations. Ed. R. Temam. Lecture Notes in Mathematics. Vol. 565, Springer-Verlag (1976).

[5]   Meneveau, C. and Nelkin, M.: Phys. Rev. A 39 , 3732 (1989).

[6]   Moiseev, S.S., Petviashvili, V.I., Toor, A. V. and Yanovskii, V.V.: Physica D 2, 218 (1981).

[7]   Passot, T. and Pouquet, A.: Euro. J. Mech. B 10, 377 (1991).

[8]   Shivamoggi, B.K.: Ann. Phys. (N.Y.) 243 , 169, (1995a).

[9]   Shivamoggi, B.K.: Phys. Lett. A 166, 243 (1992).

[10] Shivamoggi, B.K.: Ann. Phys. (N.Y.) 243, 177, (1995b).

FINE STRUCTURE OF TURBULENCE IN A BOUNDARY LAYER WITH STRONG DENSITY DIFFERENCES

A. SOUDANI, M. FAVRE MARINET, S. TARDU, J. L. HARION
Laboratoire des Ecoulements Géophysiques et Industriels
B.P. 53 X 38041 Grenoble Cédex France UJF-INPG-CNRS

1. INTRODUCTION

Turbulent flows with strong density differences occur in a number of practical situations including premixed combustion over a strongly heated wall , the re-entry phase of a space shuttle into the atmosphere or the cooling of turbine blades by tangential injection. Although a number of well conducted studies exist on this topic (1,2,3) several questions concerning the detailed structure of the near wall turbulence remain unanswered. That is the case for instance when the mechanism generating the overall turbulent heat transfer is considered. Several investigations have converged to the observation that the heat transfer coefficient is independent of wall flux, even when the former is quite strong and enhances the ejection of low momentum fluid near the wall into the inner layer (3). The mechanism that compensates this phenomena and maintains unaffected the overall heat transfer mechanism is still not clear. The aim of the present investigation is to provide further experimental data obtained in a turbulent boundary layer subject to strong density differences $\Delta\rho$ in order to contribute to the understanding of these complex non equilibrium flows.

2. EXPERIMENTAL SET-UP and FLOW CONDITIONS

Density differences are generated by injecting different gas mixtures (air-helium) to the wall in a pressurized wind tunnel described in detail in (4). The present results were obtained at atmospheric pressure. Simultaneous measurements of the density $\rho(t)$ and the velocity $u(t)$ are performed by means of an interfering probe. The calibration procedure as well as the details concerning the frequency response of the probe together with first order statistics of fluctuating streamwise velocity and density may be found in (3). The measurements were performed at several distances x from the injection slot in the entire boundary layer. The closest point to the wall was $y^+=20$ ($^+$ indicates variables non dimensionalized with local shear velocity u_τ and viscosity v). The accent here is mainly put on the characteristics obtained in the inner layer. The experimental conditions are:
External flow: $\rho_\infty = 1$ kg/m^3, $U_\infty = 6$ m/s; Injection of pure air ($\rho_{inj} = 1.2$ kg/m^3) or pure helium (He, $\rho_{inj} = 0.16$ kg/m^3) through a slot of height 3mm. with a bulk velocity $U_{inj} = 2$ m/s . At x = 100 mm from the injection slot: Boundary layer thickness

S. Gavrilakis et al. (eds.), Advances in Turbulence VI, 295-298.
© *1996 Kluwer Academic Publishers.*

$\delta = 20.5$ mm., $Re_\delta = 6000$, $u_{\tau\,air} = 0.24$ m/s, $u_{\tau\,He} = 0.29$ m/s. Record length $T_r = 3500\;\delta/U_\infty$.

3. RESULTS

3.1. HIGH ORDER STATISTICS

Previous results have shown that the rms of the instantaneous streamwise velocity is not significantly affected by the presence of $\Delta\rho = \rho_{inj} - \rho_\infty$. This is no more the case when the higher order statistics are considered. Fig. 1a compares the distribution of the skewness of u' obtained with injection of He and air at $x/\delta = 5$ downstream of the injection slot, with the DNS data of (5) in the canonical $\Delta\rho = 0$ turbulent boundary layer. It is seen that $\overline{u'^3} / \overline{u'^2}^{\,3/2}$ is significantly lower and negative in the low log layer ($y^+ < 100$) when $\Delta\rho < 0$ (injection of He) while the skewness factor is quite similar to the results given by DNS in the case of small $\Delta\rho > 0$. This shows that the ejection like motions with u'<0 are intensified when $\Delta\rho < 0$. The skewness of the density fluctuations is even more interesting (Fig. 1b). The structural parameter $\overline{\rho'^3} / \overline{\rho'^2}^{\,3/2}$ is strongly negative once $y^+ > 50$, i.e in the constant shear region when He is injected at the wall. The distributions of the flatness of both u' and ρ' (not shown here) in the inner layer are found similar suggesting that the intermittency mechanism is not significantly altered by $\Delta\rho$.

3.2. CONDITIONAL ANALYSIS

Fig. 2a shows the quadrant distributions of ρ' and u' obtained at $y/\delta = 0.25$ and $x/\delta = 5$. It is seen that ρ'-u' is oriented from quadrants 2-4 when $\Delta\rho>0$ (air injection) and from quadrants 1-3 when $\Delta\rho <0$. This is in agreement with the distribution of the correlation coefficient $\overline{\rho'u'} / \sqrt{\overline{\rho'^2}}\,\sqrt{\overline{u'^2}}$ which is positive or negative depending upon the injection of lighter or heavier fluid at the wall (4). The data is further analyzed by means of a conditional analysis similar to that given by (3). The large excursions of a quantity q' are determined by

$$q'_{+,-} = \frac{\sqrt{\sum_i (S_{+,-})_i \left[q'_i - <q'_{i+,->}\right]^2}}{\sum_i (S_{+,-})_i}$$

where the subindices + or - correspond respectively to the ejections and sweeps, < > denotes the conditional average and S is a detector function based both on u' and ρ'. The ejection phase is defined for He as: $(S_+)_i = 1$ if $u'_i < -\sqrt{\overline{u'u'}}$ and $\rho'_i < -\sqrt{\overline{\rho'\rho'}}$ and $(S_+)_i = 0$ otherwise. The data obtained this way (Fig. 2b) compares fairly well with the measurements over a strongly heated wall reported in (3). Close inspection of the results shows that the ejection type flow is intensified when lighter fluid (He) is injected into the boundary layer and that, in return, the sweeps are weakened. This, too, is in agreement with (3). Although the correlation coefficient between velocity and density fluctuations is significantly larger than the $-\overline{u'v'}$ correlation in the inner layer, it was thought first, that $\overline{u'\rho'}$ could be dynamically similar to $-\overline{u'v'}$ through the distributions of the quadrant contributions. A detailed analysis conducted for the reference case (injection of air with small $\Delta\rho > 0$)

has, revealed that this is not exactly the case. Qualitative agreement has been found between the contributions of the ejection and sweep type events ($\overline{C_{ej}}$, $\overline{C_{sw}}$) when compared with the canonical boundary layer. However, some quantitative differences have been observed at $y^+ > 30$, in the ratio $\overline{C_{ej}} / \overline{C_{sw}}$ (see (6) for further details). The hole analysis as described in (2) has also been applied to the $\overline{u'\rho'}$ quadrants. The main result is that the ejections occupy a larger area in the quadrant in the case of strong $\Delta\rho < 0$ in particular in the constant shear region. Fig. 3 recapitulates the distributions of $\overline{C_{ej}}$ vs. the hole size for the reference case and the injection of He at $y^+ = 80$.

4. DISCUSSION

The energetic events in the inner layer are intimately related to the presence of the quasi-streamwise vortices. The rate of change of ω_x vorticity reads,

$$\frac{D\omega_x}{Dt} = \vec{\omega}.\nabla u - \frac{\omega_x}{\rho}\frac{D\rho}{Dt} +.... $$ The term $-u\frac{\omega_x}{\rho}\frac{\partial\rho'}{\partial x}$ of this equation plays a role which

is similar to the stretching of vorticity via $\omega_x\frac{\partial u}{\partial x}$. With He, and during the ejections it

is likely that the stretching of the streamwise vorticity is intermittently enhanced by this term and that the quasi-streamwise vortices are consequently reinforced. Results

concerning the quantity $S=\overline{\left(\frac{\partial u'}{\partial t}\right)^2\frac{\partial\rho'}{\partial t}} / \left[\overline{\left(\frac{\partial u'}{\partial t}\right)^2}\ \overline{\left(\frac{\partial\rho'}{\partial t}\right)^2}\right]^{1/2}$ (which is equivalent to the

skewness of the velocity time derivatives related to the vorticity stretching strength) are further analyzed in (6) and seem to strengthen this assumption. Further results concerning the conditional averages related to the shear layer events determined by an adapted version of VITA may also be found in (6).

References

1-Larue J.C, Libby P.A.,1980 Phys. Fluids, 23(6) p. 1111
2- Cheng R.K, T.T., Ng, 1984 Phys. Fluids, 28 (2) p. 473
3- Wardana I.N.G, Ueda T., Mizomoto M., 1995 Exp. Fluids, 18, p. 454
4- Harion J.L., 1994 PhD Thesis Grenoble
5- Kim J., Moin P., Moser R., 1987 J. Fluid Mech. 177, 1081
6-Tardu S., Soudani A., Favre Marinet M., Harion J.L. (1996) Response of near wall turbulence to strong density differences, In preparation.

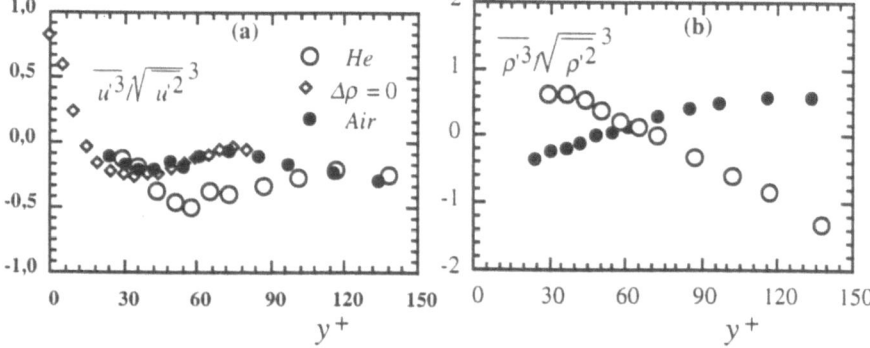

Figure 1 - Skewness of the velocity (a) and density fluctuations (b) at $x/\delta=5$

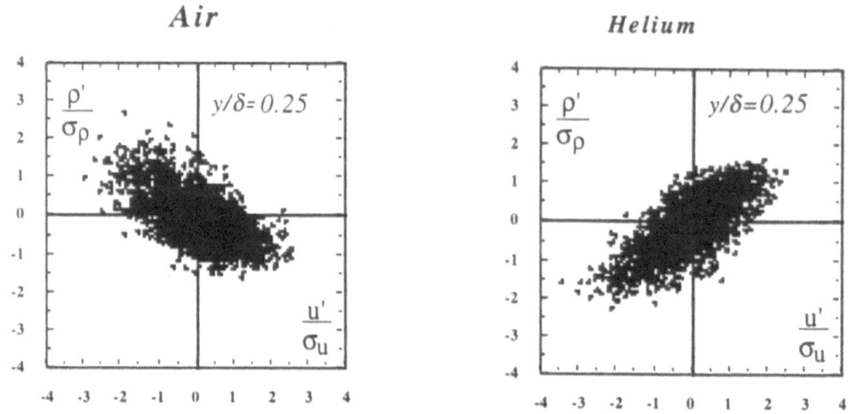

Air

Helium

Figure 2a - Quadrant distribution of the density and velocity distributions at y/δ=0,25.

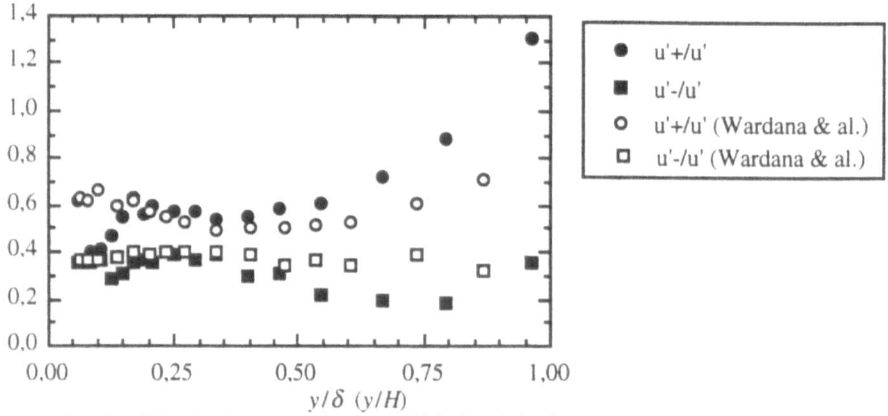

- ● u'+/u'
- ■ u'-/u'
- ○ u'+/u' (Wardana & al.)
- □ u'-/u' (Wardana & al.)

y/δ (y/H)

Figure 2b - Conditional velocity contributions with helium injection.

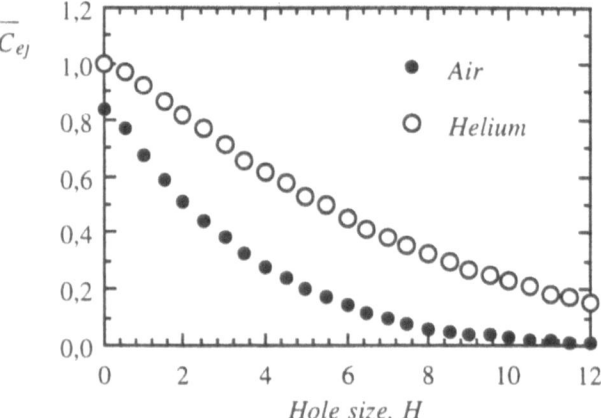

- ● Air
- ○ Helium

Hole size, H

Figure 3 - Hole analysis at x/δ≈2,75 and y⁺=80.

SECOND ORDER ENTROPY CONSISTENT MODELLING OF TURBULENT COMPRESSIBLE FLOWS

G. BRUN[2], J.M. HÉRARD[1], L. LEAL DE SOUSA[1], M. UHLMANN[2]

[1] *EDF/DER/AEE/LNH/GRET. 6, quai Watier. 78400. Chatou. France.*
[2] *METRAFLU. 64, chemin des Mouilles. 69134. Ecully. France.*

Abstract

We examine herein the suitability of some second order closures to describe turbulent compressible flows with shocks, applying the standard Favre averaging technique. The basic set of equations reads :

$$(\rho)_{,t} + (\rho U_i)_{,i} = 0 \tag{1}$$

$$(\rho U_i)_{,t} + (\rho U_i U_j)_{,j} + (p\, \delta_{ij})_{,j} + (R_{ij})_{,j} = -(\Sigma_{ij}^{visc})_{,j} \tag{2}$$

$$(E)_{,t} + (E\, U_j)_{,j} + (U_i(p\, \delta_{ij}+R_{ij}))_{,j} = -(U_i(\Sigma_{ij}^{visc}))_{,j} + \left(\sigma_E\left(\frac{p}{\rho}\right)_{,j}\right)_{,j} \tag{3}$$

$$(R_{ij})_{,t} + (R_{ij}\, U_k)_{,k} + R_{ik}U_{j,k} + R_{jk}U_{i,k} = \Phi_{ij} - \frac{2}{3}\left(\frac{\varepsilon}{I}\right)\text{trace }(R)\, \delta_{ij} + \left(\Phi_{ij}^k\right)_{,k} \tag{4}$$

setting :

$$R_{ij} = <\rho\, u_i'' u_j''> \; ; \; 2K = I = \text{trace}(R) = R_{ii} \tag{5}$$

$$\Sigma_{ij}^v = -\mu\, (U_{i,j} + U_{i,j} -\frac{2}{3} U_{l,l}\delta_{ij}) \; ; \; p = (\gamma\text{-}1)\, (E - \frac{\rho U_j U_j}{2} - \frac{1}{2} R_{jj}) \tag{6}$$

The realisability constraints read :

$$n_i R_{ij}(x,t)n_j \geq 0 \quad ; \quad \rho(x,t) \geq 0 \quad ; \quad p(x,t) \geq 0$$

An entropy inequality, which is valid for regular enough solutions, may be derived. It requires that a few assumptions are made for the unknown terms Φ_{ij} and Φ_{ij}^k. The realisability constraints hold for C^1 solutions, if the components of the mean velocity gradient are bounded (see [3], [4]), focusing on a few models ([3], [9], [12], [13]). Dealing with statistically two-dimensional turbulence, the whole system takes the form:

$$W_{,t} + \sum_{i=1}^{3} A_i(W)\, W_{,i} = S(W) + (F_i^v(W, \nabla W))_{,i}$$

when choosing the simplest closure ([9]), and setting :

$$W^t = (\rho,\, \rho U,\, \rho V,\, R_{11}, R_{22}, R_{12}, R_{33},\, E).$$

By introducing some normal vector **n** and also :

$$R_{nn} = \dot{n}^t\, R\, n \text{ and}: U_n = U^t\, n$$

the one dimensional Riemann problem associated with the first order non conservative system :

S. Gavrilakis et al. (eds.), Advances in Turbulence VI, 299-300.
© *1996 Kluwer Academic Publishers.*

$$W_{,t} + A_n(W)\ W_{,n} = 0$$

connecting states with a linear path through discontinuities ([1], [10], [11]) may be investigated. Owing to the entropy inequality :

$$- \sigma [\eta] + [U\eta] \leq 0$$

it may be shown that the one-dimensional Riemann problem admits a unique realisable entropy consistent solution, provided that some condition holds for the initial condition on each side of the initial discontinuity. This result ([5]) is close indeed to those detailed in [2], [7,] and [8].

We then focus on numerical modelling of the previous set of partial differential equations. Owing to the above mentionned result, an exact Riemann solver, based on the choice of the linear path, may be developped ; even more, approximate Riemann solvers may be derived, which behave quite well ([14]). The most difficult point dwells in the computation of the non conservative terms which are part of the non conservative system, even when focusing on the simplest model issuing from [12]. The need for such an effort was demonstrated in [2] and [7]. In particular, the preservation of the invariance of the Riemann invariants in the Linearly Degenerate fields is essential. Due to the presence of new contact discontinuities (with regard to the standard gas-dynamics system), it may be checked that naïve solvers, which are based on Euler solvers, fail to predict shock tube simulations ; this is even more true when the turbulent Mach number is closer to 1, and when the anisotropy of the Reynolds stress tensor is dominant.

References

[1] Colombeau, J.F. (1992) Multiplication of distributions, *Springer-Verlag*.
[2] Forestier, A., Hérard, J.M. and Louis, X. (1995) Exact or approximate Riemann solvers to compute a two-equation turbulent compressible model, *Proc. of the Ninth Int. Conf. on Finite Elements in Fluids*, New Trends and Applications . Part I. pp. 677-686.
[3] Hérard, J.M. (1994) Basic analysis of some second moment closures, Part I : incompressible isothermal turbulent flows *TCFD* vol. 6, n°4, pp. 213-233.
[4] Hérard, J.M. (1994) Basic analysis of some second moment closures. Part II : incompressible turbulent flows including buoyant effects, *Coll. Notes internes DER* 94NB00053.
[5] Hérard, J.M. Basic analysis of some second moment closures. Part III : turbulent compressible flows with shocks *in preparation* .
[6] Hérard, J.M. (1995) Suitable algorithms to preserve the realisability of Reynolds stress closures *ASME FED* vol. 215, pp. 73-80.
[7] Hérard, J.M. (1995) Solveur approché pour un système hyperbolique non conservatif issu de la turbulence compressible, *EDF report* HE-41/95/009/A.
[8] Hérard, J.M., Forestier, A. and Louis, X. (1995) A non strictly hyperbolic system to describe compressible turbulence, *Coll. Notes internes DER* 95NB00003.
[9] Fu, S, Launder, B.E. and Tselepidakis, D.P. (1987) Accomodating the effects of high strain rates in modelling the pressure strain correlations *UMIST report* TFD 87/5.
[10] Le Floch, P. (1988) Entropy weak solutions to non linear hyperbolic systems in non conservative form, *Comm. in Part. Diff. Eq.* 13(6), pp. 669-727.
[11] Le Floch, P. and Liu, T. P. (1992) Existence theory for non linear hyperbolic systems in non conservative form, CMAP report 254, to appear in *Forum Mathematicum*..
[12] Lumley, J.L. (1978) Computational modelling of turbulent flows, *Advances in Applied Mechanics*, vol. 18, pp. 123-176.
[13] Shih, T. and Lumley, J.L. (1985) Modeling of pressure corelation terms in Reynolds stress and scalar flux equations, *Cornell University report* FDA-85-3.
[14] Uhlmann, M. and Brun, G. (1995) Solveur pour le système des équations hyperboliques non conservatives issu d'un modèle de transport de tensions de Reynolds, *ECL-METRAFLU report*, (unpublished).

PATTERN FORMATION IN PASSIVE SCALAR DISTRIBUTION IN COMPRESSIBLE TURBULENT FLUID FLOW

T. ELPERIN, N. KLEEORIN

The Pearlstone Center for Aeronautical Engineering Studies, Department of Mechanical Engineering, Ben-Gurion University of the Negev, Beer-Sheva 84105, P. O. Box 653, Israel

and

I. ROGACHEVSKII

Racah Institute of Physics, The Hebrew University of Jerusalem, 91904 Jerusalem, Israel

1. Introduction

The large variety of interesting phenomena related to the passive scalar transport in a random incompressible fluid flow were investigated both theoretically and experimentally. These effects include anomalous turbulent diffusion, intermittency, fractal structure of a concentration field. However the passive scalar transport by a compressible turbulent flow is a subject of a relatively few investigations and some interesting aspects of this problem were not addressed.

The main goal of this paper is to discuss new effects, turbulent barodiffusion and turbulent thermal diffusion in gases. These phenomena result in the additional fluxes of the mass of gaseous admixture advected by a compressible turbulent fluid flow with low Mach numbers. These effects caused by compressibility of turbulent fluid flow.

2. Turbulent Mass Flux of Gaseous Admixture

We consider a mixture of two gaseous component with different number densities: $n_p \ll n$. Instead of a gas component with the number density n_p light particles can be considered. Hereafter the gas component with number density $n = \rho_f / m_\mu$ is called fluid, while the component with number density n_p is called gaseous admixture (m_μ is the mass of moleculas of the surrounding fluid). The number density $n_p(t, \mathbf{r})$ of gaseous admixture in a turbulent compressible flow is determined by equation:

$$\frac{\partial n_p}{\partial t} + \nabla \cdot (n_p \mathbf{U}) = D \Delta n_p \,, \tag{1}$$

S. Gavrilakis et al. (eds.), Advances in Turbulence VI, 301-304.

where **U** is a random velocity field of the gaseous admixture which it acquires in a turbulent fluid velocity field **v**. The density ρ_f of the surrounding fluid satisfies the continuity equation

$$\frac{\partial \rho_f}{\partial t} + \nabla \cdot (\rho_f \mathbf{v}) = 0 . \tag{2}$$

In Eq. (2) we do not take into account the diffusion of the fluid in the gaseous admixture because $n_p \ll n$.

The velocity **v** of the surrounding fluid is assumed to be known, and the velocity of the gaseous admixture **U** strongly depends on the velocity of the surrounding fluid: $\mathbf{U} = \mathbf{v} + \mathbf{v}_s$. The second term in the velocity of gaseous admixture describes a sedimentation of the gaseous admixture in a gravity field with a terminal fall velocity $\mathbf{v}_s$. We consider the case of large Reynolds and Peclet numbers.

In order to derive equation for mean concentration of the gaseous admixture Eq. (1) must be averaged over an ensemble of random velocity fluctuations. To this purpose we use here the stochastic calculus [1; 2; 3]. The use of this technique allows to derive equation for the mean field $N = \langle n_p \rangle$:

$$\frac{\partial N}{\partial t} + \nabla \cdot (\mathbf{J}_M + \mathbf{J}_T) = 0 , \tag{3}$$

where the turbulent flux of the gaseous admixture is given by

$$\mathbf{J}_T = -D_T \left[\nabla N + N \frac{\nabla T}{T} - N \frac{\nabla P}{P} \right] , \tag{4}$$

$D_T = u_0 l_0 / 3$ is the coefficient of turbulent diffusion, u_0 is the characteristic velocity in the scale l_0. The molecular flux of the gaseous admixture

$$\mathbf{J}_M = -D \left[\nabla N + k_t \frac{\nabla T}{T} + k_p \frac{\nabla P}{P} \right] . \tag{5}$$

comprises three terms: molecular diffusion ($\sim \nabla N$), molecular thermal diffusion ($\sim k_t \nabla T$, where k_t is the molecular thermal diffusion ratio), and molecular barodiffusion ($\sim k_p \nabla P$, where k_p is the molecular barodiffusion ratio, see, e.g., [4]). Comparing the molecular (5) and turbulent (4) fluxes of the gaseous admixture we can interpete the new additional turbulent fluxes as fluxes caused by the effects of turbulent thermal diffusion ($\sim k_T \nabla T$, where $k_T = N$ is the turbulent thermal diffusion ratio), and turbulent barodiffusion ($\sim k_P \nabla P$, where $k_P = -N$ is the turbulent barodiffusion ratio).

3. Mechanism of large-scale pattern formation

Now we study the dynamics of the large-scale distribution of the concentration of gaseous admixture in the small-scale turbulent fluid flow. Turbulent barodiffusion and turbulent thermal diffusion may result in formation of inhomogeneous structures in large-scale distribution of gaseous admixture advected by compressible turbulent fluid flow. The mechanism of this effect is as follows. In incompressible flow at any time mass of fluid flowing into a small volume exactly equals to a mass outflow from this volume. In the limit of infinite Peclet number the light gas component is frozen into a flow of a surrounding fluid. This means that there is no accumulation of the gaseous admixture at any point of the volume.

The situation changes if $\nabla \cdot \mathbf{u} \neq 0$ in a turbulent fluid flow. In this case a mass of fluid flowing into a small volume does not equal to a mass outflow from the volume at any instance. Therefore at times smaller than a characteristic time of the turbulent velocity field there is accumulation (or outflow) of the gaseous admixture. Note that accumulation and outflow of the the gaseous admixture in a small control volume are separated in time and molecular diffusion breaks a symmetry between accumulation and outflow (i.e., it breaks a reversibility in time). The latter can cause pattern formation in concentration distribution of the gaseous admixture advected by a compressible turbulent fluid flow.

Now we analyze formation of large-scale inhomogeneous structures in spatial distribution of the gaseous admixture concentration. Equation (3) for the mean number density of the gaseous admixture is reduced to the Schrödinger equation and we can use quantum mechanics analogy for the analysis of the inhomogeneities formation in a spatial distribution of the concentration of gaseous admixture. In order to estimate the the growth rate of the instability γ_0 we use a modified variational method (e.g., a modified Ritz method) which yields

$$\gamma_0 \sim \frac{\lambda_0^2 \eta}{2} \left[1 - \left(\frac{4(1 + c_0)}{\eta^2} \right)^p \right] , \tag{6}$$

where

$$p = \frac{1}{2} \left(\frac{Y^2}{Y_{cr}^2} - 1 \right), \quad Y_{cr}^2 = \frac{2(1 + c_0)}{c_0 \lambda_0^2} \ln \left(\frac{\eta}{2\sqrt{1 + c_0}} \right) ,$$

Y is determined from an equation

$$(Y + 1)^2 = 2\eta Y \sqrt{1 - c_0} \exp \left(-\frac{c_0 \lambda_0^2 Y^2}{2(1 - c_0)} \right) ,$$

and $\eta = v_0/\lambda_0$, $v_0 = v_s \Lambda_T/D_T$. The gravity acceleration $\mathbf{g}$ and $\nabla \rho$ are directed against the Z axis. Here we used the following spatial distributions of the turbulent kinetic energy $\langle \mathbf{u}^2 \rangle$ and mean temperature $T(Z)$:

$$\langle \mathbf{u}^2 \rangle - \exp[(c_0 Z^2/2) \exp(-\beta_0 Z^2)] ,$$

$$\lambda(Z) = \frac{1}{T} \frac{dT}{dZ} = (Z - \lambda_0) \exp(-\epsilon_0 Z^2) ,$$

where $\beta_0 \ll 1$ and $\epsilon_0 \ll 1$. These distributions satisfy the necessary condition for excitation of the instability. We used dimensionless variables: coordinates are measured in units Λ_T, time t is measured in units Λ_T^2/D_T, value λ is measured in units Λ_T^{-1}, and Λ_T is the characteristic scale of the spatial mean temperature distribution, the mean temperature T is measured in units of temperature difference δT in the scale Λ_T, and concentration N is measured in units N_*. It is seen from Eq. (6) that the instability is excited when $v_0/\lambda_0 > 2\sqrt{1+c_0}$ and $Y > Y_{\mathrm{cr}}$. For example, when $c_0 \ll 1$ (i.e., the inhomogeneity of turbulence is very weak), the growth rate of the instability is given by $\gamma_0 \sim c_0 \lambda_0^4 Y_0^2/2$, where $Y_0 = \eta - 1 + \sqrt{\eta(\eta-2)}$. Thus in the homogeneous turbulence the instability is not excited. The analyzed instability results in formation of inhomogeneous distribution of concentration of the gaseous admixture. The exponential growth during the linear stage of the instability can be damped by the nonlinear effects (e.g., hydrodynamic interaction between gaseous admixture and turbulent fluid flow, a change of temperature distribution in the vicinity of temperature inversion layer). The Schrödinger equation was also solved numerically. The numerical results are in a good agreement with the analytical estimates obtained by means of the modified Ritz method. Instability is excited when $0 < c_0 < 1.14$. Note that the analytical estimate obtained by means of the modified Ritz method yields the upper bound for $c_0 = 1$ in agreement with the numerical results.

References

Zeldovich, Ya. B., Molchanov, S. A., Ruzmaikin, A. A. and Sokoloff, D. D. (1988) Intermittency, diffusion and generation in a nonstationary random medium. *Sov. Sci. Rev. C. Math Phys.*, **7**, 1-110, and references therein.

Elperin, T., Kleeorin N. and Rogachevskii, I. (1995) Dynamics of passive scalar in compressible turbulent flow: large-scale patterns and small-scale fluctuations, *Phys. Rev. E*, **52**, 2617-2634.

Elperin, T., Kleeorin N. and Rogachevskii, I. (1996) Turbulent thermal diffusion of small inertial particles, *Phys. Rev. Lett.*, **76**, 224-228.

Landau, L. D. and Lifshitz, E. M. (1987) *Fluid Dynamics*, Pergamon, Oxford.

A TWO-SCALE TURBULENCE MODEL
FOR COMPRESSIBLE MIXING FLOWS

O. GREGOIRE, D. SOUFFLAND, S. GAUTHIER

CEA/Limeil-Valenton, 94195 Villeneuve St Georges CEDEX, France

AND

R. SCHIESTEL

IRPHE, 1 Rue Honnorat, 13003 Marseille, France

We are interested in strongly unsteady, turbulent mixing flows, such as those occurring in shock tube experiments. In these situations, production bursts, inducing spectral non-equilibrium, are separated by time intervals of relaxation to an isotropic turbulence. Since classical one-point closure models do not describe these departures from equilibrium, we have developed a *two-scale* statistical turbulence model [1] for *compressible mixing* flows.

1. Features of the two-scale model

An adaptive partition of the turbulent spectrum is defined. Since specific physical processes (production, transfer and dissipation) are associated to each spectral zone, we derived the corresponding transport equations for the turbulent kinetic energies ($\tilde{k}_1$, $\tilde{k}_2$) and the energy transfer rates ($\tilde{\Gamma}, \mathcal{E}$) attached to each domain of the turbulent spectrum. It results a set of four equations. The model accounts for the density fluctuations through the Favre averaging. The modelisation of the enthalpic production, important in mixing flows, is performed. Mixing is described by mass concentrations.

The two-time-scale modeling doubles the number of unknown coefficients. However, the spectral character of this approach allows us to derive several equations for these phenomenological coefficients. The model is first applied to a free decaying grid turbulence. We also use the analytical model for large-scale turbulence established by Canuto and Goldman (CG) [2]. This model depends on the growth rate of the instability-generating turbulence. The first spectral zone of the closure model is associated to the the large-scale spectrum of the CG model, while the second one, corresponding to smaller spatial scales, is given by the Kolmogorov inertial spectrum. This method is successively applied to the weak and strong shear flows and the Rayleigh-Taylor instability. These specific cases provide several algebraic equations for the coefficients.

This two-scale model has been embedded in a 1D hydrocode.

S. Gavrilakis et al. (eds.), Advances in Turbulence VI, 305-306.

306

2. Rayleigh-Taylor Induced Turbulence

The model is first used to simulate the turbulent mixing flow at a boundary between two gases, forced by a constant acceleration. The main interest of this test is that turbulence is continuously fed as opposed to the impulsive production occuring in shock tubes. The proportion of turbulent energy contained in the large scale zone increases continuously. The mixing length scales in t^2, and so does the upper bounds of the spectral zones (Figure 1). The ratio between the transfer rate ($\widetilde{F}$), which supplied the small scale region, and the dissipation rate ($\widetilde{\mathcal{E}}$), quickly reaches an asymptotic value of 1.16, indicating a large departure from equilibrium. Thus, it turns out that the two-scale model provides a description of a Rayleigh-Taylor turbulent flow in agreement with the physical intuition.

3. Richtmyer-Meshkov Induced Turbulence

We also simulate Meshkov's experiments in a helium/air shock tube [3]. In these experiments, the turbulence is mainly provided by the enthalpic production term $\overline{u_i'' \, \nabla_i \overline{P}}$ during the travelling of shock waves through the turbulent mixing zone (TMZ). Figure 2 displays comparison between the experimental data of the turbulent mixing length and the results of the two-scale model. The effects of the main shock and the reshocks are clearly seen in this figure. Moreover, there exists a delay of $\widetilde{\mathcal{E}}$ with respect to $\widetilde{F}$. It expresses the expected departure from equilibrium, which is the main result brought by the multiple-scale concept.

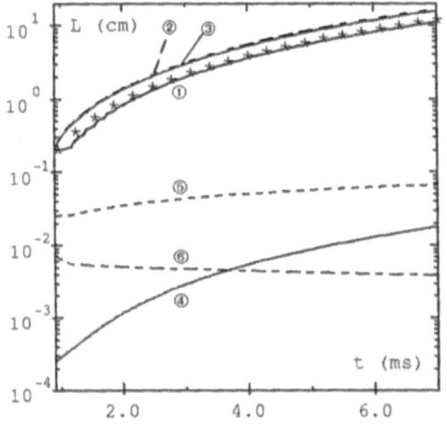
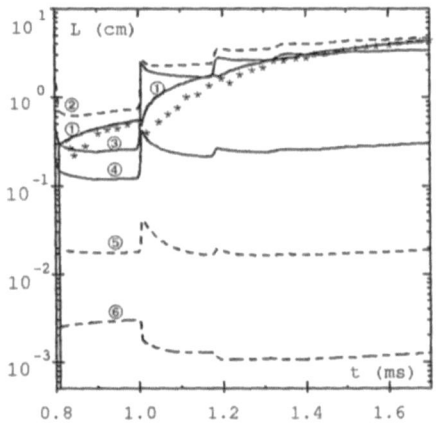

Figure 1. : Rayleigh-Taylor instability. Thickness of the TMZ : (*) analytical fit $\alpha g t^2$, (1) present numerical results. Evolution of the characteristic lengths in the two-scale simulation : (2) integral length, (3,4) upper bounds of the spectral zones, (5,6) Taylor and Kolmogorov scales. Note that curves (4) and (5) reach the same order of magnitude.

Figure 2. : Meshkov's experiment. Thickness of the TMZ : (*) experimental data of Andronov et al. [3], (1) simulation results. Evolution of the characteristic lengths in the two-scale simulation : (2) integral length, (3,4) upper bounds of the spectral zones, (5,6) Taylor and Kolmogorov scales. Note that evolution of curves (4) and (5) are very similar.

[1] R. Schiestel, J. de Mécanique Théor. et Appl. **2**, 417 (1983).
[2] V.M. Canuto and I. Goldman, Phys. Rev. Lett. **54**, 430 (1985).
[3] V.A. Andronov et al., Sov. Phys. Dokl. **44**, 424 (1976).

VI

Transition and Dynamical Systems

HIGHER BIFURCATIONS IN FLUID FLOWS AND COHERENT STRUCTURES IN THE TURBULENT STATE

F.H. BUSSE

Institute of Physics, University of Bayreuth, D-95440 Bayreuth

AND

R.M. CLEVER

Institute of Geophysics and Planetary Physics, UCLA, Los Angeles, CA 90024

1. Introduction

Many systems of fluid flow which are approximately homogeneous in two spatial dimensions and in time exhibit a sequence of supercritical bifurcations in which new dynamical mechanisms are introduced. Typically symmetries are broken in each of the bifurcations and in principle a stepwise evolution from simple to complex spatially periodic flows is obtained. The Taylor-Couette system and the Rayleigh-Bénard system are the best known examples of such fluid systems. While the primary state exhibits the full symmetries of the system, secondary states generically assume the form of two-dimensional rolls. Except for special situations in which other patterns compete with them, rolls must be considered as the universal secondary motion in the case of supercritical bifurcation. The transition from rolls to three-dimensional forms of fluid flow is not universal, however. Tertiary and quarternary states typically reflect the characteristic properties of the fluid system and of the specific region in the parameter space.

Although the spatially periodic tertiary and quarternary states are often stable with respect to infinitesimal disturbances, their basins of attraction tend to decrease with increasing control parameter. Nevertheless, experimentally realized irregular flows exhibit many features of the spatially periodic solutions and typical elements of these structures persist even in states of fully developed turbulence. Numerous examples from plane Couette flow, from thermal convection in horizontal and inclined layers and from the Taylor-Couette system have been derived in recent years. For reviews see references [1], [2]. In this short paper we intend to present some

S. Gavrilakis et al. (eds.), Advances in Turbulence VI, 309-312.
© 1996 *Kluwer Academic Publishers.*

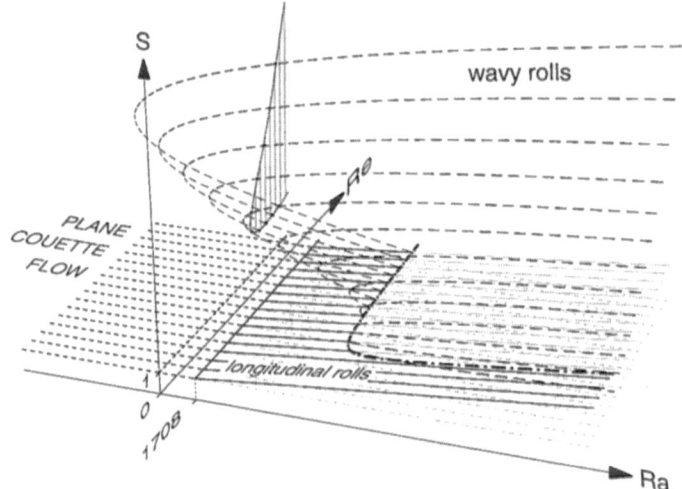

Figure 1. Sketch of positions of primary solution ($S \equiv 1$), secondary solution in the form of longitudinal rolls and tertiary solution in the form of wavy rolls in the Ra-Re-S parameter space for the Bénard-Couette problem. The shaded area indicates the existence of tertiary solutions in the isothermal case ($Ra = 0$) of plane Couette flow.

recently obtained results for tertiary and quarternary states of fluid flow in the plane Couette system.

2. Convection in a Sheared Fluid Layer

A plane Couette shear can be superimposed onto a Rayleigh-Bénard fluid layer with rigid boundaries without a change in the critical value of the Rayleigh number Ra. After introducing a Cartesian coordinate system with the horizontal x-coordinate in the direction of the shear we find that convection rolls with an axis parallel to the x-direction set in at the canonical value 1708 of Ra. Based on the relative motions $\frac{1}{2}U$ of the lower boundary and $-\frac{1}{2}U$ of the upper boundary in the x-direction the Reynolds number of the problem can be defined by $Re = Ud/\nu$ where d is the height of the fluid layer and ν is the kinematic viscosity. While the onset of the rolls and their amplitude are independent of Re, their instabilities are strongly dependent on this parameter. In figure 1 a qualitative sketch of the bifurcation diagram is shown where in addition to the parameters Ra, Re the shear Nusselt number S has been introduced. S measures the average stress exerted on the rigid boundaries divided by the stress exerted in the case of pure plane Couette flow. The main feature of interest is the extent of the tertiary fluid state in the form of steady wavy rolls into the region $Ra \leq 0$ of the Re-Ra-S parameter space. For details see [3]. The property that steady tertiary states of flow exist in the case of plane Couette flow, $Ra = 0$, was found first by Nagata [4]. Unfortunately the steady wavy rolls are unstable [3,5] and symmetry breaking instabilities lead to time dependent flows.

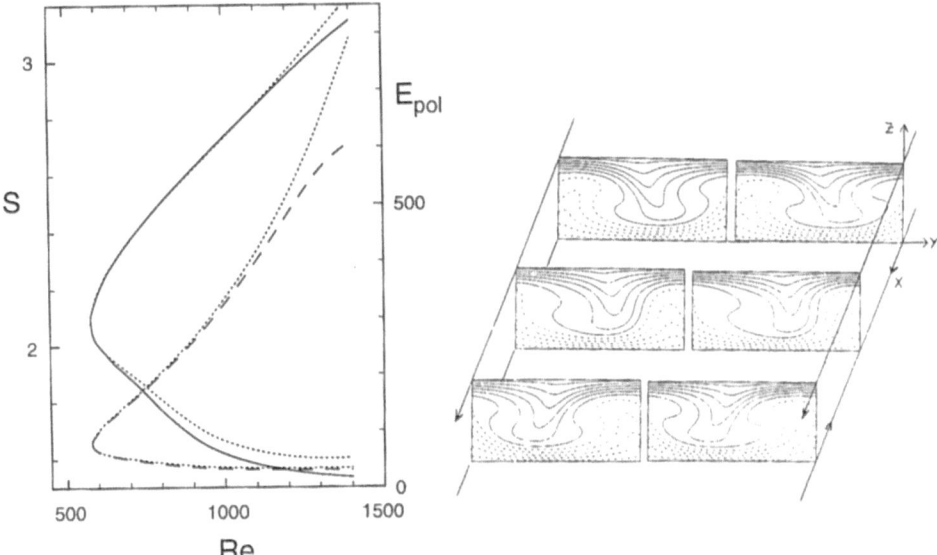

Figure 2. Shear Nusselt number S (solid line) and kinetic energy E_p of steady wavy rolls (dashed line) as a function of Re for $Ra = 0, \alpha_x = 1.5, \alpha_y = 2.5$. The dotted lines correspond to lower truncation ($N_\tau = 14$ instead of $N_\tau = 16$).

Figure 3. Lines of constant x-component of the velocity field of the steady tertiary solution with $\alpha_x = 1.5, \alpha_y = 3.0$ in the planes $x = 0, x = \pi/2\alpha_x, x = \pi/\alpha_x$ for $Re = 580$ (left plot) and $Re = 650$ (right plot).

3. Tertiary and Quarternary States of Flow in the Plane Couette Problem

In order to discuss tertiary and quarternary states of fluid flow it is convenient to separate the velocity field u into the part $U(z, t)$ that is averaged over the horizontal x-y-plane and into the fluctuating part $\check{u}$ for which the general representation in terms of poloidal and toroidal components can be used

$$u = U + \check{u} = U(z,t)i + \nabla \times (\nabla \times k\varphi) + \nabla \times k\psi. \qquad (1)$$

Here we have used the property that the mean component is always directed in the x-direction indicated by the unit vector i. The unit vector in the vertical direction is denoted by k. Without losing generality it can be assumed that the average of φ and ψ over the x-y-plane vanishes. In figure 2 the shear Nusselt number S and the kinetic energy E_p of the poloidal component of motion of steady wavy rolls are plotted as a function of Re for a particular set of wavenumbers α_x, α_y which represent typical values of wavy rolls [3]. As can be seen from this figure, there exist a lower branch with weak tertiary motion and an upper branch with a rather strong fluctuating component of motion. In figure 3 the x-component of the velocity field is shown at different positions along the x-axis for a solution on the upper branch. Both, the solutions on the upper and the lower branch, are

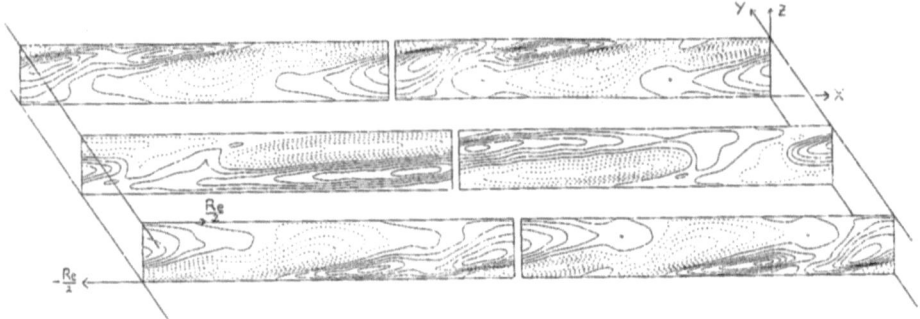

Figure 4. Lines of constant y-component of the velocity field of a time periodic quarternary solution with $\alpha_x = 1.0, \alpha_y = 3.0$ in the planes $y = 0, y = \pi/2\alpha_y, y = \pi/\alpha_y$ for $Re = 740$. Left and right plots differ by half a period, $t_p = 0.3048$.

unstable. A time periodic quarternary solution bifurcating from the upper branch is shown in figure 4.

4. Concluding Remarks

Although in experiments on plane Couette flow it is usually found that the transition from laminar flow to turbulent flow occurs at Reynolds numbers of the order 1400 (for a survey see [6]), more recent experiments [7] have demonstrated that three-dimensional states of fluid flow can be realized at much lower Reynolds numbers of the order of 640. A more detailed comparison between the streamwise vortices and the weakly time dependent wavy rolls similar to those shown in figure 4 will be of considerable interest.

References

[1] Busse, F.H. and Clever, R.M. (1994) Higher order bifurcations in fluid systems and coherent structures in turbulence, in K.-H. Spatschek and F.G. Mertens (eds.), *Nonlinear Coherent Structures in Physics and Biology*, Plenum Press, New York, pp. 405-416.

[2] Busse, F.H. and Clever, R.M. (1995) Bifurcations far from Criticality in Fluid Systems, in A. Doelman and A. van Harten (eds.), *Nonlinear dynamics and pattern formation in the natural environment*, Pitman Res. Notes in Math., Vol. 335, pp. 37-51.

[3] Clever, R.M. and Busse, F.H. (1992) Three-dimensional convection in a horizontal fluid layer subjected to a constant shear, *J. Fluid Mech.* **234**, 511-527.

[4] Nagata, M. (1990) Three-dimensional finite amplitude solutions in plane Couette flow: bifurcation from infinity, *J. Fluid Mech.* **217**, 519-527.

[5] Nagata, M. (1993) Stability of non-axisymmetric flows in the Taylor-Couette system, in S. Kida (ed.), *Unstable and Turbulent Motion in Fluid*, World Scientific, pp. 3-9.

[6] Tillmark, N. and Alfredsson, P.H. (1992) Experiments on transition in plane Couette flow, *J. Fluid Mech.* **235**, 89-102.

[7] Dauchot, D., and Daviaud, F. (1995) Streamwise vortices in plane Couette flow, *Phys. Fluids* **7**, 901-903.

THE STABILITY OF STEADY AND PERIODIC SOLUTIONS OF A LOW-DIMENSIONAL DYNAMICAL SYSTEM FOR 2D DRIVEN CAVITY FLOWS

W. CAZEMIER, R.W.C.P. VERSTAPPEN AND A.E.P. VELDMAN

Department of Mathematics, University of Groningen
P.O.Box 800, 9700 AV Groningen, The Netherlands

1. Introduction

A fully three-dimensional stability analysis of a flow governed by the Navier-Stokes equations leads to a computationally extremely expensive eigen-problem. Replacing the Navier-Stokes equations by a low-dimensional dynamical system that mimics the dynamics of the Navier-Stokes equations promises to save so many CPU-hours, that the analysis of the bifurcation scenario becomes feasible. In this paper, we will investigate how well an 80-dimensional dynamical system predicts the stability of steady and (quasi-)periodic flow in a 2D lid-driven cavity.

The dynamical system is constructed via a Galerkin projection of the Navier-Stokes equations on a set of basisfunctions. The basisfunctions are subtracted by a proper orthogonal decomposition (POD) of a (weakly) turbulent flow in a driven cavity at Reynolds number Re=22,000. POD-basisfunctions are constructed in such a way that they contain as much as possible energy (in time average) for a fixed number of basisfunctions. The flow data have been supplied by a direct numerical simulation (DNS). DNS data is also used to verify the bifurcation analysis of the dynamical system.

2. Proper Orthogonal Decomposition

POD is a technique to subtract the most energetic structures from a given flow. The basic idea is to find the "best" approximation of the fluctuating velocity field $u(x, t)$ with N structures, i.e. to determine a sequence of functions of time $a_i(t)$ and of space $\sigma^i(x)$ such that

S. Gavrilakis et al. (eds.), Advances in Turbulence VI, 313-316.
© *1996 Kluwer Academic Publishers.*

$$\int_{space} \int_{time} [u(x,t) - \sum_{i=1}^{N} a_i(t)\sigma^i(x)]^2 dt dx, \qquad (1)$$

is minimal. By calculus of variations it can be shown that the spatial structures $\sigma^i(x)$ satisfy

$$\int_{space} R(x,x')\sigma^i(x')\,dx' = \lambda_i \sigma^i(x), \qquad (2)$$

with $R(x,x') = \int_{time} u(x,t)u^*(x',t)dt$ the space-correlation tensor. The space-correlation tensor is symmetric and positive definite. Therefor, equation (2) has a set of orthogonal solutions σ^i with corresponding real and positive eigenvalues λ_i. The eigenvalues are ordered in increasing magnitude. The POD-basisfunctions $\sigma^i(x)$ are divergence-free and satisfy the boundary conditions.

The numerical computation of the POD-basisfunctions by directly solving (2) is expensive. We have used a cheaper method, the so-called snapshot method of Sirovich (1987). For a recent review of the POD, the reader is referred to Berkooz *et al.* (1993).

3. Galerkin Projection

The velocity field U is decomposed in the mean velocity $\overline{U}$ and the fluctuating velocity u: $U = \overline{U} + u$. The Galerkin projection of the Navier-Stokes equations on σ^i reads

$$\left(\sigma^i, \frac{\partial u}{\partial t}\right) = -(\sigma^i, \overline{U} \cdot \nabla u) - (\sigma^i, u \cdot \nabla \overline{U}) - (\sigma^i, \overline{U} \cdot \nabla \overline{U})$$
$$-(\sigma^i, u \cdot \nabla u) + \frac{1}{Re}(\sigma^i, \Delta u) + \frac{1}{Re}(\sigma^i, \Delta \overline{U}). \qquad (3)$$

substituting the approximation $u = \sum_{i=1}^{N} a_i(t)\sigma^i(x)$ into equation (3) yields a set of ordinary differential equations for $a_i(t)$:

$$\dot{a}_i = A_{ijk}a_j a_k + B_{ij}a_j + C_i, \qquad (4)$$

The coefficients A_{ijk}, B_{ij}, and C_i depend on $\overline{U}$ and the σ's. Now the idea is that we can take a relatively low number of σ's (80 in our case) due to the optimally fast convergence of the captured energy. A closure model for the energy exchange between the resolved and non-resolved POD-basisfunctions can be used.

We have compared the 80-dimensional dynamical system (computed at Re=22,000) with the DNS results at Re=22,000: their outcome agrees well both for short and for longer integration times (Cazemier *et al.*, 1994). For

TABLE 1. *Solution of the DNS at different Reynolds numbers*

Re	computed solution	period(s) in sec.
6,000	stationary	-
8,000	periodic	2.22
10,000	quasi-periodic	2.29 and 3.66
11,000	quasi-periodic	1.42 and 3.67
12,000	periodic	1.45

this we have used a model based on energy exchange between modes at Re=22,000. For the bifurcation analysis we do not use any model at all.

The POD-basisfunctions depend on the Reynolds number. Nevertheless, we want to use the 80-dimensional dynamical system for a whole range of Reynolds numbers to investigate the stability of steady and periodic flow in a driven cavity. Since the most energetic POD-basisfunctions correspond with large scale structures we expect these to stay important for other Reynolds numbers. Among others, Liu *et al.* (1994) have provided experimental evidence for this expectation. They have shown that, in the outer region of wall bounded turbulent flow, the 1D POD-basisfunctions are independent of the Reynolds number when scaled with the proper outer variables.

4. Results

DNS showed (see Table 1) that the flow in a 2D driven cavity converges to a stationary flow at Re=6,000, at Re=8,000 the flow is periodic with a period of 2.22 seconds, at Re=10,000 the flow quasi-periodic with periods 2.29 and 3.66 sec., at Re=11,000 the flow is also quasi-periodic, and at Re=12,000 the flow converges again to a periodic solution. These simulations used a 250x250 slightly stretched grid, a fourth-order discretization method in space, and a second-order Adams-Bashforth time integration method. With the DNS we can not calculate the exact bifurcation-point, but we can do this with the dynamical system.

The low-dimensional dynamical system has a stationary solution up to Re=7,819 (see Table 2). At this Reynolds number a Hopf bifurcation takes place. The period which belongs to this instable periodic solution is 1.63 seconds. After this first Hopf bifurcation this periodic solution stays stable till at Re=11,188 two complex conjugated Floquet multipliers cross the unit circle in the complex plane, indicating that the periodic solution

TABLE 2. *Solution of the low-dimensional dynamical system at different Reynolds numbers*

Re			computed solution	period(s) in sec.
0	-	7,819	stationary	-
7,819	-	11,188	periodic	1.63 at Re=7.819
11,188	-	$\pm 11,500$	quasi-periodic	1.67 and 2.70 at Re=11,188
$\pm 11,500$	-	$\pm 11,900$	periodic	1.67 at Re=11,500
$\pm 11,900$	-	?	quasi-periodic	1.66 and 4.65 at Re=12,000

becomes unstable. The second period of the instable solution is 2.70 sec., the ratio of the two periods 2.70/1.67=1.62 is similar to the ratio of the two periods of the DNS at Re=10,000, 3.66/2.29=1.59. After the second bifurcation the Floquet multipliers re-enter the unit circle at approximately Re=11,500 indicating that the 2-periodic solution becomes 1-periodic again. At Re=11,900 again an eigenvalue leaves the unit circle. The ratio of the 2 periods at Re=12,000, 4.65/1.66=2.80, is similar to the ratio of the two periods of the DNS at Re=11,000, 3.67/1.42=2.58.

So, the results show that the behavior of the low-dimensional dynamical system and the DNS is similar, with the difference that the Reynolds numbers at which the bifurcations take place and the periods are not the same. The ratio's of the periods, however, are similar. More solutions at the same Reynolds number are possible, therefor more DNS and dynamical system computations have to be done.

Acknowledgements. Both the Stichting Nationale Computerfaciliteiten (National Computing Facilities Foundations, NCF) with financial support from the Nederlandse Organisatie voor Wetenschappelijk Onderzoek (Netherlands Organization for Scientific Research, NWO) and the National Aerospace Laboratory NLR are gratefully acknowledged for the use of supercomputer facilities.

References

Berkooz, G., Holmes, P. and Lumley, J.L. (1993) The Proper Orthogonal Decomposition in the Analysis of Turbulent Flows,*Annu. Rev. Fluid Mech.* **25**, pp. 539-575

Cazemier, W, Verstappen, R.W.C.P., Veldman, A.E.P. (1994) DNS of Turbulent Flow in a Driven Cavity and their Analysis Using Proper Orthogonal Decomposition,*AGARD conference proceedings 551: Application of Direct and Large Eddy Simulation to Transition and Turbulence*, pp. 36-1-36-11

Sirovich, L. (1987) Turbulence and the Dynamics of Coherent Structures, *Quart. Appl. Math.* **45**, pp. 561-590

Liu, Z.C., Adrian, R.J., Hanratty, T.J. (1994) Reynold's Number Similarity of Orthogonal Decomposition of the Outer Layer of Turbulent Wall Flow, *Phys. Fluids* **6**, pp. 2815-2819

EXISTENCE AND STABILITY OF FINITE-AMPLITUDE STATES IN PLANE COUETTE FLOW

A. CHERHABILI and U. EHRENSTEIN
Université Lille I, Laboratoire de Mécanique de Lille, URA CNRS 1441
Bd. P. Langevin, F - 59655 Villeneuve d'Ascq Cedex, France

ABSTRACT

To elucidate the transition mechanism in plane Couette flow we compute finite-amplitude equilibrium solutions by extending numerically 2D nonlinear waves in plane Poiseuille flow to the plane Couette flow limit. The 2D nonlinear states in plane Couette flow take the form of spatially localized (solitary-like) stationary waves, they represent a new basic state for a secondary stability analysis. Secondary stability characteristics are computed as well as secondary bifurcation branches leading to 3D nonlinear states spatially localized in the streamwise direction and periodic in the spanwise direction.

1. Introduction

Plane Couette flow (PCF), that is the flow in a channel induced by the relative motion of two infinite parallel walls, can be considered as a kind of prototype for shear flows with a subcritical stability behaviour, due to the absence of a linear critical Reynolds number (cf. Romanov, 1973). Therefore the subcritical transition from laminar to turbulent PCF must be abrupt in contrast to a so-called supercritical transition scenario where disorder appears progressively. Transition in PCF must be driven by finite-amplitude disturbances and a related question is the existence of nonlinear states, solution of the Navier-Stokes equations. The onset of nonlinearities in PCF cannot be described in a classical frame of bifurcation theory due to the absence of a primary instability. Nagata (1990) provided three-dimensional nonlinear states in PCF by extending finite-amplitude solutions for a circular system between co-rotating cylinders with a narrow-gap to the case with zero average rotation rate. Bifurcation sequences in a Bénard-Couette problem have been produced by Busse and Clever (1992) leading to PCF as limiting case. In the present study we try to connect nonlinear 2D equilibrium states which exist in the pressure driven plane Poiseuille flow (PPF) to the PCF-limit, by continuously increasing the wall velocity. The travelling waves present in PPF evolve into solitary-like stationary states for PCF, solutions of the Navier-Stokes equations. Secondary stability results with respect to 2D and 3D perturbations of these new nonlinear equilibrium states are performed. Global numerical continuation analyses for bifurcating equilibrated 3D solutions are shown as well.

2. 2D nonlinear states in PCF

The flow is that of an incompressible viscous fluid of viscosity μ^* and constant density ρ^* between two parallel plates. The fluid motion results from the combined effects of the

S. Gavrilakis et al. (eds.), Advances in Turbulence VI, 317-320.

relative motion of the walls, the wall velocities being $V*$ and $-V*$ respectively, and an imposed pressure gradient $P*$. We introduce a continuation parameter η and a reference velocity $U*$ to define a (dimensionless) basic Poiseuille-Couette flow profile

$$U(y,\eta) = (1 - \eta)(1 - y^2) + \eta\, y \qquad (1)$$

$\eta = 0$ and $\eta = 1$ being the PPF- and the PCF-limits respectively. The Reynolds number $Re = \rho* U* h*/\mu*$ is defined with half the channel width $h*$, for $\eta = 1$ the pressure gradient vanishes and $U*$ is equal to half the velocity difference between the walls $V*$. A velocity disturbance $\vec{v} = (u, v, w)$ is superimposed to the laminar basic flow $(U(y,\eta),0,0)$ and the solution is supposed to be periodic in the streamwise x-direction and spanwise z-direction. The normal velocity v and normal vorticity $\omega = \partial u/\partial z - \partial w/\partial x$ are used as independent variables, they are expressed as truncated Fourier series in the directions of periodicity

$$\begin{Bmatrix} v(x,y,z,t) \\ \omega(x,y,z,t) \end{Bmatrix} = \sum_{n=-N}^{N} \sum_{m=-M}^{M} \begin{Bmatrix} \hat{v}_{nm}(y,t) \\ \hat{\omega}_{nm}(y,t) \end{Bmatrix} \exp\left[i(n\alpha x + m\beta z)\right] . \qquad (2)$$

The modes $(\hat{v}_{nm}(y,t), \hat{\omega}_{nm}(y,t))$ are solution of nonlinear modal equations by injecting the above expansion into the Navier-Stokes equations. Starting point for our computations are nonlinear 2D travelling wave solutions in PPF ($\omega \equiv 0$ in (2)) which emerge from neutrally stable points in streamwise wavenumber α and Reynolds number Re. Those solutions are stationary in a frame of reference $x' = x - ct$ (moving with the unknown wave speed c) they can be computed by use of Newton-Raphson iteration as solutions of a large nonlinear system obtained after discretization of the modal equations by Chebyshev-collocation.

To recover nonlinear states for PCF we increase (starting from the equilibrium surface in PPF) the previously defined parameter η and it is only after a sufficiently high Fourier truncation ($N \geq 10$ in (2)) that we indeed succeeded in computing 2D nonlinear states for $\eta = 1$, that is for PCF (Cherhabili and Ehrenstein, 1995). During this continuation procedure the nonlinear travelling waves present in PPF (with $c \neq 0$ and $\eta = 0$) evolve into spatially localized stationary waves for $\eta = 1$. This gives some explanation why previous attempts to compute 2D nonlinear waves in PCF failed. Convergence of these nonlinear states in terms of the modal expansion are discussed in (C. & E., 1995). A typical structure of the nonlinear 2D state is shown in Fig.1, where the solitary-like profiles of the streamwise component of the disturbance with 2 humps at the center of the channel $y = 0$ and 3 humps near the walls $y = \pm 1$ are depicted. The lowest Reynolds number for existence of these states is close to 1500 (C. & E. 1995).

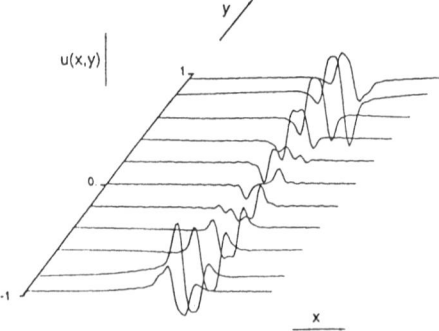

Fig. 1 - Perspective view of evolution over one wavelength in x and across the channel y of the streamwise component u of the disturbance velocity in PCF, at Re = 2200, with N = 25, α = 0.11.

3. The stability of 2D states and global 3D computations

The computed 2D nonlinear equilibrium states form a new basic state $\vec{u}_{2D}(\alpha, Re)$ for a secondary stability analysis. The idea is that the transition is related to the appearance of successive bifurcations leading to solutions with increasing spatial or temporal complexity (Saffman, 1983). Superimposing the secondary perturbation

$$\vec{v} = e^{\sigma t} e^{i\beta z} \sum_{n=-N}^{N} \vec{\hat{v}}_n(y) e^{in\alpha x} \tag{3}$$

to the new basic state and injecting into the Navier-Stokes equations the linearization around $\vec{u}_{2D}(\alpha, Re)$ leads to an eigenvalue problem for the temporal eigenvalue $\sigma = \sigma_r + i\sigma_i$ in (3) (the basic state being unstable if $\sigma_r > 0$). The stability analysis reveals that 3D disturbances have amplification rates σ_r of an order of magnitude higher than 2D secondary disturbances similar to what one observes in PPF (cf. Ehrenstein and Koch, 1991). Concentrating on 3D secondary instabilities Fig. 2 shows the amplification rate of the most amplified (zero-frequency) mode as function of β, for a 2D nonlinear state located at $Re = 2200$. First σ_r increases to reach a maximum at $\beta \approx 3$, for greater β-values σ_r decreases continuously to a point of neutral stability at $\beta_c \approx 23$. This critical spanwise wavenumber β_c corresponds to a (stationary) bifurcation point for the emergence of 3D nonlinear stationary states, solution of the full Navier-Stokes system. We define an amplitude ε for the nonlinear solution by a suitable normalization condition (cf. C. & E., 1995), and a bifurcating branch in the (ε, Re) -plane is depicted in Fig. 3. The cut through the 2D equilibrium surface is shown as the dashed lines. The computed 3D solution branch is represented as the solid line which connects the upper and the lower branch of the 2D equilibrium surface through bifurcation points. The structure of the 3D equilibrium state is illustrated in Fig. 4, where isolines for the wall-normal vorticity ω are plotted. Fig. 4a shows ω in the (x,y)-plane for a fixed spanwise location $z = 1$. Fig. 4b is a cut through the center of the structure in Fig. 4a in the (y,z) -plane. The solution is again spatially localized in the streamwise direction, and it exhibits counter-rotating cells in the periodic spanwise z-direction, which is a consequence of our symmetry assumption with respect to $z \rightarrow -z$.

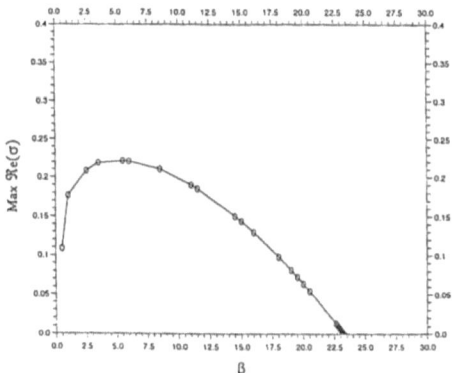

Fig. 2 - Highest amplification rate Max ($\Re e(\sigma)$) for 3D secondary instability, as function of the spanwise wavenumber β, for the 2D equilibrium state at Re = 2200, $\alpha = 0.132$.

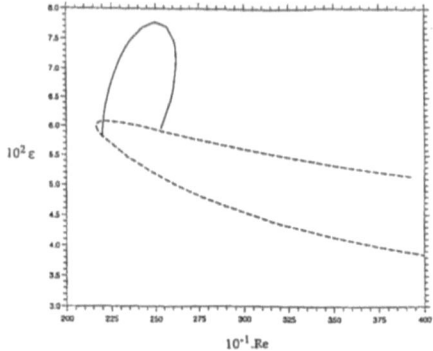

$10^{-1}.Re$

Fig. 3 - 2D and 3D equilibrium states for PCF in the (ε, Re) - plane. ---- : 2D state at $\alpha = 0.132$; —— : 3D state bifurcating from 2D - solution, for $\beta = 23.2406$ (with N = 13, M =1, 29 Chebyshev polyn.)

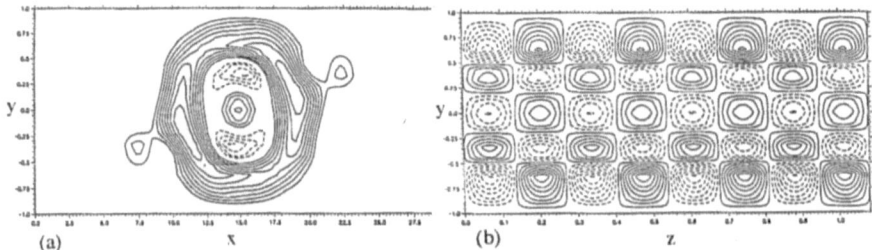

(a)   x   (b)   z

Fig. 4 - Normal vorticity of the 3D equilibrium state at Re = 2355, $\alpha = 0.132$, $\beta = 23.24$. (a) Spatially localized structure in the (x,y)-plane at z=1. (b) x = constant-cut in the (y,z)-plane over 4 wavelengthes.

4. Discussion

Our numerical bifurcation study of finite amplitude solutions in PCF led to the discovery of spatially localized 2D stationary nonlinear states in PCF. Stationary bifurcation points on the 2D equilibrium surface are starting points for the computation of 3D nonlinear states spatially localized in the streamwise direction and periodic in the spanwise direction. Recent experimental results reported by Dauchot and Daviaud (1995) show that perturbed regions with a qualitatively similar structure exist during transition in PCF. Exploring the equilibrium surface of 3D nonlinear stationary states we hopefully can extend the solutions reported in this work to transitional Reynolds numbers observed in experiments (D. & D., 1995; Tillmark and Alfredsson, 1992).

5. References

Busse F.H, Clever R.M., 1992, *J. Fluid Mech.*, **234**, 511-527.
Cherhabili A., Ehrenstein U., 1995,*Eur. J. Mech. B/Fluids*, **14**, 667-696.
Dauchot O., Daviaud F., 1995, *Phys. Fluids*, **7**, 901-903.
Ehrenstein U., Koch W., 1991, *J. Fluid Mech.*, **228**, 111-148.
Milinazzo F.A., Saffman P.G., 1985, *J. Fluid Mech.*, **160**, 281-295.
Nagata M., 1990, *J. Fluid Mech.*, **217**, 519-527.
Romanov V.A., 1973, *Funct. Anal. Applics*, **7**, 137-146.
Saffman P.G., 1983, *Ann. N. Y. Acad. Sci.*, **404**, 12-24.
Tillmark N., Alfredsson P.H., 1992, *J. Fluid Mech.*, **235**, 89-102.

ON THE RELATION BETWEEN LOW-DIMENSIONAL MODELS BASED ON POD AND DYNAMICS OF COHERENT STRUCTURES IN A TURBULENT PLANE MIXING LAYER

L. CORDIER, R. MANCEAU, J. DELVILLE and J.-P. BONNET

CEAT, Laboratoire d'Etudes Aérodynamiques Poitiers (France)

The organized character of turbulent shear flow is now well admitted from experimental and numerical evidence. These organized, often called <u>coherent</u>, large scale <u>structures</u> are responsible for important characteristics of the flows, such as vibrations, mixing, drag, sound emission, etc ... Such aspects are of great practical importance for fundamental research as well as from an industrial point of view. The long-term goal is to achieve active control in industrial flows using coherent structures dynamics.

In this paper, coherent structures in a turbulent plane mixing layer are studied using the notion of *deterministic chaos* originally introduced in 1971 by Ruelle and Takens [7]. Following their approach, we assume that we can build a low-order dynamical system able to describe the major features of the dynamical behavior of the flow. The basic idea is to introduce through this dynamical system a way to take into account the instantaneous state of the flow to predict its temporal evolution. Up to now, the methods of control in aerodynamics were based only on the mean characteristics of the flow. Similar approaches have been recently used with success in the case of open flow systems : Aubry et al. [1] studied the dynamics of coherent structures in the wall region of a turbulent boundary layer and Corke et al. [4] utilized low-dimensional dynamical systems models to guide control experiments in the case of a jet.

Here, large scale structures of the flow are studied through the use of Proper Orthogonal Decomposition (POD), a method introduced in Turbulence by Lumley (1967) [5]. With this technique, a coherent structure $\sigma(\underline{X})$ with $\underline{X} = (x_1, x_2, x_3, t)$ is specified by requiring that its projection on the flow realizations $u(\underline{X})$ shall be maximum in quadratic mean :

$$\frac{\left\langle (\sigma, u) \right\rangle}{\|\sigma\|^2} \overset{def}{=} Max \tag{1}$$

where $\left\langle \ \right\rangle$ is the ensemble average operator and (,) the complex inner product

S. Gavrilakis et al. (eds.), Advances in Turbulence VI, 321-324.
© *1996 Kluwer Academic Publishers.*

operator defined as :

$$(a, b) = \int_{\mathcal{D}} a^*(\underline{X}).b(\underline{X})d\underline{X} \tag{2}$$

where * is the complex conjugate. Using the calculus of variations, this maximization problem can be reduced to a Fredholm integral value problem, where the kernel is the correlation tensor. In the present study, POD is applied only in the direction of the mean velocity gradient x_2 assumed inhomogeneous. In the two other directions, streamwise x_1 and spanwise x_3 (nearly homogeneous), we applied a Fourier transform. Following the approach of Lumley [5], the dominant structure of the flow can be determined from the following equation :

$$\sum_{j=1}^{3} \int_{\mathcal{D}} \Psi_{ij}(x_2, x_2'; k_{x_1}, k_{x_3})\Phi_j^{(n)}(x_2'; k_{x_1}, k_{x_3})dx_2' = \lambda^{(n)}(k_{x_1}, k_{x_3})\Phi_i^{(n)}(x_2; k_{x_1}, k_{x_3}) \quad n = 1, \ldots, +\infty$$

$$\tag{3}$$

where n is the mode number of the POD and where the cross-spectrum $\Psi_{ij}(x_2, x_2'; k_{x_1}, k_{x_3})$ is the Fourier transform in the two homogeneous directions x_1 and x_3 of the two point space correlations $R_{ij}(x_2, x_2'; \delta_x, \delta_z) = \left\langle u_i'(x_2, x_1 = X_0, x_3 = 0)u_j'(x_2', X_0 + \delta_x, \delta_z) \right\rangle$. The averaging process involved $\langle \; \rangle$ is a classical time average.

The set of eigenfunctions $\Phi_i^{(n)}(x_2; k_{x_1}, k_{x_3})$ is complete in the sense that the flow fields of the given ensemble can be expanded in terms of the eigenfunctions via :

$$\hat{u}_i(x_2; k_{x_1}, k_{x_3}, t) = \sum_{n=1}^{+\infty} a_{k_{x_1}, k_{x_3}}^{(n)}(t)\Phi_i^{(n)}(x_2; k_{x_1}, k_{x_3}) \tag{4}$$

where $\hat{u}_i(x_2; k_{x_1}, k_{x_3}, t)$ is the two dimensional Fourier transform of $u_i'(x_1, x_2, x_3, t)$ the fluctuation velocity.

The global energy contained within the flow can be defined by :

$$E_T = \sum_{n=1}^{+\infty} \int_{-\infty}^{+\infty} \int_{-\infty}^{+\infty} \lambda^{(n)}(k_{x_1}, k_{x_3})dk_{x_1}dk_{x_3} \tag{5}$$

The problem 3 was solved using the experimental data of Delville [3]. The energy contained in the POD modes converge rapidly with the first mode being dominant (49% of the turbulent kinetic energy), indicating that a low-order dynamical systems approach would be fruitful. Moreover, this first mode contains evidence of both known types of flow organization in the mixing layer, i.e quasi-two dimensional spanwise structures and streamwise aligned vortices.

The next step is to introduce into the Navier Stokes equations written for an incompressible and inviscid flow, the Reynolds decompositions for both velocity $u_i = \overline{u_i} + u_i'$ $(i = 1, 2, 3)$ and pressure $p = \overline{p} + p'$ where the ensemble average $\overline{(\;)}$ is evaluated by spatial integration over the two homogeneous directions x_1 and x_3.

Adding classical mixing layer assumptions like the thin-layer hypothesis ($\overline{u_2} = \overline{u_3} = 0$), homogeneity in the x_3 direction ($\dfrac{\partial \overline{()}}{\partial x_3} = 0$), and using the continuity equation for the mean velocity we arrive at $\overline{u_1} = \overline{u_1}(x_2, t)$. Finally, we derive the Navier Stokes equations for the turbulent components :

$$\frac{\partial u_i'}{\partial t} + u_j' \frac{\partial u_i'}{\partial x_j} - \overline{u_j' \frac{\partial u_i'}{\partial x_j}} + u_2' \frac{\partial \overline{u_1}}{\partial x_2} \delta_{i1} + \overline{u_1} \frac{\partial u_i'}{\partial x_1} + \frac{1}{\rho} \frac{\partial p'}{\partial x_i} = 0 \qquad (6)$$

where δ_{i1} is the Kronecker symbol. Then, the method consists in projecting the Fourier transform in x_1 and x_3 of equations (6) onto the empirical basis set using a Galerkin method. This projection yields a set of time dependent ODEs for the so-called random coefficients of the POD : $a_{k_{x_1}, k_{x_3}}(t)$.

A truncation is then performed to include only the <u>first POD mode</u> for selected streamwise/spanwise modes in order to achieve the low dimensional set of equations. As a first step, a truncation including only $k_{x_3} = 0$ and some non-zero streamwise wavenumbers k_{x_1} is studied. As a closure hypothesis of equation 6, we chose to use a velocity profile based on the Boussinesq equation [2] and estimated with the first POD mode only :

$$\overline{u_1}(x_2) = -\frac{1}{\nu_t} \int_0^{x_2} \overline{u_1' u_2'}(x_2') dx_2' + U_m \qquad (7)$$

with

$$\overline{u_1' u_2'}(x_2) = \overline{u_1' u_2'}^{(1)}(x_2) = \int_{-\infty}^{+\infty} \lambda_{k_{x_1}, 0}^{(1)} \Phi_{1, k_{x_1}, 0}^{(1)}(x_2) \Phi_{2, k_{x_1}, 0}^{(1)*}(x_2) \, dk_{x_1} \qquad (8)$$

where U_m is the mean velocity of the flow and where ν_t is estimated such that the boundary conditions $\overline{u_1}(-L) = U_b$ and $\overline{u_1}(L) = U_a$ are satisfied.

Finally, the set of derived equations is written :

$$\frac{da_{k_{x_1}, 0}}{dt}(t) = L_{k_{x_1}, 0} \, a_{k_{x_1}, 0}(t) + \sum_{k_{x_1}'} Q_{k_{x_1}', 0, k_{x_1}, 0} \, a_{k_{x_1}', 0}(t) \, a_{k_{x_1} - k_{x_1}', 0}(t) \qquad (9)$$

We can note that with this closure hypothesis, no cubic term appear in 9.

This set of derived equations, is analysed through the use of dynamical systems techniques and special attention is paid to the linear instability result obtained. In figure 1, we compare the linear growth rate obtained for an inviscid mixing layer by Michalke [6] to the one obtained by our dynamical system. We find that very good agreement is obtained. We retrieve the most linear unstable mode α^{max} and the value of its linear growth rate. This mode is essential to model the dynamical behavior of the system because it forces the system with the wavenumber of the first instability of the plane mixing layer. Moreover, with our low-dimensional system, the parabolic shape of the theoretical curve is conserved, which indicates

324

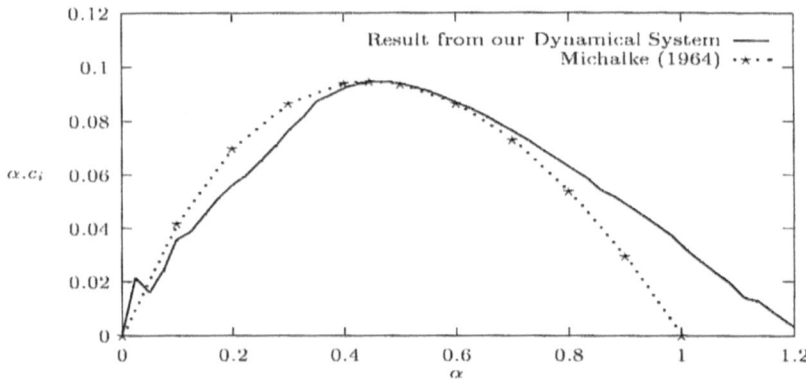

Figure 1: Linear growth rate of the inviscid turbulent plane mixing layer.

that in our system, α^{max} dominates clearly the other modes.

To conclude, we find that the first POD mode is not solely linked to energy but to linear stability theory too. Thus, we can hope that a low-order dynamical system approach, constructed with the first POD mode only, would be fruitful to describe the dynamical behavior of the flow.

References

[1] Aubry N., Holmes P., Lumley J.L. & Stone E. : 1988. "The dynamics of coherent structures in the wall region of a turbulent boundary layer" *J. Fluid Mech.*, **192**, 115.

[2] Boussinesq T. V. : 1877. "Théorie de l'écoulement tourbillonnant" *Mém. Pré. par div. Sav. Paris*, **23**.

[3] Delville J. : 1995 "La décomposition orthogonale aux valeurs propres et l'analyse de l'organisation tridimensionnelle des écoulements turbulents ci-saillés libres" *Thèse Université de Poitiers*, in french.

[4] Corke T. C., Glauser M. N., & Berkooz G. : 1994. "Utilizing low-dimensional dynamical systems models to guide control experiments" *Applied Mechanics Reviews*, **6, part 2**, S132.

[5] Lumley, J. L. : 1967. *Atm. Turb. and RadioWave Prop.*. Yaglom and Tatarsky eds. Nauka, Moscow, 166-178.

[6] Michalke A. : 1964 "On the inviscid instability of the hyperbolic-tangent velocity profile" *J. Fluid Mech.*, **19**, part 4, 543.

[7] Ruelle D. & Takens F. : 1971. "On the Nature of Turbulence" *communs. Math. Phys.*, **Vol. 82**, 137.

SPATIAL NUMERICAL SIMULATIONS AND EXPERIMENTS ON THE BREAKDOWN OF STREAMWISE VORTICES IN A BLASIUS BOUNDARY LAYER

Y. CUI, B. TANGUAY, A. BOTTARO AND P.A. MONKEWITZ

IMHEF - DGM, École Polytechnique Fédérale de Lausanne, CH-1015 Lausanne, Switzerland

Hairpin vortices have been found to be one of the predominant coherent structures in both transitional and turbulent wall-bounded shear flows. Here, we study, both experimentally and numerically, the evolution of a hairpin vortex from its birth, through its growth as it interacts with the mean shear, to its breakdown. The vortices are introduced in a subcritical laminar boundary layer by fluid injection through a long and thin slot aligned with the free stream direction. With the Reynolds number and slot geometry fixed, the control parameter of this experiment is the ratio of the injection velocity to the free stream velocity. This investigation is relevant to both the study of near-wall turbulent coherent structures and by-pass transition.

NUMERICAL STUDY

The effect of the localized perturbations at the wall on the generation and amplification of vortices is studied with spatially developing numerical simulations of the full incompressible Navier-Stokes equations. The finite volume approach is adopted, with centred schemes for convective and diffusive fluxes, and a Crank-Nicolson scheme for the time evolution. At low injection ratios no oscillations are observed downstream of the slot and the perturbation kinetic energy decays exponentially along the streamwise direction. When the amplitude of the localized disturbance is past a critical value, the three-dimensional shear layers created by the interaction of the jet with the oncoming boundary layer roll up to form hairpin-shaped vortices. These vortices are shed periodically behind the slot, convected downstream and their legs oscillate in a varicose fashion (see figure 1). The evolution of the hairpins downstream proceeds via lateral spreading and

325

S. Gavrilakis et al. (eds.), Advances in Turbulence VI, 325-328.
© 1996 *Kluwer Academic Publishers.*

the appearance of inclined wave fronts, until a stage is reached where an incipient turbulent spot appears.

EXPERIMENTAL INVESTIGATION

The experiments were carried out in the low-speed wind tunnel of the IMHEF. As shown in figure 2a, the 1 mm x 40 mm slot was located 250 mm from the flat plate leading edge. At a freestream velocity of 4 m/s, this corresponds to a displacement thickness Reynolds number of 430 at the slot center. The freestream turbulence level was 0.25 %.

Figure 2b shows a phase-referenced slice through two successive hairpin vortices. The visualization was performed by seeding the injection flow with smoke. It depicts the main features of the flow and clearly shows the growth (in size) of the structures with downstream distance.

The observed extreme sensitivity of the vortices to any external disturbance indicates that the instabilities inherent to this flow are of a convective nature. In order to phase-lock the (primary) instability, small-amplitude (typically 0.005 m/s) periodic fluctuations were superimposed on the mean injection velocity . The "preferred mode" was then determined by measuring the system response to constant amplitude perturbations over an appropriate frequency band (figure 2c). The streamwise growth rate and wavelength of the instability as a function of the control parameter will be presented.

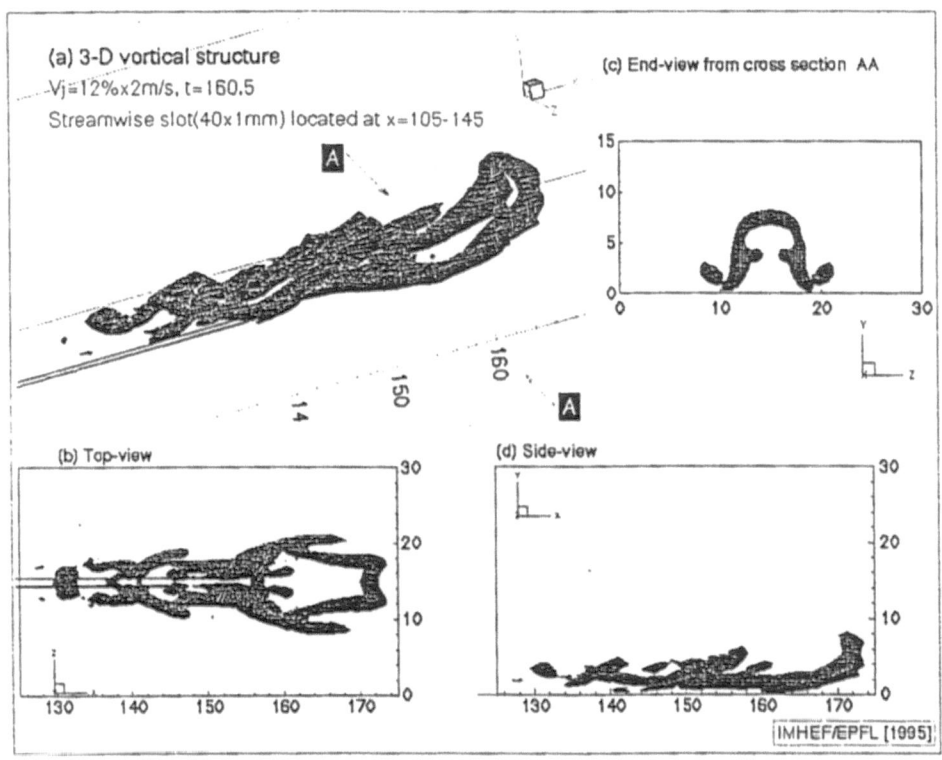

Figure 1. Visualization of a hairpin-shaped vortex in a Blasius boundary layer by isosurface of negative λ_2 at a fixed nondimensional time $t = 160.5$; λ_2 is the intermediate eigenvalue of the symmetric tensor $S^2 + \Omega^2$, where S and Ω are the symmetric and antisymmetric part of the velocity gradient tensor, respectively. (a) Three-dimensional view of the vortical structure. The free-stream direction is from lower left to upper right. The streamwise slot of dimensions 40 mm x 1 mm is located at $x = 105 - 145mm$. (b) Top view (c) End view of hairpin structure downstream of cross-section AA (d) Side view.

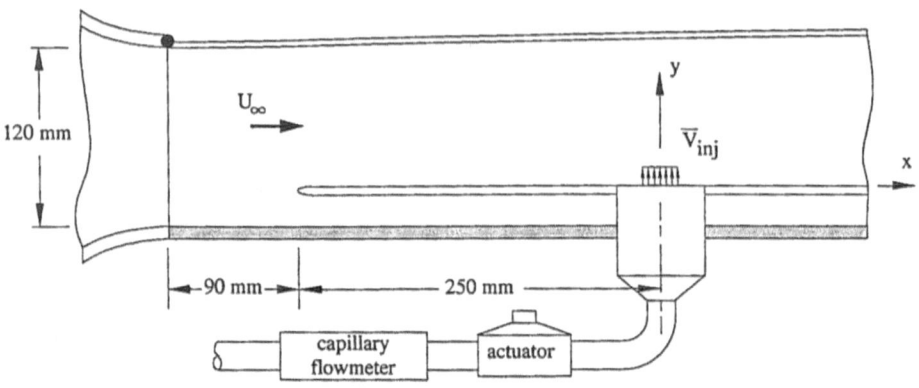

Figure 2 a) Test section and injection system

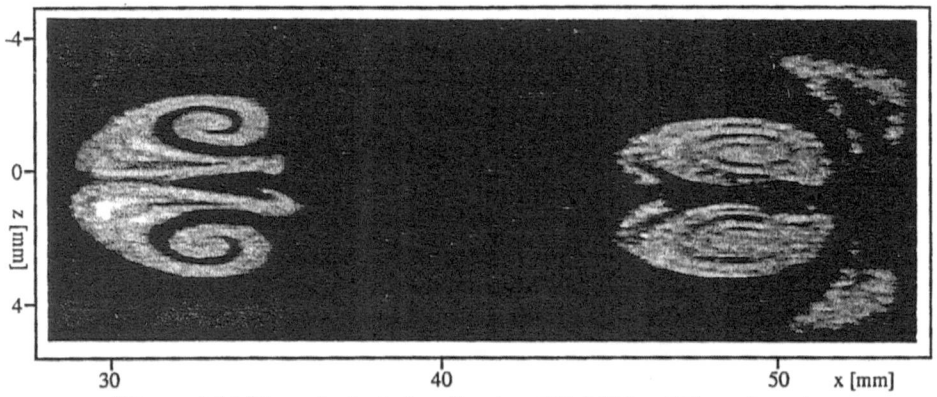

Figure 2 b) Phase-locked visualization (Vinj / U_∞= 7%; y=2 mm)

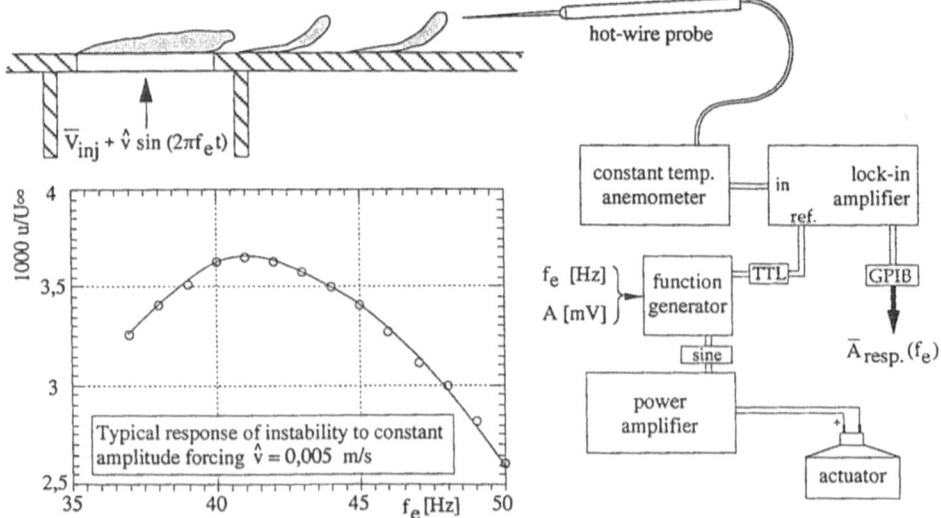

Figure 2 c) determination of the preferred mode of the instability

Pairs of counter-rotating streamwise vortices : A finite amplitude solution in plane Couette flow.

O. Dauchot, S. Bottin and F. Daviaud
Groupe Instabilités et Turbulence, CEA Saclay, SPEC.
F-91191 Gif-sur-Yvette CEDEX, France

Abstract

New experimental observations of the transition to turbulence in the plane Couette flow (PCF) allow us to identify the streaky structures previously observed on the turbulent spots border as pairs of streamwise counter-rotating vortices. The PCF is slightly modified by introducing a wire in its central plane, parallel to the spanwise direction. A destabilization then occurs and similar pairs of streamwise vortices are stabilized. Comparing the modified flow with analytical calculation and interpolating our results to the limit of an infinitesimal wire diameter, leads us to consider these structures as finite amplitude solutions in PCF.

When the transition to turbulence occurs because of infinitesimal perturbations, it is possible to determine the bifurcate solution which branches off the basic state continuously. But in many flows, the transition is subcritical and caused by finite amplitude perturbations. The plane Couette flow does not face primary linear instability and can not be studied by weakly nonlinear analysis. Numerically, several studies[1][2][3] have tried to overpass this difficulty by simulating the continous deformation of finite amplitude solutions in other flows towards the PCF. Experimentally[4], [5] the PCF has been destabilized when $R \approx 370$, while Dauchot et Daviaud[6] revealed the existence of a critical amplitude of perturbation and its dependence on the Reynolds number.

Two of the authors had already noticed the existence of coherent structures on the edges of the turbulent spots[6]. Furthermore correlation measurements[7] have revealed the existence of streamwise turbulent structures with a periodicity in the spanwise direction. More generally streamwise vortices appear to be present in many process of transition to turbulence. In this study, we succeed in stabilizing streamwise vortices and in giving a precise description of their behaviour in the PCF, the simplest shear flow. Their independence on the forcing process leads us to conclude to their existence in the unmodified PCF.

The PCF set up[6] is created by the opposite motion of two parallel walls. The Reynolds number is defined by $R = Uh/\nu$, where U is the speed of either wall, $h = d/2$ is half the gap and ν is the fluid kinematic viscosity. The accuracy of R is within 3%. The flow is perturbated by a thin wire of diameter 2ρ ($\rho/h \in$

S. Gavrilakis et al. (eds.), Advances in Turbulence VI, 329-332.

$([0.01, 0.15])$ attached in the central plane parallely to the spanwise direction. We define as x (resp. y and z) the streamwise (resp. the normal to the wall and the spanwise) direction. The origin in x and y is fixed at the wire position.
A laser plane illuminates the flow seeded with yellow colouring or Iriodin. A back scattering laser doppler velocimetry allows streamwise velocity profiles $U_x(y)$ measurement, with a spatial resolution of 0.1 mm. Measuring laminar profiles, the accuracy of U_x, given by U_x^{rms}/U_x is within 3%.

The flow is first disturbed by an instantaneous and localized finite amplitude perturbation in order to create a turbulent spot[6]. Figure 1 reveals that the streaky structures observed in turbulent spots are produced by pairs of streamwise counter-rotating vortices. It is particularly obvious on the right picture. On the central picture, one can distinguish four or five pairs of vortices. Even closer to the center, they destabilize. These structures have also been observed by mean of spatiotemporal diagrams[6]. They appear transciently during spot relaxation for very low Reynolds number such as 50.
The wire is now introduced in the central plane. For $R < R_0$ the flow is, close to the wire, different from the PCF but remains laminar, stationary, and invariant in the spanwise direction. Figure 2(a) displays velocity profiles for two different diameters wires. U_x is normalized by $U_x(y = \rho)$ in the unmodified PCF and y is normalized by ρ. The wire induces a significant deformation of the flow, even if very locally. Note that the dependence on the wire radius is well described by the chosen scaling. Figure 2(b) displays the streamwise velocity profiles measured and analytically calculated[8].

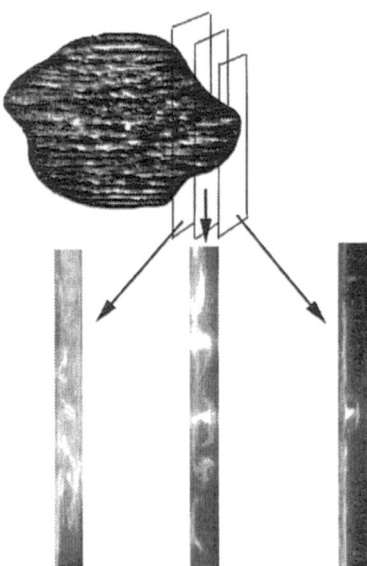

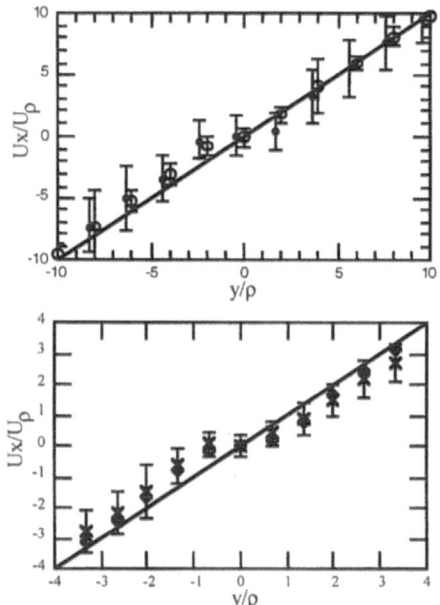

Figure 2: Streamwise velocity profile close to the central plane. (a): $\bullet$:$\rho/h = 0.0143$, o:$\rho/h = 0.0429$, $x/\rho = 1.33$. (b): $\times$: Measured profile, $\diamond$: Analytical calculation. $x/\rho = 1.33$, $\rho/h = 0.0429$

Figure 1: A (y, z) view of the streamwise vortices on the border of the turbulent spots.

A very good agreement is observed between the measurements and the calculations. Measurements for further position from the wire show that the experimental profile has recover the PCF one when $x = 15\rho$.

When $R = R_0$ a transition occurs[9]. One can see on figure 3(a) that the transition is characterized by the apparition of localized coherent structures occupying a limited extention Δ in the streamwise direction, much larger ($30h$) than the wire influence area. The transition appears to be subcritical: the arising structures already have a finite streamwise extension, and an hysteresis cycle even if very small is observed. Figure 3(b) displays a visualization of the flow in a (y.z) plane, where one can observe the high velocity streams origin: a lift up mechanism induced by pairs of streamwise counter-rotating vortices, which are very similar to those observed on the border of the turbulent spot. Each vortex, centered in the gap, has a ellipsoidal section, slightly tilted in opposite direction for the vortices of a same pair. The large diameter is about twice the small one. The vortices angular velocity grows from π to 2π rad/s, with the Reynolds number. The characteristic spanwise spacing between the pairs of vortices, is badly selected. Its mean value $\lambda/h \simeq 5 \pm 3\%$, independently from the Reynolds number. When $R = R_1$, the vortices are destabilized and the flow become turbulent. The destabilization is a continuous process, which the authors describe elsewhere[10]. It suggests that a detailed study of the streamwise vortices destabilization could lead to a better understanding of the turbulent state in the PCF.

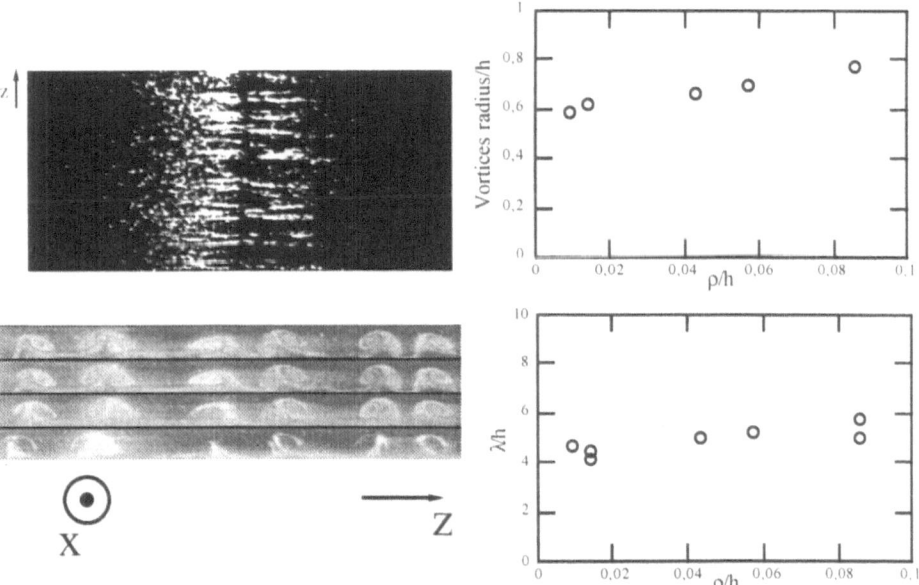

Figure 3: Streamwise vortices around the wire. $\rho/h = 0.0857$, $R = 250$. (a): An (x, z) view. (b): An (y, z) view.

Figure 4: Dependance on the wire radius. (a): vortices radius in function of ρ/h. (b): spanwise lengthscale λ in function of ρ/h.

Let us now describe the influence of ρ/h on the apparition and the stabilisation of the streamwise vortices. First, the thiner the wire is, the higher R_0 is. It also appears that the transition becomes more and more subcritical when ρ/h decreases. On the contrary, as shown on figure 4(a), the wire radius has little influence on the streamwise vortices radius. Similarly, λ remains constant in the studied range of wire radius (figure 4(b)).

In this study, we slightly modified the PCF by introcucing a thin wire in its central plane. Pairs of counter-rotating streamwise vortices are observed in the vicinity of the wire and precisely described. Several observations lead us to conjecture that these coherent structures are a finite amplitude solution, which play a central role in the subcritical transition to turbulence. First, the same structures are observed on the border of turbulent spots, as well as transciently during the relaxation of a turbulent spot, in the unmodified PCF. Second the transition leading to the streamwise vortices stabilization appears to be subcritical. From this point of view, λ bad selection is the sign of a nonlinear characteristic spanwise lengthscale and not of a linearly bifurcated solution wavelength. Finally the adequacy between our profiles measurement and the analytical calculation[8] indicates that the modified PCF around the wire can be considered as an infinite shear flow past a cylindrical cylinder. Consequently ρ should be the only lengthscale in any solutions related to the wire. However our observations reveal the independence of the counter-rotating vortices on ρ and further observations, to be published elsewhere[10], indicate that h is the dominant geometrical factor. These results lead us to conclude that the streamwise vortices observed in this study are intrinsic to the PCF, the wire playing the role of a forcing or a noise generator.

References

[1] M. Nagata, (1990). "Three-dimensional finite-amplitude solutions in plane Couette flow: bifurcation from infinity,"J. Fluid Mech. **217**, pp 519

[2] F. H. Busse, R. M. Clever, (1995). "Bifurcation sequences in problems of thermal convection and of plane Couette flow,"B. Gjévik, J. Grue, and J. E. Weber eds., Nluwer Acad. Publ.

[3] A. Cherhabili, U. Ehrenstein, (1995). "Spatially localized two-dimensional finite amplitude states in plane Couette flow"Eur. J. Mech. B/ Fluids, **14**, n, pp 677-696

[4] F. Daviaud, J. Hegseth and P. Bergé, (1992). "Subcritical transition to turbulence in plane Couette flow,"Phys.Rev.Lett. **69**, pp 2511

[5] N. Tillmark and P. H. Alfredsson, (1992). "Experiments on transition in plane Couette flow", J. Fluid Mech. **235**, pp 89

[6] O. Dauchot and F. Daviaud, (1994) . "Finite-amplitude perturbation and spots growth mechanism in plane Couette flow,"Phys. Fluids, **7** (2), pp. 335-343

[7] K. H. Bech, N. Tillmark, P. H. Alfredsson, H. I. Andersson, (1995). "An investigation of turbulent plane Couette flow at low Reynolds numbers"J. Fluid Mech. **286**, pp. 291-325

[8] A. T. Chwang, T. Y-T Wu, (1975). "Hydromechanics of low Reynolds-number flow. Part 2. Singularity method for Stokes flows,"J. Fluid Mech. **67**, pp 787

[9] O. Dauchot and F. Daviaud, (1994) . "Streamwise vortices in plane Couette flow,"Phys. Fluids **7** (5), pp. 901-903

[10] S. Bottin, O. Dauchot and F. Daviaud, (1996) . "Transtion to turbulence in plane Couette flow: The streamwise vortices role,"Submitted to J. Fluid Mech.

NONLINEAR INTERACTION OF LAMINAR THREE-DIMENSIONAL DISTURBANCES IN PLANE POISEUILLE FLOW.

BEN DIEDRICHS

Department of Fluid Mechanics, Luleå University of Technology,

S-971 87 Luleå, Sweden

PER A. ELOFSSON

Department of Mechanics, Royal Institute of Technology,

S-100 44 Stockholm, Sweden

An experimental study of localized disturbances in plane Poiseuille flow is presented and the main focus is on the nonlinearityies of localized (laminar) disturbances. Measurements of the streamwise disturbance velocity are carried out with a hot-wire technique at a Reynolds number of 1600. The set-up is arranged so that two localized disturbances are interacting to form a composite disturbance (*CD*). They are generated a distance of 50 half channel heights apart so that the second disturbance strikes the primary one when this has reached its peak energy. The second disturbance is introduced with a smaller amplitude. Subtracting the individual disturbances from the *CD* the nonlinear interaction (*NI*) is extracted, directly relating the amplitudes between the two of them. Perturbing the front of the primary disturbance keeps the motion lasting for longer times than perturbing somewhere else on the centreline. In addition, it is found that the *NI* of the centre of the structure in general appears in connection with large velocity *gradients* in the streamwise and spanwise directions of the disturbances. At the outer regions of the *CD* the *NI* followed the negative *amplitudes* of the velocity, shown in Fig. 1. The relatively large amplitudes of the *NI* suggests that the disturbances are of a nonlinear type rather than linear, but with laminar characteristics (i.e. deterministic behaviour). Moreover, the findings reveal that there is a threshold value (for the jet-velocity emanating from the wall) for the

S. Gavrilakis et al. (eds.), Advances in Turbulence VI, 333-334.
© 1996 *Kluwer Academic Publishers.*

developing disturbances. Exceeding the pivotal jet-velocity the disturbances evolve entirely from the strongly nonlinear front. Else, they die out relatively quickly due to the lack of the strongly nonlinear initial stage (triggering). From a theoretical viewpoint it is noted that the least damped oblique modes are among the ones travelling with the greatest phase speeds, but with a damping rate larger than the spanwise modes (fully oblique, i.e. with the spanwise wavenumber $\alpha=0$), c.f. the P-branch of the eigenvalue spectra of the Orr-Sommerfeld equation. Nonlinearly, spanwise disturbances couple only with oblique waves, suggesting that the incipient nonlinear stage, or additional triggering by the second disturbance, is responsible for the non-proportional sustained disturbance motion and growth of the spanwise structures.

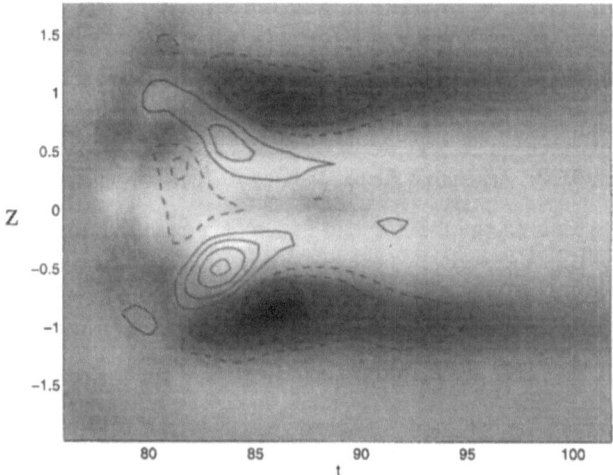

Fig. 1. *Nonlinear interaction (NI)* plotted with iso-contours on top of the *composite disturbance* represented as a shaded area ranging from white to black (i.e. maximum positive disturbance velocity to maximum negative disturbance velocity). Solid and dashed lines denote positive and negative *NI*, respectively. The disturbance is measured 85 channel half heights downstream the first loudspeaker. Contours of the nonlinear interaction are 2% of the centreline velocity indicating significant nonlinear amplitudes. z, indicates the spanwise coordinate.

LARGE-SCALE KOLMOGOROV FLOW ON THE BETA-PLANE, RESONANT WAVE INTERACTIONS AND SCALE SELECTION

U. FRISCH

CNRS, Observatoire de Nice, BP 229, 06304 Nice Cedex 4, France.

B. LEGRAS

CNRS/LMD/ENS, 24 rue Lhomond, 75231 Paris Cedex 5, France.

AND

B. VILLONE

Istituto di Cosmogeofisica, CNR, C. Fiume 4, 10133 Torino, Italy.

The large-scale dynamics of the Kolmogorov flow near its threshold of instability is studied in the presence of the β-effect (Rossby waves). The governing equation, obtained by a multiscale technique, fails the Painlevé test of integrability when $\beta \neq 0$. This "β-Cahn–Hilliard" equation with cubic nonlinearity is simulated numerically in various régimes. The dispersive action of the waves modifies the inverse cascade associated with the Kolmogorov flow [1]. For small values of β the inverse cascade is interrupted at a wavenumber which increases with β. For large values of β only resonant wave interactions (RWI) survive. An original approach to RWI is developed, based on a reduction to normal form, of the sort used in celestial mechanics [2]. Otherwise, wavenumber discreteness effects, which are dramatic in the present case, are not captured. The method is extendable to arbitrary RWI problems of the kind encountered in plasma physics, spin waves, oceanography, etc. (see, e.g., Ref. [3]). The only four-wave resonances present involve two pairs of opposite wavenumbers. This allows leading-order decoupling of moduli and phases of the various Fourier modes, so that an exact kinetic equation is obtained for the energies of the modes. It has a Lyapunov functional (gradient) formulation and multiple attracting steady-states, each with a single mode excited. The final state depends thus on the initial con-

S. Gavrilakis et al. (eds.), Advances in Turbulence VI, 335-336.
© *1996 Kluwer Academic Publishers.*

dition chosen, as illustrated in Fig. 1. A detailed presentation of this work may be found in Ref. [4].

Recent calculations, suggested by V. Yakhot, have been performed with a large number n of linearly unstable modes and a small value of β, such that the β-effect is important only up to some wavenumber $1 < k_\beta \ll n$, thereby selecting a preferred scale $\sim k_\beta^{-1}$. After relaxation of transients the solution becomes a slowly traveling wave; the only wavenumbers excited are then (odd) multiples of k_β. The energies of the modes are very close to those obtained for *unstable* steady-state solutions of the Cahn–Hilliard equation with k_β pairs of kinks/antikinks [5], which are thus stabilized by the combined effect of the Rossby waves and the slow drift.

References

1. She, Z.S. (1987) Metastability and vortex pairing in the Kolmogorov flow, *Phys. Lett.* A124, pp. 161–164.
2. Arnold, V.I., Kozlov, V.V. & Neishtadt, A. (1988) Mathematical aspects of classical and celestial mechanics, in *Dynamical Systems III*, ed. V.I. Arnold, pp. 1–291, Encyclopaedia of Mathematical Sciences, vol. 3, Springer Verlag.
3. Zakharov, V.E., L'vov, V.S. & Falkovich, G. (1992) *Kolmogorov Spectra of Turbulence I*, Springer Verlag.
4. Frisch, U., Legras, B. and Villone, B. (1996) Large-scale Kolmogorov flow on the beta-plane and resonant wave interactions, *Physica D*, in press.
5. Kawasaki, K. & Ohta, T. (1982) Kink dynamics in one-dimensional nonlinear systems, *Physica* A116, 573–593.

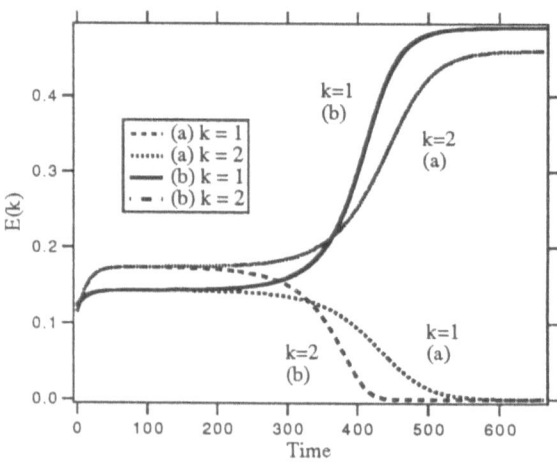

Figure 1. Simulation of the β-Cahn–Hilliard equation with strong Rossby waves when the dynamics are dominated by resonant wave interactions. At long times the Fourier amplitudes go to a steady state with a single Fourier mode excited, as predicted by the asymptotic theory. Several single–Fourier mode attractors are competing, as indicated by (a) and (b) which correspond to two slightly different initial conditions.

EXPERIMENTS ON A 2:1 RESONANCE IN THE BLASIUS BOUNDARY LAYER

J.J. HEALEY
Brunel University
Department of Mathematics and Statistics
Uxbridge, Middlesex UB8 3PH, UK
email: jonathan.healey@brunel.ac.uk

1. Introduction

This paper is concerned with the evolution of small unstable wavy disturbances in a Blasius boundary layer. When the amplitude is sufficiently small the Orr–Sommerfeld equation for linearized disturbances (when the base flow is assumed parallel) gives good predictions for the disturbance's evolution. The lower the frequency, the greater the eventual growth of waves as they propagate downstream and cross the unstable parameter regime. Linear theory predicts exponential growth but experiment shows that when the amplitudes are large enough a much stronger nonlinear growth takes place and leads ultimately to turbulence. A particular scenario for the onset of nonlinearity is examined here in a set of windtunnel experiments.

If a pair of waves satisfy a 2:1 resonance, i.e. the frequency and streamwise wavenumber of one wave are twice those of the other wave, the non linear terms enter at quadratic order rather than the cubic order of nonresonant waves. Therefore, resonant waves become nonlinear at lower amplitudes than nonresonant waves and so may play a particularly important role in the transition process. The resonance leads to a dependence on the relative phase of the two waves and this will be tested for in the experiment.

Craik (1971) has shown that a 2:1 resonance can occur provided that three-dimensional waves are present that can interact with two-dimensional waves. The resulting 'Craik resonant triad' is believed to play an important role in the rapid development of three-dimensionality observed in nominally two-dimensional experiments. Healey (1994) found a new resonance for purely two-dimensional waves by allowing both frequency and wavenumber to be complex (as a model for spatially evolving waves whose input amplitude varies with time). The resonance can be followed onto the real

337

S. Gavrilakis et al. (eds.), Advances in Turbulence VI, 337-340.

frequency axis, giving a classical spatial analysis, and is then found to relate to two different branches of the dispersion relation, see Healey (1995a). The results presented here are a summary of those given in that latter paper.

In the experiments, resonant phase-dependence and nonlinear break-down were found to be strongest for disturbances close to the resonant frequencies predicted by the linear theory.

2. The resonant frequency

The Craik resonant triad selects a particular pair of oblique waves for any given planar wave. The 'resonant dyad' discussed here selects a particular frequency for a planar wave that can resonate with its harmonic at any given Reynolds number. The lower frequency wave (fundamental) is found to lie near the most unstable Orr–Sommerfeld (OS) frequency and the higher frequency wave (its harmonic) lies on a damped disconnected branch ('disconnected' refers to the behaviour at real frequencies, all the branches are connected by branch points in the complex frequency plane).

This resonant frequency can also be located on the triple-deck dispersion relationship, and so, in an asymptotic sense, it lies close to the lower branch rather than the upper branch of the neutral curve. To leading order the resonant waves are

$$\omega = 2.85 R^{-1/2}, \quad \alpha = (1.14 - 0.149\mathrm{i}) R^{-1/4} \tag{1}$$

and
$$\omega = 5.70 R^{-1/2}, \quad \alpha = (2.28 + 0.865\mathrm{i}) R^{-1/4}$$

where the frequency ω, wavenumber α and Reynolds number R, are based on displacement thickness and freestream velocity.

If it is assumed that the imaginary parts of the wavenumbers are small compared with the real parts then the amplitude equations for the resonant modes can be derived using the method of multiple scales. Strictly speaking, this requires the resonance to occur for neutral waves and then a perturbation analysis gives the nonlinear behaviour nearby, however, as argued in Healey (1995b, 1996) this is unlikely to be the case during transition. Therefore, here we will treat the 'numerically small' imaginary parts as if they were asymptotically small, and derive the following amplitude equations:

$$\dot{A}_1 = A_1 + a A_1^* A_2 \tag{2}$$
$$\dot{A}_2 = -(\gamma^{-1} - \Delta\mathrm{i}) A_2 + b A_1^2 \tag{3}$$

where A_1 and A_2 are the complex amplitudes of the two waves, the dots represent differentiation w.r.t. a slow spatial variable, the star denotes complex conjugate, γ is the ratio of the magnitude of the growth to decaying rates of the two waves, Δ is a detuning parameter for the resonance and a and b are the nonlinear coefficients obtained from the multiple-scales calculation.

The analysis can also be carried out with complex ω. Figure 1 shows how the nonlinear coefficients vary when ω becomes complex. When ω_i is sufficiently negative, i.e. for sufficiently strong decaying temporal modulation of the wave, there is a very large increase in the strength of the coefficients. The rapid change in the coefficients can be related to the resonance passing close to the branch point that joins the two branches.

An important feature of these equations can be seen by letting $A_1 = |ab|^{-1/2}B_1$, $A_2 = B_2/a$ and $\Phi = \arg(-ab)$ then the rate of change of the 'energy' is given by

$$\frac{\partial}{\partial X}\left(|B_1|^2 + |B_2|^2\right)/2 = |B_1|^2 - \gamma^{-1}|B_2|^2 + \Re\left\{B_1^2 B_2^*(1 - e^{i\Phi})\right\} \quad (4)$$

where $\Re\{\cdot\}$ denotes the real part. The case $\Phi = 0$ means that the nonlinear terms do not contribute to the disturbance growth (conservative coupling). However, if $B_1 \mapsto -B_1$ and $B_2 \mapsto -B_2$ then the nonlinear contribution changes sign, i.e. the nonlinearity could change from stabilizing to destabilizing. This does not happen for nonresonant waves.

3. Experiments

A series of experiments have been carried out in the low-turbulence wind tunnel in the Engineering Department at Cambridge University. The tunnel ran at close to 18ms^{-1} and hot-wire measurements were taken around 1m downstream from the leading of the flat plate, which was mounted at zero incidence, and at one displacement thickness from the surface of the plate. Reproducible disturbances were introduced via small a loudspeaker which acted essentially like a point source.

Figure 2 shows a set of four input disturbances (left hand column) with increasingly strong modulation and their corresponding responses measured by the hot-wire (right hand column). At this amplitude the unmodulated section of the input shows no tendency to break down into spikes, neither does the temporally growing parts of the disturbance no matter how strong the modulation. However, when the temporally damped part is sufficiently strongly modulated there is a dramatic localised break down. This might be caused by the rapid strengthing of the nonlinear coefficients found in figure 1 when ω_i is sufficiently negative.

The effect of changing the sign described in §2 was also tested for. Figure 3 shows the dramatic change in the nonlinear behaviour when the input disturbance was multiplied by -1. There is an almost complete suppresion of the breakdown for this different initial phase relationship even though the amplitudes and frequencies of the disturbances are identical.

340

However, this strong phase dependence was only found for a narrow range of frequencies. Figure 4 shows how these frequencies found in the experiment compare with the resonant frequencies predicted by the theory. Although the spanwise measurements presented in Healey (1996a) show that the spike is a strongly three-dimensional event in the boundary layer the agreement between the experimental resonant frequencies and the theory suggests that there may, nonetheless, be a two-dimensional resonant phenomenon that initiates the breakdown.

References

Craik, A.D.D. (1971) *J. Fluid Mech.* **50**, 393-413.
Healey, J.J. (1994) *Phys. Rev. Letts.* **73**, 1107-1109.
Healey, J.J. (1995a) *J. Fluid Mech.* **304**, 231–262.
Healey, J.J. (1995b) *J. Fluid Mech.* **288**, 59-73.
Healey, J.J. (1996) Characterizing boundary layer instabilities at finite Reynolds numbers. *J. Fluid Mech.* (submitted).

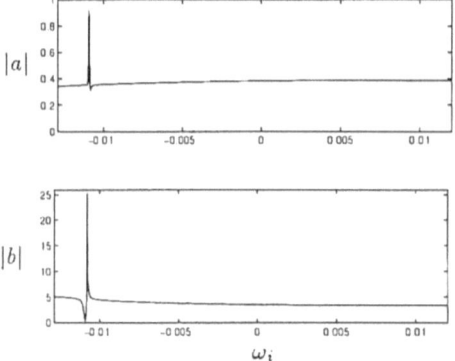

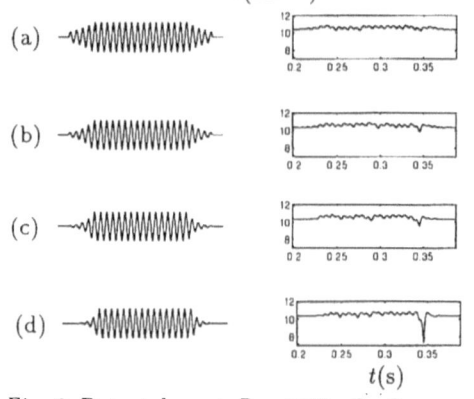

Fig. 1. Variation of the nonlinear coefficients with imaginary part of frequency ω_i.

Fig. 2. Data taken at $R = 1980$, the frequency is $\omega_r = 0.0805$ and the modulations are $|\omega_i| = 0.0037, 0.0065, 0.0102$ and 0.0153 for (a-d) respectively.

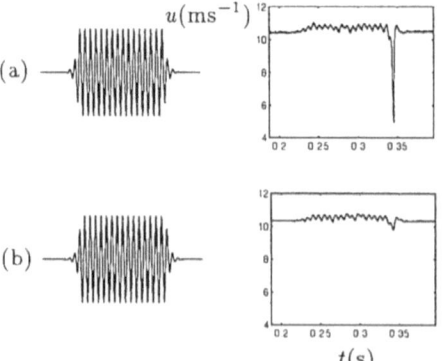

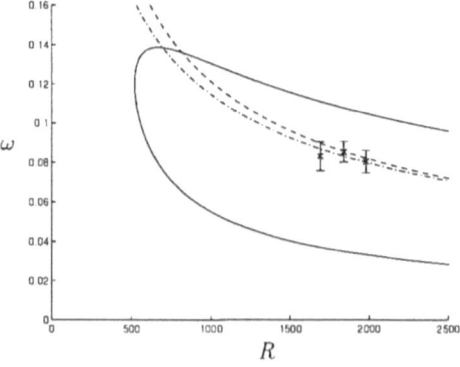

Fig. 3. (a) Same input disturbance as that in figure 2 but with $|\omega_i| = 0.021$. (b) same as (a) but with input multiplied by -1.

Fig. 4. — OS neutral curve, - - - OS resonant frequency (RF), -.-.- triple-deck RF and crosses: experimentally determined RF.

SOME STABILITY-PROPERTIES OF COUETTE-EKMAN FLOW

N. P. HOFFMANN, F. H. BUSSE
Physikalisches Institut, Universität Bayreuth
D-95440 Bayreuth

1. Introduction

The study of bifurcation-sequences leading from basic laminar fluid flow to more complex states of flow and finally to turbulence has given well acknowledged insights into transition phenomena of simple flow configurations such as Rayleigh-Bénard convection or Taylor-Couette flow.

One of the still less well understood phenomena is the stability and the transition in non-planar shear flows which are encountered in engineering applications such as rotating disk flows or in geophysical Ekman- or planetary boundary-layers. The aim of the present work is to investigate linear and nonlinear stability properties of a non-planar shear flow in a simple configuration.

2. Model and Mathematical Formulation

Couette-Ekman flow appears when an infinitely extended horizontal fluid layer is subjected to a constant shear and a constant rotation rate around a vertical axis. The model seems especially attractive, as it contains asymptotically plane-parallel Couette flow (in the limit of vanishing rotation rate) as well as Ekman-boundary-layer flow (in the limit of large rotation rate). Moreover results for rotation rates between these asymptotic limits will be of interest for the understanding of two-disk flow.

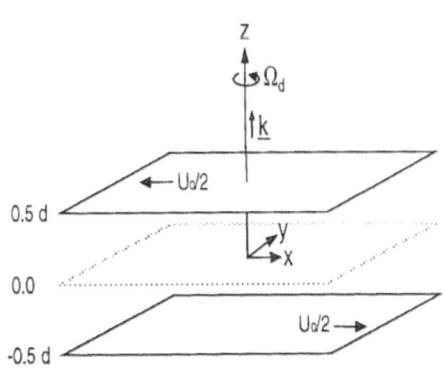

Figure 1. Geometrical configuration.

Using d as lengthscale, d^2/ν as timescale where ν is the kinematic viscosity, and ν/d as scale for velocity we write the incompressible Navier-Stokes equations for the velocity vector $\vec{u}$ and the boundary conditions:

S. Gavrilakis et al. (eds.), Advances in Turbulence VI, 341-344.

$$\nabla \cdot \vec{u} = 0, \quad \frac{\partial \vec{u}}{\partial t} + \vec{u}\nabla\vec{u} + 2\Omega \vec{k} \times \vec{u} = -\nabla\pi + \nabla^2\vec{u},$$

$$u_x = \mp Re/2 \quad \text{and} \quad u_y = u_z = 0 \quad \text{for} \quad z = \pm 1/2,$$

where $\vec{k}$ is the unit vector in the vertical direction and the dimensionless rotation rate and Reynolds number are defined by

$$\Omega = \frac{\Omega_d d^2}{\nu}, \quad Re = \frac{U_0 d}{\nu}.$$

The basic flow profile is time-independent and horizontally homogeneous,

$$U_x^0(z) = -A\sin(\sqrt{\Omega}\,z)\cosh(\sqrt{\Omega}\,z) + B\cos(\sqrt{\Omega}\,z)\sinh(\sqrt{\Omega}\,z),$$
$$U_y^0(z) = +A\cos(\sqrt{\Omega}\,z)\sinh(\sqrt{\Omega}\,z) + B\sin(\sqrt{\Omega}\,z)\cosh(\sqrt{\Omega}\,z).$$

A and B are determined by the boundary conditions.

In the following we make use of a rotated cartesian coordinate system:

$$\xi = x\cos\epsilon - y\sin\epsilon, \quad \eta = x\sin\epsilon + y\cos\epsilon, \quad z = z.$$

For the solenoidal vector field $\vec{u}$ it is possible to introduce the representation

$$\vec{u} = \vec{U}^0 + \vec{\hat{U}} + \nabla \times (\nabla \times \vec{k}\phi) + \nabla \times \vec{k}\psi = \vec{U} + \vec{\delta}\phi + \vec{\epsilon}\psi \quad \text{with} \quad \vec{\bar{u}} = \vec{U},$$

where the bar indicates the average over the (x, y)-plane. By averaging the Navier-Stokes equations over the (x, y)-plane and by operating with $\vec{\delta}$ and $\vec{\epsilon}$ onto them we obtain equations for $\hat{U}_\xi$, $\hat{U}_\eta$, ϕ and ψ:

$$(\partial_z^2 - \partial_t)\hat{U}_\xi + 2\Omega\hat{U}_\eta = -\partial_z \overline{\Delta_2\phi(\partial_{\xi z}^2\phi + \partial_\eta\psi)},$$

$$(\partial_z^2 - \partial_t)\hat{U}_\eta - 2\Omega\hat{U}_\xi = -\partial_z \overline{\Delta_2\phi(\partial_{\eta z}^2\phi + \partial_\xi\psi)},$$

$$(\nabla^2 - \vec{U}\cdot\nabla)\Delta_2\psi + 2\Omega\partial_z\Delta_2\phi + \partial_z\vec{U}\cdot\underline{\epsilon}(\Delta_2\phi) =$$
$$\partial_t\Delta_2\psi + \vec{\epsilon}\cdot[(\vec{\delta}\phi + \vec{\epsilon}\psi)\cdot\nabla(\vec{\delta}\phi + \vec{\epsilon}\psi)],$$

$$(\nabla^2 - \vec{U}\cdot\nabla)\nabla^2\Delta_2\phi - 2\Omega\partial_z\Delta_2\psi + (\partial_{zz}^2\vec{U})\cdot\nabla\Delta_2\phi =$$
$$\partial_t\nabla^2\Delta_2\phi + \vec{\delta}\cdot[(\vec{\delta}\phi + \vec{\epsilon}\psi)\cdot\nabla(\vec{\delta}\phi + \vec{\epsilon}\psi)],$$

where the symbol Δ_2 (the horizontal Laplacian) stands for $\nabla^2 - \partial_{zz}^2$.

3. Linear Stability and Energy Stability

The linear stability of the basic flow has been investigated by a normal mode ansatz and use of a shooting-method (see fig. 2 for results). It shows that at low rotation rates stationary, two-dimensional roll-like instabilities appear, whose origin is the inflection point in the middle of the fluid layer.

At larger rotation rates two-dimensional travelling-wave modes appear, that – in the asymptotic limit – may easily be identified with the well known Ekman-layer instabilities as described by Lilly [3]. It is interesting to note that at intermediate rotation rates the critical type I solutions (for notation see [2]) lie well below the critical type II solutions, while in the asymptotic limit of high rotation rate the relationship is reversed.

Furthermore an energy analysis was done to find the Energy Reynolds number as a function of the rotation rate. The results show a smooth transition between the limiting cases of plane Couette flow and Ekman-boundary-layer flow.

4. Roll-like Solutions of Finite Amplitude

The nonlinear equations are numerically solved with the Galerkin method and a Newton-Raphson iteration scheme (cf. [1]).

There are not only supercritically bifurcating solutions owing to the onset of instability, but there also exist states that do not seem to be connected with the basic state through a bifurcation. Moreover for some rotation rates these isolated states exist below the critical Reynolds number for the onset of infinitesimal disturbances and are "subcritical" in this sense.

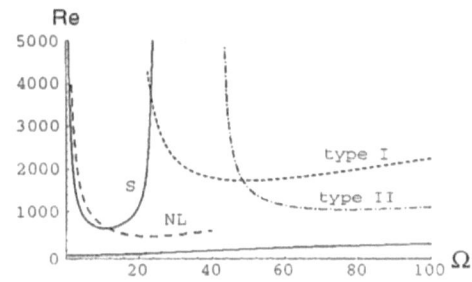

Figure 2. Critical Reynolds numbers for stationary rolls (S) and type I/II disturbances. The critical Reynolds-numbers for isolated states are also shown (NL). The lowest lying line depicts the energy Reynolds numbers.

The appearance of these subcritical isolated states is an interesting feature of our model, since finite amplitude solutions have also been found for plane, non-rotating Couette flow (see [1], [5]), where a linear stability analysis indicates stability for all Reynolds numbers, as is well known.

5. Secondary Instability

Finally a stability analysis of the secondary states has been done for selected parameters.

At sufficiently high Reynolds numbers the stationary modes that exist at low rotation rates show instability against disturbances with long wavelength (fig. 3). The region of stable rolls is bounded by a mode with vanishing Floquet parameter for small wavenumbers and by a subharmonic mode for large Reynolds numbers. All instabilities are stationary, no time dependence is introduced.

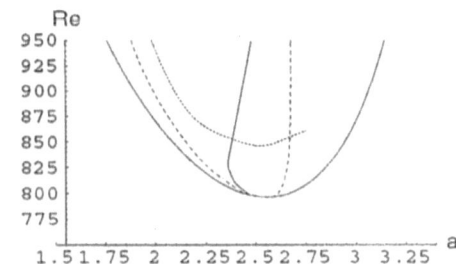

Figure 3. Results of the Floquet stability analysis for stationary rolls ($\Omega = 17$, $\epsilon = 0.122$). The domain of stable rolls is bounded by an Eckhaus instability (dashed), a zig-zag type of instability (solid) and a subharmonic instability (dotted).

Secondary states of type II show several instabilities (fig. 4) with moderate growthrates and a short-wavelength, high frequency instability (fig. 5) with large growthrate. This short-wavelength instability shows remarkable similarity with

344

observed instabilities in rotating disk flows and other three-dimensional boundary layers (cf. [2], [4]), hinting at the universal character of this kind of instability.

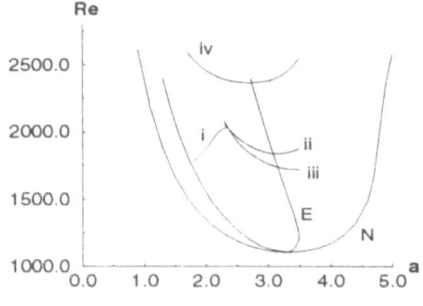

Figure 4. Results of Floquet stability analysis for type II Ekman boundary-layer flow ($\Omega = 100$, $\epsilon = 0.396$). Neutral curve (N), Eckhaus instability (E), low growthrate instabilities (i, ii, iii) and short-wavelength instability (iv) are shown.

Figure 5. Contours of constant vertical velocity at $z = -0.4$ for eigenmode of short-wavelength instability (eigenmode added to secondary state with arbitrary amplitude). The region of positive vertical velocity (solid lines) shows marked short-wavelength disturbances.

6. Conclusion

In spite of the simplicity of the model there is a rich variety of dynamical structures in the system under consideration. Linear stability calculations show stationary modes for small and traveling wave modes for large rotation rates. For higher Reynolds numbers the bifurcating stationary roll solutions are unstable against disturbances with long wavelength. The traveling wave modes have an instability with short wavelength.

Nonlinear calculations also show states that do not bifurcate from the basic state. For intermediate rotation rates these states exist for Reynolds numbers below the critical Reynolds number. The relevance of these isolated states for the transition process has to be examined in further work.

7. References

[1] Clever, R. M. and Busse, F. H., Three-dimensional convection in a horizontal fluid layer subjected to a constant shear. J. Fluid Mech. **234**, 511 (1992).

[2] Faller, A. J., Instability and transition of disturbed flow over a rotating disk. J. Fluid Mech. **230**, 245 (1991).

[3] Lilly, D. K., On the Instability of Ekman Boundary Flow. J. Atmos. Sci. **23**, 481 (1966).

[4] Kohama, Y., Study on Boundary Layer Transition of a Rotating Disk. Acta Mech. **50**, 193 (1984).

[5] Nagata, M., On wavy instabilities of the Taylor-vortex flow between corotating cylinders. J. Fluid Mech. **188**, 585 (1988).

[6] Zandbergen, P. J. and Dijkstra, D., Von Karman swirling flows. Ann. Rev. Fluid Mech. **19**, 465 (1987).

ON THE BOUNDARY LAYER TRANSITION IN TURBULENT FLOWS WITH VARIOUS LENGTH SCALES

P. JONÁŠ, O. MAZUR, V. URUBA
Institute of Thermomechanics AS CR,
Dolejškova 5, Prague 8, Czech Republic

Boundary layer transition is influenced by many factors, that cannot be described entirely and precisely in all particular cases. This fact is also the very reason for the scatter of various experimental results and computations published especially in connection with bypass-transition. Investigations of the bypass-transition consider mainly the effect of the intensity of the free-stream disturbances, almost neglecting the length scale of the disturbances. This is a little surprising in a view of the outstanding effect of an impermeable wall on the turbulent structure of the external flow (e.g. Thomas & Hancock [1977] and Hunt & Graham [1978]). Investigation of grid-turbulence passing a wall moving at the mean free-stream velocity documents the occurrence of two layers at the surface that results from the wall constraint of the flow and from the viscous sticking of the fluid to the surface. The first layer exhibits an outer kinematics region with the thickness about $\delta_S \approx 2\Lambda_e$. The second layer governed by viscosity has the thickness $\delta_W \approx \sqrt{6/\mathrm{Re}_T} \cdot \Lambda_e$. Here Λ_e and Re_T are the turbulence length scale and the turbulence Reynolds number of the outer-flow. In both these layers significant changes of the turbulence structure occur . Usually during investigations of the by-pass transition, the outer-stream length scale exceeds the boundary layer thickness. Along with the effect of the wall, the outer-stream turbulence currently decays through viscous dissipation, so that its characteristics are changing in the flow direction; e.g. in grid turbulence the intensity of longitudinal fluctuations Iu decreases according to the formula $Iu^{-2} \approx Const.(x - x_0)^m$, $m > 0$. From this the dissipation length scale $L_e = -Iu^{-1.5}/\left[d(Iu^2)/dx\right]$ can be easily derived. Hancock & Bradshaw [1983] proposed this length scale as a convenient characteristics of turbulence affecting a boundary layer. We can thus conclude, that the turbulence decisive for the transition process is not identical to the outer-stream turbulence, or turbulence as measured in the incoming stream at the leading edge. From this follows: the length scale must influence the transition.

The main aim of the contribution is to clarify the effect of the turbulence length scale on the onset and on the course of laminar-turbulent transition of a flat-plate boundary layer,

S. Gavrilakis et al. (eds.), Advances in Turbulence VI, 345-346.

and to compare the relative significance of the intensity and the length scale at the leading edge and in some other important sections. In forecasting transition we would like to find a turbulence characteristics more relevant than the intensity of turbulence in the incoming flow. At the same time the results of the experiments should serve as the Test Case T3A+ of the COST/ERCOFTAC SIG on Transition.

The investigated boundary layer develops on a wooden flat plate in the close circuit wind tunnel $(0.5 \times 0.9)m^2$. The mean velocity of the outer stream is 5m/s eventually 10m/s. The turbulence level of the flow is controlled by means of turbulence generating plane grids or screens of various geometries. Every grid is placed across the flow in the proper distance upstream from the leading edge of the plate. Thus, there has been achieved homogeneous turbulence with the turbulence level 3% at the leading edge plane downstream from every grid. The CTA measurement system is used connected to a personal calculator equipped with the Data Acquisition System controlled by the software LabVIEW. The data are digitally evaluated.

So far we have investigated transitional boundary layers in turbulence produced by plane square mesh (M) grids of cylindrical rods (d): M/d= 35mm/10mm (GT5) and 20mm/3mm (GT1). The mean velocity being U_e =5m/s. In the leading edge plane the intensity Iu is 3 percent in the both cases and the scale L_e is either 4mm or 23mm. Due to the small boundary layer thickness, it has been necessary to apply the CTA single-wire technique inside the layer only. The profiles of the longitudinal velocity characteristics have been measured: mean value, intensity, skewness and flatness factor and the skin-friction distributions in the region 0,05m up to 1,2m downstream from the leading edge. In this region the transition process has not been fully completed as it is obvious from Figure 1. At the present, the reconstruction of the working section that enable us to measure to the distance 1,6m from the leading edge has been finished. The courses of the fundamental boundary layer characteristics e.g. shape parameters and skin friction coefficient confirm the effect of the length scale on the transition in the investigated configuration. This effect is evident also from the results received by

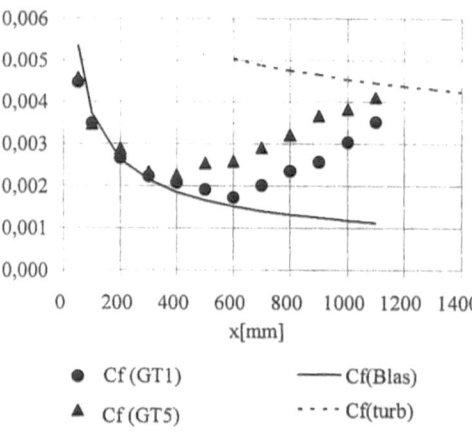

Figure 1.

means of the pattern recognition technique (TPAV see: Wallace & Brodkey & Eckelmann [1977]); e.g. nondimensional mean period of bursts. It seems evident that the larger scale turbulence causes a sooner onset of transition process.

ON TRANSITION TO INSTABILITY IN WEAKLY

INHOMOGENEOUS FLOWS WITHOUT DISSIPATION

A.G.KULIKOVSKII
Steklov Mathematical Institute,
42 Vavilova Street, Moscow, 117966, Russia

AND

I.S.SHIKINA
Department of Mathematics and Mechanics,
Moscow State University, Moscow, 119899, Russia

Abstract. The criterion is found to determine in which cases the violence of local stability conditions leads to instability of weakly inhomogeneous flow without dissipation.

1. Formulation of the problem

The loss of stedy-state flows stability is considered in the linear approximation. The basic flow depends on the slow space variable $X = \varepsilon x$, where $\varepsilon = \lambda/L \ll 1$, λ is the typical wavelength of disturbances, L is the scale of inhomogeneity, and x denotes the stream direction. Infinitesimal disturbances of such a flow are looked for in the form $f(x)exp(-i\omega t)$, where ω and $f(x)$ are the eigenfrequency and eigenfunction, respectively. If there exists an eigenfrequency ω, such that $Im\ \omega > 0$, then inhomogeneous flow is unstable.

The eigenfunction $f(x)$ can be represented as a linear combination of elementary WKB-solutions or WKB-waves

$$f(x) = \sum_n C_n A_n(X) exp\left(\frac{i}{\varepsilon}\int_{X_n}^X k_n(\omega,\xi)d\xi\right),$$

where $k_n(\omega, X)$ are the branches of the function $k(\omega, X)$, representing the local wavenumber satisfying the local dispersion relation $D(\omega, k, X) = 0$,

347

and the amplitudes $A_n(X)$ can be determined in the form of asymptotic series. In the complex X - plane the coefficients C_n can change on some lines called the Stokes lines. The latter outgo from turning points $X = X_t$, at which two or more wavenumbers coincide for a given ω.

Let us denote a frequency set by Ω, the set being such one that for each $\omega \in \Omega$ there exists a cyclic sequence of elementary WKB-waves which are subsequently transformed from one into another at turning points, or the Stokes lines, and the total product of their space amplification coefficients and their conversion coefficients is equal to unity.

In general case the following statement is valid [1]:

There exists an eigenfunction with the frequency $\omega_e = \omega_* + O(\varepsilon)$, where $\omega_* \in \Omega$ and $Im\,\omega_* = \max_\Omega(Im\omega)$. Among all eigenfrequences of the problem ω_e has the greatest imaginary part. The sequence is a main part ("skeleton") of an eigenfuncion and can involve only a part of those elementary waves which form the eigenfunction. For example, in typical cases of unbounded flows some waves forming the eigenfunction are not involved in the sequence, but propagate to infinite without interacting with other waves.

The local dispersion relation determines $\omega(k, X)$ as a multivalued analytical function of k, X. It is assumed that for real k and X all branches of ω determined by dispersion relation are real or complex conjugate. For example, $D(\omega, k, X)$ can be a polynomial in ω and k with real coefficients. This is typical for flows without dissipation. In this case the flow is locally stable, if for any given X all branches $\omega(k)$ are real for real k. There are no other restriction on the dispersion relation in our analysis.

Let the flow depend on a parameter R such that for $R \leq R_*$ the flow is locally stable everywhere and for $R > R_*$ there exists an interval (X_1, X_2) of the local instability. For any $X \in (X_1, X_2)$ there exists an interval (k_1, k_2) of real k, such that $Im\,\omega(k, X) \neq 0$ for $k \in (k_1, k_2)$. If $\Delta R \equiv R - R_* > 0$ is small, then both intervals (X_1, X_2) and (k_1, k_2) are also small.

Thus, for $R = R_*$ and for a value $X = X_*$ two branches of real curve $\omega(k)$ intersect at $k = k_*$, $\omega = \omega_*$. For small ΔR and small $\tilde{X} = X - X_*$, $\tilde{k} = k - k_*$ the dispersion relation can be written by expanding in power series and retaining the principal terms as follows:

$$(\tilde{\omega} - U\tilde{k})^2 = A\tilde{k}^2 + \beta\tilde{X}^2 - \alpha\Delta R$$

where $U, A > 0, \beta > 0, \alpha > 0$ are constants. For $\Delta R > 0$ and $\tilde{X}^2 < (\alpha\Delta R)/\beta$ we have the local instability. It is known the criterion to distinguish the absolute and convective instabilities [2]. The local instability is absolute for $U^2 < A$ and convective for $U^2 > A$.

2. The case of local absolute instability

In the case of local absolute instability the wave numbers k_1 and k_2 correspond to two waves propagating in the opposite directions. For $\Delta R > 0$ and sufficiently small $\tilde{X}$ there are two turning points at which two branches of the function $k(\omega, X)$ coincide. The two turning points coincide if the relations $\partial\omega/\partial k = 0$ and $\partial\omega/\partial X = 0$ hold simultaneously. The corresponding value $\tilde{\omega} = \tilde{\omega}_s$ has $Re\,\tilde{\omega}_s = 0$, $Im\,\tilde{\omega}_s > 0$. One can readily check, that for imaginary $\tilde{\omega}$ such that $Im\,\tilde{\omega} < Im\,\tilde{\omega}_s$ the turning points $\tilde{X} = \pm\tilde{X}_t$ lie on the real X - axis and k_1 and k_2 between them are real. For this case the eigenfrequencies should satisfy the equation [3] :

$$\frac{1}{\varepsilon} \int_{-X_t}^{X_t} [k_1(\omega, X) - k_2(\omega, X)]\, dX = \pi(n + \frac{1}{2})$$

where n is an integer. The roots of this equation are $\omega_n = \tilde{\omega}_s + O(n\varepsilon)$, $Re\,\tilde{\omega}_n = 0$ as $\varepsilon \to 0$. Hence, there exist $\tilde{\omega}_n$ for which $Im\,\tilde{\omega}_n > 0$ and the inhomogeneous flow is unstable. In this case the main part ("skeleton") of the eigenfunction is a cyclic sequence of two WKB-waves corresponding to k_1 and k_2, which are enclosed between two points of inner reflection. For $\tilde{X}$ lying beyond the interval $(-\tilde{X}_t, \tilde{X}_t)$ the eigenfunction tends to zero as $\varepsilon \to 0$. In this case the transition to instability is similar to that described in [4, 5].

3. The case of local convective instability

In this case the wave numbers k_1 and k_2 correspond to waves propagating in the same direction (for example, to the right). If $\tilde{\omega}$ is real and sufficiently small then there are two real turning points $\tilde{X} = \pm\tilde{X}_t$ in the neighbourhood of $\tilde{X} = 0$. Outside the interval $(-\tilde{X}_t, \tilde{X}_t)$ both $\tilde{k}$ are real, while inside it they are complex.

Assume that one (or both) of waves corresponding to k_1 or k_2 arrive at the turning point $\tilde{X} = -\tilde{X}_t$ from the left. At the turning point $\tilde{X} = -\tilde{X}_t$ this wave is transformed (in a standard way) in two waves corresponding to complex k_1 and k_2 propagating to the right as well. One of these waves is amplified while the other is damped. If ε is sufficiently small we can retain only the first of the waves. At the turning point $\tilde{X} = \tilde{X}_t$ this wave is transformed into two waves of equal amplitudes, corresponding to real k_1 and k_2, and propagating to the right. Thus, the interval $(-\tilde{X}_t, \tilde{X}_t)$ is an amplifier of the waves corresponding to k_1 and k_2.

However, for the instability to exist the presence of amplifier is not sufficient, a feedback should exist. The effective feedback exists, if in addition the turning points $\pm\tilde{X}_t$ there are for real $\tilde{\omega}$, at least, two other real turning points outside the local instability interval.

Let, for example, at the turning point $\tilde{X} = \tilde{X}_{2t} > \tilde{X}_t$ the wave corresponding to k_1 is transformed into the wave with real k_3 propagating to the left, while at the turning point $\tilde{X} = \tilde{X}_{1t} < -\tilde{X}_t$ the wave with k_3 is transformed into the wave corresponding to k_1. This means that for $R < R_*$ and for real frequency ω the sequence of the waves, corresponding to real k_1 and k_3 exists. Then for small $\Delta R > 0$ one can choose a small complex correction to this frequency, $\Delta\omega$, $Im\,\Delta\omega > 0$, such that there exists the sequence of waves as well.

Indeed, for complex ω with $Im\,\omega > 0$ waves, corresponding to real k for real ω, are damped in space in the direction of their propagation and $Im\,\Delta\omega$ can be chosen such that for $R > R_*$ the total wave damping outside of the interval $(-\tilde{X}_t, \tilde{X}_t)$ is compensated by the amplification of the wave on the interval $(-\tilde{X}_t, \tilde{X}_t)$. The condition of the phase coincidence can be satisfied by a choice of $Re\,\Delta\omega$. The correction $Im\,\Delta\omega$ depends on the length of the interval $(-\tilde{X}_t, \tilde{X}_t)$ and does not depend on ε. Hence, for $\Delta R > 0$ there exists the eigenfunction with eigenfrequency ω_e, $Im\,\omega_e \geq Im\,\Delta\omega - O(\varepsilon) > 0$, and the inhomogeneous flow is unstable.

The particular example of such a transition has been considered in [6].

4. Conclusions

Thus, the loss a local stability of flows without dissipation leads to instability of inhomogeneous flows in two cases:

a) local instability is absolute,

b) local instability is convective and for $R < R_*$ there exists the cyclic sequence of WKB-waves corresponding to real eigenfrequency. For $R > R_*$ the sequence involves more than two WKB-waves.

In all other cases for small $\Delta R > 0$ the flow remains stable in the linear approximation.

References

1. Kulikovskii, A.G.: On the stability condition for stationary states or flows in regions extended in one direction. *J. Appl. Math. Mech.* **49**, (1985), 316-321.
2. Lifshitz, E.M. and Pitaevskii, L.P.: *Physical Kinetics*, Pergamon, London, 1981.
3. Silin, V.P.: Oscillation of weakly non-homogeneous plasma. *Sov. Phys. JETP* **17**, (1963), 857-867.
4. Chomaz, J.M., Huerre, P., and Redekopp, L,G.: A frequency selection criterion in spatially developing flows. *Studies Appl. Math.* **84**, (1991), 119-144.
5. Huerre, P. and Monkewitz, P.A.: Local and global instabilities in spatially developing flows. *Annu. Rev. Fluid Mech.* **22**, (1990), 473-537.
6. Kulikovskii, A.G.: On the stability loss of weakly non-uniform flows in extended regions. The formation of transverse oscillations of a tube conveying a fluid. *J. Appl. Math. Mech.* **57**, (1993), 851-856.

STABILITY OF THE STUART VORTICES IN A ROTATING FRAME

S. LEBLANC AND C. CAMBON
LMFA, Ecole Centrale de Lyon
BP 163, 69131 Ecully Cedex, France

Abstract. Linear stability of the two-dimensional Stuart array of co-rotating vortices is investigated in a rotating frame. The effect of the Coriolis force, that acts only on three-dimensional perturbations, is in partial agreement with the Bradshaw-Richardson criterion, that is to say that cyclonic or strong anticyclonic rotation is stabilizing by cut-off of the spanwise wave number, whereas weak anticyclonic rotation enhances non-rotating growth rate of both fundamental and subharmonic modes. The results are compared to simpler model problems, *i.e.* parallel flow and elliptic vortex in a rotating frame.

1. Instabilities in rotating systems

1.1. PARALLEL FLOWS

When the frame rotates about an axis (say z) perpendicular to the plane of the basic flow (x, y), it is well known that two-dimensional perturbations are not influenced by rotation: *the Coriolis force acts only on three-dimensional modes*. For plane parallel shear flows in a rotating frame, the Orr-Sommerfeld and Squire equations are no longer decoupled and Squire theorem does not hold. But the effect of rotation is stabilizing or destabilizing following the semi-empirical *Bradshaw-Richardson criterion* (Tritton & Davies, 1981), that compares the mean shear of the flow to the rotation rate. For pure spanwise perturbations $e^{\sigma t} e^{i\beta z} \tilde{u}(y)$, the criterion is exact and states that, in a rotating frame with spanwise angular velocity Ωk, the inviscid basic flow $(U(y), 0, 0)$ is unstable if

$$2\Omega(2\Omega + W) < 0 \text{ for some } y \tag{1}$$

where $W(y) = -dU/dy$ is the basic relative vorticity (Pedley, 1969).

1.2. TWO-DIMENSIONAL VORTICES

In the case of two-dimensional basic flows, the role of the Coriolis force is well understood for the linear stability of the unbounded elliptic flow (Craik, 1989, Cambon *et. al.*, 1994), that is to say a stabilizing or destabilizing influence, described by the shift of the angular band of unstable oblique wave-vectors, in partial agreement with Bradshaw-Richardson criterion based on the uniform relative vorticity W of the basic flow: the effect of background rotation is destabilizing (in comparison with the non-rotating case) if $2\Omega(2\Omega + W) < 0$

S. Gavrilakis et al. (eds.), Advances in Turbulence VI, 351-354.

with maximum destabilization for zero *tilting* vorticity $(2\Omega + \frac{1}{2}W = 0)$ and stability for zero *absolute* vorticity $(2\Omega + W = 0)$.

However, the vorticity distribution of the basic flow plays an important role in the stabilizing or destabilizing effect of the Coriolis force. Thus for circularly symmetric vortices such as $(0, V(r), 0)$, Kloosterziel & van Heijst (1991) have shown that the inviscid Rayleigh criterion may be extended in order to include the Coriolis force: axisymmetric disturbances $e^{\sigma t}e^{i\beta z}\tilde{u}(r)$ are amplified if $2(\Omega + V/r)(2\Omega + W) < 0$ for some r, where the basic relative vorticity reads $W(r) = r^{-1}d(rV)/dr$.

For more complex flows exhibiting both *mean shear and concentrated vortices*, the choice of a characteristic value of the vorticity W to take into account in a "generalized" Bradshaw-Richardson stability criterion is not obvious (local vorticity in the vortex core, mean shear of the basic flow, circulation). The Stuart flow is thus an excellent candidate for exploring these issues.

2. Stability of an array of co-rotating vortices

2.1. STUART STREAM FUNCTION

Exact solution of Euler equations, the stream function discovered by Stuart (1967)

$$\Psi(x, y) = \log(\cosh y - \rho \cos x) \tag{2}$$

resembles a two-dimensional mixing layer with periodically spaced Kelvin-Helmholtz roll-ups with smooth vorticity distribution when $0 < \rho < 1$, and it recovers the tanh-mixing layer for $\rho = 0$ and the point-vortex solution for $\rho = 1$.

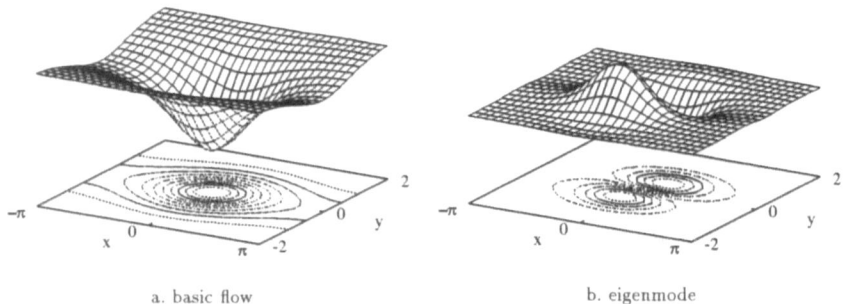

	a. basic flow	b. eigenmode

Figure 1. Basic relative vorticity of the Stuart flow ($\rho = 1/3$) and vertical vorticity component of the most unstable fundamental ($\gamma = 0$) mode in a rotating frame ($\Omega = 0.2$), for spanwise wave number $\beta = 2$.

For $\rho = 1/3$ the vortices are well shaped (figure 1.a) and the maximum vorticity (in the cores) is $W_0 = -2$, whereas the maximum value of the "mean shear" $-d\bar{U}/dy$ is -1.06 close to the maximum shear $S_0 = -1$ of the parallel mixing layer ($\rho = 0$). As noted in the previous section and in order to try to extend the Bradshaw-Richardson criterion, it seemed important that W_0 and S_0 had not closed values, a condition verified with $\rho = 1/3$ for which all the stability calculations presented in this paper have been performed.

2.2. FLOQUET ANALYSIS

In a rotating frame at constant angular velocity $\Omega\boldsymbol{k}$, a three-dimensional infinitesimal disturbance $\boldsymbol{u}(\boldsymbol{x}, t)$ is added to the two-dimensional basic flow. After linearization of Euler equations, the perturbation is solution of the system $\partial_t \mathcal{L}_1 \boldsymbol{u}(\boldsymbol{x}, t) = \mathcal{L}_2 \boldsymbol{u}(\boldsymbol{x}, t)$, where $\mathcal{L}_i = \mathcal{L}_i(\boldsymbol{\Psi}; \Omega)$; $i = 1, 2$ are linear operators involving the spatial derivatives of the basic stream function $\boldsymbol{\Psi}(x, y)$, 2π-periodic in (x).

Following previous works on instabilities of non-parallel shear flows in an inertial frame (Herbert, 1988), Floquet analysis on the linear system allows to seek modes of the form

$$\boldsymbol{u}(\boldsymbol{x}, t) = e^{\sigma t} e^{i\gamma x} e^{i\beta z} \tilde{\boldsymbol{u}}(x, y) \tag{3}$$

where $s = \mathrm{Re}(\sigma)$ is the temporal growth rate of the perturbation, β the spanwise wave number, γ a Floquet exponent (real for temporal analysis) and $\tilde{\boldsymbol{u}}(x, y)$ the eigenfunction, which is 2π-periodic in (x). σ is the eigenvalue of the system $\sigma \mathcal{L}_1 \tilde{\boldsymbol{u}}(x, y) = \mathcal{L}_2 \tilde{\boldsymbol{u}}(x, y)$, with now $\mathcal{L}_i = \mathcal{L}_i(\boldsymbol{\Psi}; \gamma, \beta, \Omega)$; $i = 1, 2$ and by symmetry considerations, it may be shown that $0 \leq \gamma \leq 1/2$ and $0 < \beta$ (if $\beta = 0$, the perturbation is two-dimensional and the Coriolis force plays no role). The two main classes of modes are:

- *fundamental modes* ($\gamma = 0$): $\boldsymbol{u}(\boldsymbol{x}, t)$ are 2π-periodic as the basic flow;
- *subharmonic modes* ($\gamma = 1/2$): $\boldsymbol{u}(\boldsymbol{x}, t)$ are 4π-periodic and excite pairing.

In a rotating frame, it may be shown that the eigenvalue problem may be reduced to a system settled by two equations that are *non-parallel* versions of Orr-Sommerfeld and Squire equations including the Coriolis force. The problem is solved numerically by a spectral collocation method. Formulation of the governing equations and details on numerical procedure would be expressed in a subsequent paper.

2.3. RESULTS

In the early 1980's, research on laminar-turbulent transition in incompressible plane shear flows (without solid body rotation) brought to the fore that two-dimensional basic states or finite amplitude traveling waves may exhibit three-dimensional linear instabilities with common generic features, namely short wave length fast growing modes resulting from an inertial mechanism: the ellipticity of the streamlines in vortical regions of the non-parallel basic flow is taken to be responsible, and the broadband instability of unbounded elliptic flow is often seen to be a generic model of those instabilities (see Baily *et al.*, 1988). For Stuart vortices, Pierrehumbert & Widnall (1982) have shown that three-dimensional fundamental modes are always unstable, whereas subharmonic ones become stable above a critical spanwise wave number β (cut-off).

On figure 2 is represented the higher growth rate of fundamental and subharmonic three-dimensional modes in function of the parameters (β, Ω). Except for short spanwise wave number (β near 0), and contrary to the non-rotating case, the two classes of modes exhibit a comparable behavior: a strong peak of unstable modes is observed for weak anticyclonic rotation (between 0 and 0.5 approximatively), and higher growth rates are obtained for large β without cut-off (because of the inviscid nature of the instability). Stronger anticyclonic and cyclonic rotations tend to stabilize the basic for long spanwise wave number β, and in fact, stronger rotation (cyclonic or anticyclonic) narrows the band of instability down to the $\beta = 0$ axis.

Except a viscous cut-off, similar behaviors are observed for the parallel mixing layer (Yanase *et al.*, 1993). As for the non-rotating case, the physical mechanisms involved in

354

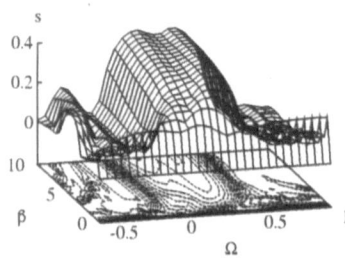
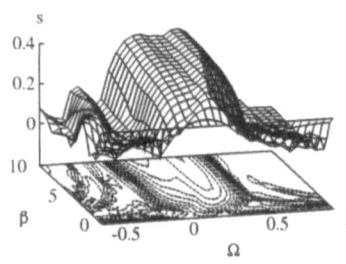

a. fundamental modes   b. subharmonic modes

Figure 2. Surfaces representing the temporal growth rate $s = \mathrm{Re}(\sigma)$ of the most unstable fundamental ($\gamma = 0$) and subharmonic ($\gamma = 1/2$) Floquet modes in function of the spanwise wave number β and of the non-dimensional angular velocity Ω of the rotating frame; isolevels are shown on the base.

the stability of the elliptic unbounded vortex show common generic features with the present results; thus the orientation of the wave vector of most unstable modes for weak anticyclonic rotation tends to be aligned with the axis of rotation, whereas they lie near the (x, y) plane for stabilizing rotation (Cambon *et al.*, 1994). The behavior of spanwise wave number β of the Floquet mode (3) is comparable to the projection of the wave vector on the rotation axis for amplified perturbations.

Trying to apply the Bradshaw-Richardson criterion to the Stuart vortices, it seems that the relevant value is the maximal mean shear S_0 (close to -1 as said before), because the band of instability is approximatively such as $2\Omega(2\Omega + S_0) < 0$ (figure 2). Note that in this unstable band, the modes are more amplified than in the non-rotating case. Now looking at the topology of an unstable fundamental mode (same periodicity as the basic flow), figure 1.b shows the dipole structure of the perturbation, for a destabilizing rotation ($\Omega = 0.2$). Similar eigenfunctions were observed without Coriolis force for the Stuart vortices and for the elliptic flow.

References

Bayly, B.J., Orszag, S.A. and Herbert, T. (1988) Instability mechanisms in shear-flow transition, *Ann. Rev. Fluid Mech.* **20**, 359–391.

Cambon, C., Benoît, J.-P., Shao, L. and Jacquin, L. (1994) Stability analysis and large eddy simulation of rotating turbulence with organized eddies, *J. Fluid Mech.* **278**, 175–200.

Craik, A.D.D. (1989) The stability of unbounded two- and three-dimensional flows subject to body forces: some exact solutions, *J. Fluid Mech.* **198**, 275–292.

Herbert, T. (1988) Secondary instability of boundary layers. *Ann. Rev. Fluid Mech.* **20**, 359–391.

Kloosterziel, R.C. and van Heijst, G.J.F. (1991) An experimental study of unstable barotropic vortices in a rotating fluid, *J. Fluid Mech.* **239**, 607–629

Pedley, T.J. (1969) On the stability of viscous flow in a rapidly rotating pipe, *J. Fluid Mech.* **35**, 97–115.

Pierrehumbert, R.T. and Widnall, S.E. (1982) The two- and three-dimensional instabilities of a spatially periodic shear layer, *J. Fluid Mech.* **114**, 59–82.

Stuart, J.T. (1967) On finite amplitude oscillations in laminar mixing layers, *J. Fluid Mech.* **29**, 417–440.

Tritton, S.C. and Davies, P.A. (1981) In *Hydrodynamic Instabilities and the Transition to Turbulence*, ed. H. L. Swinney & J. P. Gollub, pp. 229–270. Berlin: Springer-Verlag.

Yanase, S., Flores, C., Metais, O. and Riley, J.J. (1993) Rotating free-shear flows. I. Linear stability analysis. *Phys. Fluids* A **5** (11), 2725–2737.

ON THE THREE-DIMENSIONAL INSTABILITY OF ELLIPTICAL VORTICES SUBJECTED TO STRETCHING

S. LE DIZÈS

I.R.P.H.E., Universités d'Aix-Marseille I & II, CNRS,
12 Av. Général Leclerc, F-13003 Marseille, France.

M. ROSSI

L.M.M., Université de Paris VI,
4 Place Jussieu, F-75252 Paris cedex 05, France.

AND

H. K. MOFFATT

D.A.M.T.P., University of Cambridge,
Silver street, Cambridge CB3 9EW, England.

The presence of organized structures in turbulent flows has been recently emphasized by physical and numerical experiments. In particular, vorticity is mainly concentrated in localized tube-like regions: the so called "worms" or "sinews" (Moffatt *et al.*, 1994). If one agrees that such local structures are important for the global turbulent field, it is certainly worth studying their elementary dynamical behavior. Furthermore, it is known that two-dimensional vortices are subjected to generic three-dimensional instabilities (Waleffe, 1990). This phenomenon, located near the core of vortices, depends on the eccentricity of their streamlines. Since vortices present in turbulent flows are not purely two-dimensional and experience a three-dimensional strain arising from the mean field, it is natural to ask how the elliptical instability carries on when a velocity component is added along the span of the previously two-dimensional vortex.

In this work, we specifically consider the stability of the core of a three-dimensional elliptical vortex subjected to an axial strain. For this case, an exact time-dependent solution can be found. The stability of such a flow is analyzed by linearizing the Navier-Stokes equations around this basic state. The complete system for the perturbations is reduced to a single equation for the perturbed velocity along the vortex span. This equation is unfortunately non-separable in space and time variables: standard Fourier analysis is hence of no use and one should resort to a new method to solve the

S. Gavrilakis et al. (eds.), Advances in Turbulence VI, 355-356.
© 1996 *Kluwer Academic Publishers.*

problem. Fortunately enough, one is guided by the specific case of a pure solid body rotation. It is known that this flow supports neutral waves called inertial waves for which analytical expressions are available. In this work, we construct the equivalent of these solutions for the general case under study. Time-dependent equations for both amplitude and wavenumber are thus obtained. It is shown that the wavenumber modulus and angle with respect to the span increase while this vector rotates more and more rapidly around the same axis. In the limit of weak stretching and weak ellipticity, a perturbation theory can be performed and leads to a WKBJ approximation for the solution. This procedure shows that the linearized Navier-Stokes equations are transformed into a Hill equation with slowly evolving coefficients. According to this slow variation, it is possible to understand the perturbation evolution as follows. When no stretching is applied, the elliptical instability is active: an initial perturbation is unstable when its angle with the span is located inside an interval around $\pi/3$. When stretching is present, the wavector always crosses this unstable region during a finite period of time. In that area, the perturbation amplitude hence grows and an exponential factor can be expected for the global amplification. When the effect of stretching becomes important compared to the destabilizing effect of eccentricity, the wavevector remains in the unstable area during a period of time which is too short to let the perturbation reach a sufficiently large amplitude. As a result, the vortex structure is not affected by the three-dimensional elliptical instability.

The global amplification factor can indeed be computed which demonstrates that a small amount of stretching is capable to prevent the appearance of three-dimensional instabilities for vortices with a low enough eccentricity. Since most vortices are slightly elliptical (Moffatt et al., 1994) in turbulent flows, the above computations are expected to cover a wide range of experimental cases. In particular, it is tentatively argued that this mechanism may explain recent experimental observations (Cadot et al., 1995).

References

O. Cadot, S. Douady and Y. Couder (1995), Characterization of the low pressure filaments in three-dimensional turbulent shear flow, *Phys. Fluids* **7**, (3), 630-646.

H. K. Moffatt, S. Kida and K. Ohkitani (1994), Stretched vortices—the sinews of turbulence; large-Reynolds-number asymptotics, *J. Fluid Mech.* **259**, 241-264.

F. Waleffe (1990), On the three-dimensional instability of strained vortices, *Phys. Fluids* A **2**, 76-80.

EXPERIMENTAL INVESTIGATION OF NONLINEAR WAVE INTERACTIONS AND SECONDARY INSTABILITY IN THREE-DIMENSIONAL BOUNDARY-LAYER FLOW

T. LERCHE
DLR Institut für Strömungsmechanik
Bunsenstraße 10, D-37073 Göttingen, Germany

1. Experimental Setup

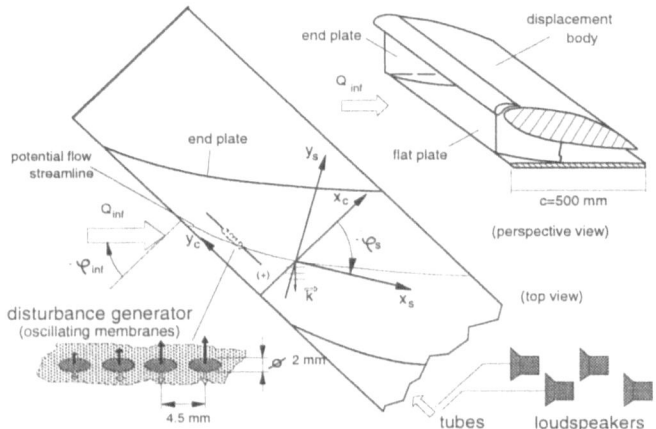

Figure 1. Experimental setup.

The present DLR transition experiment was designed to study cross-flow (CF) instability development in the presence of controlled disturbance excitation. Therefore, a new type of disturbance generator has been developed, which basically consists of an array of 40 oscillating membranes depicted in figure 1. The membranes are flush mounted and oscillate sinusoidally with adjustable phase-lag by means of pressure fluctuations produced by loudspeakers. The whole generator system has been integrated in a flat plate model (figure 1). Previous extended measurements (Lerche and Bippes, 1995) have shown that this disturbance generator excites single oblique traveling waves in agreement with local linear stability theory predictions. All subsequently presented results were obtained for an effective sweep angle of $\varphi_\infty = 43.5°$ and a freestream velocity of $Q_\infty = 16.3$ m/s by hotwire anemometry using a sub-miniature V-type probe. The mean velocity

S. Gavrilakis et al. (eds.), Advances in Turbulence VI, 357-360.

358

components U,V and W correspond with x,y and z of the correspondingly denoted coordinate system.

2. Observation of a high frequency secondary instability

In low or moderate turbulence environment streamwise vortices are observed as dominating CF instability. Following predictions of secondary stability analysis (Fischer *et al.*, 1993) and direct numerical simulation (Müller *et al.*, 1994) the stationary cross-flow vortices alter the stability characteristics of the boundary-layer flow field and thus lead to nonlinear generation of additionally amplified traveling modes.

In the experiment a single traveling mode was initiated by the newly developed disturbance generator. This case is denoted by case (1,1), where the numbers in parenthesis denote multiple integers of the fundamental wave frequency and spanwise wavenumber, respectively. Case (1,1) was compared with case (0,1)+(1,1) where additionally a dominant steady mode was initiated by a spanwise array of roughness elements. The steady mode saturation amplitude $(U_{s,max} - U_{s,min})/(2Q_e)$ differs between 2 % for case (1,1) and 10 % for case (0,1)+(1,1) while the maximum amplitude u_{rms}/Q_e of the fundamental frequency saturates at 10 % in both cases. The superposition of (n,n±m) modes in case (0,1)+(1,1) with n=1,2,3... and m=0,1,2... was found (Lerche and Bippes, 1995) to result in a spanwise modulation of unsteady disturbance amplitudes.

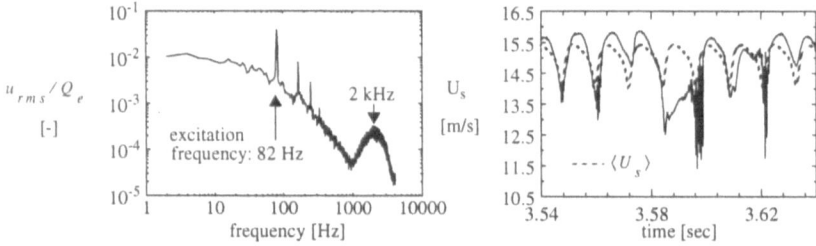

Figure 2. Left: typical Fourier spectrum of streamwise velocity component showing high-frequency secondary instability centered around 2000 Hz; Right: part of corresponding time signal and its phase average (dashed curve).

A secondary high-frequency instability as shown in figure 2 was detected in the nonlinear saturation region for both cases. In addition to earlier observations (Poll, 1985) (Kohama *et al.*, 1991) figure 3 shows the distribution of filtered (1600-2400 Hz) u_{rms} values in a wall-normal plane for case (1,1) and case (0,1)+(1,1) and a plane at $z = 2.25$mm ($z/\delta = 0.6$, $\delta = 3.75$mm) for case (0,1)+(1,1). It shows that the high-frequency instability resides in the upper part of the boundary-layer at $z/\delta \approx 0.6$. Note that the filtered u_{rms}/Q_e values differ between approx. 0.1 % for case (1,1) and 1.0 % for case (0,1)+(1,1), while the time signal reveals that the momentary high-frequency motion is much more violent.

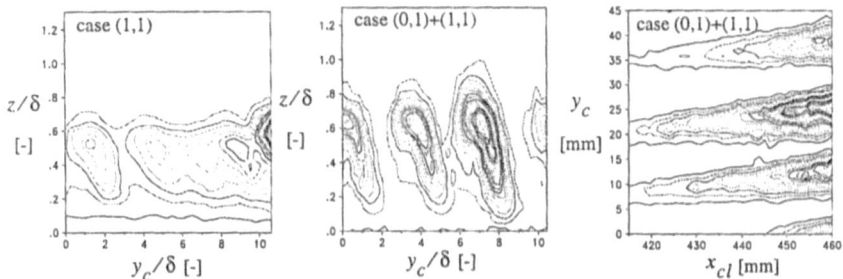

Figure 3. Left: iso-contours of secondary fluctuating velocity in spanwise and wall-normal plane (RMS values inside frequency band 1600 to 2400 Hz) for case (1,1) at $x_c/c = 0.93$; Middle: as before for case (0,1)+(1,1) at $x_c/c = 0.93$ Right: as before but in spanwise and wind-tunnel centerline plane at $z/\delta = 0.6$.

2.1. ANALYSIS OF SPATIO-TEMPORAL EVOLUTION

The experiment indicated (figure 4) the existance of a distict phase-relationship between high-frequency breakdown occurence and primary wave action at each spatial location. In particular, it was found that the phase-averaged mean velocity profiles exhibit a significant variation during one primary wave cycle T with a coincidence of secondary breakdown and highly inflexional profiles. This finding was analyzed in more detail by determination of

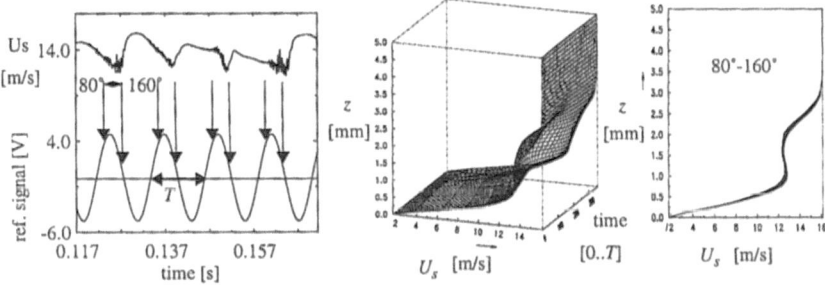

Figure 4. Left: relation of measurement signal high-frequency parts to additionally recorded reference signal of excitation device; Middle: variation of mean velocity U_s profiles during one primary wave cycle; Right: U_s profiles during secondary breakdown.

the absolute values of the second temporal derivatives of U_s (Arnal et al., 1978) to yield the spatio-temporal distribution of high-frequency activity. In figure 5 the iso-surface of high-frequency activity (regions with more than 20 % of respective maximum activity inside black mesh) is plotted with the deviation from the local mean velocity u'_s (left) and the second wall-normal derivative of U_s, U_s^{zz} (right), for case (1,1) (top) and case (0,1)+(1,1). Whereas both cases reveal that the occurrence of high-frequency secondary instability is linked to local, momentary inflexion points when at the same time the deviation from the local mean velocity is negative, i.e. the flow is decelerated, case (0,1)+(1,1) additionally shows space-time focusing of

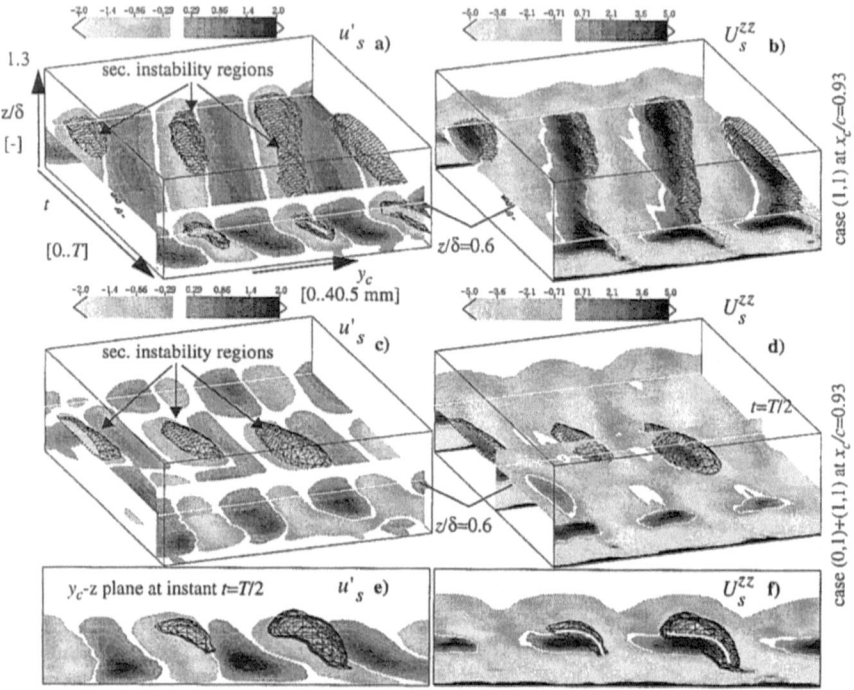

Figure 5. a) fluctuating velocity u'_s and regions of high-frequency activity for case $(1,1)$; b) as before for U_s^{zz}; c) fluctuating velocity u'_s and regions of high-frequency activity for case $(0,1)+(1,1)$; d) as before for U_s^{zz}; e) and f) $y_c - z$ planes at $t = T/2$ of c) and d), respectively.

mode action which leads to a stronger and spatially and temporally more restricted appearance of high-frequency, secondary instability.

References

Arnal, D., Juillen, J.-C. (1978) , Hot-wire conditional sampling techniques for transition intermittency studies, *ONERA paper no. 80-22226*

Fischer, T.M., Hein, S. & Dallmann, U. (1993) , A theoretical approach for describing secondary instability features in three-dimensional boundary-layer flows, *AIAA paper no. 93-0080*

Kohama, Y., Saric, W.S., Hoos, J.A. (1991) , A high-frequency, secondary instability of crossflow vortices that leads to transition, *Boundary-Layer Transition and Control Conference, Cambridge, UK*

Lerche, T., Bippes H. (1995) , Experimental investigation of cross-flow instability under the influence of controlled disturbance generation, *Colloquium on Transitional Boundary Layers in Aeronautics of the Royal Netherlands Academy of Arts and Science*

Müller, W., Bestek, H. & Fasel, H. (1994) , Spatial direct numerical simulation of transition in a three-dimensional boundary-layer, *IUTAM Symposium on Laminar-Turbulent Transition, Sendai, Japan*

Poll, D.I.A (1985) , Some observations of the transition process on the windward face of a long yawed cylinder, *J. Fluid Mech.* **Vol. 150**, pp. 329–356

INSTABILITIES IN TRAILING VORTICES AND IN THE TEMPORAL DEVELOPMENT OF VORTEX PAIRS

T. LEWEKE, G.D. MILLER & C.H.K. WILLIAMSON
Sibley School of Mechanical & Aerospace Engineering
Cornell University, Ithaca, NY 14853, USA

1. Introduction

We present new experimental results concerning the dynamics, and the development of turbulence, in a pair of counterrotating vortices in two situations: 1) the downstream evolution of the trailing vortex pair wake behind a delta wing, and 2) the more 'ideal' case of an initially uniform pair of vortices evolving in time. In addition to the scientific interest, both studies are also motivated by the need to understand the behaviour of 'real' aircraft wakes with regard to flight safety and mixing in the atmosphere.

Although the aerodynamics of lifting delta wings has been studied extensively, remarkably little attention has been devoted to their downstream wake. In the present work we study the large- and small-scale turbulent structure found in this wake, with a view to understanding the basic instabilities, mixing properties, and decay rates of the primary vortex pair. We also investigate the possibilities of a control of this decay. In our experimental work we use both a water towing tank and a wind tunnel. Our approach involves flow visualization, including Particle-Image Velocimetry, and point-velocity measurements, as well as analytical studies. Initial experiments employed a novel technique of flying delta wings in free flight through water to eliminate any interference from struts or disturbances from the towing mechanism.

The study of the more fundamental case of a uniform vortex pair allows us to identify basic mechanisms of instability in this system. Although analytical studies of vortex pairs have suggested the existence of long- and short-wavelength instabilities (Crow 1970, Widnall *et al.* 1973), it is quite surprising that there exists, in the literature, no clear evidence of such structure in laboratory experiments, beyond the suggestion that these length scales exist. Despite the fact that such experiments are often conceptually simple, they are remarkably difficult to carry out effectively. Our own approach has been to pay extreme care to the experimental set up, including the mechanical arrangement, vibrations, convection currents, visualization technique, control of end conditions, etc. The vortex pair is generated in water at the sharpened edges of two flat plates, hinged to a common base and moved in a prescribed symmetric way. Visualization is achieved using fluorescent dye. Photographs were taken from two perpendicular directions. We have varied the vortex circulation (Γ) to yield Reynolds numbers $Re=\Gamma/v$ in the range 2000-3000. The evolution of the vortex pair was found to depend strongly on the motion history of the vortex-generating plates, i.e., on the velocity profiles of the vortices. We will discuss the relation between plate motion and vortex structure in detail in our presentation.

2. Flow Behind a Delta Wing

We have made clear observations of the downstream development of the wake of a delta wing, for what appears to be the first time, in laboratory conditions. Our visualizations and measurements have led us to discover a direct link between the structures in the "braid wake" and in the primary vortex pair. This result corresponds well with turbulence measurements in a single vortex of Devenport *et al.* (1994). The visualizations in Fig. 1, which show an extremely periodic structure in the braid wake, demonstrate this link.

We have discovered a surprisingly large vertical extent of the delta wing wake, due to the distribution of vortical structures within and outside the separatrix of the vortex pair. This "curtain" of fluid remnants left behind the descending vortex pair is evident in Fig. 2. We propose that the mechanism which is responsible for the curtain involves three-dimensional dynamics of vorticity in countersigned secondary vortices which are shed from the wing surface, underneath the primary vortices.

We have found discrete wavelengths of instability, which increase as the trailing wake travels downstream. An example of the development of the large-scale instability of the vortex pair far behind the wing is provided in Fig. 2. Ultimately, we find an interconnected vortex ring-like structure, which is of a

S. Gavrilakis et al. (eds.), Advances in Turbulence VI, 361-364.

scale markedly smaller than the classical Crow (1970) instability wavelength.

Finally, we were interested in controlling the rate of decay and break-up of the coherent vortex pair wake by influencing the development of instabilities in these flows. We show that a small forcing can have strong control over the lifetime of such vortex wakes. The forcing consists in a perturbation of the flight velocity U_∞ of the wing with different amplitudes (u') and streamwise wavelengths (λ). A novel technique using image processing to measure the amplitude of the large-scale instability of the vortex pair has enabled us to measure its wavelength and growth rate, which we found to compare very well with the analysis of Crow (1970). The effect of forcing on the wake behaviour is illustrated in Fig. 3.

3. Temporal Development of a Vortex Pair

We have identified at least three different instability length scales that lead to turbulent structures in a vortex pair from our experiments:

1. *Long-wave instability.* In this case the vortices develop a waviness whose axial wavelength λ_L is several times the vortex separation b. It is a similar mechanism as the one found in the delta wing wake. This waviness is amplified in time until they touch, break up and reconnect to form periodic vortex rings. Figure 4 shows this flow at a later time from both perpendicular directions. The rings are now elongated in the transverse direction, and the side view reveals that dyed fluid is left behind at the locations where they have reconnected. The late-time evolution of this instability has not been observed or studied previously in the laboratory.

2. *Short-wave instability.* Figure 5 shows the development of a short-wave instability (wavelength $\lambda_S < b$) at the same time as the long waves ($\lambda_L \gg b$) appear. The remarkably clear visualizations of the vortex core have yielded some new results. The simultaneous perpendicular views of the instability allow us to reconstruct the core flow in three dimensions and to clearly identify the mutual interactions between the waves in each vortex core. In particular, the phase relationships observed in Fig. 5 show that the symmetry of the flow with respect to the mid-plane between the vortices is broken.

3. *Fine-scale instability.* This instability occurs during or immediately after the roll-up of the vortices and is believed to be caused by a centrifugally unstable vortex velocity profile. This rapidly leads to turbulent motion and a much faster decay of the coherent vortex pair structure than in the preceding cases.

4. Conclusions

We have made what appear to be the first clear observations of the downstream wake of a delta wing in laboratory conditions. We have discovered a direct link between turbulent stucture between the primary vortices and the structure on the primary vortices. We have found a very periodic structure in the "braid wake". We have found a remarkably large vertical extent to the wake of the delta wing. We have observed the large-scale instablity of the vortex pair and vortex ring-like structures which result from it, and we have demonstrated control over the instability wavelength and lifetime of the vortex pair.

In the more fundamental study of an initially uniform vortex pairs we have discovered a number of new phenomena concerning their dynamics. We have found clear evidence for the existence of several different long- and short-wave instabilities, which had not previously been shown experimentally for the case of a symmetric pair of straight vortices, and we were able to observe for the first time their long-time evolution, far beyond the theoretically treated linear state. We have investigated the relation between the initial vorticity distribution and the type(s) of instabilities subsequently observed on the vortex pair. Using a technique of simultaneous perpendicular views we obtained precise information about the spatial structure of the instabilities, which lead to the discovery of a symmetry breaking for the case of the short-wave instability. In addition to the material shown here, we intend to give a comprehensive presentation of these new findings, including experimental as well as analytical results.

ACKNOWLEDGEMENTS

We are grateful for discussions with Michel Provansal and Pierre Albarède. This work was supported by the Office of Naval Research under Contract Nos. N00014-90-J-1686, N00014-94-1-1197 and N00014-95-1-0332. T. L. acknowledges the support from the Deutsche Forschungsgemeinschaft under Grant No. Le 972/1-1.

References

S.C. Crow 1970, *AIAA J.* **8**, 2172.

W.J. Devenport, M.C. Rife, S.I. Liapas, and J. Miranda 1994, AIAA Paper 94-0404.

S.E. Widnall, D.B. Bliss, and C.-Y. Tsai 1974, *J. Fluid Mech.* **66**, 35.

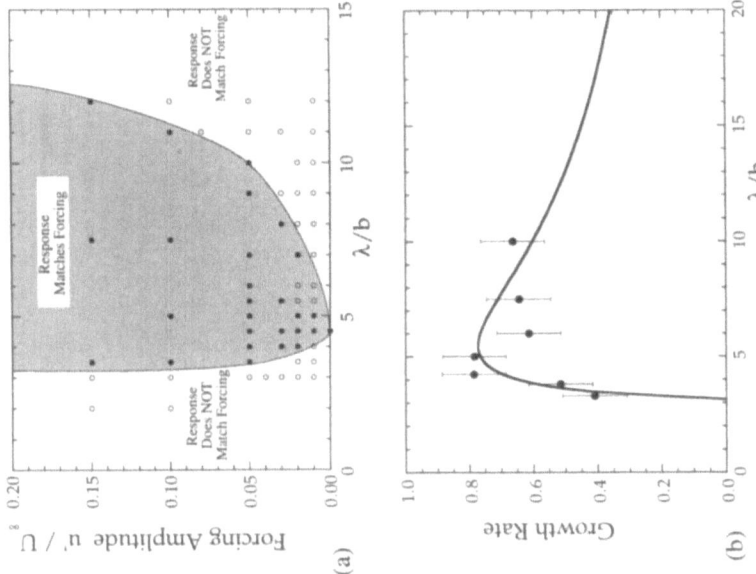

Figure 1. The near wake of a delta wing in free flight (top and side views).

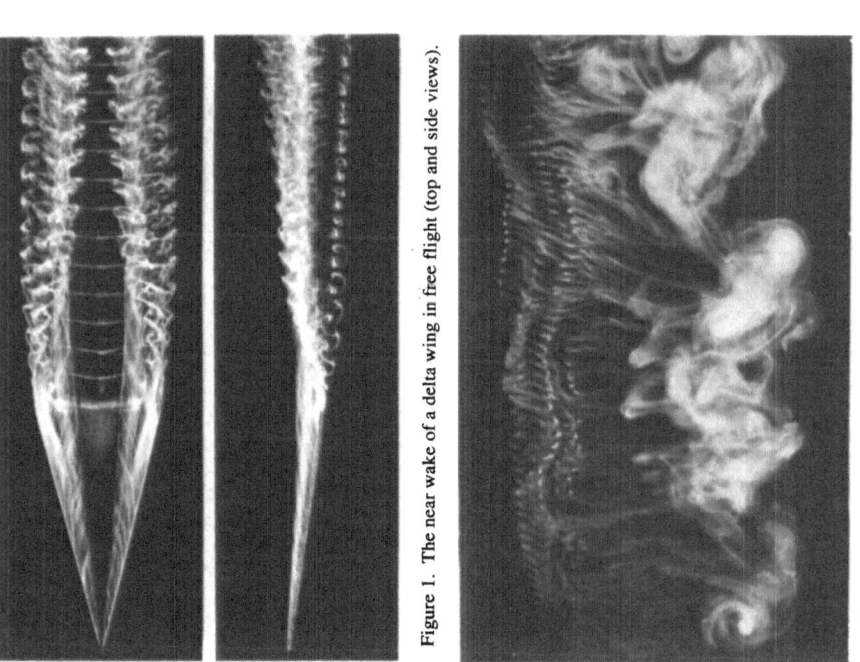

Figure 2. The far wake of a delta wing in free flight (side view).

Figure 3. (a) Different regimes of the wake's response to forcing, as function of the forcing amplitude and wavelength. b is the initial vortex spacing. (b) Growth rate of large-scale (Crow) instability as function of streamwise wavelength.

364

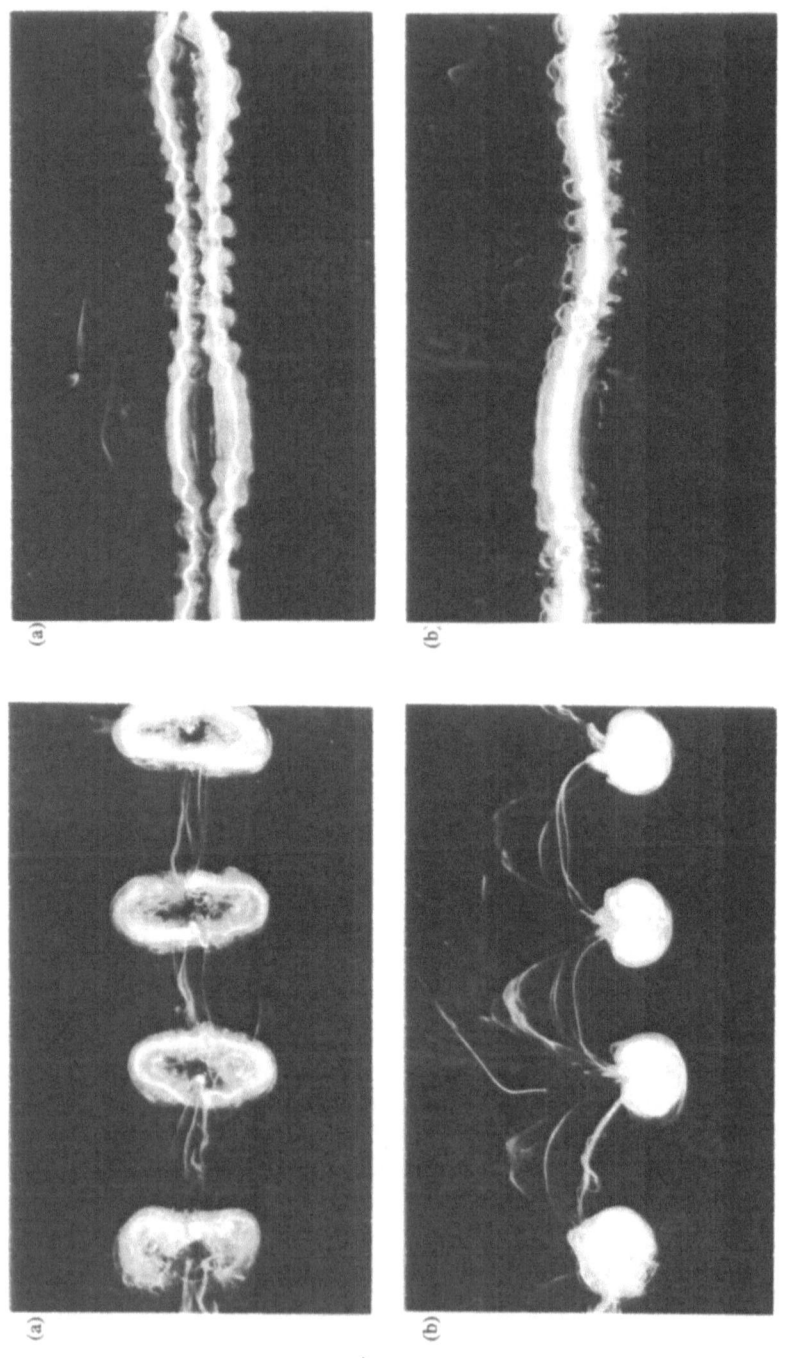

Figure 5. Short-wave instability of a vortex pair, developing simultaneously with a long-wave instability. (a) Front view. (b) Side view. $Re=2750$, $t^*=6.4$. Instability wavelengths: $\lambda_L/b=7.9$ and $\lambda_S/b=0.7$.

Figure 4. Long-wave vortex pair instability, seen at a late stage. (a) Front view, vortices are moving towards observer. (b) Side view, vortices are moving down. $Re=2100$, $t^*=t\cdot(\Gamma/2\pi b^2)=6.9$. Instability wavelength: $\lambda_L/b=3.9$.

ABSOLUTE INSTABILITY OF THE ROTATING-DISK BOUNDARY LAYER — THEORY AND EXPERIMENT

R.J. LINGWOOD

Cambridge University, Engineering Department
Trumpington Street, Cambridge CB2 1PZ, U.K.
email: rjl2@eng.cam.ac.uk

Abstract.

The rotating-disk boundary layer is studied. The inviscid stability of the flow and the stability with viscous, Coriolis and streamline curvature effects included have been studied. In both cases, the flow is radially absolutely unstable for certain parameters. The critical Reynolds number for absolute instability is $R \approx 510$. In the experimental study, the flow was perturbed impulsively at a value of R below that at which laminar–turbulent transition is observed. Convectively unstable modes were excited, which form a wave packet that initially convects away from the source. However, the radial propagation of the trailing edge of the wave packet tended towards zero as it approached R_c; the predicted critical R for the onset of absolute instability. The accumulation of energy at a well-defined radius, may cause the onset of transition, which has been consistently observed at an average value of 513 with only a small scatter. Here, transition was observed at about this value, with and without artificial excitation of the boundary layer. This lack of sensitivity to the exact form of the disturbance environment is characteristic of an absolutely unstable flow, because absolute growth of disturbances can start from either noise or artificial sources to reach the same final state, which is determined by nonlinear effects.

1. Introduction

The rotating-disk boundary layer is three-dimensional and is susceptible to crossflow instability. There is an exact similarity solution of the Navier Stokes equations for the base flow and R is equivalent to a non-dimensional radius r. The experimentally observed onset of transition (see Malik. Wilkinson & Orszag 1981) is at $R \approx 513$, with a scatter of less than 3%. This contrasts with the flow on, say, a flat plate, where transition is sudden but its location is highly dependent on the disturbance environment. This contrast

S. Gavrilakis et al. (eds.), Advances in Turbulence VI, 365-368.
© *1996 Kluwer Academic Publishers.*

led to the suggestion that radial absolute instability may be triggering the nonlinear behaviour characteristic of the onset of transition.

As discussed by Huerre & Monkewitz (1990), the response of the flow to impulsive forcing shows whether it is convectively or absolutely unstable. If the response to the transient disturbance grows with time at a fixed location in space, then the flow is absolutely unstable. Briggs (1964) showed that absolute instability can be identified by singularities in the dispersion relationship that occur when modes associated with waves propagating in opposite directions coalesce. Such points have become known as pinch points. Variation of a parameter, such as R, can cause such points, so changing the behaviour of a flow from being convectively unstable to absolutely unstable.

Large roughnesses on the disk can cause stationary crossflow vortices to grow sufficiently to distort the mean velocity profiles, causing secondary instabilities. However, it is assumed here that roughness and the turbulence levels are small, so transition is controlled by the stability of the mean profiles rather than secondary instabilities. Travelling waves are considered as well as stationary waves, and the parallel-flow approximation is used, which restricts the analysis to the local stability characteristics of the flow.

2. Results

Details of this problem, including a description of the three important spatial branches of the dispersion relation, can be found in Lingwood (1995, 1996b). A normal branch point (between branches 1 and 2) is compared with a pinch point (between branches 1 and 3) in figures 1(a) and (b). Branches 1 and 2 both originate in the upper half α-plane for $\omega_i \to \infty$ (where α is the radial wavenumber, ω is the frequency and subscripts r and i denote real and imaginary parts, respectively) and therefore branch points between these two modes do not satisfy the criterion for a pinch point and do not lead to absolute instability. A comparison between absolutely unstable pinch points ($\omega_i^o > 0$) from the viscous analysis (which only occur for $R > 510.625$) and from the inviscid analysis is shown in figures 1(b) and (c). In both cases, $\bar{\beta} = \beta/R$ (where β is the circumferential wavenumber) is 0.126. Solid lines in one plane map through the dispersion relation to solid lines in the other plane; similarly for the dashed lines. Both pinch points show a characteristic cusp at ω_i^o and, apart from the magnitude of ω_i^o which gives the degree of absolute instability, the parameter values at the pinch points are very similar. The inviscid spatial branches are the asymptotic limits of branches 1 and 3 at infinite R. Branch 2 does not appear in inviscid analyses; it is caused by a balance of viscous and Coriolis forces.

Details of the experimental procedure and results are given in Lingwood (1996a). Hot-wire measurements taken of the boundary layer without any

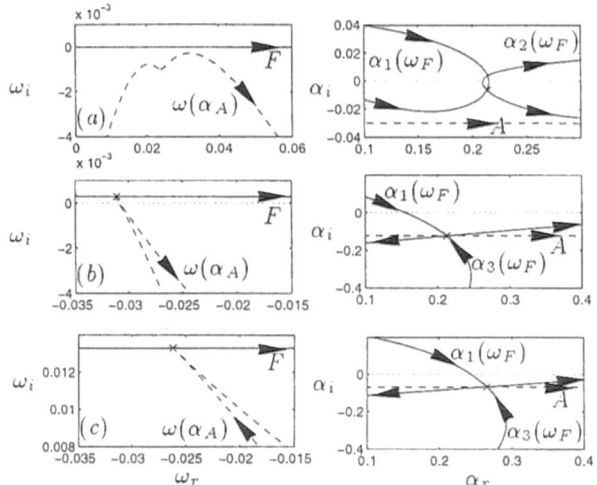

Figure 1. (a) A viscous branch point between branches 1 and 2 for $R = 515$ and $\beta = 6$. (b) A viscous pinch point between branches 1 and 3 for $R = 530$ and $\beta = 67$ ($\beta \approx 0.126$). (c) An inviscid pinch point between branches 1 and 3 for $\beta = 0.126$.

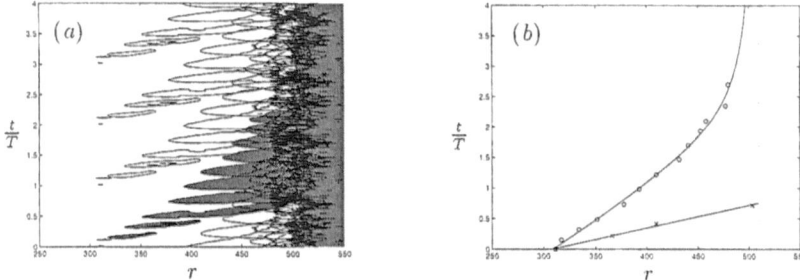

Figure 2. (a) Contour plot (0.008 level), of the wave-packet envelope. T is the time period of one revolution. (b) Leading, $\times$, and trailing, $\circ$, edges of the wave-packet trajectory and least-squares fits to the experimental data, ——. $R = 510$ is denoted by $\cdots\cdots$.

artificial excitation showed some repeatable stationary disturbances created by unavoidable roughnesses. The peak disturbance amplitude was about 3% of the local disk speed at $R \approx 500$. An analysis by Balachandar, Streett & Malik (1990) showed that at $R = 500$ a primary disturbance r.m.s. amplitude of 9% is needed for the onset of secondary instabilities. This suggests that, even close to the onset of transition, the stability is governed by the mean velocity profiles rather than secondary instabilities. With the boundary layer excited by an impulsive point source, the development of the generated wave packet can be seen in figure 2(a) in (r, t)-space. Figure 2(b) shows the leading and trailing edges of the wave-packet trajectory derived from figure 2(a). Both fitted lines were constrained to pass through the source position at $R \approx 311$ and $t = 0$. The trailing edge is modelled as an asymptote to 510, and is given by $t/T \approx 0.011r - 3.57 + 27.08/(510 - r)$.

This choice was influenced by stability theory. At the radius where the flow becomes absolutely unstable, r_c, there will be temporal growth of disturbances along the ray $(r - r_c)/t = 0$. To match this behaviour to the preceding convectively unstable behaviour, the trailing edge must tends towards the vertical as it approaches $(r - r_c)/t = 0$. The experimental points on the trailing edge closely follow this predicted behaviour and the packet disintegrates at R above about 510. Note that the stationary disturbances do not show this type of behaviour; their bounding rays have approximately constant gradient with increasing R, implying that the stationary disturbances remain convective up to the onset of transition.

3. Conclusions

Absolute instability has been found for $R > 510.625$. Neglecting all terms of $O(R^{-1})$ gives a consistent set of perturbation equations, and the absolute instability persists over a range of $\bar{\beta}$. In the experiment, the stationary disturbances have a maximum amplitude of less than a third of the threshold predicted for the onset of secondary instabilities. This suggests that the assumption made that the linear stability of the flow is determined by the mean profiles rather than secondary instabilities of modified profiles is justified. This will be true in general provided the stationary disturbance field is not too large. The progression of the wave packet in the radial direction with increasing time is shown; the radial propagation of the trailing edge of the wave packet tends towards zero as the packet approaches R_c. The accumulation of energy at this radius causes the well-defined structure of the wave packet to disintegrate there. Absolute instability may be a better explanation for transition than the convective radial growth of disturbances, leading to nonlinearity. Moreover, this mechanism may be relevant to transition on highly swept wings.

References

BALACHANDAR, S., STREETT, C. L. & MALIK, M. R. 1990 Secondary instability in rotating disk flow. *AIAA Paper* 90–1527.

BRIGGS, R. J. 1964 *Electron-stream interaction with plasmas.* MIT Press, chap. 2.

HUERRE, P. & MONKEWITZ, P. A. 1990 Local and global instabilities in spatially developing flows. *Ann. Rev. Fluid Mech.* **22**, 473–537.

LINGWOOD, R. J. 1995 Absolute instability of the boundary layer on a rotating disk. *J. Fluid Mech.* **299**, 17–33.

LINGWOOD, R. J. 1996a An experimental study of absolute instability of the rotating-disk boundary-layer flow. *J. Fluid Mech.* **314**, 373–405.

LINGWOOD, R. J. 1996b On the application of the Briggs' and steepest-descent methods to a boundary-layer flow. *Stud. Appl. Maths* (in press).

MALIK, M. R., WILKINSON, S. P. & ORSZAG, S. A. 1981 Instability and transition in rotating disk flow. *AIAA J.* **19**, 1131–1138.

A TIME-REVERSED APPROACH TO THE STUDY OF GÖRTLER INSTABILITIES

PAOLO LUCHINI

Dip. Ingegneria Aerospaziale, Politecnico di Milano,
Via Golgi 40, 20133 Milano, Italy

AND

ALESSANDRO BOTTARO

IMHEF - DGM, École Polytechnique Fédérale de Lausanne,
CH-1015 Lausanne, Switzerland

The original analysis (by Görtler himself and others) of the curvature-excited streamwise vortices that have become known as Görtler vortices was based on a quasi-parallel extension of the theory of parallel centrifugal instabilities of the Taylor-Couette and Dean type. However, the possibility of applying a quasi-parallel analysis to Görtler vortices has always been questioned because of the difficulties in identifying a suitable scaling parameter that could allow the theory to be cast in the form of a proper asymptotic expansion continuable to all orders.

Actually, there is more than one way in which such an expansion can be formalized for large values of the longitudinal coordinate x. Let us introduce a local Görtler number $G = (U/\nu)^{1/4}x^{3/4}R^{-1/2}$ and dimensionless wavenumber $b = (x\nu/U)^{1/2}\beta$, where U is the free-stream velocity, ν the kinematic viscosity, R the radius of curvature and β the spanwise wavenumber of the perturbation. Hall (*JFM* **124**, 1982) obtained the result that when G and b tend simultaneously to infinity in such a way that $G = O(b^2)$, a multi-layered asymptotic expansion exists, and thus showed that at the abscissa x where $G \approx 1.7b^2$ there is a transition from amplification to damping of the leading Görtler mode. Bottaro and Luchini (in *Math. and Physical Modeling in Hydro. Stability*, D.N. Riahi ed., World Sci. 1996) showed that, when G tends to infinity independently of b, a multiple scale asymptotic expansion can be set up that has the classical theory as its leading term and permits an extension to higher orders, thus allowing a theoretically justified description of the amplification phase in terms of mode analysis.

S. Gavrilakis et al. (eds.), Advances in Turbulence VI, 369-370.
© *1996 Kluwer Academic Publishers.*

Nevertheless, no asymptotic analysis is available for the initial region of G of order unity where the Görtler instability originates. Indeed, the concept of mode loses meaning in this region. That is why the study of the generation of Görtler vortices today does not employ mode analysis but rather a numerical solution of the linearized boundary layer equations, according to the method started by Hall (JFM **130**, 1983). There is, however, the drawback that the numerical calculation must be repeated for every different initial condition, thus making it very difficult to answer general questions such as which kind of disturbance is most effective in generating Görtler vortices and what thresholds must be imposed on ambient noise if one wants to avoid them.

The present work originated from the observation that if one accepts the limitation of studying small perturbations such that the problem can be linearized, the above questions do possess a general answer. For, in a linearized setting, the final mode amplitude in the large G-range of x (where modes distinguished from each other do exist) must be a linear functional of the initial conditions, expressed by the integral of the product of the initial conditions times a suitable Green function.

We thus set up to calculate the Green function numerically. The difference with the previous approach is that the numerical calculation needs to be performed just once; after that, the influence of any initial condition on the final mode amplitude can be directly expressed through the Green function, and the analysis of the effects of given, deterministic or random, noise sources becomes very easy. In a terminology that has recently become quite common, the Green function expresses the *receptivity* of Görtler vortices to free-stream and wall disturbances (represented, respectively, by initial and wall boundary conditions).

It turns out that, for a parabolic problem like the one governed by the linearized boundary layer equations, the Green function can be determined through the solution of an adjoint system of partial differential equations which is parabolic too, but with a reversed natural direction of evolution, from downstream towards upstream locations. Therefore this is in a sense a time-reversed approach to instability analysis (that can in pronciple be applied to a much wider range of problems), in which the disturbance is followed, in adjoint function space, from the time when it is observed back to the time when it is generated. From the numerical point of view this is a problem of the same order of difficulty as a single numerical integration of the forward system of equations. From a conceptual point of view, however, a careful analysis is needed of the singularities that can arise, under general boundary conditions, at the leading edge of the boundary layer.

LAMINAR–TURBULENT TRANSITION IN PIPE FLOW

F.T.M. NIEUWSTADT, A.A. DRAAD AND G.D.C. KUIKEN
Lab. for Aero- & Hydrodynamics
TU-Delft, 2628 AL Delft, The Netherlands

1. Experimental facility

Recently, a horizontal pipe facility has been completed in our laboratory. It has a length of 34m and an inner diameter of 40mm. A sketch of the facility is given in Fig. 1. The working fluid is water. The facility has been designed for transition studies. So, in order to maintain laminar flow up to very high Reynolds numbers, much care has been devoted to avoid any disturbances which may trigger transition. A settling chamber is used to eliminate swirl and suppress turbulence in the entrance flow by means of several honeycombs and screens. In the design of the shape of the contraction, which connects the settling chamber to the pipe, attention has been given to minimise adverse pressure gradients and to suppress the generation of Görtler vortices.

The pipe itself is made up of a number of Plexiglas sections, each with a length of 2m. The ends of these sections are made slightly conical to eliminate in the connection of two sections any steps in inner pipe diameter that might trigger transition. The flow facility is of the re-circulatory type. Water is pumped from storage vessels, through the settling chamber in the pipe and back to the vessels. A flowmeter is coupled to the pump so that a constant flow rate can be obtained even in a transitional flow. A large part of the facility, i.e. pipe and settling chamber, has been insulated and the water temperature is thermostatically controlled in order to avoid any secondary circulations due to free convection following from temperature differences between the water and the air outside the pipe.

As a result of our careful design, we have able to maintain laminar flow up to a Reynolds number of $Re=60\,000$. It should be noted here that only a few facilities exist(ed) that sustain laminar pipe flow at such high Reynolds numbers, none of which are re-circulatory systems.

S. Gavrilakis et al. (eds.), Advances in Turbulence VI, 371-374.
© 1996 Kluwer Academic Publishers.

372

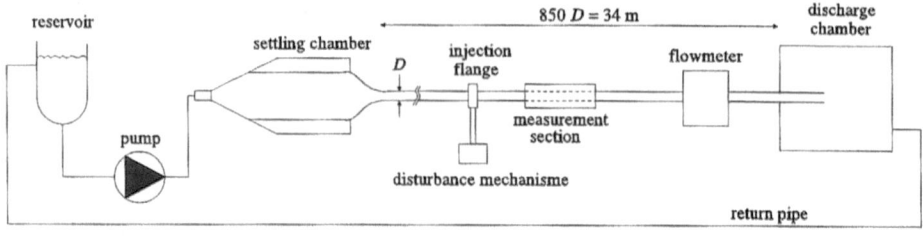

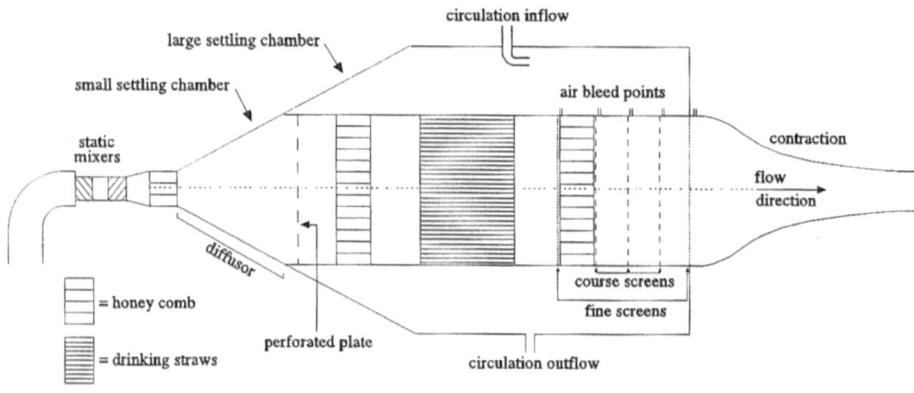

Figure 1. Sketch of the flow facility (upper figure) and settling chamber (lower figure).

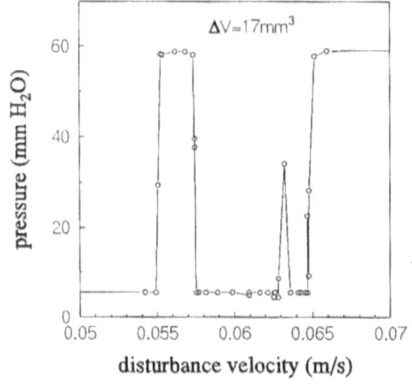

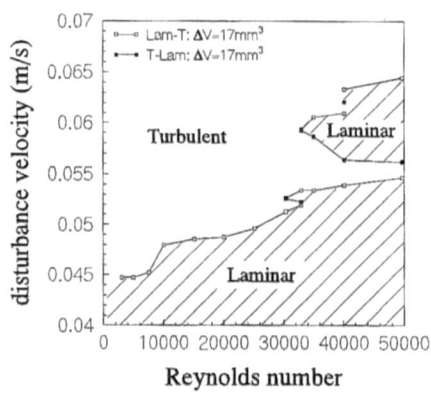

Figure 2. Pressure drop (a high value means turbulent flow and low value laminar flow) as a function of the frequency of the disturbance for a Reynolds number of $Re=40\,000$.

Figure 3. Disturbance velocity at which transition occurs, as a function of the Reynolds number for a given volume displacement of the forcing system.

In order to trigger the laminar flow in a controlled way, the flow is excited with a disturbance consisting of fluid injection/extraction through a slit over the entire circumference of the pipe. This set-up results in a smooth disturbance as a function of the angular coordinate. No mass is added to the flow and only a local disturbance is generated. The frequency of the disturbance can be varied between 0 and 40Hz. As the disturbance mechanism is constructed to displace a constant volume, increasing the frequency leads also to an increase in disturbance velocity. The frequency is determined by a pulse counter and has a resolution of 0.02Hz. This accuracy is necessary since the transition from laminar to turbulence appears to depend very sensitively on the settings of the disturbance mechanism.

STABILITY MEASUREMENTS

We have found the laminar pipe flow to be extremely sensitive to the magnitude of the disturbance generated with the forcing system described above. A clear and distinct threshold value at which the flow changes almost discontinuously from laminar to turbulent, is present over the entire range of frequencies and Reynolds numbers studied. An example is given in Fig. 2 where we show the pressure drop behind the forcing device as a function of the disturbance frequency at a fixed Reynolds number and a given displacement volume (ΔV) of the disturbance mechanism (one should remember that in this case increase in frequency also implies increase in disturbance velocity). Note that the large pressure drop signifies turbulent and the low pressure drop laminar flow. It is clear that the transition is very sharp. We also find the perhaps surprising result that transition is not a monotonous function of the disturbance velocity. Fig. 2 shows that increasing the frequency not only results in transition but it can also result in relaminarization of the flow. We return to this behaviour below.

At Reynolds numbers close to the minimum transition Reynolds number of approximately $Re \simeq 2\,000$, the generation of so called puffs is observed which have been also reported in the literature (Wygnanski and Mullin, 1973; Darbyshire and Mullin, 1995). The puffs were most effectively generated with a disturbance parameter setting which also triggers so-called turbulent slugs above $Re \simeq 3\,000$.

For high frequencies, above $Re = 30\,000$, bifurcations have been found in the transition behaviour, as already mentioned in relation to Fig. 2. Another illustration of this behaviour is given in Fig. 3 where the disturbance velocity at which transition occurs is plotted as a function of Reynolds number for a fixed volume displacement of the forcing device. We see that when increasing the disturbance velocity, transition to turbulence occurs above a certain threshold level. Increasing the disturbance velocity further, the flow relaminarizes to become turbulent again at higher velocities. To our knowledge such bifurcations have not been reported before.

Another way to analyze our experimental results is by plotting the disturbance velocity at which transition occurs as function of the Reynolds

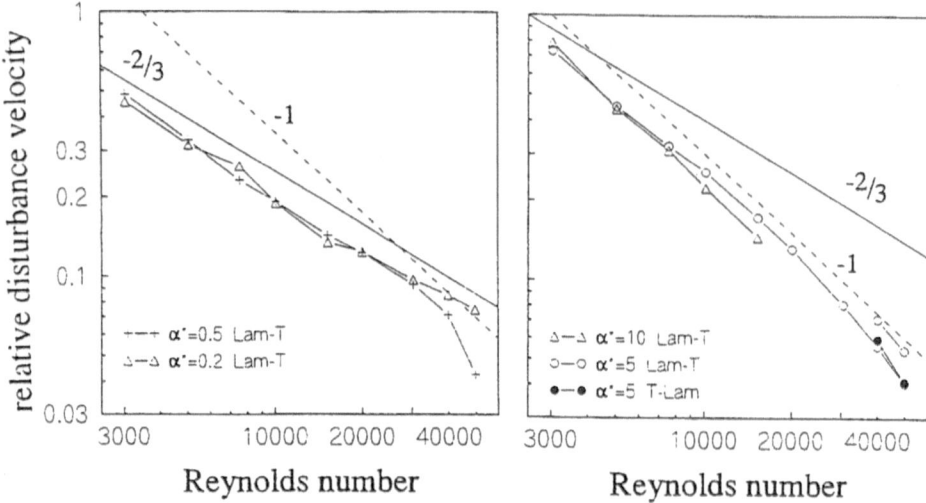

Figure 4. Relative critical disturbance velocity versus Reynolds number for large dimensionless wave numbers.

Figure 5. Relative critical disturbance velocity versus Reynolds number for small dimensionless wave numbers.

number. In Figs. 4 and 5 we have plotted the results for large and small wave numbers. The curves are found to scale with the Reynolds number according to a power law with the exponents -1 and $-2/3$, respectively.

These results have important implications for theory and will be object of further study, both experimentally and theoretically.

References

Darbyshire, A.G.& Mullin, T. (1995) Transition to turbulence in constant-mass-flux pipe flow, *J. of Fluid Mech.* **289**, pp. 83–114.

Wygnanski, I.J., & Champagne, F.H. (1973) On transition in a pipe. Part 1. The origin of puffs and slugs and the flow in a turbulent slug. *J. of Fluid Mech.*, **59**, part 2, pp. 281–335.

PHYSICAL ANALYSIS OF THE THREE-DIMENSIONAL TRANSITION TO TURBULENCE IN THE FLOW AROUND A CIRCULAR CYLINDER BY MEANS OF DIRECT NUMERICAL SIMULATION

H. PERSILLON AND M. BRAZA

Institut de Mécanique des Fluides, URA D0005 CNRS.
Avenue du Professeur Camille Soula,
31400 Toulouse Cedex, France

1. Introduction

The onset of turbulence in separated unsteady flows around bodies is a topic of major interest in the domain of fundamental research. The circular cylinder configuration offers a precious test-case, owing to recently observed fascinating phenomena concerning the transition to turbulence (Williamson (1988) (1989), Eisenlohr *et al* (1989)). Especially, the establishment of a "universal" Strouhal-Reynolds number relationship within 1 percent of accuracy, by removing the disparity of previous experiments due to end-plate conditions, offers a challenging phenomenon to investigate physically by means of Direct Numerical Simulations. Actually, the increased capacity of super computers offers the possibility to DNS approaches to start the investigation of these phenomena, concerning non-Cartesian body geometries. This study is a continuation of our studies developed during the last fifteen years on the numerical simulation of transition to turbulence phenomena in the wake past a circular cylinder, firstly devoted to analyse 2D transition features, among which the onset and interaction of the shear-layer instability with the vortex shedding mechanism (Braza *et al*(1984)-(1990)). Our more recent studies since 1990's in this field, have been devoted to the analysis of three-dimensional transition to turbulence phenomena in the context of PhD thesis of H. Persillon (1995). Since '91's we have simulated firstly, under the hypothesis of parallel shedding, the frequency modulation of the Bénard-Kàrmàn instability within the discontinuity region in the Strouhal-Reynolds relation, results reported in a review paper by Braza (1995). In the present paper, we analyse in details this transition to turbulence process, over a wide Reynolds number range, with an emphasis to the space variation of the instability mode in the discontinuity region.

S. Gavrilakis et al. (eds.), Advances in Turbulence VI, 375-378.
© *1996 Kluwer Academic Publishers.*

2. Results

The direct numerical simulation of the full Navier-Stokes equations is performed for Reynolds number values 100, 150, 180, 185, 190, 200, 220, 250 and 300 over long physical time values, to obtain an established flow process. The code ICARE developed in the context of this study is used, based on a predictor-corrector pressure scheme, on a finite-volume approach and on second order of accuracy in space and time. The boundary conditions in the spanwise direction are chosen to correspond to the parallel shedding phenomenon. The boundary conditions at the outlet boundary are non-reflecting ones Jin & Braza (1993) in order to inhibit feedback effects and to allow the vortices to travel without confinement through the outlet boundary. The spanwise size of the domain is 2.25 diameters. Figure 1 shows the variation of the Strouhal number versus Reynolds number. The present DNS shows clearly the formation of a discontinuity region in the Reynolds number range 190 to 250 in agreement with the physical experiment. Furthermore, by using the complete DNS approach, able to take into account the non-linearity of the physical process, we show that the appearance of the discontinuity occurrs beyond Re=185 and before Re=190. As it can be seeing in the same figure, the 2D numerical simulation, carried out with the 2D version of the code ICARE, shows a continuous curve, which follows smoothly the rising part of the curve (up to Re=170) and the almost constant part of the curve (beyond Re=250).

Therefore, the discontinuity region is due to strictly three-dimensional mechanisms. We can analyse these mechanisms through the three-component vorticity field. In fact, streamwise vorticity component is progressively formed from Re=100 to Re=300. At Re=190 a contra-rotating coherent vortex structure is formed along a plane normal to the x axis, at x/D=3.05. This phenomenon 'pumps' a part of energy from the externally supplied one. The energy attributed to the main vortices decreases and the increasing rate of the Strouhal number is reduced, causing the discontinuity appearance. This structure is found to be fragmented to a three-lobe vortex structure at Re=300. These new structures are smaller and less inertial. We can suppose that they have a shorter life than the structures of the second part of the curve. Hence, the energy consumed by this configuration is lower than the amount of energy used by the previous configuration. For this reason, the von Kàrmàn vortex pattern earns an amount of energy and hence the Strouhal number increases and leads to the second discontinuity of the curve. In addition to the aforementioned classes of vortices, the present DNS has allowed the spontaneous generation of shear-layer vortices already at the present Reynolds number range. The frequency of these eddies is found to be very close to the Strouhal number one. For this reason, this phenomenon is difficult to obtain through a signal processing of experimental time-dependent signals, because of the non-distinction of the shear-layer frequency to the Strouhal number. The DNS simulation offers the possibility to detect this class of eddies in this situation, because of the access to the unsteady vorticity fields. Therefore, the present result allows the "extrapolation" of the law $F_{shear-layer}/F_{Strouhal} \sim Re^n$ (n=0.5, suggested by Bloor (1964) either n=0.67, according to recent studies of Prasad and Williamson (1996), towards the low Reynolds number range. The number of shear-layer vortices is found to increase as a function of Reynolds number. The present study furnishes the way of amplification of the von Kàrmàn modes in the near wake, according to the aforementioned three-dimensional and non-linear process. Figure 2 shows the u component amplitude of the fundamental frequency versus Reynolds number, as a function of x/D along the rear axis. The region of the maximum amplitude is more pronounced in the 3D case, indicating a higher coherent character of the process, than under the simplifying hypothesis of two-dimensionality. All the amplitude variations form a plateau beyond x/D=5, a characteristic which is strengthened in the 3D case. This illustrates a saturation effect due to the persistence of the coherent character upstream the decay region. The increasing part of the curve, in the very near wake, towards the maximum amplification point, corresponds to the establishment of the organized features of the process, in respect to the von Kàrmàn mode. This process is governed by the same law for all Reynolds numbers investigated. In this study, the relation between the maximum amplitude location versus Reynolds number is established, as a universal feature of the present transition to turbulence process. We find that this variation is described by the law $X_{max} \sim Re^{-1/2}$. In the same way, the variation of A_{max} versus Reynolds number is quantified by the law: $A_{max} \sim Re^{1/2}$. We also provide the space-variations of the mean velocity components and of the complete turbulent stress tensor in the near-wake (figure 3), a precious information to quantify the onset of turbulent motion in the present wake flow.

References

M. S. BLOOR (1964) - *The transition to turbulence in the wake of a circular cylinder.* J. Fluid Mechanics, **19**, 290.

M. BRAZA, P. CHASSAING & H. HaMINH (1984) - *Numerical simulation of the vortex shedding past a circular cylinder using a pressure-velocity formulation.* Numerical Methods for Transient and Coupled Problems, pp. 673-687, Venice, Italy, 9-13 July 1984.

M. BRAZA, P. CHASSAING & H. HaMINH (1986) - *Numerical study and physical analysis of the pressure and velocity fields in the near wake of a circular cylinder.* J. Fluid Mechanics, **165**, 79-130.

M. BRAZA, P. CHASSAING & H. HaMINH (1990) - *Prediction of large-scale transition features in the wake of a circular cylinder.* Physics of Fluid, **A2**, 1461.

M. BRAZA (1995) - *Transition features in wake flows by means of numerical analysis.* Book chapter review, Current Topics in the Physics of Fluids, Ed. Research Trends.

H. EISENLOHR & H. ECKELMAN(1989) - *Vortex splitting and its consequence in the vortex street wake of cylinders at low Reynolds numbe.* Physics of Fluids, **A1**, 189.

G. JIN & M. BRAZA (1993) - *A Non-reflective outlet boundary condition for incompressible unsteady Navier-Stokes calculations.* J. Computational Physics, **107**, 239-253.

H. PERSILLON (1995) - *Analyse Physique par simulation numérique bi- et tri-dimensionnelle de la transition laminaire-turbulente dans l'écoulement autour d'un cylindre.* Thèse de Doctorat, I.N.P.T., Toulouse, France.

A. PRASAD and C.H.K. WILLIAMSON (1996) - *The instability of the separated shear layer from a bluff body.* Physics of Fluids, (Letter), to appear, 1996.

C.H.K. WILLIAMSON (1988) - *Defining a universal and continuous Strouhal-Reynolds number relationship for the laminar vortex shedding of a circular cylinder.* Physics of Fluids, **31**, 2742.

C.H.K. WILLIAMSON (1988) - *Oblique and parallel modes of vortex shedding in the wake of a circular cylinder at low Reynolds numbers.* J. Fluid Mechanics, **206**, 579.

378

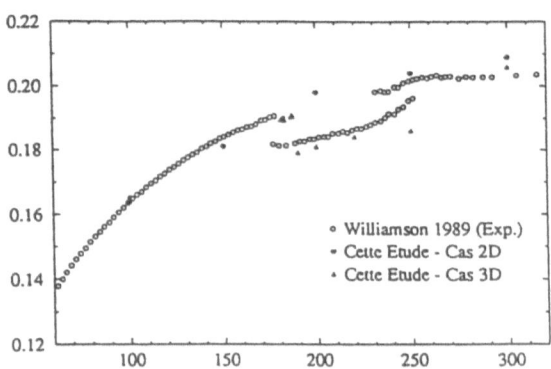

Figure 1. Strouhal versus Reynolds number.

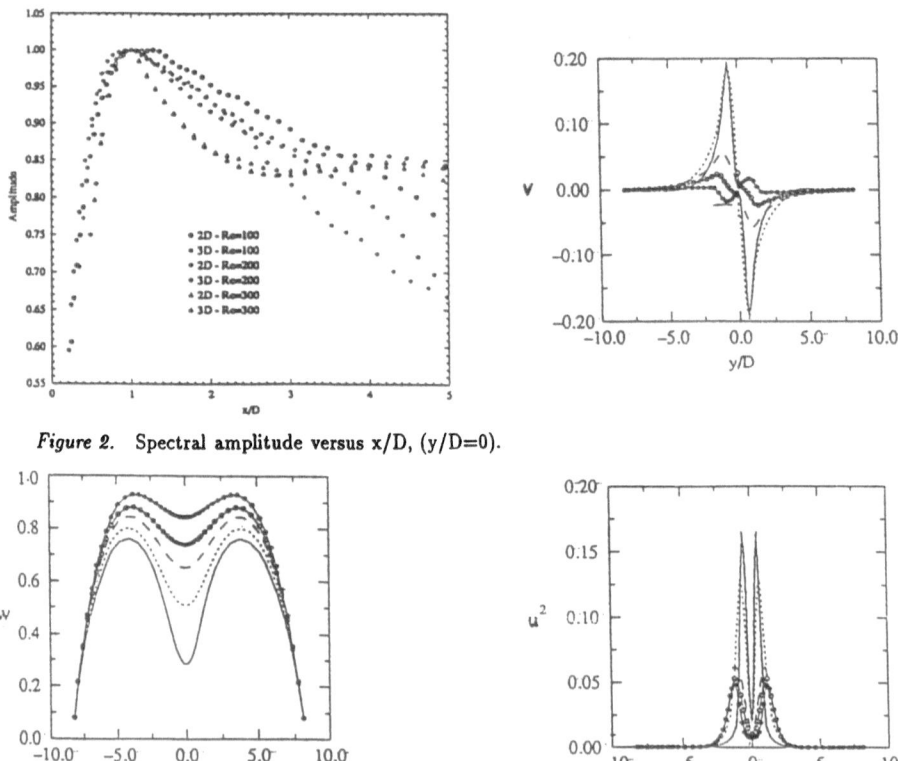

Figure 2. Spectral amplitude versus x/D, (y/D=0).

Figure 3. Mean v, w velocities and u^2 Reynolds stress.

WAKE TRANSITION IN COAXIAL JETS

H. REHAB, E. VILLERMAUX AND E. J. HOPFINGER
LEGI/IMG - CNRS
BP 53, 38041 Grenoble Cedex 9, France

A new flow pattern transition is observed in the near field of high Reynolds number coaxial jets. The configuration consists of a central jet of velocity u_1 surrounded by an annular jet of velocity u_2, and the study focuses on the limit of large velocity ratios $r_u = u_2/u_1$. For moderate velocity ratios (i.e. for $r_u < r_{uc} = 5$ typically), the fast annular jet which dominates the near flow field development pinches the central slow jet at the end of the inner potential core at a frequency corresponding to the outer Strouhal jet mode ($St = fD_2/u_2 = 0.35$). The structure of the near flow field is strongly dependent on r_u. It is observed experimentally that above the critical velocity ratio r_{uc}, the inner core breaks down into an unsteady recirculation bubble. The size of the recirculation region corresponds to the inner jet diameter D_1 and the velocity of the reverse flow is proportional to u_2 (figure 1). The transition mechanism is explained by a simple model whose ingredients are the turbulent entrainment rate, governed by the outer jet, and mass conservation. This model satisfactorily predicts r_{uc}. The transition to this recirculating regime results a wake type instability. The recirculation bubble oscillates periodically at a low frequency ($St = fD_1/u_2 = 0.035$) and a large amplitude relatively to Strouhal fundamental mode (figure 2). This low frequency mode persists until $x/D_1 \approx 3$ downstream. Angular cross-correlations in the plane parallel to the jet outlets show moreover that this oscillation diplays an azimuthal precession such that the rotation time of the phase of the oscillations is equal to the oscillation period.

S. Gavrilakis et al. (eds.), Advances in Turbulence VI, 379-380.
© 1996 *Kluwer Academic Publishers.*

380

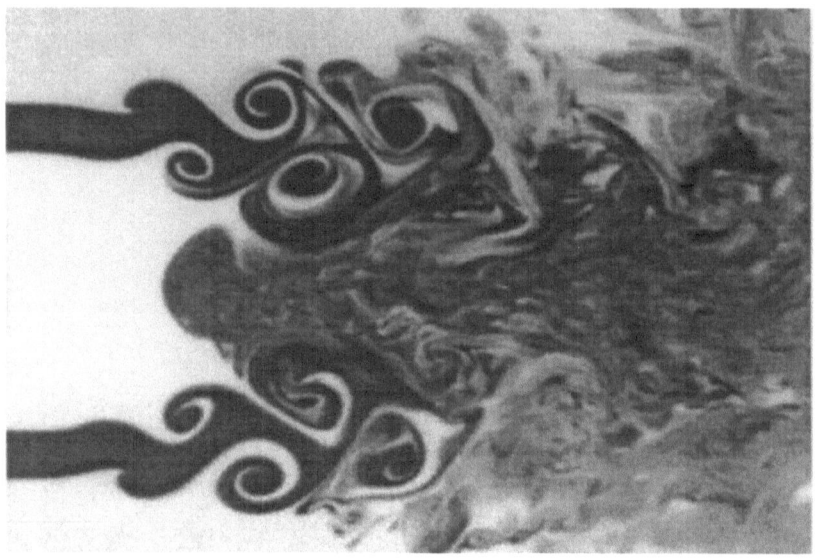

Figure 1. Plane cut along the jet axis. Instantaneous picture of the flow at $r_u \approx 8$.

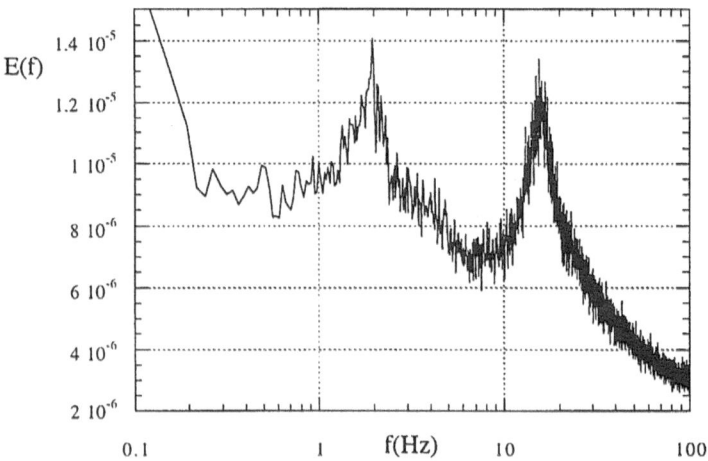

Figure 2. Velocity fluctuations spectrum measured at $x/D_1 \approx 2$ for $r_u = \infty$.

TRANSITION THRESHOLDS
IN BOUNDARY LAYER AND CHANNEL FLOWS

P.J. SCHMID
Department of Applied Mathematics, Box 352420
University of Washington, Seattle, WA 98195, U.S.A.

S.C. REDDY
Department of Mathematics
Oregon State University, Corvallis, OR 97331, U.S.A.

AND

D.S. HENNINGSON
Aeronautical Research Institute of Sweden (FFA)
Box 11021, S-16111 Bromma, Sweden

1. Introduction

Despite more than a century of work, transition to turbulence in boundary layer and channel flows is still not completely understood.

A transition scenario studied in detail is the secondary instability theory (Klebanoff, Tidstrom, & Sargent, 1962; Orszag & Patera 1983; Bayly, Orszag & Herbert, 1988):

$$\text{TS-wave} \rightarrow \text{nonlinear 2D-state} \rightarrow \text{secondary instability of 2D-state} \quad \textbf{(A)}$$

The initial disturbance is a finite-amplitude 2D Tollmien-Schlicting wave, the least stable mode of the linearized Navier-Stokes equations, plus noise. This wave evolves into a equilibrium or quasi-equilibrium state, which is unstable to three-dimensional disturbances. The theory agrees qualitatively and quantitately with experiments where a 2D disturbance is created using the vibrating ribbon technique. On the other hand, the theory may not explain natural transition, which is inherently three-dimensional.

In recent work on transition in channel and boundary layer flows the following scenarios have been investigated (Schmid & Henningson 1992; Kreiss, Lundbladh & Henningson 1994; Lundbladh, Reddy & Henningson, 1994; Berlin, Lundbladh & Henningson 1994):

$$\text{vortices} \rightarrow \text{streaks} \rightarrow \text{secondary instability of streaks} \quad \textbf{(B)}$$

S. Gavrilakis et al. (eds.), Advances in Turbulence VI, 381-384.
© *1996 Kluwer Academic Publishers.*

oblique waves → vortices → streaks → secondary instability of streaks

(C)

Scenario (B) is initiated by an array of streamwise vortices, periodic in the spanwise direction, plus noise. This disturbance has the greatest potential for linear transient growth in channel and boundary layer flows (Butler & Farrell 1992; Reddy & Henningson 1993). Streamwise vortices generate streaks by the lift-up mechanism. Streaks break down due to a spanwise inflectional instability (Yu & Liu, 1994; Waleffe 1995). In scenario (C) oblique waves interact nonlinearly to create streamwise vortices. The rest of the scenario is similar to (B). Noise is added to the initial disturbances to break symmetry, and is required for (B) for transition.

The purpose of this paper is to compare the threshold energy and time for transition to turbulence for the above three scenarios in the *temporal* case for plane Poiseuille flow (PPF) and the Blasius boundary layer (BL).

2. Numerical Simulations

Computations are done using a spectral Fourier-Chebyshev algorithm (Lundbladh, Henningson & Johansson 1992). The computational domain is assumed to be periodic in the streamwise and spanwise directions. In the boundary layer case, the growth of the boundary layer is modeled by translating the computational domain downstream with a representative velocity and adjusting the boundary layer thickness accordingly.

We define disturbance energy by $E = \frac{1}{2V} \int_{\text{period}} (u^2 + v^2 + w^2) d\mathbf{x}$, where the region of integration is one period and V is the volume of the periodic box. We define the time for transition T as the time at which the friction coefficient c_f reaches its mean value. If a turbulent state is not achieved, we define $T = \infty$. For PPF the Reynolds number is based on the half-channel height and the centerline velocity. For BL flow the Reynolds number is based on the displacement thickness and the freestream velocity.

We compute the initial TS waves by solving the Orr-Sommerfeld equation by a spectral method. The initial vortices and obliques waves in (B) and (C) are optimal disturbances (Butler & Farrell 1992; Reddy & Henningson 1993). Noise of 1% of the disturbance energy in the form of Stokes modes is added to the initial disturbance.

3. Results and Discussion

The results are shown in Figure 1. For both flows the lowest energy for transition is achieved by (C). For plane Poiseuille flow our results indicate that the threshold energy for transition for (A) is about 2 orders of greater for (A) than for (B) and (C). Scenarios (B) and (C) are three-dimensional from the outset and make use of linear transient growth mechanisms. In

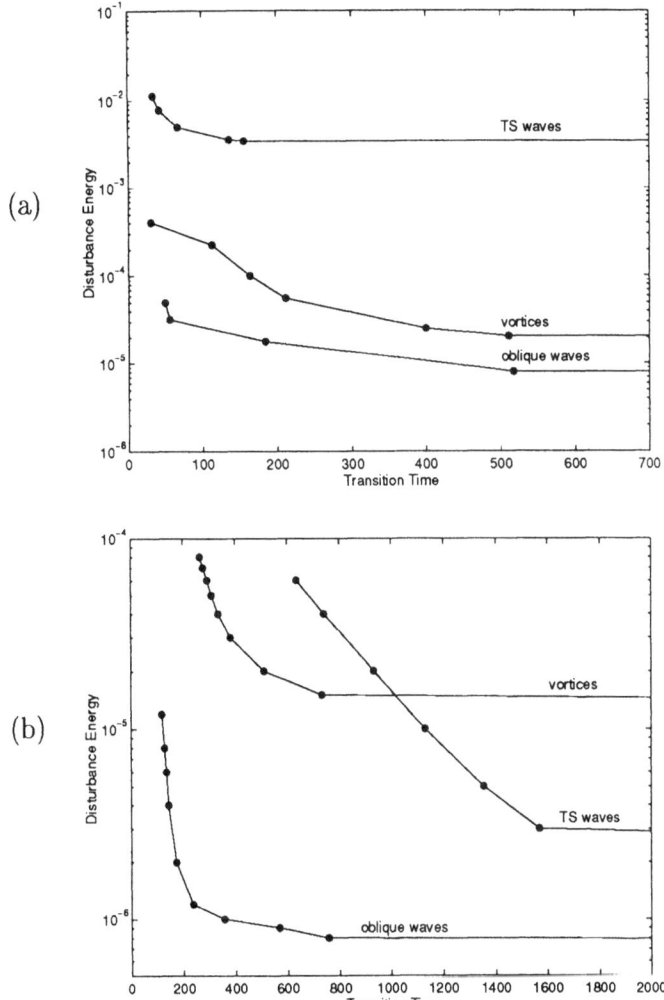

Figure 1. Initial disturbance energy versus transition time. (a) PPF with $R = 1500$. For (A) and (C), the fundamental streamwise and spanwise wavenumbers are $\alpha_0 = 1$ and $\beta_0 = 1$, respectively, and $\alpha_0 = 1$ and $\beta_0 = 2$ for (B). The initial disturbance has $\alpha = 1$ and $\beta = 0$ for (A), $\alpha = 0$ and $\beta = 2$ for (B), and $\alpha = 1$ and $\beta = \pm 1$ for (C). The results have been confirmed using three grids (highest resolution $32 \times 81 \times 64$.) (b) Temporally growing BL with initial Reynolds number 500 and frequency $F = 160$. For (B) and (C), the fundamental streamwise and spanwise wavenumbers are $\alpha_0 = 0.3$ and $\beta_0 = 0.325$, respectively, and $\alpha_0 = 0.2$ and $\beta_0 = 0.2$ for (A). The initial disturbance has $\alpha = 1$ and $\beta = 0$ for (A), $\alpha = 0$ and $\beta = 2$ for (B), and $\alpha = 1$ and $\beta = \pm 1$ for (C). The resolution has been chosen as $32 \times 65 \times 32$, and the advection speed of the computational domain has been 0.4.

the Blasius boundary layer, the secondary instability scenario is more competitive. The Reynolds number increases in the downstream direction. It is expected that scenario (A) can be even more competitive for lower frequen-

384

cies since the distance between branches I and II in the stability diagram increases as the frequency decreases.

We plan on investigating these scenarios in greater detail in future work.

Acknowledgements

PJS has been supported in part by NSF grant DMS-9406636. PJS and SCR gratefully acknowledge the hospitality of the Department of Mechanics, Royal Institute of Technology, Stockholm, Sweden, where part of this work was done. Some computations were done at the Pittsburgh Supercomputing Center.

References

Bayly, B.J., Orszag, S.A., Herbert, T. (1988) Instability mechanisms in shear flow transition, *Ann. Rev. Fluid Mech.*, **20**, pp. 359-391.

Berlin, S., Lundbladh, A. and Henningson, D.S. (1994) Spatial simulations of oblique transition in a boundary layer, *Phys. Fluids A*, **6**, pp. 1949-1951.

Butler, K.M. and Farrell, B.F. (1992) Three-dimensional optimal perturbations in viscous flows, *Phy. Fluids A*, **4**, pp. 1637-1650.

Klebanoff, P.S., Tidstrom, K.D. & Sargent, L.M. (1962) The three-dimensional nature of boundary-layer instability, *J. Fluid Mech.*, **12**, pp. 1-34.

Kreiss, G., Lundbladh, A. and Henningson, D.S. (1994) Bounds for threshold amplitudes in subcritical shear flows. *J. Fluid Mech.*, **270**, 175-198.

Lundbladh, A., Henningson, D.S. and Johansson, A.V. (1992) An efficient spectral integration method for the solution of the Navier-Stokes equations. *FFA Technical Report, FFA-TN 1992-28, Aeronautical Research Institute of Sweden, Bromma, Sweden.*

Lundbladh, A., Henningson, D.S. and Reddy, S.C. (1994) Threshold amplitudes for transition in channel flows, in: Hussaini, M.Y., Gatski, T.B. and Jackson, T.L. (eds.) *Transition ,Turbulence, and Combustion, Volume I*, pp. 309-318, Kluwer, Dordrecht, Holland.

Orszag, S.A. & Patera, A.T. (1983) Secondary instability of wall-bounded shear flows, *J. Fluid Mech.*, **128**, pp. 347-385.

Reddy, S.C. and Henningson, D.S. (1993) Energy growth in viscous channel flows, *J. Fluid Mech.*, **252**, pp. 209-238.

Schmid, P.J. and Henningson, D.S. (1992) A new mechanism for rapid transition involving a pair of oblique waves *Phys. Fluids A*, **4**, 1986-1989.

Spalart, P.R. and Yang, K.-S. (1987) Numerical study of ribbon-induced transition in Blasius flow, *J. Fluid Mech.*, **178**, pp. 345-365.

Yu, X & Liu, J.T.C. (1994) On the mechanism of sinuous and varicose modes in three-dimensional viscous secondary instability of nonlinear Görter vortices, *Phys. Fluids A*, **6**, pp. 736-750.

Waleffe, F. (1995) Hydrodynamic stability and turbulence: Beyond transients to a self-sustaining process, *Stud. Appl. Math.*, **95**, pp. 319-343.

EXPERIMENTAL STUDY OF THE STABILITY OF THE FLOW BETWEEN A ROTATING AND A STATIONARY DISK

L. SCHOUVEILER, P. LE GAL, M.P. CHAUVE and Y.TAKEDA*

Institut de Recherche sur les Phénomènes Hors Equilibre
UM 138, CNRS - Universités d'Aix-Marseille I & II
12, Avenue Général Leclerc,13003 Marseille, France

*Paul Scherrer Institute, CH-5232, Villigen PSI, Switzerland

1. Introduction

Different processes of laminar to turbulent transition are exhibited by the viscous flow between a rotating and a stationary disk according to the values of the aspect ratio and rotation speed. The various flow regimes, resulting from the successive losses of stability, are described here, using experimental techniques. Their domain of existence and stability are displayed in a transition diagram which completes the previous studies of San'kov et al. [1] and Sirivat [2].

2. Experimental details

2.1 APPARATUS

The experimental setup (Fig. 1.) consists of a horizontal rotating disk (radius $R = 140$ mm) set in a water-filled cylindrical housing with a sliding fit. The angular velocity of the disk ω can be continuously varied in the range $0 - 4\pi$ rd / s.

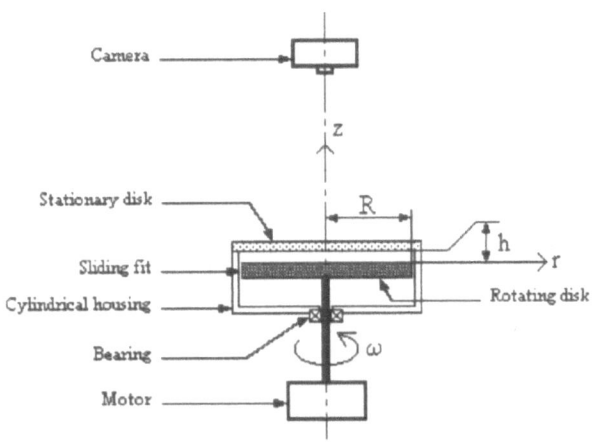

Fig. 1. Experimental apparatus

385

S. Gavrilakis et al. (eds.), Advances in Turbulence VI, 385-388.
© 1996 *Kluwer Academic Publishers.*

The top of the housing constitutes the stationary disk. The height of the rotating disk inside the housing is adjustable in such a way that the axial distance h between the two disks can be continuously varied from 0.125 to 20 mm.

The control parameters of the flow are the aspect ratio h/R, the Ekman number $Ek^{-1} = \omega.h^2/\nu$ (where ν is the kinematic viscosity of the working fluid) and the Reynolds number $Re = \omega.R^2/\nu$.

2.2 FLOW VISUALIZATION AND UDA TECHNIQUES

A visualization technique is used to observe the different patterns of the flow. Visualization is performed with a small amount of flake particles added to the working fluid. These particles have a high reflective index and they tend to align themselves along stream surfaces, thus becoming visible under appropriate lighting conditions (for further details see Savas [3]). The images are captured, through the stationary disk (made of Plexiglas), with a CCD camera (see Fig. 1.) and recorded. Then, they are digitized and processed on a micro-computer through a standard library of image processing and graphics functions.

The velocity field is characterized using an Ultrasonic Doppler Anemometer working with a pulsed ultrasonic emission. This device allows the measurement of an instantaneous profiles of the velocity component in the ultrasound beam direction, by detecting the Doppler shift frequency of ultrasound as a function of time. Details on this UDA method can be found in [4].

3. Results and discussion

3.1 BASIC FLOW

At low angular velocity, the laminar basic flow is axisymmetrical. Two types of basic flows are distingued [5]. When the aspect ratio is large ($h/R > 1.75 \ 10^{-2}$), the basic flow is of Batchelor type with a boundary layer on each disk separated by an inviscid rotating core in which the velocity is only tangential (Fig. 2.a). For the smaller $h/R < 1.75 \ 10^{-2}$, the two boundary layers are merged, the basic flow is purely viscous (Fig. 2.b) and tends, for $Ek^{-1} \rightarrow 0$, to the torsional Couette flow [1].

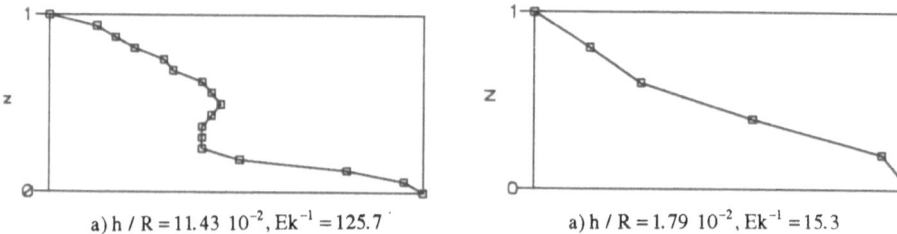

a) $h/R = 11.43 \ 10^{-2}, Ek^{-1} = 125.7$ a) $h/R = 1.79 \ 10^{-2}, Ek^{-1} = 15.3$

Fig. 2. Mean profile (UDA measures) of the velocity component onto a direction perpendicular to the radius with an angle of 20° versus the vertical at $r/R = 0.57$ (rotating and stationary disk resp. at $z=0$ and $z=1$)

3.2 FLOWS REGIMES AND STABILITY

Some of the flow patterns observed are constituted of well organized systems of vortices which appear like alternate dark and light bands on the visualizations. These vortices are corotating, their sense of rotation being imposed by the horizontal mean flow. Their axes form spirals in a plane parallel to the disks. The orientation of these wave systems is defined with the local angle ε between the vortex axis and the tangential direction, positive in the same rotation sense as the disk.

Mainly, two different processes of transition, associated with the two basic flow types, are observed.

For large aspect ratios $(h / R > 1.75 \ 10^{-2})$, the basic flow becomes unstable for $\omega_0 = f(h / R)$. From ω_0, we observe a regular system of travelling circular waves (Fig. 3.a). The circular vortices $(\varepsilon = 0)$ appear on the periphery and move towards the center. This mode of instability seems to be correlated to the disk rotation, since their frequency of formation is three times the rotation frequency. At a higher velocity $\omega_1 = f(h / R) > \omega_0$, a system of spiral waves $(\varepsilon > 0)$ forms in a peripherical region of the disk. The two systems exist together (Fig. 3.b).

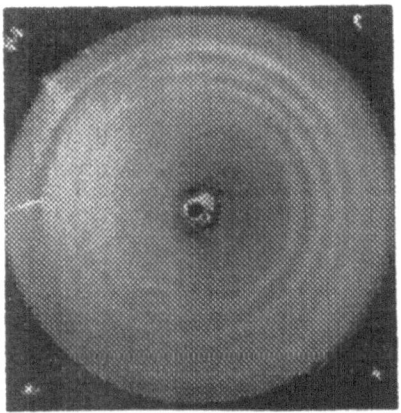 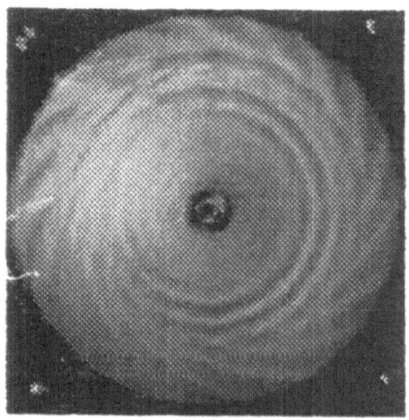

a) Re = 17241   b) Re = 23398

Fig. 3. Wave systems for h / R = 11.43 10^{-2} (Clockwise rotation)

We have observed, through Fourier analysis, non linear interactions between the two wave systems. When increasing the rotating velocity, more and more complex wave patterns appear and lead to turbulence.

For smaller aspect ratios $(h / R < 1.75 \ 10^{-2})$, the destabilization of the basic flow (at $\omega'_0 = f(h / R)$) occurs when the basic flow is purely viscous (see 3.1) and conducts to the formation of a regular array of spiral vortices with negative angle ε (Fig. 4.a). Then, a secondary instability arises when the angular velocity is further increased to $\omega'_1 = f(h / R)$. It appears in the form of travelling spiral-shaped solitary waves (to keep the terminology of [1]) constituted of a narrow band of turbulent flow (Fig. 4.b). These waves cross the primary system of vortices and their number and size increase with ω, until the flow is entirely invaded.

A last kind of structures is observed for intermediate values of the aspect ratio $(1.43\,10^{-2} < h/R < 2.5\,10^{-2})$, from $\omega'_3 = f(h/R)$. They consist in spots (called "rollers" in [1]) which are formed on the periphery and move towards the center, along spiral trajectories, crossing the different systems previously described.

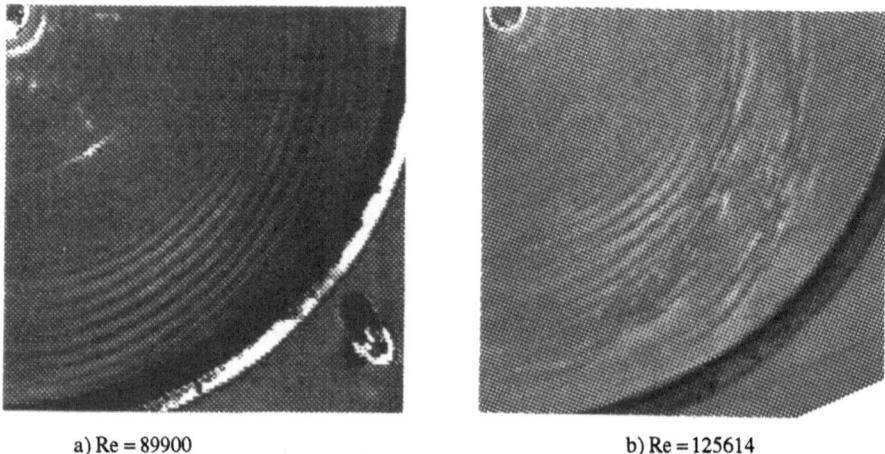

a) Re = 89900   b) Re = 125614

Fig. 4. Flow patterns $h/R = 1.43\ 10^{-2}$, center at the left hand top corner (clockwise rotation)

The boundaries of the flow regimes are displayed in the original transition diagram (Fig. 5.).

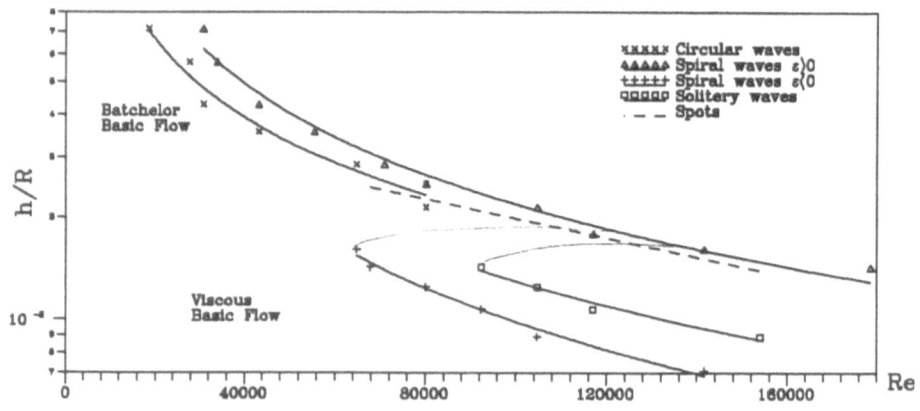

Fig. 5. Transition diagram

4. References

[1] San'kov, P. L. and Smirnov, E. M., Bifurcation and transition to turbulence in the gap between rotating and stationary parallel disks, *Fluid Dyn.* **19** (5) (1985) 695

[2] Sirivat, A., Stability experiment of flow between a stationary and a rotating disk, *Phys. Fluids A* **3** (11) (1991) 2664

[3] Savas, Ö., On flow visualization using reflective flakes, *J. Fluid Mech.* **152** (1985) 235

[4] Takeda, Y., Fischer, W. E., Sakakibara, J. and Ohmura, K., Experimental observation of the quasiperiodic modes in a rotating Couette system, *Phys. Rev. E* **47** (1993) 4130

[5] Daily, J. W. and Nece, R. E., Chamber dimension effects on induced flow and frictional resistance of enclosed rotating disk, *J. Basic Eng.* **82** (1960) 217

DIRECT NUMERICAL SIMULATION OF HIGH RAYLEIGH NUMBER CONVECTION IN A ROTATING AND NON-ROTATING SPHERICAL SHELL: THE PRANDTL NUMBER DEPENDENCE

A. TILGNER AND F.H. BUSSE

Physikalisches Institut, Universität Bayreuth, 95440 Bayreuth, Germany

Thermal convection in a self gravitating spherical fluid shell with a ratio of inner to outer radius of 0.4 is investigated by direct numerical simulation with a pseudospectral method. In the non-rotating case the Rayleigh number (Ra) has been varied for the Prandtl numbers $Pr=0.1, 1, 10$ over nearly 3 decades. At $Pr = 1$ the Reynolds number of the flow varies as $Ra^{1/2}$ and the Nusselt number Nu as $Ra^{0.23}$ for $Ra > 10^4$. The former scaling thus agrees with the one observed in experiments on low temperature gaseous helium, whereas the latter differs from the 2/7 law [1]. The square root of the horizontal area at the mean radius divided by the thickness of the shell is 4.1 in the simulation, so that the numerical results should be compared to experimental data obtained in large aspect ratio cells. The 2/7 scaling was reported for $Ra > 4 \cdot 10^7$ in cells of aspect ratio 0.5 and 1, but could be observed at values of Ra as low as $5 \cdot 10^3$ in a cell with aspect ratio 6.7 [2].

The Pr dependence is studied in more detail at a fixed Ra which lies well inside the scaling regime. The Pr dependence of the kinetic energy shows a kink at $Pr = 1$. Nu rises as a function of Pr for $Pr < 1$, goes through a maximum at $Pr \approx 1$ and reaches a constant level at higher Pr (see fig. 1). For $Pr < 1$, the viscous boundary layer lies within the thermal boundary layer. The mixing length theory by Kraichnan [3] takes the ordering of the two layers explicitly into account. The behavior shown in fig. 1 is qualitatively reproduced by this theory. More recent theories aimed at explaining the 2/7 law predict different Pr dependences [4, 5]. The Pr dependence is therefore a sensitive property for the test of different models of turbulent convection. Numerical simulation can play an important role in this respect since Pr can easily be varied in numerical work whereas

S. Gavrilakis et al. (eds.), Advances in Turbulence VI, 389-390.

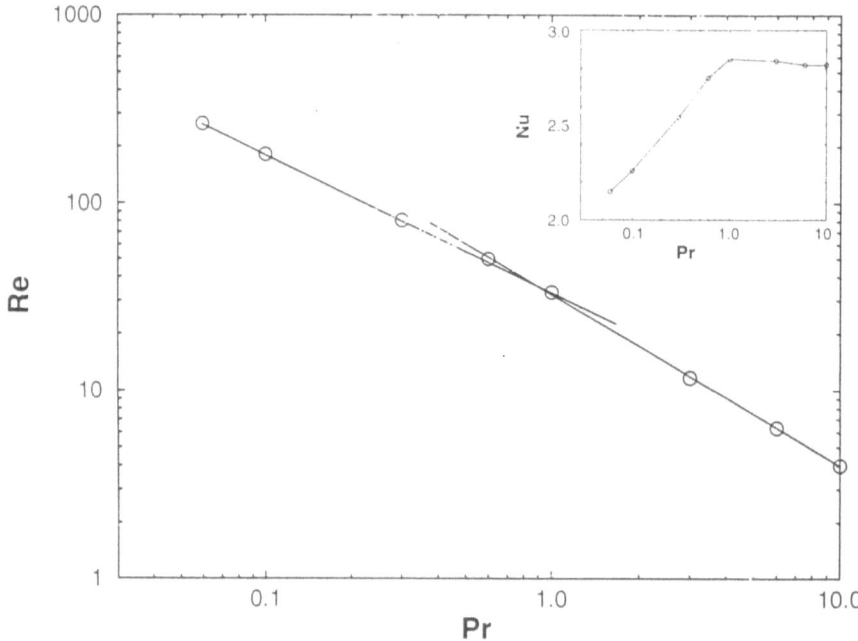

Figure 1. Re as a function of Pr at $Ra = 8 \cdot 10^4$. The straight lines correspond to exponents -0.73 (low Pr) and -0.92 (high Pr). The inset shows the variation of Nu.

experiments have to rely on available fluids. Existing experimental data is compatible with the general picture provided by fig. 1 [6].

Thermal convection has also been studied for a rotating shell at Taylor numbers 10^6 and 10^7 with free slip boundary conditions and $Pr = 0.01, 0.1, 1,$ and 10. The dominance of the Coriolis force in the rotating system modifies the scaling laws observed in the non rotating shell. For instance, a sharp transition is observed in the $Nu(Ra)$ dependence at all Pr. Changes in the flow structure occur around this transition which however vary for different Pr.

References

1. M. Sano, X.Z. Wu and A. Libchaber, Phys. Rev. A **40**, 6421 (1989)
2. X.Z. Wu and A. Libchaber, Phys. Rev. A **45**, 842 (1992)
3. R.H. Kraichnan, Phys. Fluids **5**, 1374 (1962)
4. B. Castaing, G. Gunaratne, F. Heslot, L. Kadanoff, A. Libchaber, S. Thomae, X.Z. Wu, S. Zaleski and G. Zanetti, J. Fluid Mech. **204**, 1 (1989)
5. B.I. Shraiman and E.D. Siggia, Phys. Rev. A **42**, 3650 (1990)
6. A. Belmonte, A. Tilgner and A. Libchaber, Phys. Rev. E **50**, 269 (1994); S. Globe and D. Dropkin, J. Heat Transfer **81**, 24 (1959)

EXPERIMENTS ON ROTATING PLANE COUETTE FLOW

N. TILLMARK & P.H. ALFREDSSON
Department of Mechanics, Royal Institute of Technology
S-100 44 Stockholm, Sweden

1. Introduction

System rotation may drastically change the flow behavior both for laminar and turbulent shear flows due to the effect of the Coriolis force. For plane Poiseuille flow the Coriolis force acts in such a way that in the part of the channel where the system rotation has opposite sign compared to the mean flow vorticity the flow becomes destabilized whereas in the other part of the channel it becomes stabilized. Alfredsson & Persson (1989) showed that the instability takes the form of streamwise oriented roll cells and demonstrated excellent agreement between linear theory and experiments.

For plane Couette flow with system rotation, the Coriolis force will either be stabilizing or destabilizing across the full channel width (figure 1). Linear stability theory (Lezius & Johnston, 1971) shows that the critical Re is 20.65 ($Re = U_w h/\nu$) and the related critical rotation number $Ro = 0.5$ ($Ro = 2\Omega h/U_w$). The corresponding critical wavenumber is $\beta = 1.56 h^{-1}$. It can also be shown that the flow becomes stabilized for $Ro > 1$. Figure 2 shows the region of destabilization for laminar Couette flow.

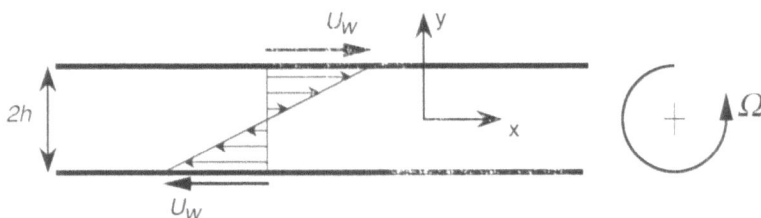

Figure 1. Laminar rotating plane Couette flow with destabilizing rotation

For non-rotating plane Couette flow the transitional Reynolds number, i.e. the Reynolds number below which there is no self sustained turbulence,

S. Gavrilakis et al. (eds.), Advances in Turbulence VI, 391-394.
© 1996 *Kluwer Academic Publishers.*

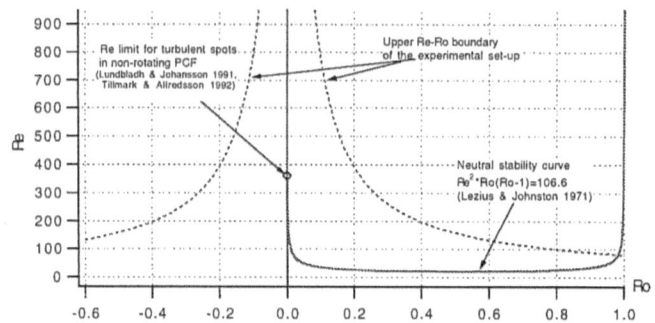

Figure 2. Neutral stability curve and upper limit for present experimental set-up.

has been shown to be around 360 both experimentally (Tillmark & Alfredsson, 1992) and numerically (Lundbladh & Johansson, 1991). Komminaho, Lundbladh & Johansson (1996) showed that turbulent plane Couette flow can be stabilized by negative rotation. For $Re = 750$ they found the fully turbulent flow to relaminarise at $Ro = -0.060$.

2. Experimental apparatus

In the present study we have used the same equipment as in our previous experiments though this time the apparatus is placed on a rotating turntable. In short, the equipment is a transparent 1.5 meter long plane Couette channel with vertical glass walls and a moving plastic belt covering the facing surfaces, i.e. both walls are moving in opposite direction. The distance between the glass walls is 20 mm. The working fluid is water, seeded with light reflecting platelets.

3. Results

A major objective of the present study was to get a general overview of the different flow regimes within the $Re - Ro$ region covered by the apparatus. These include flow behaviour close to the neutral curve (both at high and low Ro), secondary instabilities of the roll cells, the region of self sustained turbulence as well as the effect of rotation on fully developed turbulence.

3.1. EFFECTS OF ROTATION ON LAMINAR FLOW

At Reynolds numbers close to the neutral stability curve primary instabilities in the form of stationary elongated streamwise roll-cells with a spanwise width of approximately $2h$ appear in the flow. When Ro and Re are slightly increased secondary instabilities set in. Figure 3a,b shows, at two different times, the flow at a fairly low Re ($Re = 200$) as the channel is starting to rotate at $Ro = 0.01$. The flow sets up almost stationary elongated streamwise roll-cells, figure 3a, with a spanwise width of approximately $2h$. However, for this Re the roll-cells eventually become wavy, figure 3b, and finally break down. The formation of roll-cells and their subsequent break down is cyclic with a period time of approximately $3\Omega^{-1}$.

As the rotation rate increases the sinusoidal undulation of the roll-cells ceases. The roll-cells become more pronounced and merging and splitting occur in more or less stationary dislocations. The average spanwise width is $2h$ but with large variations. Further increase of the rotation rate causes a secondary instability in the form of a spiral pattern on the roll-cells. As the rotation is increased above $Ro = 1$ linear theory predicts stabilization which is verified in the experiments as both the primary and secondary instabilities vanish and the flow becames fully laminar (for all accessible Reynolds numbers, $Re < 80$).

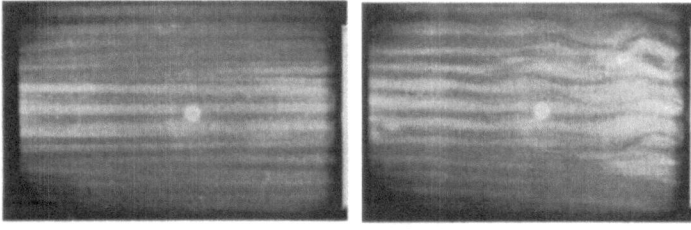

Figure 3.   a,b: $Re \approx 200$, $Ro \approx 0.01$

3.2. EFFECTS OF ROTATION ON TURBULENT FLOW

For $Ro < 0$ the Coriolis force stabilises the flow and the transition Reynolds number increase as the rotation becomes more negative. To determine the transition Re we start from a fully turbulent flow at a fixed Re, gradually increasing the Ro until the turbulence in the flow has vanished except in the inlet regions. Figure 4 shows the relation between transition Re and Ro for Reynolds numbers up to 900 indicating a linear relation between these quantities. Experiments also confirm the transition Re found in the numerical simulation by Komminaho et al. (1996). When the rotation is reversed the Coriolis force destabilizes the flow and the transition Reynolds number decreases further but for $Re \leq 200$ turbulence is not sustainable at any rotation rate.

At Reynolds numbers well above 200 and at moderate rotation rates as in figure 5 ($Re \approx 700, Ro \approx 0.1$) the turbulent flow contains regularly spaced streamwise roll-cells extending all over the entire channel length. The cellular structure becomes more pronounced the stronger the rotation.

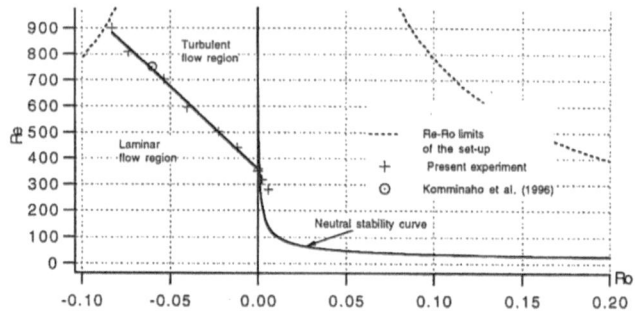

Figure 4. Experimental data points demarking regions of laminar flow and self sustained turbulence

4. Final remarks

The results described show that rotating plane Couette flow exhibits a rich variety of flow phenomena, some of which has not been observed in other flow situations, such as the relaminarization for stabilizing rotation. This and other phenomena will be reported in the near future.

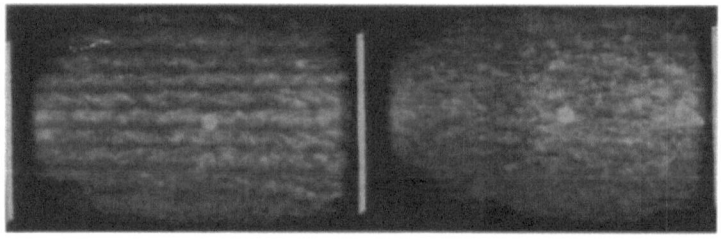

Figure 5. a: $Re \approx 700$, $Ro \approx -0.1$; b: $Re \approx 700$, $Ro = 0$.

Financial support from the Swedish Research Council for Engineering Sciences is gratefully acknowledged.

References

Alfredsson, P.H. & Persson, H. 1989 Instabilities in channel flow with system rotation. *J. Fluid Mech.* **202**, 543-557.

Komminaho, J., Lundbladh, A. & Johansson, A. 1996 Very large structures in plane turbulent Couette flow. (Accepted for publication in *J. Fluid Mech.*)

Lundbladh, A. & Johansson, A. 1991 Direct simulation of turbulent spots in plane Couette flow. *J. Fluid Mech.* **229**, 499-516.

Lezius, D.K. & Johnston, J.P. 1971 The structure and stability of turbulent wall layers in rotating channel flow. Report MD-29 Thermoscience Division, Dept. Mech. Engineering, Stanford University.

Tillmark, N. & Alfredsson, P.H. 1992 Experiments on transition in plane Couette flow. *J. Fluid Mech.* **235**, 89-102.

NATURAL TRANSITION IN AN ADVERSE PRESSURE GRADIENT BOUNDARY LAYER

B.F.A. VAN HEST, D.M. PASSCHIER AND R.A.W.M. HENKES
Faculty of Aerospace Engineering
Delft University of Technology
Delft, the Netherlands

1. Introduction

The paper is devoted to an experimental study of the *natural* transition process in boundary layers subjected to an *adverse* pressure gradient and low free stream turbulence intensity. The study was performed to obtain several goals. The first was to extend the available data-set of transitional boundary layers under adverse pressure gradient with detailed measurements of the streamwise and wall–normal velocity components (mean values and Reynolds stresses), velocity spectra, and measurements of the intermittency across the boundary layer and in the streamwise direction. The velocity spectra were used to verify the applicability of linear stability theory, (especially with respect to the parallel-flow assumption with fast boundary layer growth) and the possible effects of nonlinear interaction (because of the large amplification rates) on transition prediction. The measurements of the intermittency were used to verify available empirical relations, like Klebanoff's (1956) and Narasimha's (1957) intermittency distributions, which are often used in models for the flow in the transition region.

2. Experimental setup

The measurements were conducted on the flat test wall of a closed return wind tunnel. The flexible wall opposite to the test wall was set to impose a near–neutral pressure gradient over the first $750\,mm$ to delay transition onset and obtain a thick boundary layer, followed by an adverse pressure gradient close to Hartree $\beta = -0.18$. The unit Reynolds number was $0.67 \times 10^6\,m^{-1}$ ($U_{x=170} \approx 10.4\,m/s$) and free stream turbulence $Tu = 0.1\,\%$. A miniature cross hotwire ($l = 0.5\,mm$, $l/d = 200$) connected to a constant temperature anemometer was used to measure the streamwise and wall–normal components of the velocity.

3. Tollmien-Schlichting wave development

The TS–wave amplitude distribution across the boundary layer (shown for the frequency band of the most amplified TS–wave in figure 1) shows two near wall maxima. The second maximum is not predicted by linear stability theory (solid lines). Because for natural transition not only 2D but also 3D disturbances enter the boundary layer through the

395

S. Gavrilakis et al. (eds.), Advances in Turbulence VI, 395-398.

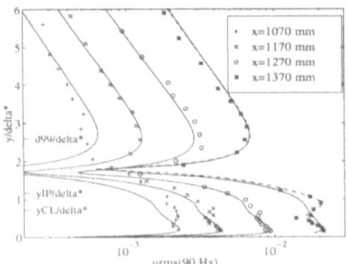

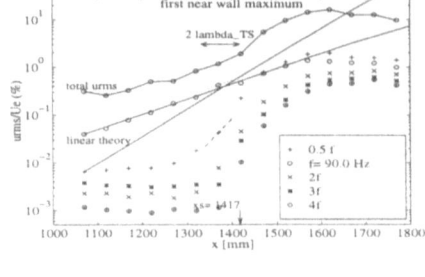

Figure 1. Measured and calculated TS–wave amplitude distributions.

Figure 2. Streamwise growth of TS–waves and harmonics at first near–wall maximum.

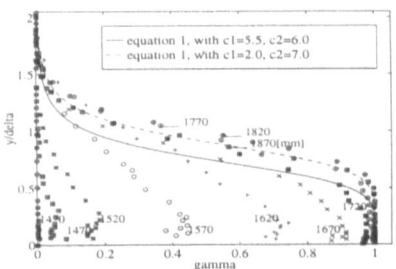

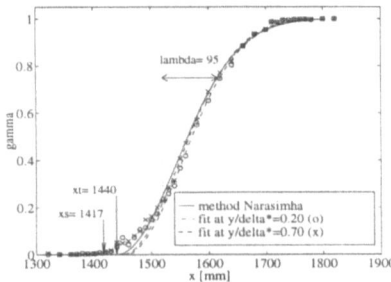

Figure 3. Intermittency distribution across the boundary layer in the transition region and initial turbulent boundary layer.

Figure 4. Intermittency in streamwise direction, measured at the wall distance of both near–wall maxima.

receptivity mechanism, the measured amplitude distribution is a combination of these disturbances. The solution of the Orr–Sommerfeld equation for 3D–waves indeed shows that waves with propagation angles up to $\phi \approx 40°$ are amplified in the present case (calculations by Rist (1996)). Figure 1 also shows the 3D–eigenfunction with $\phi = 19°$ (dotted line), and the combination of this mode with the 2D eigenfunction (dashed line), which closely resembles the experimental amplitude distribution. Figure 2 shows the growth of the most amplified Tollmien–Schlichting wave in streamwise direction. The agreement with linear stability theory is fair, which indicates that non–parallel effects can be neglected even for a flow with an adverse pressure gradient. It can be observed that approximately two TS–wavelengths upstream of transition onset the subharmonic and higher harmonic frequencies start to grow rapidly. Because the subharmonic amplitude becomes larger than that of the fundamental frequency ($f = 90\,Hz$) only after transition onset (at $x_s = 1417\,mm$), it is concluded that empirical methods of transition prediction (e.g. the e^n–method) based on linear theory can be used to predict transition onset. From a bispectral analysis (see e.g. (Corke , 1987)) it has been shown ((Hest et al., 1996)) that at the first near–wall maximum a strong phaselocking exists between two fundamental frequencies, with the appearance of higher harmonics at subsequent downstream positions, while at the second near–wall maximum phaselocking is found between the fundamental frequency and its subharmonic.

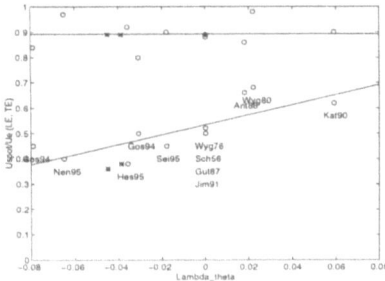 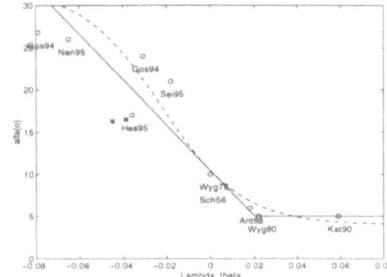

Figure 5. Spot leading and trailing edge propagation velocities (left graph) and spanwise growth rate (right graph) as a function of pressure gradient. The asterix ($*$) denotes values obtained from the present natural transition experiments.

From these results it is assumed that in the present case the nonlinear development of the Tollmien-Schlichting waves results in a combined resonance, i.e. both a subharmonic and fundamental secondary instability take place.

4. The intermittency distribution

Breakdown of the Tollmien-Schlichting waves to turbulence occurs at the second near-wall maximum. The turbulent spots were detected by comparing a detection function, taken as the short time average of $|d^2u/dt^2|$, with a threshold level. Next, the velocity record was divided into laminar and turbulent regions. The intermittency, i.e. the fraction of time the flow is turbulent, across the boundary layer (figure 3) remains maximum at this wall distance at the different streamwise positions in the transitional boundary layer. The intermittency distribution in the turbulent boundary layer does not follow Klebanoff's (1956) relation, given by

$$\gamma(y) = \left[1 + c_1 \left(\frac{y}{\delta}\right)^{c_2}\right]^{-1}$$
(1)

with $c_1 = 5.5$ and $c_2 = 6$. These constants are valid for turbulent boundary layer flow with zero pressure gradient. The difference between the experimental values and this relation may be due to the fact that the turbulent boundary layer in the present case is not completely developed as it has become completely turbulent just a short distance upstream. When the discrepancy is due to the effect of the adverse pressure gradient, a change in constants is proposed for the present case, with new parameters given by $c_1 = 2.0$ and $c_2 = 7.0$. Through the transition region, the intermittency increases from zero at the onset of transition to one in the turbulent boundary layer. The streamwise intermittency (figure 4) is found to follow Narasimha's (1957) universal intermittency distribution, given by

$$\gamma(x) = 1 - \exp\left[-n_t \sigma \frac{(x - x_t)^2}{U_e}\right]$$
(2)

with $n_t \sigma = 0.411 \frac{U_e}{\lambda^2}$, x_t the onset of transition and $\lambda = x_{\gamma=0.75} - x_{\gamma=0.25}$ is a characteristic parameter for the length of the transition region. A fit of equation 2 to the experimental data with $n_t \sigma$ given by a relation for the spot formation and growth, $n_t \sigma =$

$n_t \tan \alpha (U_e/U_{te} - U_e/U_{le})$ (see Chen & Thyson, 1971), results in a spot formation rate $n_t = 1.05 \times 10^3 \, m^{-1} s^{-1}$, a spanwise growth $\alpha = 16.5°$ and trailing edge velocity[1] $U_{te}/U_e = 0.38$. These values agree well with results from triggered turbulent spot experiments for the same local pressure gradient $\lambda_\theta = \theta^2/\nu \, dU/dx$ (see figure 5 (Hest *et al.*, 1995)). This result gives confidence in models for the length of the transition region based on triggered spot experiments.

5. Conclusions

Considering the development of the Tollmien–Schlichting waves and the transition process, the following conclusions can be made

- linear stability theory provides a good prediction of the TS–wave growth.
- the second near–wall maximum in the TS–wave amplitude distribution seems to be caused by a combination of 2D and 3D TS–waves.
- the secondary instability is of a combined resonance type, with fundamental resonance found at the first near–wall maximum and subharmonic resonance at the second maximum.
- breakdown to turbulence first occurs at the second near–wall maximum.
- the normal intermittency distribution needs a modification ($c_1 = 2.0, c_2 = 7.0$ for present case), which is maybe due to the non–equilibrium state of the turbulent boundary layer.
- the streamwise intermittency distribution agrees with Narasimha's universal distribution, and provides spot propagation parameters in agreement with triggered spot experiments.

References

Corke, T.C. (1987) Measurements of resonant phase locking in unstable axisymetric jets and boundary layers, In *Non-linear Wave Interactions in Fluids*, editors Miksad, R.W., Akylas, T.R., and Herbert, T., editors, **Vol. no. 87**, pp. 37–65.

Hest, B.F.A. van, Groenen, H.F., and Passchier, D.M. (1996) Non-linear development and breakdown of Tollmien-Schlichting waves in an adverse pressure gradient boundary layer, In *Transitional Boundary Layers in Aeronautics*, editors Henkes, R.A.W.M. and van Ingen, J.L., K.N.A.W colloquium Amsterdam, the Netherlands.

Hest, B.F.A. van, Passchier, D.M., and Ingen, J.L. van (1995) The development of a Turbulent Spot in an Adverse Pressure Gradient Boundary Layer, In IUTAM symposium on *Laminar-Turbulent Transition*, editors Kobayashi, R., editor, pp. 255–262.

Klebanoff, P.S. (1956) Characteristics of Turbulence in a Boundary Layer with Zero Pressure Gradient, *Technical Report* **TN 3178**, NACA.

Narasimha, R. (1957) On the distribution of intermittency in the transition region of a boundary layer, *J. Aero. Sci.*, **24**, pp. 711–712.

Rist, U. (1996) I.A.G., Universität Stuttgart, Germany (private communication).

[1]It has been shown that the leading edge velocity ($U_{le}/U_e = 0.89$) is almost constant.

THREE-DIMENSIONAL WAKE TRANSITION.

C.H.K.WILLIAMSON

Mechanical and Aerospace Engineering, Cornell University
Ithaca, NY 14853, U.S.A.

1.Introduction

It is now well-known that the wake transition regime for a circular cylinder involves two modes of small-scale three-dimensional instability (modes "A" and "B"; Williamson, 1988), depending on the regime of Reynolds number (Re), although almost no understanding of the physical origins of these instabilities, or indeed their effects on near wake formation, have hitherto been made clear. There is now some strong interest in this problem, coming not only from experiment, but also from Direct Numerical Simulation, where, in some cases, these modes A and B have been found clearly (Thompson & Hourigan, 1996; Zhang et al., 1995; Henderson, 1995; Mittal & Balachandar, 1996). Much of the recent surge of activity concerning the wake transition and development of turbulence in wakes has been addressed comprehensively in a review paper. Williamson (1996a).

2.   Wake transition

The wake transition regime (Re=190-260) is characterised by two distinct three-dimensional modes of instability, as shown experimentally (Williamson, 1988, 1992, 1996b, 1996c; Wu et al., 1994; Zhang et al. 1995; Mansy et al., 1994). "Mode A" instability has a spanwise wavelength of 3-4 diameters, as visualized in Figure 1, whereas "Mode B" instability has a wavelength of close to 1 diameter. These data are indicated in Figure 2(a). In the presentation, we show that both modes involve the generation of streamwise vortex pairs in the wake, which reside and are stretched in the streamwise direction in the "braid" regions between primary Karman vortex structures. Such stretching of streamwise vorticity appears to be a generic (and well-known) feature in transition and development of turbulence in free shear layers.

We demonstrate that the incipience of wake transition is triggered early by contamination in the form of vortex dislocations coming from the ends of the body (Miller & Williamson, 1994). Over the last forty years a range of critical Reynolds numbers of Re=140-190 has been reported. Noack & Eckelmann (1994), using an approximate Floquet stability analysis, also find a critical Re_{cri}=170. However, by minimising any end contamination in the experiments, one finds Re_{cri}=194. which is remarkably close to the results from the recent Floquet stability analysis of Barkley & Henderson (1996) and Henderson & Barkley (1996), which find Re_{cri}=189. In this presentation, we find that the presence of interfering vortex dislocations (_not_ caused by end effects) causes the large scatter in previous measurements of spanwise wavelength for the vortex loops of mode A instability. It is only by careful and accurate measurements,_without_ the presence of dislocations, that one finds a clear trend of decreasing wavelength with Re, which is quite distinct from the assumptions of constant wavelength in previous studies. The present data yield an excellent agreement with the curve of maximum growth rate from the secondary stability analysis of Barkley & Henderson, as shown in Figure 2(b). The marked disparity in spanwise wavelength, and

S. Gavrilakis et al. (eds.), Advances in Turbulence VI, 399-402.
© 1996 _Kluwer Academic Publishers._

visual appearance, between modes A and B suggests that the modes are due to distinct physical instabilities, which are demonstrated comprehensively in Williamson (1996b).

The wake transition regime can also be characterised by velocity and pressure measurements, and the inception of the different modes of instability, along with the presence of dislocations yield discontinuities in the S-Re relationship, the first of which is hysteretic while the second involves a gradual transfer of energy between modes. The presence of the dislocations has an important impact on the flow measurements; it causes a discontinuous reduction in Strouhal frequency, and large levels of turbulent fluctuations, which decay only slowly downstream. The time traces and spectra of wake fluctuations, shown in Figure 3, show that most of the downstream energy resides in the low-frequency fluctuations, coming from the large-scale dislocations. *Energy levels downstream can differ by a factor of 100 depending on whether dislocations are present or not!* These dislocations are two-sided, in the manner described in Williamson (1992), rather than the one-sided structures caused by end effects. One might question from where do these "two-sided" dislocations originate ? We show here that they are triggered at the sites of vortex loops (see Figure 4), and are not a manifestation of end effects or of experimental artifact.

3.   Conclusions

It is deduced in this work that the large variation in previous measurements concerning mode A secondary instability are due to the presence of vortex dislocations.   In the absence of such dislocations, we find an excellent agreement of the critical Re as well as spanwise wavelength of mode A with the linear secondary stability analysis of Henderson & Barkley. We further demonstrate that these large-scale dislocations in wake transition are triggered at the sites of some of the vortex loops for mode A; they are an intrinsic feature of transition, independant of end conditions.   These studies lead us to a new clarification of the possible flow states through wake transition, as follows.   If one defines a Mode A* as (Mode A + Dislocations), then the route through transition appears to follow the scenario of wake modes:

$$(2D \rightarrow A \rightarrow A^* \rightarrow B)$$

Acknowledgements: The support from the Ocean Engineering Division of the O.N.R., monitored by Tom Swean, is gratefully acknowledged. (O.N.R.Contract No. N00014-94-1-1197 and N00014-95-1-0332).

4.   References

Barkley, D. & Henderson, R.D.   1996 Three-dimensional Floquet stability analysis of the wake of a circular cylinder. Accepted for *J. Fluid Mech.*.

Henderson, R.D. 1995   Private communication.

Henderson, R.D. & Barkley, D. 1996 Secondary instability in the wake of a cylinder. Accd for *Phys.Fluids*.

Mansy, H., Yang, P. & Williams, D.R.   1994   Quantitative measurements of spanwise-periodic three-dimensional structures in the wake of a circular cylinder. *J. Fluid Mech..*, **270**, 277.

Miller, G.D. & Williamson, C.H.K. 1994   Control of three-dimensional phase dynamics in a cylinder wake. *Experiments in Fluids.*, **18**, 26.

Mittal, R. & Balachandar, S. 1996   Generation of streamwise vortical structures in wakes. *Phys. Rev. Lett.*.

Noack, B.N. & Eckelmann, H. 1994 A global stability analysis of the periodic wake. *J. Fluid Mech.* **270**, 297.

Thompson, M., Hourigan, K. & Sheridan, J. 1996   Three-dimensional instabilities in the wake of a circular cylinder.   To appear in *Experimental, Thermal and Fluid Science*.

Williamson, C.H.K. 1988 The existence of two stages in the transition to three-dimensionality of a cylinder wake. *Phys. Fluids*, **31**, 3165.

Williamson, C.H.K. 1992 The natural and forced formation of spot-like "vortex dislocations" in the transition of a wake. *J. Fluid Mech.* **243**, 393.

Williamson. C.H.K. 1996a Vortex dynamics in the cylinder wake. *Annual Review of Fluid Mech.*, **28**, 477.

Williamson, C.H.K. 1996b Three-dimensional wake transition. Accepted for *J. Fluid Mech.*

Williamson. C.H.K. 1996c Mode A secondary 3D wake instability. Accepted for *Phys. Fluids*.

Wu, J., Sheridan, J., Welsh, M.C., Hourigan, K. & Thompson, M. 1994 Longitudinal vortex structures in a cylinder wake. *Phys. Fluids*, **6**, 2883.

Yang, P., Mansy, H. & Williams, D.R. 1993 Oblique and parallel wave interaction in the near wake of a circular cylinder. *Phys. Fluids*, A **5**, 1657.

Zhang, H., Fey, U., Noack, B.R., Koenig, M. & Eckelmann, H. 1995 On the transition of the cylinder wake. *Phys. Fluids*, **7**, 1.

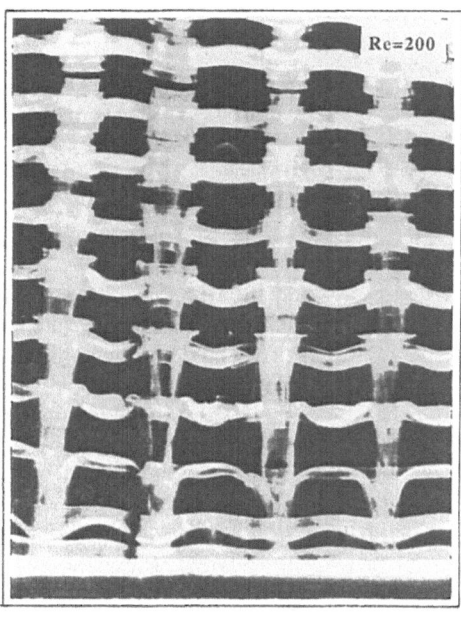

Figure 1.   Mode "A" three-dimensional instability.
This mode is associated with the inception of streamwise vortex loops. This specific example for Re=200, corresponds with a spanwise wavelength: λ/D=4.01, which is remarkably close to the maximum growth rate from Floquet analysis, Barkley & Henderson (1995).

Figure 2.   Spanwise instability wavelengths of the two 3-D instabilities.
Normalised spanwise wavelength of streamwise vortex structures (λ_Z/D), versus Re. It can be seen that there are two distinct wavelengths for modes A and B instabilities. The lower plot comprises only part of the collected data of the upper plot, and compares experimental data for mode A instability wavelengths with the Floquet analysis of Barkley & Henderson (1995). It is significant that in the latter more-accurate experimental data, measurements are made for purely mode A, in the absence of interfering dislocations.

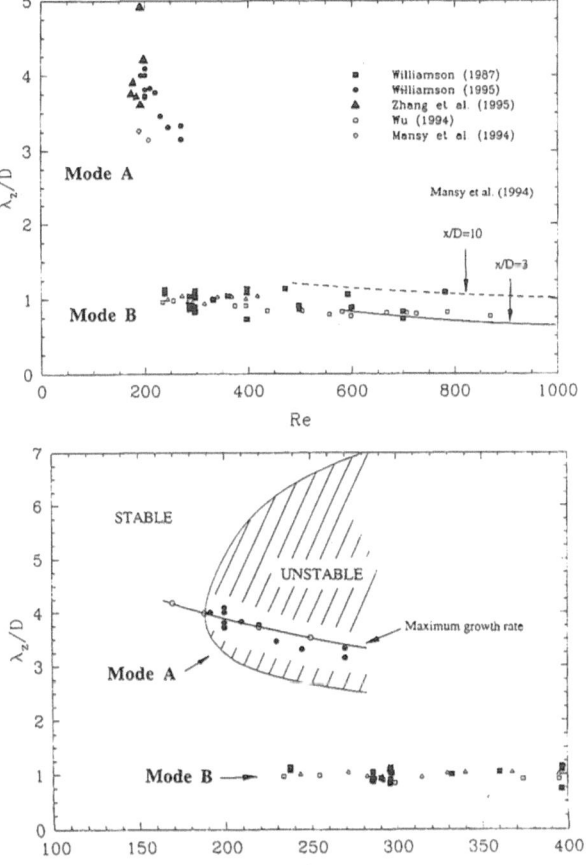

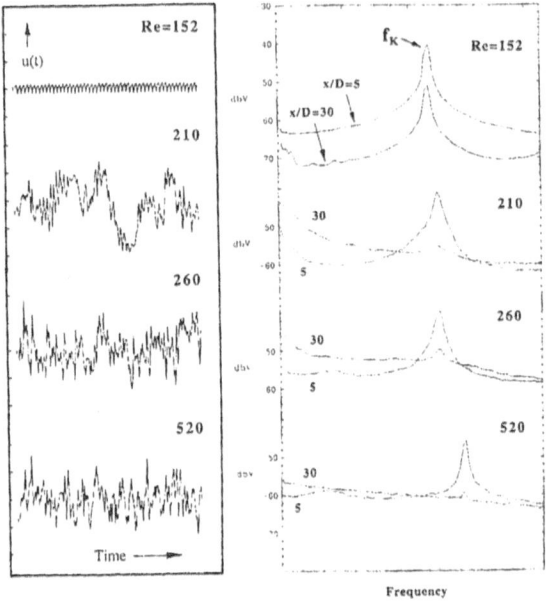

Figure 3. Large low-frequency fluctuations and corresponding spectra, through the transition regime.

(a) Low-frequency intermittent velocity fluctuations in the transition regime explain the large fluctuation energy measured in the profile and downstream decay plots of Figure 21. **(b)** Spectra at $x/D = 5$ and 30 for a selection of Re show the dominance of energy at low frequencies through transition, although by Re=520, the broad low-frequency peak has diminshed.

Figure 4. Evolution of a "vortex dislocation" at the site of a vortex loop of mode A.

It can be seen, from this sequence, that large-scale "vortex dislocations" can evolve at the sites of some of the vortex loops of mode A. These are the two-sided dislocations that were forced to occur in the case of Williamson (1992), but it is clear that in the case of wake transition, they can form independantly of the end conditions.

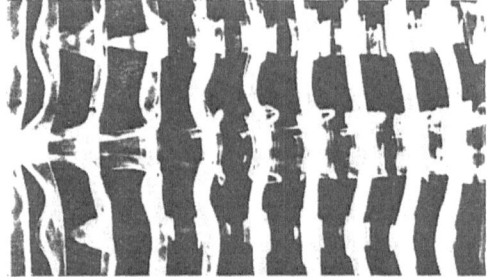

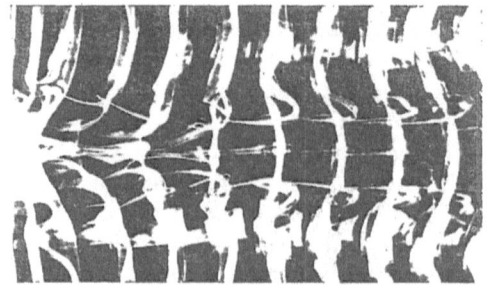

Stability of the Spiral Flow in the Annulus between Concentric Cylinders

Michael KLIKA, Gerta MARLIANI and Venkatesa I. VASANTA RAM

Institut für Thermo- und Fluiddynamik, Ruhr Universität, D 44780 Bochum, Germany

Abstract

Subject of this paper is the stability of the spiral flow (flow with helical streamlines) in the annulus between concentric circular cylinders (see figure below). This is the fully developed flow in the annulus when both axial pressure gradient and rotation of one of the cylinders, either the inner or outer one, are simultaneously present. Its stability is studied through the linearised equations for small disturbances. The dispersion relation of the eigenvalue problem contains as parameters of the basic flow the Reynolds number of the through flow, the ratio of the gap width to cylinder radius and the ratio of the circumferential to through flow (axial) velocities. The *temporal and spatial* eigenvalue problem has been solved numerically by the spectral collocation method to obtain the growth/decay rates of small disturbances and thus the wave numbers of neutrally stable disturbances. The focus of attention in this work is the region on the surface of neutral stability over which transition changes character from one close to Taylor instability (axial pressure gradient is zero) to the one closer to the instability of the channel flow (no rotation of either cylinder, gap width small compared to radius).

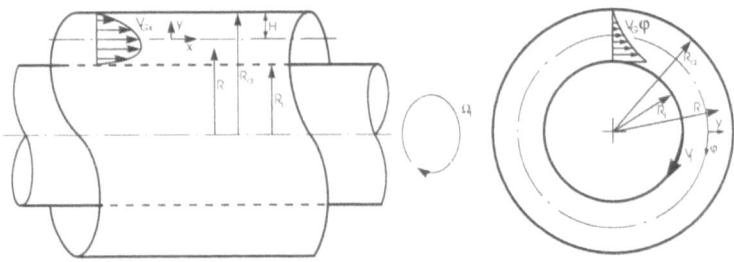

Introduction

It is known that transition in fluid flows is affected by body forces acting on the flow. A body force that is inevitably present in a flow with curved streamlines is the centrifugal force and this is associated with transition caused by Taylor and Görtler instabilities. The flow with helical streamlines is susceptible to transition from both mechanisms of instability, that due to waves of the Tollmien-Schlichting type as well as that in the Taylor kind of instability. An outstanding difference in character between the two is that while the latter exhibits "exchange of stabilities" (i.e. the neutrally stable disturbance has the real part of the period also equal to zero) the former

S. Gavrilakis et al. (eds.), Advances in Turbulence VI, 403-404.

does not belong to this category. The phenomenon of transition in the spiral flow has therefore to cross over in character from one to the other, the cross-over position in the parameter space depending upon the ratio of circumferential to axial (through flow) velocities and gap width to cylinder radius. The resolution of details in this cross-over region is the subject of the paper.

The basic flow

The basic flow is indicated schematically in the figure above. It is the fully developed flow in the annulus between the concentric circular cylinders one of which, the inner or the outer, is rotating. The basic flow may be written as follows:

$$V_{Gx} = V_{Gx}^*/U_{Ref} = \frac{V_{Gx}^*}{\frac{H^2}{2\mu}\frac{dp_G^*}{dx^*}} = \frac{1}{2\epsilon_R}\left(\frac{A\ln(1-\epsilon_R) + B\ln(1+\epsilon_R) + 4\ln(1+\epsilon_R y)}{\ln(1-\epsilon_R) - \ln(1+\epsilon_R)}\right)$$

$$V_{G\varphi} = \left[V_a\left(1 + \frac{(1-\epsilon_R)^2}{(1+\epsilon_R)^2 - (1-\epsilon_R)^2}\right) - V_i\frac{(1-\epsilon_R)(1+\epsilon_R)}{(1+\epsilon_R)^2 - (1-\epsilon_R)^2}\right]\frac{1}{1+\epsilon_R}(1+\epsilon_R y)$$
$$+ \left[V_i(1-\epsilon_R) - V_a\frac{(1-\epsilon_R)^2}{(1+\epsilon_R)}\right]\frac{(1+\epsilon_R)^2}{(1+\epsilon_R)^2 - (1-\epsilon_R)^2}\frac{1}{(1+\epsilon_R y)}$$

with

$$A = (-2+2y) + (-1+y^2)\epsilon_R, \quad B = (-2-2y) + (1-y^2)\epsilon_R,$$

where $\epsilon_R := \frac{R_a-R_i}{R_a+R_i}$ is the ratio of the semi gap width to the mean of the cylinder radii,

The disturbance The equations for small amplitude disturbances of the above basic flow are derivable by methods described in standard works, see eg. Drazin and Reid (1981). Writing the disturbances in wave form as

$$(v_{sr}, v_{s\varphi}, v_{sx}, p_s) = (A_r(y), A_\varphi(y), A_x(y), A_p(y)) \cdot e^{i(\lambda_x x + n\varphi - \omega t)} + c.c.$$

the equations governing the disturbances can be brought into a form identifiable as extensions of the Orr-Sommerfeld and Squire equations.

Results

The resulting eigenvalue problem for the behaviour of small disturbances was solved by the spectral collocation method. They contain as limiting cases: 1. the equations for the Taylor instability between concentric cylinders when the inner cylinder is rotating and the axial pressure gradient is zero, and 2. the classical Orr-Sommerfeld equations for plane channel flow when the gap width is small. A parameter study was conducted to obtain the shape of the surface of neutral stability in the parameter range of interest, i.e. when the character of the neutrally stable disturbance changes from one to the other.

References

DRAZIN, P.G.; REID W.H.: *Hydrodynamic Stability*, Cambridge University Press, Cambridge, (1981).

SPATIAL NUMERICAL SIMULATIONS OF TRANSITION TO TURBULENCE IN A CURVED CHANNEL ROTATING AROUND THE SPANWISE AXIS

T. RANDRIARIFARA,
*IMHEF - DGM, École Polytechnique Fédérale de Lausanne,
CH-1015 Lausanne, Switzerland*

A. BOTTARO
*IMHEF - DGM, École Polytechnique Fédérale de Lausanne,
CH-1015 Lausanne, Switzerland*

AND

W. COUZY
*IMHEF - DGM, École Polytechnique Fédérale de Lausanne,
CH-1015 Lausanne, Switzerland*

Our concern is with the transition to turbulence of spatially develop-ing flows in curved, rotating channels. The flow in a curved channel is susceptible to a primary instability, the so-called Dean instability, when the critical parameter, the Dean number - an appropriate combination of the Reynolds number and a curvature parameter - exceeds a critical value. This instability manifests itself with the appearance of streamwise vortices. When spanwise system rotation is applied, the growth of the vortices can be enhanced or damped, depending on the direction of the rotation vector. Interestingly, the primary instability can be of oscillatory type for some choices of the rotation number. After the primary instability, a number of secondary instabilities can occur, preceding the transition to turbulence: the most common such secondary instability is of travelling wave type.

We pursue the task of exploring the instability and the breakdown to turbulence of rotating Dean vortices with direct numerical simulations of the governing Navier-Stokes equations. High order spatial and temporal ac-curacy are key issues for our analysis. Therefore, we use a Legendre spectral element method based on modern velocity-pressure decoupling techniques and stable, up to third order accurate, time integration schemes. The it-

S. Gavrilakis et al. (eds.), Advances in Turbulence VI, 405-406.
© 1996 *Kluwer Academic Publishers.*

erative nature of the solver supplies the right framework for a parallel implementation. The code developed runs with a high parallel efficiency and good single-node performance on the Cray T3D.

Since the first instabilities in a channel are of convective type, they are normally triggered by disturbances at the boundaries: typically the inlet boundary is the most important, but the outlet and the solid boundaries can be influential as well. If no perturbations are supplied (except, clearly, for the computer round-off errors), the development of the instabilities is so slow that their effect can not be felt in a computational box of reasonable length. To trigger the primary instability, steady inlet perturbations of small amplitude are applied; they are taken from the eigenfunctions of the linear stability theory. This way the vortical flow can evolve from the initial stage where nonlinear effects are negligible, through the nonlinear interaction stage and up to saturation; at this point, typically, secondary instabilities become operational. To excite the secondary, wavy instability and the subsequent rapid transition to turbulence, small amplitude fluctuating disturbances are introduced. These disturbances consist of the eigenfunctions of a linear, inviscid secondary stability analysis of Floquet-type carried out locally on the vortical profiles previously computed.

Numerical results will be compared to experiments carried out at KTH.

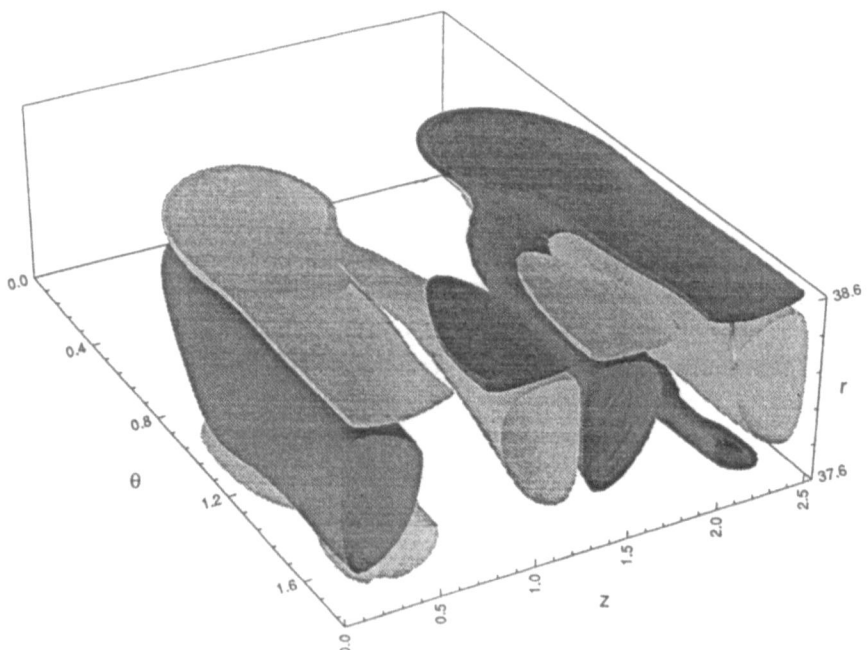

Visualization through isosurfaces of the streamwise vorticity of the vortical structures that appear in a curved channel, for a case of positive rotation is shown below. The curved channel has been straightened for graphical purposes.

DYNAMICS OF A CHAIN OF COUPLED WAKES

J.F. RAVOUX, P. LE GAL, M.P. CHAUVE and Y. TAKEDA*
Institut de Recherche sur les Phénomènes Hors Equilibre
UM 138, CNRS - Universités d'Aix-Marseille I & II
12, Avenue Général Leclerc, 13003 Marseille, France

Laboratory for Spallation Neutron Source
Paul Scherrer Institute
CH-5232, Villigen PSI, Switzerland

1. Introduction

This paper is devoted to experimental and analytical investigations concerning a one dimensionnal assembly of coupled oscillators, namely Bénard-Von Karman vortex wakes. The latter are shed by parallel identical and equidistant cylinders placed side by side in a row perpendicular to a uniform flow. Our experimental facility consists of a horizontal water tunnel with a section of 20 mm high and 128 mm wide. Each cylinder being 4 mm diameter and 20 mm length, the physical situation permits us to assume a 1D dynamics along the row axis. Quantitative transversal velocity measurements are obtained by Ultrasound Doppler Anemometry (for details on the method, see [1]) performed on a line parallel to the cylinders row. Thus, we obtain spatiotemporal profiles of the wakes oscillations.

2. Dynamics

Each wake is coupled to its neighbours and as expected, global behaviour of the one dimensionnal array of oscillators is observed. Different dynamical regimes are obtained by controlling the distance (g = D/d, where D is the distance between two cylinder axes and d their diameter) separating the cylinders and the Reynolds number of the flow. Two typical dynamics have been studied : situations with a weak or a strong coupling.

In the first case, a set of experiments has been done where g = 3. At a Reynolds number Re_{u}, a Hopf bifurcation appears and the Bénard-Von Karman streets are shed with first neighbours oscillating in phase opposition, this is analog to the optical mode of phonon propagation. Figure 1.a) presents a space time diagram corresponding to the latter, it consists in a part of the data from an experiment with 11 wakes.

407

S. Gavrilakis et al. (eds.), Advances in Turbulence VI, 407-408.
© 1996 *Kluwer Academic Publishers.*

408

For stronger coupling, the spatiotemporal diagram (-see figure 1.b)- for $g = 2$, 14 wakes) shows an acoustical mode of oscillation, i.e. with all the oscillations in phase. At very strong coupling, a first stationary bifurcation arises for a Reynolds number Re_C, it consists in a spatial symetry breaking. Due to the Coanda effect, each wake can be deviated towards one side or the other. Thus several groups of merged wakes are created. An analogy with magnetic domains separated by Ising walls can be made. When increasing the Reynolds number to Re'_H, some of the wakes exhibit an oscillatory behaviour. These wakes are confined in some regions of the flow and are locked in phase.

Figure 1 : Space time diagrams of optical a) and acoustical b) modes obtained by UDA.

Besides these experimental investigations, analytical and numerical studies of a coupled oscillators model have been realized. This model which is based on the diffusive coupling of Hopf bifurcations leads to a dicrete form of the Ginzburg-Landau equation (GLE) [2]. Stable states and transition to chaos in this model present strong similarity with the experimental facts. In particular the destabilization of the optical mode of oscillation is exhibited by numerical simulations : a weak coupling corresponds to the optical mode while the strong one to the acoustical mode. These two extreme situations are shown in figures 2.a) and 2.b) ; they are separated by a chaotic behaviour, which is also observed experimentally.

Figure 2 : Numerical simulations of the GLE : respectively a) optical and acoustical b) modes.

3. References

[1] Takeda, Y., Fischer, W. E., Sakakibara, J. and Ohmura, K. (1993) Experimental observation of the quasiperiodic modes in a rotating Couette system, *Phys. Rev. E* **47** 4130
[2] Cardoso, O., Willaime, H., Tabeling, P. (1990) Chaos in a linear array of vortices, *Phys. Rev. Lett.* **65** 1869

EXPERIMENTAL STUDY OF ARTIFICIAL HAIRPIN VORTICES

H.A. ZONDAG AND J.H. VOSKAMP
Eindhoven University of Technology
Department of Physics
Postbox 513, 5600 MB Eindhoven, the Netherlands

Background of the research   Several features of turbulence cannot be addressed properly by statistics: the dynamics remain out of reach. These dynamics are necessary to understand aspects like: *'How are new vortices generated near the wall?'* or *'Why do longitudinal riblets cause drag reduction?'* Therefore, coherent structures are attracting interest, especially hairpin vortices. A problem in the experimental study of coherent structures in a turbulent background is the fact that these structures do not reproduce in a turbulent flow nor is it easy to detect and define them. To overcome this, artificially created coherent structures in a laminar environment have been studied by several researchers. Special interest was concentrated on the interaction with the wall. It was found that new hairpin vortices could be generated by the interaction of the vortex and the wall (Haji-Haidari, 1990). In addition, Walker made simulations indicating the effect of the velocity profile on the shape of the vortex (Hon & Walker, 1987).

Experimental set-up   Our experimental set-up consists of both a water channel and a low-speed wind tunnel. The water channel is used for flow visualization by means of a bubble wire. In the wind tunnel measurements were done using a linear array of 9 single hot-wires. The hairpin vortices are shedded periodically downstream of a hemisphere or a half teardrop (diameter $D = 16$ mm) on a wall in a laminar boundary layer ($Re_D \approx 2000$).

Results   If the obstacle has a blunt upstream part (hemisphere), a standing vortex is trailing alongside of it. This standing vortex draws high speed fluid towards the wall at $|z/D| < 0.5$ and pushes the legs of the hairpin vortex down, while at $|z/D| = 1$ low speed fluid is drawn up from the wall and a low-speed streak is created. This standing vortex influences the development of the hairpin vortex: in the case of a hemisphere the head is located

S. Gavrilakis et al. (eds.), Advances in Turbulence VI, 409-410.
© 1996 *Kluwer Academic Publishers.*

410

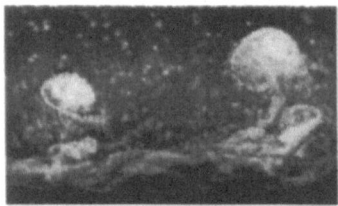 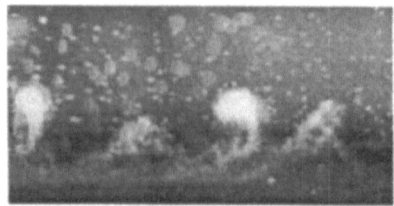

Figure 1. The influence of the shape of the obstruction (side view, the flow direction (u) is from left to right). Vortices shedded by **left:** a hemisphere, **right:** a half teardrop.

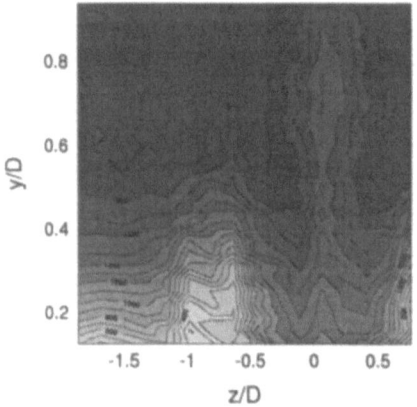

Figure 2. The distorted average u-profile: darker color implies higher velocity. The vortex head passes at $z = 0$. At z/D=-1 there is a low speed streak generated by the standing vortex. In the region $-0.5 < z/D < 0.5$ the near wall velocity is much higher than in the undisturbed flow at z/D=-1.5. The flow direction is into the paper.

higher and the upflow of fluid from the wall by the legs is reduced (figure 1). This might influence the ability of the vortex to generate secondary structures. The velocity profile is deformed by the vortices (figure 2): the average flow velocity near the wall (and thus the wall friction) increases for $|z/D| < 0.5$. The low speed streak that is created by the standing vortex was found to be destabilized by the passage of the hairpin head, growing into new vortices having a streamwise, as well as a spanwise part. In our visualizations these new vortices did not grow into isolated hairpin vortices.

References

Acarlar, M.S. & C.R. Smith (1987), A study of hairpin vortices in a laminar boundary layer. *J. Fluid Mech.* **Vol. no. 175**, pp. 1–41.

Haji-Haidari, A. (1990), Generation and growth of single hairpin vortices (*thesis, Lehigh University*).

Hon, T.L. & J.D.A. Walker (1987), An analysis of the motion and effects of hairpin vortices (*report FM-11, Lehigh University*).

VII

Experiments and Novel Experimental Techniques

INTERACTIVE CONTROL OF WALL STRUCTURES BY MEMS-BASED TRANSDUCERS

CHIH-MING HO AND STEVE TUNG
Mechanical and Aerospace Department
University of California, Los Angeles
Los Angeles, CA 90095-1597
USA

YU-CHONG TAI
Department of Electrical Engineering
California Institute of Technology
Pasadena, CA 91106
USA

1. Introduction

In turbulent boundary layers, the near-wall streamwise vortices contribute to a significant part of the turbulence production. Successful control of the streamwise vortices can lead to the desirable skin-friction drag reduction. In the past, passive devices such as longitudinal riblets were used to achieve a 6% reduction (Bruse et al. 1993). Recent numerical simulations (Choi et al. 1994) indicate that an active control scheme based on surface shear-stress measurements can increase the drag reduction to almost 30%. We are currently developing a large-scale integrated systems of micromachined sensors, micromachined actuators, and neural network VLSI circuits for interactive control of the wall structures. The newly developed MEMS technology allows us to, for the first time, fabricate micro sensors and actuators that can match both the temporal and spatial resolutions of the wall structures (Ho and Tai 1994). Individual components of the integrated system have been developed and tested. The following is a discussion of their design and capabilities.

2. The Micro Shear-Stress Sensors

A micro shear-stress imaging chip with arrays of micro shear-stress sensors has been designed and fabricated by surface micromachining technology (figure 1a). Each chip consists of about 100 micro sensors within a 1cm x 1cm area. The design of the micro shear-stress sensor is similar to that of the traditional hot-film anemometry. Each sensor consists of a 150µm x 3µm polysilicon hot wire deposited on a 200µm x 200µm

413

S. Gavrilakis et al. (eds.), Advances in Turbulence VI, 413-416.
© 1996 *Kluwer Academic Publishers.*

414

diaphragm that seals a 2μm deep vacuum cavity underneath. The presence of the vacuum chamber reduces the heat transfer from the hot wire to the substrate and increases the sensitivity of the sensor by almost 2 order of magnitude over the one without the vacuum chamber.

The imaging chip has been successfully used to map out the instantaneous surface shear-stress distribution of a turbulent channel flow (figure 1b). The streaky structures in the contour plot represent high shear stress regions produced by the random near-wall flow structures. The statistics based on our experiments at different freestream velocities show that the average length, width, and spanwise spacing of the structures agree very well with that of previous flow visualization results.

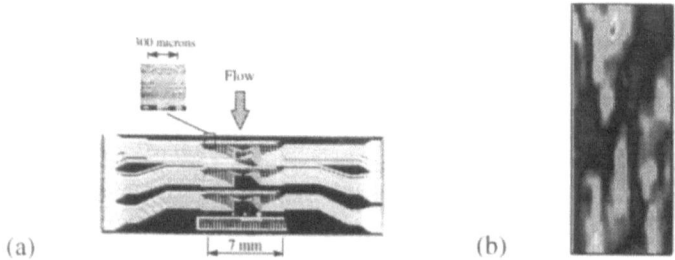

(a)   (b)

Figure 1. (a) micro shear-stress imaging chip; (b) iso-contours of instantaneous turbulent surface shear stress.

3.   The Micro Actuators

An individually addressable magnetostatic actuator has been fabricated by micromachining technique (figure 2). The active flap is a 4mm by 4mm single-crystalline silicon plate (40μm thick) supported by two serpentine springs on the substrate bulk silicon. The electromagnetic components of the actuator consist of a 30-turn planar copper coil and an insulated triangular permalloy layer on top of the coil. When activated, the actuator rotates from the substrate to a prescribed angle. Static (dc) rotation is achieved by interacting the actuator permalloy with an external magnetic field. Dynamic (ac) rotation or oscillation is accomplished by supplying an ac current to the copper coils.

The interaction between an active flap and a stationary streamwise vortex in a laminar boundary layer has been investigated. The actuator is flush mounted on the wall where the streamwise vortex induces a local peak in the surface shear stress distribution. With the micro actuator activated at a prescribed oscillation, the velocity distribution in a plane normal to the mean flow and slightly downstream from the actuator is recorded by hot-wire anemometry. Different actuator rotation angles and oscillating frequencies are experimented. From the phase averaged streamwise velocity measurement, the drag coefficient for each case is computed. Figure 3 shows the result of the 30^0 rotation case. At 20Hz, the maximum reduction of 25 % in the drag coefficient is achieved. Since 20Hz is neither the highest nor the lowest frequency tested, our result suggestes the existence of an optimum frequency (for drag reduction) for a particular maximum off-plane movement. We are currently conducting more experiments with different maximum vertical displacement to determine the relationship between the actuator operation conditions and the resulting drag reduction.

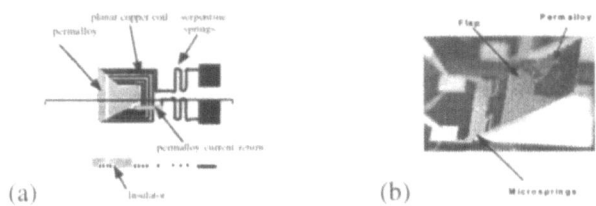

Figure 2. (a) Schematics of the active flap; (b) a micrograph of current activated active flap.

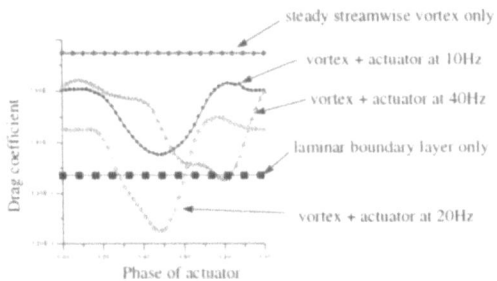

Figure 3. Drag reduction by a micro actuator at different actuating frequencies.

4. Control Circuits

The control circuits act as the interface between the upstream sensors and the downstream actuators. They process the real-time signals from the micro shear stress sensors to detect regions of high shear stress and output control signal to the micro actuators accordingly. An analog CMOS VLSI system has been designed and fabricated for this purpose. Figure 4 shows a diagram of how the information is processed in the system.

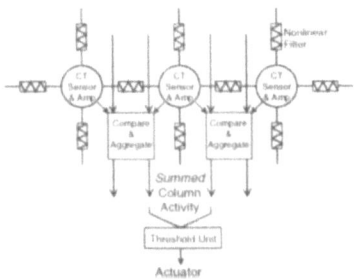

Figure 4. Block diagram of the complete detection/control chip.

A non-linear filtering network connects together the amplified outputs of the shear stress sensors. The filtering preserves large differences between adjacent sensors while smoothing small differences. The comparison and aggregation is organized by columns corresponding to different actuators. Once the signal is aggregated and exceeds a threshold then the actuator is activated. Figure 5 shows the results of real-time processing of the instantaneous shear-stress signals measured by the imaging chip. The close matching between the shear stress distribution and the control signal confirms the capability of the control circuits.

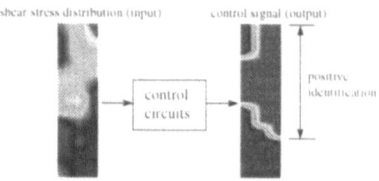

Figure 5. A schematic of the real-time processing of the instantaneous shear-stress distributions by the control circuits.

5. Conclusions

A large-scale integrated system of MEMS-based micro sensors, micro actuators, and neural network VLSI is proposed for distributed control of turbulent surface shear stress. Individual component of the system has been fabricated and tested. An imaging chip with arrays of micro shear-stress sensors successfully captures the near-wall streamwise streaks of a turbulent boundary layer. A CMOS VLSI control chip identifies high shear-stress regions from the real-time sensor signals of the imaging chip. Finally, wind tunnel studies shows a 25% reduction in surface shear stress when an oscillating micro actuator is applied to a steady streamwise vortex.

6. Acknowledgment

The authors would like to acknowledge Professor Rodney Goodman, Bhusan Gupta, Fukang Jiang, and Raanan Miller for providing some of the results. This work is supported by Air Force Office of Scientific Research through the URI Project.

7. References

Choi, H., Moin, P. and Kim, J. (1994) Active turbulence control for drag reduction in wall-bounded flows, *J. Fluid Mechanics* **262**, 75-110.

Ho, C.M. and Tai, Y.C. (1994) MEMS - Science and Technology, *ASME Winter Annual Meeting*, Chicago,Ill.

Bruse, M., Bechert, D.W., van der Hoeven, J.G.Th., Hage, W. and Hoppe, G. (1993) Experiments with conventional and with novel adjustable drag-reducing surfaces, *Near-Wall Turbulence Flows*, Elsevier Science Publishers B.V.

AN INVESTIGATION OF THE TURBULENCE FIELD
IN A THREE-DIMENSIONAL WALL JET

H. ABRAHAMSSON, B. JOHANSSON AND L. LÖFDAHL

Chalmers University of Technology
Thermo- and Fluid Dynamics, 41296 Göteborg, Sweden

1. Introduction

A three-dimensional wall jet (3DWJ) is created when a jet of finite extension is ejected close to a wall, forming a complex and highly turbulent flow field, which spreads rapidly in the lateral direction. It is difficult to capture the 3DWJ in numerical simulations, especially the large lateral spreading rate, and as this flow case is of significant importance in many industrial applications, e.g. film cooling of gas turbine blades, a better understanding of the flow is required in order to improve its modelling.

Sforza & Herbst (1970) studied the plane 3DWJ in a quiescent surrounding , and their work has been followed by e.g. Newman et al. (1972), Davis & Winarto (1980), Koso & Ohashi (1982), Padmanabham & Lakshmana Gowda (1991), LeBlanc (1995), and review articles by Launder & Rodi (1981, 1983). However, large deviations are observed when different experiments are compared, and most investigations are devoted to the turbulence on the symmetry line close to the inlet. Almost no information is available on the turbulence off the symmetry line in the fully developed region of the flow.

The objective of the present investigation is to study the turbulent field in the fully developed and self-preserving region of an isothermal, three-dimensional wall jet in a quiescent surrounding, in order to determine the turbulent kinetic energy budget on the symmetry line, and to increase the knowledge about the large lateral spreading.

2. The experiment

All measurements were performed in a large scale wall jet facility at Chalmers, where the 3DWJ is formed by a tangential injection of air (top hat profile with a low turbulence intensity) through a circular orifice, diameter, d= 20 [mm]. The wall jet develops over a large horizontal flat plate, length 2.1 [m] and width 3.2 [m]. A vertical back wall, height 1.2 [m], is placed over the inlet in order to give well-defined boundary conditions, Eriksson et al. (1991). A stationary cross as well as a rotated single wire were employed for the measurements, and they were focused on the region 50 to 90 x/d. Here, vertical velocity profiles on the symmetry line and lateral profiles at the level of the maximum mean velocity were carried out. In the position 80 x/d several profiles in the lateral plane were measured to determine the gradients needed in the turbulent kinetic energy equation. Further details on the experiment are found in Abrahamsson et al. (1995), Fransson & Götberg (1995) and Auckenthaler (1996).

S. Gavrilakis et al. (eds.), Advances in Turbulence VI, 417-420.
© 1996 *Kluwer Academic Publishers.*

418

3. Results and Discussion

Figure 1a shows the streamwise mean velocity. It can be noted that the profiles at different streamwise positions exhibit similarity when they are scaled by the maximum mean velocity and the half width. Furthermore, the measurements in a lateral plane at 80 x/d reveal a lateral twisting of the mean velocity vector when the wall is approached, and also a negative normal mean velocity in large regions despite the normal growth (the angles are positive when flow is directed from the wall and the symmetry line). This observation supports the idea of a secondary flow, e.g. a streamwise vorticity, prevailing in the 3DWJ, causing the rapid lateral spreading, see e.g. Launder & Rodi (1983) and Matsuda et al. (1990).

The streamwise development of the 3DWJ is found in Figure 1b, where it is observed that the lateral spreading $(z_{1/2})$ is much larger as compared to the normal $(y_{1/2})$, and that the growth is approximately linear. In agreement with other investigations, the ratio of the lateral and normal growth rates is 4.9, see e.g. Padmanabham & Lakshmana Gowda (1991). However, the level of the present growth rates $(dy_{1/2}/dx=0.065$ and $dz_{1/2}/dx=0.32)$ are slightly larger as compared to other investigations, which might be connected to a slightly faster decay of the maximum mean velocity $(x^{-1.29})$ in the present study as compared to earlier investigations.

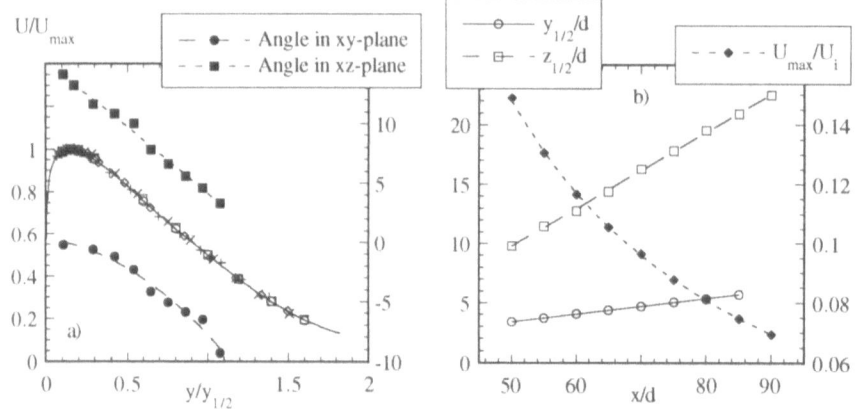

Figure 1: a) Left axis: Similarity of the streamwise mean velocity profiles on the symmetry line. (Legends see Figure 2a). Right axis: Mean velocity direction in position x/d=80 z/d=10. b) The streamwise development of the maximum mean velocity (U_{max}) and the normal $(y_{1/2})$ and lateral $(z_{1/2})$ half widths. U_i is the mean velocity in the inlet (equal to 60 m/s).

An interpretation of the turbulent structure in the lateral direction, Figure 2a, reveals a self-preserving behaviour at the level of y_{max}, and it is also clear that the shear stress is changing sign on the symmetry line. The scatter in the shear stress seems to be larger as compared to the normal stresses, which might due to the scaling used. See a recent similarity theory by George et al. (1996) showing that the shear stress in the outer region of a two-dimensional wall jet scales with the friction velocity. In agreement with the two-dimensional wall jet, the location of the change of sign in the shear stress is closer to the wall as compared to the location of the zero mean velocity gradient. It is noticed in Figure

2b, showing the turbulent structure on the symmetry line, that the turbulence level is substantially larger in the 3DWJ as compared to the two-dimensional wall jet.

Recalculating the turbulence levels in Figure 2 shows a considerably high local turbulence intensity, having the minimum value (approximately 25%) on the symmetry line at y_{max}, and the level increases off the symmetry line and from the wall. Those high local turbulence intensities increases the error in the outer parts of the 3DWJ.

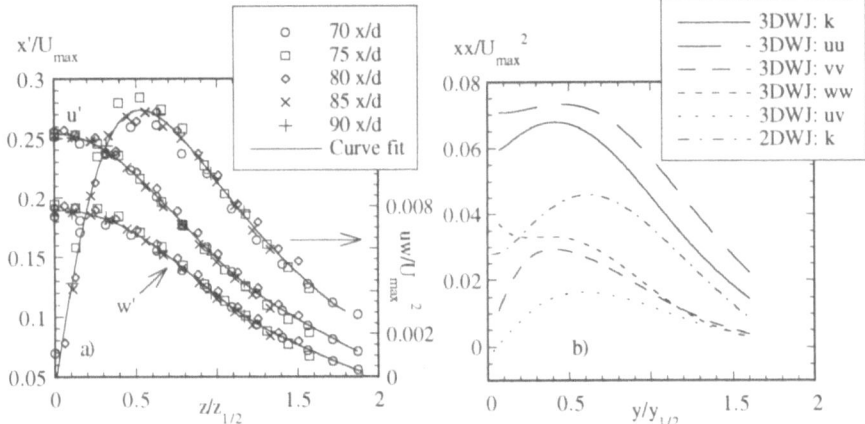

Figure 2: a) Self-preserving profiles (u', w' and uw) in the lateral direction at the level of maximum mean velocity. b) The turbulent structure on the symmetry line. The lines represent curve fits from the self-preserving profiles. "2DWJ" represent measurements by Abrahamsson et al. (1994).

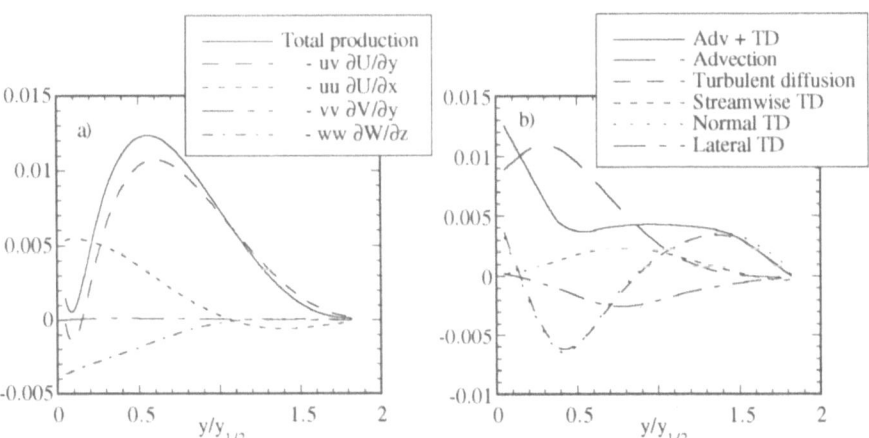

Figure 3: a) The production of turbulent kinetic energy in position 80 x/d on the symmetry line. b) The transport of turbulent kinetic energy by advection (Adv) and turbulent diffusion (TD). Scaled by $y_{1/2}$ and U_{max}. Positive numbers indicate gain of energy, negative numbers indicate a loss.

Figure 3a shows that the total turbulent production has a maximum in the outer free shear layer, a minimum in the region around y_{max} and again increases closer to the wall.

The component uv dU/dy is negative in the region around y_{max} due to the displacement in the position of zero gradient and shear stress, all in agreement with observations by Padmanabham & Lakshmana Gowda (1991b). Another important observation is the non zero production in the lateral component, a difference from the two-dimensional wall jet, see Vicari & Abrahamsson (1994), a phenomenon that might be responsible for the difference in the lateral Reynolds stress close to the wall as compared to the two-dimensional wall jet, see Abrahamsson et al. (1994).

In Figure 3b the transport terms in the turbulent kinetic energy budget are displayed. The turbulent diffusion, integrating to zero within 7%, indicates a turbulent kinetic energy redistribution from the central part of the flow, where the highest turbulence levels prevail, towards the wall and the outer region. An energy diffusion from the symmetry line into the lateral direction is indicated by the lateral component, and the advection level is higher than in the two-dimensional wall-jet, see Vicari & Abrahamsson (1994). The large contribution of the advection term, i.e. a transport of energy from the upstream regions, is connected with the large retardation of the flow and is a plausible interpretation of the large turbulence activity in the 3DWJ.

4. References

Abrahamsson, H., Johansson, B. and Löfdahl, L., 1994, "A plane two-dimensional wall-jet in a quiescent surrounding", European Journal of Mechanics B/Fluids, Vol. 13, No. 5, pp. 533-556.

Abrahamsson, H., Chevalier, P.Y. and Lebars, C., 1995, "Turbulence measurements in a three-dimensional wall jet", Report 95/19, Thermo- and Fluid Dynamics, Chalmers University of Technology, Göteborg, Sweden.

Auckenthaler, T., 1996, "An experimental investigation of the three-dimensional turbulent wall jet", Thermo- and Fluid Dynamics, Chalmers University of Technology, Göteborg, Sweden.

Davis, M.R. and Winarto H., 1980, "Jet diffusion from a circular nozzle above a solid surface", Journal of Fluid Mechanics, Vol. 101, Part 1, pp. 201-221.

Eriksson, M., Johansson, B. and Löfdahl, L., 1991, "Flow visualisations and turbulence measurements in a three-dimensional wall jet", First European Fluid Mechanics Conference, Cambridge, England.

Fransson, Ö. and Götberg, J., 1995, "Measurements with a rotated single hot wire in a three dimensional wall jet", Diploma work 95/12, Thermo- and Fluid Dynamics, Chalmers University of Technology, Göteborg, Sweden.

George, W.K., Abrahamsson, H. and Löfdahl, L., 1996, "A similarity theory for the plane wall jet", 3rd International Symposium on Engineering Turbulence Modelling and Measurements, Crete, Greece.

Koso, T. and Ohashi, H., 1982, "Turbulent diffusion of a three-dimensional wall jet", Bulletin of the JSME, Vol. 25, No. 200, pp. 173-181.

Launder, B. and Rodi, W., 1981, "The turbulent wall jet", Prog. Aerospace Sci., Vol. 19, pp. 81-128.

Launder, B. and Rodi, W., 1983: "The turbulent wall jet, measurements and modelling", Ann. Rev. Fluid Mech., pp. 429-459.

LeBlanc, D.D., 1995, "Reynolds stress measurements in a three dimensional turbulent wall jet using hot wire anemometry", Department of Mechanical Engineering, McGill University, Montreal, Canada.

Matsuda, H., Iida, S. and Hayakawa, M., 1990, "Coherent structures in a three dimensional wall jet", Journal of Fluids Engineering ASME Transactions, Vol. 112, pp. 693-698.

Newman, B.G., Patel, R.P., Savage, S.B. and Tjio, H.K., 1972, "Three-dimensional wall jet orginating from a circular orifice", Aeronautical Quartely, pp .188-200.

Padmanabham, G. and Lakshmana Gowda, B.H., 1991, a) "Mean and turbulence characteristics of a class of three-dimensional wall-jets - Part 1: Mean flow characteristics", Journal of Fluids Engineering, Vol. 113, pp. 620-628. b) "Mean and turbulence characteristics of a class of three-dimensional wall jets - Part 2: Turbulence characteristics", Journal of Fluids Engineering, Vol. 113, pp. 629-634.

Sforsa, P.M. and Herbst, G., 1970, "A study of three-dimensional, incompressible, turbulent wall- jets", AIAA Journal, Vol.8, No. 2, pp. 276-283.

Vicari, K.F.F. and Abrahamsson, H., 1994, "An investigation of the production, advection and turbulent diffusion in a plane two-dimensional wall jet", Intern skrift 94/8, Department of Thermo- and Fluid Dynamics, Chalmers University of Technology, Göteborg, Sweden.

SPATIAL ENSTROPHY SPECTRUM

IN A FULLY TURBULENT JET

Fully developed turbulence

C. BAUDET AND R.H. HERNANDEZ
Laboratoire de Physique
Ecole Normale Supérieure de Lyon (URA 1325 CNRS)
46, Allée d'Italie 69364 Lyon Cedex 07 France

1. Introduction

The scattering of sound by a turbulent medium has been investigated theoretically by many authors [1, 2]. In particular, Chu *et al.* have shown that in a viscous heat-conducting compressible medium three modes of fluctuation exist, namely the entropy mode, the vorticity mode and the sound mode. In a linear approximation these three modes are decoupled from each other, but taking into account the non-linearities of the equations governing the evolution of a turbulent flow (mass, heat and momentum conservation) they have shown that sound waves can be scattered by either temperature fluctuations or vorticity fluctuations. More recently, Lund *et al.* [3] have derived a linear relation between the scattered ultrasound intensity and space-time Fourier transform of either the vorticity field or the temperature field. The physical mecanism at the origin of the acoustic scattering by vorticity can be thought as follows : an acoustic wave incident on a vorticity distribution induces fluctuations of the vorticity at the incoming sound frequency (by virtue of Kelvin circulation theorem). In turn the flucuating vorticity field radiates a sound wave which is the scattered acoustic wave. Using a Born approximation Lund *et al.* obtain the following relation between the amplitude of the scattered pressure p_s and the time-spatial Fourier transform of the vorticity distribution $\Omega(q_s, \nu)$:

$$\frac{p_{scat}(\nu)}{p_{inc}} = \pi^2 i \frac{-cos(\theta_s)}{1 - cos(\theta_s)} \frac{\nu e^{i\nu D/c}}{c^2 D} (\vec{n} \wedge \vec{r})\vec{\Omega}(q_{scat}^{\rightarrow}, \nu - \nu_o) \qquad (1)$$

where $\vec{q_s} = 4\pi \frac{\nu_o sin(\theta_s/2)}{c}$ is the scattering wave vector, ν_o is the incoming sound frequency and θ_s the scattering angle. Unit vectors $\vec{n}$ (*resp.* $\vec{r}$) is in

S. Gavrilakis et al. (eds.), Advances in Turbulence VI, 421-424.
© 1996 *Kluwer Academic Publishers.*

the direction of the incoming sound wave (*resp. scattered*). Equation (1) shows that, using acoustic scattering, it is possible to probe one component (perpendicular to the scattering plane defined by the vectors $\vec{n}$ and $\vec{r}$) at a chosen length scale through the selection of a known scattering wavevector $q_{\vec{scat}}$. Moreover it is a non perturbating technique. Using, as a test flow, the von Kármán vortex street behind a cylinder at a low Reynolds number we have confirmed [4] experimentally the validity and main features of equation (1). The purpose of the present work is to apply the acoustic scattering technique to the characterization of the vorticity in a fully turbulent flow. More precisely, we were interesting in the determination of the exponent of the power law (if any) of the enstrophy $\Omega(\vec{k}) \equiv \langle \frac{1}{2}|\vec{rot}(\vec{v})|^2 \rangle$ [6].

2. Experimental Setup

We have performed an acoustic scattering experiment on an axisymetric turbulent jet flow. The jet flow emerged from a circular nozzle 5 cm in diameter, at a Reynolds nunber of about 10^5. The measurements have been performed 60 diameters downstream of the nozzle where the jet flow is expected to be reasonably self-preserving [5]. We have performed a statistical analysis of the longitudinal velocity component (along the jet axis) using classical hot film anemometry. The statistics of the longitudinal is found to be gaussian to a very good approximation , with mean velocity about 5.3 m/s and standard deviation about 1.5 m/s. From the computation of the third order structure function, we determined the mean energy transfer rate $\epsilon = 13 m^2/s^3$, the Taylor microscale $\lambda = 6.3 mm$ and the Reynolds number $R_\lambda \approx 600$. The inertial range extends over more than one decade of wavevectors. The acoustic scattering device consists in two large ultrasound transducers ($50 cm X 30 cm$) of the Sell type (in order to ensure a precise selection of the scattering angle and thus of the direction and modulus of the scattering wavevector [4]). The scattering wavevector is then adjusted by tuning precisely the incoming sound frequency ν_o between 2 kHz and 250 kHz (the scattering angle is held at a constant value of 60 degrees). For each frequency ν_o the scattered pressure signal is digitally sampled with a 16 bits resolution, and numerically demodulated resulting in a complex time signal. The average time spectra are then computed using a Fast Fourier Transform algorithm. We have represented on Figure 1.a, the spectrum of the scattered pressure for the incoming sound frequency $\nu_o = 32 kHz$. One clearly observes a scattered peak centered around a frequency shift $F_{dop} = 424 Hz$ as a consequence of the doppler effect related to the mean velocity of the flow : $F_{dop} = \langle \vec{q_s}.\vec{V} \rangle$ (in this case the direction of $\vec{q_s}$ was parallel to the jet axis). Indeed, we have plotted the evolution of F_{dop} *versus* ν_o, the slope of the best linear fit gives a good estimation of the mean flow velocity (5.35

m/s) close to the value obtained with the hot-film anemometry. Moreover, the shape of the scattered peak is very close to a gaussian, like the PDF of the longitudinal velocity. The best gauss fit, enables us to determine the standard deviation of the doppler frequency shift. We find that the ratio of the standard deviation of the Doppler frequency over its mean value is nearly independant of the incoming sound frequency (*i.e the probed length scale of the flow*) and slightly larger than the turbulence intensity (37 %). We interpret these last results as a consequence of the advection of the vorticty field by the flow velocity field.

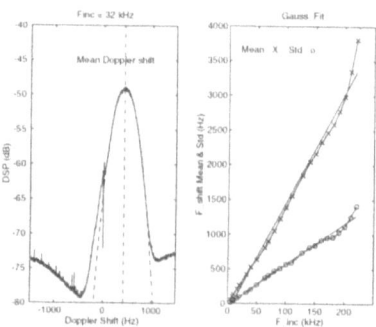

Figure 1. a. DSP of the demodulated scattered pressure signal *versus* doppler shift frequency (– best gaussian fit), the incoming sound frequency is 32 kHz. b. Evolution of the mean doppler shift (x) and the standard deviation of the doppler shift (o) *versus* incoming sound frequency ν_o.

On the figure 2, we have also reported, the evolution of the total scattered pressure corresponding to the area under the scattered peak. By virtue of the Bessel-Parseval theorem (identifying averages of the squared amplitude in the time space and the Fourier space), we obtain the spatial enstrophy spectrum $\Omega(\vec{k} = (q_s, 0, 0))$. Assuming that the flow is isotropic, we can compute the enstrophy spectrum $\Omega(K)$. Reproducing the argument of Kolmogorov [7], one expects a scaling law for $\Omega(K)$ with a scaling exponent $+1/3 \equiv 2 - 5/3$. Indeed, by representing on the same plot the compensated spectra of the energy (obtained by hot-film anemometry) and that of the enstrophy, one observes a good agreement, over more than one decade of wavectors, with the Kolmogorov theory (K41).

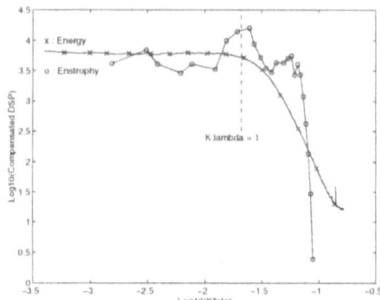

Figure 2. Log-log plot of the Enstrophie DSP (o) end the Energy DSP (x) *versus* the spatial wavevector K normalized by the Kolmogorov length η. The vertical dashed line indicates the Taylor microscale λ. The DSP are compensated by their expected K41 power laws (1/3 for the enstrophy and -5/3 for the energy).

3. Conclusion

We have analysed the scattering of acoustic wave, with frequencies between 2 kHz and 200 kHz, incident on a fully turbulent jet flow. The large frequency response of our transducers, enables us to probe a wide range of length scales, covering the inertial range and the dissipation range. The analysis of the scattered time spectrum, at a fixed frequency of the incoming sound wave gives access to the temporal dynamic of the vorticity field at a well defined length scale. When the incoming sound frequency is varied, the spatial dynamic of the vorticity field is analysed. Our measurements show that the enstrophy spectrum follows, in the inertial range, a power law scaling with an exponent close to the value 1/3, expected from a dimensional argument similar to the argument proposed by Kolmogorov in 1941.

References

1. Kraichnan R.H., 1953, The Scattering of Sound in a Turbulent Medium. *J. Acoust. Soc. Am.* **25-6** , 1096-1104.
2. Chu B-T. & Kovasznay L. S. G., 1969, Non-Linear Interactions in a Viscous Heat-Conducting Compressible Gas. *J. Fluid Mech.* **3**.
3. Lund F. & Rojas C., 1989, Ultrasound as a Probe of Turbulence. *Physica D* **37**, 508-514.
4. Baudet C., Ciliberto S. & Pinton J-F., 1991, Spectral Analysis of the von Kármán Flow Using Ultrasound Scattering. *Phys. Rev. Lett.* **67-2**, 193-195.
5. Wygnanski I. & Fiedler H., 1969, Some Measurements in the Self-Preserving jet. *J. Fluid Mech.* **38-3**, 577-612.
6. Frisch U., 1995, Turbulence. *Cambridge University Press*.
7. Kolmogorov A.N., 1941, The Local Structure of Turbulence in Incompressible Viscous Fluid for Very large Reynolds Number. *Dokl. Akad. Nauk SSSR* **32** , 9-13.

ON TURBULIZATION OF THE TORNADO-LIKE VORTICES
AND TRANSITION TO A SYSTEM OF VORTICES

B.M.BOUBNOV
INSTITUTE OF ATMOSPHERIC PHYSICS RAS,
109017. MOSCOW, RUSSIA

1. Introduction

Inhomogeneous heating of a fluid is a major source of motions in geophysical and other natural processes. In many cases the system is rotating as a whole, or in part, forming a vortex, or vortices. The role of rotation is usually determined by the Rossby number Ro=$U/L\Omega$, where U is a characteristic velocity of motions and could be determined by the other parameters of the problems, Ω is the angular rate of rotation and L is a characteristic size of the system. In the case of global rotation the value of $\Omega=\Omega_{gl}$ is fixed while in the case of local or differential (vorticity) its value may vary in broad limits depending both on Ω_{gl} and on heating inhomogeneity and its gradient ∇T =$\Delta T/D$. Here ΔT is a characteristic value of the temperature difference within the system and D is spatial scale of the inhomogeneity.

The most studied are the cases of convection when one of the spatial scales is very large and can be excluded from consideration and the heating is homogeneous: "horizontal" and "vertical" convection (Boubnov and Golitsyn, 1995). The "horizontal" convection caused by horizontally inhomogeneous heating and leads to a formation of baroclinic waves and vortices (a laboratory analogue are the experiments on baroclinic instability in rotating cylinders (Hide,1953 , Fultz 1951)). The "vertical" convection caused by vertical temperature gradient in plane rotating layer of fluids heated from below and cooled from above and a vortex grid or geostrophic vortices can arise in fluid layer.

Convection from the local heating is the case where all the main regimes of "vertical" and "horizontal" convection can exist and also the additional regimes with the intensive solitary vortex can be developed. Here we will consider the experiments results of the convection from the rotating disk with diameter D and the influence of the aspect ratio δ-D/H (where H is the neight of the fluid layer) to the structure of the intensive vortex. The intensive vortex is developed near the heating disk due to the influence of the rotation to the upflow from the disk and is defined by the processes in the boundary layers (thermal and Ekman). When $\delta\ll1$ (and there is a space for the propagation of the vortex from the disk) the vortex is laminar and have a cylindrical

425

S. Gavrilakis et al. (eds.), Advances in Turbulence VI, 425-428.
© 1996 *Kluwer Academic Publishers.*

form, but when δ>1 the disturbances from the upper boundary are propagated on a cylindrical vortex and destroy it.

2. Experimental results

The experiments are carried out with the heating disk with a diameter $D=15$ cm which rotate with a constant angular velocity Ω around a vertical axis of symmetry. The temperature of the disk T_i changes from 320 K to 720 K, and the temperature difference between the disk and an ambient non-rotating air ΔT changes from 27 K to 427 K. The height of the air layer H is defined by the distance between the disk and the upper boundary. Two cases of the upper boundary are used: rigid boundary where upflow changes its direction, and the linear grid, through which the upflow can propagates, but the grid stops the rotation of the tornado-like vortex.

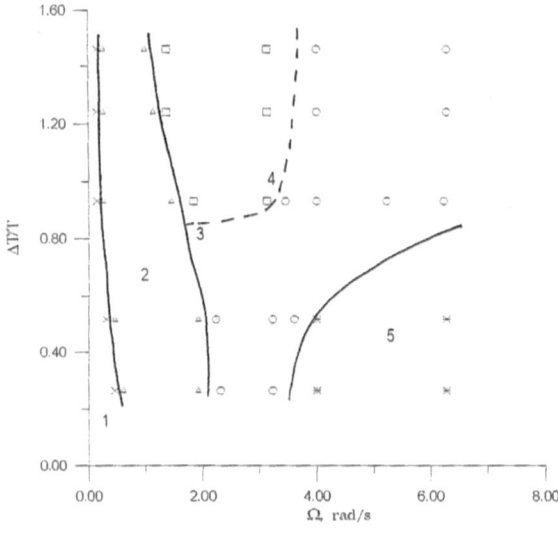

Fig.1

The structure of the flow above the disk is defined by a processes at the boundary layer , and if the height of the layer is large enough, the upper boundary do not have any influence to these processes. Regime diagram for a large height $H=2$ m ($\delta=0.075<<1$) is presented in figure 1. Depend from the temperature difference ΔT and the rotation rate Ω there are five different regimes : *1* - turbulent convection, which is negligible differ from the non-rotating case $\Omega=0$ (Boubnov and Heijst,1994), *2* - "spiral-like" flow, when the hot air from the disk propagate as a spiral in non-rotating air (Fitzjarrald, 1973), *3* - stable quasi-laminar cylindrical vortex (see Fig. 2a), *4*-unstable cylindrical vortex (periodically a cylindrical vortex is arisen above the disk, but after some time it moves out from the disk), *5*- convective vortices (more than one), which are propagated out of disk (Boubnov and Golitsyn,1990, Maxworthy and Narimosa ,1991).

Regime of the stable quasi-laminar cylindrical vortex is defined by the following stage: intensive horizontal flow from the edge of the disk moves to the centre in a limit of Ekman layer, a change of the direction of the flow from horizontal to the vertical through a break point and than the vertical cylindrical vortex is established above the disk at some distance r from its centre (Fig. 2a). The distance $r=2$-5 cm is changed due to the complicated motion of the vortex above the disk, the horizontal scale of the vortex d_v changes from 1 to 3 cm and weakly depends from ΔT and Ω.

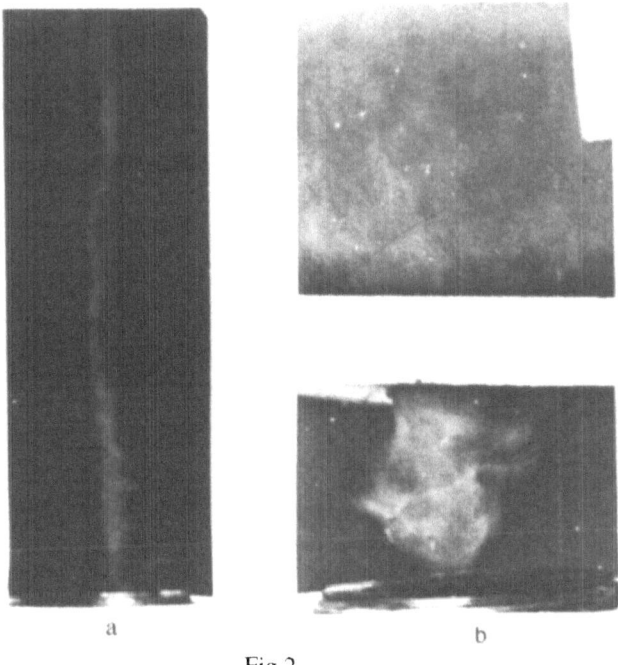

a   b

Fig.2

The mean temperature above the disk have a sharp changes near the disk at the regimes of a single intenive vortex (*3* and *4)*, and much less at the regimes of the multiple vortices (*1* and *5*). The maximum temperature of the vortex is changed linear with the height and does not depend from the temperature of the disk and rotation rate. Note that the linear dependence also is a regular regimes of a convection in rotating plane layer, where the vertical gradient is proportional to the Ω.

When the height of the layer H is the order of the diameter of disk D (or $\delta \cong 1$), the laminar vortex is destroyed and only turbulent vortex can exist (Fig.2b, where the linear grid is the upper boundary). This destruction of the laminar cylindrical vortex and transitions to the turbulent vortex is defined by the propagation of the disturbances from the upper boundaries (grid) in both directions of the vortex: above the disk a rotation of the vortex is completely stopped and picture of the flow does not differ from the turbulent convection in non-rotating fluids, while below a grid a turbulent vortex with a conical shape is rotated with large angular velocity. A wave propagation of the disturbances in both direction of the vortex was studied by Hopfinger *et al* , (1982) at the case of vortices produced by the oscillating grid in rotating tank. The conical shape of the turbulent vortex is defined by the different processes at the lower (heating disk) boundary , where the breaking point is the base of the vortex, and upper boundary, where there are the disturbances and mixing from the grid. When the upper boundary is rigid the upflow will moved horizontally in all direction on this boundary and the conical form is expressed much better than in the case of a grid boundary. The rotating turbulent vortex is much more stable to side disturbances than the conical vortex, and mostly its position above the disk is symmetric.

When the height of the layer is much less than diameter of disk $H<<D$ ($\delta>>1$) the conical vortex transform to the turbulent toroidal vortex with the different rotations at the upper and lower boundaries and the well pronounced eye at the centre.

If we will applied the result of these experiments to the geophysics, we can classified the intensive vortices using the aspect ratio parameter δ, which is the ratio of the scale of the vortex (diameter) or the heating region for the origin of this vortex to the height of the layer, where this vortex can propagate (atmosphere or the distance between clouds and ground in the case of tornado). When $\delta>>1$, the hurricane or toroidal turbulent vortex can exists, when $\delta \cong 1$, the tornado or conical turbulent vortex, and $\delta<<1$ the dust devils, or cylindrical vortices can be be developed.

References

Boubnov B.M. and Golitsyn G.S. (1995) *Convection in Rotating Fluids*, Kluwer Academic Publishers, Dordrecht.

Boubnov B.M. and Golitsyn G.S. (1990) Temperature and velocity field regimes of convective motions in a rotating plane fluid layer, *J. Fluid Mech.*, **219**, 215-239.

Boubnov B.M. and van Heijst G.J.F. (1994) Experiments on convection from a horizontal plate with and without background rotation, *Experiments in Fluids* **16**, 155-164.

Fitzjarrald D.E.(1973) A laboratory simulation of convective vortices, *J. Atmos.Sci.*, **30**, 894-902.

Fultz D. (1951) Experimental analogies to atmospheric motions, in Malone T.F. (ed.), *Compendium of Meteorology*, Boston, pp. 1235-1248.

Hide R. (1953) Some experiments on thermal convection in rotating fluids, *Quart.J. Roy. Soc.* **78**, No. 339, 161.

Hopfinger E.J., Browand F.K. and Gange Y. (1982) Turbulence and waves in rotating tank, *J. Fluid Mech.*, **125**, 505-534.

Maxworthy T. and Narimosa S. (1991) Vortex generation by convection in a rotating fluid, *Ocean Modelling*, 1037-1040.

THREE-DIMENSIONAL TURBULENT BOUNDARY LAYER IN AN "S"-SHAPED DUCT

J.M. BRUNS, H.H. FERNHOLZ
Hermann-Föttinger-Institut für Strömungsmechanik, TU Berlin
T.V. TRUONG
Institut de Machines Hydrauliques et de Mécanique des Fluides, EPFL

1. Introduction

A three-dimensional turbulent boundary layer in an "S"-shaped duct has been investigated experimentally. The quasi two-dimensional boundary layer in the straight entry section turns three-dimensional in the "S"-section, driven by a strong lateral pressure gradient, and finally recovers back to a two-dimensional flow in the straight part at the outlet. The boundary layer downstream of the curved section has "cross-over" velocity profiles with a change of sign in the W-component. Earlier measurements with a smaller curvature and thus a smaller lateral pressure gradient have been reported by Truong and Brunet (1992) and Löfdahl et al. (1993). First results of the present flow configuration were presented by Bruns and Truong (1994).

2. Experimental Arrangement

The test section and its coordinate system are presented in Fig. 1. Internal sidewalls are installed to remove the sidewall boundary layers, thereby reducing their influence. The Reynolds number at the entrance of the test section was set to 10^6 per meter. Approximately 1200 pressure taps in the test plate allow detailed measurements of the static pressure. A traverse mechanism enables the probe to move in the two lateral and the vertical directions and to rotate around the vertical axis as well as around the probe axis (Truong and Brunet, 1992). Measurements were carried out at 33 locations along three geometrical lines in the three-dimensional "S"-section, but only results at selected positions along the center line (P1-P6) will be presented.

The skin friction was measured by means of a Preston tube at position P1 and by a surface fence at positions P2-P6. The velocity field was investigated utilizing single hot-wire and triple hot-wire probes. Because probes and traverse mechanisms moving in the flow field severely disturb the flow close to the wall, near wall measurements were performed with a special hot-wire probe, where the prongs extended through the wall. All hot-wire probes have 2.5 μm tungsten wires with a sensor length of 0.5 mm and gold plated ends. The local flow angle β of the velocity vector in planes parallel to the

S. Gavrilakis et al. (eds.), Advances in Turbulence VI, 429-432.
© 1996 *Kluwer Academic Publishers.*

wall was determined by the bi-sector method with a rotating single-wire probe and the wall angle by rotating the surface fence. Miniature triple hot-wire probes made at TU Berlin were used to measure the Reynolds stresses which can measure all components of the instantaneous velocity vector. The three wires are positioned orthogonally on the edges of a cube. During the measurements the triple hot-wire probe was always aligned with the local mean flow direction.

3. Results

Fig. 2 shows the development of the pressure and skin-friction coefficient along three geometrical lines indicated in Fig. 1 as "Up, Center, and Down". The location of the minima of the C_p distributions corresponds to maxima of C_f.

Fig. 3 presents profiles of the flow angle $\beta - \alpha$ with respect to the external streamline against the wall distance y^+ for 6 positions along the center line. The cross flow angle decreases in the streamwise direction from P1 to the typical three-dimensional profiles (P2-P4) to a minimum of -18 degrees at P3. Downstream of the inflection point of the "S"-shape $\beta - \alpha$ increases up to 4 degrees at P5 and P6. The flow angle changes sign at the "cross-over" profiles (P5 and P6). The wall streamline angle corresponds with the symbol at $y^+ = 2$ and represents independent measurements with the surface fence.

The velocity magnitude profiles (wall units, Fig. 4) approximate the behavior of two-dimensional boundary layers in that they coincide with the logarithmic law of the wall.

Profiles of twice the turbulent kinetic energy, scaled by the skin friction velocity, are presented in Fig. 5. The profiles show their peak values close to the wall where the production of the turbulent energy is largest. The typical three-dimensional profiles show q^2 distributions similar to those in two-dimensional boundary layers, whereas the "cross-over" profiles have much higher values due to the increased production by the cross flow velocity gradient $\partial W / \partial y$. The uv Reynolds shear-stress profiles are affected by both the streamwise and lateral pressure gradient with peak values moving away from the wall for the "cross-over" profiles (Fig. 6). The vw component of the Reynolds shear-stress reaches peak values of 30% of uv (Fig. 7). It is zero for a 2-D boundary layer and changes sign from negative to positive when the "cross-over" occurs. uw is the second largest component with peak values of 60% of uv (Fig. 8). Again a change of sign, now from positive to negative, occurs in correspondence with the "cross-over".

The influence of the lateral pressure gradient on the mean flow and the turbulence structure can also be characterized by the direction of the shear-stress vector γ_τ in relation to the direction of the velocity gradient vector γ_g which are defined by

$$\gamma_\tau = \arctan\left(\frac{vw}{uv}\right), \quad \gamma_g = \arctan\left(\frac{\partial W / \partial y}{\partial U / \partial y}\right).$$

Fig. 9 and 10 show γ_τ and γ_g for positions P3 (typical three-dimensional profile) and P6 ("cross-over" profile). The shear-stress vector lags behind the velocity gradient vector for the three-dimensional as well as for the "cross-over" profile. Note that the "cross-over" profile in Fig. 10 shows maximum values of γ_τ, γ_g and $\gamma_\tau - \gamma_g$ where the cross flow velocity changes sign.

4. References

Truong T.V. and Brunet M. (1992) Test Case T1: Boundary layer in a "S" shaped channel, *Proceedings Ercoftac Workshop on Numerical Simulation of Unsteady Flows, Transition to Turbulence*, Cambridge University Press

Löfdahl L., Truong T.V., Johansson B. Bruns J. and Olsson C.O. (1993) Influence of different sensor Configurations on the Turbulent Quantities in a Complex Three-dimensional Flow Field, *ASME-Fluid Engineering Conference*, Washington DC, USA

Bruns, J. and Truong, T.V.(1994) Flow in an "S"-Shaped Duct, *Proceedings of the 2nd International Conference on Experimental Fluid Mechanics*, July 4-8, 1994, Torino, Italy.

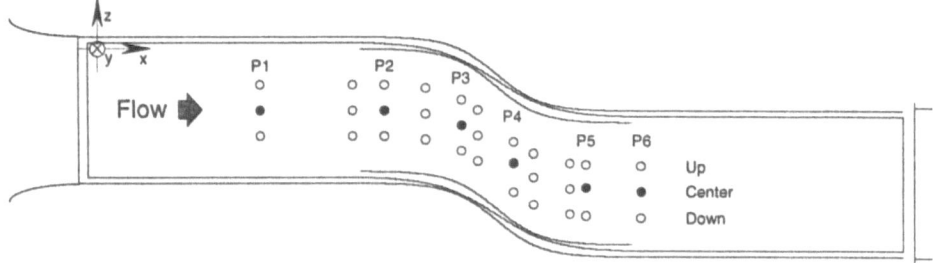

Figure 1: Location of the measured stations in the "S"-shaped channel

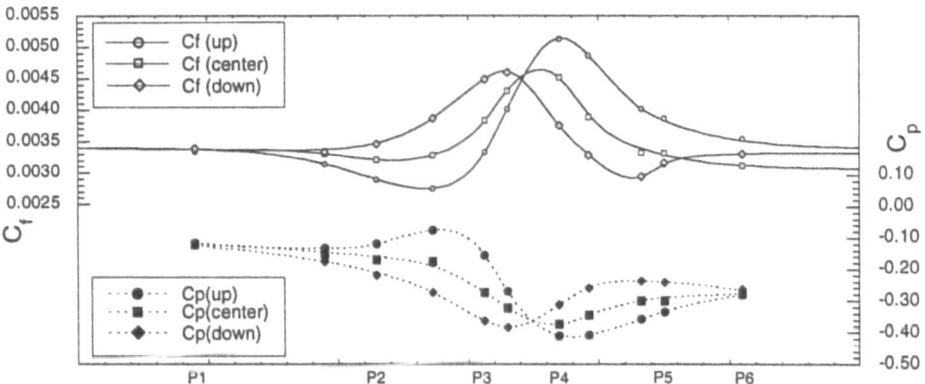

Figure 2: Evolution of the C_f, C_p along the "S"-shaped channel

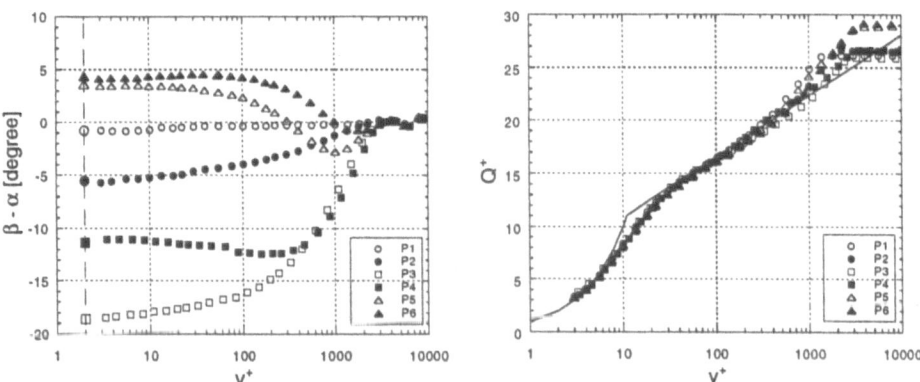

Figure 3: Flow angle profiles   Figure 4: Velocity magnitude profiles

432

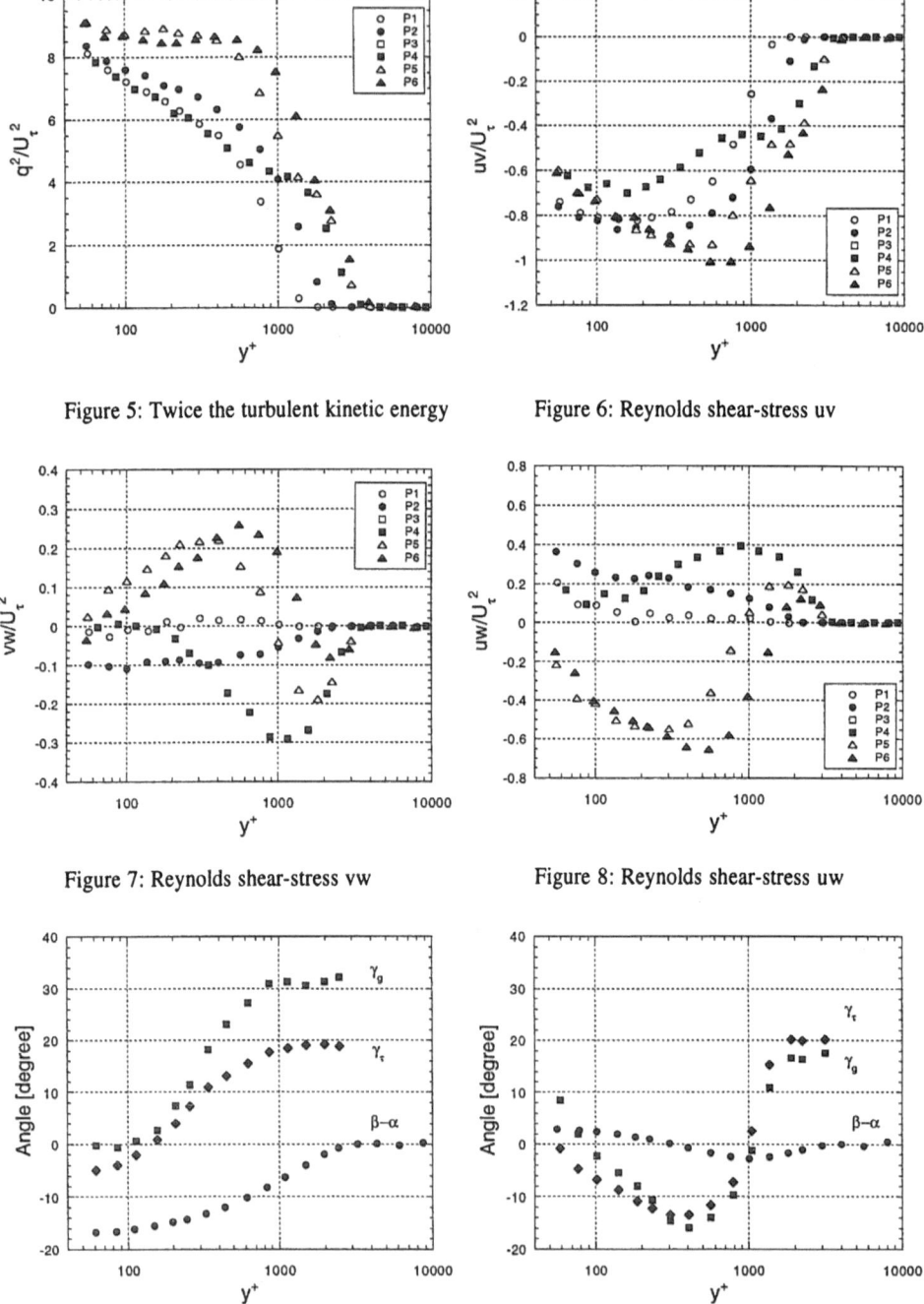

Figure 5: Twice the turbulent kinetic energy

Figure 6: Reynolds shear-stress uv

Figure 7: Reynolds shear-stress vw

Figure 8: Reynolds shear-stress uw

Figure 9: Velocity (β-α), velocity gradient γ_g and shear-stress γ_τ vectors at P3

Figure 10: Velocity (β-α), velocity gradient γ_g and shear-stress γ_τ vectors at P5

SOME TURBULENCE CHARACTERISTICS DOWNSTREAM A SHOCK WAVE - BOUNDARY LAYER INTERACTION

J. DELEUZE AND M. ELENA
Institut de Recherche sur les Phénomènes Hors Équilibre
(Ex. Institut de Mécanique Statistique de la Turbulence)
12, avenue du Général Leclerc
13003 Marseille France

1. Introduction

For many years, shock wave - boundary layer interactions have been studied in different cases. Nevertheless, turbulence measurements are not much numerous, especially when the interaction is produced by an incident shock wave (Délery, 1992).

The experimental results, which are presented here, have been obtained in a turbulent boundary layer downstream its interaction with an oblique shock wave inducing a separation.

2. Experimental conditions and method of measurement

The experiment was performed in the S8 nozzle of the IRPHE (ex-IMST) supersonic wind tunnel. The studied boundary layer developed on the lower quasi-adiabatic wall and was characterized as being reasonably two dimensional and fully turbulent when it was touched by an incident shock wave. This shock wave was created by a shock generator fixed on the upper wall and inclined at 8°. The penetration of this shock wave down to the subsonic region of the boundary layer creates a separation, generates a reflected shock upstream the point of impact and induces an expansion fan.

Upstream the interaction, the nominal test conditions and the parameters of the boundary layer were as follow:
- freestream Mach number: $M_e = 2.32$,
- boundary layer thickness relative to $U = 0.99 \, U_e$: $(\delta_{99})_0 = \delta_0 = 11$ mm,
- Reynolds number: $R_{\delta_{2i}} = 3400$, where $R_{\delta_{2i}} = \dfrac{\rho_e U_e \delta_{2i}}{\mu_w}$, and δ_{2i} the incompressible momentum thickness,
- skin friction coefficient: $C_f = 2 \cdot 10^{-3}$.

The length of the central region of the interaction constituted by the shocks - expansion system, corresponds to $4 \, \delta_0$. The results presented here concern the relaxation region of

433

S. Gavrilakis et al. (eds.), Advances in Turbulence VI, 433-436.
© 1996 *Kluwer Academic Publishers.*

434

the boundary layer which is located between the abscisses $X = 8\ \delta_0$ and $X = 16\ \delta_0$ counted from the beginning of the interaction.
In the experiments, the velocity was measured with a two components laser Doppler velocimeter in the forward scattering mode. The laser velocimetry set up was already described by Elena et al. (1985).

3. Results

In the relaxation region, downstream the interaction, the main experimental results are (Deleuze, 1995):
- The evolution of the streamwise mean velocity shows an amplification in the wake region (fig. 1).
- An amplification of the second-order statistical moments of streamwise and vertical velocity fluctuations is shows (fig. 2); it follows a maximum which is detached from the wall. Downstream the interaction, these evolutions tend to get closer to the evolutions of a fully turbulent boundary layer.
- The main term of turbulence production $(-\overline{u'v'}\ \frac{\partial U}{\partial y})$ presents a second peak in the middle of the boundary layer (fig. 3).
- Quadrant analysis using the distribution of the instantaneous Reynolds stresses $(\overline{u'v'})_i$ in each of the four quadrants of the (u',v') plane is made in the relaxation region. It gives informations on the structure of the boundary layer. The instantaneous stresses distributions are deeply modified: in the first half part of the boundary layer ($y/\delta < 0.5$), the well-known "sweep events" are predominant (fig. 4).

So, some great similarities in the relaxation mechanisms are observed between the results of this study and those of Skäre & Krögstad (1994) obtained in a subsonic boundary layer submitted to an adverse pressure gradient.

4. Conclusion

In the relaxation region ($4.5\ \delta_0 < X < 16\ \delta_0$), the studied boundary layer seems to be submitted to perturbations essentially governed by the pressure gradient represented by the shocks - expansion system in the interaction. For the investigated longitudinal distances, the separation does not seem to act on the boundary layer structure. Only the effect of compression through the shocks propagates to the relaxation region.

References

Délery J. (1992): Etude expérimentale de la réflexion d'une onde de choc sur une paroi chauffée en présence d'une couche limite turbulente, *La recherche aérospatiale*, n°1, 1-23.
Deleuze J.(1995): Structure d'une couche limite turbulente soumise à une onde de choc incidente, *Thèse de l'Université d'Aix-Marseille II*.

Elena M., Lacharme J.P., Gaviglio J.(1985): Comparison of hot-wire and laser Doppler anemometry methods in supersonic turbulent boundary layer, *International symposium on laser anemometry*, FED vol 33, Miami.

Skäre P.E., Krögstad P.A.(1994): A turbulent equilibrium boundary layer near separation", *J. Fluid Mech.*, vol. 272, 319-348.

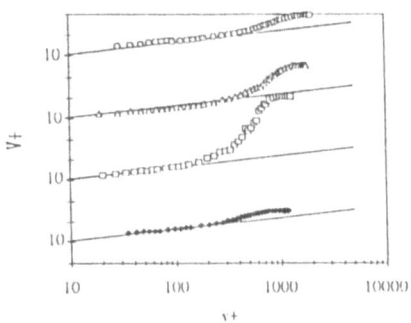

Figure 1 : Streamwise mean velocity

◆ : X=-4.2 δ_0, □ : X=8.5 δ_0, △ : X=12.2 δ_0, ○ : X=15.8 δ_0

——— : $V^+ = \frac{1}{K}\ln(y^+) + C_1$

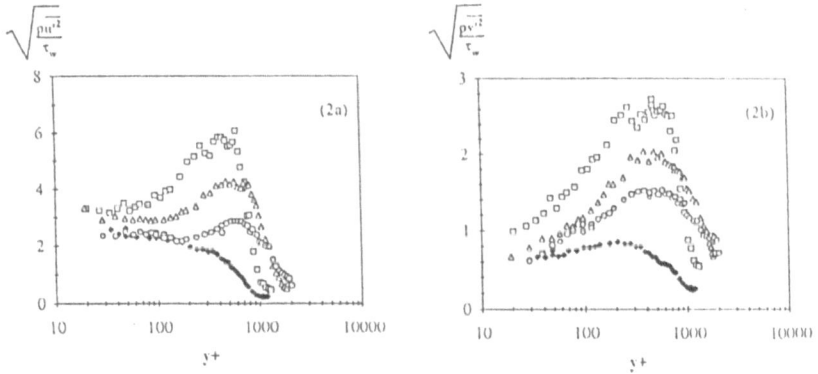

Figure 2. (a): Fluctuations of streamwise velocity,
(b): Fluctuation of vertical velocity

◆ : X = -4.2 δ_0, □ : X = 8.5 δ_0,
△ . X = 12.2 δ_0, ○ . X = 15.8 δ_0

(a) (b)

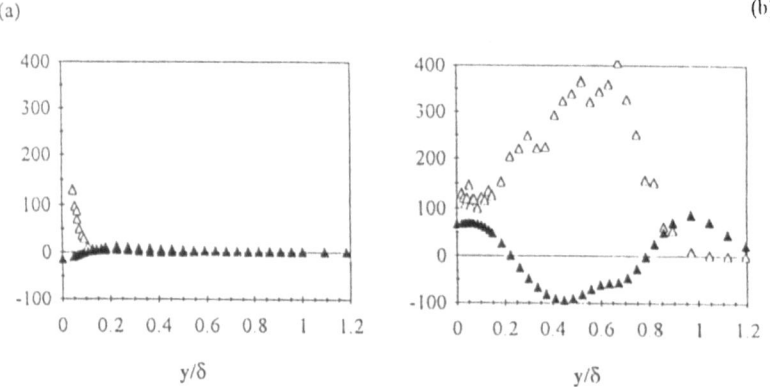

Figure 3: **Production and turbulence diffusion**

Δ : Production, ▲ : Diffusion

(a) : X=-4.2 δ_0, (b) : X=8.5 δ_0

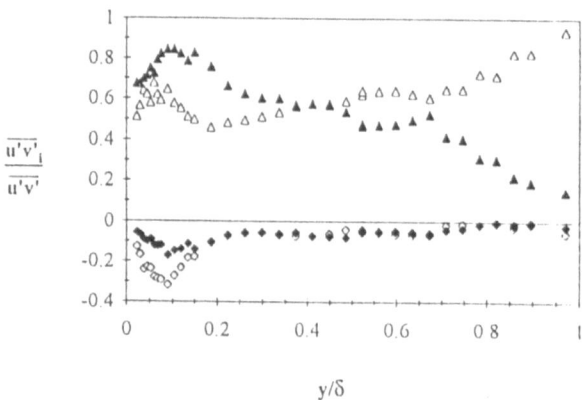

Figure 4: **Contributions of $\overline{u'v'}_i$ to $\overline{u'v'}$ (X=8.5 δ_0)**

◇ : $\overline{u'v'}_1$, Δ : $\overline{u'v'}_2$, ◆ : $\overline{u'v'}_3$, ▲ : $\overline{u'v'}_4$

SCALING OF VORTICITY FILAMENTS IN A TURBULENT SWIRLING FLOW

B. DERNONCOURT, J.-F. PINTON, S. FAUVE

École Normale Supérieure de Lyon
Laboratoire de Physique CNRS, 69364 Lyon France

Abstract. We study vorticity fluctuations in a turbulent swirling flow generated in the gap between coaxial contra-rotating disks. The flow is probed acoustically by an ultrasonic incident wave and the scattered pressure reveals intense vorticity concentrations in the bulk of the flow. They are compatible with vorticity filaments observed in numerical simulations and their core size scales like the Taylor microscale [1].

1. Introduction

Intense vorticity concentrations have been first observed in direct numerical simulations of fully developed turbulence and their characteristics have since been widely studied [2,3,4,5,6]. The existence of "vorticity filaments" with length l_f corresponding to a large spatial scale, a core size r_f corresponding to a small spatial scale and velocity increment δu_f (on their core size) in the large scale range (probably u_{rms}) is fairly well established. However, the exact scalings of r_f, l_f and δu_f versus the Reynolds number are not known due to the finite Reynolds number range of the numerical simulations and to possible problems in the filament detection algorithm.

In experiments, coherent vorticity structures in the form of filaments have been visualized in the von Kármán swirling flow [7]. Studies of their dynamical and statistical properties have been performed via pressure measurements [8,9] and pressure-velocity correlations [10]. These measurements, while yielding quantitative results, rely on *indirect* detection of the filaments and were performed at the flow boundaries. In contrast, we use an ultrasound scattering technique that allows a *direct* measurement of the vorticity in the *bulk* of the flow.

S. Gavrilakis et al. (eds.), Advances in Turbulence VI, 437-440.

2. Experimental set-up and measurement technique

A turbulent flow is produced inside a closed vessel, in the gap between two coaxial contra-rotating disks – the experimental apparatus is described in [9]. The rotation frequency $(1/T_{rot})$ is varied from 650 to 1500 rpm, and 3 different fluids are used (F1: pure water), (F2: mixture water/glycerol 28% mass fraction), (F3: mixture water/glycerol 48% mass fraction). Reynolds numbers – based on disk radius and rotation speed – ranging from $1.5\,10^5$ to $1.5\,10^6$ are achieved. We use an ultrasound scattering technique to probe the Fourier components of the vorticity field parallel to the rotation axis [12,13,14]. The sound emitter and receiver are reversible piezoelectric broad band transducers located in a plane perpendicular to the rotation axis. The scattering angle is $\theta = 60°$ in the forward direction.

The scattering amplitude in the case of a localized vortex has been obtained by Ferziger [10]. For a rectilinear Gaussian vortex of core size a and intensity Γ_0, he showed that the scattering cross-section $S(q, \theta)$ is peaked at $q_{max} = 1/2a \sin(\theta/2)$ (q is the scattering wave-vector). A maximum of scattered amplitude is thus observed for a wavelength of the order of the vortex core size ($q_{max} = 1/a$, in our experimental set-up).

Because the transducers allow the detection of the sound refracted by the flow and reflected on the boundaries, the acoustic signal always displays a peak at the frequency of the incoming sound wave. In order to obtain the scattered contribution alone it is filtered out using a short-term Fourier analysis. In this process, Fourier transforms are calculated with a 20 kHz span, giving an effective integration time $\sim$ 1ms. This time is small compared to the advection time of the filaments across the measurement volume ($\sim$ 10ms), but it is long compared to the filaments turnover time ($\sim$ 100μs, of the order of the Kolmogorov time scale). Therefore, the process of short-time Fourier transform and filtering of the scattered pressure signal allows the detection of large vorticity concentrations on small spatial scales but not the study of their detailed internal dynamics.

3. Results

Figure 1a shows a time evolution of the axial vorticity field, measured using the short time Fourier transform method for an incoming sound frequency equal to 10MHz, i.e. $\lambda = 150\mu$m, close to the Taylor microscale of the flow. It is readily observed that the scattered sound amplitude has a very intermittent behavior. Sharp rare events, with magnitude up to 10 standard deviations, occur over a more regular background.

If one isolates only the largest events (for example with $|\omega(q, t)| > 4$ standard deviations), the corresponding time spectra are similar to those in figure 1b. They display very sharp peaks at well defined, but not constant,

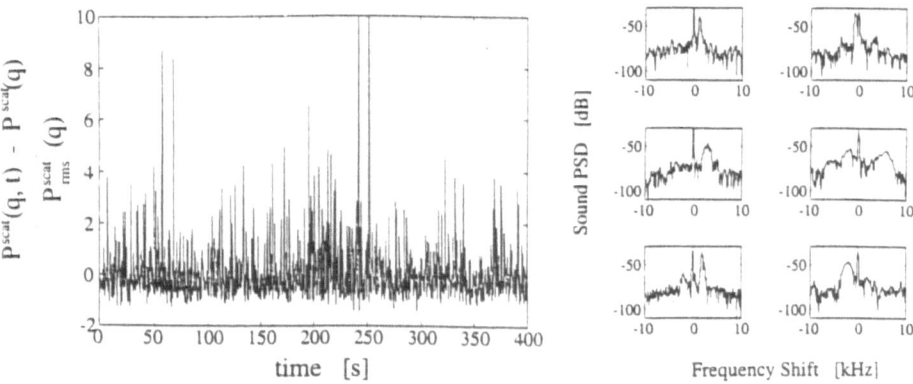

Figure 1. Time evolution of the axial vorticity field (a) and examples of corresponding individual power spectra (b). The Reynolds number is $Re = 10^6$.

Doppler shifts. Therefore in each case the scattered sound field is due to a structure that has an advection speed constant across the measurement volume. This is coherent with the existence of worm-like structures formed from interactions of large eddies.

This intermittent behavior of the scattered pressure is observed only in a certain range of incoming sound frequencies, that is when the vorticity field is probed in a given range of scales. Figure 2a shows the fluctuations in time outside that range. Compared with figure 1a, one clearly observes an absence of isolated large events and a more Gaussian behavior of the vorticity fluctuations.

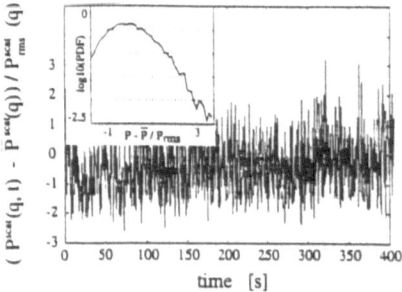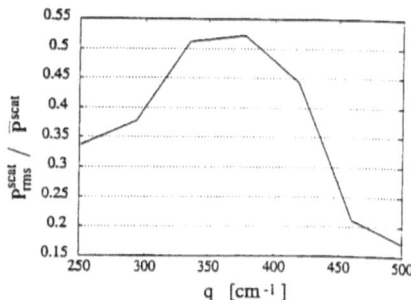

Figure 2. (a) Time evolution of the axial vorticity field for a scattering wave vector q corresponding to a scale of 250 μm. (b) Axial vorticity fluctuations *vs.* q, at ($Re = 10^6$).

Indeed, when the fluctuation level $- P_{rms}^{\text{scat}}(q)/\overline{P}^{\text{scat}}(q)$, where rms refers to the standard deviation and the overbar denotes time averaging – is computed for varying scattering wavenumbers (figure 2b), one observes a characteristic scale for the detection of the filaments: $\sim 180\mu m$ for a integral Reynolds number $Re = 10^6$.

The variation of the filaments core size with the integral Reynolds number is displayed in figure 3. For a relevant scaling for the characteristic size

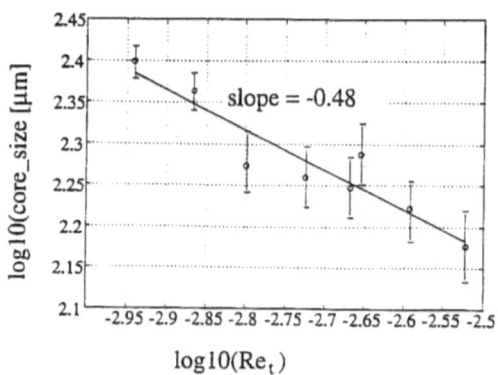

$\log 10(Re_t)$

Figure 3. Scaling of the core size with the turbulent Reynolds number – Re_t is known up to a proportionality constant, since u_{rms} is computed from pressure measurements at the wall, using $P_{rms} \propto \rho u_{rms}^2$.

of these vorticity concentrations to be obtain, one must use the *turbulent* Reynolds number $Re_t = u_{rms}R/\nu_v$, which in our experimental set-up using smooth disks is not proportional to the *integral* Reynolds number $Re = R^2 f_{rot}/\nu_v$ in the explored range of rotation frequency. With this correction taken into account, our results indicate a $Re_t^{-0.48}$ scaling for the filaments core size which confirms the relevance of the Taylor microscale as the characteristic size of the core of the vorticity filaments.

References

1. B. Dernoncourt, J.-F. Pinton and S. Fauve, submitted to *J. Phys. II France*
2. E. D. Siggia (1981), *J. Fluid Mech.*, Vol. no. **107**, pp. 375-406.
3. M. Brachet (1990), *C. R. Acad. Sci. (Paris)*, Vol. no. **311**(II), pp. 775-780.
4. Z.S. She, E. Jackson and S. A. Orszag (1990), *Nature*, Vol. no. **344**, pp. 226-228.
5. A. Vincent and M. Menneguzi (1991), *J. Fluid Mech.*, Vol. no. **225**, pp. 1-20.
6. J. Jiménez *et al.* (1993), *J. Fluid Mech.*, **Vol. no. 255**, pp. 65-90.
7. S. Douady *et al.* (1991), *Phys. Rev. Lett.*, Vol. no. **67**, pp. 983-986.
8. S. Fauve *et al.* (1993), *J. Phys. II France*, Vol. no. **3**, 271-278.
9. P. Abry *et al.* (1994), *J. Phys. II France*, Vol. no. **4**, pp. 725-733.
10. O. Cadot, S. Douady and Y. Couder (1995), *Phys. Fluids* A, Vol. no. **7**, 630-646.
11. J. H. Ferziger (1974), *J. Acoust. Soc. Am.*, Vol. no. **56**(6), pp. 1705-1707.
12. P. Gromov *et al.* (1982), *Sov. Phys. Acoust.*, Vol. no. **28**, pp. 452-455.
13. C. Baudet, S. Ciliberto, JF. Pinton (1991), *Phys. Rev. Lett.*, Vol. no. **76**(2), pp. 193.
14. JF. Pinton, C. Baudet (1993), in *Turbulence in Spatially Extended Systems*, Nova Science.

DYNAMICAL FLOW TOMOGRAPHY BY LASER INDUCED FLUORESCENCE

S.DEUSCH[1], T.DRACOS[2], P.RYS[1]
Swiss Federal Institute of Technology, Zurich, Switzerland,
[1] Dept. of Industrial Chemistry and Chemical Engineering,
[2] Prof. em.

1. Introduction

Laser induced fluorescence (LIF) imaging is a well developed method to directly obtain the 2D structure of a flow by imaging chemical species which emit a sufficient signal to be detected. The principles of LIF are also applicable to a 3D-imaging technique. The first one to put this into practice was Dahm et al.[1], others are Merkel[5] and Dracos et al.[2]. This work presents the development of a new 3D-LIF imaging method where the illuminating laser beam is swept parallel to the imaging sensor and a volume of 45x45x18 mm dimension is recorded with a resolution of 256x256x90 pixels. The obtained 3D images are evaluated to extract the velocity, vorticity, and rate-of-strain tensor fields from the grey level distribution by an alternative technique to [1]. Also some experimental evaluations are presented briefly.

2. Experiment

The turbulent flow studied is a circular free jet. The flow facility has been described by Merkel [5]. The turbulent free jet issues from a d=5mm orifice with a jet Reynolds number of 4300. The investigated flow is situated at x/d=255 from the orifice, but 10 cm off the axis where the steepest mean velocity gradient is expected. The discharging jet water is homogeneously diluted with fluorescent tracer which is excited to fluorescence emission by irradiation of an Ar-ion laser at the described location. During the recording time the camera and the laser excitation optics move with the local mean velocity of the flow for a Lagrangian view. Fig. 1 shows an example of two tomographic volumes with turbulent mixing. The volumes consist each of 90 layers of 256x256 pixels each, representing 45x45x18mm^3 of physical space. In contrast to former work, the laser sheet for the excitation illumination of one layer is obtained dynamically by passing the laser beam through a rapidly rotating plexiglass prism at right angle to its spinning axis. This type of scanner produces a completely parallel shift of the laser beam and a nearly uniform plane of illumination. The prism rotates at 3750 rpm thus illuminating 250 planes/s. The time increment for recording two consecutive flow volumes is 0.4s. The illuminating

441

S. Gavrilakis et al. (eds.), Advances in Turbulence VI, 441-444.

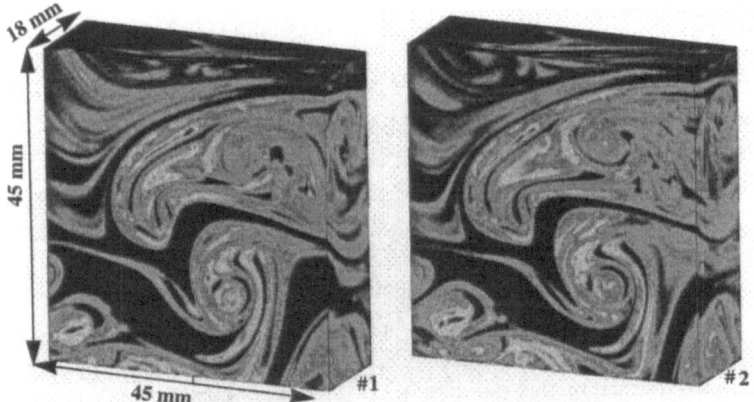

Fig. 1. :Two consecutive tomographic flow volumes, 45x45x18mm³, Δt=0.4s

laser sheet is scanned in depth direction parallel to itself by a small deflection mirror which is mechanically synchronized to the prism. The boundary conditions for this tomographic method are satisfied since the time for recording a volume of 90 layers is 0.36s, considerably shorter than the found Kolmogorov time scale of 1.5s, and the spatial resolution of 0.2mm is also well below the calculated Kolmogorov length scale of 1.3mm.

3. Determination of Velocity and Deformation Tensor for each Fluid Particle of the Tomographic Volume by Adaptive Least Square Correlation (ALSC)[4]

The theoretical fundament of the combined LIF-Tomography-ALSC technique is that the calculated displacement of a patch of dyed voxels can be equalled to the displacement of the corresponding fluid particle itself as long as the molecular diffusion of the fluorescent tracer is negligible compared to the momentum diffusion of the fluid. The equation of the scalar isosurface velocity $u_c = u - D((\nabla^2 c)/|\nabla c|)e_g = u - \nu/(Sc)((\nabla^2 c)/|\nabla c|)e_g$ can be approximated by $u_c = u$, if the second term on the right hand side is small. c is the scalar field, D is the scalar diffusivity, and e_g is the unit vector of the gradient direction. The fluorescent tracer is highly diluted, $\sim 1.10^{-7}$mol/l disodium fluorescein in water, the molecular diffusion coefficient is $3,7.10^{-6}$ cm²/s [3]. The Schmidt number, Sc = ν/D, is 3200 thus indicating the dominating role of advective transport over diffusion. The ALSC algorithm finds correlated patches of pixels in subsequent flow volumes. The original patch will be found displaced and distorted in the subsequent configuration by a displacement vector and a homogeneous deformation tensor. Introducing the experimental time Δt, one obtains the velocity vector and the velocity gradient tensor $Y_{ij} = \partial v_i / \partial x_j$. The latter tensor can be split up into its symmetric part and its skew-symmetric part according to $Y_{ij} = 1/2(Y_{ij}+Y_{ji})+1/2(Y_{ij}-Y_{ji}) = D_{ij}+R_{ij}$, with D_{ij}, the rate-of-strain tensor, and R_{ij}, the vorticity tensor.

4. Results

The discussion of the flow geometry and topology is given in terms of tensor invariants. Soria[6] applied this method to the study of incompressible mixing layers of shear flows generated by direct numerical simulations. The eigenvalues of the velocity gradient tensor Y_{ij} satisfy the characteristic equation $\lambda^3 - I\lambda^2 + II\lambda - III = 0$,

where $I = Y_{ii}$, $II = Y_{ii}Y_{jj} - Y_{ij}Y_{ji}$, and $III = \det(Y_{ij})$.

The first invariant, $I = Y_{ii} = D_{ii}$, is the continuity and therefore should be zero in water. Fig. 2. shows a correlation of 0.9988 between $|u_{1,x} + u_{2,y}|$ and $|u_{3,z}|$. With $I = 0$, the local flow geometry is completely described by the second and third invariants. Of interest are also the invariants of the rate-of-strain tensor D_{ij} , and of the rate of rotation tensor R_{ij}. Figures 3-5 show scatter diagrams of the invariants, cf.[6]. In the $I = 0$ plane of invariants, figs. 3-4, the solid line separates the regions of real λ from complex values. Fig. 3 shows that the observed data points cluster near the origin of the plot with a slight orientation into the lower right quadrant where unstable node saddle-saddle prevails. Unlike [6], not many points are found at high gradients in the upper left quadrant (stable focus stretching). Possibly, this is because the results are only a small portion of the entire mixing layer, see fig.1. Fig. 4 shows the rate-of-strain invariants, from where it can be seen that the velocity gradient tensor admits all possible incompressible topologies. The more negative values of II_D are points of higher dissipation. These points tend to lie in the right half plane, fig. 4, depicting a rate-of-strain tensor of the topology of saddle-saddle-unstable node. However, these points of highest dissipation constitute only a small fraction of the total dissipation, i.e. the bulk of the dissipation is contributed by points belonging to small velocity gradients and intermediate to large scales. A magnification around II_D close to zero, also reveals a tendency for the intermediate and large scales to lie in the right half-plane of topology unstable-node-saddle-saddle. Fig.5 is most interesting since it tells about the roles of strain and enstrophy. The mechanical dissipation of kinetic energy due to viscous friction, $\varepsilon = 2\nu D_{ij}D_{ij} = -4\nu II_D$, is proportional to $- II_D$. Thus large negative values of II_D correspond to high rates of kinetic energy dissipation. Because of the $I = 0$ case, II may be decomposed to give $II = 1/2(R_{ij}R_{ij} - D_{ij}D_{ij})$. Large values of II indicate regions with strain dominating over the enstrophy, large positive values of II indicate regions with enstrophy dominating over the strain. Fig.5 confirms the finding of fig.3 that little rotational dissipation is found, instead more irrotational dissipation is found, i.e. points above the 45° line, and a slight accumulation along the 45° line corresponding to a vortex sheet-like structure of the free shear layer. Fig. 6 shows the tendency of the enstrophy generation term, normalized by ω', to have the same signum function as the intermediate eigenvalue of the rate of-strain-tensor. Fig.7 shows the typical probability density function of the three eigenvalues of the rate-of-strain tensor (normalized by $z/\sqrt{6}$, $z^2 = \alpha^2 + \beta^2 + \gamma^2$) with the most compressive value, $\gamma < 0$, balancing the two other stresses, α and β. Also could be verified the well known result of alignment of β with vorticity ω, both being even more aligned with the vorticity magnitude increasing.

444

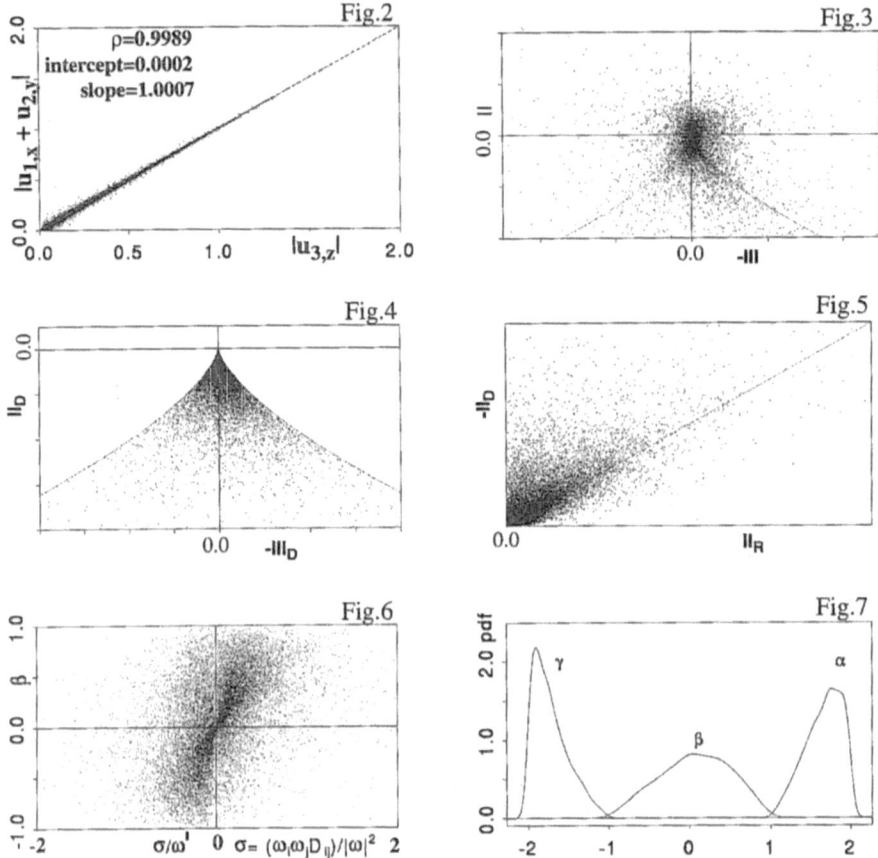

Fig.2-7: Some kinematic results of 3D LIF-ALSC method

5. References

1. Dahm, W.J.A., Lester, K.S., Southerland, K.B.; A scalar imaging velocimetry technique for fully resolved four-dimensional vector velocity field measurements in turbulent flows, *Phys.Fluids* A 4 (10), 2191 (1992)

2. Dracos, T., Merkel, G.J., Rys, F.S., Rys, P., Maas, H.G., Grün, A.W.; Velocity field measurements using Laser Induced Fluorescence Tomography, *Proc. Int. Workshop on PIV*, Fukui 95, July 2-5, Japan

3. Hodges, K.C., La Mer, V.K, J. *Am.Chem.Soc.* **70**, 724 (1948)

4. Maas, H.G.; Determination of velocity fields in flow tomography sequences by 3D least squares matching. Optical 3-D Measurement techniques II, (ed. Grün, A., Kahmen, H.), Wichmann Verlag, Karlsruhe, (1993)

5. Merkel, G.J.; Ph.D. thesis ETH No. 11174, (1995), see also Merkel,G.J., Dracos, T., Rys, P.; Two-dimensional and three-dimensional imaging of passive scalar fields in a turbulent jet, *Progress in Visualization*, **Vol. 2**, CRC Press Inc., (1996), in print

6. Soria, J., Sondergaard, R., Cantwell, B.J., Chong, M.S., Perry, A.E.; A study of the fine-scale motions of incompressible time-developing mixing layers, *Phys.Fluids* **6** (2), 871 (1994)

LOW REYNOLDS NUMBER TURBULENT NEAR WAKES

T. D. GOUGH and P. E. HANCOCK
Department of Mechanical Engineering
University of Surrey,
Guildford, Surrey, UK

1. Introduction

Accurate calculation of flow around an aerofoil or blade requires accurate prediction of the flow near the trailing edge and in the near wake. The near wake is primarily characterised by two changes in boundary condition, namely the sudden removal of the no-slip and impermeability conditions.

Measurements have been made in two low-Reynolds number wakes, one symmetrical and one asymmetrical with a boundary layer thickness ratio of 1.9. The geometry and flow conditions were matched as closely as possible to a large eddy simulation study using a rectangular variable density mesh. If the smallest mesh size is to remain an acceptable multiple of the smallest scales the computation time will increase with Reynolds number roughly as $(U_e\theta/v)^4$, where θ is the momentum thickness and U_e the free-stream velocity. So as to keep the computation times acceptably short, and to keep the effect of the sub-grid scale model acceptably small, it was decided that $U_e\theta/v$ should not exceed about 600. Given that turbulence is only naturally sustainable for Reynolds numbers greater than about 300 this Reynolds number range imposed severe constraints on the experiments, in particular the tripping of the boundary layer so as to be free of residual trip effects.

The simulations were performed in two parts, the first being of the boundary layer, and the second of the wake using the boundary layer data to provide inflow data at a position sufficiently far upstream of the trailing edge, so that the expected not insignificant upstream effects on the boundary layer flow were properly taken into account. The simulations (Gao and Voke, unpublished) will be reported separately.

The trailing edge thickness, t, was about 0.1 of the boundary layer thickness, δ, at the trailing edge. This is larger than typical of an aircraft wing but comparable to that of a wing element or blade.

2. Experimental arrangement and techniques

The wake was developed from the trailing edge of a flat plate, 300mm long, 600mm span and 0.96mm thick, the largeness of the span permitting the measurements to extend in to the far wake sufficiently free from end-flow effects. The plate length and free-stream velocity is a compromise; at 10m/s the boundary-layer thickness is about

445

S. Gavrilakis et al. (eds.), Advances in Turbulence VI, 445-448.

9mm, for $U_e\theta/v = 600$. Decreasing the free-stream velocity (inversely) increases the boundary layer thickness, thereby reducing probe resolution volume and position errors, but requires probe calibration over a lower velocity range and measurement of still smaller (inversely squared) wall shear stress. The free-stream velocity was 9.6m/s, and the pressure gradient was negligible. The boundary layer was tripped by means of a round wire finely glued to the plate surface. Its size was found to be crucial in terms of providing the correct amount of stimulation and, contrary to the suggestions in the literature, strongly dependent on the upstream laminar boundary layer thickness when this was relatively thick.

Hot-wire measurements were made by means of single-wire and sub-miniature x-wire probes. No wire-length effects were observed for single-wire measurements for lengths less than 1.0mm. The sensors of the cross-wire probes were nominally 0.45mm long separated by about 0.5mm. Probe resolution errors were shown to have been small, even in the regions of the large stress gradients immediately downstream of the trailing edge. Wall shear stress was measured using Preston tubes, and by fitting the velocity profile to the logarithmic law. It is now well demonstrated (eg Erm and Joubert, 1991) that the logarithmic law is not Reynolds number dependent at low Reynolds numbers. The Preston tube technique requires the existence of an inner-layer law, and the calibration is implied by this law; self consistency from tubes of differing diameters and with that inferred from the velocity profile therefore implies the existence of the inner-layer form. Agreement in friction velocity u_τ was within ±1.6%.

3. Results and Discussion

Integral parameters such as boundary layer and momentum thicknesses show that the effects of the trailing edge extend no further than about $10t$ - ie about 1δ - upstream of the trailing edge. Beyond this point (where δ_{995} was 9.1mm) there is a marked decrease in thickness parameters and inward movement of streamlines arising from the positive $\partial U/\partial x$ in the wake, requiring negative $\partial V/\partial y$, and the 'in filling' behind the plate. Figure 1 shows the mean velocity profile in law-of-the-wall coordinates at 2, 5, 10 and 20mm upstream of the trailing, ie at $x/t = -2, -5, -10$ and -20, where x is the streamwise distance from the trailing edge. In what follows y is the distance perpendicular to the plate and wake, with the origin taken either on the plate surface or on the wake centre-line. Measurements at and upstream of $-10t$ show good agreement with the standard logarithmic law, and with wall shear stress inferred from Preston tubes. A logarithmic region clearly exists between y^+ of 30 and 100, the outer limit corresponding to y/δ_{995} of 0.3, which is larger than at high Reynolds numbers where the extent is about 0.2, and in good agreement with Erm and Joubert. A Reynolds-number dependence in the outer edge is acceptable and plausible given the dependence of the outer layer structure. Nearer to the trailing edge the velocity profile departs from the inner layer law, and does so in a manner consistent with the expected acceleration. There is a clear departure at $-2t$, exhibiting a larger U. Here it has been assumed that, by contrast, the flow in the outer layer will have had little time to adjust. This implies an effective u_τ for the outer layer, which we have taken here to be equal to that at $-10t$ - ie about one boundary layer thickness upstream. Of course, this u_τ no longer corresponds to the shear stress at the wall or in the inner layer, which one would anticipate to be higher. The station at $-2t$ is about 0.2 boundary layer thicknesses upstream, or 0.7 inner-layer thicknesses, taking this as the point at which $y^+ = 100$. It is unlikely therefore that the inner layer will be in local equilibrium in this region.

Figure 2 shows $\overline{v^2}$ and $\overline{uv}$ at four stations, the first being 2mm upstream of the trailing edge, and the second 2mm downstream - the latter case is very similar to the first, as might be expected. (The slight non-antisymmetry in $\overline{uv}$ is because of wire-angle errors in the miniaturised x-wire probe.) Dramatic peaks in $\overline{v^2}$ and $\overline{uv}$ develop, and are maximum at about x = 5mm. These are largely (and probably entirely) due to vortex shedding, and develop in an inner layer. $\overline{u^2}$ (not shown) indicates two peaks slightly outside those in the shear stress, but these are much less noticeable, probably because of the higher level of $\overline{u^2}$. Peaks have also been found by Hayakawa and Iida (1992) and others, and likewise attributed to vortex shedding. Figure 3 shows the spectrum of the u-fluctuations near the centre-line at $x = 5$mm. The two distinct but broad peaks are evidence of alternate shedding. The width of each is about $\pm 0.3f$, where f is the corresponding centre frequency. Broad peaks presumedly arise because the rate at which vorticity is shed at the trailing edge fluctuates according to the streamwise fluctuation u near each surface. In that the velocity gradient in a turbulent boundary layer changes drastically outside the viscous sublayer - the ratio of $\partial U/\partial y$ in a logarithmic region to that at the wall is equal to $1/y^+$ - it seems reasonable to suppose to a first approximation that flow outside this inner layer behaves in effect like a free stream. In the case of figure 2 this edge is at 1.5mm, ie $y^+ = 50$, which is just beyond the 'knee' in the velocity profile $U(y)$, and based on the velocity at this point the Strouhal number is 0.16, remarkably similar to that of bluff bodies. Furthermore, the intensity, u'/U_e, at this position is about 0.1, and supposing the fluctuations to be Gaussian would imply a quasi-steady frequency spread of $\pm 0.3f$ - as indeed is observed. No vortex shedding was detected at $x = 40t$, ie at 4δ, and the shedding was weak at $10t$ - though the bandwidth was no different - nor much outside $y^+ = 50$. Persistence in v appears longer than in u because v is much smaller than u in the boundary layer; presumedly, the vortices become submerged within the surrounding turbulence and are eventually destroyed by it.

The inner layer of the wake as inferred from the centre-line velocity scales on $u_\tau x/v$, where u_τ is that at separation, whilst the streamwise extent to which this scaling applies appears to scale on outer-layer scales (eg θ), as might be anticipated. At about $3t$ there is a mean free stagnation point ending a recirculation region, downstream of which the centre-line velocity increases rapidly (and roughly linearly) so that at $10t$ it has already reached $0.5U_e$; this variation resembles that found in the wakes of bluff bodies. The extra-strain rate $\partial U/\partial x$ on the centre-line reaches about 0.23 of the maximum in $\partial U/\partial y$; the high $\partial U/\partial x$ presumably intensifies the streamwise vorticity streaks from the sub/buffer layers while the decreasing $\partial U/\partial y$ will tend to do the opposite.

4. References

Erm, L.P. and Joubert P.N. (1991) Low-Reynolds-number turbulent boundary layers,
  J. Fluid Mech. **230**, 1-44.
Hayakawa, M. and Iida, S. Behavior of turbulence in the near wake of a thin flat plate at low Reynolds
  numbers, Phys. Fluids A **4**, 2282-2291.

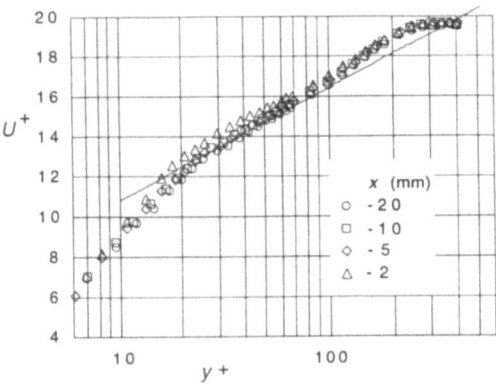

Figure 1   Effect of trailing edge on mean velocity .
Line is with k = 0.41 and C = 5.2.

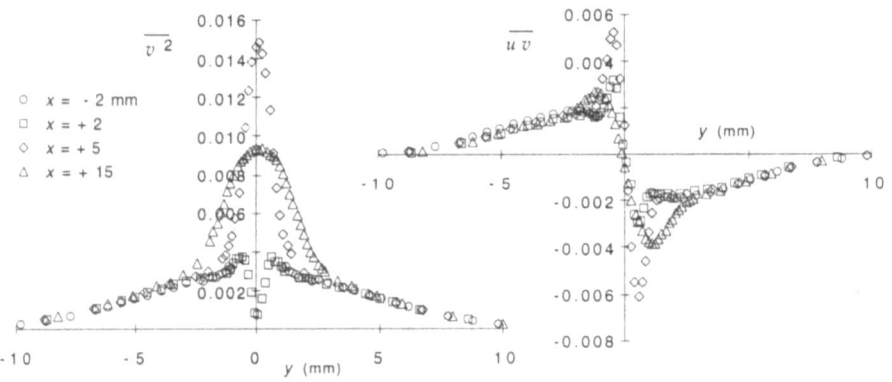

Figure 2   Reynolds stresses in near wake, $\overline{v^2}$ and $\overline{uv}$.

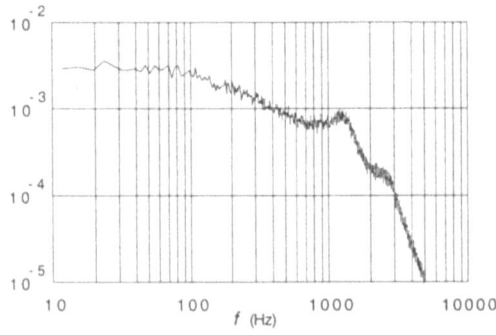

Figure 3   Spectrum of *u* at *x*, *y* = 5, 0.27mm.

ON SOME LOCAL ASPECTS OF TURBULENT DRAG REDUCING FLOWS OF DILUTE POLYMERS AND SURFACTANTS

A. GYR
Inst. of Hydromechanics and Water Resources Management
Swiss Federal Institute of Technology, CH-8093 Zurich
A. TSINOBER
Department of Fluid Mechanics and Heat Transfer,Faculty of Engineering
Tel-Aviv University, 69978 Tel-Aviv, Israel

Abstract

Local aspects of turbulent drag reducing flows were studied via evaluating the surrogate dissipation, the third order velocity structure function, the third order moment of the derivative of the velocity fluctuations of a flow in water and in drag reducing solutions (polymer and surfactants) at the same flow conditions. While both the dissipation and the enstrophy production are reduced strongly for both drag reducing flows, they exibit some qualitative differences. Namley, in a polymer solution flow the disispation is increasing with the distance from the wall, whereas it is decrasing for the flow of surfactant solution. The inhibition of enstrophy production is stronger in the near wall region in the polymer solution, while for the surfactant the tendency is opposite.Thus there are clear indications on the qualitative differences in mechanisms in the two drag reducing flows.

1. Introduction

The phenomenon of drag reduction is known mostly via its global manifestation in the dependence of the overall pressure drop versus mean flow rate. In spite of a great number of speculations the physical mechanism/s behind the phenomenon remain not well understood, Gyr and Bewersdorff (1995). Therefore it is useful to obtain some local information related to basic physical processes characteristic of turbulent flows with and without drag reducing additives under the same flow conditions.

For this purpose we made a comparison of the mean surrogate dissipation $(\partial u'/\partial t)^2$, the structure function F_3 eq. (3) and the skewness factor s eq. (4) at two

449

S. Gavrilakis et al. (eds.), Advances in Turbulence VI, 449-452.
© 1996 *Kluwer Academic Publishers.*

locations in a turbulent flow at $Re_D = 9600$ in a pipe of square cross section for water, dilute polymer solution (Separan AP 30 of 20 ppm) and surfactant ($C_{14}TASal$ 1.9 mMol/l) using the experimental velocity measurements of Stüer and Gyr (1995). The term 'surrogates' stands for the fact that ε is estimated via the steamwise derivative of the velocity component in flow direction only.

2. Theoretical preliminaries

2.1 MEAN SURROGATE DISSIPATION

The mean surrogate dissipation was estimated via the relation between the Taylor micro-scale, $\lambda_{(\tau)}$, and the Kolmogorov length-scale, η_k , given by

$$\eta_k = \left(\frac{v^3}{\varepsilon}\right)^{1/4} = \left(\frac{v^2 \lambda_{(\tau)}^2}{15\sigma_{(u)}^2}\right)^{1/4} \Rightarrow \varepsilon = \frac{15 v \sigma_{(u)}^2}{\lambda_{(\tau)}^2} \tag{1}$$

with the standard deviation $\sigma_{(u)}$ of u and $\lambda_{(\tau)}$, the Taylor micro scale.

2.2 STRUCTURE FUNCTION AND LOCAL DISSIPATION.

Since Kolmogorov (1941) it is known that the third order velocity structure function defined as

$$F_3(r,t) = \left\langle \left[u(x,t) - u(x+r,t) \right]^3 \right\rangle \tag{2}$$

is proportional to εr

$$F_3(r) \propto \varepsilon r \tag{3}$$

with the prefactor precisely equal to -4/5 in the inertial range of r.

Thus F_3 can be used as another direct, at least qualitative, estimate of reduction of dissipation in drag reducing flows.

2.3 THE SKEWNESS OF THE VELOCITY DERIVATIVES AND THE ENSTROPHY PRODUCTION.

The skewness factor, s, of the velocity derivative in flow direction is

$$s = - \left\langle \left(\frac{\partial u}{\partial x}\right)^3 \right\rangle \Big/ \left\langle \left(\frac{\partial u}{\partial x}\right)^2 \right\rangle^{2/3} \tag{4}$$

For a Gaussian velocity field s is identically zero. It is well known (but not well understood) that in turbulent flows s>0 explained by a continuous enstrophy production as a result of prevalent vortex stretching. It is natural to expect that the enstrophy production in non-Newtonian drag reducing flows should be strongly reduced as compared to the water flow under the same flow conditions, since the vortex stretching process is intimately related to the dissipative properties of turbulent flows (Taylor 1938). It should be emphasized that in order to investigate this hypothesis it is necessary to compare not the skewness factor but the

$$<(\partial u/\partial x)^3>_w/<(\partial u/\partial x)^3>_s, \tag{5}$$

since it is this quantity which is, at least approximately, proportional to the enstrophy generation (e. g. Batchelor and Townsend 1949). The skewness is not good for comparison since it has in the denominator the mean of the squared velocity derivative to the power 3/2, which is reduced too in drag reducing flows. There is no such a problem in water flow.

3. Results

The results for the mean surrogate dissipation for the above mentioned flows at the two wall distances representative for the near wall layer and the edge of the buffer layer are shown in Table I

TABLE 1 The mean surrogate dissipation $<\varepsilon>$ [m^2/s^3]

$y^+ \approx$	Water	Dilute polymer	Surfactant
24	0.0152	0.0004	0.0013
80	0.0100	0.0009	0.0008

In the near wall region the energy dissipation is much higher for water than for the two drag reducing solutions. At the edge of the buffer zone this is not the case anymore, i.e. for all three fluids the dissipation becomes almost equal. These results confirm the view that the major contribution to drag reduction process comes from the near wall region.

Another interesting aspect seen from table 1 is an anomalous behaviour of dissipation for a dilute polymer solution: it increases with the distance from the wall.

In order to use the structure function to evaluate the dissipation one has to choose values of r belonging to the inertial subrange.

We recognized that the inertial subrange was too small to evaluate the energy dissipation via relation (3). Therefore we used the sequence for which the structure function exhibited a power law behaviour, with the assumption that the prefactor is proportional to ε, i.e.

$$F_3 \propto \beta \varepsilon r^\alpha \tag{6}$$

The values for β are shown in table 2 and the one for the corresponding α evaluated in this range in table 3. Therefore the data can be used as an indication of the behavior and are showing the same trends as the mean surrogate dissipation.

TABLE 2 The slope $\beta\epsilon$ of the structure function $F_3(r)$

$y^+ \approx$	Water	Dilute polymer	Surfactant
24	-0.090	-0.016	-0.009
80	-0.015	-0.008	-0.001

TABLE 3 The exponent α of the structure function $F_3(r)$

$y^+ \approx$	Water	Dilute polymer	Surfactant
24	2.00	1.86	1.64
80	1.85	1.86	1.40

The inhibition of the enstrophy production in drag reducing flows can be seen from the comparison of the values of $<(\partial u/\partial x)^3>_w$ and $<(\partial u/\partial x)^3>_s$ for water and drag reducing fluids respectively (see table 4). One can see that the ratio $<(\partial u/\partial x)^3>_w/<(\partial u/\partial x)^3>_s$ is really much larger than unity, i.e. enstrophy production in drag reducing flows is strongly inhibited. An interesting aspect is that it is stronger in the near wall region in polymer solution, whereas for surfactants the tendency is opposite. This is a clear indication of the qualitative differences in the mechanisms in the two drag reducing flows (see also table 1).

TABLE 4 The skewness and the ratios between the water and the solutions for $<(\partial u/\partial x)^3>_w/<(\partial u/\partial x)^3>_s$

	Water		Dilute polymer		Surfactant	
	near-wall	outer flow	near-wall	outer flow	near-wall	outer flow
$<(\partial u/\partial x)^2>$	1873	711	173	80.6	98.6	38.7
$<(\partial u/\partial x)^3>$	47708	5360	2503	765	427	5.55
ratio w/s			17.29	6.52	45.8	384
s	.589	.282	1.101	.983	.436	.023

The skewness s exhibits a qualitative difference in the drag reducing flows too, namely the polymer solutions deviate more from a Gaussian state, whereas the flow of surfactant is rather close to Gaussian in the outer flow.

4. References

G.K. Batchelor & A.A. Townsend (1949) The nature of turbulent motion at large wave-numbers. Proc. Roy. Soc. A **199**, 238-255.

A. Gyr & H.-W. Bewersdorff (1995) Drag reduction of turbulent flows by additives. Kluwer Academic Publishers, Dordrecht, Boston, London.

A.N. Kolmogorov (1941) Dissipation of energy in locally isotropic turbulence. Dokl. Akad. Nauk SSSR **32**, 16-18.

H. Stüer & A. Gyr (1996) Fractal dimension and intermittency coefficients for various flow variables in Newtonian and non-Newtonian turbulent flows. J. Non Newtonian Fluid Mech.**62**, 207-224.

G.I. Taylor (1938) Production and dissipation of vorticity in a turbulent fluid. Proc. Roy. Soc. Lond. A **164**, 15-23.

MEASUREMENTS IN A THREE-DIMENSIONAL SEPARATION

J. R. HARDMAN and P. E. HANCOCK
Department of Mechanical Engineering
University of Surrey,
Guildford, Surrey, UK

1. Introduction

Most detailed measurements of the turbulence structure in separated flow have been made using a number of two-dimensional geometries, the most common being the backward-facing step. In these flows the separation is nominally both invariant with spanwise position and co-planar. By 'sweeping' the separation line with respect to the flow direction a cross-flow is in effect superimposed with the result that the flow is still spanwise-invariant (provided the flow width is sufficient to avoid end effects) but is now non-coplanar, where here the spanwise direction is taken along the separation line. While the swept case is an obvious extension of the original two-dimensional case a more general but systematic set of flows is less obvious. The current study is part of a systematic extension outlined by McCluskey et al. (1991).

Figure 1 shows the attachment and surface streamlines in a mildly three-dimensional separation generated downstream of a symmetrical v-shaped separation line, angled at $\pm 10°$. The separation was generated behind a vertical fence fixed to the front of a horizontal splitter plate - see figure 2. The fence height, h_f, above the plate surface was 10.0mm, the plate thickness 3.0mm, and the overall flow width 1500mm. The wind tunnel working section height was 500mm. In all the measurements made so far it appears that the separation is also symmetrical, a significant finding in itself. While not necessary, symmetry has the advantage that it provides some additional (time-mean) checks on the measurements and convenient boundary conditions for computation. In this type of flow there is an inwards lateral flow inside the separation from the sides with the result that the separating streamlines are *not* joined to the attachment line. At the attachment position the separating streamlines remain above the surface allowing a mass outflow (equal to the side inflow), as illustrated in figure 2. Only when the separation is spanwise invariant is it correct to describe separation and attachment lines as joined by *re*attaching streamlines. The width of the flow in figure 1 is sufficient for the side flow regions to be spanwise invariant, thus providing well defined boundary conditions and a progressive departure from spanwise invariance.

2. Results

The mean velocity and Reynolds stress measurements reported here have been made using pulsed-wire anemometry. All velocities are normalised by the free-stream reference velocity, U_{ref}, measured upstream of the flow rig.

S. Gavrilakis et al. (eds.), Advances in Turbulence VI, 453-456.

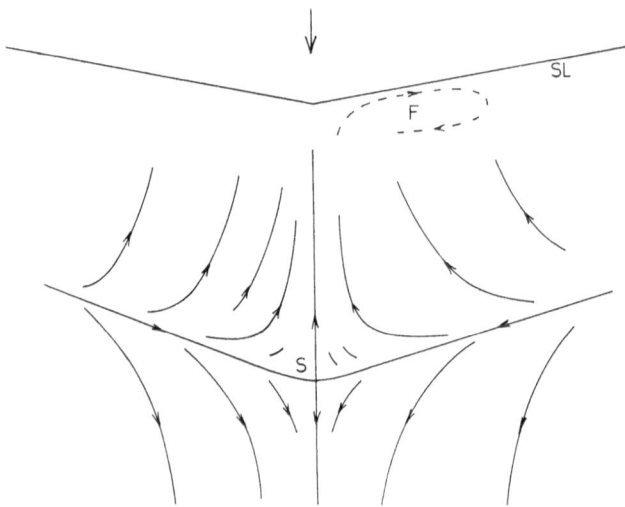

Figure 1 Surface streamlines on splitter plate.
Free-stream flow is from top to bottom. SL denotes (fence)
separation line; S and F denote saddle and foci singularities.

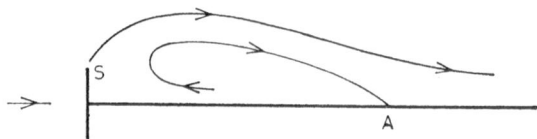

Figure 2 Separating (S) and attaching (A) streamlines on centre plane.

Figure 1 is a scaled plot of the surface streamlines, and figure 3 shows the variation of the surface pressure, where x is in the free-stream direction measured (locally) from the fence, and z is in the lateral direction measured from the centreline. y is perpendicular to the splitter plate, measured from the plate surface. From figure 3 it is clear that the lateral extent is in excess of one attachment length before the flow becomes spanwise invariant. Surface streamlines (beyond figure 1) indicate about two attachment lengths. The pressure downstream of attachment of this central region is distinctly slightly lower than that downstream of the spanwise invariant flow, and is probably caused by streamwise vorticity originating within the separation which is then convected out of the separation by the flow between the separation and attachment streamlines (figure 2).

Figure 4 shows the x-direction velocity $U(y)$ at $x/X_A = 0.5$ and 1.0 for four lateral positions, where that at $z \doteq 385$mm is in the spanwise-invariant region. The distance to attachment, X_A, is shown in figure 5. The shapes of $U(y)$ are fairly similar with z, but extend substantially further over a fairly narrow central region about ±half an attachment length in width. This can be seen also from δ_{95} in figure 5, where δ_{95} is the

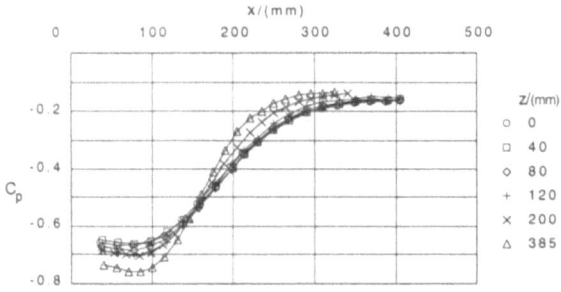

Figure 3 Pressure coefficient $C_p = (p-p_{ref})/0.5\rho U_{ref}^2$.

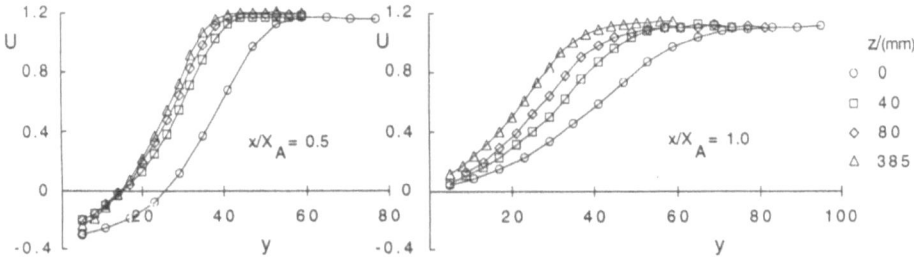

Figure 4 Mean velocity $U(y)$ at $x/X_A = 0.5$ and 1.0.

position at which $U(y)$ is 0.95 of the local maximum. At attachment δ_{95} is about 70% larger than it is in the spanwise-invariant flow, while at $x/X_A = 0.5$ it is larger by about 40%. As a fraction of X_A, δ_{95} is larger in the central region by a factor of about 1.4, probably at least partly as a result of the displacing effect of the 'outflow' between the separating and attaching streamlines illustrated in figure 2.

The z-direction mean velocity $W(y)$ is shown in figure 6a at $x/X_A = 1$, and as a polar plot $(U{\sim}W)$ in figure 6b. The scatter in the case $z = 0$ indicates the level of precision possible at the relatively low level of W that occurs for the present small sweep angle. Figure 6b indicates a fairly linear variation between U and W. Hancock and McCluskey (1995) found U and W to be closely related by velocity triangles in the spanwise-invariant case, including the flow in the reverse-flow region, where nearer to the surface W is proportional to U such that $tan(W/U)$ is equal to the limiting streamline direction - ie the flow very near the surface is co-planar. They concluded that except near the surface the response to the cross flow is essentially inviscid, the linear variation between U and W concurring with that implied by inviscid secondary-flow theory. There is no obvious reason why the same form of relationship should not apply more generally.

Lastly, figure 7 shows $\overline{u^2}$ at x/X_A of 0.5 and 1.0 at the same lateral positions as in figure 5. The profiles of $\overline{w^2}$ are of vary comparable shape at the respective stations, but

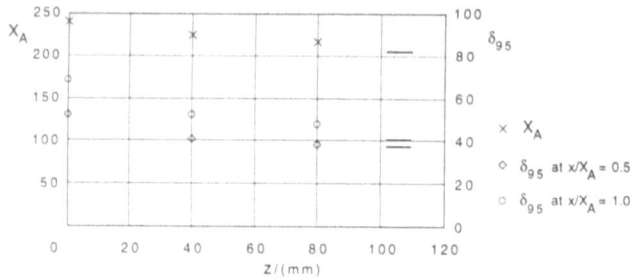

Figure 5 Attachment length, X_A, and bubble height, δ_{95}.
Horizontal bars indicate levels in spanwise-invariant case.

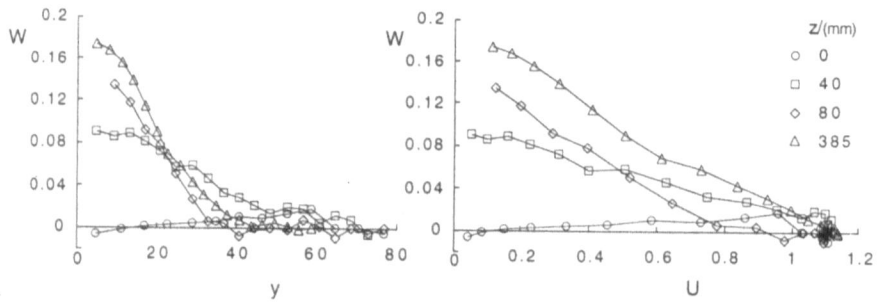

Figure 6 Mean velocity, W; a) $W(y)$, b) W against U.

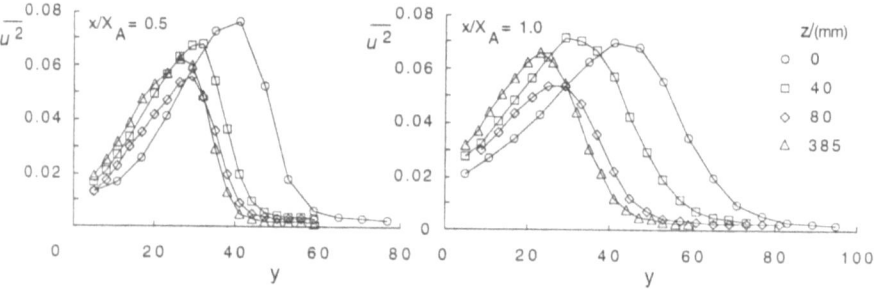

Figure 7 Reynolds stresses $\overline{u^2}$ at $x/X_A = 0.5$ and 1.0.

smaller; $\overline{w^2}/\overline{u^2}$ is about 0.65. The levels of stress - at least as indicated by these stresses - do not vary much, though at $z = 80$ the maxima are below those either side, by roughly 25% at both x/X_A.

3. References

McCluskey , F.M., Hancock, P.E. and Castro, I.P. (1991) Three-dimensional separated fows, Eighth Turb. Shear Flows Symp., Munich, 9-5-1, September.
Hancock, P.E. and McCluskey, F.M. (1995) Three-dimensional separated flow, Tenth Turb. Shear Flows Symp., Penn State Univ., 18-7, August.

THE MEASUREMENT OF TURBULENT STRUCTURE USING PIV

T.R.HAYDON*, C. A.GREATED[‡].

* Research Student, ‡ Professor.

Fluid Dynamics Unit, Department of Physics and Astronomy,
The University of Edinburgh, Edinburgh, EH9 3JZ, U.K.

1. Introduction

Particle Image Velocimetry (PIV) is an optical technique for obtaining a spatial record of the velocities, at a point in time in a plane of a flow (Adrian, 1991). The application of PIV to turbulent flows is problematical due to the relatively large dynamic range of velocities and the 180° directional ambiguity inherent in the analysis using autocorrelation techniques. The use of image shifting overcomes these difficulties in essence, although the measurement of turbulent structures is still non-trivial. The values chosen for experimental parameters such as image-shift velocity, exposure, scan rate and seeding particle density affect the range of turbulent eddies that are resolved.

The limitations of PIV with respect to turbulent flows and the experimental methods used to maximise the range of eddy sizes resolvable in the images, are reviewed. Work done on the estimation of errors involved when measuring turbulent parameters using PIV maps (Dam, 1995) is used to estimate the relative merits of increasing the frame size and increasing the number of pictures. The application of PIV to turbulence produced after water waves break and the processing of data to yield quantatative turbulent information is subsequently discussed.

2. Practical Implementation of PIV in the study of turbulence

The use of a Kodak Megaplus 4.2 CCD camera enabled high-resolution, digital images to be grabed and stored to the hard disc of a PC. By upgrading the memory on the frame-grabber board, a succession of these images can

457

S. Gavrilakis et al. (eds.), Advances in Turbulence VI, 457-460.

be obtained, giving a high quality record of the evolution of the vorticity in a particular region. The time interval between these frames is limited by the frame rate of the Megaplus and the reset time of the image shifter device. The images recorded in these experiments were analysed digitally, considerably speeding up the analysis compared to previously used photographic techniques.

The camera exposure and seperation of laser pulses can place limitations on the timescale of turbulent fluctuations that can be measured. Furthermore, to resolve the complete range of turbulence structures the separation between seeding particles must be of the order of the smallest size of eddy that might be expected. If the size of the interrogation area used in the analysis is too large, small eddy sizes may be excluded..

3. Relative merits of increasing frame size or number of pictures

With the most recent CCD-based PIV systems it is possible to capture a rapid sequence of complete frames, thus combining the advantages of temporal as well as spatial resolution. However, ultimately the amount of imformation that can be recorded and analysed is limited and, in designing a system, a compromise has to be made between the number of pixels used per frame and the number of pictures recorded. In order to examine this dilema a little more quantitatively, consider the very simplest case of measuring the mean velocity in a statistically stationary and homogeneous turbulent flow. We will assume that the distance between pixels on each frame and the time interval between exposures has been fixed; determined by the smallest eddy size of interest. Thus the size of each frame and the total number of frames are variables.

For a single frame, of a small size compared to the integral length scale (L) then the mean square error is

$$\epsilon^2 = \sigma^2 \tag{1}$$

where σ^2 is the mean square velocity fluctuation. For an N^2 size frame, where $N \geq L$, this is reduced to (Dam 1995)

$$\epsilon^2 = (\frac{L}{N})^2 \sigma^2 \tag{2}$$

If we now take an evenly spaced sequence of frames over time T then, since the integral time scale is $L/\bar{v}$ where $\bar{v} = $ mean velocity,

$$\epsilon^2 = (\frac{L}{N})^2 \sigma^2 \frac{2L}{T\bar{v}} \tag{3}$$

where T is long compared to $L/\bar{v}$. Again if $T \geq L/\bar{v}$ then $\epsilon^2 = \sigma^2$. Thus for $T \geq L/\bar{v}$, the rms error is proportional to $1/N$ and $1/\sqrt{T}$. Noting that the number of pixels used (say n) is proportional to N^2 and to T, it appears that $\epsilon \sim n$.

This rather simplistic result indicates that increasing spatial averaging and temporal averaging are given equal weight in the overall averaging process although, in practice, cost considerations may favour increasing the number of frames.

4. Response time of seeding

As with LDA, errors can arise in the measurement of turbulence parameters if the seeding is not chosen appropriately. These errors can be very significant for PIV due to the optical requirement for relatively large seeding particles. At large wave numbers the relaxation time of the particles may be of the same order as the turbulence periods. This results in either damping or enhancement of these frequency components, dependent on the relative density of the seeding and flow medium, with corresponding errors in the estimation of rms turbulence levels. This form of error will be assessed in the paper with reference to the breaking wave problem.

5. Turbulence generated after breaking waves

Breaking water waves are an interesting example of turbulence in that the structure generated is initially two-dimensional but becomes three-dimensional as it develops. The turbulence generated by a single breaking wave is non-homogeneous manifesting itself in the form of a patch or layer near the surface which is mixed downwards and forwards. The form of the patch and it's development with time is highly dependent on the type and strength of the wave. The use of PIV in the manner described earlier in this paper enables an excellent visual analysis of the turbulent mixing and leads on to the measurement of associated parameters. Comparison with LDA studies by researchers such as Rapp and Melville (1990) enable us to conclude if PIV can be used on these complicated turbulent flows to yield useful and accurate quantatative data.

Velocity fields were obtained for instances beginning just after breaking and extending up to 10 seconds after breaking for several types of wave; these vector maps generally show a range of turbulent structures. Figure 1 shows an example of a PIV map of a large vortex created by a steep plunging breaker. This is part of a sequence taken where this vortex is shown to travel accross the field of view. The visual record in itself is very useful, providing an idea of the size and velocities associated with the vortex.

The rate at which the energy is dissipated from the surface waves by the process of breaking is important in the identification of the turbulent mixing layer in which this dissipation is enhanced. Dissipation has been estimated

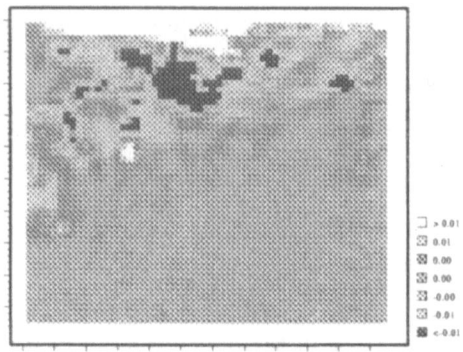

Figure 1. typical post breaking vorticity field (Flow direction to the right).

from the form of the wave-number spectrum by many experimenters producing reasonable agreement with each other and giving conclusive evidence of the existence of the layer. Wave number spectra were calculated from PIV flow maps and dissipation rates estimated. The transition from 2-d to 3-d flow was studied by observing the wave number spectra and the two-dimensional continuity. Comparisons were made with both theoretical and experimental work in these areas (George *et al.* 1994, Agrawal *et al.* 1992, Montgomery and Kraichnen 1979, and Battjes and Sakai 1981).

References

Adrian, R.J. (1991) *Particle Imaging Techniques for Experimental Fluid Dynamics.* Ann. Rev. Fluid Mech., 23:261-304.

Elgard Dam, C. (1995) Ph.D thesis *Particle Image Velocimetry, Accuracy of the Method with Particular Reference to Turbulent Flows*

George, R., Flick, R.E., Guza, R.T. (1994) *Observations of turbulence in the surf zone.* J. Geophys. Research, 99:801-810.

Agrawal, Y. C., Terray, E. A.,Donelan, M. A., Hwang, P. A., Williams III, A. J., Drennan, W. M., Kahma, K. K. and Kitaigorogskii, S. A. (1992) *Enhanced dissipation of kinetic energy beneath surface waves* Nature 359:219-20.

Kraichnan, R.H., Montgomery, D. (1979) *Two-dimensional turbulence* Reports on Progress in Physics.

Battjes, J.A., Sakai, T. (1981) 111:421-437. *Velocity field in a steady breaker*

Rapp, R.J., Melville, W.K. (1990) *Laboratory Measurements of Deep Water Breaking Waves.* Phil. Trans. R. Soc. Lond. A 331:735-800.

PRELIMINARY EXPERIMENTS
ON THE CONTROL OF THREE-DIMENSIONAL MODES
IN THE FLOW OVER A BACKWARD-FACING STEP

A. HUPPERTZ AND G. JANKE

Hermann-Föttinger-Institut für Strömungsmechanik,
Technische Universität Berlin,
Straße des 17. Juni 135, 10623 Berlin, Germany

1. Introduction

Flows with separation and reattachment occur in many practical engineering devices [1]. Among separating and reattaching flows, the two-dimensional backward-facing step flow with a fixed separation line is the simplest configuration which shows most of the significant features of separated flow.

The experiment shown in Figure 1 was designed to study the effect of three-dimensional forcing on the size of the separation bubble.

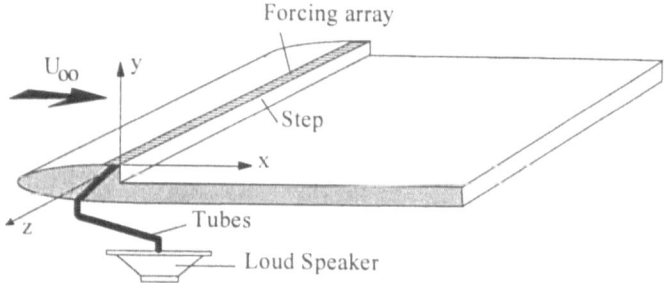

Figure 1. Schematic experimental configuration

The model consisted of a 5:1 elliptical nose followed by a straight-wall section, the step and an off-center splitter plate. The step height was H = 20 mm, the aspect ratio was 20 and the expansion ratio was 1.09. A forcing array was located at the trailing edge of the step. It consisted of 105 small holes (cross section 2 x 2 mm) which were inclined under 45° and connected via tubes with loud speakers. The loud speakers were driven by a multi-channel arbitrary-signal generator, which permits to study a wide variety of forcing modes. In the preliminary experiments reported here two time-harmonic types of spatial excitations were studied: 1. spanwise-uniform excitation and 2. spanwise-

S. Gavrilakis et al. (eds.), Advances in Turbulence VI, 461-464.
© *1996 Kluwer Academic Publishers.*

standing square-waves excitation produced by a phase shift between groups of array elements. The excitation frequency, the phase shift and the spanwise wavelength were varied. The integral fluctuation energy of the disturbed flow, 1 mm behind the step edge, was kept constant. Its maximum fluctuation amplitude was less than 1% of the free-stream velocity.

The experiments were conducted at a Reynolds number of $(U_\infty \cdot H/\nu) = 2900$, i.e., at a free stream velocity $U_\infty \approx 2.2$ m/s. The separating boundary layer was laminar and a Blasius profile with displacement thickness of $\delta = 1.365$ mm and momentum thickness of $\theta = 0.535$ mm was measured. The flow underwent transition in the separated shear layer and was turbulent at reattachment.

A companion DNS of this flow was undertaken by Bärwolff et al. [2].

2. Results and Discussion

First we discuss the spanwise-uniform forcing. For this case the open circles in Figure 2 show the mean reattachment length (measured with oil-film interferometry) as a

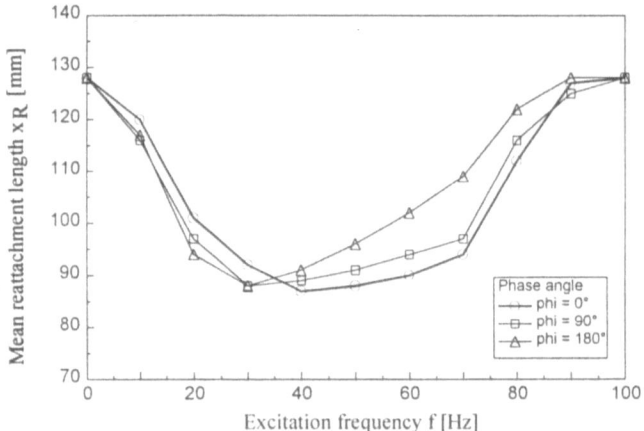

Figure 2. Mean reattachment length x_r vs. excitation frequency
Spanwise wavelength $\lambda_z/2 = 40$ mm

function of the excitation frequency. Zero Hz represents the unforced case. The shortest separation bubble, 32% shorter than in the non-forced case, is obtained at $f \approx 40$ to 50 Hz. This trend has also been observed in the DNS [2]. The corresponding Strouhal number formed with the momentum thickness at separation $St \equiv (\theta \cdot f)/U_\infty \approx 0.010$ to 0.012 is slightly below the most unstable Strouhal number of the shear layer bounding the separation region. This supports the finding of Hasan [3] that, at least for our δ/H, the instability of the step flow is mainly one of the separated shear layer and the step height is only of secondary importance.

According to the model of Chapman et al. [4] the shortening is a result of the increased shear-layer entrainment which in turn is a consequence of the shear-layer transition moving upstream towards the step edge. This is clearly visible in the smoke-wire visualizations in Figure 3, which shows the natural flow and how it changes under uniform excitation near the most effective frequency. Obviously, the forcing strongly organizes the large-scale vortex structures. The forcing utilizes the instability of the laminar shear layer to induce an early vortex roll-up followed by vortex pairing, turbulent vortex break-up and turbulent reattachment of the shear layer.

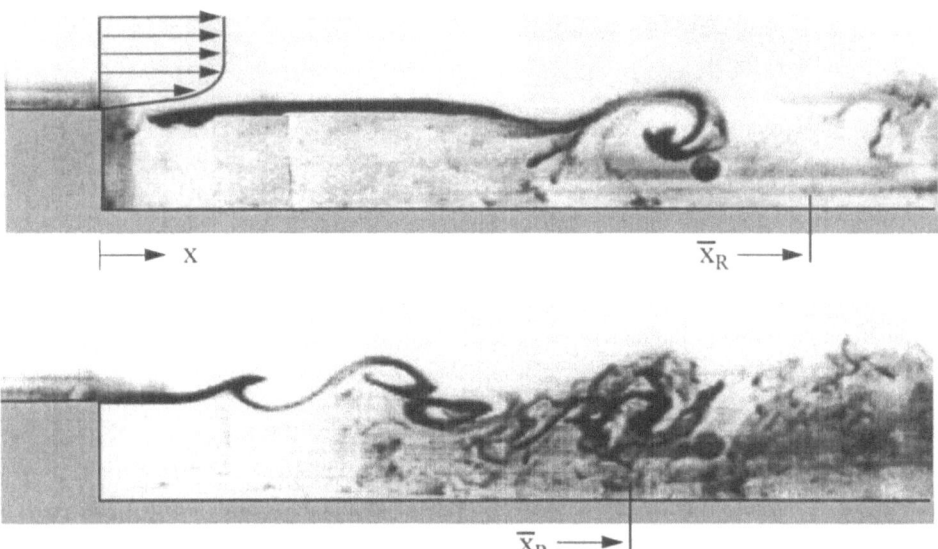

Figure 3. Spanwise structures of the shear layer: a) unforced flow; b) flow forced with 70 Hz

A test (at the most effective frequency) showed that increasing the forcing amplitude will decrease the reattachment length. However, the flow reacts most sensitively at very low amplitudes. This is readily understood from the transitional nature of the flow.

Next we discuss the three-dimensional excitation. Three spanwise wavelengths normalized by the Kelvin-Helmholtz wavelength ($\lambda_{KH} \approx 22$ mm) with nodal distances $\lambda_z/(2\lambda_{KH}) = 0.45$, 0.91, and 1.82 and two phase shifts $\varphi = 90°$ and $180°$ were tested. In Figure 2 only the results for the largest wavelength are shown. The results for $\lambda_z/(2\lambda_{KH}) = 0.91$ are very similar to the ones shown, which indicates the well-known broadband response of the shear layer [5,6]. The smallest wavelength (not shown) is substantially less effective than the other two. Figure 2 shows that compared with the 2D excitation the 3D excitation *increases* the reattachment length over a wide frequency range. Only in a small frequency range around 30 Hz a slightly shorter reattachment length is observed. This shorter reattachment is associated with the occurrence of large

464

three-dimensional structures in the shear layer and the reattachment region. Figure 4 shows a flow visualization of these structures viewed at an angle of 45°.

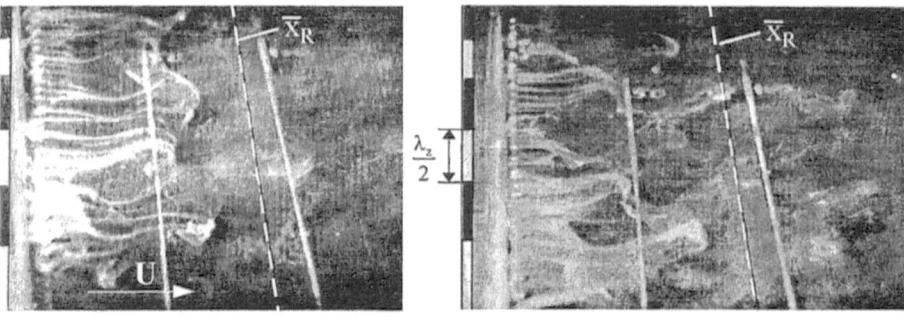

Figure 4. 3D forced flow (f = 30 Hz, $\lambda_z/2$= 20 mm): a) phase shift φ=90°; b) phase shift φ=180°

Different flow formations are observed for different phase shifts. For a phase angle of 90° undulated rollers are formed, but for a phase angle of 180° the rollers break off and form a staggered pattern of "tongues". This is associated with a distinct three-dimensional pattern of the wall streamlines in the reattachment zone and with an undulated reattachment line.

3. Conclusions

The flow over a backward-facing step can be controlled with two- and three-dimensional forcing. The largest reduction of the reattachment length with two-dimensional time-harmonic forcing is obtained in the vicinity of the most unstable frequency of the shear layer bounding the separation region. The three-dimensional harmonic excitations, tried so far, shorten the separation bubble only in a limited frequency range more than the two-dimensional ones.

4. References

1. Eck, B.: Technische Strömungslehre (in German), Springer-Verlag, Berlin, 1966
2. Bärwolff, G., Wengle, H. & Jeggle, H.: Direct numerical simulation of transitional backward-facing step flow manipulated by oscillating blowing/suction. 3rd International Symposium on Engineering Turbulence Modeling and Measurements, Crete, 1996
3. Hasan, M.A.Z.: The flow over a backward-facing step under controlled perturbation: laminar separation. *J. Fluid Mech.* **238,** (1992), 73-96
4. Chapman, D.R., Kuehn, D.M. & Larson, H.K.: Investigation of separated flow in supersonic and subsonic streams with emphasis on the effect of transition. *NACA Report* **1356,** 1958
5. Pierrehumbert, R.T. & Widnall, S.E.: The two and three-dimensional instabilities of a spatially periodic shear layer, *J. Fluid Mech.* **114** (1982), 59-82
6. Nygaard, K.J. & Glezer, A.: Evolution of streamwise vortices and generation of small-scale motion in a plane mixing layer. *J. Fluid Mech.* **231** (1991), 257-301

AN INTEGRATED SILICON BASED WALL PRESSURE-SHEAR STRESS SENSOR FOR MEASUREMENTS IN TURBULENT FLOWS

L. Löfdahl*, E. Kälvesten**, T. Hadzianagnostakis* & G. Stemme**

Thermo & Fluid Dynamics, Chalmers University of Technology,
S-412 96 Göteborg, Sweden
*** Department of Signals, Sensors & Systems, Royal Institute of Technology,*
S-10044 Stockholm, Sweden

1. Introduction

By the introduction of silicon micro machining into fluid mechanics the experimentalists have been given a new incentive to carry out measurements of fundamental parameters in turbulent flows. One such unknown parameter is the pressure-velocity correlation (PVC) which turns out to be a key term in the kinetic energy budget as well as in the transport equations for the Reynolds stresses.

By the integration of the pressure and shear stress sensors on one single chip we have now achieved a PVC sensor with an edge-to-edge distance between the two active areas, x_1, of 100 µm, which corresponds to a dimensionless distance in the range $5.0 < x_1^+ < 11.6$, see Kälvesten et al. (1996a). In this paper we present some turbulence measurements conducted with this new PVC sensor in the Reynolds number range of $4.9 \times 10^3 < Re_\theta < 1.0 \times 10^4$.

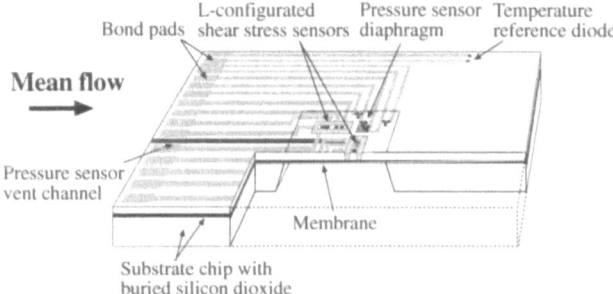

Figure 1. A schematic cut-out view of the integrated pressure-shear stress sensor. The $6 \times 4 \times 0.5$ mm silicon substrate chip has a $1.5 \times 1.5 \times 0.03$ mm membrane on which the pressure and shear stress sensing functions are located.

2. Experimental Arrangements

As shown in Figure 1 the integrated sensor consists of a $6 \times 4 \times 0.5$ mm substrate chip with a $1.5 \times 1.5 \times 0.03$ mm membrane where the pressure and shear stress sensing functions are located. To conduct simultaneous measurements of the fluctuating normal stress (pressure) and shear stress (friction velocity) the pressure sensor is located at the corner of the two L-configured shear stress sensors. This L-configuration makes it possible to simultaneously

465

S. Gavrilakis et al. (eds.), Advances in Turbulence VI, 465-469.
© *1996 Kluwer Academic Publishers.*

measure and distinguish the two orthogonal, in-plane components of the friction velocity. In our design the pressure sensitive diaphragm area was chosen to 100×100 μm and the flow sensing hot-chip top areas are 300×60 μm. The two sensors are located to give a centre-to-centre distance between the pressure sensing diaphragm and the shear stress sensing hot chip areas of about 300 μm. To minimize the flow induced disturbances which originates from the bonding wires, the substrate chip is relatively large so that the bonding pads at the substrate chip edge are located far away from the sensing areas. A more comprehensive description of the pressure and shear stress sensors can be found in Löfdahl et al. (1992) (1994) and (1996).

All measurements were conducted in a closed low-speed wind tunnel. To generate the turbulent boundary layer, a flat plate of 2.5 m in length was positioned in the horizontal symmetry plane of the test section ($1.25 \times 1.80 \times 3.00$ m). The boundary layer parameters were calculated from velocity data which was obtained using standard hot-wires mounted on a traverse system. Our turbulence measurements were conducted for free-stream velocities in the range $20\ \text{m/s} < U_\infty < 50\ \text{m/s}$ corresponding to Reynolds numbers $4.9 \times 10^3 < Re_\theta < 1.0 \times 10^4$. We have in earlier works shown that the current flat plate boundary layer agree very well with "bench mark" cases from the literature, see e.g. Löfdahl et al. (1992).

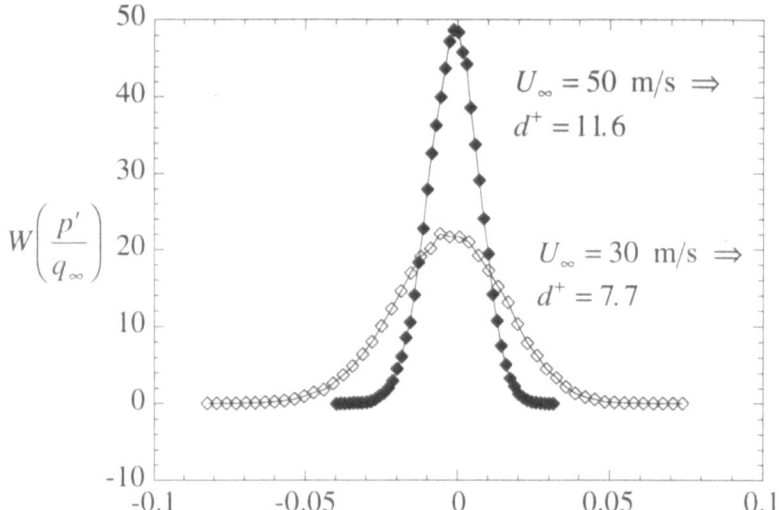

Figure 2. Probability density distribution of the pressure fluctuations measured at free stream velocities of 30 m/s ($Re_\theta = 6.6 \times 10^3$) and 50 m/s ($Re_\theta = 1.0 \times 10^4$) scaled in (a) outer and (b) inner variables.

3. Results and discussion

To evaluate the performance of the integrated pressure-shear stress sensor the probability density distributions, the rms values and the power spectra for both the pressure and wall shear stress fluctuations were measured.

Figure 2 shows the measured probability density distributions of the fluctuating pressure amplitudes for free stream velocities of 30 and 50 m/s scaled in outer variables. The corresponding dimensionless diaphragm sizes, $d^+ = du_\tau/v$, are 7.7 and 11.6 (wall units) with $d = a = 100$ μm as the diaphragm side length. The pressure amplitudes are scaled with the dynamic pressure q_∞. All distributions are normalized so that the area under each curve is equal to unity. The scaling with the outer parameter, q_∞, shows that a decreasing diaphragm size is associated with a shrinking peak in the probability distribution of the fluctuating pressure. The same trend has been found earlier by Schewe (1983) and Löfdahl et. al. (1994).

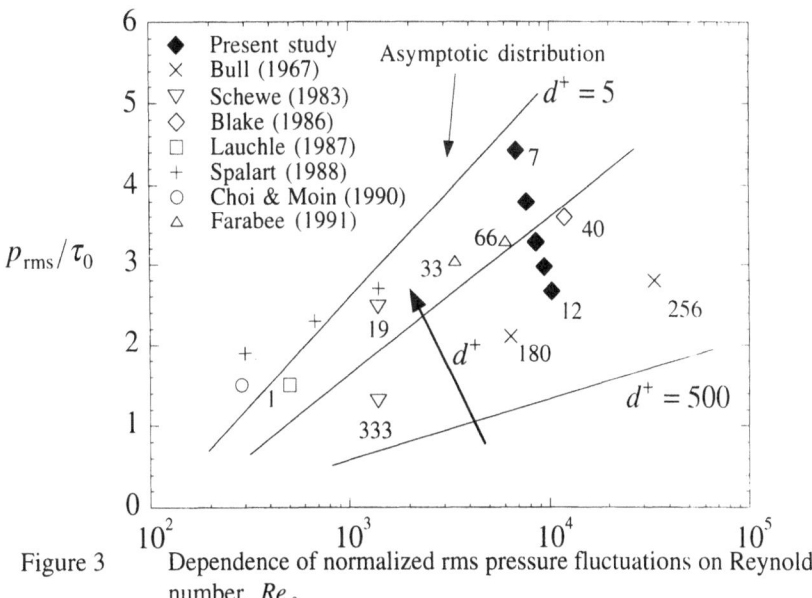

Figure 3   Dependence of normalized rms pressure fluctuations on Reynolds number, Re_θ

In Figure 3 the rms values of the wall pressure fluctuation, p_{rms}/τ_0, are shown as a function of the Reynolds number Re_θ. We have also included some measurements of Bull (1967), Schewe (1983), Blake (1986), Lauchle & Daniels (1987) and Farabee & Casarella (1991) together with direct numerical simulations (DNS) of Spalart (1988) and Choi & Moin (1990). (The numbers indicated in the figure correspond to the transducer sizes expressed in terms of d^+ units). As proposed by Bandyopadhyay (1995) an decreasing sensor sensitivity for Reynolds number effects for increasing d^+ is obtained. This is schematically shown with the three d^+ iso-lines in the figure. Moreover, an asymptotic limit for the rms values for $d^+=5$ is suggested. The increase of p_{rms}/τ_0 as the diaphragm size decreases has already been shown by e.g. Schewe (1986). Our measurements underline that a smaller d^+ value implies higher rms pressures due to better spatial resolution. In our work the d^+ range, $7.7 \leq d^+ \leq 11.6$, was considerably smaller than in the other works, where d^+ varied from 19 to 330. An exception is the measurements by Lauchle & Daniels (1987) who had glycerine as the working fluid but considered an entirely different experimental

468

arrangement. Noteworthy from Figure 3 is also that the pinhole measurements of Farabee & Casarella (1991) gives lower values than ours, in spite of the fact that these measurements were taken in approximately the same Reynolds number range. These deviations are discussed in Löfdahl et al. (1996). It is worthwhile to point out that the uncertainties of the data in Figure 3 are large since they come from different sources with varying background noise, different signal processing and for some cases corrections have been applied.

In Figure 4 the correlation coefficients for the pressure and wall shear stress fluctuations are shown as function of the Reynolds number, Re_θ. For Reynolds numbers of, $4.9 \times 10^3 < Re_\theta < 1.0 \times 10^4$, these coefficients were in the range 0.40 to 0.50 for the parallel and 0.20 to 0.25 for the perpendicular configuration to the mean flow. As shown in the figure, the correlation is higher for the parallel than for the perpendicular sensor configuration. This is the same trend obtained in earlier investigations for the pressure-pressure correlation coefficient, see e.g. Löfdahl et al. (1996). Increasing correlation for decreasing Reynolds numbers can be explained by the relatively better sensor resolution (smaller d^+ and l^+) and smaller dimensionless sensor separation (x_i^+).

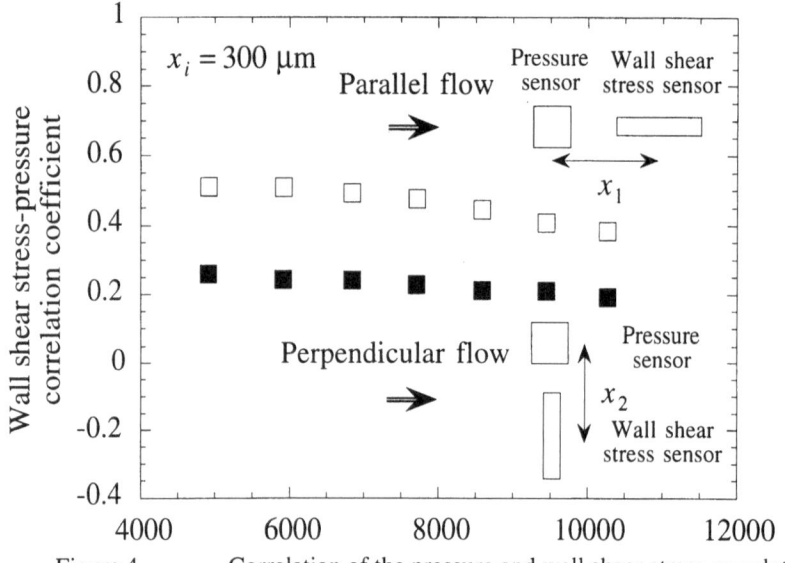

Figure 4.   Correlation of the pressure and wall shear stress correlation plotted against the Reynolds number, Re_θ.

Support from the Swedish Research Council for Engineering Sciences (TFR) is gratefully acknowledged.

4. References

Bandyopadhyay, R., 1995, Discussion on "Small Silicon Pressure Transducers for Space-time Correlation Measurements in a Flat Plate Boundary Layer", Accepted for J. Fluid Engineering

Blake, W. K., 1986, "Mechanics of Flow-Induced Sound and Vibration," Applied Mathematics and Mechanics, Vol. 17-1, Academic Press, New York.

Bull, M.K., 1967, "Wall Pressure Fluctuations Associated with Subsonic Turbulent Boundary Layer Flow", J. Fluid Mech. Vol. 29, pp 597-599.

Choi, H. & Moin P., 1990, "On the space-time characteristics of wall-pressure fluctuations," Phys. Fluids, Vol. 2, pp. 1450-1460

.Farabee, T. M. & Casarella, M. 1991, "Spectral features of wall pressure fluctuations beneath turbulent boundary layers," Phys. Fluids A, Vol. 3, No. 10, pp. 2410-2419.

Kim, J., 1989, "On the structure of pressure fluctuations in simulated turbulent channel flow", J. Fluid Mech., Vol. 205, pp. 421-451.

Kälvesten, E., Löfdahl, L. & Stemme, G., 1996a, "Analytical characterization of piezoresistive square-diaphragm silicon microphone, Sensors and Materials, Vol. 8, No. 2, pp. 113-136.

Kälvesten, E., Löfdahl, L. & Stemme, G., 1996b, "An integrated silicon based wall pressure-shear stress sensor for measurement in turbulent flows", Sensors and Actuators, accepted.

Lauchle, G. C. & Daniels, M. A., 1984, "Wall-pressure fluctuations in turbulent pipe flow". Physics of Fluids, Vol. 30, (10), pp. 3019-3024.

Löfdahl, L., Stemme, G. & Johansson, B., 1992, "Silicon based flow sensors used for mean velocity and turbulence measurements", Exp. in Fluids, Vol. 12, pp. 270-276.

Löfdahl, L., Kälvesten, E. & Stemme, G., 1994, "Small silicon based pressure transducers for measurements in turbulent boundary layers," Exp. in Fluids., Vol. 17, pp 24-31.

Löfdahl, L., Kälvesten, E. & Stemme, G., 1994, "Small Silicon Pressure Transducers for Space-time Correlation Measurements in a Flat Plate Boundary Layer", To appear in the Sept. isuue of J. Fluid Engineering

Panton, R. L., 1984, "Incompressible flow", John Wiley & sons.

Schewe, G., 1983, "On the structure and resolution of wall-pressure fluctuations associated with turbulent boundary-layer flow," J. Fluid Mech., Vol. 134, pp. 311-328.

Spalart, P., 1988, " Direct simulation of a turbulent boundary layer up to Re= 1410", J. Fluid Mech., Vol 187, pp 61-98.

SEPARATION CONTROL ON A GLIDER WING
WITH ARTIFICIAL BIRD'S FEATHERS

R. MEYER[1], D.W. BECHERT[2] and W. HAGE[1]

[1] Hermann-Föttinger-Institut für Strömungsmechanik,
   TU Berlin, Strasse des 17. Juni 135, 10623 Berlin, Germany
[2] DLR, Abteilung Turbulenzforschung,
   Mueller-Breslau-Str. 8, 10623 Berlin, Germany

The covering feathers on the upper side of bird's wings tend to pop up during the landing approach of birds and in flight through very turbulent air. Liebe´s (1938) explanation of this observation has been that flow separation can be limited by these covering feathers. An initial flight experiment with a piece of leather on one wing of a Messerschmitt Me 109 fighter aircraft (Liebe 1938) and previous experiments by Nachtigall, Wedekind & Dreher (1985) with bird's wings do hint at an appreciable effect, but were somewhat preliminary. In the present experimental research project being carried out together with the Institute for Biophysics (Bionik und Evolutionstechnik) of the Technical University of Berlin and with the STEMME Aircraft Company, Strausberg, we investigate this particular effect, supported by the German Federal Ministry of Education, Science and Technology (BMBF). Our experimental setup consists of a real wing section of a STEMME S10 motor glider aircraft, inserted into our wind tunnel and attached to a force balance. The flow around the wing is two-dimensional. The wing section is a so-called "laminar" airfoil, type HQ 41, developed by Horstmann & Quast, DLR Braunschweig. The Reynolds number range of our experiments is $1 \div 2 \times 10^6$, so that the lower flight velocities of the real aircraft are covered. This particular regime is relevant for the high-lift condition considered here.

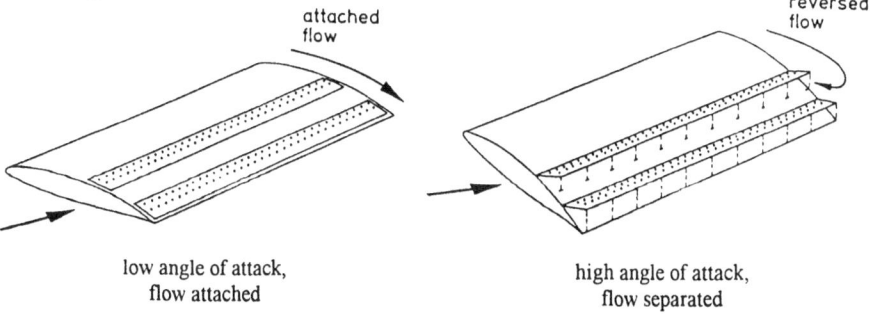

low angle of attack,
flow attached

high angle of attack,
flow separated

Fig. 1. Operation of the "separation controllers".

471

S. Gavrilakis et al. (eds.), Advances in Turbulence VI, 471-472.

In our experiments, the bird's feathers are not emulated in their original biological appearance, but as thin plastic or metal sheets attached to the upper surface of the airfoil (see Fig. 1). We call these devices "separation controllers". They are actuated by the flow itself and are in operation (i.e., bristled) only if flow separation on the wing starts or takes place and they stabilize and limit the separation. Under attached flow conditions, they are attached to the airfoil and thus the production of parasitic drag is avoided. We have tested about 150 different configurations of our "separation controllers" and have made the following observations: A design which permits pivoting of the movable sheets on their leading edges (see Fig. 1) and with tether threads (to prevent tipping over) seems to be easier to operate than any devices which emulate real bird's feathers. A perforation of the sheets is necessary in order to prevent lifting of the devices by the pressure distribution of attached flow. When flow separation starts to occur, the local flow reversal lifts the devices more and more for increasing angles of attack. It turned out that a soft trailing edge of the perforated sheets of our devices helps both for lifting and reattaching at the required angles of attack. Thus, in a transformed way, the soft feather tips and the porosity of real bird's feathers reappear as a necessary requirement for the design of our "separation controllers".

In Fig. 2 we provide data at $Re = 10^6$ for one device on the airfoil. The increase in maximum lift is 11 %. The gain in lift is increased beyond that value for more than one device as, e.g., with the configuration shown in Fig. 1. The rear device in Fig. 1 produces a direct extension of the lift curve as can be seen in Fig. 2. Additional devices attached farther upstream on the wing do produce additional beneficial effects for higher angles of attack. We have tested up to three devices which produce as much as 23 % increase of maximum lift. In our paper we also will give data on the interaction of "separation controllers" with conventional mechanical flaps. The increase of lift also persists there.

Besides measuring lift and drag with a balance, we have documented transition and separation patterns on the wing. This was carried out with white TiO_2-oil-benzine paint on the black wing surface. It turned out that our control devices do not only limit the separation. The occurring paint patterns suggest a dramatic stabilization of the separation wake. A wonderfully steady and optically impressive secondary flow pattern with fractal structure emerges on the wing surface, which is not observed without our devices.

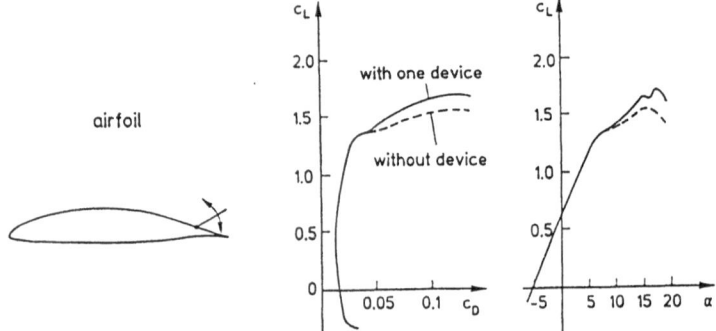

Fig. 2   Drag polar and lift coefficient C_L versus angle of attack α for our airfoil with one "separation controller".

EXPERIMENTAL INVESTIGATION OF THE AXISYM-METRIC SHEAR LAYER BETWEEN COAXIAL JETS WITH SWIRL

C. NAYERI[1], S. BÉHARELLE[2],J. DELVILLE[2], J.P. BONNET[2] & H.E. FIEDLER[1]

[1]TU-Berlin, Hermann-Föttinger-Institut Berlin,
Müller-BreslauStr.8, D-10623 Berlin
[2]CEAT/LEA , 43 Route de l'Aerodrome, F-86036 Poitiers

Introduction

Swirling motion is of practical interest as it can be found in a variety of industrial applications such as combustors, cyclonic separators or in trailing vortices of aircraft. Hence, the effect of swirl in various flow configurations has been studied intensively during the past two decades. The major interest was to determine its influence on mean flow quantities and the onset of the vortex breakdown.

In the present work swirl is used to create three-dimensional axisymmetric turbulent shear layers. The axisymmetric configuration was motivated by previous studies in a plane configuration with cross shear [1] which indicated that side wall effects can be problematic.

The three-dimensional shear layer is generated at the interface between swirling coaxial jets and is characterized by additional cross shear when the jets swirl at different magnitudes or directions. Of particular interest is the influence of the cross shear on the evolution of Kelvin-Helmholtz instability and the question as to how far flow control is applicable and useful.

The Wind Tunnel

The wind tunnel (Fig.1) consists of two coaxial plenum chambers with attached swirl generators and contractions. For the design of the wind tunnel a low turbulence level was of major interest. Therefore the swirl generators are far upstream of the jet exits as

473

S. Gavrilakis et al. (eds.), Advances in Turbulence VI, 473-476.

recommended in [2]. Plenum chambers and swirl generators are made of steel and the contractions of fiber glass. Each swirl generator has 36 adjustible radial inlet vanes and is enclosed in a separate cabin. Both cabins are supplied with air by means of two independent centrifugal blowers. Air filters mats mounted around the inlet of the swirl generators improve the uniformity of the entering radial flow. For reduction of turbulence and further improvement of flow uniformity three screens with successively derceasing mesh size are installed in the plenum chambers. Each plenum chamber is composed of three cylindrical sections, between which screens are mounted.

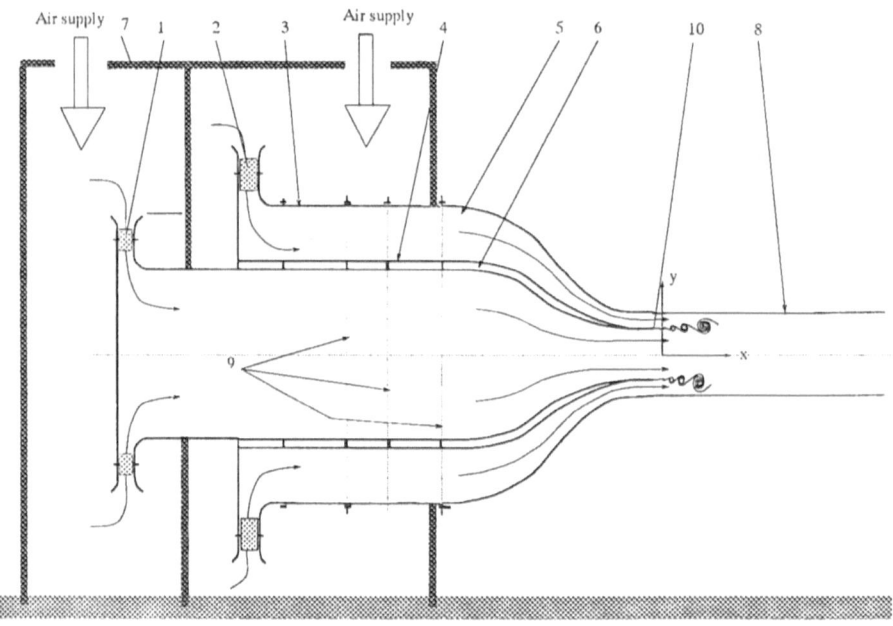

Figure 1: The wind tunnel: 1, 2 radial inlet with vanes; 3 annular plenum chamber; 4 central plenum chamber; 5, 6 contractions; 7 cabine; 8 test section; 9 screens; 10 axisymmetric trailing edge and coordinate system

The maximal free stream velocity of either jet is $U=30$ m/s . For the inner stream the contraction ratio is 1:9.8 and for the annular stream 1:12. The contraction contours consist of two cubic arcs matched at a position x= -0.3 L where L is the contraction length. Effects of flow separation have not been observed.

To allow modification of the exit velocity profile of the inner jet an axial inlet at the upstream end of the inner plenum chamber can be

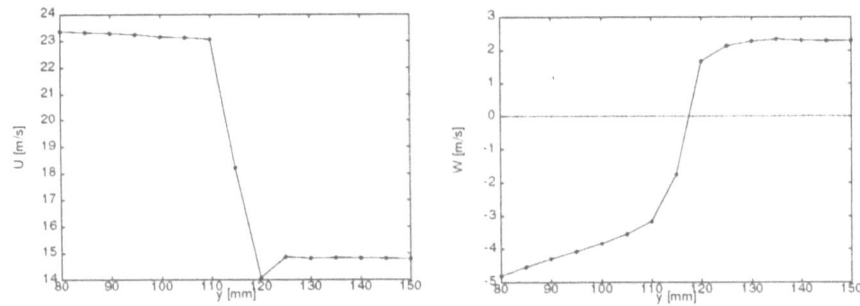

Figure 2: Velocity profiles at $x/R=0.2$

opened.

It is neccessary to shield the flow to be investigated from perturbations originating in the shear layer between the external jet and
the environment. Therefore a test section of diameter 400 mm made
of plexiglass is mounted at the exit of the external jet.

Experiments

To test the performance of the tunnel experiments with mean
axial velocities $U_1=25$ m/s (central jet)and $U_2=15$ m/s (annular jet)
as obtained from averaging over the cross section were carried out.
Three cases were studied: no-swirl, counter-swirl and co-swirl. Mean
axial and mean tangential velocity profiles (obtained from LDA) as
function of r for the shear layer region in the counter-swirl case at
$x/R=0.2$ are shown in Fig.2. The free stream turbulence level was
found to be less than 0.4 %. All data evaluated so far are based on
the classical coordinate system where x is parallel to the axis of the
jet, y denotes the radial direction, and z the circumferential direction.

Growth rates of the momentum thickness $d\Theta/dx$ as computed
from hot wire measurements at several downstream positions are
0.0068, 0.0092 and 0.0085 for no swirl, counter-swirl and co-swirl ,
respectively.

The streamwise turbulent energy $\overline{u'^2}$ for the three cases shows
largest values in the counter swirl case. Smallest values are in the

no swirl case.

Spectra measured at various radial positions show a maximum arounf f=200 Hz for all cases considered (Fig.3).

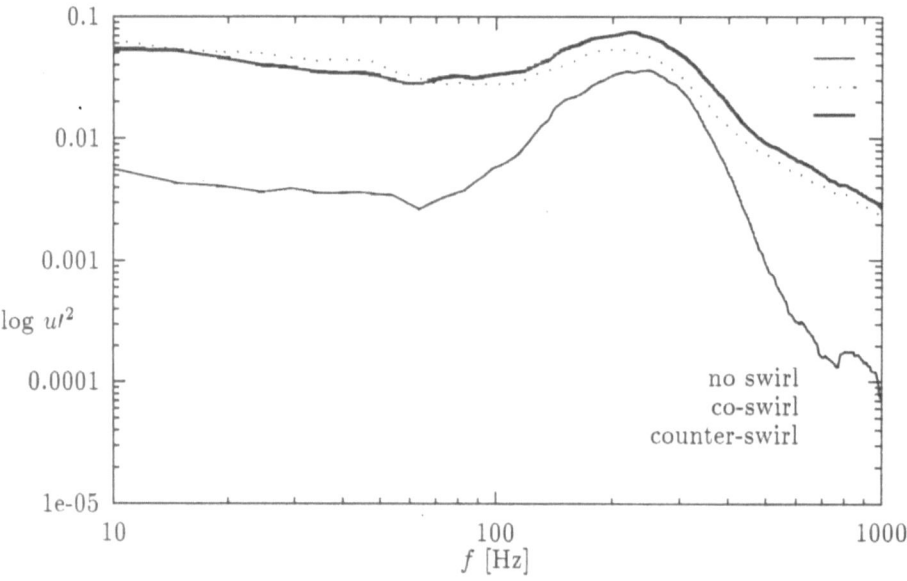

Figure 3: Figure 3: Spectra of $\overline{u'^2}$ at x=500 mm

This observation is corroborated by eigenvalue spectra from a scalar proper orthogonal decompostion (POD) of the velocity signals of a rake of 23 normal hot wires. The rake was aligned in the radial direction and the wires in the tangential direction. The first mode λ^1 of the no-swirling case shows a typical maximum found in plane shear layers [3]. In the swirling cases no such modal maximum seems to exist.

Whether these structures are a genuine feature of the flow or a facility feature remains at this point undetermined.

References

[1] H. Gründel. Strukturen in symmetrischen und asymmetrischen scherschichten. Doctoral thesis, Technische Universität Berlin.

[2] F.C. Gouldin, J.S. Depsky, and S.-L. Lee. Velocity field characteristics of a swirling flow combuster. *AIAA journal*, 23, no. 1:95–102, 1985.

[3] J. Delville, S. Bellin, and J.P. Bonnet. Use of the proper orthogonal decomposition in a plane turbulent mixing layer. In M. Lesieur and O. Metais, editors, *Turbulence and Coherent Structures*, pages 75–90. Kluwer Academic Publisher, 1990.

NEAR-WALL STRONG SWEEPS AND HIGH KURTOSIS LEVELS

Direct simulation and experiment

F.T.M. NIEUWSTADT AND J.M.J. DEN TOONDER
Laboratory of Aero- and Hydrodynamics
TU-Delft, Delft, The Netherlands

AND

Z. ZHANG AND C. XU
Engineering Mechanics
Tsinghua University, Beijing, China

In direct numerical simulations of wall turbulence, i.e. channel, pipe and boundary layers, in general a very high near-wall value is found for the kurtosis of the wall-normal velocity fluctuations. For instance values up to $\sim$25 have been reported, see e.g. (Kim *et al.*, 1987). In contrast, the experimental data up to now have not confirmed such high kurtosis levels and give considerable lower values, say around $\sim$5 (Durst *et al.*, 1995). This has led to some controversy between numerical simulators and experimentalists to explain the background of this discrepancy. We aim to resolve this problem with help of data from a direct simulation together with experimental data.

The direct numerical simulation (DNS) has been carried for a channel flow at a Reynolds number $Re_m = 2666$ (based on the channel half width H and the bulk mean velocity U_m). We have used a spectral code and a resolution of $128 \times 129 \times 128$ in the streamwise, normal and spanwise directions, respectively. The experimental data have been obtained in a pipe flow with Laser-Doppler Velocimetry (LDV). The Reynolds number is $\text{Re}_m = U_m D/\nu = 10018$ (U_m being the bulk mean velocity and D the diameter). For further details regarding the direct simulation and the experiments we refer to (Xu *et al.*, 1996).

The analysis of the DNS data shows that the high kurtosis levels are caused by streamwise vortices in the lowest part of the buffer layer. Very occasionally, i.e. at about 0.01 % of the time, these vortices cause a very strong sweep with a value of velocity fluctuation (in the direction of the wall) in excess of $|v'/v_{rms}| > 10$. This sweep is transported downstream

S. Gavrilakis et al. (eds.), Advances in Turbulence VI, 477-478.

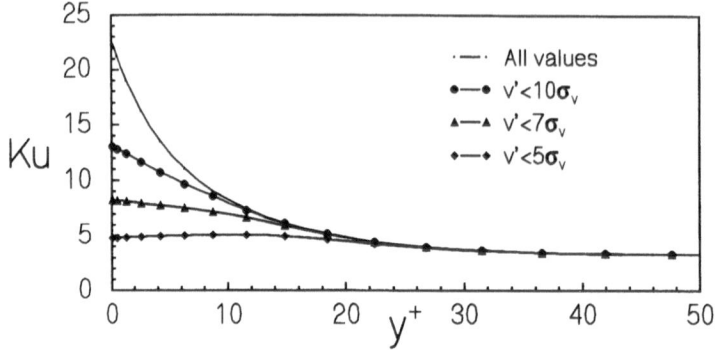

Figure 1. Kurtosis of the wall-normal velocity fluctuations in a channel flow where the highest values (as indicated by the symbols) have been removed from the time series.

together with the streamwise vortex, i.e. with the mean velocity at the height of the vortex. This means that at a fixed point in space, e.g. a LDV measuring point, the wall-normal velocity signal exhibits a very strong spike with a rather short duration due to the large transport velocity of the spike. Based on this information from the DNS data, we searched in our experimental data for such spikes and their existence could be indeed confirmed, i.e. large wall-normal velocity fluctuations were observed at a height of $y^+ = 5$ with $|v'/v_{rms}| > 8$. The time trace of these fluctuations resemble closely the 'spike'-structure found in the DNS data.

That these velocity spikes have not been observedup to now is due to the fact that velocity fluctuations with such a large magnitude as found above, are usually treated in experiments as outliers and consequently they are removed from the measuring series. This removal of high values has a large effect on the value of the kurtosis value as is illustrated in Fig. 1. In this figure we show the kurtosis values obtained from DNS data where the different curves give the kurtosis when values beyond a certain threshold (in terms of the standard deviation) have been removed from the data set. It is clear that removal of fluctuations of say $|v'/v_{rms}| > 7$ considerably reduces the kurtosis with respect to the value for the full data set.

Based on these results, we tend to conclude that the high kurtosis values found in the direct simulation are indeed realistic.

References

Kim, J., Moin P. and Moser R.. (1987) Turbulence statistics in fully developed channel flow at low Reynolds number, *J. Fluid Mech.* **177**, pp. 133–166.,

Durst F., Jovanovic J. and Sender J. (1995) LDA measurements in the near-wall region of turbulent pipe flow, *J. Fluid Mech.* **295**, pp 305–335.

Xu C., den Toonder J.M.J., Nieuwstadt F.T.M. and Zhang Z. (1996) Origin of High Kurtosis Levels in the Viscous Sublayer; Direct Numerical Simulation & Experiment, *Physics of Fluids* in press

INVESTIGATION OF THE DYNAMICS OF NEAR-WALL TURBULENCE USING NONLINEAR TIME SEQUENCE ANALYSIS

Amilcare Porporato and Luca Ridolfi
Department of Hydraulics, Transports and Civil Infrastructures
Polytechnic of Turin

1. Introduction

It is widely accepted that the wall turbulent flows are globally high dimensional (e.g. Keefe, Kim & Moin 1992). Therefore the system is not globally treatable either theoretically or in experiments and, at best, we can only focus our study on the main features of the dynamics. Fortunately this approach is not in actual fact restrictive, since it is also naturally suggested by the presence of coherent structures which, in a certain sense, constitute the skeleton of the near-wall turbulence. Because of their coherency and simplicity in relation to the rest of the dynamics, it is likely that in phase space such organized flow structures correspond to distinct and fairly simple orbits that, at random intervals, leave and suddenly reconnect to the large dimensional part of the attractor (Newell et al., 1988). In particular, in the low dimensional model of Aubry et al. (1988), the bursting phenomenon is shown to correspond in phase space to heteroclinic excursions (see also Sanghi & Aubry, 1993, and references therein). These theoretical studies, as well as certain ad hoc simplified numerical simulations (Hamilton et al. 1995), can capture the main features of the near-wall dynamics and are very important for highlighting the physical origin of the bursting process. However, they still have too little contact with the real turbulence. From the experimental and numerical point of view, the bursting process is quantitatively studied by means of conditional sampling techniques (e.g. Luchik & Tiederman, 1987), that unfortunately suffer from a certain subjectivity in the choice of the threshold values.

In this work we study the link between experimental measurements and low dimensional models and, more generally, we check the various potentials of nonlinear time sequence analysis (e.g. Grassberger et al. 1991) for studying the bursting process. For the sake of brevity many details cannot be reported here, so for further details the reader is referred to another work by one of the authors (Porporato, 1996).

2. Measurements and Preliminary Analysis

The turbulence time series was obtained by using a Laser Doppler Anemometer to measure the longitudinal velocity component in the near-wall region of a

S. Gavrilakis et al. (eds.), Advances in Turbulence VI, 479-482.
© *1996 Kluwer Academic Publishers.*

480

hydraulically-smooth pipe flow. The Reynolds number, based on the mean global velocity and the pipe diameter, was 7000 and the measurement point was at y^+= 15. The measurement was extended for 999 s and the mean sampling frequency was 108 Hz, well beyond the Kolmogorov frequency. In order to be able to use the estimators for equispaced signals, the non-uniformly sampled signal was linearly interpolated and equispaced at the mean sampling frequency. Since the power spectrum (not reported here) shows a clear distinction between turbulence and noise, we could eliminate this latter by low-pass filtering the signal (Butterworth, 48 dB/oct) at 3 Hz; this distinction in the frequency ensures that the signal is not significantly altered by the filtering, as this would dramatically bias the results of the nonlinear analysis.

The statistics of the filtered signal are in complete agreement with the results given in literature.

3. Conditional Sampling

In order to have a reference point for the non linear analysis we subjected the signal to two traditional conditional sampling techniques, namely VITA and MULEVEL (Luchik & Tiederman, 1987). Because it has been shown that, with suitable grouping, MULEVEL provides reliable results, we decided to set its thresholds to the values usually employed, and to adjust the VITA thresholds up to obtain the same mean bursting time as MULEVEL. We therefore adopted the values of k=1 and C=0.25 for MULEVEL and then, to avoid multiple detections during a single burst, we grouped the detections, according to Luchik and Tiederman (1987). The mean time between bursting was T_B^+=125, and this agrees well with the classical results. As far as the threshold values of VITA are concerned, the value of T^+ was set at 10 (Luchik & Tiederman, 1987), whilst the value of the other threshold k was tuned by trial and error to 0.6. Fig. 1a shows a tract of the signal together with the detections of the two techniques, and fig. 1b-c show the ensemble averaged velocity profiles during MULEVEL and VITA detections.

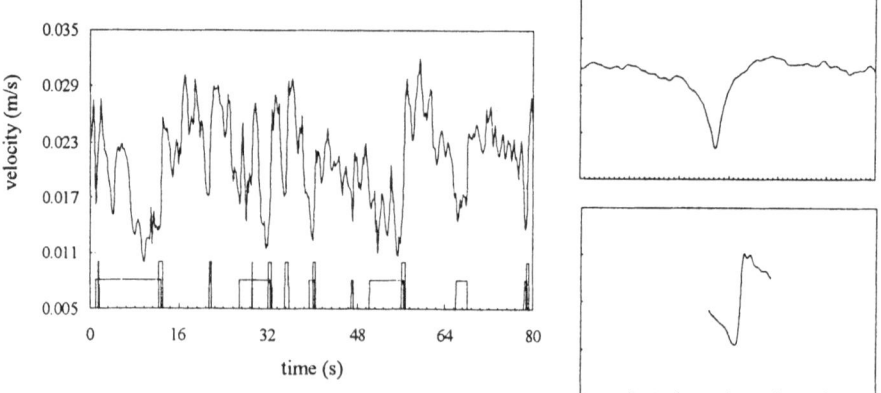

Fig. 1a-c (a) Tract of the measured turbulent signal and related bursting detections (higher steps refer to VITA and the lower ones to grouped MULEVEL) and ensemble average velocity profile during grouped MULEVEL (b) and VITA (c) detections.

4. Nonlinear Analysis

We applied the delay time method of Takens, choosing the optimum delay using the methods of Buzug and Pfister (1992). Even if the globally underlying dynamics is high-dimensional, a correct reconstruction may give interesting insights into possible low dimensional elements of the attractor. Fig. 2a shows a two dimensional projection of the reconstruction, compared with the analogue projections of the velocity mean profiles (fig. 2b-c) of fig. 1b-c. In particular the mark of the large positive gradients, which correspond to the regular orbits in the upper-left part of the projection and which are typical of the bursting process phases detected by VITA, can be clearly seen. Such observation might provide experimental evidence for the heteroclinic excursions theoretically expected in presence of a wall for Navier Stokes equations or its simplifications (Newell et al., 1988; Aubry et al., 1988). Moreover, since it is quite easy to distinguish when the system leaves the high dimensional attractor, it also suggests that it might be possible to find a more objective, physically based criterion for conditional sampling to detect bursting from the measurement of only one velocity component. The Poincarè sections of the three-dimensional reconstruction (not shown here) reveal that the attractor is globally flattened in the direction normal to the plane of the two-dimensional projection of fig. 2a. This assures that the large scale orbits, which we are interested in, are not biased by the projection in such a representation.

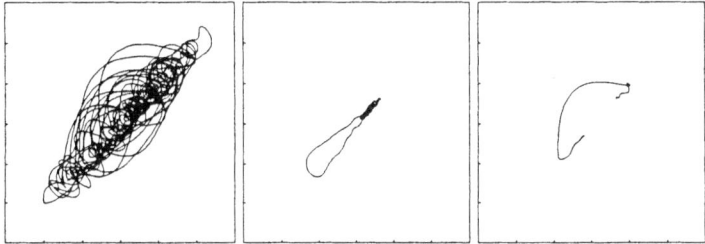

Fig. 2a-c. Projections of the attractor (a) and of the profile of fig. 1b (b) and of fig. 1c (c).

The second main part of nonlinear analysis deals with the possibility of investigating the signal using nonlinear prediction (Farmer & Sidorovich, 1987). This technique may be used both to evidence low-dimensional nonlinear interactions and therefore to detect periods of lower or higher complexity as well as (from a practical point of view) to try to perform short-term forecasts of the bursting, useful for drag reduction by active walls. Fig. 3a shows an example of prediction for a tract of the turbulent signal using the forecasting interval $\Delta t=0.290$ s (i.e. slightly over the estimated Kolmogorov time scale). The forecasts are made with the direct technique, using embedding dimension 11 and 50 neighbours, values that preliminary trials had shown to give the best results. The figures show that the large scale dynamics is quite well captured, while the small scales are not because of the method's low dimensional character. The prediction errors for $\Delta t=0.160$ s (fig. 3b) reveal a certain decrease in the predictability at the beginning of the bursting. We do not consider the errors for $\Delta t=0.290$ s because the local linear approximation used may cause an increase of errors which have nothing to do with the dynamic errors of interest. Such error behaviour

482

might be due to the influence of the outer region acting as an external quasi-random excitation. This conjecture is reinforced by the results (not shown here) of the noise reduction performed using the algorithm of Grassberger et al. (1991). The resulting corrections, which may be considered as errors of a nonlinear prediction performed simultaneously both forward and backward, are no longer concentrated at the beginning of the bursting and this confirms that it is difficult to predict the start of the bursting.

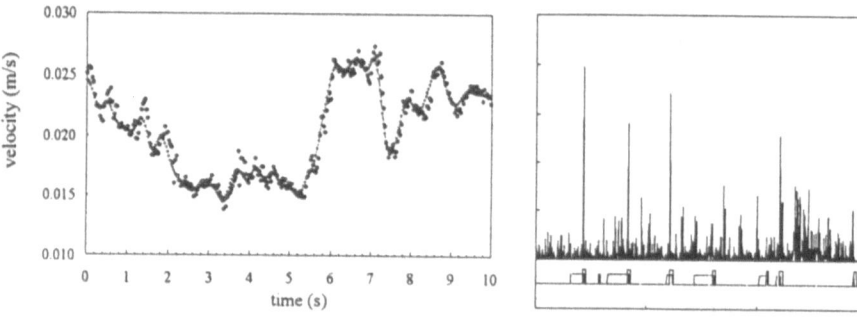

Fig. 3a-b Predictions for $\Delta t=0.290$ s (a); Comparison of errors ($\Delta t=0.160$ s) and detections (b).

5. Conclusions

We have shown some results of the nonlinear analysis of a near-wall turbulence velocity signal. Although our work is still at a preliminary stage, quite interesting aspects of the bursting process seem to have emerged. It is now progressing in all the directions indicated above, in particular in developing new criteria for the choice of VITA thresholds, in trying the possible improvements of nonlinear prediction and in finding possible traces of the regeneration mechanisms of coherent structures (Hamilton et al., 1995) in a real near-wall turbulent signal.

6. References

Aubry, N., Holmes, P., Lumley, J. L. & Stone, E. (1988) *The dynamics of coherent structures in the wall region of a turbulent boundary layer.* J. Fluid Mech. **192**, 115.

Buzug, Th. & Pfister, P. (1992) *Comparison of algorithms calculating optimal embedding parameters for delay time coordinates.* Physica D **58**, 128.

Farmer, D. J. & Sidorovich, J. J (1987) *Predicting chaotic time series.* Phys Rev. Lett. **59**(8), 845.

Grassberger, P. Schreiber, Th. & Schaffrath, C. (1991) *Nonlinear time sequence analysis.* Int. J. Bif. & Chaos. **1**(3), 521.

Hamilton, J. M., Kim, J. & Waleffe, F. (1995) *Regeneration mechanism of near-wall turbulent structures.* J. Fluid. Mech. **287**, 317.

Keefe, L., Moin, P. & Kim, J. (1992) *The dymension of attractors underlying periodic turbulent Poiseuille flow.* J. Fluid Mech. **242**, 1.

Luchik, T. S. & Tiederman, W. G. (1987) *Timescale and structure of ejections and burst in turbulent channel flow.* J. Fluid Mech. **174**, 529.

Newell, A. C., Rand, D. A. & Russel, D. (1988) *Turbulent transport and the random occurrence of coherent events.* Phisica D **33**, 281.

Porporato, A. (1996) *Ricerca di elementi di bassa dimensione nella turbolenza di parete.* Ph.D. Thesis, Dept. of Hydraulics-Polytechnic of Turin. (in Italian)

Sanghi, S. & Aubry, N. (1993) *Mode interaction models for near-wall turbulence.* J. Fluid Mech. **247**, 455.

Takens, F. (1980) *Detecting strange attractors in turbulence.* In Lectures Notes in Math. vol. 898. Springer.

A TRISTATIC DOPPLER VELOCITY PROFILER AND ITS APPLICATION TO TURBULENT OPEN-CHANNEL FLOW

SHEN C. and U. LEMMIN
Laboratoire de Recherches Hydrauliques
DGC-EPFL, 1015 Lausanne, Switzerland
Fax: +41 21 693 6767
E-mail: weidong.shen@lrh.dgc.epfl.ch

Abstract: A tristatic acoustic Doppler velocity profiler (ADVP) has been developed to measure instantaneous two dimensional velocity profiles non-intrusively. The measuring volume is about $\phi 20 \times 4.5mm$, and the frequency response is better than 20Hz. Applications of the ADVP to clear-water open-channel flow are presented. The tristatic ADVP measurements show a strong organization of the turbulent two component velocity field over the total water column which is correlated to the Reynolds stress intermittency.

1. Introduction

A number of measuring techniques and instruments have been developed to study turbulent flow. A Laser Doppler velocimeter can measure small scale turbulence non-intrusively, but is limited by its inability to simultaneously measure profiles. Hot-wire anemometers have the same problem and perturb the flow. Therefore, the possibility of probing turbulence with an acoustic Doppler velocity profiler (ADVP) has been investigated.[1-3] Monostatic ADVPs which have been used for more than two decades, have recently been applied to clear water flow.[4,5] The ADVP can measure instant profiles of velocity, turbulence intensity and other turbulence parameters with a resolution on turbulence scales. However, a monostatic ADVP can only work in stationary steady flow. To investigate temporarily and spatially varying turbulent flow as well, we developed a tristatic ADVP system for open-channel application. This paper presents the details of this technique and some results.

2. Principle of a tristatic ADVP

As shown in Fig. 1, a tristatic ADVP system consists of one ultrasonic emitter and two wide-angle receivers. The central transducer TRE emits a pulse of acoustic waves with frequency f_0 and a pulse duration τ. When propagating through turbulent flow, this wave pulse is scattered by the targets located in the insonified water volumes. Two symmetrically installed transducers TRP (Positive Transducer) and TRN (Negative Transducer) will receive the signals scattered in the two directions simultaneously. The frequencies of the two signals are different from the one emitted by f_{d+} and f_{d-} respectively, where f_{d+} or f_{d-} is the Doppler frequency shift. When the projections of

S. Gavrilakis et al. (eds.), Advances in Turbulence VI, 483-486.
© 1996 *Kluwer Academic Publishers.*

velocity vector $\mathbf{V}_j$ on the two Bragg wavenumber vectors $\mathbf{B}_{j+}$ and $\mathbf{B}_{j-}$ are V_{j+} and V_{j-} (Fig. 1), one finds $f_{dj\pm} = \pm 2V_{j\pm} f_0/c$. The horizontal and vertical velocities, u_j and w_j, in the volume R_j can be therefore obtained by

$$u_j = c\left(f_{d+} - f_{d-}\right) / 4f_0 \sin\left(\alpha_j/2\right) \tag{1}$$

$$w_j = -c\left(f_{d+} + f_{d-}\right) / 4f_0 \cos\left(\alpha_j/2\right). \tag{2}$$

The emitter TRE sends out a sequence of acoustic wave pulses with a repetition frequency PRF. To improve the reliability of the results and to eliminate the fixed echoes which often occur in boundary layer flow, the phase of each emitted pulse is changed randomly between o and π, so that the fixed echoes become white noise. Based on a so-called "pulse-pair" (PP) algorithm, which is a correlation-based estimation procedure, one can extract the velocity information from these coded Doppler signals.[6]

A recent study[5] has found that for a high frequency sonar system, the scattering targets are scalar microstructures which are formed by the concentration variation of fine air bubbles. Due to their small size these follow the turbulent motion of the water well. Therefore, the velocity measured by the ADVP is the velocity of fluid, rather than that of suspended particles, which other instrument depends on.

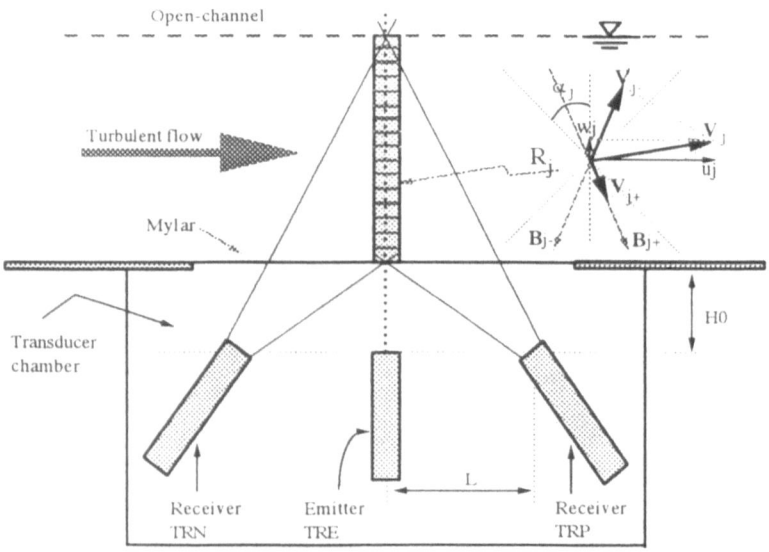

Fig. 1 Configuration of tristatic ADVP

3. Measurement results

The recirculating open channel has a dimension of $43\times2\times1.5$ m^3 and a smooth bed. A tristatic ADVP is installed at about one half of the total length of the flume where the turbulence is well developed. The working fluid is the normal tap water held in a large reservoir. It is constantly filtered, hence free of sediments.

The present tristatic ADVP works at $1MHz$, the pulse duration is chosen to be $6\mu s$. hence, the dimension of insonified region is about $\phi20\times4.5mm$. This spatial resolution is in the order of the microscales of open-channel turbulence when the water depth is at least $10cm$. The measurements were made in uniform flow with a water depth $10cm$ and a Reynolds number of $R_e\approx3.6\times10^5$. The instant velocity vector was calculated by the PP-algorithm using 32 data samples, providing a maximum detectable velocity frequency about $33Hz$. The results coincide well with the known laws for the profiles of mean velocities and variances for the two velocity components.[3] The time series of the profiles of the two velocity components allow for velocity imaging. Fig. 2 shows an example of a velocity image after subtracting the long term mean profiles. Large scale organization becomes evident over a large portion of the flow field when the instantaneous vectors are averaged over a short time scales.

For the same data set, images of the Reynolds stress distribution were obtained by calculating the instantaneous $u\circ v$ profiles, which were again time averaged. Fig. 3 shows an intermittence in the distribution in the near wall layer where maxima coincide with strong, jet-like features in the velocity field in the outer region (Fig. 2).

4. Conclusions

The main advantage of the tristatic acoustic Doppler velocity profiler which has been presented in this paper is its ability to non-intrusively measure time series of 2-D instant velocity profiles, which are difficult to observe by other methods. Applications to turbulent open-channel flow has demonstrated the capacity of the tristatic ADVP to provide new insight into the turbulence structure in this flow field. An extension to time series of profiles of the complete three-dimensional velocity vector is only a matter of adding two more wide angle receiver transducers in a plane perpendicular to the present one.

References

[1] J. C. Schuster, "Measuring water velocity by ultrasonic flowmeter," J. Hydr. Div., ASCE **Vol. 101**(HY12), pp 1503-1517 (1975).

[2] J. L. Garbini, F. K. Forster, and J. E. Jorgensen, "Measurement of fluid turbulence based on pulsed ultrasound techniques (part I and II)," J. Fluid Mech. **Vol. 118**, pp 445-505 (1982).

[3] T. Rolland, *Développement d'une instrumentation Doppler ultrasonore adaptée à l'étude hydraulique de la turbulence dans les canaux*, Thèse NO. 1281, EPFL (Lausanne, 1994).

[4] R. Lhermitte, and U. Lemmin, "Probing water turbulence by high frequency Doppler sonar," Geophys. Res. Letters **17**(10), 1549 - 1552 (1990).

[5] C. Shen, and U. Lemmin, "The scattering of high-frequency sound in open-channel turbulent flow," RAPPORT ANNUEL, Laboratoire de Recherches Hydrauliques, EPFL, 1993.

[6] R. Lhermitte, and R. Serafin, "Pulse-to-pulse coherent Doppler sonar signal processing techniques," Journal of Atmospheric and Oceanic Technology **1**(4), (1984).

486

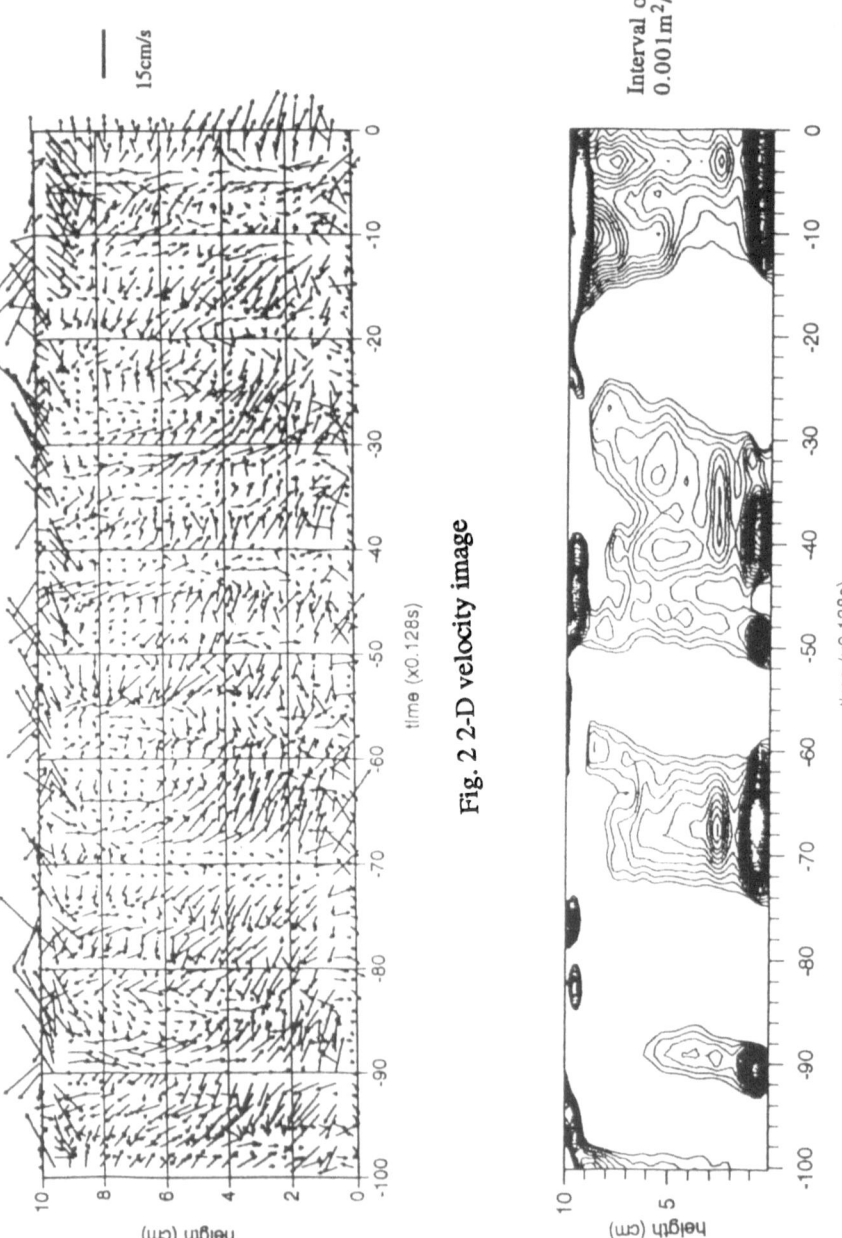

Fig. 2 2-D velocity image

Fig. 3 Temporal Reynolds stress image

TURBULENT SEPARATION REGIONS IN FRONT AND DOWNSTREAM OF A FENCE

H.A. SILLER AND H.H. FERNHOLZ
Hermann-Föttinger-Institut für Strömungsmechanik
Technische Universität Berlin,
Straße des 17. Juni 135, 10623 Berlin, FRG

Abstract. In the turbulent separation region located upstream of a two-dimensional fence measurements of the pressure and wall shear-stress distributions in the main flow direction were taken along with measurements of boundary layer profiles. The flow was then manipulated by a periodic disturbance which was located upstream of the forward separation region. Two different disturbances were tested: an oscillating spoiler and a two-dimensional oscillating jet with zero mass flow, driven by a loudspeaker. Both manipulators were oriented parallel with the fence. With proper tuning of the parameters, the re-attachment length behind the fence could be reduced by 50%. This project is performed in cooperation with H. Wengle in Munich, who works on a large-eddy-simulation of the flow.

1. The undisturbed flow over a fence

The flow over a two-dimensional fence on a flat plate (Fig. 1) is very complex: there are two reverse flow regions up- and downstream of the fence, which interact with one another, and the shear layer which separates from the sharp edge of the fence. Great care has been taken to generate a two-dimensional boundary layer approaching the fence. The optimisation process resulted in a span-wise skin-friction variation of ±3% and a two-dimensional flow over about two thirds of the span downstream of the fence.

The flow regime upstream of the fence (Fig. 2) is characterised by the distributions of the pressure c_p, the skin-friction coefficient c_f, and the reverse flow parameter χ_w (where $\chi_w = 50\%$ denotes separation or re-attachment). The Reynolds number at the fence position is $\mathrm{Re}_{\delta_2} = 1300$

S. Gavrilakis et al. (eds.), Advances in Turbulence VI, 487-490.
© *1996 Kluwer Academic Publishers.*

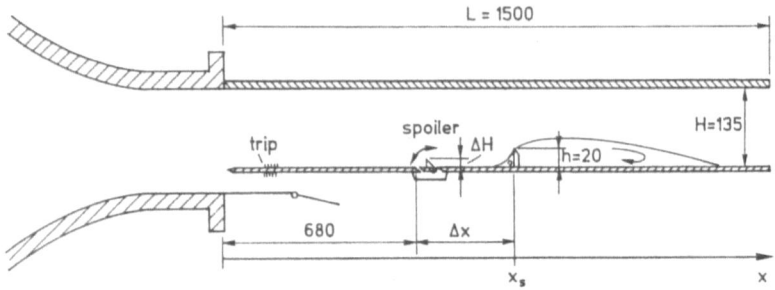

Figure 1. Experimental setup

and the ratio of boundary layer thickness to fence height $\delta/h \approx 1$ (both measured with the fence removed).

The lengths of the upstream and downstream separation regions increase with the blocking ratio H/h (here $H/h = 6.75$). separation in front of the fence occurs at $x_f/h = 0.66 \pm 0.2$, re-attachment behind the fence at $x_r/h = 14.15 \pm 0.2$.

The boundary layer profiles upstream of the fence (Fig. 3) were measured with a traversable pulsed wire probe, which extended up to 13 mm, i.e. $y/h = 0.65$, above the wall.

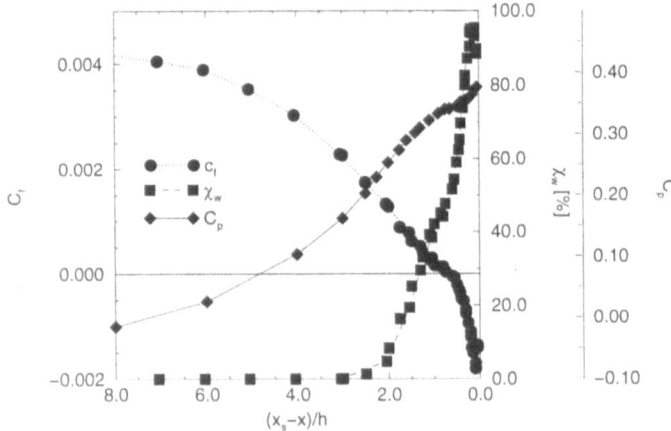

Figure 2. Undisturbed flow upstream of the fence

2. Manipulation of the flow

For the two manipulation devices used, three critical parameters have been identified:

(1) the frequency f of the spoiler or jet oscillation,
(2) the amplitude of the spoiler ΔH and the sound pressure p_c in the settling chamber under the slot, respectively, and
(3) the position x relative to the upstream reverse-flow region.

Both manipulators influence x_r in a similar way. Figure 4 shows x_r, scaled with the fence height h, as a function of the Strouhal number $St = (fh)/U_\infty$ for different excitation amplitudes ($\Delta H/h$ in the spoiler experiments and p_c for the oscillating jet). The experiments with the spoiler and the oscillating jet had slightly different Reynolds numbers Re_h (10300 and 10700) and different blockage ratios H/h (6.75 and 9.50).

The upstream position $\Delta x/h$ of the manipulator relative to the fence influences the maximum reduction of x_r. For the oscillating jet, best results are obtained, if it is placed two fence heights in front of the fence.

For the oscillating spoiler, the optimum position is between $3 \leq x/h \leq 4$. For $x/h \leq 2$ and $x/h > 5$ the reduction of x_r is diminished.

The largest reductions of x_r were achieved with maximum spoiler amplitude or sound pressure. Flow visualisations with a smoke wire (Fig. 5) show that the separation bubble oscillates in time with the excitation. The visualisations and spectral measurements taken in the shear layer separating from the sharp tip of the fence indicate that vortex pairing is one of the mechanisms to increase mixing and thus to reduce the re-attachment length.

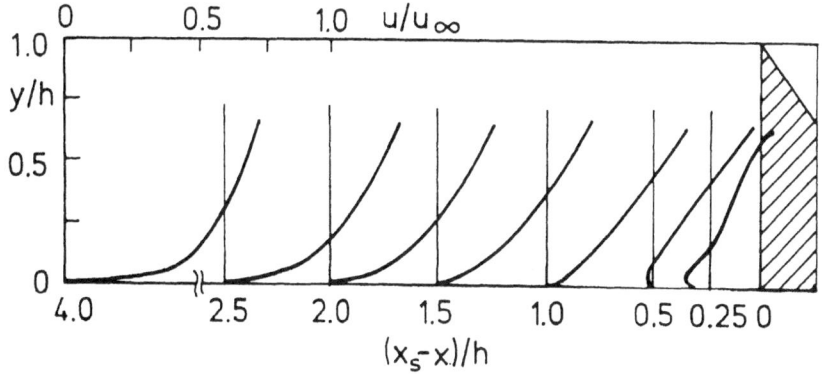

Figure 3. Boundary layer profiles upstream of the fence

490

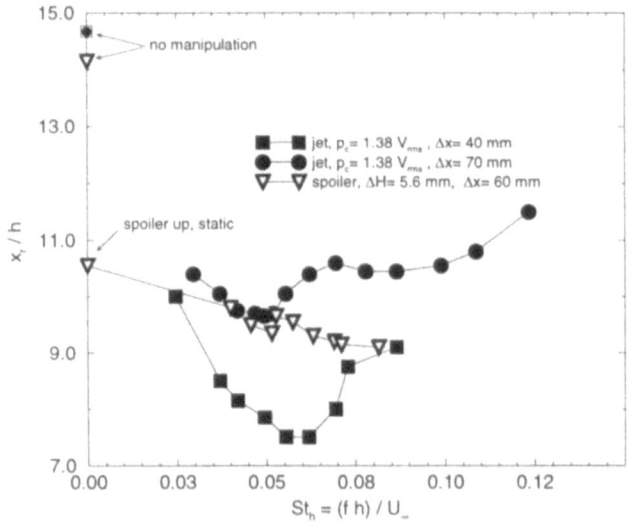

Figure 4. Effect of the manipulation

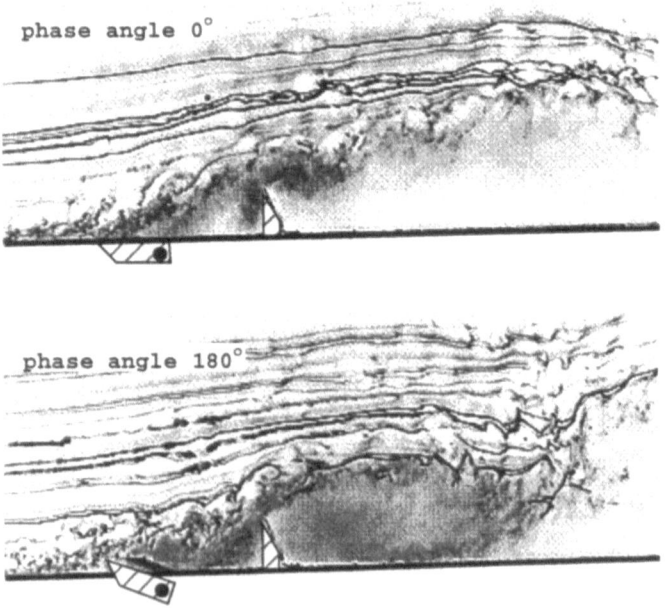

Figure 5. Smoke wire flow visualisation

MEASUREMENT OF PRESSURE-STRAIN THROUGH TWO-POINT VELOCITY CORRELATIONS IN HOMOGENEOUS AXISYMMETRIC TURBULENCE

T.I.Å. SJÖGREN & A.V. JOHANSSON

Department of mechanics, KTH, S-100 44 Stockholm, Sweden

Abstract. A new method for determining the pressure-strain directly from wind-tunnel experiments has been developed with the aid of a newly developed theoretical description of the kinematics of axisymmetric homogeneous turbulence. In the special case of axisymmetric turbulence one can measure the pressure strain correlation without measuring the pressure.

1. Introduction

Reynolds stress closures based on the modeling of the transport equations for the kinetic energy, K, the dissipation, ε, and the Reynolds stress anisotropy, $a_{ij} = \overline{u_i u_j}/K - 2/3\delta_{ij}$, are considered in the situation of homogeneous turbulence. For an infinitely rapid strain the anisotropy state is determined by the action of the production term P_{ij}^a and the rapid pressure-strain rate $\Pi_{ij}^{(r)}$. In the absence of mean strain, on the other hand, the return-to-isotropy process is governed by the slow pressure-strain rate, $\Pi_{ij}^{(s)}$, and the anisotropy of the dissipation rate tensor, e_{ij}. In order to separate these effects in a situation with finite strain one has to be able to measure all of these different quantities. The transport equations for the anisotropy tensor and the turbulent energy read

$$U\frac{da_{ij}}{dx} = P_{ij}^a + \frac{1}{K}(\Pi_{ij}^{(r)} + \Pi_{ij}^{(s)}) - \frac{\varepsilon}{K}(e_{ij} - a_{ij}) \qquad (1)$$

$$U\frac{dK}{dx} = P - \varepsilon \qquad (2)$$

S. Gavrilakis et al. (eds.), Advances in Turbulence VI, 491-494.
© 1996 *Kluwer Academic Publishers.*

2. Axisymmetric Homogeneous Turbulence

The kinematics of axisymmetric turbulence have been analyzed by,*e.g.*, Batchelor (1953), and recently by Lindborg (1995). For this situation Lindborg derived the following relation for the pressure-strain correlations

$$\lambda_i\lambda_j\Pi_{ij}^{(s)} = \int_0^\infty\int_0^\infty [6\rho z(2z^2 - 3\rho^2)r^{-7}(M_1 - M_2) + 6\rho zr^{-5}(M_2 - M_3)$$
$$+6\rho^2(4z^2 - \rho^2)r^{-7}M_7]\,dzd\rho \tag{3}$$

$$\lambda_i\lambda_j\Pi_{ij}^{(r)} = \int_0^\infty\int_0^\infty 18\sigma\rho^2 zr^{-5}R_4\,dzd\rho \tag{4}$$

where $r_i = z\lambda_i + \rho e_i^\rho$ is the separation vector between the two points and λ_i is the unit vector defining the axis of symmetry, and ρ is the radial coordinate in this cylindrical coordinate system. The axisymmetric straining parameter is here denoted $\sigma = \frac{\partial U}{\partial x}$. The R and M-functions in (3,4) are the measurable two-point velocity-correlations

$$R_4 = \langle uv'\rangle \tag{5}$$
$$M_1 = \langle uuu'\rangle, \quad M_2 = \langle vvu'\rangle, \quad M_3 = \langle wwu'\rangle, \quad M_7 = \langle uvu'\rangle \tag{6}$$

where $'$ denotes a quantity evaluated at a separation $\bar{r}$ from the unprimed quantities. With two X-probes separated on a line perpendicular to the mean flow one can measure these functions (using Taylor's hypothesis in the symmetry direction) and thereby also the pressure-strain term.

3. Experimental Setup

The experiments have been carried out in the MTL low turbulence, low speed wind tunnel at KTH. Turbulence was generated by a monoplane square rod grid with 10 cm mesh width positioned in the beginning of contraction, which has an area ratio of nine. The test section is 7 m long and has a rectangular cross-section of 0.8m × 1.2m. In the contraction the turbulence is strongly distorted by the accelerating mean flow producing a turbulence state at the beginning of the test section that is highly anisotropic with relative turbulence levels around 1%. In the test section the turbulence is allowed to relax towards isotropy in the absence of any mean strain. The turbulence Reynolds number in the experiments is typically about $Re_T \equiv 4K^2/\nu\varepsilon \approx 5000$, and the maximum nondimensional strain $\sigma_{max}K/\varepsilon \approx 6$. A double X-probe configuration was used for all the velocity correlation measurements. The hot-wire X-probes had 2.5μm platinum wire sensors and a measurement volume of $(0.75\text{mm})^3$. An angular calibration procedure was used where third order polynomials were fitted to the calibration data.

4. Results

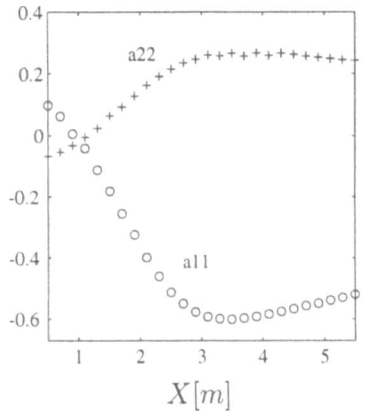

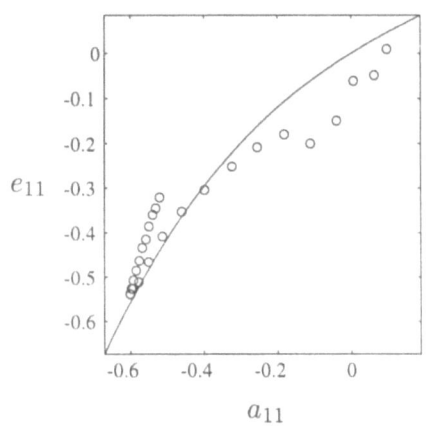

Figure 1. Reynold stress anisotropy a_{11} and a_{22}. The contraction ends at $X = 4m$.

Figure 2. The anisotropy of the dissipation. The solid curve is the model of Halläck *et al.*(1990).

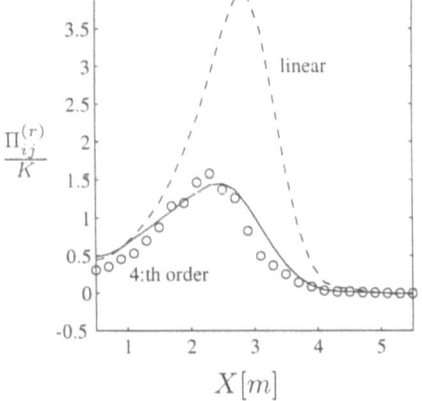

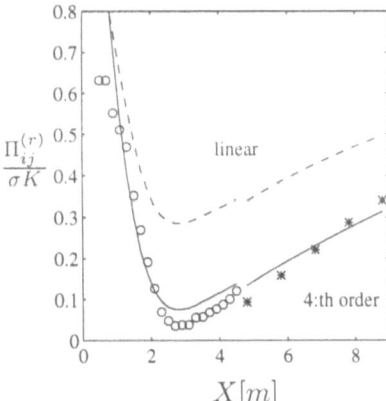

Figure 3. Rapid pressure-strain vs. downstream distance.

Figure 4. Rapid pressure-strain normalized with strain rate vs. downstream distance.

Figure 1 shows the Reynolds stress anisotropy development in the contraction and the subsequent short relaxation part. Note that $a_{33} = a_{22}$ due to axisymmetry, and that a state quite close to the two-component limit (where $a_{11} = -2/3$) is reached in the latter part of the contraction. Even though the turbulence Reynolds number here is as high as 5000 the dissipation rate anisotropy becomes quite large in the straining phase of the

494

contraction. It is also seen in figure 2 that it reasonably well adheres to the behaviour predicted by the model of Hallbäck *et al.* (1990). After the straining the return towards isotropy is somewhat faster than predicted by the model.

Equation (4) was used to compute the rapid pressure strain rate term. It was shown that this could be done with a good accuracy. The results shown in figure 3 were found to be highly repeatable between different experiments. One should stress here that these are the first measurements presented in the literature to this date where the different parts of the pressure strain have been obtained separately. A comparison is also shown in figure 3 between the experimental results and the predictions of the (general) linear Launder, Reece & Rodi (1975) model and the nonlinear (fourth order) model of Johansson & Hallbäck (1994). The latter is seen to give an excellent agreement with the experimental results.

The fourth order tensor term, that multiplied by the mean velocity gradient tensor and the kinetic energy, can be determined also in the absence of a mean strain. In axisymmetry this element is given by eq. (4). Figure 4 shows this quantity in the contraction and the subsequent relaxation ($x > 4$ m). The fourth order model is seen to give a good description of its behaviour in both parts.

Figure 5 shows the budget of eq. (1) where all terms have been determined experimentally. The slow part of the pressure-strain could also be determined directly with the aid of eq. (3), but is in figure 5 determined through balance of eq. (1). We note that although the strain is quite rapid in the central part of the contraction the slow pressure-strain is comparable in magnitude to the rapid part. The effects caused by the difference in anisotropy between the stress and the dissipation rate are quite small.

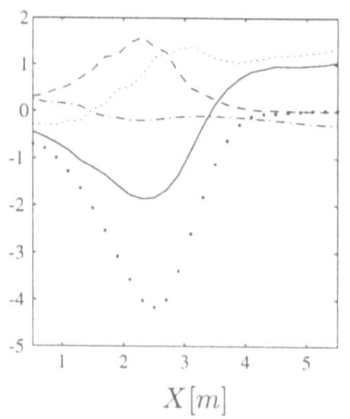

$X\,[m]$

Figure 5. Reynold stress anisotropy (a_{11}) budget. The full line represents left hand side of eq. (1). The right hand side containing production (coarse dotted), rapid pressure-strain (dashed), slow pressure-strain (dotted) and anisotropic dissipation (chain-dashed).

Financial support from TFR is gratefully acknowledged.

CONDITIONAL SAMPLING OF DISSIPATION IN LARGE-SCALE ROTATIONAL REGIONS OF GRID TURBULENCE

S. T. THORODDSEN

Affiliation
Department of Theoretical and Applied Mechanics
University of Illinois at Urbana-Champaign
Urbana, IL 61801-2935, USA.

1. Introduction

Multi-point hot-wire measurements in grid turbulence were used to conditionally average the dissipation of turbulent kinetic energy over inertial-scale features of a turbulent velocity field. Velocities were measured at 5 separate spatial points arranged in the cross-configuration shown in Figure 1. The probes were all located in a cross-stream plane. There are 4 single hot-wires at the edges, with an ×-wire at the center. Six separate velocity components were thus measured simultaneously. The widest spacing between any two probes ($=2L$ in Fig. 1) spans 0.65 of the integral length scale Λ. The flow features used in the conditioning described below are thereby inside the inertial range and much larger than the the worms observed at the relatively small scales (see e.g. Jimenez et al., 1993). The wind-tunnel turbulence was generated with a grid of a very large mesh-size ($M = 19$ cm), producing turbulent Reynolds number Re_λ of about 200, while retaining good transverse homogeneity of the turbulence. The dissipation was approximated with the socalled pseudo- or surrogate dissipation ϵ'_r based on one or two velocity gradients, to avoid the kinematic constraints previously described by Thoroddsen (1995).

2. Conditional Sampling

The conditional sampling which was applied to the data to identify the large-scale rotational regions was the following. The large-scale rotation

495

S. Gavrilakis et al. (eds.), Advances in Turbulence VI, 495-498.
© 1996 *Kluwer Academic Publishers.*

496

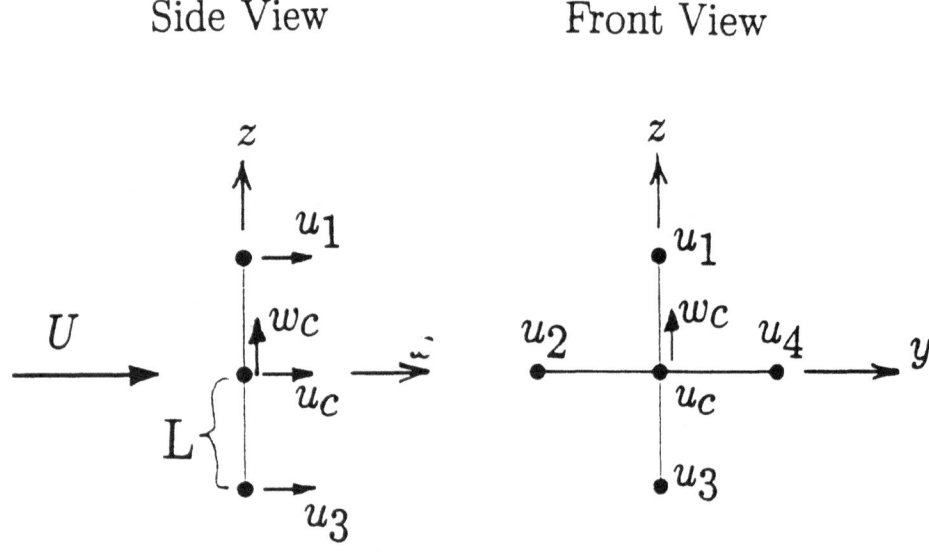

Figure 1. The configuration of the hot-wire probes. The wire separation L was 3.55 cm. Both u and w were measured at the center point.

about the transverse y-axis is defined as

$$\Omega_y(x,t) = \frac{\Delta u}{\Delta z} - \frac{\Delta w}{\Delta x} \tag{1}$$

where $\Delta u/\Delta z = (u_1 - u_3)/2L$ and $\Delta w/\Delta x = [w_c(x + L) - w_c(x - L)]/2L$. Taylor's hypothesis was used to shift in the x-direction by $L = U\Delta t$. The wire numbering and locations are defined in Figure 1. Additional details of the experimental approach are given by Thoroddsen (1996).

3. Results

The conditional averaging was carried out around peaks in the Ω_y signal where $|\Omega_y| > \beta(\Omega_y)_{rms}$. Figure 2 shows the average dissipation rate in the conditioned rotational regions for many different values of the parameter β. The conditioned dissipation ϵ'_r inside the most intense large-scale rotational regions of the flow is more than twice as large as the expected value of ϵ'_r over the whole domain. The dissipation was here averaged over segments of length $r = L/2$, where L is defined in Figure 1. These segments were aligned on both sides of the peaks in $|\Omega_y|$.

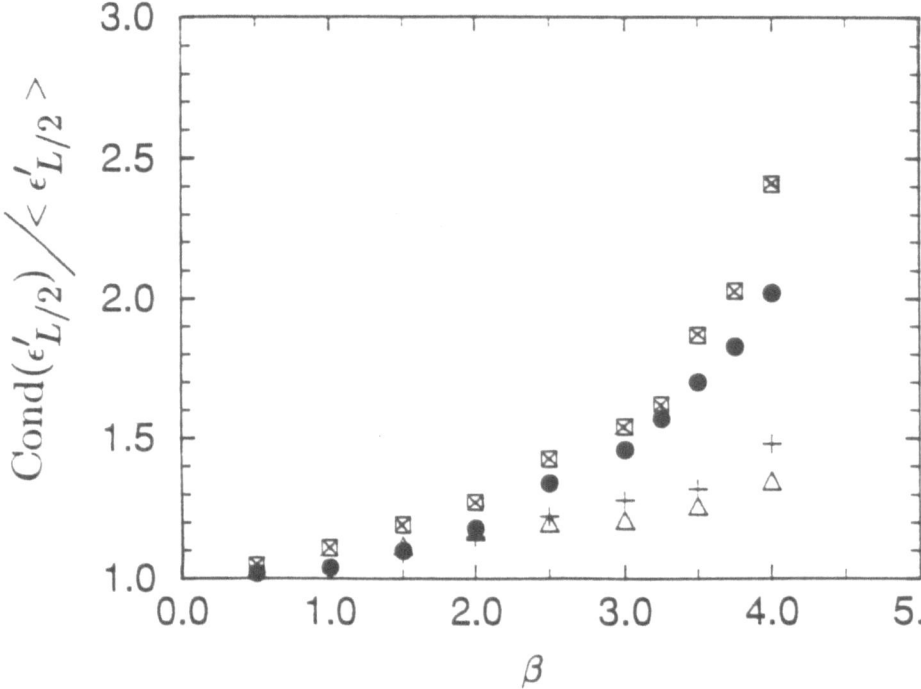

Figure 2. Excess conditioned dissipation $\epsilon'_{L/2}$ vs the parameter β used in the conditioning; Results from conditioning on Ω_y, ($\boxtimes$) when $(\partial w/\partial x)^2$ is used to estimate the dissipation; $\bullet$ using $(\partial u/\partial x)^2$. Results from conditioning on $\Delta u/\Delta x$, ($\triangle$) for $\epsilon'_{L/2}$ based on $\approx (\partial w/\partial x)^2$ and $(+)$ when using $(\partial u/\partial x)^2$.

For comparison with the above results, conditional averaging of the dissipation was also carried out over large-scale fronts (or streamwise velocity increment, following the Kolmogorov Refined Similarity Hypothesis **RSH** (Kolmogorov, 1962; Obukhov, 1962)). This conditioning showed smaller excess in dissipation. These results are included in Figure 2.

4. Discussion

These results show that the local large-scale rotation is more effective in predicting large values of the turbulent dissipation than the streamwise velocity increments. It may thereby be argued that the Kolmogorov **RSH** may not be the optimal way of describing the spatial structure of turbulent dissipation. Models based on the large-scale rotation might give better results. This could be of importance for models used in *Large-Eddy Simulations.*

References

Jimenez, J., Wray, A. A., Saffman, P. G. and Rogallo, R. S. (1993) The structure of intense vorticity in isotropic turbulence, *J. Fluid Mech.*, **Vol. 255**, pp. 65.

Kolmogorov, A. N. (1962) A refinement of previous hypotheses concerning the local structure of turbulence in a viscous incompressible fluid at high Reynolds number, *J. Fluid Mech.*, **Vol. 13**, pp. 82.

Obukhov, A. M. (1962) Some specific features of atmospheric turbulence, *J. Fluid Mech.*, **Vol. 13**, pp. 77.

Thoroddsen, S. T. (1995) Reevaluation of the Experimental Support for the Kolmogorov Refined Similarity Hypothesis, *Phys. Fluids*, **Vol. 7**, pp. 691–693.

Thoroddsen, S. T. (1996) Conditional sampling of dissipation in moderate Reynolds number grid turbulence, *Phys. Fluids*, in press.

AN APPLICATION OF 3D-PTV ON THE MEASUREMENT OF TURBULENT QUANTITIES AND PARTICLE DISPERSION IN TURBULENT CHANNEL FLOW

M. VIRANT, TH. DRACOS
Institute of Hydromechanics and Water Resources Management
Swiss Federal Institute of Technology, ETH Hönggerberg,
CH-8093 Zürich, Switzerland

1. Introduction

The 3D-Particle-Tracking-Velocimetry (3D-PTV) technique developed at the ETH Zürich is capable of measuring three-dimensional Eulerian and Lagrangian properties in three-dimensional fields of turbulent flows simultaneously. This technique is based on imaging illuminated flow markers with four steroscopically arranged CCD cameras. Digital image processing, in combination with methods of digital photogrammetry and object tracking allows a reconstruction of the flow marker movement in an observation volume. The photogrammetric and physical basis of these techniques are presented comprehensively in Maas et al. [1993] and Malik et al. [1993]. We are reporting on the application of 3D-PTV on the measurement of dispersion in a fully developed turbulent channel flow.

2. Experimental performance

The experiments were performed in a 24 m long open duct (width 60 cm. adjustment of constant flow depth h by tilting the channel above angle α).

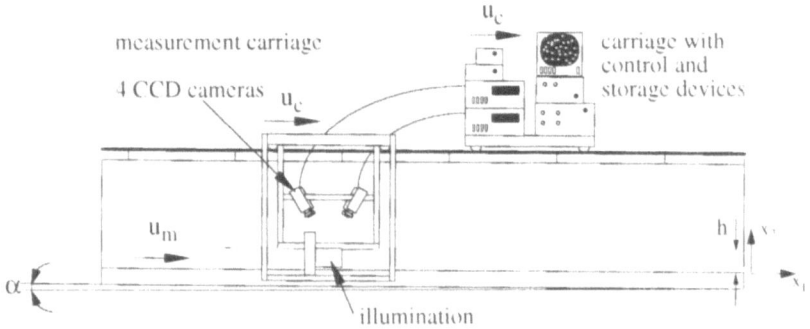

Figure 1 Side view on the open channel.

The flow was marked with plastic spheres of an average diameter of 50μm and a specific density of 1.02. Due to the frame repetition rate of the used european video standard

S. Gavrilakis et al. (eds.), Advances in Turbulence VI, 499-502.
© 1996 *Kluwer Academic Publishers.*

CCD cameras the temporal resolution was 40 ms. Lagrangian analysis requires long trajectories, therefore the carriages - one with the cameras and one with the control and storage devices - were driven by a stepper motor each. The carriage velocity u_c was independently measured by 12 light diods.

With this set-up it was possible to measure velocity vectors of up to 1300 particles. As due to ambiguities the lenght of the measured trajectories decreases with an increasing particle number density, for Lagrangian measurements less particles were used. Figure 2 shows trajectories of 450 particles over 100 time steps.

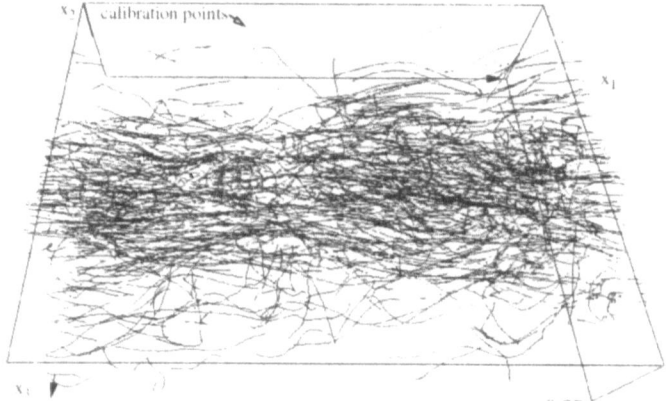

Figure 2. Observation volume size 43x25x9cm3, 450 vectors per time step over 100 time steps, reference velocity eq. carriage velocity uc=14.8 cm/sec. Reynolds number Reh=11570.

The extracted streamwise velocity vectors of a whole run over 1200 time steps are displayed on Figure 3. Assuming ergodicity in horizontal layers, the velocity was calculated in 1/2 mm intervals. This figure combines Eulerian and Lagrangian information, as in addition the velocity of one particle along its trajectory is displayed.

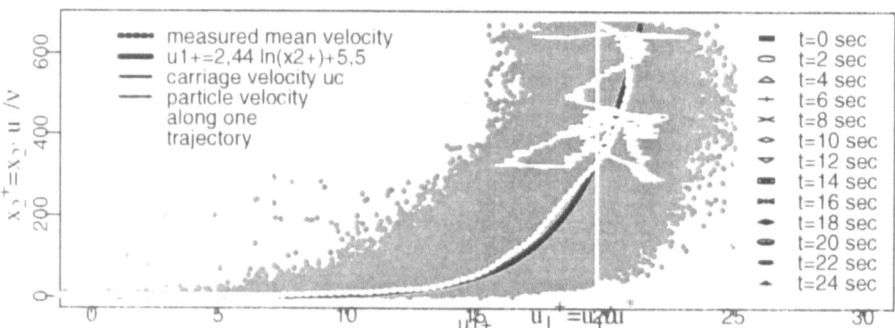

Figure 3 550'000 streamwise velocity components over a flow depth of h=9 cm of the statistics over flow shown in Fig. 2, normalized with friction velocity u*=7.5 mm/sec.

The agreement of the PTV measurements with the LDA measurements and with the direct numerical simulations is well within an error band of 1%.

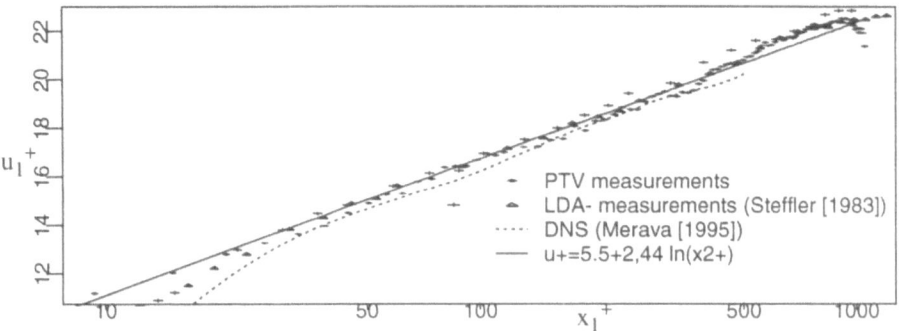

Figure 4. Streamwise mean velocities, normalized by friction velocity u* in semi-log scale.

The Lagrangian velocity autocorrelation and crosscorrelation coefficients $R_{ij}(t)$ displayed in Figure 5 are computed along the trajectories:

$$R_{ij}(\tau) = \frac{\overline{u_i'(t=0) \cdot u_j'(t=\tau)}}{\sqrt{\overline{u_i'^2(t=0) \cdot u_j'^2(t=\tau)}}} \ .$$

The Lagrangian integral length scales are $T_{1L}=3.2$, $T_{2L}=0.35$ and $T_{3L}=0.48$. The R_{13} and R_{23} crosscorrelations are zero. Due to the shear in the x_1-x_2-direction the streamwise and depthwise velocities are correlated ($R_{12}(0)=0.31$).

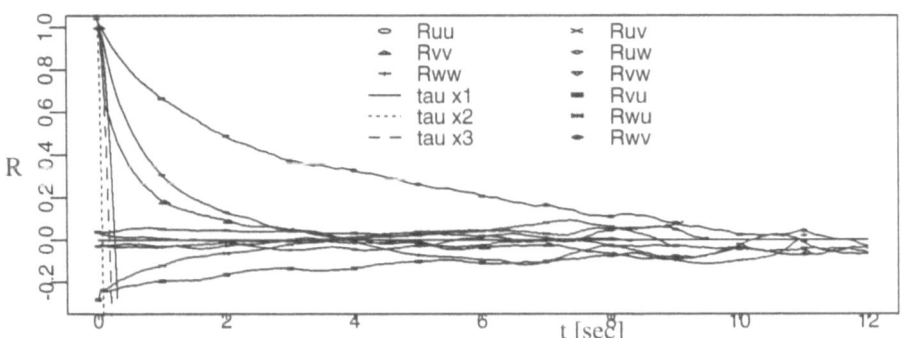

Figure 5. Lagrangian auto- and cross-correlations

According to Taylor's turbulent diffusion theory [1921], the lateral diffusion of fluid particles at time elapsed t due to fluctuation velocities u for homogeneous isotropic turbulence is expressed as a mean-square value:

$$\overline{y_3^2}(t) = 2u_3'^2 \int_0^t \int_0^{t'} R_{3L}(\tau) \, 'd\tau'dt' \ .$$

For diffusion times larger than the integral length scale Taylor predicted a linear rela-

502

tion between the variance of the particle distance and the diffusion time t: $\overline{y_3^2}(t) \approx 2u_3'^2 \cdot T_{3L} \cdot t$. This particle distance variances, made dimensionless with the Lagrangian integral length scales and the RMS of the fluctuation velocity for the lateral direction, are plotted in Figure 6 against the diffusion time, made dimensionless with Lagrangian integral length scales.

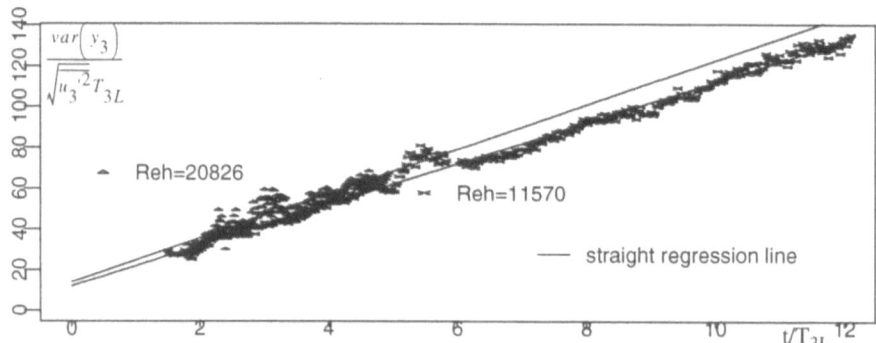

Figure 6. Variance of particle distance over time.

The straight behaviour of the variance agrees well with the theoretical prediction and the coincidence of the lines for the two Reynoldsnumbers corroborates the hypothesis that the diffusion process is mostly governed by the turbulence intensity and the scales of turbulence.

3. Conclusion

Three-dimensional velocity vectors, trajectories and particle distribution in a turbulent channel flow have been measured using PTV. From the Lagrangian autocorrelation coefficients the Lagrangian integral length scales could be computed and Taylor's diffusion theory verified.

4. References

Maas, H.-G., Gruen, A., Papantoniou, D. A., 1993, Particle tracking velocimetry in three-dimensional flows, Part I: Photogrammetric Determination of Particle Coordinates, Exp. in Fluids 15, 279-294.

Malik, N. A., Dracos, Th., Papantoniou, D. A., 1993, Particle tracking velocimetry in three-dimensional flows, Part II: Particle Tracking, Exp. in Fluids 15, 279-294.

Steffler, P. M., Rajaratnam, N., Peterson, A. W., 1983, LDA Measurements of mean velocity and turbulence distribution in a smooth rectangular open channel, Water Resources Eng. Rep. WRE 83-4, University of Alberta, Edmonton, Canada.

Taylor, G. I., 1921, Diffusion by continous movements, Proc. Roy. Soc. London, Vol. 20, Series 2, Part 1, pp. 196-212.

This research was supported by the Swiss National Science Foundation throught Grant No. 20-29852.90 and by the ETH Zürich.

MEASUREMENT OF FLOW STRUCTURES
IN A TURBULENT BOUNDARY LAYER

J. WESTERWEEL, J.G.TH. VAN DER HOEVEN AND M.M. VAN AARSSEN
Delft University of Technology
Laboratory for Aero and Hydrodynamics
Rotterdamseweg 145, 2628 AL Delft, The Netherlands

It is conjectured that riblet surfaces effect a drag reduction by organizing the structure of a turbulent flow in the near-wall region of a turbulent boundary layer. However, direct experimental observation is inconclusive, as these are based on single-point measurement probes (like LDV) or qualitative observations with conventional flow visualization. Digital particle image velocimetry (DPIV) is a measurement technique that yields quantitative information of the instantaneous velocity field in a planar cross-section of the flow, and it is especially useful for the investigation of flow structures. We therefore decided to use this technique to make a quantitative investigation of the flow structures in a boundary layer over smooth and riblet surfaces.

The measurements were carried out in a boundary layer over a flat plate in a free-surface water flow facility. The Reynolds number was about 800 (based on the free-stream velocity and the momentum thickness). In a previous investigation [1] it was demonstrated that DPIV can provide accurate measurements of the velocity in a turbulent pipe flow. We used an image acquisition system that can record high-resolution images (1000×1016 pixels) at a rate of 10 images per second, with a real-time storage capacity of 262 images. An overview of the experimental conditions is presented in Table 1.

In this experiment we focus on the so-called low-speed streaks, so we choose our measurement planes parallel to the wall, with a view area of approximately 40×40 mm^2. This configuration yields the streamwise and spanwise velocity components with a spatial resolution of 0.56 mm (i.e., 3782 velocity measurements per image), and with an accuracy of about 1% of the mean streamwise velocity. In order to obtain accurate measurements of the flow statistics we also carried out measurements with LDV; this also allows us to validate the DPIV results.

S. Gavrilakis et al. (eds.), Advances in Turbulence VI, 503-506.
© 1996 Kluwer Academic Publishers.

TABLE 1. An overview of relevant experimental conditions.

Flow:		
fluid	water	
free-stream velocity	144	mm/s
boundary layer thickness	44	mm
momentum thickness	6	mm
Reynolds number	800	
Seeding:		
type	Sphericel	
nominal diameter	12	μm
Light sheet:		
type	scanning beam	
source	cw Ar$^+$ laser	
power	2	W
thickness	0.4	mm
exposure time delay	2.7	ms
number of exposures	5	
Recording:		
type	electronic (CCD)	
resolution	1000×1016	pixels
lens focal length	50	mm
numerical aperture	4	
image magnification	0.3	
viewing area	40×40	mm^2
Interrogation:		
area	1.12×1.12	mm^2
resolution	32×32	px
Data set:		
vectors/image	3,782	
images/data set	262	

Several time series have been recorded over different surfaces at various distances from the surface, ranging from the buffer layer to the outer edge of the boundary layer. Here we present preliminary results obtained from the measurements of the flow over a smooth surface. The velocity statistics are homogeneous within each measurement plane. To determine the turbulent flow statistics from the DPIV measurements we ensemble averaged the data over all frames of each time series. In Figure 1 we compare the mean streamwise velocity profiles from the DPIV and LDV measurements. Fig-

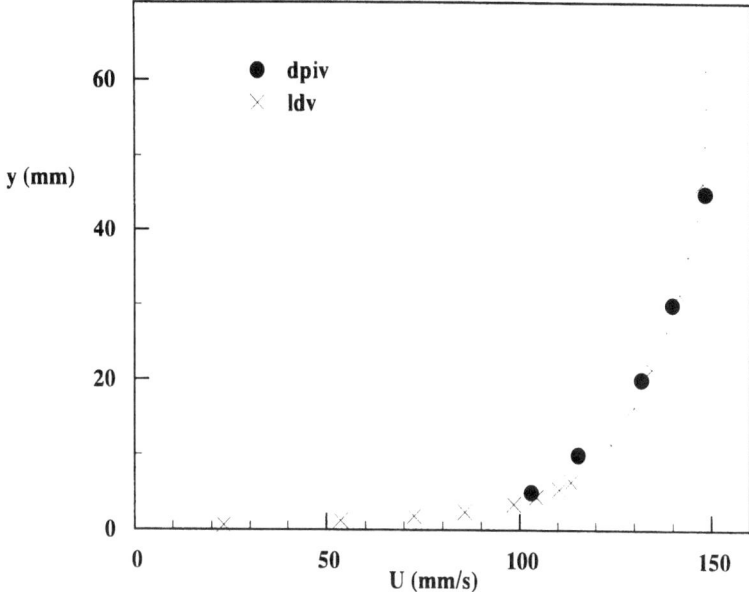

Figure 1. Streamwise velocity profile as a function of distance from the wall. LDV ($\times$) versus DPIV ($\bullet$).

ure 2 shows four vector maps of the instantaneous velocity relative to the mean streamwise velocity in the measurement plane (U(y)). The maps in Fig. 2 are representative for the measured series at the corresponding distance. Clearly, as we approach the wall the structure of the instantaneous flow changes from small structures (Fig. 2-a) in the outer region of the boundary layer, to elongated patterns close to the buffer layer (Fig. 2-d).

From these preliminary results we can already conclude that the results for the mean velocity profile obtained with DPIV agree with those obtained with LDV. The vector maps show how the structure of the flow depends on the distance from the wall. We are currently investigating the series recorded with different riblet surfaces; the results will be included in the final presentation.

Acknowledgements

The research of dr.ir. J. Westerweel has been made possible by a fellowship of the Royal Netherlands Academy of Arts and Sciences.

References

1. Westerweel, J., Draad, A.A., van der Hoeven, J.G.Th. and van Oord, J. (1996) Measurement of fully-developed turbulent pipe flow with digital particle image velocimetry, *Exp. Fluids*, **Vol. no. 20**, pp. 165–177

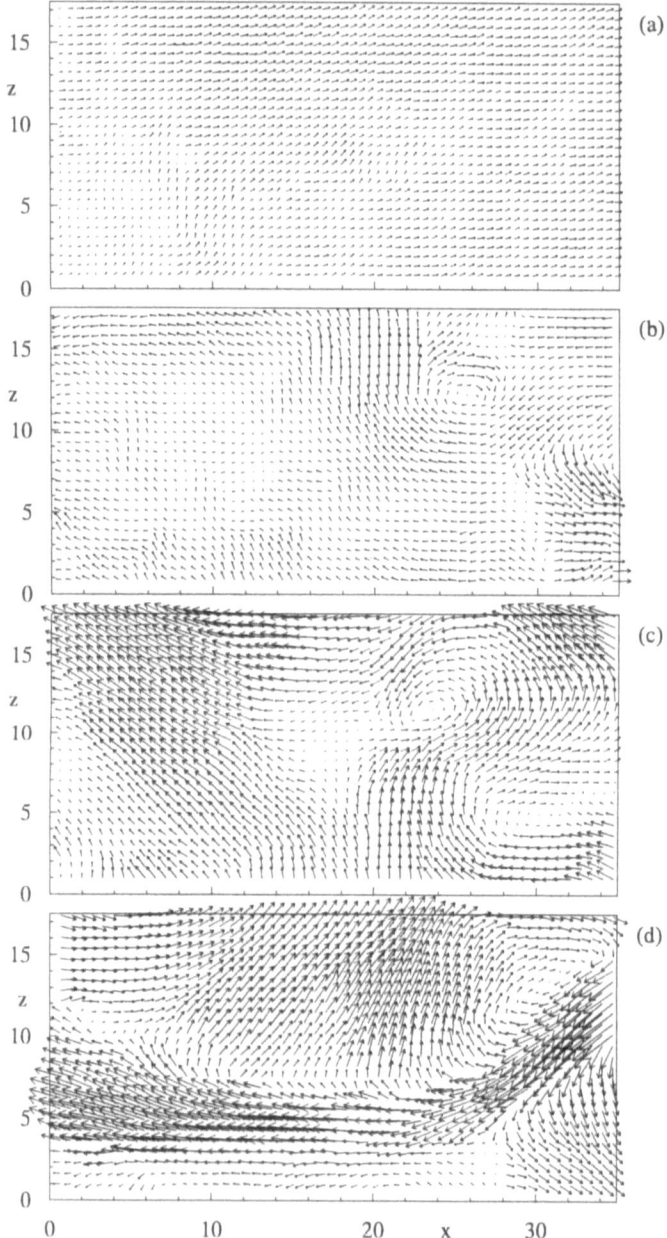

Figure 2. Instantaneous relative velocity maps at different distances from the wall: (a) at y = 45 mm (at the edge of the boundary layer), (b) at y = 30 mm (in the logarithmic layer), (c) at y = 10 mm and (d) at y = 5 mm (in the logarithmic layer, close to the buffer layer). The flow direction is from left to right, and the scales of the axes are in mm.

Correlation Between the Enstrophy and the Energy Dissipation Rate in a Turbulent Wake

Y. Zhu and R. A. Antonia

Department of Mechanical Engineering
University of Newcastle, N.S.W., 2308, Australia

Abstract

All three components of the vorticity fluctuation have been measured simultaneously in a turbulent wake using a new eight-sensor vorticity probe. The correlation between the enstrophy and the energy dissipation rate has been examined in some detail. The homogeneous value of ϵ is strongly correlated with ω^2. The full value of ϵ and, more especially its isotropic value, are less well correlated with the enstrophy.

1 Introduction

The enstrophy ω^2 and the turbulent energy dissipation rate ϵ are two important characteristics of small scale turbulence. While these quantities are available in direct numerical simulations of turbulent flows (e.g. Kim et al., 1987), few measurements have been made (e.g. Vukoslavčević et al., 1991, Balint et al., 1991; Tsinober et al., 1992; Marasli et al., 1993; see also Wallace and Foss, 1995). Such measurements are useful for testing intermittency models, assessing the adequacy of $(\partial u/\partial x)^2$ as a surrogate for ϵ and quantifying the correlation that exists between the dissipation and vorticity fields. However, the latter quantity has received little attention even though there have been investigations of the properties of ϵ and ω^2 (Browne et al., 1987; Antonia et al., 1988; Meneveau et al., 1990; Bershadskii et al., 1993) and of the correlation between vorticity and velocity derivatives (e.g. Tsinober, 1990; Tsinober et al., 1992). The main aim of this paper is to quantify the correlation between ω^2 and ϵ.

2 Experimental Details

Measurements were made in the self-preserving turbulent wake of a circular cylinder at $x_1/d = 240$, where x_1 is measured from the cylinder and d is the cylinder diameter (6.35 mm). The free stream velocity is 3.6 m/s. At the measurement

S. Gavrilakis et al. (eds.), Advances in Turbulence VI, 507-510.

508

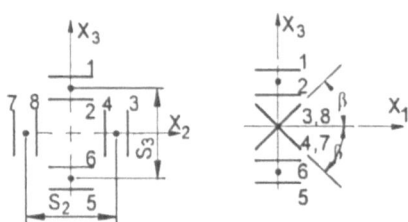

Figure 1: Configuration of the vorticity probe.

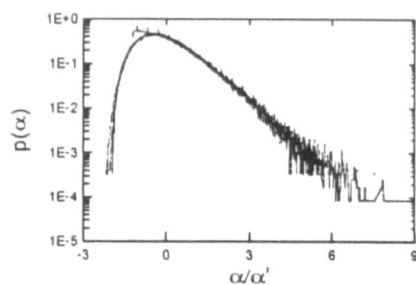

Figure 2: Probability density function of α. —, $\alpha = |\omega|$; $\cdots$, $\epsilon_{iso}^{1/2}$; $--$, $\epsilon_{hom}^{1/2}$; $-\cdot-$, $\epsilon^{1/2}$.

position, the wake half-width L is about 26 mm; the turbulence Reynolds number R_λ is about 40 and the Kolmogorov length scale is 0.64 mm on the centreline.

The three components of the vorticity fluctuation were measured simultaneously with a new eight-sensor vorticity probe. The probe (Figure 1) consists of four X-wires; two are in the $x_1 - x_2$ plane and are separated in the x_3 direction ($s_3 = 2.76$ mm); the other two are in the $x_1 - x_3$ plane and are separated in the x_2 direction($s_2 = 2.5$ mm). The lateral separation between the inclined wires in each X-probe is about 1 mm.

3   Results

The measured velocity and vorticity spectra were compared with those obtained by direct numerical simulation (DNS) [Kim et al., 1987] on the centreline of a fully developed turbulent channel flow (figure not shown). There is evidence (e.g. Antonia and Kim, 1994) that the high wavenumber part of the spectra exhibit universality in support of the Kolmogorov (1941) hypothesis. The measured Kolmogorov normalized spectra agreed quite well with the corresponding DNS data at high wavenumbers suggesting an adequate performance of the new probe.

Figure 2 shows the pdfs of $|\omega|$ and $\epsilon^{1/2}$. ϵ is estimated using full, homogeneous and isotropic expressions. For the full dissipation, the sum of the unmeasured quantities $u_{2,2}^2$ and $u_{3,3}^2$ is approximated by $(u_{1,1}^2 - 2u_{2,3}u_{3,2})$ using continuity and homogeneity (strictly, only the mean values of $u_{2,3}u_{3,2}$ and $u_{2,2}u_{3,3}$ are equal by homogeneity). The isotropic dissipation rate ϵ_{iso} is assumed to be $15\nu u_{1,1}^2$. These quantities show an exponential distribution at large α and are almost identical for $\alpha/\alpha' \geq -0.5$ (a prime denotes a rms value). Such an agreement is not observed in the pdfs of $|\omega|$ and ϵ_{iso} measured in a grid flow and in a circular jet (Bershadskii et al., 1993). The close similarity between pdfs of $|\omega|$ and $\epsilon^{1/2}$ implies that ω^2 may be a good substitute for the total energy dissipation rate.

The correlation between ϵ and ω^2 can be quantified by examining their joint pdf. Figure 3 shows the joint pdf measured on the centreline. There is a strong correlation between ϵ_{hom} and ω^2. This is not surprising since ϵ_{hom} and ω^2 have

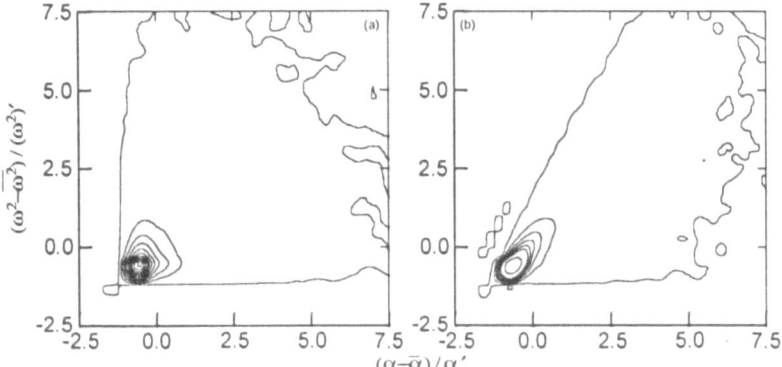

Figure 3: Joint pdfs between ω^2 and α. (a) $\alpha = \epsilon_{iso}$. (b) $\alpha = \epsilon_{hom}$.

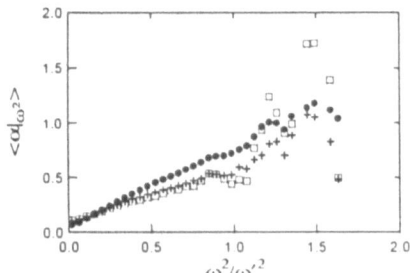

Figure 4: Distribution of mean dissipation rate conditioned on the enstrophy. ●, $\alpha = \epsilon_{hom}$; □, ϵ_{iso}; +, ϵ.

similar expressions and $\bar{\epsilon}_{hom} = \nu\overline{\omega^2}$. The correlation between either ϵ and ω^2 or ϵ_{iso} and ω^2 is relatively smaller (the joint pdfs for the latter two pairs are similar to each other and therefore only one is shown). The correlation coefficients $\rho_{\omega^2\epsilon}$, $\rho_{\omega^2\epsilon_{iso}}$ and $\rho_{\omega^2\epsilon_{hom}}$ are 0.4, 0.22 and 0.81 respectively, on the centreline. As x_2 increases, $\rho_{\omega^2\epsilon}$ and $\rho_{\omega^2\epsilon_{iso}}$ increase while $\rho_{\omega^2\epsilon_{hom}}$ remains almost constant. For example, at $x_2/L = 1.6$, these coefficients are 0.62, 0.38 and 0.88 respectively. It is evident that ϵ_{iso} does not represent ϵ adequately. This casts some doubt on the use of $(\partial u/\partial x)^2$ as a surrogate for ϵ.

The large value of $\rho_{\omega^2\epsilon}$ is consistent with the observation that strong alignment exists between the vorticity and the strain rate (e.g. Kerr, 1985; Tsinober et al., 1992). Due to the alignment, high vorticity regions may be associated with a high energy dissipation rate. This association may be examined by conditional averaging. Figure 4 shows the distribution, at the wake centreline, of the average value of the dissipation rate conditioned on ω^2. At small values of ω^2, all conditional dissipation rates vary linearly with ω^2 and the rate of increase is largest for ϵ_{hom}, indicating that a strong correlation exists between the enstrophy and the dissipation rates. For large ω^2, there is a large scatter in the data, especially for ϵ_{iso}, probably due to an insufficient number of samples. Nevertheless, the general trend indicates that large values of ω^2 are indeed correlated with high energy dissipation rates. A similar behaviour is also observed off the centreline.

510

4 Conclusions

A new vorticity probe has been used to measure the three components of the
vorticity fluctuation in a turbulent wake flow. Although the pdfs of ϵ, ϵ_{iso} and
ϵ_{hom} seem to follow closely the pdf of ω^2, the conventional correlation between
ω^2 and ϵ is smaller than that between ω^2 and ϵ_{hom}. The correlation between the
enstrophy and the isotropic dissipation rate is the smallest of the three measured
correlations. Conditional correlations suggest that regions of high enstrophy are
associated with a high energy dissipation rate.

Acknowledgements

The support of ARC is gratefully acknowledged.

References

Antonia, R. A. and Kim, J. (1994). *Phys. Fluids A*, **6**, 834-841.

Antonia, R. A., Shah, D. A. and Browne, L. W. B. (1988). *Phys. Fluids*, **31**, 1805–1807.

Balint, J.-L., Wallace, J. M. and Vukoslavčević, P. (1991). *J. Fluid Mech.*, **228**, 53–86.

Bershadskii, A., Kit, E. and Tsinober, A. (1993). *Phys. Fluids A*, **5**, 1523–1525.

Browne, L. W. B., Antonia, R. A. and Shah, D. A. (1987). *J. Fluid Mech.*, **179**, 307–326.

Kerr, R. M. (1985). *J. Fluid Mech.*, **153**, 31-58.

Kim, J., Moin, P. and Moser, R. (1987). *J. Fluid Mech.*, **177**, 133–166.

Kolmogorov, A. N. (1941). *Dokl. Akad. Naud. SSSR*, **30**, 301-305.

Marasli, B., Nguyen, P. and Wallace, J. M. (1993). *Expts. in Fluids*, **15**, 209–2183.

Meneveau, C., Sreenivasan, K. R., Kailasnath, P. and Fan, M. (1990) *Phys. Rev. A.*, **41**, 894–913.

Tsinober, A. (1990). *Phys. Fluids A*, **2**, 484–486.

Tsinober, A., Kit, E. and Dracos, T. (1992). *J. Fluid Mech.*, **242**, 169–192.

Vukoslavčević, P., Wallace, J. M. and Balint, J.-L. (1991). *J. Fluid Mech.*, **228**, 25–51.

Wallace, J. M. and Foss, J. F. (1995). *Ann. Rev. Fluid Mech.*, **27**, 467–514.

INFLUENCE OF THE TRANSVERSE SHEAR ON THE DEVELOPMENT OF WAKE FLOWS

S. BEHARELLE [†], C. N. NAYERI [††], J. DELVILLE [†],
J.-P. BONNET [†], H. E. FIEDLER[††]
[†] CEAT, Laboratoire d'Etudes Aérodynamiques Poitiers (France)
[††] Hermann-Föttinger-Institut Berlin (Germany)

Investigation of coaxial jets with weak counter-swirl can allow to understand the influence of the transverse instabilities on the wake or mixing layer development. The understanding of this phenomena is of primordial importance for improvement of many industrial applications such a combustors, industrial burners, trailing vortices of aircraft....

In order to estimate the influence of the transverse shear on the wake flows, several experiments were carried out in a wind tunnel which allows to generate two coaxial jets with counter-swirl (diameter of the inner jet 24 cm, diameter of the external jet 40cm). This facility allows independent variation of the swirl number S (hence transverse shear) and the exit velocity for each jet.

The diameter of the inner jet is large compared to the momentum thicknesses of the boundary layers at the trailing edge ($\theta/D \sim 0.004$). The axisymmetric shear layer can then be compared to a plane configuration.

A wake flow configuration is studied for various transverse shears, the exit velocity of each jet being 30 m/s. The configuration without rotation is compared to the plane wake case and then to the counter-rotating cases.

The effect of the additional transverse shear is analyzed by means of the Proper Orthogonal Decomposition (POD Lumley [1]). POD is applied to the velocity data obtained by hot wire anemometry (rake of 23 probes) at various downstream locations from the nozzle exit, following the procedure used by Delville & al.[2][3].

In this study, the eigenvalue problem which is solved is :

$$\int_D \Psi_{uu}(y, y'; f)\Phi_u^{(n)}(y'; f)dy' = \lambda^{(n)}(f)\Phi_u^{(n)}(y; f). \qquad (1)$$

where Ψ_{uu} is the Fourier transform of the two-point correlation tensor $< u(y, t)\ u(y, t + \tau) >$.

In terms of POD modes, the axisymmetric no-rotating case yields eigenvalue spectra very close to the ones that can be obtained in the plane case. In both configurations, two maxima can be noticed for $\lambda^{(1)}$ at frequencies $fb/U_c \sim 0.15$ and 0.35 (Figures 1.a and 2), corresponding to the varicose and alternate organization of the wake.

S. Gavrilakis et al. (eds.), Advances in Turbulence VI, 511-512.
© 1996 Kluwer Academic Publishers.

512

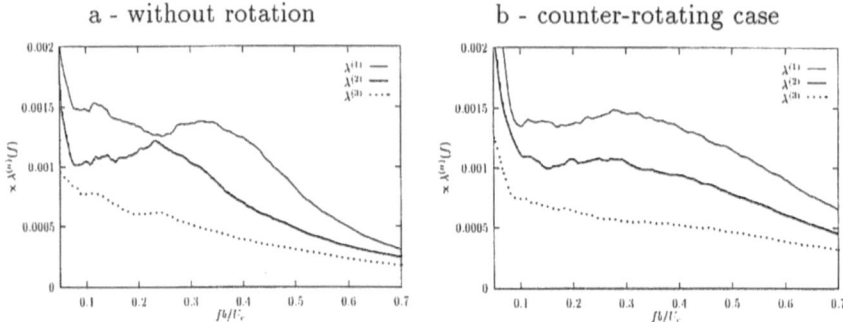

a - without rotation   b - counter-rotating case

Figure 1: Evolution of the first three eigenspectra for the axisymmetric case with and without rotation

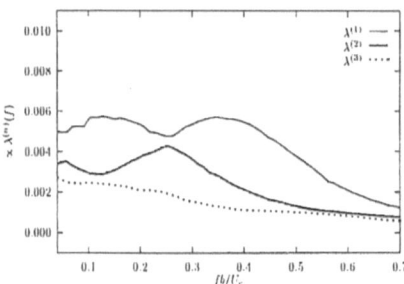

Figure 2: Evolution of the first three eigenspectra for the plane wake

When transverse shear is added (counter-rotating case), the three dimensionalization of the flow is increased as can be seen from the spreading of the first mode $\lambda^{(1)}$ and from the increase of the contribution of the higher frequencies to the modes of the POD.

Complementary measurements, carried out with a four-hotwire probe, allow to obtain similar conclusions about the increase of the spreading rates of the wake with addition of transverse shear.

References

[1] J.L. Lumley. The structure of inhomogeneous turbulent flows. In *Atm. Turb. and Ratio Wave Prop.*, pages 166–178. Yaglom and Tatarsky eds., Nauka, Moscow, 1967.

[2] J. Delville, S. Bellin, and J.P. Bonnet. Use of the proper orthogonal decomposition in a plane turbulent mixing layer. In M. Lesieur and O. Metais, editors, *Turbulence and Coherent Structures*, pages 75–90. Kluwer Academic Publisher, 1990.

[3] J. Delville. Characterization of the organization in shear layers via the proper orthogonal decomposition. *Kluwer Academic Publishers*, 53:263–281, August 1994.

THE UNSTEADY SPECTRAL PROPERTIES
OF TURBULENT CHANNEL FLOW
AT TIME VARIANT BOUNDARY CONDITIONS

M. BEYKIRCH, K. HESSE AND D. RONNEBERGER
Drittes Physikalisches Institut der Universität Göttingen
Bürgerstraße 42 - 44, D - 37073 Göttingen

The experiment reported here is conducted in a 2D channel at a Reynoldsnumber of 10^4 (based on the height of the channel). One of the channel walls homogeneously oscillates in the direction of the mean flow producing a shear wave propagating normal to the wall into the flow. The frequency of the oscillation $(0.5 \cdot 10^{-3} \leq f_w^+ \leq 3 \cdot 10^{-3})$ covers the range between quasi–steady and quasi–laminar behaviour of the flow (the $^+$ denotes the normalization with wall units). The streamwise component of the flow velocity $u(t)$ has been measured simultaneously at four different spanwise (z) locations by means of hot–film probes, i. e., the streamwise (x) and the normal (y) coordinates of the probes are identical. Besides the frequency of the oscillation the normal coordinate, i. e. the distance of the probe assembly from the wall has been varied.

The measured signals are decomposed into a phase averaged and a turbulent part: $u(t) = \langle u(t) \rangle + u'(t)$. The spatial and temporal correlation of $u'(t)$ contains information on the turbulent flow field in the x-z-plane. Therefore, we are interested in the temporal modulation of this correlation due to the imposed unsteadiness of the flow. The temporal modulation of a temporal correlation is not unambiguously defined. So we have used the Wigner distribution which can be considered as the basis of all other possible definitions.

As an example figure (1a) shows the modulation of the power auto–spectral density for various wall frequencies f_w^+ and for a fixed distance from the wall $(y^+ = 10)$. The presentation of the modulation factors in the complex plane needs some explanation. First of all, the amplitude of the sinusoidal oscillation of the wall is so small that the response of the flow contains practically no higher harmonics. So the response can be described by a set of complex amplitudes comprising the relative amplitudes and the phase lags of the various quantities reacting to the oscillating wall shear

S. Gavrilakis et al. (eds.), Advances in Turbulence VI, 513-514.
© 1996 *Kluwer Academic Publishers.*

stress. Secondly, the time variant power of the band–pass filtered $u(t)$ in each of the frequency bands (octave bands) is regarded as the response of a linear system that is described by a transfer function depending on f_w^+. Two of these transfer functions for $f_u^+ = 0.033$ and $f_u^+ = 0.26$ are depicted on figure (1a) by the dashed lines. These transfer functions exhibit a dependency on the wall frequency that is typical of a low pass filter responding with some time delay. It is remarkable that the amplitude of the frequency band with the low mid frequency $f_u^+ = 0.033$ goes through zero at the highest wall frequency, and it is anticipated that the amplitude will increase again if the wall frequency is further increased.

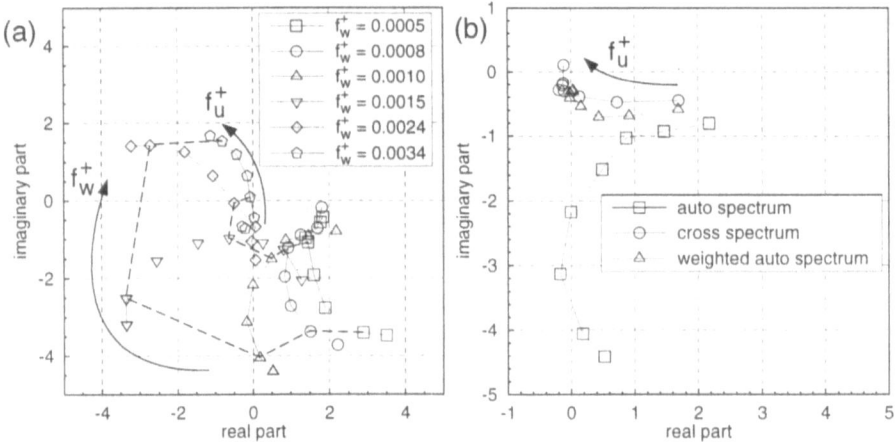

Figure 1. Modulation factors of the power spectral density in the complex plane at a wall distance of $y^+ = 10$. (a) Modulation of the auto spectrum for various modulation frequencies. (b) Modulation of the cross spectrum at a sensor spacing of $\Delta z^+ = 17$ and a modulation frequency of $f_w^+ = 0.0010$ (circles). The triangles show the modulation of the auto spectrum weighted with the normalized mean cross-power spectral density (coherence factor).

The set of modulation factors describing the modulation of the power spectral density at a fixed wall frequency form a characteristic curve in the complex plane (solid lines in figure 1) which is nearly independent of the wall frequency. However, the position of the curves relative to zero depends strongly on the wall frequency. So we speculate that two different mechanisms are responsible for the modulation of the form of the spectrum and of the total power of $u'(t)$, respectively. Figure (1b) shows a typical example of the fact that the modulation of the cross spectrum is caused mainly by the modulation of the auto spectrum. The much smaller modulation of the coherence factor is proportional to the difference between the circles and the triangles in figure (1b). So we conclude that the spatial structure of the turbulence is much less modulated than its intensity. The stationary coherence factors further suggest the existence of structures which are comparatively narrow with respect to their streamwise dimensions.

MEASUREMENTS OF A STRONGLY SWIRLING PIPE FLOW

M. DIRKZWAGER AND F.T.M. NIEUWSTADT
Laboratory for Aero and Hydrodynamics
TU-Delft, 2628 AL Delft, the Netherlands.

EXPERIMENTAL SET-UP

Two-component Laser-Doppler Velocimtry (LDV) measurements have been carried out in a swirling pipe flow in order study the turbulent structure of this flow. The pipe set-up is vertical with a total length of 10 m and a diameter of 50 mm. The working fluid is water which flows upwards through the pipe. The swirl is generated by means of vanes, which are carefully designed to induce a flow deflection of about 60° without separation. The pipe section with the constant diameter behind the swirl element has an adjustable length of 40, 60 or 80 diameters. This section is followed by a settling chamber in which the swirl is damped and a symmetric outflow is ensured. Measurements of the axial and tangential velocities have been carried out at several positions along the pipe section behind the swirl element, separated by 5 pipe diameters.

RESULTS

The flow rotation can be characterised by the swirl number which is defined as:

$$S = \frac{R \int_0^R U W r^2 dr}{2(\int_0^R r U dr)^2},\tag{1}$$

with R the pipe radius, $U(r)$ the mean axial velocity and $W(r)$ the mean tangential velocity. The first measurement of this parameter is located at 10 diameters behind the swirl element and we observe there $S = 3$. For this value of the swirl number, measurements are performed at two Reynolds numbers, $2.5\ 10^4$ and $5.0\ 10^4$, respectively, where the Reynolds number is based on the mean axial velocity and pipe diameter.

S. Gavrilakis et al. (eds.), Advances in Turbulence VI, 515-516.

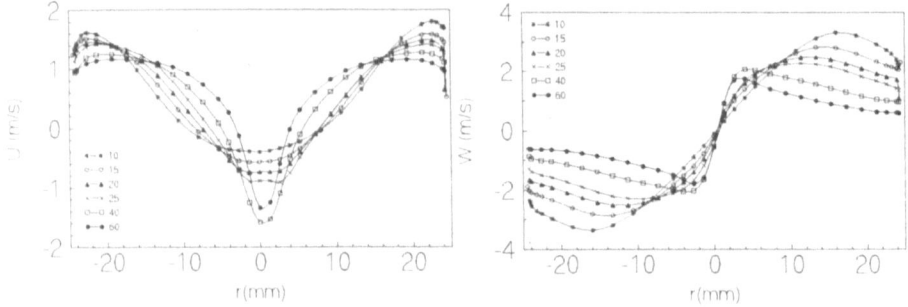

Figure 1. The mean axial (left) and tangential (right) velocity, measured at 10, 15, 20, 25, 40 and 60 diameters after the swirl element. $Re = 5.0 \cdot 10^4$.

In Fig. 1 profiles of the mean axial and tangential velocity are shown. A reverse axial flow at the centreline ia apparent because damping of the swirl leads to an adverse pressure gradient at the centreline. At a larger distance behind the swirl element, the reverse flow region becomes smaller and its intensity stronger. The tangential velocity profile is dominated by a solid body rotation at 10 diameters behind the swirl element and transforms later on to a so-called free vortex profile, i.e. the vorticity is concentrated at the centre of the pipe. The gradient of the tangential velocity at the centre increases until the swirl is damped to a value of $S \approx 1$. Results of velocity fluctuations in all three directions will be discussed during the presentation. These fluctuations show a minimum at the centre of the pipe for the high swirl numbers at which we have done our observations. At all measurement positions we have also measured the static wall pressure. The pressure profiles throughout the flow can be determined by using the measured velocity profiles. The pressure profiles confirm the adverse pressure gradient at the centre of the pipe.

At large distances behind the swirl element the centre of the vortex starts to move (vortex precessing). This influences the velocity profiles measured by the LDA by introducing low frequency oscillations. A triggering method has been investigated to correct for this effect. This method results in an adjustment of the mean velocity profiles.

With this experimental data set we feel that we have the data available which can be used to improve the turbulence models for this case. This will be the objective of further study.

THE BOUNDARY LAYER APPROXIMATION: AN APPLICATION TO ENERGY TRANSPORT IN WAKES

Thierry M. FAURE
Laboratoire de Mécanique des Fluides et d'Acoustique, UMR CNRS 5509
École Centrale de Lyon, 69 131 Écully, France

1. Introduction

The boundary layer approximation is an useful tool for the reduction of turbulence equations. This approximation, valid for a developed flow, is based on an analysis of the order of magnitude for all these terms (Corrsin 1963). In the present paper, we measure the Reynolds stress transport equations in the far wake of an axisymmetric streamlined body. Then, one can obtain the pressure-strain distributions and compare them with the evolutions predicted by second order closure models.

2. Results

The experiment is realised in a wind tunnel where is placed the model, 50 cm in length and with a diameter $D = 8$ cm (Faure 1995). The Reynolds number based on this diameter and the freestream velocity is $Re_D = 5.8 \times 10^4$. An automated triple hot-film anemometry system is used, that gives all three components of instantaneous velocity.

The Reynolds stress transport equations, can be written as the balance:

$$C_{ij} = P_{ij} + T_{ij} + \Pi_{ij} - \varepsilon_{ij} \qquad (1)$$

where the terms denote respectively : C_{ij} convection, P_{ij} production, T_{ij} kinetic and pressure transfer, Π_{ij} pressure-strain and $-\varepsilon_{ij}$ dissipation. Measurements are made of all the terms of these equations except for the pressure-strain correlation, which is determined from the overall balance. The dissipation rate is evaluated using the isotropic expression and Taylor's hypothesis. In the self-similar wake, the axial equation is given in figure 1, where the radial direction is divided by the radius of the wake $r*$. In the centreline, the gain of turbulent energy is mainly realised with convection, while losses result from dissipation and pressure transfers. As radial distance increases, production grow and reaches a maximum around $r / r* = 0.5$.

We present results of the comparison between the pressure-strain term balanced from the measurements and the corresponding distributions given by second-order closure models. For its linear part, the expression proposed by Rotta (1951) is used. Two

517

S. Gavrilakis et al. (eds.), Advances in Turbulence VI, 517-518.

518

models, given by Launder *et al.* (1975), are tested for the rapid part of the pressure-strain: the quasi-isotropic model referred as QI model, and the isotropization of production model, referred as IP model. The third part, associated to the wall effects, will be zero in the present case. Two sets of values are tested for the IP model, $c_1 = 1.5$, $\gamma = 0.6$ (denoted IP 1) which are the values published in Launder *et al.* (1975) and $c_1 = 1.8$, $\gamma = 0.6$ (denoted IP 2) which were later adjusted by Launder (1989) where the relation $(1 - \gamma)/c_1 = 0.23$ is validated. For the QI model, the single set $c_1 = 1.5$, $c_2 = 0.4$ is used. Results from comparisons of the above pressure-strain models with the distributions determined from the balance of Reynolds stress transport equations are presented in figure 2. We may note that the three models are relatively close, and predict well the amplitude of the peak value, even if its radial position is slightly underestimated.

3. References

Corrsin, S. (1963) Turbulence: experimental methods, *Handbuch der Physik*, Vol. 8, Part 2, Springer-Verlag, Berlin, pp. 524-590.

Faure, T. (1995) *Étude expérimentale du sillage turbulent d'un corps à symétrie de révolution autopropulsé par hélice*, PhD Thesis, No 95-01, École Centrale de Lyon.

Launder, B. E. (1989) Second-moment closure and its use in modelling turbulent industrial flows, *International Journal for Numerical Methods in Fluids*, **9**, 963-985.

Launder, B. E., Reece, G. J. and Rodi, W. (1975) Progress in the development of a Reynolds-stress turbulence closure, *Journal of Fluid Mechanics*, **68**, (3), 537-566.

Rotta, J. C. (1951) Statistiche Theorie nichthomogener Turbulenz, *Zeitschrift für Physik*, **129**, 547-572.

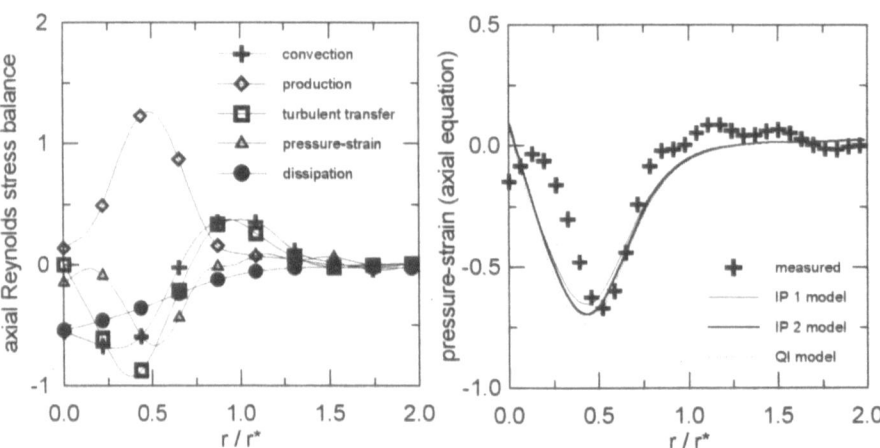

Figure 1. Self-similar axial Reynolds stress balance in the wake.

Figure 2. Comparison between the measured and modelled pressure-strain for the axial equation.

SPECTRAL ANALYSIS OF NEAR WALL TURBULENT FLOW

G. IUSO, M. ONORATO
Dipartimento di Ingegneria Aerospaziale, Politecnico di Torino
Corso Duca degli Abruzzi 24, 10129 Torino
M. ONORATO Jr.
Dipartimento di Ingegneria Aerospaziale, Università di Pisa
Via Diotisalvi 2 , 56100 Pisa

A research on wall turbulence control is in progress at the 'Modesto Panetti' Laboratory of the Politecnico di Torino. A part of this research is focused on the identification of the shear-stress producing mechanisms and on the identification of the scales and regions in which the contribution to shear stress is large.

The results shown in this short paper concern with the spectral analysis of the u-v fluctuating velocity correlation, contributing to the production of the Reynolds stress in a near wall flow.

The Wavelet transform [1] of the velocity signals is used as the analysis tool, being this technique especially suitable for performing conditional spectra.

For the case of the Mexican Hat wavelet, the tangential Reynolds stress component -ρ uv may be expressed as [2] :

$$-\rho\overline{uv} = -\rho \int\!\!\int \overline{Au(k_u, t)Av(k_v, t)}\, dk_u dk_v$$

The bar indicates time averaged quantities, $1/k$ is the duration, Au and Av are respectively the 'Amplitude Transform' of u and v, defined as $A(k,t) = Tw(k,t)/\sqrt{k}$, where $Tw(k,t)$ are the coefficients of the wavelet transform.

The hot wire signals here analysed are taken in a fully developed turbulent channel flow. The Reynolds number based on the maximum mean velocity and on the channel height is 21000; the ratio between the skin friction velocity and the maximum mean velocity is 0.047.

The amplitude transforms $Au(k,t)$ and $Av(k,t)$ of 1024 sampling points, taken at $y^+=103$, are shown respectively in figures 1a and 1b, where the quantities reported on the vertical axis are simply ordering numbers. Summation over the duration gives the original signals u(t) and v(t). For clarity, only negative values contour lines in figure 1a and positive contour lines in figure 1b, corresponding respectively to local minima and maxima, are plotted. The correlation between the two maps gives the contribution to the Reynolds stress in the second quadrant. The range of durations involved in the correlation $\overline{uv}$ is evident from the maps. It is also evident the spectral contents for the u and v velocity fluctuations: figure 1a shows peaks for duration ranging from 7 ($T^+= 6$) to 15 ($T^+= 30$), while figure 1b shows peaks ranging from 5 ($T^+= 4$) to 12 ($T^+= 20$). T^+ is the duration normalized respect to the viscous characteristic time.

S. Gavrilakis et al. (eds.), Advances in Turbulence VI, 519-520.
© *1996 Kluwer Academic Publishers.*

520

In figure 2a the spectral stress distribution map $\overline{A_u(k_u,t)A_v(k_v,t)}$ as a function of $1/k_u$ and $1/k_v$ is shown. Summation over all points yields to the value of the correlation $\overline{uv}$. The diagonal structure of this plot confirms that velocity components must be at similar scales in order to contribute to the Reynolds stress. The correlation $\overline{uv}$ is concentrated in a region having a peak at $T^+ = 30$ corresponding to 10ms. In figures 2b and 2c the conditional spectral contributions to $\overline{uv}$, respectively from A_uA_v positive values and from A_uA_v negative values, are shown. The conditional spectra associated with quadrants 1 and 3, figure 2b, and with quadrants 2 and 4, figure 2c, show also off-diagonal values, whose contribution to Reynolds stress is null.

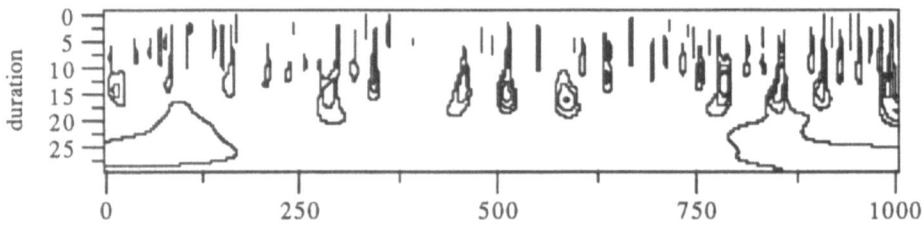

Figure 1a. Amplitude transform: Au-negative contours lines.

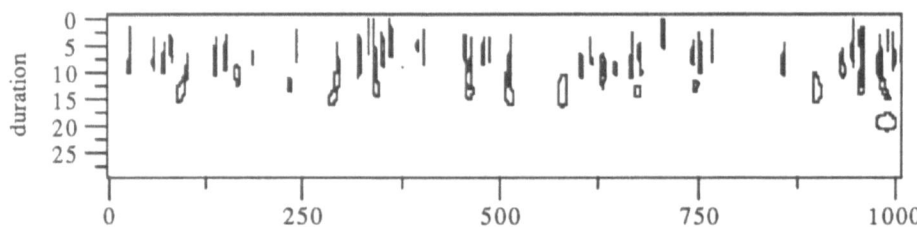

Figure 1b. Amplitude transform: Av-positive contours lines.

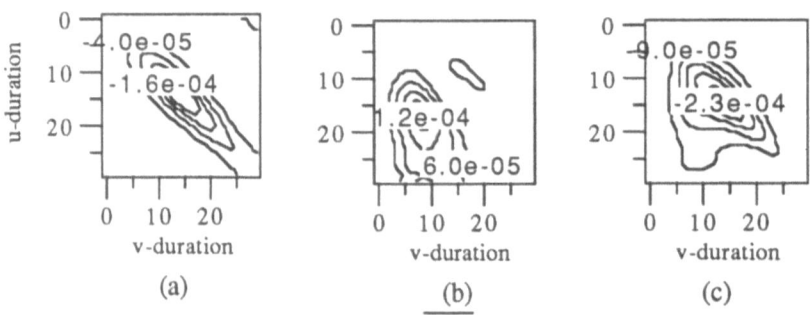

Figure 2. Spectral stress distribution maps AuAv. (a) mean spectrum; (b) conditional spectrum: AuAv>0; (c) conditional spectrum: AuAv<0.

References

1. Farge, M.: Wavelet Transforms and their Applications to Turbulence, *Ann. Rev. Fluid. Mech.* **24** (1992), 395-457.

2. Lewalle,J.: Wavelet Analysis of Experimental Data: Some Methods and the Underlying Physics, *AIAAA paper* **94-2281**, *25th AIAA Fluid Dynamics Conference*, June 20-23, 1994 Colorado Springs.

INVESTIGATION OF THE EFFECTS OF BASE INJECTION ON THE FLOW FIELD PAST A SQUARE CYLINDER

PAPAILIOU, D., KOUTMOS, P., and BAKROZIS, A.
Laboratory of Applied Thermodynamics. Department of Mechanical Engineering. University of Patras. Greece.

The complex turbulent flow produced downstream of a cylindrical or square bar has frequently been considered as a model problem due to its fundamental physical importance as it encompasses interactions among anisotropic regions, large scale recirculations, unstable shear layers and periodic or quasi-periodic energetic shedding of vortices (Bearman, 1984; Roshko, 1993). Fundamental understanding of the turbulent processes involved in a range of generic bluff-body configurations as well as other complex flow/geometry derivative combinations of practical importance is necessary to improve methods of prediction and allow better control or exploitation of these flow features.

The present work describes LDV measurements of a 2-D square cylinder wake flow under 19% confinement with a low aspect ratio 2-D air-jet injected along the symmetry plane into the wake formation region. This set up, sometimes also referred to as 'base bleed'. involves a flow/geometry combination associated with diverse engineering applications such as aerodynamics, marine hydrodynamics and flame stabilisation. Only a limited amount of information has been reported for such a configuration so far (Bearman, 1966; Woods, 1967).

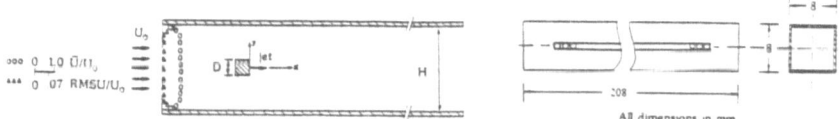

Figure 1. Flow configuration and square cylinder model geometry

The experimental facility and a sketch of the square hollow cylinder model are shown in figure 1. The 8mm diameter model was inserted into a rectangular cross section duct of 42 x 208 x 810 mm. Air was fed through the cylinder ends and then injected into the wake through 125 holes of 1mm diameter spaced 0.25mm apart over 75% of the symmetry plane. LDV measurements suggested that the discrete jet system (necessary for maintaining spanwise uniformity) due to its small aspect ratio produced a spanwise planar jet within a distance of 0.28D when operated without crossflow; this becomes even shorter with backflow. Spanwise uniformity of injection was achieved by suitably tapering the inside hollow of the cylinder to adjust the injection pressure according to inviscid manifold theory.

Parametric LDV measurements of mean and turbulence quantities and related

S. Gavrilakis et al. (eds.), Advances in Turbulence VI, 521-522.
© 1996 *Kluwer Academic Publishers.*

522

statistics were conducted for a range of conditions ranging from uninjected to fully penetrated wakes throughout the wake field at Re=8520. Careful and separate seeding of the jet and crossflow air supplies with similar concentrations was employed to minimise bias errors due to unequal particle number density.

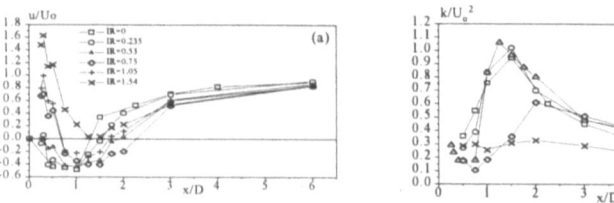

Figure 2. Development of (a) time-mean streamwise velocity and (b) turbulent kinetic energy $(k = \overline{u'^2} + \overline{v'^2})$ along the symmetry plane of square cylinder for various IR, Re=8520.

Figures 2a and b portray the streamwise velocity and the total fluctuating periodic turbulent energy development along the symmetry plane for the range of injection ratios (IR=jet to cross-flow blocked velocity). Two stagnation points can be identified, one of the jet and one of the primary air prior to jet penetration through the main recirculation. As a result of jet impingement and deflection by the primary backflow, a system of four counter-rotating recirculation regions (two on each side of the symmetry plane) is established behind the cylinder. The length, strength, width and turbulence generation characteristics of the vortex formation region seem relatively unaffected by the injection process up to IR=0.53 (corresponding to a percentage of injected mass of 0.07%). Thereafter the jet penetrates further, interacting directly with the rear end of the formation region and the development of the periodic vortex street and alters drastically the behaviour of the near wake region. This jet interference process with the basic vortex street clearly emerges in the u and v velocity power spectra as illustrated in figure 3. As the injection ratio increases, the u and v shedding frequencies continuously shift to higher values and are displaced from each other with an increasing frequency difference. The energy content of the u pronounced frequency gradually diminishes and its peak is completely suppressed after penetration. On the contrary a corresponding distinct peak of increasing energy content persists in the v-component spectra even after penetration a matter, which is currently under further investigation.

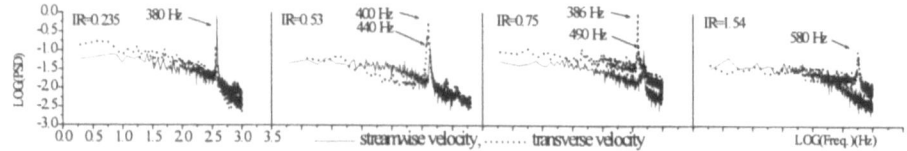

Figure 3. Unormalized power spectra at x/D=6, y/H=-0.1428, for various IR, Re=8520.

References

Bearman, P.W. (1984) Annual Review of Fluid Mechanics, 16, pp. 195-222
Bearman, P.W. (1966) AGARD CP4, pp. 497-507
Roshko, A. (1993) Jnl. Wind Eng. and Ind. Aerodynamics, 49, pp. 75-100
Woods, C.J. (1967) JFM, 29, pp. 259-272

SIMULTANEOUS MEASUREMENTS OF TEMPERATURE AND VELOCITY COMBINING COLD WIRE AND LASER DOPPLER VELOCIMETRY IN A SLIGHTLY HEATED TURBULENT JET

L. PIETRI, T. DJERIDANE, M. AMIELH, L. FULACHIER

I.R.P.H.E., 12 Av. Général Leclerc, 13003 Marseille

The present paper is concerned with simultaneous measurements of temperature and velocity in the near-field region of a turbulent air jet where the temperature acts as a passive contaminant. The difficulties encountered for such measurements when usual hot-wire techniques are used appear in particular at the jet interface where some flow reversal and high turbulence intensity are present. To circumvent this problem, a technique based on laser Doppler velocimetry and cold wire thermometry was worked out with a refined procedure compared to similar techniques used in heated boundary layers (Thole and Bogard, 1994) (Wardana *et al.*, 1995). The set-up consists in a fully developed turbulent vertical pipe flow (mean velocity $U_j = 12$m/s, Reynolds number $Re_j = 21000$) discharging into ambient air (mean velocity $U_e = 0.9$m/s) in a slightly confined configuration (jet diameter $D_j = 26$mm, enclosure section $285 \times 285mm^2$). The jet is heated at $20K$ above the ambient temperature.

Velocity measurements are performed with a two-component laser Doppler system (Argon 4W) fitted with fiber optics and two BSA's. The probe measuring volume is $0.12 \times 0.12 \times 1mm^3$ and backscatter detection is used. A very fine cold wire (CW, $\phi = 0.6\mu m$) is used to measure temperature fluctuations (heating current $I = 0.2mA$). After systematic studies, it appears that the optimum distance between these two probes is about 3η (η, Kolmogorov length scale). The temperature sampling is triggered by the envelope of the burst signal issued from the BSA. With this method, only useful and simultaneous data (around 7000) issued from the cold wire are stored in order to preserve the memory space of the computer. The coincidence interval between a velocity and a temperature acquisition is estimated at $2\mu s$. In order to restore the frequency response of the cold wire,

S. Gavrilakis et al. (eds.), Advances in Turbulence VI, 523-524.
© 1996 *Kluwer Academic Publishers.*

524

dirtied by the seeding silicone oil particles, a regular and careful cleaning is realized. So no compensation is needed, contrary to Neveu's experiment (1994) using for similar measurements a fine thermocouple instead of a cold wire. The whole system then enables simultaneous and instantaneous measurements of two velocity components (U axial component, V radial component) and temperature θ.

Only some typical results are reported in this abstract (fig. 1). They are given from the exit section up to $20D_j$. The turbulent intensity of temperature is first measured with a cold wire alone. The effects of the coflow and the confinement are discussed. These measurements are also used to check the validity of the temperature measurements when the flow is seeded. Some axial and radial profiles of the turbulent fluxes $< u\theta >$ are given. Figure 1 presents the radial profile of the correlation coefficient $R_{u\theta}$ at $15D_j$. Some profiles of $< v\theta >$ are also given and compared to measurements obtained by hot wires (Chevray and Tutu, 1978); (Chua and Antonia, 1990) and by balance of the enthalpy equation (Djeridane, 1994).

Financial support from EDF, GDF, INERIS, SNECMA and the PACA Regional Council is gratefully acknowledged. F. Anselmet provided considerable help for the development of this research at IRPHE.

References

Chevray, C.R. and Tutu, W. K. (1978) "Intermittency and preferential transport of heat in a round jet", *J. Fluid Mech* **88**, 133-160.

Chua, L.P. and Antonia, R. A. (1990) "Turbulent Prandtl number in a circular jet", *Int. J. Heat Mass Transfer* **33** , n° 2, 331-339.

Djeridane, T (1994) "Contribution à l' étude expérimentale de jets turbulents à densité variable", *Thèse I.M.S.T.* , Université d'Aix-Marseille II.

Neveu, F. (1994) "Mesures simultanées de la température et de la vitesse dans une flamme non-prémélangée méthane-air stabilisée par un brûleur de type "bluff-body", *Thèse* , Université de Rouen.

Thole, K.A. and Bogard, B. G. (1994) "Simultaneous temperature and velocity measurements", *Meas. Sci. Technol.* 5, 435-439.

Wardana, I.N.G., Ueda, T., and Mizomoto, M. (1995) "Velocity-temperature correlation in strongly heated channel flow", *Exp. in fluids* **18**, 454-461.

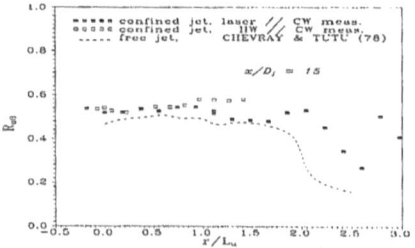 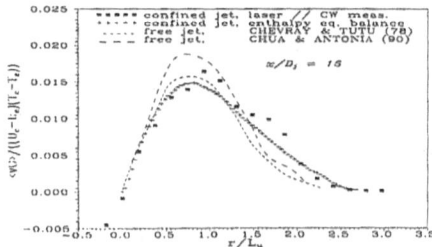

Figure 1. $X/D_j = 15$. Left: correlation coefficient between the axial velocity and temperature. Right: radial turbulent flux of temperature.

THE MEASUREMENT OF VELOCITY AND VELOCITY GRADIENTS IN A TURBULENT CHANNEL BY USING DIFFERENT PARTICLE TRACERS

G.P. ROMANO, F. BAGNOLI

Dept. Mechanics and Aeronautics, Via Eudossiana 18 - Roma- ITALY

Nowadays the statistics of velocity and velocity gradients in a turbulent channel flow at low Reynolds numbers are well established following several numerical simulations and experiments. These measurements are usually referred to the flow phase, while the effect of particle tracers is disregarded: the velocity of tracers is just what is measured in experiments. In recent years, investigations on the role of particles in the near-wall region of turbulent channel flows have been performed. A useful parameter to be considered is the particle response time: $\tau_p = \rho_p d^2/18\rho_f \nu$, where $\rho_{p,f}$ are particle and fluid density, d is the particle diameter and ν the kinematic viscosity. This time scale must be compared with Kolmogorov time scale (τ_η). For values of the ratio τ_p/τ_η less than 100, particles are expected to enhance dissipation. Hereafter the attention is focused on velocity gradients.

The effects of three different kind of particles on the evaluation of velocity gradients in the x,y plane are investigated (x and y are the streamwise and wall normal directions): air bubbles, polymer (styrene) and glass particles. The particle concentration is always less than 10^{-3} in volume. Particle characteristics are summarized in table 1: the flow is water at a temperature of about 20°C. The superscript + is used for wall variables.

Table 1. Particle characteristic parameters: $\tau_\eta \approx 4.5$ms, $\eta \approx 0.07$mm is the Kolmogorov microscale.

Seeding	ρ_n/ρ_f	$d(\mu m)$	d^+	d/η	τ_n(ms)	τ_n^+	τ_p/τ_η
Air	0.0012	40	1.36	0.57	0.00011	0.0012	0.00025
Polymer	1.07	85	2.30	1.21	0.85	0.31	0.19
Glass	2.51	75	1.95	1.07	1.50	0.53	0.31

A vertical channel flow is used (length=200cm, height=2cm, width=10cm): measurements are taken at about 180cm from the inlet ($x/\delta \approx 90$, δ being the half channel height). The Reynolds number is about 4000 (using δ and the mean flow velocity). The flow field is illuminated by a scanning beam. Images of a small region (from about 3mm^2 to 9mm^2) are recorded on a standard videorecorder using a videocamera. They are analyzed by a suitable software to validate multiple particle exposures as a trajectory (Particle Tracking Velocimetry, PTV). Once velocity vectors are validated, the two velocity components on the (x,y) plane are evaluated: velocity gradient are measured by computing the differences between the velocity components over variable distances Δx and Δy. In figure 1 the velocity gradient $\partial u/\partial y$ is shown: gradients are evaluated at a

525

S. Gavrilakis et al. (eds.), Advances in Turbulence VI, 525-526.
© 1996 *Kluwer Academic Publishers.*

526

dimensionless distance between trajectories $D=\Delta y^*=\Delta y/\eta \approx(7\div 10)$. As expected (see table 1) particles with large particle response time (polymer and glass particles) enhance the value of the velocity gradient. For this selection of Δy^*, data from air bubbles are in agreement with published DNS data (Antonia et al., 1991). Results are strongly dependent on the distance between velocity vectors. In figure 2 the gradient $(\partial u/\partial x)^+$, using air bubbles, is evaluated for different values of the distance: only for $\Delta x^*=\Delta x/\eta \approx$ 25 data are in agreement with DNS results all over the range of y^+: for $\Delta x^*<25$ there is an overestimation, whereas for $\Delta x^*>25$ an underestimation. In the same figure a similar analysis is performed for gradient along y direction: the distance which gives a good reproducibility of DNS results is $\Delta y^*\approx 10$. This difference follows that of gradient correlations and characteristic length scales over directions x and y.

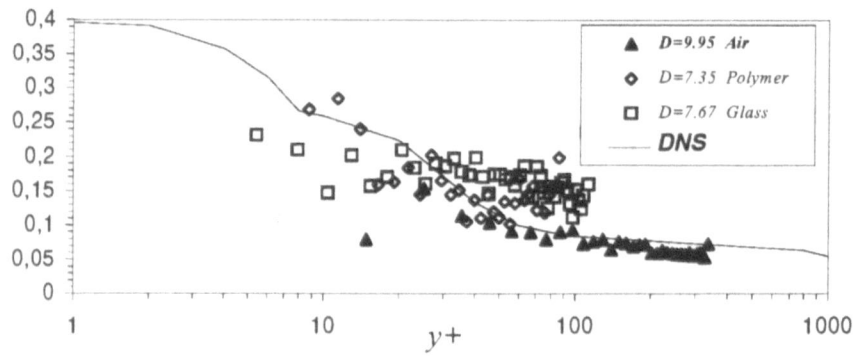

Figure 1. Normalized *rms* value of the gradient $(\partial u/\partial y)^+$ vs y^+: $D=\Delta y^*=\Delta y/\eta$.

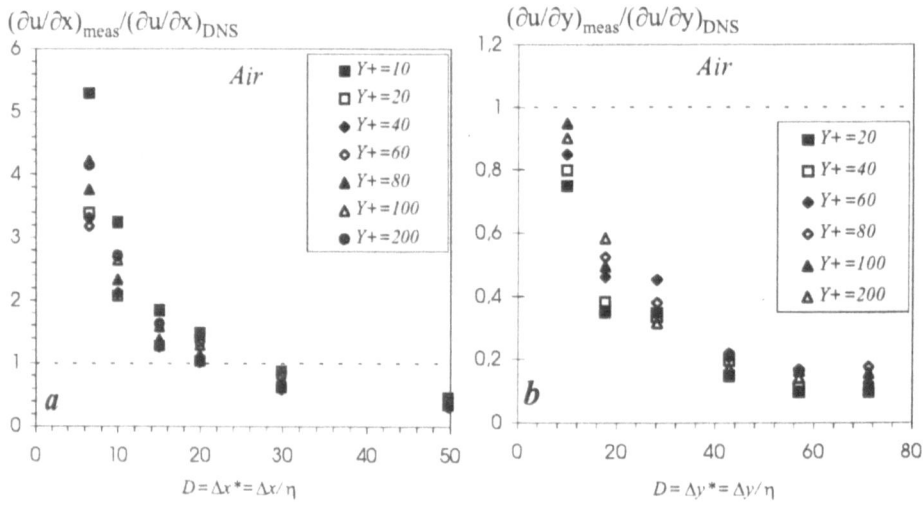

Figure 2. Ratio of measured over DNS *rms* gradients $(\partial u/\partial x)^+$ (*a*) and $(\partial u/\partial y)^+$ (*b*) vs dimensionless distance between velocity vectors for air bubbles evaluated at different distances from the wall.

FEEDBACK CONTROL OF VORTEX SHEDDING: THREE-DIMENSIONAL EFFECTS

KIMON ROUSSOPOULOS
Fluid Mechanics Laboratory (LMF)
DGM - IMHEF, EPFL, CH-1015 Lausanne, Switzerland

The use of feedback control to stabilise at low Reynolds numbers the instability behind the well-known Karman vortex street in the wake of cylindrical bodies has been studied experimentally, analytically and numerically. Most of these studies, however, consider only two-dimensional effects; i.e. they only consider effects in the plane normal to the cylinder axis, and assume that the effects along the span can be neglected.

Control is generally realised using an actuator that picks up a signal from a single location along the span, and a controller that acts equally along the span. It has been observed experimentally that when control is successful, shedding is inhibited at the spanwise location of the sensor. The control signal is therefore close to zero. In general however shedding continues or recommences along the span at locations distant from the sensor, undetected by the sensor. In these cases the sensor location is seen to act as a forced zero of shedding amplitude. In addition, control is not possible when the sensor is located too far downstream. The included figures (next page) illustrate the control configuration under consideration, and show experimental data showing the amplitude of shedding along the span when control suppresses shedding at one point.

The one-dimensional Ginzburg-Landau (G-L) equation applied to vortex shedding in the wake of bluff bodies is known to provide useful insight in two cases. One is the description of global modes for purely 2-D shedding where the spatial coordinate in the G-L equation coincides with the streamwise direction. The second is when the spanwise G-L equation is used to describe weakly 3-D effects such as oblique shedding.

This poster presentation shows experimental results demonstrating the three-dimensional effects of vortex shedding at low numbers controlled by feedback. It also shows the results of a two-dimensional numerical solution of the G-L equation, with feedback and noise added, showing that this model equation reproduces well the experimental observations. Some features that are inherently 3-dimensional (i.e. that follow from the variation of shedding amplitude along the span) are observed in both the experiments and the simulation.

S. Gavrilakis et al. (eds.), Advances in Turbulence VI, 527-528.
© 1996 *Kluwer Academic Publishers.*

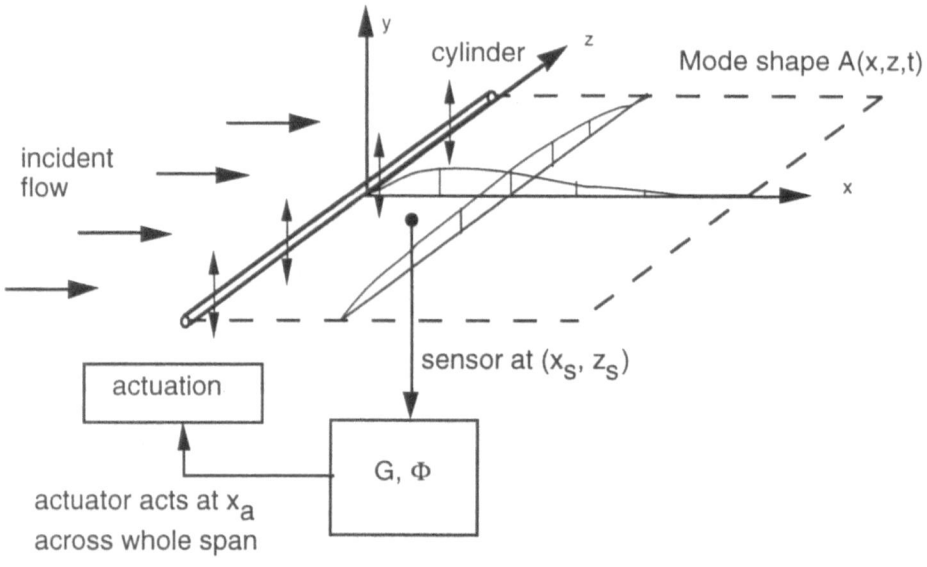

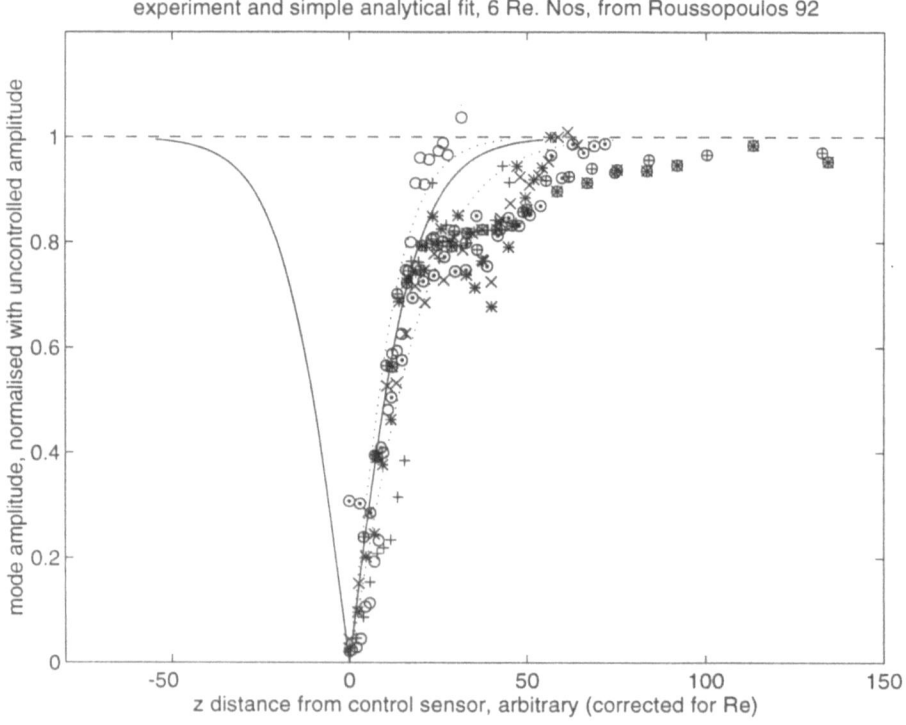

Figures illustrating the vortex shedding control configuration (top), and shedding amplitude along the span when controller acting (bottom).

TURBULENCE MANIPULATION IN WALL JETS

M. SCHOBER AND H.-H. FERNHOLZ
Hermann-Föttinger-Institut für Strömungsmechanik
Technische Universität Berlin
Straße des 17. Juni 135, 10623 Berlin, FRG

1. Introduction

A wall jet is a flow phenomenon that has features of both a boundary layer and a jet. Its main application lies with turbine blade cooling and high lift airfoils or flaps. In the former case it is desirable to prevent the cooling fluid from mixing with the ambient flow in order to keep the protective layer as far downstream as possible. Therefore the aim of this investigation is to influence the global characteristics of a turbulent wall jet, such as the spreading rate and the skin friction. This can be achieved by preventing the formation of large structures in the shear layer by disturbing the laminar shear layer in the vicinity of the nozzle of the wall jet. A thin circular cylinder in the spanwise direction can introduce such a disturbance which decreases the local half-width of the wall jet and increases the skin friction.

Hot-wire measurements and skin friction measurements (Preston tubes, surface fences and wall hot-wires) have been performed to determine the mean and fluctuating flow quantities. Flow visualisations with paraffin vapour shows the changes in the coherent structures.

2. Experiments, Results

In a first step the experiments were performed with a wall jet in a quiescent surrounding using an experimental configuration as shown in figure (1).

Figure (1) presents a comparison between the normal wall jet behaviour and that with manipulation. Figure (1a) shows the shear layer roll-up, resulting in large vortical structures extending far into the outer, quiescent surrounding. When a trip wire is placed ≈ 3 mm downstream of the lip into the shear layer (Figure 1b), the large structures do not occur. Instead, fine

S. Gavrilakis et al. (eds.), Advances in Turbulence VI, 529-530.

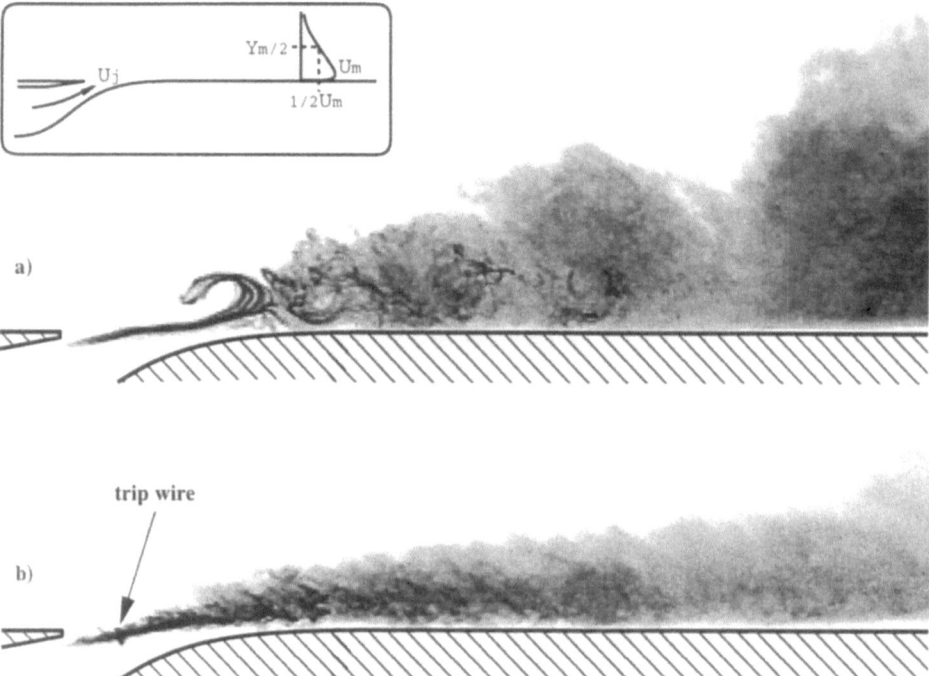

Figure 1. Shear layer structures ($Re_j = 2500$, $0 < x/b < 15$, $b = 8mm$) a) without trip wire; b) with trip wire;

scale turbulence structures are generated and the spreading of the shear layer is reduced by a factor of approximately 1.5. As the Reynolds number increases, the formation of the large structures and transition to turbulence of the flow without manipulation moves upstream and the effect of the trip wire becomes smaller but is present up to $Re_j = u_j b/\nu \approx 1.5 \times 10^4$.

The manipulated flow shows an increase of the local skin friction $\overline{\tau}_w$ and of the maximum of the local mean velocity u_m whereas the local half-width $y_{m/2}$ decreases. The ratio of the manipulated to the normal skin friction is strongly dependent on the position of the trip wire normal to the shear layer ($y_{opt} \approx y_{0.5u_j}$) and less so on the exact streamwise position.

The wire Reynolds number was varied using wires with diameters between 0.2 mm and 0.8 mm but had little influence on the manipulation. It can thus be concluded, that the shedding frequency of the Kármán vortex street plays no important role in the suppression mechanism.

In the downstream region of the wall jet $70 < x/b < 150$ both mean flow and Reynolds-stress profiles show self-similar behaviour when scaled with u_m and $y_{m/2}$.

AN EXPERIMENTAL VALIDATION OF THE "FROZEN" TURBULENCE HYPOTHESIS FOR RIVER FLOWS

B. SHTEINMAN[1], V. NIKORA[2], M. EKHNICH[3]
and A. SUKHODOLOV[4]

[1]Israel Oceanographic and Limnological Research, Yigal Allon
Kinneret Limnological Laboratory, PO Box 345, Tiberias, Israel
[2]National Institute of Water and Atmospheric Research,
PO Box 8602, Christchurch, New Zealand
[3]Odessa Hydrometeorological Institute, Odessa, Ukraine
[4]Institute of Geophysics and Geology, Kishinev, Moldova

1. Introduction

The main initial information in river turbulence investigations is velocity measurements in one or several fixed points in the flow. This information allows researchers to estimate some statistical characteristics only of time sections of the instant velocity field. To convert time turbulence characteristics into spatial ones the "frozen" turbulence hypothesis is widely used in river hydrodynamics (Grinvald and Nikora, 1988). However, this hypothesis was tested only for laboratory and atmospheric flows which are quite different from river flows. Some references as well as theoretical considerations of this problem can be found in Monin and Yaglom (1975). In this paper we present a test of the "frozen" turbulence hypothesis against field measurements for the case of river flows.

2. Method and field data

Special field experiments were designed and conducted in the Jordan (Israel), Turunchuk (Ukraine), Reut, Bik, and Ikel (Moldova) rivers with sand wave bottom and flat silty bottom. The ranges of the main hydraulic and morphometric characteristics of the investigated river reaches were: width 3.5-53 m; depth 0.2-4.5 m; water discharge 0.25-81 m³/s; water surface slope 0.00033-0.0040. The field experiments included synchronous measurements of longitudinal velocity in 5-6 points at the same level along the flow. The measurements were conducted by micropropellers with low inertia (Nikora et al., 1994). The measurements in the Jordan River were also performed by a 3D velocity meter made up of three piezo-electric slabs (Shteinman and Gutman, 1993; Shteinman et al., 1993). To test the "frozen"

S. Gavrilakis et al. (eds.), Advances in Turbulence VI, 531-532.
© 1996 Kluwer Academic Publishers.

turbulence hypothesis we used time-space correlation functions $R(\Delta x, \tau)$ and structure functions $D(\Delta x, \tau)$ of velocity fluctuations (Δx and τ are the spatial and temporal lags, respectively). This allowed us to determine empirically (1) velocity $U_t = \Delta l / \tau_{max}$ of the turbulent eddy transfer by the average flow (Δl is the distance between measuring points i and j, τ_{max} is the time lag corresponded to the maximum of the cross-correlation function $R_{ij}(\tau)$); (2) validity of the relationships $R(\Delta x = 0, \tau) = R(\Delta x, \tau = 0)$ and $D(\Delta x = 0, \tau) = D(\Delta x, \tau = 0)$; and (3) to estimate the stability of turbulent eddies.

3. Results and conclusions

For the outer flow layer ($0.10 < z/H < 1$) the distribution of $(U - U_t) / U$ value proved to be close to Gaussian with a standard deviation 7.6% which is commensurable with an accuracy of U_t (z/H is the relative distance from the bottom, U is the local mean velocity). The empirical functions $R(\Delta x = 0, \tau)$ and $R(\Delta x, \tau = 0)$, and $D(\Delta x = 0, \tau)$ and $D(\Delta x, \tau = 0)$ for this layer were close each other. The differences between them did not exceed the 95% confidence intervals. In the near bottom flow region ($z/H < 0.10$) the difference between U and U_t was more significant and in most cases $U < U_t$. According to the "frozen" turbulence hypothesis the eddies transfer occurs without any deformation of eddies. From this it follows $R(\Delta x, \tau = \tau_{max}) = 1$. However, we revealed that empirical functions $R(\Delta x, \tau = \tau_{max})$ are decreased with Δx very quickly. Values of Δx corresponding to the condition $R(\Delta x, \tau = \tau_{max}) = 0$ appeared to be 10-20 flow depths, i.e., commensurable with the scale of large eddies. Our main conclusions are: (1) the presentation of turbulent eddies as "frozen" for river flow conditions is physically unjustified; (2) the "frozen" turbulence hypothesis can be applied to mean velocity quantities (correlation functions, spectra and structure functions) in the flow region which is not very close to the bottom.

4. References

Grinvald, D. I. and Nikora, V. I. (1988) *River Turbulence* (in Russian). Hydrometeoizdat, Leningrad.
Monin, A. S. and Yaglom, A. M. (1975) *Statistical Fluid Mechanics: Mechanics of Turbulence*, vol. 2, MIT Press, Boston, Mass.
Nikora, V. I., Rowinski, P., Sukhodolov, A., and Krasuski D. (1994) Structure of river turbulence behind warm-water discharge. *J. of Hydraulic Engineering, ASCE*, **120**, 2, 191-208.
Shteinman, B. S. and Gutman, A. (1993) Flow turbulence and dispersion of different matter in river mouth. *Water Sci. Tech.*, **27**, 7-8, 397-404.
Shteinman, B. S., Mechrez, E, and Gutman, A. (1993) Spatial structure of a jet flow at a river mouth. *Boundary-Layer Meteorology*, **62**, 379-383.

VIII

Turbulence in Multi-Phase Flows

TURBULENT TWO-PHASE FLOW

by

L. van Wijngaarden

Twente University, Enschede, The Netherlands.

1. General

Turbulent two-phase flows are of great importance in industry, in particular the chemical process industry. There is a wide variety of topologies: annular flow, slug flow, dispersed flow, mist flow and so on. These flows are almost always highly turbulent.

What industry would like us to provide is a theory applicable to all turbulent two-phase flows and materialized in a users friendly CFD code. What we, fluid dynamicists, can do for them in that department is not very much. We can consider each type of turbulent two-phase flow by itself. Even then, an additional difficulty is that in most industrial turbulent two-phase flows the volume concentration of each phase is of order unity. In situations in which we have some insight this concentration is either close to zero or close to unity.

With particles suspended in a turbulent flow, for example, hydrodynamic interaction between particles can be neglected at a volume concentration of the dispersed phase well below 1% . For such very dilute suspensions it is sufficient to consider only one particle and study its behaviour in a turbulent flow.

Knowledge about the dynamics of a turbulent particle or bubble dispersion can be obtained from averaging over many trajectories of particles or bubbles. This connection between one particle behaviour and suspension hydrodynamics has proven to be very fruitful in non-turbulent two-phase flow (Biesheuvel & Van Wijngaarden 1984). We will return to this in Section 4.

With increasing concentration, hydrodynamic interaction between particles or bubbles become important. Formally, an averaged quantity is defined as ensemble average

S. Gavrilakis et al. (eds.), Advances in Turbulence VI, 535-541.
© 1996 *Kluwer Academic Publishers.*

over the ensemble of all possible configurations C_N of N particles. If the volume concentration is denoted with α, it is wellknown (e.g. Batchelor 1972) that accuracy in α is obtained when a configuration exists of only one particle, whereas accuracy in α^2 requires configurations consisting of two particles etc. In this way a theory for moderately dilute dispersions, can be obtained. In non-turbulent flow there are many examples, but for turbulent flow such calculations have not been made, as yet.

Direct numerical calculation must be attempted when multiple interactions are important. This must for solid particles be restricted to low Reynolds number flows, and to simplifying approximations for the flow along a bubble, in the case of bubbly flows.

For turbulent dispersed flow at high Reynolds numbers <u>and</u> high concentrations, the prevailing circumstances in industrial applications, DNS is beyond possibility. A time honoured way of dealing with these flows, both in the turbulent and non- or pseudo-turbulent case, is to consider a mixture as consisting of two fluids, each with its own set of conservation equations. These equations are subjected to Reynolds averaging. But then, one needs closure relations. In one-phase turbulent flow these are restricted to the Reynolds stresses (see e.g. Launder 1992). In two-phase flow many more are needed, since with the bubbles or particles new sets of scales and times appear. Usually the closure models are variatons on the k-ε theme. Good examples of what is achieved in this way are Elghobashi & About-Arab (1983) and Besnard & Harlow (1988). A recent survey can be found in Crowe, Troutt and Chung (1996).

Quite apart from the difficulties in single phase turbulent flow, there are additional ones, posed by the terms meant to describe the interaction between the phases. The great variety in scales, so typical for turbulence, is enriched or worsened, as you please, by an additional number of characteristic scales and times. Among these are particle (or bubble) size, wake size, relaxation time and so on.

2. Solid particles in turbulent flow

At low concentration by volume of particles, hydrodynamic interaction between these can be neglected. Let the particle be spherical with radius a. We denote the kinematic viscosity of the continuous phase with ν, its density with ρ, that of the particles with ρ_p and the acceleration of gravity with $\mathbf{g}$. An important parameter is the relaxation time τ, needed for the particle to adjust, at time t and position $\mathbf{X}(t)$,its velocity $\mathbf{v}$ to that of the fluid $\mathbf{u}\{\mathbf{X}(t),t\}$,

$$\tau \sim \frac{a^2\rho_p}{\rho\nu} . \tag{1}$$

Another important quantity connected with the particle, is its settling velocity $\mathbf{V}_T$ in an infinite quiescent fluid, $\mathbf{V}_T \sim \mathbf{g}\tau$.
It is usually assumed that the Reynolds number for the relative motion between particle and fluid is small with respect to unity. Then the drag is represented by Stokes's law. For particles where $a \leq l$, l being the Kolmogorov dissipation scale, the pertinent equation of motion for the particle is (Maxey & Riley 1983)

$$\frac{d\mathbf{v}}{dt} + \tau^{-1}\mathbf{v} = \tau^{-1}\mathbf{u}\{\mathbf{X}(t),t\} + \mathbf{g}. \tag{2}$$

We consider first the influence of the turbulence on the particle motion. When the influence of the initial position can be neglected, (2) has as solution

$$\mathbf{v}(t) = \tau^{-1} \int_{-\infty}^{t} \mathbf{u}\{\mathbf{X}(t')\} \exp\left(\frac{t'-t}{\tau}\right) dt + \mathbf{V}_T. \tag{3}$$

For τ small, small particles or light particles, $\mathbf{v}$ equals $\mathbf{u}\{\mathbf{X}(t),t\}$, and

$$<\mathbf{v}> = <\mathbf{u}\{\mathbf{X}(t'),t'\}> + \mathbf{V}_T. \tag{4}$$

Reeks (1977) suggested and Maxey (1987a) proved in great detail that for a completely random velocity field the mean settling speed equals $\mathbf{V}_T$. For a real turbulent $\mathbf{u}$ Maxey (1987a) concludes that the first contribution to the settling velocity $<\mathbf{v}>$ comes from third order correlations and involves

$$\left\langle \frac{\partial u_j}{\partial x_l}\frac{\partial u_l}{\partial x_j}u_i(t)\right\rangle. \tag{5}$$

For a Gaussian velocity field this three point correlation is zero. Nevertheless, for a turbulent flow the influence on the settling speed can be substantial as the numerical simulation by Wang & Maxey (1993) shows. In general the settling speed increases under the influence of the small turbulent scales. Wang & Maxey (1993) offer the explanation that particles are swept outward, when approaching a vortical region from above, toward regions where they obtain an excess settling speed.

For settling particles, the dispersion coefficient $D_p \sim <v^2>\tau$ may differ substantially from that of the fluid D, say, of order $D \sim u_0L$, where u_0 is a measure for the turbulent velocity fluctuation. This follows both from experiments (Snyder & Lumley 1971), from analytic work (Reeks 1977, Pismen & Nir 1978) and from DNS (Uijttewaal 1995). Heavy particles with a low value of u_0/V_T cross fluid particle trajectories frequently, which causes their autocorrelation to be less than that of fluid particles, leading to $D_p/D<1$. For large τ and times $t \gg \tau$, the velocity correlation seen by the particle along its trajectory nears the Eulerian velocity correlation and consequently D_p/D may be larger than one.

At a sufficiently high particle concentration, there is a feedback of the particle motion on the turbulence. With $a \leq l$, turbulence is damped, as shown in the experiments of Tsugi et al (1984) . However, when particles get larger, the relative motion becomes more important and turbulence is enhanced.

3. Bubbles in turbulent flow

Crowe et al (1996), writing on turbulent dispersed flow, exclude bubbly flow because of the totally different response of bubbles to the fluid motion, as compared with solid particles. Yet, some aspects remain, such as the "trajectory crossing " effect, and the way in which the velocity of rise in vertical flow is affected. Bubbles are drawn into eddies and display a bias towards regions with upgoing velocities, just as solid particles display preference for eddy regions with downward velocity.

Otherwise, there are important differences, indeed, as follows from the eqn. of motion (Auton et al 1988)

$$\frac{d\mathbf{v}}{dt} = 3\frac{D\mathbf{u}}{Dt} - (\mathbf{v} - \mathbf{u})\mathbf{x}\omega - 2\mathbf{g} - \tau_v^{-1}(\mathbf{v} - \mathbf{u}). \qquad (6)$$

The main differences with (2) are: fluid inertia is much more dominant, through the term $3D\mathbf{u}/Dt$ (D/Dt = material derivative in the fluid), the added mass, with associated relaxation time $\tau_v = a^2/18v$, and the lift force (second term on r.h.s.). The latter pushes bubbles towards regions where $\mathbf{u}$ is less, overshadowing the effect of eddy encounter, mentioned above. It appears (Spelt & Biesheuvel 1996) that the effect of turbulence on bubble motion depends on much more crude properties of the spectrum, as compared with solid particles. These authors by numerical simulation and by analysis, obtained results for bubble dispersion as well. As in the case of particles, the bubble dispersion coefficient D_b is less than D for small u_0/V_T, where V_T is now the speed of rise in an infinite liquid.

4. Pseudo turbulence

A rough estimate for the velocity $\mathbf{v}$ of a bubble placed in a fluid flow with local velocity $\mathbf{u}$ is (cf 5) $\mathbf{v} = \mathbf{V}_T + 3\mathbf{u}$.

This learns that random motions of bubbles may induce considerable velocity fluctuations, called pseudo-turbulence. Under the assumption of viscous potential flow around a bubble, this can be calculated for zero u_0 (Biesheuvel & Van Wijngaarden 1984), and even for arbitrary u_0. For the excess turbulent energy impaired to the the liquid, this gives as estimate

$$\alpha V_T^2 \left\{ const. + O\left(u_0^2 / V_T^2\right)\right\}. \qquad (7)$$

Measurements of pseudo turbulence have been reported by Theofanous & Sullivan (1982), Lance & Bataille (1991), Stewart (1995). All these show significantly larger excess energies.

This, presumably, is due to the flow around largely deformed bubbles, involving a complicated interplay between surface tension, vorticity accumulation, shape instability and wake instability.

5. Miscellaneous

Topics worth mentioning in the context of turbulent two-phase flow are the generation of gravity waves on the surface of a liquid by a turbulent wind flow (Belcher & Hunt 1993) and the enormous effect of bubbles on sound emission by a turbulent flow (Crighton & Ffowcs Williams 1969).

Acknowledgement

I thank Arie Biesheuvel, René Oliemans, Wim Uijttewaal and Peter Spelt for their help, in preparing this paper.

References:

Auton, T.R., Hunt, J.C.R. & Prud'homme (1988) The force exerted on a body in inviscid unsteady non-uniform rotational flow. *J. Fluid Mech.* **197**, 241-257.

Belcher, S.E. & Hunt, J.C.R.(1993) Turbulence shear flow over slowly moving waves, *J. Fluid Mech.* **251**, 109-148.

Besnard, D.C. & Hanlow, F.H. (1988) Turbulence in multiphase flow. *Int. Jnl. Multiph. Flow.* **14,6** 679-699.

Biesheuvel, A. and Wijngaarden, L van (1984) Two-phase flow equations for a dilute dispersion of gas bubbles in liquid. *J. Fluid Mech.,***148**, 301-318.

Crighton, D.G. and Ffowcs Williams (1969) Sound generation by turbulent two-phase flow. *J. Fluid Mech.***36**, 585-603.

Crowe, C.T., Troutt, T.R. and Chung, J.N.(1996) Numerical models for two-phase turbulent flows. *Ann.Rev. Fluid Mech.* **28**, 11-43.

Elghobashi, S.E. and About-Arab J.W. (1983). A two equation for two-phase flows. *Phys. Fl.* **A26**, 931-938.

Lance, M. and Bataille, J. (1991) Turbulence in the liquid phase of a uniform bubbly air-water flow *J. Fluid Mech.* **222**, 95-118.

Launder, B.E. (1992) On the modelling of turbulent industrial flows. *Comp. Meth. Appl. Sc.* 91-102

Maxey, M.R. & Riley, J.J. (1983) Equation of motion for a small rigid sphere in a nonuniform flow. *Phys. Fluids* **26**, 883-889.

Maxey, M.R. (1987a) The gravitational setting of aerosol particles in homogeneous turbulence and random fields. *J. Fluid Mech.* **174**, 441-465.

Maxey, M.R. (1987b) The motion of small sperical particles in a cellular flow field. *Phys. Fluids* **30**, 1915-1928.

Pismen, L.M. & Nir, A (1978) On the motion of suspended particles in statisionary homogeneous turbulence. *J. Fluid Mech.* **84**, 193-196.

Reeks, M.W. (1977) On the dispersion of small particles in an isotropic turbulent fluid.. *J. Fluid Mech.* **83**, 3, 529-546.

Snyder, W.H. & Lumley, J.L. (1971) Some measurements of particle velocity autocorreclation functions in a turbulent flow. *J.Fluid Mech.* **48**, 1, 41-71,

Spelt, P.D.M. & Biesheuvel, A. (1996) Numerical simulations of the motion of gas bubbles in homogeneous isotropic turbulence. Under review. with *J. Fluid Mech.*

Stewart, C.W. (1995) Bubble interaction in low-viscosity liquids. *Int. J. Multiphase Flow.* **21**, 6, 1037-1046.

Theofanous, T.G. and Sullivan, J. (1982) Turbulence in two-phase dispersed flows, *J. Fluid Mech.* **116**, 343-362.

Tsjugi, Y, Morikawa, Y and Shiomi, H. (1984) LDV measurements of an air-solid two-phase flow in a vertical pipe. *J. Fluid Mech.* **139**, 417-434.

Uijttewaal, W. (1995) Particle motion in turbulent pipe flow. Techn. Univ. Delft. **MEAH-128.**

Wang. L-P and Maxey, M.R. (1993) Settling velocity and concentration distribution of heavy particles in homogeneous isotropic turbulence. *J. Fluid Mech.* **256**, 27-68.

TURBULENT THERMAL DIFFUSION OF SMALL
INERTIAL PARTICLES

T. ELPERIN, N. KLEEORIN

The Pearlstone Center for Aeronautical Engineering Studies, Department of Mechanical Engineering, Ben-Gurion University of the Negev, Beer-Sheva 84105, P. O. Box 653, Israel

and

I. ROGACHEVSKII

Racah Institute of Physics, The Hebrew University of Jerusalem, 91904 Jerusalem, Israel

April 13, 1996

1. Introduction

A new effect of turbulent thermal diffusion of small inertial particles is discussed [1]. This phenomenon is related to the dynamics of small inertial particles in incompressible turbulent fluid flow and results in additional non-diffusive flux of particles. Inertial particles are accumulated in the vicinity of the minimum (or maximum) of the mean temperature of the surrounding fluid depending on the ratio of material particle density to that of the surrounding fluid. At large Reynolds and Peclet numbers the turbulent thermal diffusion is much stronger than the molecular thermal diffusion.

2. Mechanism of Preferential Concentration of Particles

First let us discuss the mechanism of this effect. Due to the inertia particles inside the turbulent eddy are carried out to the boundary regions between eddies by inertial force (this regions with decreased velocity of the turbulent fluid flow and maximum of pressure of the surrounding fluid). This means that in regions with maximum pressure of turbulent fluid there is accumulation of inertial particles. Similarly there is an outflow of inertial particles from regions with minimum pressure of fluid. In a homogeneous and isotropic turbulence without large-scale external gradients of temperature a drift from regions with increased (decreased) concentration of inertial particles by a turbulent flow of fluid is equiprobable in all directions. Location of these regions is not correlated with turbulent velocity field. Therefore they do not contribute to large-scale flow of inertial particles.

Situation is drastically changed when there is a large-scale inhomogeneity of the temperature of the turbulent flow. In this case the mean heat flux $\langle \tilde{\mathbf{u}}\Theta \rangle \neq 0$. Therefore fluctuations of both, temperature Θ and velocity $\tilde{\mathbf{u}}$ of fluid, are correlated. Fluctuations of temperature cause fluctuations of

543

S. Gavrilakis et al. (eds.), Advances in Turbulence VI, 543-546.
© 1996 *Kluwer Academic Publishers.*

pressure of fluid. The pressure fluctuations result in fluctuations of the concentration of inertial particles. Indeed, increase (decrease) of the pressure of surrounding fluid is accompanied by accumulation (outflow) of the particles. Therefore, the direction of the mean flux of particles coincides with that of the heat flux, i.e. $\langle \tilde{\mathbf{u}} n_p \rangle \propto \langle \tilde{\mathbf{u}} \Theta \rangle \propto -\nabla T$, where n_p is the number density of the particles, and T is the mean temperature of fluid. Therefore the mean flux of the inertial particles is directed to the minimum of the mean temperature and the inertial particles are accumulated in this region. This effect is more pronounced when turbulent fluid flow is inhomogeneous in the direction of the mean temperature gradient.

Evolution of the number density $n_p(t, \mathbf{r})$ of small particles in a turbulent flow is determined by equation:

$$\frac{\partial n_p}{\partial t} + \nabla \cdot (n_p \mathbf{v}_p) = D \Delta n_p , \tag{1}$$

where $\mathbf{v}_p$ is a random velocity field of the particles which they acquire in a turbulent fluid velocity field, D is the coefficient of molecular diffusion. We consider the case of large Reynolds and Peclet numbers. The velocity of particles $\mathbf{v}_p$ depends on the velocity of the surrounding fluid, and it can be determined from the equation of motion for a particle. Solution of the equation of motion for small particles with $\rho_p \gg \rho$ yields:

$$\mathbf{v}_p = \mathbf{v} - \tau_p \left[\frac{\partial \mathbf{v}}{\partial t} + (\mathbf{v} \cdot \nabla)\mathbf{v} \right] + O(\tau_p^2) , \tag{2}$$

where $\mathbf{v}$ is the velocity of the surrounding fluid, τ_p is the characteristic time of coupling between the particle and surrounding fluid (Stokes time), ρ_p is the material density of particles, ρ is the density of the fluid. The second term in (2) describes the difference between the local fluid velocity and particle velocity arising due to the small but finite inertia of the particle.

In this study we consider incompressible turbulent flow $\nabla \cdot \mathbf{v} = 0$. However, the velocity field of particles is assumed to be compressible, i. e. $\nabla \cdot \mathbf{v}_p \neq 0$. Indeed, Eq. (2) for the velocity of particles and Navier-Stokes equation for the fluid yield $\nabla \cdot \mathbf{v}_p = -\tau_p \nabla \cdot (d\mathbf{v}/dt) = \tau_p \Delta P / \rho$, where P is the fluid pressure. We study the large-scale dynamics of small inertial particles and average Eq. (1) over an ensemble of random velocity fluctuations. For this purpose we use the stochastic calculus [2; 3]. It yields the equation for the mean number density of particles $N = \langle n_p \rangle$ [1]:

$$\frac{\partial N}{\partial t} + \nabla \cdot (N \mathbf{V}) = -\nabla \cdot (\mathbf{J}_T + \mathbf{J}_M) , \tag{3}$$

where

$$\mathbf{J}_T = -D_T \left[\frac{k_T}{T} \nabla T + \nabla N \right], \quad k_T = N \frac{3}{\text{Pe}} \left(\frac{m_p}{m_\mu} \right) \left(\frac{T}{T_*} \right) \ln \text{Re}_* , \tag{4}$$

$D_T = u_0 l_0/3$ is the coefficient of turbulent diffusion, $\mathrm{Pe} = u_0 l_0/D_*$ is the Peclet number, the molecular diffusion coefficient $D_* = \kappa T_*/(6\pi a_* \rho \nu)$, m_μ is the mass of molecules of surrounding fluid, $T(t, \mathbf{r})$ is the mean temperature with a characteristic value T_*, $\mathrm{Re}_* = \min\{\mathrm{Re}, \mathrm{Pe_T}\}$, $\mathrm{Re} = l_0 u_0/\nu$ is the Reynolds number, $\mathrm{Pe_T} = l_0 u_0/\chi$ is the thermal Peclet number, $l_0 = k_0^{-1}$ is the maximum scale of turbulent motions, u_0 is the characteristic velocity in this scale, χ is the coefficient of molecular thermal conductivity, m_p is the mass of particles with radius a_*, $\mathbf{J}_M$ is the molecular flux of particles. Here k_T can be interpreted as turbulent thermal diffusion ratio, and $D_T k_T$ is the coefficient of turbulent thermal diffusion. Note that for $\mathrm{Re}_* \gg 1$ and $\mathrm{Pe} \gg 1$ both turbulent diffusion coefficients are much larger than the corresponding molecular coefficients.

Now we derive an equation for N^2. Multiplication of Eq. (3) by N and simple manipulations yield

$$\frac{\partial N^2}{\partial t} + (\nabla \cdot \mathbf{S}) = -N^2 (\nabla \cdot \mathbf{V}_{\mathrm{eff}}) - I_D , \qquad (5)$$

where $\mathbf{S} = N^2 \mathbf{V}_{\mathrm{eff}} - D_T \nabla N^2$, $I_D = 2D_T (\nabla N)^2$, and the effective velocity $\mathbf{V}_{\mathrm{eff}} = \mathbf{V} - \tau_p \ln(\mathrm{Re}_*)\nabla T/m_\mu$, $\mathbf{V}$ is the mean particles velocity. Equation (5) implies that if $\nabla \cdot \mathbf{V}_{\mathrm{eff}} < 0$, a perturbation of the equilibrium distribution of inertial particles can grow in time, i. e., $(\partial/\partial t) \int N^2 \, d^3 r > 0$. However, the total number of particles is conserved. Therefore the growth of N^2 when $\nabla \cdot \mathbf{V}_{\mathrm{eff}} < 0$ is accompanied by formation of an inhomogeneous spatial distribution of the inertial particles whereby regions with an increased concentration of particles coexist with regions depleted from particles.

Now we will show that turbulent thermal diffusion results in large-scale pattern formation whereby initial spatial distribution of particles in a turbulent incompressible flow of fluid evolves under certain conditions into large-scale inhomogeneous distribution due to excitation of an instability. One of the most important conditions for the instability is inhomogeneous spatial distribution of mean temperature of surrounding fluid. In particular, the instability can be excited in the vicinity of the minimum in the mean temperature distribution. The growth rate of the instability in dimensionless form is given by

$$\gamma_0 = \eta_0 + \frac{3}{2} a_0 - \left[\eta_0^2 + \frac{9}{4} a_0^2 - a_0 k^2 \right]^{\frac{1}{2}} - k^2 \qquad (6)$$

(see [1]), where

$$\eta_0 = \frac{3\alpha}{\mathrm{Pe}} \left(\frac{m_p}{m_\mu} \right) \left(\frac{\delta T}{T_*} \right) \ln \mathrm{Re}_*, \quad a_0 = \frac{1}{2} \frac{d^2}{dZ^2} \ln\langle \mathbf{u}^2 \rangle ,$$

the axis Z is directed along mean temperature gradient, the wave vector $\mathbf{k}$ is perpendicular to the axis Z. Equation (6) is written in dimensionless form where coordinate is measured in units Λ_T, time t is measured in units Λ_T^2/D_T, wave number k is measured in units Λ_T^{-1}, the temperature T is measured in units of temperature difference δT in the scale Λ_T, and concentration N is measured in units N_*. Thus the initial spatial distribution of the concentration of the inertial particles evolves into a pattern containing regions with increased (decreased) concentration of particles. Characteristic vertical size of the inhomogeneity when $\eta_0 \geq a_0$ is of the order of

$$l_z \sim \Lambda_T \left[\frac{3\alpha}{\mathrm{Pe}} \left(\frac{m_p}{m_\mu} \right) \left(\frac{\delta T}{T_*} \right) \ln \mathrm{Re}_* \right]^{-\frac{1}{2}}.$$

Remarkably $l_z \to \infty$ when $\mathrm{Pe} \to \infty$, i. e., this effect exists for large but finite Peclet numbers.

3. Discussion

The effect of turbulent thermal diffusion is important in combustion and some atmospheric phenomena (e. g., atmospheric aerosols, cloud formation and smog formation). Observations of the vertical distributions of aerosols in the atmosphere show that maximum concentrations can occur within the temperature inversion layers. Using the characteristic parameters of the atmospheric turbulent boundary layer: maximum scale of turbulent flow $l_0 \sim 10^3 - 10^4$ cm; velocity in the scale l_0 : $u_0 \sim 30 - 100$ cm/s; Reynolds number $\mathrm{Re} \sim 10^6$ and of the temperature inversion: scale $\Lambda_T \sim 3 \times 10^4$ cm and dimensionless mean spatial temperature variation $\delta T/T_0 \sim (1 - 3) \times 10^{-2}$, we obtain that the characteristic time of excitation of the instability of concentration distribution of aerosols with material density $\rho_p \sim 2$ g / cm^3 and radius $a_* = 10\mu$m varies in the range from 0.3 to 3 hours. This value is in compliance with the characteristic time of growth of inhomogeneous structures in atmosphere. It is essential that this time strongly depends on the aerosol size, i.e., $\sim a_*^{-2}$.

References

Elperin, T., Kleeorin N. and Rogachevskii, I. (1996) Turbulent thermal diffusion of small inertial particles, *Phys. Rev. Lett.*, **76**, 224-228.

Zeldovich, Ya. B., Molchanov, S. A., Ruzmaikin, A. A. and Sokoloff, D. D. (1988) Intermittency, diffusion and generation in a nonstationary random medium. *Sov. Sci. Rev. C. Math Phys.*, **7**, 1-110, and references therein.

Elperin, T., Kleeorin N. and Rogachevskii, I. (1995) Dynamics of passive scalar in compressible turbulent flow: large-scale patterns and small-scale fluctuations, *Phys. Rev. E*, **52**, 2617-2634.

NUMERICAL STUDY OF BUBBLE AND PARTICLE MOTION IN A TURBULENT BOUNDARY LAYER USING PROPER ORTHOGONAL DECOMPOSITION

I.A. JOIA, T. USHIJIMA, M.R. ELSDEN AND R.J. PERKINS

Laboratoire de Mecanique des Fluides et d'Acoustique,
Ecole Centrale de Lyon
36 av Guy de Collongue, 69131 Ecully, FRANCE

1. Introduction

There are many practical problems in environmental and industrial flows which involve the transport of bubbles or particles in a turbulent boundary layer. Recent experimental work (Ushijima & Perkins 1995) has suggested that the instantaneous structure of the flow in the boundary layer plays a dominant role in determining the motion of particles and bubbles.

In order to investigate these processes in detail, we have developed a simulation of the unsteady, three–dimensional flow in a turbulent boundary layer over a flat plate, using the method of Proper Orthogonal Decomposition. We have used this model to compute the trajectories of bubbles and particles.

2. The POD Method

POD was used by Aubry *et al* (1988) and Sanghi & Aubry (1993) to model flow in a turbulent boundary layer, and our model closely follows their work. They provide a full description of the method and the properties of the velocity fields which it generates, so we only provide a brief summary here.

The basic principle of the method is that the velocity field can be expressed in terms of the sum of a limited number of eigenfunctions, $\phi_i(\mathbf{x})$, which can be computed from the velocity correlation matrix $R_{ij}(\mathbf{x}, \mathbf{x}') = \langle u_i(\mathbf{x}) u_j(\mathbf{x}') \rangle$ by solving the eigenvalue problem:

$$\int R_{ij}(\mathbf{x}, \mathbf{x}') \phi_j(\mathbf{x}') \, d\mathbf{x}' = \lambda \phi_i(\mathbf{x}) \tag{1}$$

547

S. Gavrilakis et al. (eds.), Advances in Turbulence VI, 547-550.

The velocity field $u_i(\mathbf{x}, t)$ is given by a sum over the eigenfunctions:

$$u_i(\mathbf{x}, t) = \sum_n a^n(t)\phi_i^n(\mathbf{x}). \tag{2}$$

The coefficients $a^n(t)$ are determined by solving a set of ODE's which are obtained by Galerkin projection of the eigenfunctions onto the Navier–Stokes equations.

An advantage of POD over other numerical techniques such as LES or DNS is that the modes which are used to approximate the turbulent velocity field are optimal at capturing the energy in the field, so it is possible to model the evolution of the velocity field using a relatively low–order model. This means that the technique requires only moderate computing resources. POD is a model of a turbulent velocity field (rather than a full solution of the Navier-Stokes equations) so it is easier to identify and investigate the influence of particular processes. It is also easier to extract information about the underlying turbulent structure of the velocity field.

A disadvantage of the technique is that it is necessary to specify the full correlation matrix $R_{ij}(\mathbf{x}, \mathbf{x}')$, and this requires either detailed experimental measurements or results from numerical simulations such as LES or DNS.

The results presented here are based on the correlation matrix obtained from measurements in pipe flow, at a Reynolds number of 8750, made by Herzog (1986). The results are for a 32–D simulation which contains 6 (Fourier) modes in the spanwise direction, 2 (Fourier) modes in the streamwise direction and a single (Proper) mode in the wall normal direction. The size of the computational domain is 666 wall units in the streamwise direction, 333 in the spanwise direction and 40 wall units in wall–normal direction.

The model contains one free parameter, α, which controls the rate of transfer of energy from large scales to the small, unresolved, scales. We have chosen a value of α for which the statistics of our velocity field are close to those measured in turbulent boundary layers.

3. Results

Figures 1a and 1b show respectively the mean velocity profile and turbulence intensities predicted by the model (all velocities have been normalised by the wall shear velocity u_*). The mean velocity profile is in good agreement with experimental results, both in the near–wall, linear, and in the outer, logarithmic, region. The profiles of $\langle u'/u_* \rangle$ and $\langle u'v'/u_*^2 \rangle$ are in close agreement with experimental measurements, but the model slightly under-predicts the profiles of $\langle v'/u_* \rangle$ and $\langle w'/u_* \rangle$.

Figure 2 shows the structure in the velocity field at two different instants in time. The velocity vector is projected onto cross–sections perpendicular

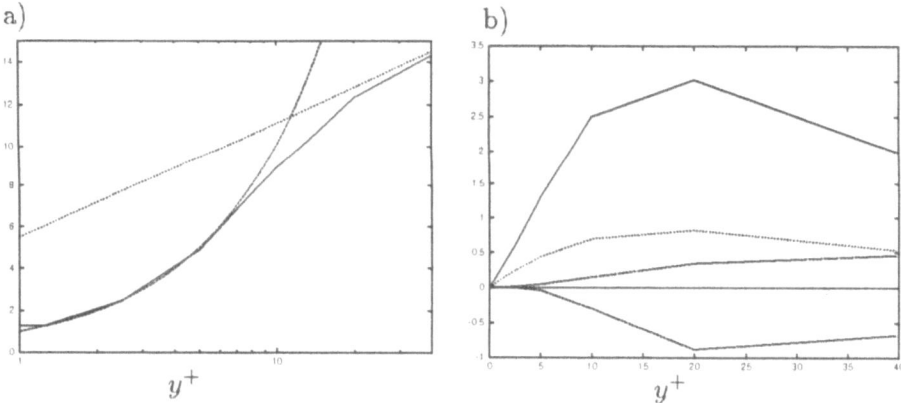

Figure 1. a) The mean velocity profile, U/u_*, on a log–linear graph, included is the linear velocity profile valid for $y^+ < 5$ and the logarithmic profile valid for $y^+ > 30$. b) Turbulence intensities in the boundary layer, from top downwards, $\langle u'/u_* \rangle$, $\langle w'/u_* \rangle$, $\langle v'/u_* \rangle$, $\langle u'v'/u_*^2 \rangle$.

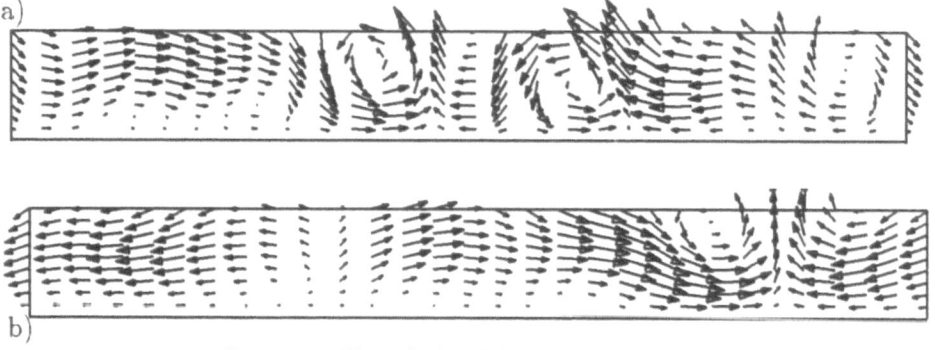

Figure 2. The velocity field at different times.

to the streamwise direction. Fig (2a) shows two, co–rotating, vortex tubes, these vortices often persist in the streamwise direction up to distances of 400 wall units. In fig (2b) we note a region where (slow moving) fluid is being ejected from the wall region, these regions are associated with counter–rotating vortex tube pairs.

We have used these velocity fields to compute particle and bubble trajectories in the near wall region, by integrating the relevant equation of motion. Some examples of the trajectories of heavy particles are shown in fig 3. The relative density of the particles is 2000, and the particles are characterised by two dimensionless parameters – V_T/v' and τ_p/T_L, V_T being the particle terminal velocity, τ_p the particle response time and T_L the Lagrangian timescale of the turbulence.

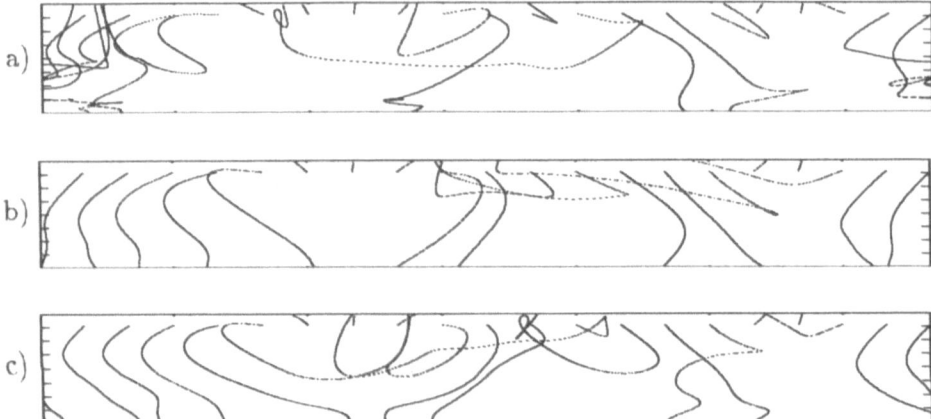

Figure 3. Particle trajectories projected onto a plane perpendicular to the mean flow direction. a) $V_T/v' = 0.1$, $\tau_p/T_L = 1.0$, b) $V_T/v' = 1.0$, $\tau_p/T_L = 1.0$, c) $V_T/v' = 1.0$, $\tau_p/T_L = 0.1$.

4. Conclusions

As well as providing a promising method for investigating the role of near–wall turbulent structure in the transport of particles and bubbles, this approach also enables us to estimate quantities (such as 'trapping distances') which can be used directly in practical engineering calculations.

5. Acknowledgments

The authors would like to express their gratitude for the help provided by Dr. Nadine Aubry, without which the project would have proceeded at a much slower rate.

6. References

Aubry, N., Holmes, P., Lumley, J.L. & Stone, E. 1988 The dynamics of coherent structures in the wall region of a turbulent boundary layer. *J. Fluid Mech.* **192**, pp 115–173.

Herzog, S. 1986 The large scale structure in the near–wall region of turbulent pipe flow. *Ph.D. thesis, Cornell University.*

Sanghi, S. & Aubry, N. 1993 Mode interaction models for near–wall turbulence. *J. Fluid Mech.* **247**, pp 455–488.

Ushijima, T. & Perkins, R.J. 1995 The Dispersion of Heavy Particles in a Turbulent Boundary Layer. *Second International Conference on Multiphase Flow '95-Kyoto* pp PT31–PT37.

THE EFFECTS OF SHEAR AND SPIN ON PARTICLE LIFT AND DRAG IN A SHEAR FLOW AT HIGH REYNOLDS NUMBERS

S. KOMORI AND R. KUROSE

Department of Chemical Engineering, Kyushu University
Hakozaki, Fukuoka 812-81, JAPAN

1. Introduction

Particle motions in the turbulent boundary layer are often seen in significant environmental problems such as desertification and air pollution, and they also occur in many industrial processes. It is, therefore, of great practical interest to investigate particle motions both in settling environmental problems and in designing industrial equipment. Particle motions in turbulent flows have often been investigated using numerical simulations. However, most of the numerical simulations were limited to small nonspinning particles at low particle Reynolds numbers of $Re < 1$, since drag and lift acting on a small particle comparable to or less than Kolmogorov scale can easily be given by analytical formulas based on the Stokes's assumption. On the other hand, motions of large particles at high particle Reynolds numbers of $Re > 100$ have not been simulated, because of the difficulty for estimating the lift and drag acting on large particles.

The purpose of this study is to numerically investigate the effects of fluid shear and particle spin on particle lift and drag in a linear shear flow at high particle Reynolds numbers. Three-dimensional numerical solutions were obtained for a steady, linear shear flow past a spinning or nonspinning spherical particle over a wide range of particle Reynolds number ($0.5 < Re < 500$), and the effects of shear and spin rate on lift and drag coefficients were investigated .

2. Numerical Simulation

A three-dimensional numerical simulation was used to estimate particle lift and drag induced by fluid shear and particle spin (Fig.1). The imposed flow was a linear shear flow without turbulence. The three-dimensional Navier-Stokes equations were directly solved using a third-order finite difference scheme based on the marker and cell method and the cylindrical coordinates were used[1]. Three parameters; the particle Reynolds number $Re(= 2aU_c/\nu)$, the dimensionless fluid shear rate $\alpha^*(= a/U_c(\partial U/\partial y)$ and the dimensionless particle spin rate $\Omega^*(= \Omega a/U_c)$, were defined and they ranged from 0.5 to 500, 0 to 0.4 and 0 to 0.25, respectively.

S. Gavrilakis et al. (eds.), Advances in Turbulence VI, 551-554.

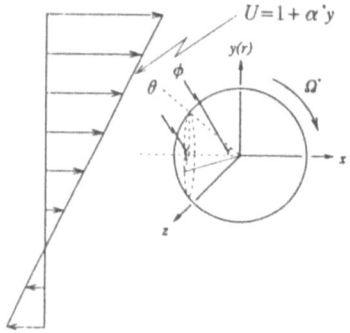

Figure 1. Coordinate system for a spinning sphere in a linear shear flow.

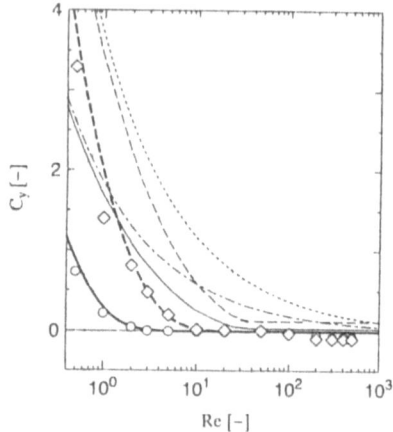

Symbol		
$\alpha^{\cdot}$	0.1	0.4
Saffman	———	— ·· — ·· —
McLaughlin	———	— — — —
Mei	— · — · —	···········
Present computation	○	◇

Figure 2. Comparison of the shear lift coefficient C_y for a stationary sphere in a linear shear flow between the present and previous[2,3,4] studies $(0.5 < Re < 500)$.

Figure 3. Shear lift coefficient C_y for a stationary sphere in a linear shear flow $(10 < Re < 500)$.

Here U_c is the fluid velocity at the central point of a spherical particle, a the radius of a spherical particle and Ω the particle spin rate.

3. Results and Discussion

Figure 2 shows the variations of the shear lift coefficient C_y with Re for a stationary sphere in a linear shear flow. C_y rapidly decreases with increasing Re in the low particle Reynolds number region $(Re < 10)$. Although the present C_y deviates from the predictions by Saffman[2] and Mei[3], it is in good agreement with the prediction by McLaughlin[4] who extended the analytical solutions of Saffman[2] to higher Re. However, in the region of $Re > 100$, the present C_y shows the negative values in contrast with the previous results. The negative values of C_y are magnified in Fig.3. C_y becomes negative in the region of $Re > 50$, and the negative values increase with increasing α^*. In previous studies, the lift force has been considered to act towards the high speed region $(C_y > 0)$. Only Jordan and

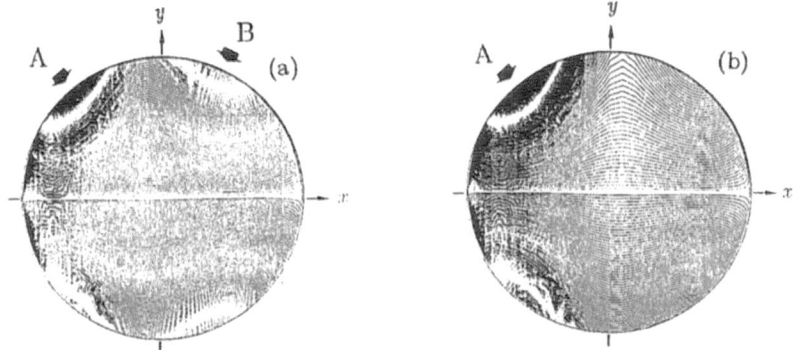

Figure 4. Surface contours of y-component of pressure on the surface of a stationary sphere; (a) for $Re = 200$ and $\alpha^* = 0.2$; (b) for $Re = 50$ and $\alpha^* = 0.2$.

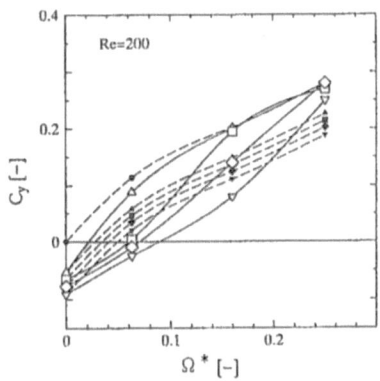

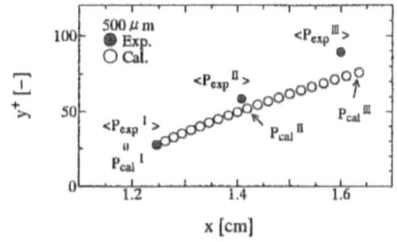

Figure 6. Comparison of the particle trajectory predicted under the mean velocity field with the trajectory measured in the turbulent boundary layer.

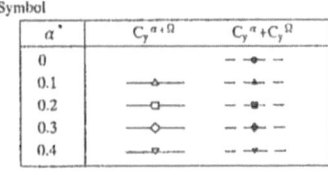

Symbol		
α^*	$C_y{}^{\alpha,\Omega}$	$C_y{}^\alpha + C_y{}^\Omega$
0		$--\bullet--$
0.1	$--\triangle--$	$--\blacktriangle--$
0.2	$--\square--$	$--\blacksquare--$
0.3	$--\diamond--$	$--\blacklozenge--$
0.4	$--\nabla--$	$--\blacktriangledown--$

Figure 5. Lift coefficient C_y on a spinning sphere in a linear shear flow.

Fromm[5] numerically showed the negative C_y for a cylinder with $Re = 400$ in a linear shear flow, but they did not discuss why C_y becomes negative.

To investigate the negative lift, the effects of pressure and viscous forces on C_y were estimated. The results showed that the pressure force significantly contributes to the negative C_y but the viscous force has no effect on the negative C_y. In fact, the instantaneous pressure distributions on the surface of a stationary sphere show that the pressure force acts on the rear part of the sphere in the negative y direction as indicated by an arrow B in Fig.4a for a high particle Reynolds number of $Re = 200$, in contrast with the low Reynolds number case of $Re = 50$ (Fig.4b).

The negative lift coefficient was also confirmed by carrying out an experiment of a falling iron-particle in a linear high-viscosity shear flow produced between two belts moving in the counter direction.

For a spinning particle, particle spin promotes the particle drag and lift as well as fluid shear. However, the sign of the lift coefficient does not change even in the high particle Reynolds number region. Figure 5 shows the distributions of the lift coefficient against the dimensionless spin rate Ω^* at $Re = 200$ for a spinning sphere in a linear shear flow. The solid lines indicate the lift coefficient $C_y^{\alpha+\Omega}$ at $Re = 200$ for a spinning particle in a linear shear flow, and the dashed lines show the sum of the lift coefficient C_y^{α} at $Re = 200$ for a stationary sphere in a linear shear flow and the lift coefficient C_y^{Ω} for a spinning sphere in a uniform unsheared flow. Although the values of $C_y^{\alpha+\Omega}$ are close to those of $C_y^{\alpha} + C_y^{\Omega}$, $C_y^{\alpha+\Omega}$ does not strictly coincide with $C_y^{\alpha} + C_y^{\Omega}$. This means that the effects of fluid shear and particle spin rate cannot be independently treated for a spinning particle in a shear flow.

By using the predictions of the particle lift and drag coefficients, a trajectory of a spinning particle at high particle numbers was simulated under the mean velocity field in the turbulent boundary layer by solving a Lagrangian equation of particle motion. The predicted trajectory is compared with the measured trajectory [6] in Fig.6. The result shows that the organized motion, which cannot be given only by the mean velocity field, strongly affects the particle motion in the turbulent boundary layer.

4. Acknowledgments

This work was supported by the Japan Ministry of Education, Science and Culture through Grants-in-Aid (No.072274). The computation was carried out by the super computer of the Center for the Global Environmental Research, National Instiue for Environmental Studies.

5. References

1. Hanazaki, H. : A numerical study of three-dimentional stratified flow past a sphere, *J. Fluid Mech.* **192** (1988), 393-419.
2. Saffman, P.G. : The lift on a small sphere in a slow shear flow, *J. Fluid Mech.* **22** (1965), 385-400.
3. Mei, R. : An approximate expression for the shear lift force on a spherical particle at finite Reynolds number, *Int. Multiphase Flow* **18** (1992), 145-147.
4. McLaughlin, J.B. : Inertial migration of a small sphere in linear shear flows, *J. Fluid Mech.* **224** (1991), 261-274.
5. Jordan, S.K. and Fromm, J.E. : Laminar flow past a circle in a shear flow, *Phys. of Fluid* **15** (1972), 972-976.
6. Kurose, R. and Komori, S. : Relationship between particle motion and turbulence structure in turbulent boundary layer flow over particle-rough wall, *J. JSME (in Japanese)* **61** (1995), 1693-1700.

TURBULENCE STRUCTURE NEAR A SHEAR-FREE GAS-LIQUID INTERFACE IN STABLY STRATIFIED OPEN-CHANNEL FLOWS

RYUICHI NAGAOSA AND TAKAYUKI SAITO

National Institute for Resources and Environment
Onogawa 16-3, Tsukuba, Ibaragi 305, JAPAN

1. Introduction

Understanding of the turbulence structure near a gas-liquid interface in open-channel flows is important to investigate the scalar exchange process and the turbulence scalar diffusion below the gas-liquid interface. Therefore, many researchers have paid attention to the turbulence structure near the gas-liquid interface (Rashidi *et al.*, 1991; Komori *et al.*, 1989; Komori *et al.*, 1993; Lam and Banerjee, 1992).

When the active scalar is transferred across the interface, the turbulence structure is strongly affected due to the effect of the stratification. According to Gerz *et al.*(1989) and Komori *et al.*(1982), distributions of turbulence statistics in the stratified flows depend only on the gradient Richardson number. Their conclusions show that the effect of stratification has a significant role in the scalar transfer near the gas-liquid interface. However, the relationship between the scalar transfer near the gas-liquid interface and the stable stratification still remain unknown although several remarkable studies were performed (Komori *et al.*, 1982; Gerz *et al.*, 1989).

The purpose of this study is to investigate the turbulence structure in both unstratified and stratified open-channel flows by means of the direct numerical simulations. Fully-developed turbulence statistics are estimated to clarify the effect of the stable stratification on the scalar transfer near the interface. The velocity fields on the interface in both unstratified and stratified flows are discussed to clarify the effect of the stratification on the turbulence structure near the gas-liquid interface.

S. Gavrilakis et al. (eds.), Advances in Turbulence VI, 555-558.

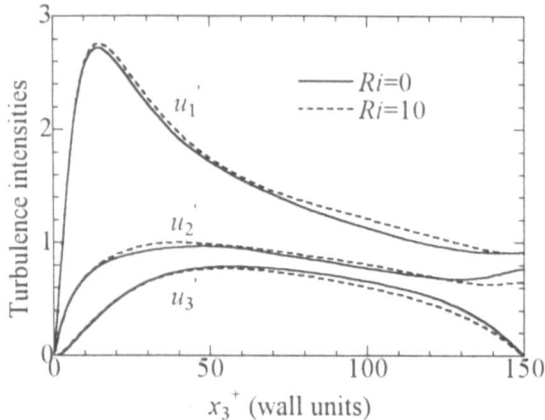

Figure 1. Vertical distributions of the turbulence intensities.

2. Numerical Method

The flows between the non-slip (bottom) and the free-slip (top) walls are simulated in this study. The sizes of the simulation box are 943, 471 and 150 wall units in streamwise (1), spanwise (2) and vertical (3) directions.

The governing equations used here are the continuity equation, the Navier-Stokes equations with Boussinesq approximation and the energy transport equation. The computations are carried out with 64^3 grid points for a Reynolds number of 150, based on the channel depth and the wall friction velocity. The Prandtl number used in this study is 1.0 and the Richardson number is varied between 0 and 10.

3. Results and Discussions

3.1. TURBULENCE STATISTICS NEAR THE INTERFACE

Figure 1 shows the vertical profiles of the turbulence intensities in both unstratified and stratified flows. The spanwise turbulence intensity slightly increases near the interterface in the unstratified flow although the streamwise turbulence intensity is not considerably increase. These profiles show that the turbulence kinetic energy in the vertical direction, which is damped near the gas-liquid interface, is mainly redistributed to the spanwise direction (Komori *et al.*, 1993; Handler *et al.*, 1993). However, the increasing trend of the spanwise turbulence intensity near the interface is not apparent when the stable stratification is imposed.

According to Perot and Moin(1995), the redistribution of the turbulence kinetic energy are induced by the impinging structures (splats) near the gas-liquid interface through the pressure-strain effect. Therefore, the velocity

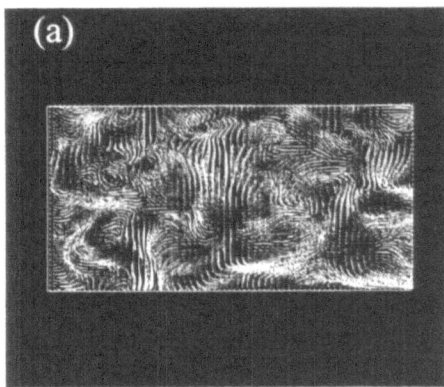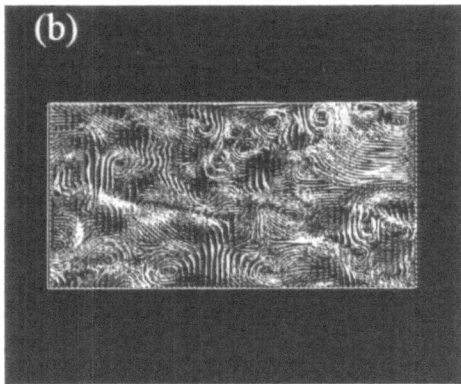

Figure 2. Velocity field on the free surface. (a)unstratified flow, (b)stratified flow

fields on the gas-liquid interface in both unstratified and stratified flows are visualized to investigate the effect of the stable stratification on the turbulence structure. Figure 2(a) show that the impinging structures are dominant near the free surface in the unstratified flow. On the other hand, these impinging structures are restricted in the stratified flow as shown in figure 2(b). Many whirlpool-like structure appears on the interface in the stratified flow. Since the impinging structures have strong vertical motion, they are highly restricted due to the stable stratification.

3.2. TURBULENCE STRUCTURE NEAR THE INTERFACE

To see the redistribution mechanism near the gas-liquid interface in both unstratified and stratified flows, the vertical distributions of the pressure-strain terms, $\Pi_{ii} = 2\overline{pu_{i,i}}$ (i=1,2,3), are estimated in figure 3. In both unstratified and stratified flows, $\Pi_{11} \approx 0$ and $\Pi_{22} + \Pi_{33} \approx 0$ are satisfied very close to the interface ($140 < x_3^+ < 150$). These distributions strongly suggest that the vertical turbulence kinetic energy is mainly redistributed to the spanwise direction, rather than the streamwise direction.

These distributions of the pressure-strain terms are almost same in the stratified flow. However, the absolute values of Π_{22} and Π_{33} very close to the interface are smaller than that in the unstratified flow. Clearly the redistribution of the turbulence kinetic energy, which is apparent in the unstratified flow, is restricted in the stratified flow. These phenomena are correspond to the restriction of the impinging structures near the interface in the stratified flow.

558

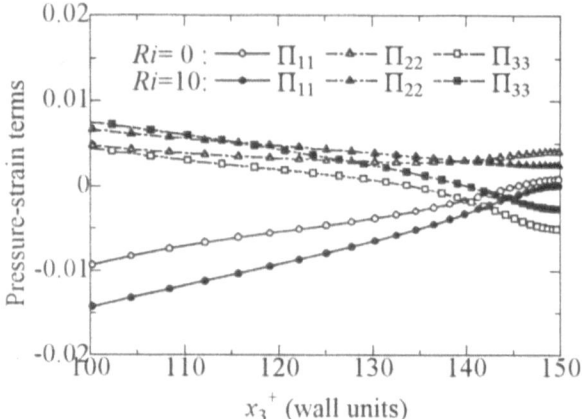

Figure 3. Vertical distributions of the pressure-strain terms

4. Conclusions

Turbulence structure near the shear-free gas-liquid interface in both un-stratified and stratified flows are investigated. When the stable stratification is not imposed, the impinging structures are dominant near the gas-liquid interface. However, these structures on the interface are restricted due to the stable stratification. As a consequence, the redistribution of the turbulence kinetic energy from vertical to horizontal direction is not apparent in stratified flow.

References

Gerz, T., Schumann, U. and Elgobashi, S.E. (1989) Direct numerical simulation of stratified homogeneous turbulent shear flows, *J. Fluid Mech.* **200**, pp.563-594.

Handler, R. A., Swean, Jr., T. F., Leighton, R. I. and Swearingen, J. D. (1993) Length scale and the energy balance for turbulence near a free surface, *AIAA J.* **31**, pp.1998-2007.

Komori, S., Ueda, H., Ogino, F. and Mizushina, T. (1982) Turbulence structure in stably stratified open-channel flow, *J. Fluid Mech.* **130**, pp.13-26.

Komori, S., Ueda, H. and Murakami, Y. (1989) Relationship between surface renewal and bursting motions in an open-channel flow *J. Fluid Mech.* **203**, pp.103-123.

Komori, S., Nagaosa, R. and Murakami, Y. (1993) Direct numerical simulation of three-dimensional open-channel flow with zero-shear gas-liquid interface, *Phys. Fluids A* **5**, pp.115-125.

Lam, K. and Banerjee, S. (1992) On condition of streak formation in a bounded turbulent flow, *Phys. Fluids A* **4**, pp.306-320.

Pan, Y. and Banerjee, S. (1995) A numerical study of free-surface turbulence in channel flow, *Phys. Fluids* **7**, pp.1649-1664.

Perot, B. and Moin, P. (1995) Shear-free boundary layers. Part 1. Physical insights into near-wall turbulence, *J. Fluid Mech.* **295**, pp.199-227.

Rashidi, M., Hetsuroni, G. and Banerjee, S. (1991) Mechanisms of heat and mass transport at gas-liquid interface, *Int. J. Heat Mass Transf.* **34**, pp.1799-1810.

LINEAR STABILITY OF TWO-PHASE FLOWS

V. YA. RUDYAK, E. B. ISAKOV AND E. G. BORD
Novosibirsk State Academy of Civil Engineering
630008, Novosibirsk, Russia

1. Introduction

The laminar–turbulent transition problem remains one of the most actual problems of the mechanics and physics during the whole century. An essential progress in understanding this phenomenon was achieved when studying the steady laminar flow stability with respect to small external disturbances. However practically all achievements of the stability theory both theoretical and experimental are relevant to the one phase fluid study.

The present paper is devoted to study of the two phase flows stability. The stability of monodispersed suspensions and gas suspensions has been investigated within the framework of a linear theory. The plane Poiseuille flow, wake and jet were analysed. It is supposed that a dispersed phase consists of the solid spherical particles and a carrier medium is an incompressible fluid. The volume concentration of the particles is small. The influence of the particles mass concentration f, the sizes of the particles, their relaxation time S and distribution through the flow have been studied.

The dynamics of the system is described in the framework of the two-fluid hydrodynamic equations in which the interface forces are proportional to the hydrodynamic velocities difference

$$\nabla \cdot \mathbf{V}_f = 0, \qquad \frac{\partial \mathbf{V}_f}{\partial t} + \mathbf{V}_f \cdot \nabla \mathbf{V}_f = -\nabla p + \frac{\Delta \mathbf{V}_f}{Re} - \frac{\rho_p}{S\,Re}(\mathbf{V}_p - \mathbf{V}_f),$$

$$\frac{\partial \rho_p}{\partial t} + \nabla \cdot (\rho_p \mathbf{V}_p) = 0, \qquad \frac{\partial \mathbf{V}_p}{\partial t} + \mathbf{V}_p \cdot \nabla \mathbf{V}_p = \frac{1}{S\,Re}(\mathbf{V}_p - \mathbf{V}_f).$$

where $\mathbf{V}_f$ and $\mathbf{V}_p$ are the carrier fluid velocity and the particles one, respectively; ρ_p is the density of the particulate phase; p is a pressure; $Re = U_0 L \rho_f / \mu$ and $S - \tau / Re$; τ is the relaxation time of the particles; U_0 and L are the characteristic values of the flow velocity and length scale. ρ_f

S. Gavrilakis et al. (eds.), Advances in Turbulence VI, 559-562.

is a density of the carrier fluid, μ is its viscosity. For the Stokeses interface force the parameter $S = 2/9(a/L)^2(\rho_p^*/\rho_p)$, where ρ_p^* is a density of the particle material.

By the standard way the linear stability problem for the two–dimensional disturbances (the highest instability as in a clean fluid case) is reduced to solution of the equation [1, 2]

$$(W - c)\Delta\psi - W''\psi + \frac{d}{dy}(\psi J f') = \frac{1}{i\alpha Re}\Delta^2\psi, \tag{1}$$

$$W(y) = U + fJ, \quad J = \frac{U - c}{1 + i\alpha SR(U - c)}, \quad \Delta = (\frac{d^2}{dy^2} - \alpha^2)$$

for the stream function ψ. Here ω and α are the frequency and wave number of the disturbances. $C = \omega/\alpha$. $f = \rho_p/\rho_f$. Eq. (1) is reduced to the Orr–Sommerfeld one only for homogeneous distribution of particles ($f =$const) and $SRe \gg 1$.

2. Poiseuille flow. $U(y) = 1 - y^2$. Uniform distribution of particles

The qualitative analysis and calculations showed that the finely dispersed fluid ($S \lesssim 10^{-6}$) was less stable then a clean one. Such a behaviour is caused by increasing the effective medium density and agrees with the Saffman's qualitative conclusion [3]. The flow stability increases with increasing the particles sizes. This effect is explained by the inertia of large particles. Such particles move at the average flow velocity and damp the high–frequencies oscillations.

At a certain value of the relaxation time S the flow stability becomes maximum and then it gradually decreases because the particles number per unit volume of the dispersed fluid is decreasing. The maximum stabilization effect is reached when $SRe \sim 1.5 \div 2$.

The flow stability characteristics and the neutral stability curves essentially depend on the value of the particles mass concentration. If the particles mass concentration is large enough the two dimensional infinitezimal disturbances can be completely suppressed in a wide range of the flow velocities. Depending on the particles size and their density sharp increasing of the critical Reynolds number Re (maybe on two or three orderes) is observed.

3. Poiseuille flow. Nonuniform distribution of particles

In present paper we supposed that the mass concentration distribution f is either the Gauss function $f(y) = f_V(\sigma)\exp(-y^2/\sigma^2)$ or the superposition of two such functions

$$f(y) = f_V(\sigma, \xi) \frac{1}{2} \left[\exp\left(-(y - \xi)^2/\sigma^2\right) + \exp\left(-(y + \xi)^2/\sigma^2\right) \right], \quad (2)$$

where the normalization factor f_V is chosen so that the total number of particles is always constant. Within the limit $\sigma \to \infty$ such a distribution is converted into a homogeneous one, but when $\sigma \ll 1$ we have a limit of thin dusty layers. In principle we can represent any distribution $f(y)$ as a set of dusty layers.

The flows with the Gauss particles distribution are less stable as compared with homogeneous dispersed fluid. The similar behaviour is typical for the fluids with the distribution (2) at small values of the parameter ξ ($\xi \lesssim 0.5$), too. However, the stability character of the flow changes when dusty layers come to the critical layers. At first, the flow stability decreases a little and then increases sharply. As it was shown in [4] such a behaviour is connected with the third term on the left hand side of Eq. 1. This term changes the sign near the critical layer. Therefore it is an additional dissipative force at first and then it come an active force generating the Reynolds stresses.

It is very interesting that the antisymmetric mode can destabilize the Poiseuille flow at certain conditions. For the flow with the particles distribution (2) the antisymmetric disturbances destabilizing the flow are mainly a large wavelength as compared with the symmetric ones. The critical Reynolds number for this mode is determined by the dusty layer thickness and increases with decreasing σ.

4. Wake and jet

In this section the stability of the plane jet velocity profile $U_j(y) = \exp(-y^2)$ and wake ($U_w(y) = 1 - \exp(-y^2)$) was studied. These problems are practically equivalent. Here as in the case of the Poiseuille flow a fine dispersed fluid is less stable then a clean one. The critical Reynolds number for dispersed fluid with homogeneous distribution of particles is simply reduced in $(1 + f)$ times. However the particles destabilize the flow only if the respective relaxation time $\tau = S Re \ll 1$. On the countrary the flow becomes more stable as compared with the clean fluid one when $\tau \sim 1$. The highest stability of the flow is reached for the relaxation times $\tau \sim 1.0 \div 8.0$.

The particles distribution nonhomogeneity can essentially change the disturbances increments. The flow stability is determined by the location of the dusty layer with respect to the critical layer.

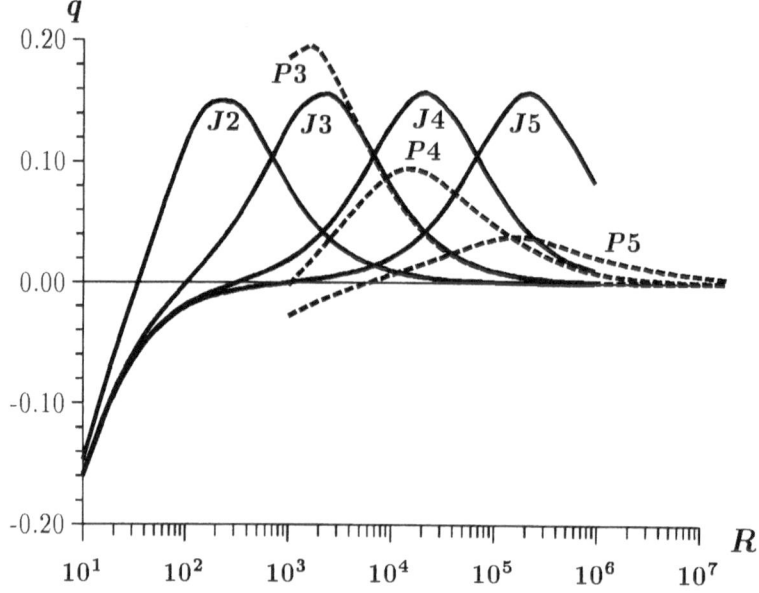

Figure 1. Relative change of the disturbances growth increments.

5. Conclusion

We see that the effect of the particles on the stability of different flows have common features. Figure illustrates this influence. Here the dependence of $q = (\omega_{i0} - \omega_i)/f$ as a function of the Reynolds number for the different values S is given (ω_{i0} and ω_i are the increments of the disturbances growth for the one–phase and two–phase flows, respectively). The curves $J2, J3, J4, J5$ and $P3, P4, P5$ correspond to the symmetric mode disturbances for the jet and Poiseuille flow when $S = 10^{-2}, 10^{-3}, 10^{-4}, 10^{-5}$. We see that the dispersed flows are less stable then the one phase flows if $SR \ll 1$. The maximal stabilization effect is reached when $SR \sim 1.5 \div 2$.

References

1. Rudyak V.Ya., Isakov E.B., Bord E.G. (1995) *Heterogeneous Media Stability. II. Poiseuille flow with suspended particles.* Preprint NSACE N1(6)-95, Novosibirsk.
2. Isakov E.B., Rudyak V.Ya. (1995) Flow Stability of Rarefied Gas Suspensions in Plane Channel, *Fluid Dynamics,* **Vol. 30, No. 5**.
3. Saffman P.G. (1962) On the stability of laminar flow of a dusty gas. *J. Fluid Mech.* **Vol. 13**, pp. 120–128.
4. Rudyak V.Ya., Isakov E.B. (1996) Poiseuille Flow Stability of Two–Phase Fluid with Inhomogeneous Distribution of Particles, *J. Applied Mech. Technical Phys.* **Vol. 37, no. 1**.

EXPERIMENTAL STUDY OF THE FLOW BEHIND AN ELONGATED BUBBLE IN A VERTICAL PIPE

L. SHEMER, S. POLONSKY and D. BARNEA
Department of Fluid Mechanics and Heat Transfer
Faculty of Engineering, Tel-Aviv University
Tel-Aviv 69978, Israel

Slug flow pattern occurs over a wide range of flow parameters in gas-liquid pipe flow. It is characterized by quasi-periodic sequences of liquid slugs and long bullet-shaped bubbles. In vertical two-phase slug flow, the liquid around the bubbles moves downstream as a thin falling film. Each slug sheds liquid in its back to the subsequent film, which is then injected into the bubble wake as a circular jet. In spite of the extreme importance of this process for understanding of two-phase slug flow [1, 6, 7], only limited data is presently available on the flow structure in the wake region [2, 3, 4, 5].

Motion of a single elongated bubble in a liquid flow can be seen as a basic phenomenon in slug flow. The sharp contrast between the two phases suggests using of video imaging and digital processing of the resulting sequence of images for the study of the wake region of such a bubble. The experimental facility is made of a 4m long transparent pipe with an inner diameter of 24 mm. Water is flowing in a closed loop. Air enters the inlet section from a settling chamber, which is separated from the inlet tube by a computer-controlled valve. The duration of the valve opening and the pressure in the air chamber determine the length of the injected bubbles.

The pictures of bubble movement are taken simultaneously at different axial locations by two interlaced video cameras and are recorded on an optical video disc. The time shift between consecutive images is 1/60s. A mirror is attached to the pipe at an angle of $45°$, providing two orthogonal views of the bubble in the same image.

Bubbles of lengths varying from 1D to 12D were studied. Time sequences of instantaneous shapes of the bubble tail were determined from the recorded images (Figure 1). The variation in time of the instantaneous velocity of the interface at various radial locations can be determined from these sequences. It

563

S. Gavrilakis et al. (eds.), Advances in Turbulence VI, 563-566.
© 1996 *Kluwer Academic Publishers.*

was found that the amplitude of the bubble tail oscillations increases with the bubble length. Frequency spectra of tail oscillations at various radial locations were obtained from the sequence of images. The dominant frequency suggests the resonant character of those oscillations. Turbulent fluctuations are superimposed on these oscillations.

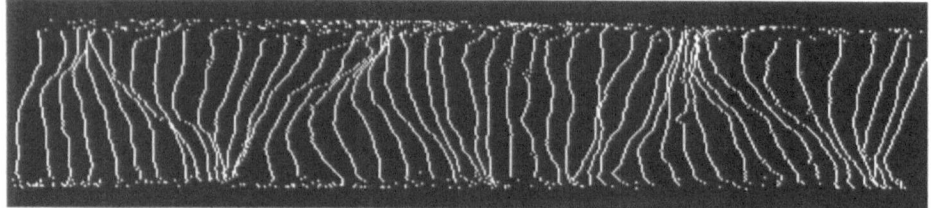

Figure 1. Tail profiles for long bubble (192 mm, or 8 D).

Interactions of the free surface oscillations with the circumferential wall jet result in small bubbles shedding from the periphery of the bubble tail. These bubbles perform vortical movement in the wake, as can be seen in the sequence of images from the two orthogonal directions depicted in Figure 2a. (Contours of the elongated bubble tail, as well as those of the following small bubbles are shown). The ejection of a single bubble from the wall region of the elongated bubble is also seen in this figure. Some of the bubbles reunite with the large elongated bubble close to the axis of the pipe (see Figure 2b). The quantitative information on the movement of those bubbles is obtained from such images.

References

1. Barnea, D. and Brauner, N. Hold-up of the liquid slug in two phase intermittent flow, *International Journal of Multiphase Flow* **11** (1985), 43-49.
2. Campos, J. B. L. M. and de Carvalho, J. R. F. An experimental study of the wake of gas slugs rising in liquids, *Journal of Fluid Mechanics*, **196** (1988), 27-37.
3. Campos, J. B. L. M. and de Carvalho, J. R. F Mixing induced by air slugs rising in narrow columns of water, *Chemical Engineering Science*, **43** (1988), 1569-1582.
4. Mao, Z.-S. and Dukler, A. The motion of Taylor bubbles in vertical tubes.: II. Experimental data and simulations for laminar and turbulent flow, *Chemical Engineering Science*, **46** (1991), 2055-2064.
5. Moissis, D. and Griffith, P. Entrance effects in a two-phase slug flow, *Journal of Heat Transfer, Transactions of ASME, Series C*, **2** (1962), 29-39.
6. Shemer, L. and Barnea, D. Visualization of the instantaneous velocity profiles in gas-liquid slug flow. *Physicochemical Hydrodynamics*, **8** (1987), 243-253.
7. Taitel, Y. Barnea, D. and Dukler, A.E. Modeling flow pattern transitions for steady upward gas-liquid flow in vertical tubes. *AIChE J.*, **26** (1980), 345-354.

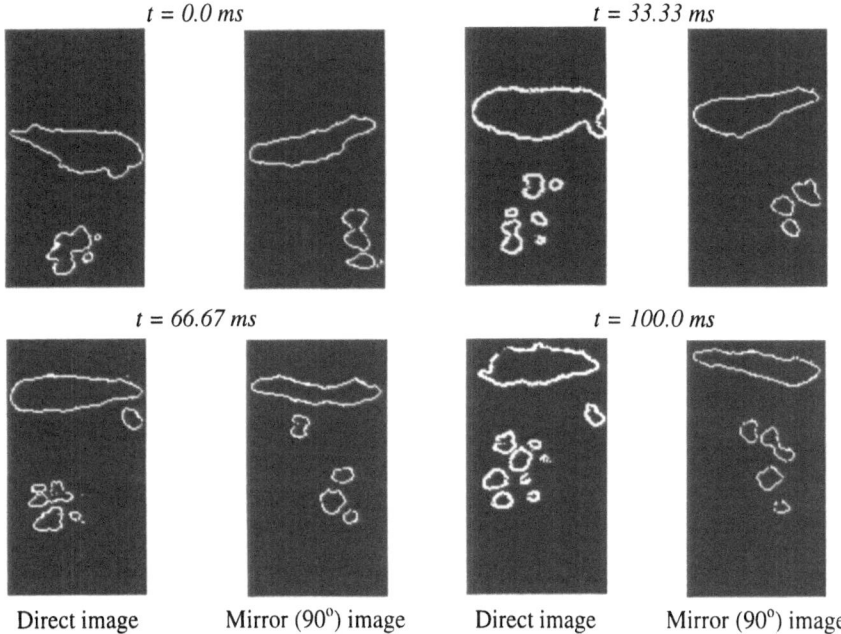

Figure 2a. Elongated bubble ejects small bubbles (L = 4D)

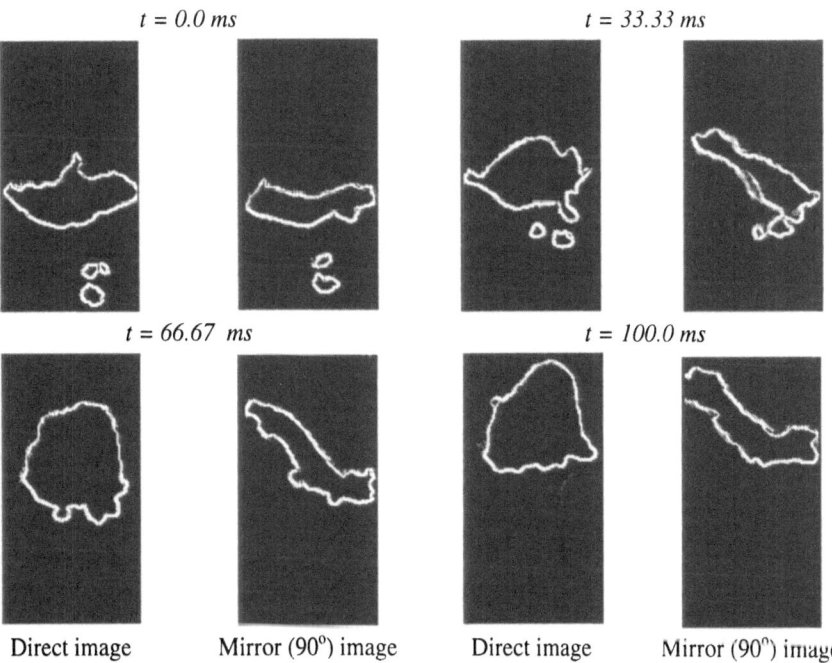

Figure 2b. Elongated bubble absorbs small bubbles (L = 8D)

566

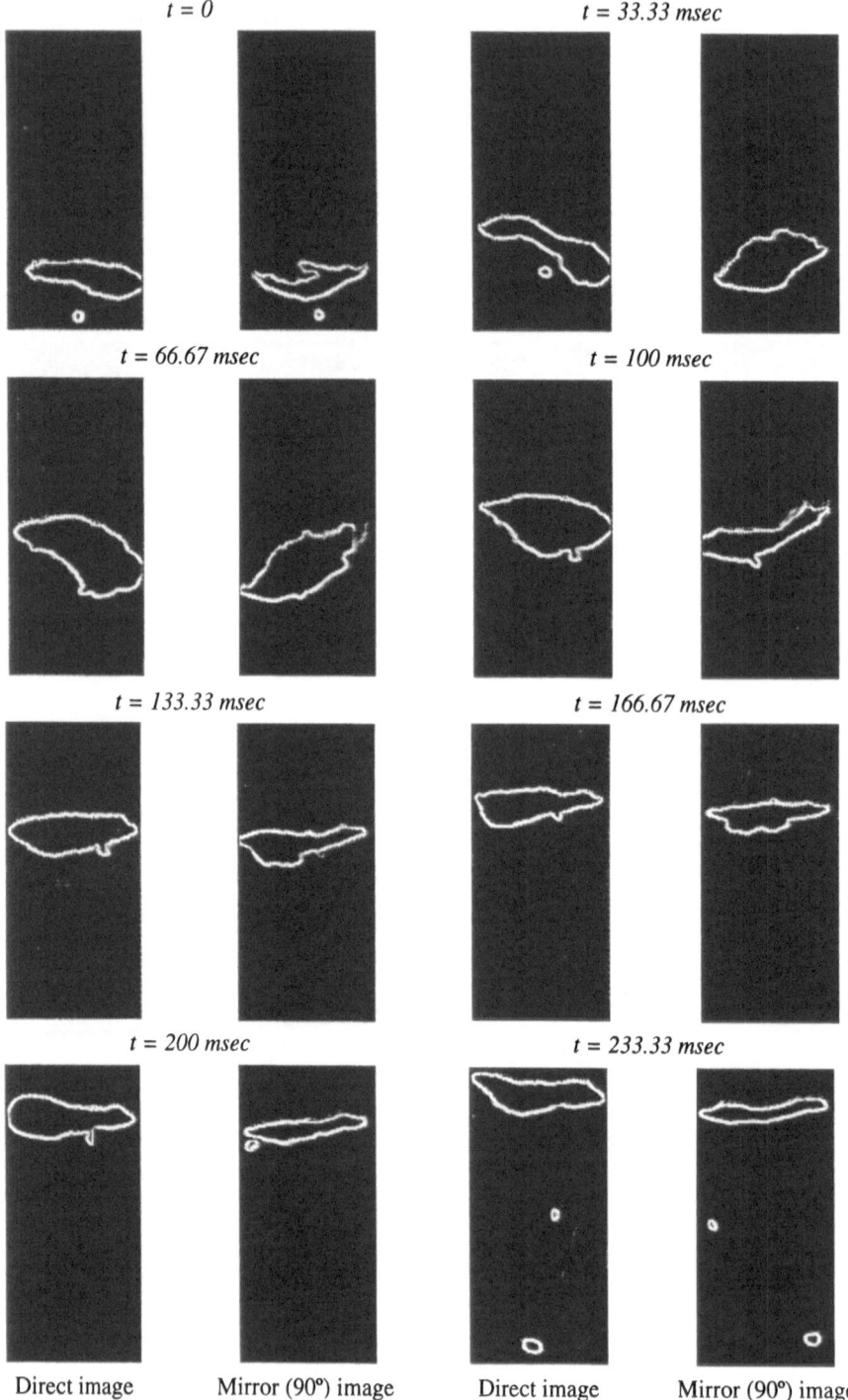

Figure 2c. Elongated bubble absorbs and ejects small bubble (L = 4 D)

BUOYANCY PROFILE EFFECTS IN INCLINED BUBBLY SHEAR FLOW

NEALE THOMAS, K. SANAULLAH, X YANG
FAST Team, School of Chemical Engineering
University of Birmingham B15 2TT, England

Summary

We describe some studies of nominally two-dimensional fully turbulent and developed duct flows (breadth/depth B/D=5; overall length/depth L/D=60; D-based Reynolds number Re>10^4) for inclination departures to 30 degrees from vertical at low voidages (<5% sectional average) representative of disperse regime using tap water bubbles (4mm) and smaller bubbles (2mm) stabilised in ionic solution. The bubbles were injected via a manifold located on the upper face at the entry station and turbulently dispersed transversely into the main body of flow. Pitot and static probe instrumentation, primitive but validated, provided adequate (10% local value) discrimination of main aspects of the mean velocity and voidage profiles at representative streamwise stations in L/D.>30. Snapshot photography with axial slit-lighting was used to capture and characterise key features of buoyancy induced transverse layering and intermittent instability seen as Kelvin Helmoltz eddies responsible for sustaining dispersion at larger inclination departures of 10+ degrees.

Our results can be divided into three categories of behaviour. For vertical flow (0 degrees) the evidence is inconclusive as to whether bubbles are preferentially trapped within the wall-layer as found in some, maybe most, earlier experimental work. Thus the 4mm bubbles showed indication of voidage retention but the 2mm bubbles did not.. For nearly vertical flow (5 degrees) there was pronounced profiling of voidage especially with 4mm bubbles but the transverse transport was not suppressed sufficiently to induce any obvious layering However, with inclined flow (10+ degrees) a distinctively layered pattern was invariably manifested in which voidage confinement increased with increasing inclination This behaviour is akin to buoyancy effects found in miscible fluids where the transverse eddy flux is sustained only through cyclic renewal of Kelvin Helmholtz instabilities by finescale turbulent erosive sharpening of the interface except with bubbles the sharpening is accelerated or dominated by the transverse component slip flux. The resulting pattern of disperse bubbly flow overlying bubble-free flow is distinctive and persistent enough to warrant discrimination as an additional regime residing between the established fully disperse and fully segregated patterns traditionally distinguished in established maps.

S. Gavrilakis et al. (eds.), Advances in Turbulence VI, 567-570.
© 1996 *Kluwer Academic Publishers.*

Main elements of the experimental set-up and findings were reported in Sanaullah & Thomas (1994, ST below; also 1996, submitted) based on Sanaullah's (1995) PhD project, here combined with an overview of simple modelling approaches for eddy dispersion of voidage reported in Yang & Thomas (1993, YT below) as backdrop to Lagrangian trajectory simulations of bubble transport in turbulent free shear flows of which main elements were reported in Yang & Thomas (1994; 1996, in preparation) and derived from Yang's (1996, submitted) PhD project.

Vertical Flow

For vertical upflows there is a broad concensus on voidage transport as drift toward the wall and retention there under the influence of vorticity lift force augmented by transverse pressure gradient due to the eddy normal stress gradient, with equilibrium profiles sustained by opposing downgradient turbulent dispersion. YT recapitulated such simple formulations and assessed the extent to which eddy diffusion models can accommodate the behaviour reported in several studies, notably Serizawa's (1976; see YT) pioneering experiments. It seems that simple models can capture the main features despite their incomplete characterisation of eddy transport inhibition by wall-layer buoyancy which is known to be crucial for representation of uni-phase mixed convection (ST refers). Practical interest here centres on reduced shear stress and hence also reduced production of turbulence as potential hazard for operation of intensive heat exchangers. With bubbly fluids the important issue is whether fine-scale "burbulence" near the wall suffices to compensate for buoyantly impaired "shurbulence", to our knowledge not yet adequately resolved despite significant safety implications. Considerable current fundamental interest is demonstrated in the ongoing work by leading groups in Europe (e.g. Marie, Ecole Centrale Lyon), Japan (e.g. Matsumoto, Tokyo University) and the USA (e.g. Maxey, Brown University).

The present study recovered only velocity profiles for vertical flow. The results with 4mm bubbles were certainly consistent with voidage excess retained in the wall zone insofar as the velocities were higher than expectation for neutral buoyancy. On the other hand, in our set-up the bubbles were introduced at the wall so this finding might equally be explained in terms of insufficient test-section length to properly accommodate evolution to fully developed conditions. Establishment of equilibrium voidage profiles is known to take considerably longer than for mean velocity and shear stress in uni-phase turbulent flows because of the weak lateral forces associated with bubble transverse fluxes. Indeed, although the mean velocity profiles obtained with 2mm bubbles showed no obvious indication of wall-layer confinement we equally cannot take this contrary result as satisfactory evidence that attraction is absent from the asymptotic approach to far downstream fully developed flow. More studies are required to resolve these uncertainties, including systematic attention to bubble size as independent parameter determining the burbulence scale and intensity for prescribed voidage independent of shurbulence. scale and intensity.

Nearly Vertical Flow

Inclination departure from vertical of just 5 degrees suffices to encourage flow evolution to a nominally fully developed state possessing distinctively skewed profiles of voidage, though not sufficiently confined to warrant description as a segregative pattern. For this "nearly-vertical" flow the transverse eddy dispersion is manifestly reduced by profile buoyancy but is certainly not completely eliminated as occurs in the layered patterns encountered at higher inclinations of 10 degrees or more; see below. The important distinction between inhibition and suppression of eddy diffusion has long been recognised in geophysical fluid dynamic representations for turbulent transport of uni-phase buoyant inhomogeneities by salt or heat where (we note) misciblilty always provides for partial irreversibility via molecular mixing on the finest scales whereas here with bubbles the transport is entirely reversible. Although less skewness is displayed in the voidage profiles with 2mm bubbles compared with 4mm bubbles it appears that both cases (i.e. for the same sectional average voidage) can be approximately accommodated in terms of eddy dispersion modelled as a downgradient flux using the local Richardson number to characterise buoyant damping of the effective diffusivity; refer to ST for an outline demonstration.

Inclined Flow

For inclinations to the vertical of 10 degrees or more, both the voidage and the velocity profiles display profound skewness increasing with increasing angle and always more pronounced for the 4mm bubbles compared with the 2mm bubbles. Whilst this trend is in line with simple qualitative expectations on growing buoyant confinement, there also emerges a marked adjustment in the transverse structure which is increasingly apparent at larger angles (20+ degrees) and with lower flowrates, especially for the larger bubbles. Although still maintaining fully disperse pattern, the voidage is strikingly restricted to an upper layer, with profiles peaking close to the upper wall and tapering well above the lower wall. The velocity profiles here display striking asymmetry being substantially bigger in the voided upper layer than in the lower layer of essentially bubble-free flow and in-between possessing a roughly uniform gradient reminiscent of free shear layers between uniform parallel uni-phase streams. Indeed, the transverse structure overall is typical of buoyantly suppressed turbulent mixing at higher Richardson numbers when eddy dispersion is restricted by a capping interface that supports only intermittent patchy activity by localised transient overturning motions associated with Kelvin Helmoltz instabilities. As confirmed by numerous snapshot photo-records, the transverse transport in these circumstances comprises the following cyclic process (refer to ST): upwards drift of bubbles resulting in sharpening of gradients in the interior shear layer; instability leading to an overturning eddy event; downwards dispersion by vortex scavenging of bubbles into the eddy; eddy weakening and decay due to entrainment exacerbated by bubble buoyancy as energy sink; collapse and subsequent renewal of the sequence.

Reflections

Our experimental study was probably conducted at sufficiently large scale to be relevant for benchmarking of practical calculation schemes, notably the two-fluid formulations built on one-point closures of the Reynolds averaged conservation equations In this regard, it may deserve attention insofar as the exercise served to distinguish and broadly characterise a "segregative-disperse" pattern that resides between the established patterns of (fully) disperse and (fully) segregated bubbly flows usually employed as adjunct to these calculation methods, for example in preliminary parameterisation of the bubble slip-speeds. Indeed, the segregative-disperse pattern described here could not be directly accommodated within the framework of earlier generation methods based on simple sectionally-averaged schemes like the drift-flux model without (at least) the introduction of voidage fractional depth as an extra degree of freedom or additional conservation quantity. In this respect, then, the study was probably worthwhile although more work is needed to consolidate the connection with considerable established knowledge on buoyancy effects in inclined two-layer flows of uni-phase fluids. The apparent significance of Richardson number as correlating parameter for the transverse profiles indicates that buoyant potential energy associated with voidage comprises a controlling sink on the shear kinetic energy associated with the velocity profiles for inclinations exceeding 10 degrees or so and is significant even for inclinations of 5 degrees to the vertical.

Another potentially important practical connection suggested by our study links to mixed convection in (nearly) vertical flows of uni-phase fluids where axial buoyancy forces due to thermally induced density reduction can attenuate the wall-zone shear stress sufficiently to cause major reduction in the turbulence energy and thus also the heat flux. For intensive heat exchangers such coupling is capable of causing catastrophic collapse corresponding to conditions of buoyancy-induced laminarisation, so it is sensible to ask whether axial buoyancy due to excess voidage in the wall-zone might perhaps act in a similar fashion. Whilst there is no room here to incorporate established and emerging fundamental literature on the behaviour and effects of bubbles in turbulent wall-layers, we must mention that bubble size appears to be a key factor insofar as microbubbles (100s microns) seemingly can reduce the wall shear stress whereas macrobubbles (mm sizes as used here) apparently act to increase it. With boiling water exchangers supporting a spectrum from micro to macro the issue is obviously of considerable practical significance.

References

Sanaullah, K. and Thomas, N.H. 1994. Velocity / voidage profiles for steeply inclined bubbly flows in "segregated-disperse" regime. ASME-FED 180, 119-128.

Yang, X. and Thomas, N.H. 1993. Void fraction profiles in bubbly upward and downward flow. 6th Julich Workshop "Two-Phase Flow Predictions", 274-282.

Yang, X. and Thomas, N.H. 1994. Simulation of particle and bubble dispersion in turbulent free shear flows. ASME-FED 185, 259-268.

CORRECTION FACTOR FOR VOID FRACTION MEASUREMENTS WITH A LINE SENSOR

TH. PANIDIS
Laboratory of Applied Thermodynamics
University of Patras, GR 265 00 Patras-Rio, Greece

It is well known that sensors measure the flow properties in a "control" volume rather than at a single point. This appears to be systematically ignored in void fraction measurements in dispersed two phase flows. In this work a correction factor is proposed to account for the mentioned inaccuracy in evaluating void fraction in two phase flows with a line sensor. The theory was developed to address the problem in the case of employing a hot film sensor in bubbly flows. In this derivation effects of deformation and deflection of the bubbles by the probe were not considered.

The general derivation assumes non deformable particles with constant orientation for each shape. We define $w(\mathbf{x})$ as the mean number of particle volume centres per unit volume at a point $\mathbf{x}$, and $p(\mathbf{x}, g, D)$ expresses the probability density function that a particle at $\mathbf{x}$ is of shape g, ($g \in G$) and equivalent volume diameter D. A point $\mathbf{r}$ will be in the particulate phase while a centre of a particle (g, D) remains in a volume $V_\alpha(\mathbf{r}, g, D)$ equal to the volume of the particle, with centre at the point $\mathbf{r}$ and having a shape symmetric to that of the particle about point $\mathbf{r}$.

The local void fraction $\alpha_L(\mathbf{r})$ is the expectation that $\mathbf{r}$ is in the dispersed phase and an estimator of $\alpha_L(\mathbf{r})$ is the time fraction that $\mathbf{r}$ is in the dispersed phase for large observation time T.

$$\alpha_L(\mathbf{r}) = \int_{G} \int_{0}^{\infty} \int_{V_\alpha(\mathbf{r},g,D)} w(\mathbf{r})p(\mathbf{r},g,D)dV\,dD\,dg = \lim_{T\to\infty} \frac{T_p}{T} \tag{1}$$

Let us suppose a sensor which is descriminating the dispersed phase while a centre of a particle (g, D) is in a volume $V_s(g, D)$. The expectation n_s that a particle centre is in $V_s(g, D)$ can be estimated, assuming that only one particle interacts with the sensor, by the time fraction that the sensor is descriminating the dispersed phase for large observation time T.

$$n_s = \int_{G} \int_{0}^{\infty} \int_{V_s(g,D)} w(\mathbf{r})p(\mathbf{r},g,D)dV\,dD\,dg = \lim_{T\to\infty} \frac{T_s}{T} \tag{2}$$

The sensor estimation n_s must be multiplied by the correction factor defined as

$$s = \frac{\alpha_L}{n_s} = \left(\int_{G} \int_{0}^{\infty} \int_{V_\alpha(g,D)} w(\mathbf{r})p(\mathbf{r},g,D)dV\,dD \right) \div \left(\int_{G} \int_{0}^{\infty} \int_{V_s(g,D)} w(\mathbf{r})p(\mathbf{r},g,D)dV\,dD \right) \tag{3}$$

S. Gavrilakis et al. (eds.), Advances in Turbulence VI, 571-572.
© 1996 *Kluwer Academic Publishers.*

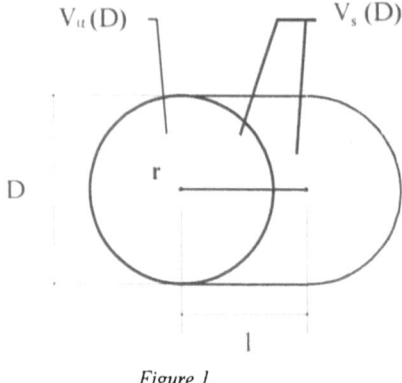

Figure 1.

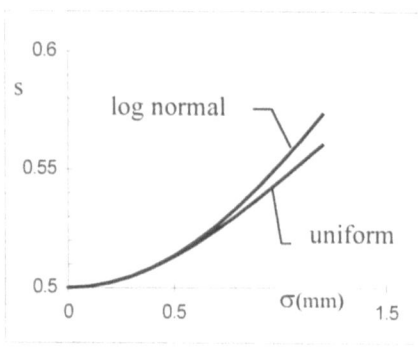

Figure 2.

to calculate the local void fraction α_L.

Let us assume that a cylindrical hot film is a line sensor whose V_s is the union of all $V_\alpha(r, g, D)$ along its length, l. For spherical particles of diameter D (see Figure 1),

$$V_\alpha(D) = \frac{\pi D^3}{6} \quad \text{and} \quad V_s(D) = \frac{\pi D^3}{6} + \frac{\pi D^2 l}{4} = \frac{\pi}{6} D^2 \left(D + \frac{3}{2} l \right) \tag{4}$$

and the correction factor can be evaluated if $p(r, D) = p(D) = ct$ in $V_s(D)$ as :

$$s = \left(\int_0^\infty D^3 p(D) dD \right) \div \left(\int_0^\infty D^2 (D + \tfrac{3}{2} l) p(D) dD \right) \tag{5}$$

The equations for s have been derived analytically for monodispersed particles as well as for uniform and logarithmic normal probability distributions of the diameter, and are presented in Table 1. The results for sensor length, $l = 2$ mm and mean bubble diameter, $\mu = 3$ mm, are shown in figure 2, as a function of the variance of the distribution. For the corresponding monodispersed case the value of s is 0.5. These values indicate that severe errors may occur if a correction is not applied in the void fraction estimation.

TABLE 1. The correction factor s for simple diameter distributions

Distribution	p(D)	s
constant $D = d$	$\delta(D-d)$ (Dirac delta)	$\dfrac{2d}{2d + 3l}$
uniform in $[(d-\varepsilon), (d+\varepsilon)]$ $\mu = d, \quad \sigma^2 = \dfrac{\varepsilon^2}{3}$	$1/(2\varepsilon)$ for $d-\varepsilon < D < d+\varepsilon$ 0 otherwise	$\dfrac{2d}{2d + \dfrac{3d^2 + \varepsilon^2}{d^2 + \varepsilon^2} l}$
logarithmic normal $\mu = e^{\lambda + \frac{1}{2}\zeta}, \quad \sigma^2 = \mu^2 \left(e^{\zeta^2} - 1 \right)$	$\dfrac{1}{\sqrt{2\pi}\zeta D} e^{-\frac{1}{2}\left(\frac{\ln D - \lambda}{\zeta}\right)^2}$	$\dfrac{e^{3\lambda + \frac{(3\zeta)^2}{2}}}{e^{3\lambda + \frac{(3\zeta)^2}{2}} + \dfrac{3l}{2} e^{2\lambda + \frac{(2\zeta)^2}{2}}}$

STATISTICAL RESULTS ON BUBBLE TRAJECTORIES
IN TURBULENT BOUNDARY LAYER

S. TRAN-CONG, J.L. MARIÉ AND R.J. PERKINS

Laboratoire de Mécanique des Fluides et d'Acoustique
Ecole Centrale de Lyon/Université Claude Bernard - Lyon 1
UMR CNRS 5509
BP 163 - 69131 Ecully CEDEX *- FRANCE*

1. Introduction

Several experimental investigations (Serizawa *et al.*, 1975, Moursali *et al.*, 1995a) have shown that a wall void-peaking distribution can be expected in upward flows, depending on the bubble size. This is due to the deceleration of the bubbles at the surface and also to the deflection of a significant number of bubbles towards the wall (Moursali *et al.*, 1995a). Two important conclusions were reached concerning this migration. Firstly, it is mainly the small bubbles ($D_B < 4mm$) which migrate and remain at the wall. Secondly, the migrations seem to be quite random and characterized by a short timescale. The latter characteristics, together with the scaling analysis of Moursali *et al.*, suggest that the motion of migrating bubbles might be the result of their interaction with coherent structures in the boundary layer.

The objective of the present work is to complete the previous investigation by providing a more detailed experimental description of these phenomena and a better understanding of the underlying mechanisms.

2. Experimental facility and instrumentation

Experiments were carried out in a vertical upflowing boundary layer on a flat plate (Moursali *et al.*, 1995a) with velocities up to 1m/s (figure 1). At a point X=1m downstream of the leading edge, the thickness of the turbulent boundary layer was of the order of 22mm and the Reynolds number 22000. Two types of mesasurement have been performed.

As a first step, the trajectories of bubbles released at a given distance from the plate were analysed as a function of the deformation of their interface, for mean equivalent diameters over the range 1- 6mm, by taking films of the flow with a high speed video camera. This was then processed using image analysis techniques (Perkins & Hunt, 1989) to yield a statistical analysis on the bubble shapes, deformations and trajectories in the wall region. The bubble migration and injection frequencies were measured using a laser beam and an optical probe.

As a second step, a visualization of both the bubble trajectories and the large scale turbulent stuctures in the boundary layer has been performed. This was done by injecting fluorescent dye from a hole at X=0.3m, and illuminating the flow with a laser sheet.

S. Gavrilakis et al. (eds.), Advances in Turbulence VI, 573-574.
© *1996 Kluwer Academic Publishers.*

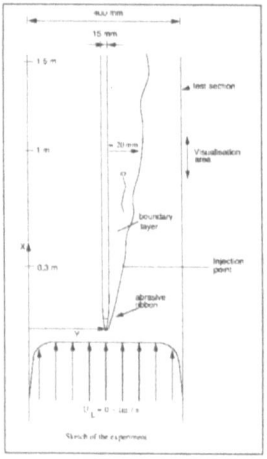

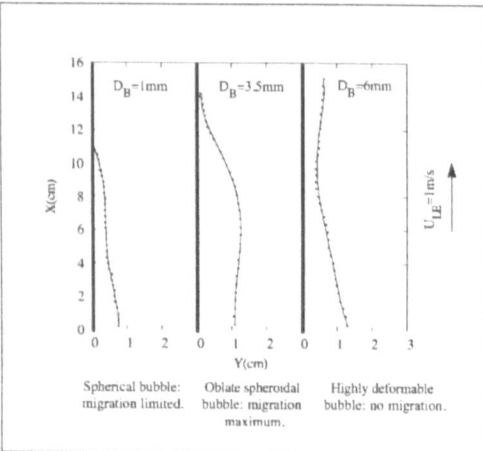

| Figure 1 | Figure 2 |

3. Qualitative results

Three distinct behaviours can be identified (figure 2). Below 1mm, the bubbles (which are roughly spherical and undeformable) scarcely migrate. Between 1 and 4mm, the migration is significant, with a maximum for 2mm diameter bubbles. This typically corresponds to bubbles having an oblate spheroidal deformable shape associated with an helicoidal trajectory. Above 4mm, deformation of the interface is important and the trajectory more complex. Indeed big bubbles are only deflected with difficulty towards the wall. As they reach the surface, they are strongly elongated by the high shear, and then finally return to the free-stream. The existence of these three behaviours has been noted by Zun, but the influence of the deformation on the bubble trajectory has been investigated only in the case of laminar shear flow (Tomiyama *et al.*, 1995).

Premilary results from the visualisation of the interaction between coherent structures and bubbles are promising, and suggest that migration takes place as the bubbles cross the turbulent bulges. But this has still to be confirmed by improving the quality of the visualisations.

References

Serizawa, A., Kataoka, I. & Michiyoshi, I. (1975) Turbulence structure of air-water bubbly flows, Part I, II & III, *Int. J. Multiphase Flow*, **Vol. 2**, pp. 221-259.

Moursali, E., Marié, J.L. & Bataille, J. (1995a) An upward turbulent bubbly boundary layer along a vertical flat plate, *Int. J. Multiphase Flow*, **Vol. 21, No 1**, pp. 107-117.

Moursali, E., Marié, J.L. & Bataille, J. (1995b) Law of the wall and turbulent intensity profiles in a bubbly boundary layer at low void fraction. Proceedings of *2nd International Conference on Multiphase Flow*, pp. I17-I21. Kyoto, Japan.

Perkins, R.J. & Hunt, J.C.R. (1989) Particle tracking in turbulent flows. In *Advanced in Turbulence 2*, (ed. Fernholz, H.H. & Fielder, H.E.), pp. 286-291. Springer-Verlag.

Zun, I. (1980) The tranverse migration of bubbles influenced by walls in vertical bubbly flow, *Int. J. of Multiphase Flow*, **Vol. 6**, pp. 583-588.

Tomiyama, A., Sou, A., Zun, I., Kanami, N. & Sakagushi, I. (1995) Effects of Eötvöes number and dimensionless volumetric flux on lateral motion of a bubble in a laminar flow, *2nd International Conference on Multiphase Flow*, **Vol. 1**, pp. PD11-PD18. Kyoto, Japan.

IX

Turbulent Mixing

ANOMALOUS SCALING EXPONENTS

OF A PASSIVE SCALAR ADVECTED BY TURBULENCE

G.FALKOVICH

Weizmann Institute

Rehovot 76100 Israel

For Kraichnan's problem of passive scalar advection by a velocity field delta-correlated in time, any simultaneous correlation function of a scalar satisfies a closed differential equation so that all common hypotheses about intermittency could, in principle, be verified by direct calculation. In an isotropic turbulence, the n-point correlation function depends on $n(n-1)/2$ distances, which makes direct solution of the respective partial differential equation quite difficult if the parameters are not at all constrained. In the limit of large space dimensionality $d \gg 1$, the anomalous exponents are small and are found by perturbation theory in $1/d$. We demonstrate how the anomalous part of the many-point correlation funstion appears as a zero mode of the operator of turbulent diffusion which exploits the the interchange symmetry between the points. We then consider passive scalar convected by multi-scale turbulent velocity field with short yet finite temporal correlations. Taking the limit of a white velocity as a zero approximation we develop perturbation theory with respect to a small velocity correlation time τ. We show that the anomalous scaling exponents of the scalar field continuously depend on τ i.e. they are nonuniversal.

The advection of a passive scalar field $\theta(t, \mathbf{r})$ by an incompressible turbulent flow is governed by the equation

$$(\partial_t + u^\alpha \nabla^\alpha - \kappa\Delta)\theta = \phi, \qquad \nabla^\alpha u^\alpha = 0 . \tag{1}$$

The external velocity $\mathbf{u}(t, \mathbf{r})$ and the source $\phi(t, \mathbf{r})$ are independent random functions of t and $\mathbf{r}$, both Gaussian and δ-correlated in time. The pair correlation function $\langle \phi(t_1, \mathbf{r}_1)\phi(t_2, \mathbf{r}_2) \rangle = \delta(t_1 - t_2)\chi(r_{12})$ as a function of the argument $r_{12} = |\mathbf{r}_1 - \mathbf{r}_2|$ decays on the scale L. The value $\chi(0) = P$ is the production rate of θ^2. The velocity field is multi-scale in space with a power spectrum. The pair correlation function $\langle u^\alpha(t_1, \mathbf{r}_1)u^\beta(t_2, \mathbf{r}_2) \rangle =$

S. Gavrilakis et al. (eds.), *Advances in Turbulence VI*, 577-580.

$\delta(t_1 - t_2)[V_0\delta^{\alpha\beta} - \mathcal{K}^{\alpha\beta}(\mathbf{r}_{12})]$ is expressed via the so-called eddy diffusivity

$$\mathcal{K}^{\alpha\beta} = \frac{D}{r^\gamma}(r^2\delta^{\alpha\beta} - r^\alpha r^\beta) + \frac{D(d-1)}{2-\gamma}\delta^{\alpha\beta}r^{2-\gamma},$$

where $0 < \gamma < 2$ and isotropy is assumed.

Considering steady state and averaging (1) over the statistics of $\mathbf{u}$ and ϕ, one gets the closed balance equation for the simultaneous correlation function of the scalar $F_{1...2n} = F(\mathbf{r}_1,\ldots,\mathbf{r}_{2n}) = \langle\theta(\mathbf{r}_1),\ldots,\theta(\mathbf{r}_{2n})\rangle$:

$$-\hat{\mathcal{L}}F_{1...2n} = F_{1...2n-2}\chi_{2n-1,2n} + \text{permutations}. \tag{2}$$

The operator $\hat{\mathcal{L}} \equiv \sum_{i,j}\mathcal{K}^{\alpha\beta}(\mathbf{r}_{ij})\nabla_i^\alpha\nabla_j^\beta/2 + \kappa\sum\Delta_i$ describes both turbulent and molecular diffusion, it may be rewritten in terms of relative distances r_{ij} [1]:

$$
\begin{aligned}
\hat{\mathcal{L}} = &\frac{D(d-1)}{2-\gamma}\sum_{i>j} r_{ij}^{1-d}\partial_{r_{ij}}(r_{ij}^{2-\gamma} + r_d^{2-\gamma})r_{ij}^{d-1}\partial_{r_{ij}} \\
&-\frac{D(d-1)}{2(2-\gamma)}\sum(r_{in}^2 - r_{ij}^2 - r_{jn}^2)\frac{r_{ij}^{1-\gamma}}{r_{jn}}\frac{\partial^2}{\partial r_{ij}\partial r_{jn}} \\
&-\frac{D}{4}\sum\frac{1}{r_{ij}^\gamma r_{im}r_{jn}}\left(\frac{d+1-\gamma}{2-\gamma}r_{ij}^2(r_{in}^2 + r_{jm}^2 - r_{ij}^2 - r_{mn}^2)\right. \\
&\left.+\frac{1}{2}(r_{ij}^2 + r_{im}^2 - r_{jm}^2)(r_{ij}^2 + r_{jn}^2 - r_{in}^2)\right)\frac{\partial^2}{\partial r_{im}\partial r_{jn}} \\
&+\kappa\sum\frac{r_{ij}^2 + r_{im}^2 - r_{mj}^2}{2r_{ij}r_{im}}\frac{\partial^2}{\partial r_{ij}\partial r_{im}}. \tag{3}
\end{aligned}
$$

Here, the summation is performed over $n(2n-1)$ independent distances (for $d > 2n - 2$) with subscripts satisfying the conditions $i \neq j$ and $m \neq i,j$, $n \neq i,j$; the diffusion scale $r_d^{2-\gamma} = 2\kappa(2-\gamma)/(D(d-1))$ has been introduced. We consider the convective interval $L \gg r_{ij} \gg r_d$ where the operator $\hat{\mathcal{L}}$ is scale invariant. The scaling exponent of $\hat{\mathcal{L}}$ is $-\gamma$; the solution of (2) may thus be presented in the form $F = F_{\text{forc}} + \mathcal{Z}$, where we separated the so-called "forced" part of the solution (with the scaling exponent $\zeta_{2n-1} + \gamma$ prescribed by the rhs) from the zero mode $\mathcal{Z}$ that may have a different scaling. They are the zero modes of the operator $\hat{\mathcal{L}}$ that are responsible for the anomalous scaling. It has been demonstrated in [1] that r_d does not appear in the leading terms of F as long as at least some of the distances r_{ij} are in the convective interval.

Let us consider now the case of large space dimensionality where the anomalous dimensions can be calculated analytically. Despite the small level of fluctuations at large d, the statistics of the scalar is substantially non-Gaussian at small scales. However, since the anomalous dimensions are small, there exists a wide interval of scales where the correlation functions are close to their Gaussian values and can be calculated by perturbation theory in $1/d$. As $d \to \infty$, the main part of $\hat{\mathcal{L}}$ is the operator of the first order $\hat{\mathcal{L}}_0 = (d^2 D/(2-\gamma)) \sum r_{ij}^{1-\gamma} \partial_{r_{ij}}$. The zeroth term in the perturbation series for $F_{1...2n}$ is given by a Gaussian reducible expression. We iterate it once by applying the operator $\hat{\mathcal{L}}_0^{-1}(\hat{\mathcal{L}} - \hat{\mathcal{L}}_0)$. In the first correction to be thus found, the logarithmic terms $\ln(L/r_{ij})$ are of interest because they appear at expanding the anomalous scaling factors $(L/r_{ij})^{\Delta_{2n}}$ over $\Delta_{2n} = n\zeta_2 - \zeta_{2n}$. The straightforward calculation shows that the main contribution to the correlation function is given by the respective zero mode with the scaling exponent $\zeta_{2n} = n\zeta_2 - 2(2-\zeta_2)n(n-1)/d$:

$$\langle (\theta_1 - \theta_2)^{2n} \rangle \sim r^{n\gamma} (L/r)^{2n(n-1)(2-\gamma)/d}. \tag{4}$$

For example, for the fourth-order correlator, the zero mode $\mathcal{Z}_4 = \sum (r_{ij}^\gamma - r_{kl}^\gamma)^2$ of the operator $\hat{\mathcal{L}}_0$ give the main contribution in the first order in $1/d$. We see that anomalous exponents are positive and grow with n so that statistics is getting more and more non-Gaussian at smaller scales.

To find the correlation functions of the scalar derivatives, one should consider some distances r_{ij} as going to zero. While some distance passes the diffusion scale the dependence on that distance changes. To describe that, we include the diffusion operator into $\hat{\mathcal{L}}$. As a result, the diffusion scale r_d appears in the correlation functions. The form of r_d dependence could be readily established for an arbitrary n, d, γ by using a straightforward perturbation expansion in the ratio between small and large distances (see Sect. III of [1]). Considering $r_{ij} \simeq R$ one gets for the correlation functions that involve the dissipation field $\epsilon = \kappa[\nabla\theta]^2$:

$$\langle \langle \epsilon_1 \theta_3 \cdots \theta_{2n} \rangle \rangle \sim R^{(n-1)\gamma} (L/R)^{\Delta_{2n}}. \tag{5}$$

In a similar way one gets [2]

$$\langle \epsilon_1^n \epsilon_2^m \rangle \sim (L/r_{12})^{\Delta_{2n+2m} - \Delta_{2n} - \Delta_{2m}} (L/r_d)^{\Delta_{2n} + \Delta_{2m}}. \tag{6}$$

The dissipation field is thus highly intermittent, the single-point means $\langle \epsilon^n \rangle \sim \langle \epsilon \rangle^n (L/r_d)^{\Delta_{2n}}$ grow unlimited when diffusivity decreases. Statistics at the convective interval is better related to local average ϵ_r over the ball with the radius r. Since spatial integration and time average commute then our knowledge of $\langle \epsilon_1 \cdots \epsilon_n \rangle$ with all distances in the convective interval allows one to obtain by spatial integration (which converges if $\Delta_4 < d$) a version of the refined similarity hypothesis valid in our case :

$$\langle (\epsilon_r)^n \rangle \sim \langle \epsilon \rangle^n (r/L)^{\mu_n}, \quad \mu_n = \zeta_{2n} - n\zeta_2 = -\Delta_{2n}. \tag{7}$$

The first $(n \ll \gamma d)$ moments of the locally averaged dissipation have $\mu_n = -n(n-1)\Delta_4/2$ so they are described by *log*-normal statistics:

$$P(\epsilon_r) \sim \exp\left(-\frac{(\ln[\epsilon_r/\langle\epsilon\rangle] - \Delta_4 \ln[L/r]/2)^2}{2\Delta_4 \ln[L/r]}\right).$$

The law (4) qualitatively corresponds to the observed behavior of ζ_n. Quantitatively, we cannot use (4) for $d = 3, \gamma = 2/3$ (which would give an overestimation of the anomalous exponents) because the validity condition of our theory $n \ll \gamma d$ will be violated already at $n = 2$. Note that quadratic dependence of ζ_n and μ_n on n (and lognormality) is violated when $n \simeq \gamma d$, while the similarity relation (7) is true for any n, γ, d.

We can now consider the velocity field

$$\langle[v^\alpha(t,\mathbf{r}) - v^\alpha(0,\mathbf{0})][v^\beta(t,\mathbf{r}) - v^\beta(0,\mathbf{0})]\rangle = 2K^{\alpha\beta}(t,r),$$

$$K^{\alpha\beta} = \frac{Dr^{2-\gamma}}{\tau_r}\left[\left(\delta^{\alpha\beta} - \frac{r^\alpha r^\beta}{r^2}\right)g_\perp(t/\tau_r) + \delta^{\alpha\beta}g_\|(t/\tau_r)\right]$$

with the finite correlation time $\tau_r = \tau_L(r/L)^z$. We assume $\epsilon = \tau_r/t_r \ll 1$ where the turnover time $t_r = (2-\gamma)r^\gamma/D\gamma d(d-1)$ so that

$$\epsilon = D\tau_L L^{-\gamma}\gamma d(d-1)/(2-\gamma) \ll 1.$$

We calculate the perturbation of the turbulent diffusion operator $\mathcal{L}$ due to ϵ and find the perturbation of the scaling exponents [3]

$$\Delta_4(\epsilon) = \Delta_4 + \frac{\epsilon(2-\gamma)}{d\gamma}(4 + 6\gamma - 2\gamma^2), \quad \Delta_{2n}(\epsilon) = n(n-1)\Delta_4(\epsilon)/2$$

Velocity's non-Gaussianity influences scalar exponents as well. We denote by ϵ_4 the ratio of the cumulant to the fourth-order correlator and consider the limit $1 \gg 1/d \gg \epsilon_4 d^3$. A direct calculation shows that the contribution to Δ_n is proportional to $n(n-1)\epsilon_4 d^2$ for $n \ll \gamma d$. Scalar exponents are thus sensitive to the details of velocity statistics.

References

1. Chertkov M., Falkovich G., Kolokolov I. and Lebedev V.: Normal and anomalous scaling of the fourth-order correlation function of a randomly advected passive scalar, *Phys. Rev. E* **52** (1995) 4924–41.
2. Chertkov M. and Falkovich G.: Anomalous scaling exponents of a white-advected passive scalar, *Phys. Rev. Lett.* **76** (1996) 2706.
3. Chertkov M., Falkovich G. and Lebedev V.: Non-universality of the scaling exponents of a passive scalar convected by a random flow, *Phys. Rev. Lett.* **76** (1996).

DIRECT NUMERICAL SIMULATION OF TURBULENT LIQUID TAYLOR VORTEX FLOW WITH AN IMMISCIBLE DROPLET

Y. HAGIWARA, M. NAKAMURA, M. TANAKA and Y. TAKASHINA

Dept. of Mech. and System Engng, Kyoto Institute of Technology
Matsugasaki, Sakyo-ku, Kyoto, JAPAN

1. Introduction

It is one of the latest arguments on turbulent dispersed multiphase flows whether turbulence is attenuated or enhanced by the dispersed phase. Various models have been proposed to explain the turbulence modulation. However, the turbulence modulation caused by the deformable dispersed phase and the deformation of the dispersed phase due to turbulence have not yet been described in these models. Therefore, none of these models are directly applicable to turbulent liquid flows with immiscible droplets.

It is crucial for the detailed description of the phase interaction to simulate the change of a specific structure of turbulence near a deforming droplet. We deal with an eddy-dominated structure of turbulence in the classification by Wray and Hunt (1990) as the specific structure, and focus on a turbulent Taylor vortex as a model of the structure.

A direct numerical simulation is at present the most appropriate method we have of simulating this kind of flow. Therefore, a direct numerical simulation is conducted for a turbulent liquid Taylor vortex flow with an immiscible droplet in the present study.

2. Computational Procedures

We dealt with the Taylor vortex flow occurring in an annulus gap between a fixed cylinder of R_O in radius and an inner cylinder of R_I ($=0.75R_O$) in radius in a stable state of rotation. The computational domain had the dimensions of $R_I \leq r \leq R_O$, $0 \leq \theta \leq 2\pi$, $0 \leq z \leq 4(R_O - R_I)$ in radial, azimuthal and axial directions, respectively.

The computational domain was divided into small cells, and the pressure and the velocity components were given at the centre and the surfaces of the cell respectively to obtain the finite difference expression for the governing equations. A total of 10 × 188 × 40 cells were used so that all the sides of the cells were almost the same dimension. The second-order consistent scheme (Suzuki and Kawamura, 1994) and the central difference scheme were applied for the finite-differencing of the convection terms and the viscous terms of the Navier-Stokes equations, respectively. The convection terms for the velocity, w, in the z direction are expressed as follows:

S. Gavrilakis et al. (eds.), Advances in Turbulence VI, 581-584.
© *1996 Kluwer Academic Publishers.*

$$u\frac{\partial w}{\partial r}\Big|_{i,j,\,k+1/2} + v\frac{\partial w}{r\partial\theta}\Big|_{i,j,\,k+1/2} + w\frac{\partial w}{\partial z}\Big|_{i,j,\,k+1/2}$$

$$=\frac{1}{2}[(u\frac{\partial w}{\partial r}\Big|_{i+1/2,\,j,\,k+1/2}+u\frac{\partial w}{\partial r}\Big|_{i-1/2,\,j,\,k+1/2})+(v\frac{\partial w}{r\partial\theta}\Big|_{i,\,j+1/2,\,k+1/2}+v\frac{\partial w}{r\partial\theta}\Big|_{i,\,j-1/2,\,k+1/2})+(w\frac{\partial w}{\partial z}\Big|_{i,\,j,\,k+1}+w\frac{\partial w}{\partial z}\Big|_{i,\,j,\,k})]$$

$$=\frac{1}{2}[(\frac{u_{i+1/2,\,j,\,k+1}+u_{i+1/2,\,j,\,k}}{2}\cdot\frac{w_{i+1,\,j,\,k+1/2}-w_{i,\,j,\,k+1/2}}{\Delta r}+\frac{u_{i-1/2,\,j,\,k+1}+u_{i-1/2,\,j,\,k}}{2}\cdot\frac{w_{i,\,j,\,k+1/2}-w_{i-1,\,j,\,k+1/2}}{\Delta r})$$

$$+(\frac{v_{i,\,j+1/2,\,k+1}+v_{i,\,j+1/2,\,k}}{2}\cdot\frac{w_{i,\,j+1,\,k+1/2}-w_{i,\,j,\,k+1/2}}{r\Delta\theta}+\frac{v_{i,\,j-1/2,\,k+1}+v_{i,\,j-1/2,\,k}}{2}\cdot\frac{w_{i,\,j,\,k+1/2}-w_{i,\,j-1,\,k+1/2}}{r\Delta\theta})$$

$$+(\frac{w_{i,\,j,\,k+3/2}+w_{i,\,j,\,k+1/2}}{2}\cdot\frac{w_{i,\,j,\,k+3/2}-w_{i,\,j,\,k+1/2}}{\Delta z}+\frac{w_{i,\,j,\,k+1/2}+w_{i,\,j,\,k-1/2}}{2}\cdot\frac{w_{i,\,j,\,k+1/2}-w_{i,\,j,\,k-1/2}}{\Delta z})] \quad (1)$$

where u and v are the velocity components in the r and θ directions, respectively. These low-order schemes are better than any other higher-order schemes for localising the effect of the velocity inside the droplet on the near-interface velocity of the continuous-phase flow. Using the successive over relaxation method, the velocity and the pressure were simultaneously adjusted so that the conservation of mass was satisfied. The third-order Adams-Bashforth scheme was used for the time integration of the discretised equations.

3. Phase Interface

The change in the droplet shape with time was obtained by tracking the position of the interface. The position of interface was determined by the fraction of the continuous-phase fluid, F ($0\leq F\leq 1$), occupying a cell. The fraction was estimated with the modified Volume-of-Fluid algorithm (Hirt and Nichols, 1981). The change in F, assigned for each cell, with time was obtained by the equation shown as follows:

$$\frac{\partial F}{\partial t} + \frac{\partial(uF)}{\partial x} + \frac{\partial(vF)}{\partial y} + \frac{\partial(wF)}{\partial z} = 0 . \quad (2)$$

The consistent scheme and the Adams-Bashforth scheme were used for the spatial derivatives and the time integration of equation (2), respectively. The interface was identified with the surface having the value of $F = 0.5$.

The average viscosity over the cell, μ_{eff}, was defined by the following equation:

$$\mu_{eff} = \mu_c F + \mu_d (1 - F) , \quad (3)$$

where subscripts c and d denote continuous-phase and dispersed-phase, respectively.

The pressure due to the interfacial tension was considered in the computational procedures. The pressure, p_σ, was determined by the product of the interfacial tension, σ, and the sum of two principal curvatures, (C_1, C_2). The curvatures were expressed by the first-order and the second-order derivatives of the distance, f, between the interface and the reference axes, ξ_1 and ξ_2, as follows:

$$p_\sigma = \sigma(C_1 + C_2) = \sigma \ [\frac{\partial^2 f_1}{\partial\xi_1^2} / \{ 1 + (\frac{\partial f_1}{\partial\xi_1})^2\} ^{3/2} + \frac{\partial^2 f_2}{\partial\xi_2^2} / \{ 1 + (\frac{\partial f_2}{\partial\xi_2})^2\} ^{3/2}] . \quad (4)$$

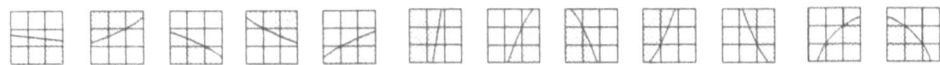

Figure 1. Twelve patterns of interface position

A 3×3 cell matrix shown in Figure 1 was utilised for deciding f in equation (4) as a quadratic function of the reference coordinate. The boundary between shaded and unshaded areas in the figure indicates the interface. For example, if the interface position corresponds to one of the five patterns from the left in Figure 1, the bottom side of the matrix is the reference axis and f is obtained as the distance from the bottom side using the values of F for the interface cells (Hagiwara et al., 1996).

4. Initial and Boundary Conditions

A database was constructed for the developed velocity field of the single-phase flow by a preliminary computation. A linear azimuthal velocity distribution in the r direction superimposed with four azimuthally-uniform vortices was given as the initial velocity field for the computation. The vortices had the velocity distribution similar to that of the Rankine vortex, and the rotation of the adjacent vortex was opposite. The results after the initial velocity field had disappeared were adopted as the database.

The nonslip condition was imposed on the cylinder walls. The periodical boundary condition was given for velocity, pressure and F in the θ and z directions.

The spherical droplet was defined by assigning the values of F to the database as the initial condition of the multiphase-flow simulation. The Reynolds number based on the annulus gap width and the azimuthal velocity of the inner cylinder, V_w, was 1500.

5. Results and Discussion

Figure 2 shows the instantaneous velocity field on the (r, z)-plane including the centre of the droplet. The contour of the energy obtained by subtracting the mean azimuthal velocity from the total energy, $k[=\{u^2 + (v - V)^2 + w^2\}/(2V_w^2)] = 0.005$, and the cross section of the droplet are shown by the fine and bold lines, respectively. Figure 2(a) indicates the result for the case where the fluid properties of the droplet are the same as those of the Taylor vortex flow at the non-dimensional time of $t^*[=tV_w/(R_O - R_I)]$ $= 0.5$. It is found in the figure that the droplet whose cross section is almost a round shape exists inside the central Taylor vortex. Figures 2(b) shows the counterpart of Fig. 2(a) at $t^*=1.5$. The deformation of the interface due to the Taylor vortex is found.

The result is shown in Figure 2(c) for the case where the viscosity of the droplet is 500 times larger than that of Taylor vortex flow. The region of $k < 0.005$ near the interface is found to become larger compared with the counterpart of Fig. 2(b). This tendency was observed throughout the present computation range. This suggests that the viscous droplet attenuates the energy of the vortex near the interface. The effect of the vortex on the deformation of the viscous droplet is found not to be noticeable.

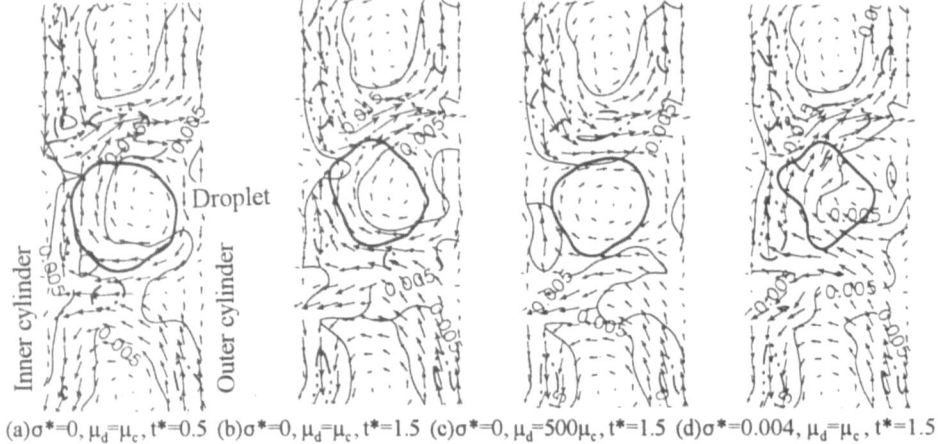

(a)$\sigma^*=0, \mu_d=\mu_c, t^*=0.5$ (b)$\sigma^*=0, \mu_d=\mu_c, t^*=1.5$ (c)$\sigma^*=0, \mu_d=500\mu_c, t^*=1.5$ (d)$\sigma^*=0.004, \mu_d=\mu_c, t^*=1.5$

Figure 2. Velocity field, contour of kinetic energy and cross section of droplet

It is found by comparing Fig. 2(d) with Fig. 2(b) that the interfacial tension, $\sigma^*[=\sigma/\{\rho V_w^2(R_O - R_I)\}] = 0.004$, causes an increase in the energy of k near the interface and a noticeable deformation of the droplet.

6. Conclusions

A direct numerical simulation was conducted for the turbulent liquid Taylor vortex flow with an immiscible droplet. The main conclusions obtained are as follows.
(1) The consistent scheme and the central difference scheme are effective for the direct numerical simulation of the turbulent Taylor vortex flow considered.
(2) The Volume-of-Fluid method is applicable to the prediction of the droplet shape.
(3) The computational results show that the energy of the vortex near the interface is attenuated by the viscous droplet.
(4) The interfacial tension is found to cause an increase in the energy near the interface and a noticeable deformation of the droplet in the present computation.

The authors acknowledge the support of the present study given through the Grant-in-Aid by the Ministry of Education, Science and Culture, Japan, No.05240212.

7. References

Hagiwara, Y., et al. (1996) Direct numerical simulation of liquid turbulent plane Couette flow with an immiscible droplet, *Proc. 2nd Int. Symp. on Numerical Methods for Multiphase Flows* (in print).
Hirt, C.W. and Nichols, B.D. (1981) Volume of fluid method for the dynamics of free boundaries, *J. Comp. Physics* **39**, 201-225.
Suzuki, T. and Kawamura, H. (1994) Consistency of finite-difference scheme in direct numerical simulation of turbulence (in Japanese), *Trans. Japan Soc. Mech. Engrs.*, **60B**, 3280-3286.
Wray, A.A. and Hunt, J.C.R. (1990) Algorithms for classification of turbulent Structures, in H.K. Moffatt and A. Tsinober (eds.), *Topological Fluid Mechanics*, Cambridge Univ. Press, Cambridge, pp.95-104.

DIRECT NUMERICAL SIMULATION OF HEAVY-GAS DISPERSION IN A PLANE CHANNEL

C. HÄRTEL AND M. MICHAUD
Institute of Fluid Dynamics, ETHZ
CH-8092 Zürich, Switzerland

AND

C. F. STEIN
Dept. of Mathematics, Chalmers University of Technology
S-412 96 Göteborg, Sweden

Abstract. A direct numerical simulation study of two-dimensional lock-exchange flows in a plane channel is conducted. The simulations are based on the Boussinesq equations where density differences are assumed to be small. The influence of the Reynolds number on the propagation speed of the gravity currents is examined. The simulation results are in good agreement with experimental reference data and theoretical findings.

1. Introduction

The dispersion of heavy-gas clouds in an environment of lighter fluid is a problem of considerable interest in natural science and engineering (see Simpson 1987). In the past numerous experimental studies were devoted to this problem which aimed at clarifying issues like the inner structure of the heavy-gas cloud, their propagation speed or the mixing with ambient fluid. In recent years also numerical simulation studies were conducted in this field, but the computational grids employed were too coarse to capture the small-scale structures of the flow (Droegemeier and Wilhelmson 1987; Klempp *et al.* 1994). Direct numerical simulations (DNS) of this problem, where all relevant physical phenomena are thoroughly resolved in space and time, have not been presented yet.

In the present project the DNS approach is for the first time applied to study fundamental physical properties of the dispersion of heavy gases. The flow under consideration is the mutual intrusion of two gas clouds of different density in a plane channel. Initially the two gases are separated by a vertical membrane. After the release of the gases a heavy-gas front

S. Gavrilakis et al. (eds.), Advances in Turbulence VI, 585-588.

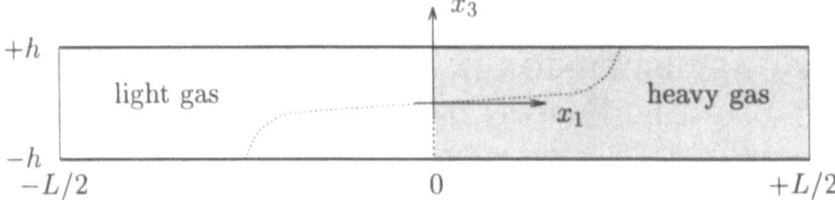

Figure 1. Principle sketch of lock-exchange flow in a channel of length L and height $2h$. The dotted line gives the interface between the two gases some time after the release.

and a light-gas front develop and propagate along the lower and the upper channel wall, respectively (see figure 1). This type of flow has extensively been studied in experiments and is often called the lock-exchange problem.

In the initial stage of our study only two-dimensional computations were conducted in order to gain a first insight into the prevailing physical mechanisms and to clarify numerical issues like the appropriate length of the computational domain and the necessary numerical resolution. The next stage of the project will be concerned with three-dimensional simulations.

2. Numerical Method, Initial and Boundary Conditions

The present simulations are based on the Boussinesq equations. In the longitudinal direction x_1 (see figure 1) the flow is assumed to be symmetric with respect to $x_1 = \pm L/2$ which allows to employ a Fourier spectral method for the spatial discretization. In the normal direction x_3 a spectral-element technique is utilized and no-slip boundary conditions are applied at the walls. For the temporal discretization of the nonlinear terms a third-order Runge-Kutta method is used together with a second-order accurate Crank-Nicolson scheme for the viscous terms and the pressure. At each discrete time step of the simulation the resulting system of coupled implicit equations is solved by an influence-matrix technique (see Canuto *et al.* 1988).

In the lock-exchange simulations the flow field is initialized with a fluid at rest. The initial temperatures of the fluid in the left and right part of the channel differ by a prescribed temperature difference ΔT, corresponding to a relative density difference between the cold (heavy) and warm (light) fluid of $\Delta \varrho / \varrho = -\alpha \Delta T$ (α being the heat-expansion coefficient). Since a spectral method is used in the longitudinal direction, a smooth transition has to be provided at the interface at $x_1 = 0$. The transition region, however, extends over a few grid points only. In all simulations the walls are assumed to be adiabatic and the Prandtl number is set to unity.

For the validation of the code we performed simulations of Rayleigh-Bénard convection (which differs from the lock-exchange problem with respect to the boundary and initial conditions for the temperature only) and compared the results with both linear-stability theory and reference data from the literature. In all cases an excellent agreement was obtained.

3. Results

In figure 2 the temporal evolution of a lock-exchange flow at a Reynolds number of $Re_b = 1118$ is illustrated by isocontours of the temperature for four successive time instants of the simulation. The Reynolds number $Re_b = U_b h/\nu$ (ν denotes the kinematic viscosity of the fluid and h is the channel half width) is the ratio of buoyancy forces and viscous forces. The reference velocity U_b is defined as $U_b = \sqrt{g\,\alpha\,\Delta T\,h}$, g being the gravitational acceleration. The simulation was performed on a computational mesh with 768 and 64 grid points in longitudinal and normal direction, respectively.

From figure 2 the steep density gradients across the heads of the fronts and the thin boundary layers that form at the walls become clearly visible. The interface between the light and heavy gas is subject to a Kelvin-Helmholtz type instability and features large vortices at later times of the simulation. Note that the light-gas and heavy-gas front are symmetrical which is due to symmetry properties of the initial conditions.

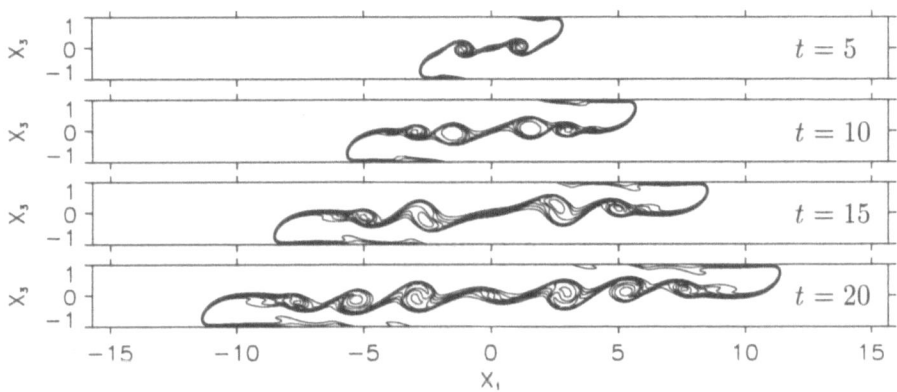

Figure 2. Isocontours of the temperature for four successive time instants t after the initial release of the gases (time non-dimensionalized by h/U_b). Results for $Re_b = 1118$.

In order to examine how the propagation speed of the front (defined as the speed of the foremost point of the front) depends on the Reynolds number, we conducted a number of simulations where Re_b was varied systematically. In the individual simulations the front speed could be observed to remain essentially constant after an initial transient of about 5 time units h/U_b. From this constant front speed U_F, say, a Froude number $Fr = U_F/U_b$ of the gravity current can be computed. The dependence of the Froude number on Re_b is shown in figure 3 where it is seen that Fr increases continuously with Reynolds number over the whole range examined. For low Reynolds numbers the front is considerably retarded by viscous forces, while for the highest Reynolds number ($Re_b = 10^4$) the Froude number is already close to the experimentally established limit of about 0.67 (Simpson 1987). Experimental results of Müller and Fanneløp (1996)

588

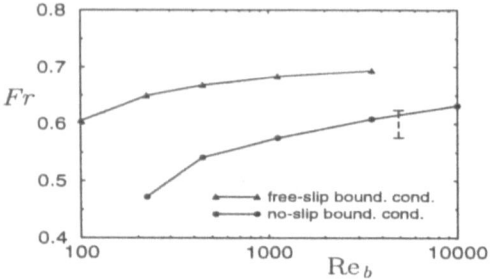

Figure 3. Froude number Fr as a function of Reynolds number Re_b (symbols identify the individual simulations). The experimental results of Müller and Fanneløp (1996) are included as a dashed vertical bar.

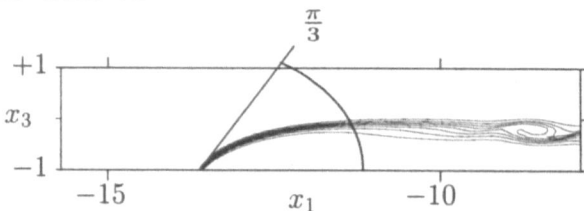

Figure 4. Isocontours of the temperature for $Re_b = 1118$. Simulation with free-slip boundary conditions. $\pi/3$ is a theoretical result for the front angle (see Benjamin, 1968).

are included in figure 3 for comparison and a good agreement between their data and ours is seen.

In addition to the simulations with no-slip boundary conditions at the walls we performed a number of computations where the walls were frictionless (free-slip boundary conditions). In this case no boundary layer develops and, as can be seen from figure 4, the foremost point of the front is located on the wall which is in agreement with established theoretical and experimental findings (Simpson 1987). The Froude numbers obtained from these simulations are included in figure 3. A comparison with the results obtained with no-slip boundary conditions reveals that the wall friction may exert a significant influence on the propagation speed of the front.

References

Benjamin, T. B. (1968) Gravity currents and related phenomena, *J. Fluid Mech.*, **31**, pp. 209–248.

Canuto, C., Hussaini, M. Y., Quarteroni, A. and Zang, T. A. (1988) *Spectral Methods in Fluid Dynamics*, Springer Verlag, Berlin.

Droegemeier, K. K. and Wilhelmson, R. B. (1987) Numerical simulation of thunderstorm outflow dynamics. Part I: Outflow sensitivity experiments and turbulence dynamics, *J. Atmos. Sci.*, **44**, pp. 1180–1210.

Klempp, J., Rotunno, R. and Skamarock, W. (1994) On the dynamics of gravity currents in a channel, *J. Fluid Mech.*, **269**, pp. 169–198.

Müller, J. and Fanneløp, T. K. (1996) Private Communication.

Simpson, J.-E. (1987) *Gravity Currents: In the Environment and the Laboratory*, Ellis Horwood Limited, Chichester.

STOCHASTIC MODELS OF DISPERSION AND MIXING IN HOMOGENEOUS TURBULENCE

B. M. O. HEPPE

DAMTP, Silver Street, Cambridge CB3 9EW, U.K.

1. Introduction

A novel approach to stochastic modelling of two-particle relative dispersion and mixing in stationary homogeneous and isotropic turbulence is proposed. The nonlinear Mori-Zwanzig projector formalism [1] exactly transforms the Navier-Stokes and passive scalar field equations in form of generalized Langevin type which might be more suitable for approximations. The forces on a fluid particle are divided into a deterministic and a random part, the former of which can be evaluated systematically whereas only little is known about the latter. Assuming the random part to be Gaussian white noise yields a much reduced description of truly non-Gaussian and non-Markovian turbulent transport.

The theory sketched in the following offers a systematic approach to statistical closure of conditional averages which are of great importance for probability density function (pdf) methods of turbulent reacting flows [2] and scalar intermittency models [3].

2. The nonlinear Mori-Langevin equation

We consider two fluid particles with relative position $\mathbf{r}(t) = \mathbf{r}(\mathbf{r}_0, t)$ initially at time $t=0$ separated by $\mathbf{r}_0$. The Lagrangian joint pdf P_L at time t of the relative particle property $\mathbf{a}(t)$ (i.e. velocity or concentration difference) conditional upon the initial separation reads

$$P_L(\mathbf{a}, \mathbf{r}, t|\mathbf{r}_0) = \langle \delta(\mathbf{a}(\mathbf{r}_0, t) - \mathbf{a})\delta(\mathbf{r}(\mathbf{r}_0, t) - \mathbf{r}) \rangle = \langle p_L(\mathbf{a}, \mathbf{r}, t|\mathbf{r}_0) \rangle \quad (1)$$

where p_L denotes the Lagrangian fine grained pdf [2] and $\langle . \rangle$ is the average over realizations of the initial velocity field. In a homogeneous and incom-

S. Gavrilakis et al. (eds.), Advances in Turbulence VI, 589-592.

pressible flow the Eulerian pdf P_E of the difference $\mathbf{a}$ between two fixed points at distance $\mathbf{r}$ is related to the Lagrangian pdf by [4]

$$P_E(\mathbf{a}|\mathbf{r}, t) = \int d\mathbf{r}_0 P_L(\mathbf{a}, \mathbf{r}, t|\mathbf{r}_0), \qquad (2)$$

with initial condition $P_L(\mathbf{a}, \mathbf{r}, 0|\mathbf{r}_0) = P_E(\mathbf{a}|\mathbf{r})\delta(\mathbf{r}-\mathbf{r}_0)$ where in a stationary state P_E is independent of time. Motivated by (2) we introduce a scalar product $(\mathbf{a}(t), \mathbf{b}(t)) = \int d\mathbf{r}_0 \langle \mathbf{a}(t)\mathbf{b}(t) \rangle$ and the conditional average

$$\langle \mathbf{b}(t)|\mathbf{a}, \mathbf{r} \rangle = \frac{(\mathbf{b}(t), p_L(\mathbf{a}, \mathbf{r}, 0|\mathbf{r}_0))}{P_E(\mathbf{a}|\mathbf{r})}. \qquad (3)$$

In order to apply the Mori-Zwanzig formalism we define a projector

$$\mathcal{P}\mathbf{b}(t) = \int d\mathbf{a} \int d\mathbf{r} \langle \mathbf{b}(t)|\mathbf{a}, \mathbf{r} \rangle p_L(\mathbf{a}, \mathbf{r}, 0|\mathbf{r}_0) \qquad (4)$$

which projects out the initial fine grained Lagrangian pdf $\mathcal{P}p_L(0) = p_L(0)$. The Mori algorithm [1] reformulates the equations of motion written in effective Liouville form (collecting $\mathbf{r}(t)$ and $\mathbf{a}(t)$ in a single vector $\mathbf{X}(t)$)

$$\frac{d}{dt}\mathbf{X}(t) = \mathcal{L}\mathbf{X}(t) = e^{\mathcal{L}t}\mathcal{L}\mathbf{X}_0 \qquad (5)$$

as generalized non-Markovian Langevin equations

$$\frac{d}{dt}X_i(t) = V_i(\mathbf{X}(t)) + \int_0^t d\tau P_E^{-1}(\mathbf{x})\partial_{x_j}m_{ij}(\tau, \mathbf{x})P_E(\mathbf{x})\Big|_{\mathbf{x}=\mathbf{X}(t-\tau)} + f_i(t). \qquad (6)$$

The three terms are, respectively, the nonlinear streaming term

$$\mathbf{V}(\mathbf{X}(t)) = e^{\mathcal{L}t}\mathcal{P}\mathcal{L}\mathbf{X}_0 = \int d\mathbf{x}\langle \mathcal{L}\mathbf{X}_0|\mathbf{x} \rangle p_L(\mathbf{x}, t|\mathbf{r}_0), \qquad (7)$$

a memory term, and a random forcing orthogonal to $p_L(0)$ at all times

$$\mathbf{f}(t) = e^{(1-\mathcal{P})\mathcal{L}t}(1 - \mathcal{P})\mathcal{L}\mathbf{X}_0 \qquad (8)$$

with vanishing mean $\langle \mathbf{f}(t)|\mathbf{x} \rangle = 0$. In a stationary state the memory $\mathbf{m}$ is related to the random forces by the fluctuation-dissipation theorem [5]

$$m_{ij}(t, \mathbf{x}) = \langle f_i(t)f_j(0)|\mathbf{x} \rangle. \qquad (9)$$

There has been no approximation so far, but in general only little is known about the stochastic force $\mathbf{f}(t)$ representing the rapid acceleration due to pressure and viscous forces.

3. The Lagrangian velocity field

We specify (6) to describe the Lagrangian velocity differences and write $\mathbf{v}(t)$ instead of $\mathbf{a}(t)$. The random force (8) is assumed to be isotropic Gaussian white noise, independent of $\mathbf{r}$ and $\mathbf{v}$ with $m_{ij}(\tau) = 2m\delta_{ij}\delta(\tau)$ which might be a reasonable assumption in the inertial range of high Reynolds number homogeneous and isotropic turbulence. The Fokker-Planck equation (FPE) for $P_L(\mathbf{v}, \mathbf{r}, t|\mathbf{r}_0)$ corresponding to (6) will also be satisfied by $P_E(\mathbf{v}|\mathbf{r})$ since the model coefficients are independent of $\mathbf{r}_0$. The FPE reduces to

$$v_i \partial_{r_i} P_E + \partial_{v_i} V_i(\mathbf{v}, \mathbf{r}) P_E = 0 \qquad (10)$$

which is known as the *well-mixed* condition for the streaming term $\mathbf{V}$. Thomson [6] determines $\mathbf{V}$ from (10) by assuming P_E to be multivariate Gaussian which leads to a dispersion model quadratic in the velocities. Pedrizzetti and Novikov [4] instead make no assumption on P_E. but derive a hierarchy of moment equations from (10) constraining their stochastic model which they assume to be linear in the velocities.

We evaluate the streaming term directly from (7). Expanding the delta function $\delta(\mathbf{v} - \mathbf{v}_0)$ in (3) in Hermite polynomials up to second order in $\mathbf{v}_0$ and using the skewness property of the Liouvillian $\mathcal{L}$ [7] we find

$$V_i = \frac{1}{2}\partial_{r_m} D_{mij} D_{jk}^{-1} v_k - \frac{1}{6}\partial_{r_m} D_{mijk} D_{jk}^{-1} + \frac{1}{6}\partial_{r_m} D_{mijk} D_{jl}^{-1} D_{kn}^{-1} v_l v_n + \dots$$

$$(11)$$

The leading order is determined by the third order structure function $D_{ijk}(\mathbf{r})$ known from the Kolmogorov equation to represent energy transfer to the small scales. The model linear in $\mathbf{v}$ has been extended to include the viscous scales. Neglecting intermittency we found qualitative agreement with direct numerical simulation (DNS) data [8]. The last two terms in (11) survive in a Gaussian velocity field and are essentially the one obtained by Thomson [6]. In conclusion, the nonlinearity entering via the Lagrangian paths is handled in a natural way. The lack of time-reversal symmetry relates relative dispersion to the skewness of the velocity field.

4. The passive scalar field

We now consider the concentration differences of a passive scalar $(\mathbf{a}(t) = \Delta\phi(t))$ with molecular diffusivity κ. For the sake of simplicity the velocity field is assumed to be delta correlated in time, which could be realized by adiabatic elimination of $\mathbf{v}$ from our dispersion model. The effect of the velocity field on the scalar is then reduced to an eddy diffusivity in the Fokker-Planck equation for $P_L(\Delta\phi, \mathbf{r}, t|\mathbf{r}_0)$ or $P_E(\Delta\phi|\mathbf{r})$. respectively [3].

592

Evaluation of the streaming term yields up to $O((\Delta\phi)^3)$

$$V_\phi(\mathbf{r}, \Delta\phi) = \frac{1}{2} \frac{\partial_{r_m}\langle v_m(\Delta\phi)^2\rangle}{\langle(\Delta\phi)^2\rangle} \Delta\phi \equiv \gamma_\phi(r)\Delta\phi \qquad (12)$$

which by use of the Yaglom equation, the equivalent of the Kolmogorov equation for a passive scalar, leads to $\gamma_\phi = \kappa(\nabla_r^2 S_2(r) - \nabla_r^2 S_2(0))/S_2(r)$. Kraichnan et al. [3] consider a similar expansion, however, without prescribing how to evaluate higher orders. They claim that in the inertial range for $\kappa \to 0$ and $r \to \infty$ only the term linear in $\Delta\phi$ survives.

The main difference between our approach and the Kraichnan theory results from the random force, which they do not consider at all. Whereas the amplitude of the scalar force $f_\phi(t)$ proportional to the viscous acceleration is likely to decrease with increasing Peclet number, it surely contributes on viscous scales where the Kraichnan theory is known to fail. Note that an external driving force sustaining stationarity can be neglected on the small scales considered here.

Allowing for a zero mean Gaussian force with correlation $\langle f_\phi(t)f_\phi(0)\rangle = 2m_\phi(r)S_2(r)\delta(t)$ the FPE reduces to a differential equation for even order scalar structure functions $S_{2n}(r) = \langle(\Delta\phi)^{2n}\rangle$

$$1/r^2\partial_r(r^2\eta_{LL}(r)\partial_r S_{2n}) = -2n(\gamma_\phi + m_\phi)S_{2n} + 2n(2n-1)m_\phi S_2 S_{2n-2}, \quad (13)$$

where $\eta_{LL}(r)$ denotes the longitudinal eddy diffusivity. If $m_\phi = 0$ (13) is exactly the Kraichnan closure predicting anomalous inertial range scaling $S_{2n}(r) \propto r^{\zeta_{2n}}$ with exponents $\zeta_{2n} \propto \sqrt{n}$ which has been supported by DNS [3]. Our generalization including the memory is still exact at second order ($n=1$) and for a Gaussian $\Delta\phi$. In case the memory contributes in the inertial range, regular scaling is restored. In the viscous convective range for large Peclet and Prantl numbers m_ϕ is still r-independent but S_2 has a plateau leading to $\zeta_{2n} = 0$ for all n. In the far viscous range, m_ϕ is constant again and the expected regular scaling follows $\zeta_{2n} = 2n$. Note that the far viscous behaviour cannot be represented by a model neglecting the memory.

References

1. Lindenberg, K. & West, B.J. (1990) *The non-equilibrium statistical mechanics of open and closed systems.* VCH Publishers, New York.
2. Pope, S.B. (1994) *Ann. Rev. Fluid Mech.* **26**, 23-63
3. Kraichnan, R. H., Yakhot, V., and Chen, S. (1995) *Phys. Rev. Lett.* **75**, 240.
4. Pedrizzetti, G. & Novikov, E.A. (1994) *J. Fluid Mech.* **280**, 69-93.
5. Grabert, H., Hanggi, P. & Talkner, P. (1980) *J. Stat. Phys.* **22**, 537-552.
6. Thomson, D.J. (1990) *J. Fluid Mech.* **210**, 113.
7. Grossmann, S. & Thomae, S. (1982) *Z. Phys. B - Condensed Matter* **49**, 253-261.
8. Yeung, P.K. (1994) *Phys. Fluids* **6**, 3416.

THE EARLY STAGES OF TURBULENT MICROMIXING

C.INNOCENTI AND E.VILLERMAUX

LEGI/IMG-CNRS, PB 53X, 38041 Grenoble Cedex, France.

The aim of this work is to investigate some fundamental aspects of small scales scalar mixing in high Reynolds numbers ($Re = u'L/\nu = 5 \times 10^3 - 1 \times 10^4$) turbulence, in the generic situation where the injection scale is small compared to the stirring (integral) scale, L. We observe the first steps of the time and space evolution of a small quantity of a scalar introduced in the far field of a water turbulent jet, where the velocity fluctuations are known to be nearly homogeneous, isotropic and constant. We inject the scalar on the centerline of the jet, 20 diameters downstream the outlet (see Fig.1) throught a small tube, whose diameter d, can be varied ($0.04 < d/L < 0.125$). In order to investigate the role played by molecular diffusion in the mixing evolution, two different types of scalar are used: disodium fluorescein in water ($Pr = 2000$) and temperature in water ($Pr = 7$). The local concentration fluctuations of the scalar are detected by an optical probe (spatial resolution $150\mu m$) for the fluorescein and by a cold film (spatial resolution $250\mu m$) for the temperature. The probe is placed at different distances x, $0 < x < L$, from the injection-tube outlet and so, using Taylor hypotesis, we can have informations about the time evolution of the mixing for $0 < t < t(L) = L/u'$.

We first focus on the concentration power spectra, $E(k)$, for wave number k lying in the inertial range $1/L < k < 1/\eta$, ($\eta \sim 100\mu m$). A quite unexpected result concerns the behavior of the hight Prandtl number case (2000), i.e. $E(k) \sim k^{-1}$. By contrast, the temperature field ($Pr = 7$) is characterized by a steeper power law, close to $k^{-5/3}$ (the same spectral law as for the velocity field, as theoretically expected from a scalar in equilibrium with the straining field, see Fig. 2). No sensitive changes in these two spectral laws are noticed by varing the parameters Re, d and the initial dye or temperature concentration, c_o.

The probability density functions of the scalar concentration c exhibit an exponential decrease, $P(c/c_o) \sim e^{-\alpha(c/c_o)}$, where $\alpha = \alpha(t) \sim t^\gamma$ and

S. Gavrilakis et al. (eds.), Advances in Turbulence VI, 593-594.
© *1996 Kluwer Academic Publishers.*

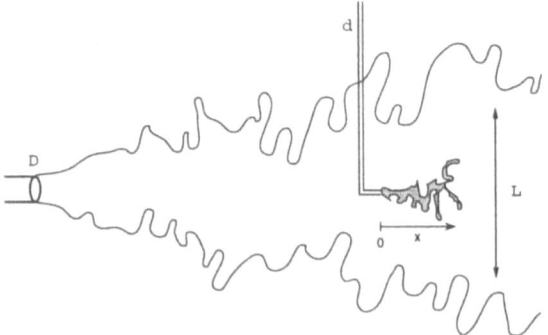

Figure 1. Sketch of the experimental device: the scalar injection tube is located at 20 D downstream of the outlet of the jet (D being the initial diameter of the jet).

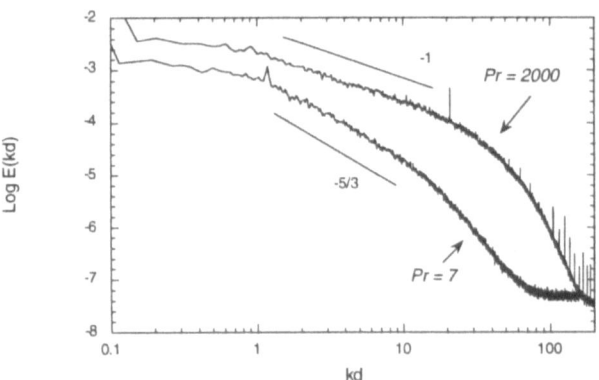

Figure 2. Scalar concentration power spectra at $x = 9d$ (corresponding to $t/t(L) = 0.3$) on the jet axis ($d/L = 0.125$, $Re = 5 \times 10^3$): above, fluorescein ($Pr = 2000$); below, temperature ($Pr = 7$).

smaller values of γ are found by increasing Pr, or decreasing d.

The use of a fluorescent dye also allows us to visualize the time evolution of the spatial organization of the scalar at $Pr = 2000$, as revealed by a planar cut of the medium. In particular, we investigate the evolution of the dye-clear water interface. Following a "blob" of dye from its formation at the outlet of the tube injector, we qualitatively show that its boundary, initially smooth, is rapidly distorted and elongated, with a subsequent increase of the surface of its interface. This evolution is quantified computing the fractal dimension d_f of the interface and the blob area to volume ratio r at different instants of time. After an initial rapid increase, d_f reaches a saturation value, while r increases continuously.

FREQUENCY SPECTRA OF SCALAR AND VELOCITY FLUCTUATIONS AT ENTRAINING STRATIFIED INTERFACES

E. KIT
Department of Fluid Mechanics and Heat Transfer,
Tel-Aviv University, Ramat-Aviv 69978, ISRAEL
H.J.S. FERNANDO and C.Y. CHING
Department of Mechanical and Aerospace Engineering,
Arizona State University, Tempe, AZ 85287-6106, USA

The present work deals with the measurement of scalar fluctuations in zero-mean-flow turbulence, so that the Eulerian spectra of scalar and velocity fluctuations, $E_b(\omega)$ and $E_w(\omega)$ can be measured. The measurements are made at and away from entraining shear-free density interface, so that scalar fluctuations in the presence and absence of buoyancy effects can be delineated. The measurements are helpful in delineating how the entrained dense fluid particles at an interface are broken down and diluted (thus loosing buoyancy) by the background turbulence. The rapidness with which the fluid particles are molecularly mixed determines the effectiveness of entrainment and overall mixing.

To achieve the above goals, we employed an oscillating grid in a two fluid system with the turbulence in the dense layer. The turbulent velocity fluctuations were measured using a two-component fiber optic laser-Doppler velocimeter (LDV) and by x-type hot film probes. The salinity fluctuations were measured by a microscale conductivity probe. The results indicate that the spectra in the region contiguous to the entraining density interface are of transitionary nature, although stationary. Near the interface beyond the layer dominated by waves, $(say, z < 0.2 L_H)$ the flow is replcted with newly entrained buoyant fluid particles of different scales which still retain their buoyancy. This region is characterized by a ω^{-3} spectra, which is characteristics of turbulence in the presence of buoyancy effects.

As the entrained fluid parcels are advected by the turbulent eddies, the scalar inhomogeneities are rapidly broken up and loose their buoyancy, thus developing a spectrum similar to that of the inertial convective region ($E_b \propto \omega^{-5/3}$) around $z / L_H \sim 0.5$; yet the viscous-convective mixing region is not fully developed and, hence, there is no ω^{-1} type spectra can be seen. As the fluid parcels move further upward, $z / L_H \sim 0.7$, the viscous-convective scales become distinct and developed, and a passive scalar spectra containing both inertial-convective ($\omega^{-5/3}$) and viscous-convective (ω^{-1}) can be noticed. At still larger $z / L_H \geq 0.8$, the molecular mixing causes the scalar inhomogeneities to disappear, and the scalar fluctuations cascading

595

S. Gavrilakis et al. (eds.), Advances in Turbulence VI, 595-596.

down the inertial-convective range are not replenished at larger scales. Thus ($\omega^{-5/3}$) region gradually disappear, but retaining a (ω^{-1}) region; it is interesting that this ω^{-1} region protrudes into wave numbers that are much smaller than the Kolmogorov frequency ω_k, and hence it should not be considered merely as a viscous-convective subrange. Hence, the present spectral measurements indicate that the buoyancy effects of entrained fluid parcels persist only up to $z/L_H \sim 0.5$.

The case of $z \sim L_H$ is shown in Fig. 1, where $E_b(\omega)$ and $E_w(\omega)$ are presented. Since $h_{pi} \sim L_H$, the probe can be considered as in a region beyond which there is no a noticeable influence of buoyancy. Since the Schmidt number was very high in the present experiments (~ 500) one can expect the existence of a ω^{-1} subrange at large ω ($> \omega_k$). However, the experiments did not show a $\omega^{-5/3}$ subrange at moderate values of ω; this is somewhat surprising since the vertical velocity spectra, measured by LDV (Fig. 1b) clearly indicates such a subrange. Interestingly enough, the direct numerical simulations of Metais and Lesieur (1992) and L.E.S. studies of Lesieur and Rogallo (1989) also indicate the same behavior as found here. In their study of turbulence in a stably stratified shear flow, Rohr et al. (1988) showed that at low Richardson numbers, such that salinity or temperature behave as a passive scalar, their spectra are proportional to $k^{-1.2}$ while velocity spectra clearly indicate a slope of -5/3.

References

Lesieur, M. and Rogallo, R. S. (1989) Large-eddy simulation of passive scalar diffusion in isotropic turbulence. *Physics Fluids A* **1**, 718-722.

Metais, O. and Lesieur, M. (1992) Spectral large-eddy simulation of isotropic and stably stratified turbulence. *J. Fluid Mech.* **239**, 157-194.

Rohr, J.J., Itsweire, E.C., Helland, K.N. and Van Atta, C.W. (1988) Growth and decay of turbulence in a stably stratified shear flow. *J. Fluid Mech.* **195**, 77-111.

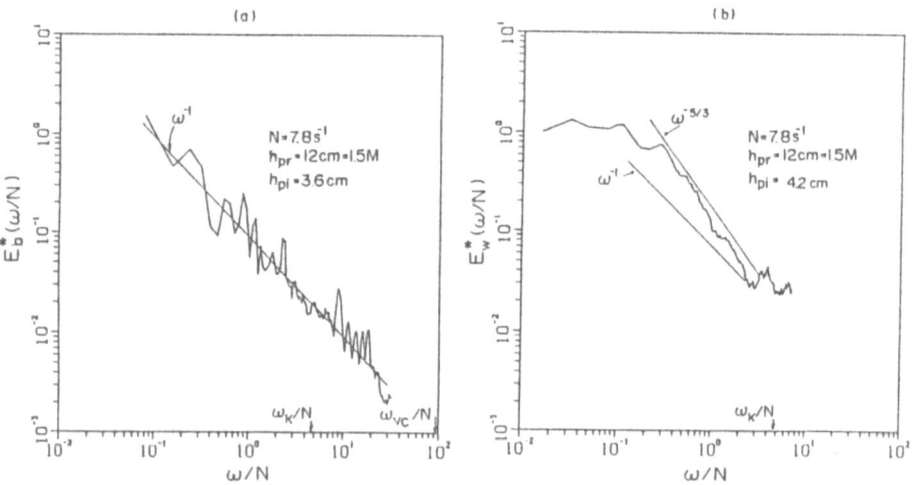

Fig. 1   Normalized salinity (a) and vertical velocity (b) spectra. h_{pr} and h_{pi} are the distances between the probe and oscillating grid and the density interface and the probe. N is the buoyancy frequency.

DISPERSION IN HOMOGENEOUS STRATIFIED FLOWS USING KINEMATIC SIMULATIONS

NADEEM A. MALIK, F.S. GODEFERD & C. CAMBON
L.M.F.A., École Centrale de Lyon, 69131 Ecully, FRANCE

1. Introduction

To overcome the low Reynolds number limitation of DNS, various simplifying models have been proposed, such as *Eulerian* closure models which yield Eulerian statisics of the model flow, such as the time-dependent second-order energy tensor $U_{ij}(\boldsymbol{k}, t)$. However, dispersion studies are easier to model in a *Lagrangian* framework. Kinematic Simulation (KS), Fung, Hunt, Malik & perkins 1992, (*J. Fluid Mech.*, **236**, 281), is a non-local Lagrangian based model in which entire flow fields with prescribed energy spectrum are generated which has the advantage of containing genuine flow structure. KS has been used to study *structural diffusion* of 1- and 2-particle relative diffusion, (Malik 1996, present conf. proc.), and to model dispersion of contaminants in marine environments, (Perkins, Malik & Fung 1993, *J. Applied. Sci. Res.*, **51**, 539). In this paper a KS model for the homogeneous stratified turbulence for the study of particle dispersion is described.

2. The Linearised Model Equations

The Boussinesque equations for the fluctuating velocity $\boldsymbol{u}$ and temperature τ fields with vertical gradient γ of mean temperature (stable for $\gamma < 0$), are

$$\left(\frac{\partial}{\partial t} + u_j \frac{\partial}{\partial x_j} - \nu\nabla^2\right) u_i + \frac{\partial p}{\partial x_i} = \beta g \tau \delta_{i3}$$

$$\left(\frac{\partial}{\partial t} + u_j \frac{\partial}{\partial x_j} - \nu\nabla^2\right) \tau = -\gamma u_3 \qquad (1)$$

The temperature diffusivity is assumed equal to the kinematic viscosity ν (Prandtl number $= 1$). β is the thermometric expansivity, and g is the acceleration due to gravity. The characteristic frequency of oscillation of the internal gavity waves that arise due to the stratification is the Brunt-Vaisala frequency $N = (\beta\gamma g)^{1/2}$.

   The treatment of these equations is made easier by reference to the Cray-Herring frame, in which an orthonormal eigenframe ($e^i(\boldsymbol{\kappa}, \theta, t)$, $i = 1, 2, 3$) associated with the *linearised* regime is readily derived, see Godeferd & Cambon (1994, *Phys. Fluids*, **6**(6), 2084).

S. Gavrilakis et al. (eds.), Advances in Turbulence VI, 597-600.
© 1996 *Kluwer Academic Publishers.*

By adding the temperature field, normalised by $\beta g/N$, parallel to the third eigen-vector $e^3 = \kappa$, we can define a complex velocity field v whose Fourier component $\hat{v}(\kappa, t)$ is given by

$$\hat{v}(\kappa, t) = \phi^1(\kappa, t)e_i^1(\kappa) + \phi^2(\kappa, t)e_i^2(\kappa) + I\phi^3(\kappa, t)e_i^3(\kappa) , \qquad (2)$$

where and $I^2 = -1$, and ϕ^1 and ϕ^3 are the energy spectral densities in their respective directions. The solenoidal part of v gives the actual velocity field u. The spectral density of the total energy (kinetic + potential) is equal to

$$\frac{1}{2}\hat{v}_i^*\hat{v}_i = \left(\frac{1}{2}\hat{u}_i^*\hat{u}_i + \frac{1}{2}\left(\frac{\beta g}{N}\right)^2 \hat{T}^*\hat{T}\right).$$

The two-point velocity-temperature correlation tensor Φ_{ij} is such that $\Phi_i(\kappa, t) = \langle\hat{\phi}_i^*\hat{\phi}_i\rangle/2$, and $\Psi(\kappa, t) = \Phi_{23}^* = \Phi_{32} = \langle\hat{\phi}_2^*\hat{\phi}_3\rangle/2$, whose integration in the vertical direction gives the vertical heat flux $\frac{1}{2}\langle u_3\tau\rangle = -\frac{N}{\beta g}\int \Psi_R(\kappa) \sin\theta_\kappa \, d^3\kappa$ – the 'R' denotes the real part. Φ_1 is the spectrum of kinetic energy contained in the horizontal vortical motions; Φ_2 is that of the wave motions, and Φ_3 is that of the potential energy of the wave motions.

In the linearised limit, the dynamical equations of motion yield an explicit expression for the Fourier components $\hat{v}(\kappa, t)$ from an initial $\hat{v}(\kappa, t_0)$ at time t_0 via a Greens' operator $G(\kappa, t, t_0)$, (Godeferd & Cambon 1994), viz

$$\hat{v}_i(\kappa, t) = G_{ij}(\kappa, t, t_0)\hat{v}_j, \qquad (3)$$

$$G_{ij}^\epsilon(\kappa, t) = \sum_{\epsilon\in\{0,\pm1\}} N_i^\epsilon N_j^\epsilon e^{I\epsilon\sin\theta_k Nt} . \qquad (4)$$

where $N_i^\epsilon(\kappa) = \left(e_i^{|\epsilon|}(\kappa) + \epsilon e_i^3(\kappa)\right)/\sqrt{|\epsilon|+1}$ is the eigenmode associated with the eigenvalue $-\epsilon I N e_3^2(\kappa) = \epsilon I N \sin(\theta_k)$. In equation (4) there appears the phase (when multiplied by the Fourier mode $e^{I\kappa\cdot x}$) $\phi_w = \kappa \cdot x \pm N \sin\theta_k t$ of the gravity wave of frequency $\omega_k = N \sin\theta_k$.

3. The KS Model

The mathematical formulation of the KS flow field for isotropic turbulence is given in detail in Fung *et. al.* 1992. The stratified KS model solves first for the *complex* velocity velocity field whose Fourier modes, $\hat{v}_{mn}(x, 0)$ at initial time $t = 0$ are

$$\hat{v}_{nm}(x, 0) = \sum_{j=1}^3 \left(A_{nm}^j e^{I\chi_{nm}^j}e_{nm}^j\right)\exp(I\kappa_{nm}\cdot x + I\omega_{nm}t) \qquad (5)$$

where ω_{nm} are the frequencies due to the intrinsic unsteadiness of eddying motions. χ_{nm}^j, $j = 1, 2, 3$, are random phases selected between $[0, 2\pi]$. $\hat{v}_{nm}(x, t)$, at later time t, is obtained from $v_{nm}(x, 0)$ from equations (3) and (4) using the Green's operator G_{ij}^{nm}. Thus, the phase in the Fourier modes contains an additional time-varying term $\exp(\pm IN \sin\theta_k t)$ due to the internal waves in the linearised regime. The actual velocity field $u(x, t)$ is then obtained by a summation of each of the Fourier components $(\hat{v}_{mn})$ projected onto the plane orthogonal to the vector e_3^{mn}, and the normalised temperature is the sum of components projected parallel to e_3^{mn}.

The spectra that determine the A_{nm}^j initially must be specified. Physically meaningful spectra must come from established theory or from computed spectra from more fundamental models such as DNS or closure models such as EDQNM2 (Orsag 1970, Godeferd & Cambon 1994). The random phases χ_{mn}^2 and χ_{mn}^3 are adjusted in the KS model so that it yields the correct value for the vertical heat flux(which must also be given at the initial time $t = 0$) so that the velocity-temperature fields are correctly correlated.

Fluid particle trajectories are computed by integrating the particle's Lagrangian velocity $u^L(t) = u(x, t)$ using a fourth order Adams-Bashforth predictor-corrector method. Lagrangian statistics of particle motion are obtained from an ensemble of trajectories from different realisations of the flow field.

4. Results

We present results from a set of data obtained from the EDQNM2 closure model of Godeferd & Cambon 1994 for a stratification of $N = 16\pi \ s^{-1}$. The spectral data is in the form of $N_k = 37$ wavenumber shells, distributed on $N_\theta = 19$ discrete angles between $[0, \pi/2]$, figure 1. A consistency check of the effects of the stratification on the model is a plot of the energy in the energy modes as a function of time; figure 2 shows that the potential and kinetic energies oscillating in anti-phase as they exchange energy.

Figure 3 shows the log-log plot of one-particle diffusion against time t. The solid line is the horizontal dispersion $\langle (x - x_0)^2 + (y - y_0)^2 \rangle / 2$, and the dotted line is the vertical dispersion $\langle (z - z_0)^2 \rangle$. The dispersion follows the Taylor diffusion laws; the $\langle X^2 \rangle \sim \langle u^2 \rangle t^2$ law for small times $t \ll T^L$; and $\langle X^2 \rangle \sim Dt$ for long times $t \gg T^L$, where T^L is the Lagrangian (decorrelation) time scale, and the diffusivity $D = \langle u^2 \rangle T^L$. The difference in the horizontal and vertical diffusion reflects the anisotropy in the flow: $\langle (u_z)^2 \rangle / \langle (u_x)^2 \rangle \approx 0.86$; $D_z / D_x \approx 0.46$. This gives $T_z^L / T_x^L \approx 0.5$. Pearson, Puttock & Hunt, (1983, J. Fluid Mech. **129**, 219), have conjectured on the basis of a stochastic model that if the stratification is sufficiently high, i.e. $NT \gg 1$, where T is the Eulerian time scale of the flow, then the

600

vertical dispersion will be confined by buoyancy forces to a finite spread, i.e. $\langle z^2 \rangle / L_z^2 \to$ constant as $t \to \infty$. Our KS model suggests that we need $NT > NT^L > 16$ for this to happen.

By the time of the conference we hope to present results for the dispersion of inertial particles.

This work was financied by the E.E.C. under contract number ERBCGBI-CT94-1230. The present address for NAM is D.A.M.T.P., University of Cambridge, CB3 9EW, England. E-mail N.A.Malik@damtp.cam.ac.uk.

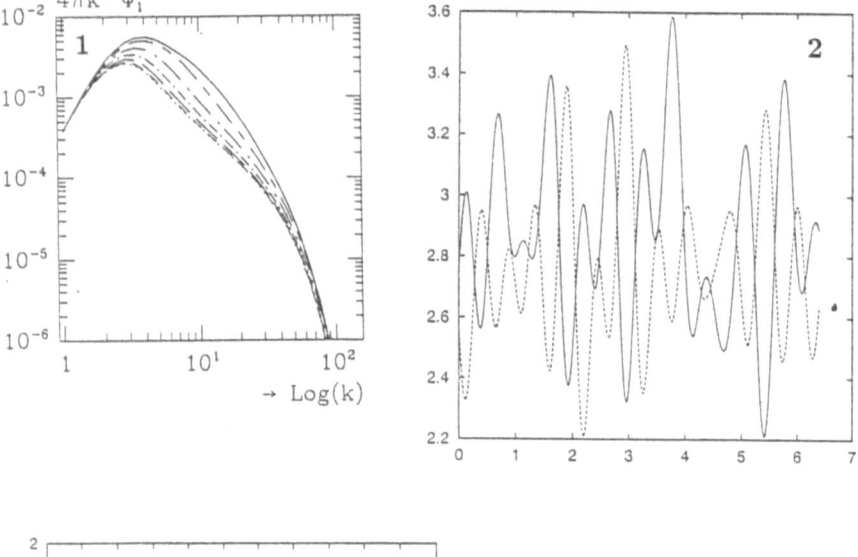

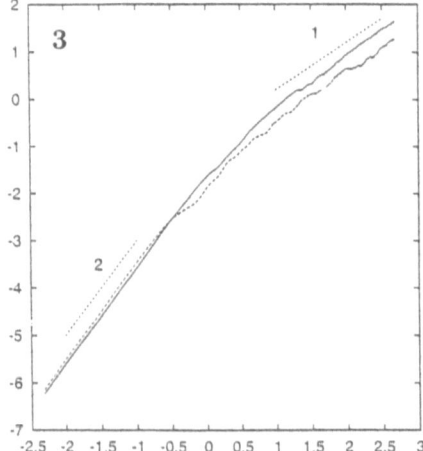

Figure 1, the spectra for Φ_1 given by EDQNM2 and used in the KS model. **Figure 2**, Oscillating exchange between the kinetic (solid line) and potential (dashed) energyies. **Figure 3**, Log-log of $\langle x^2 + y^2 \rangle / 2$ (solid line) and $\langle z^2$ against thew time Nt, $N = 16\pi$.

COHERENT STRUCTURES AND SCALES OF LAGRANGIAN TURBULENCE

V. MELESHKO, T. KRASNOPOLSKAYA, G.W.M. PETERS AND
H.E.H. MEIJER
Eindhoven University of Technology,
Postbus 513, 5600 MB Eindhoven, The Netherlands

1. Introduction

The objective of this paper is to provide qualitative and quantitative understanding of coherent structure formation in Lagrangian turbulence for time periodic two-dimensional motion of incompressible viscous fluid in a bounded domain, and study their scales using the basic notions of statistical mechanics. Our approach to investigate the coherent structures and scales of Lagrangian turbulence is based on an analysis of Eulerian and Lagrangian two-dimensional topology of creeping flow.

For this purpose an analytical solution for the Eulerian velocity field is constructed providing an accurate description of the steady streamline patterns in the cavity. Based upon the analytical solution we developed a numerical algorithm of a contour line tracking which permits us to follow the deformation process of an initial circular dye blob under considerable (up to 10^3 times) increase in the blob boundary. When the flow is such that one viscous fluid is spreading inside a cavity occupied by another viscous fluid, without destroying the continuity of the two fluid domains, the contour line plays an important role as the interface between the two fluids. Our algorithm preserves the basic Lagrangian topological properties of the flow, such as connectedness, compactness, nonselfintersection, Poincaré – Bendixon index. It also conserves the area (volume) of the blob during any topological transformations, while the traditional technique based upon representation of the blob interface as an accumulation of uniformly distributed points, due to exponential divergence of neighbouring points, can provide only qualitative picture of mixing after few periods and do not conserve its topological properties.

S. Gavrilakis et al. (eds.), Advances in Turbulence VI, 601-604.

2. Mathematical Formulation and an Analytical Solution

The model of two-dimensional Stokes flow in a curved wedge cavity $a \leq r \leq b$, $|\theta| \leq \theta_0$ is applied to studying the motion of viscous fluid under generation by a "blinking" periodical tangential movement of the curved walls. The velocity field is given in terms of the streamfunction $\Psi(r, \theta, t)$ by

$$u_r = \frac{1}{r}\frac{\partial \Psi}{\partial \theta}, \qquad u_\theta = -\frac{\partial \Psi}{\partial r}, \tag{1}$$

where Ψ satisfies the biharmonic equation

$$\nabla^2 \nabla^2 \Psi = 0. \tag{2}$$

The boundary conditions are written as:

$$u_r = 0, \quad u_\theta = V_{bot}(\theta, t) = V_a(t), \quad r = a, \quad -\theta_0 \leq \theta \leq \theta_0, \tag{3}$$

$$u_r = 0, \quad u_\theta = V_{top}(\theta, t) = V_b(t), \quad r = b, \quad -\theta_0 \leq \theta \leq \theta_0, \tag{4}$$

$$u_r = 0, \quad u_\theta = 0, \quad \theta = \pm\theta_0, \quad a \leq r \leq b. \tag{5}$$

The basic idea of the analytical method used here, namely a superposition method, is to use the sum of two ordinary Fourier series on the complete systems of trigonometric function on r and θ coordinates for the stream function Ψ:

$$\Psi = r \sum_{n=1}^{\infty} \frac{1}{\beta_n}\left(G_n \frac{\sinh \beta_n \theta}{\sinh \beta_n \theta_0} \sin \theta + H_n \frac{\cosh \beta_n \theta}{\cosh \beta_n \theta_0} \cos \theta \right) \sin \beta_n \tau + \tag{6}$$

$$\sum_{m=1}^{\infty} \left[A_m \left(\frac{r}{b}\right)^{\alpha_m} + B_m \left(\frac{r}{b}\right)^{\alpha_m+2} + C_m \left(\frac{a}{r}\right)^{\alpha_m} + D_m \left(\frac{a}{r}\right)^{\alpha_m-2} \right] \cos \alpha_m \theta$$

with

$$\alpha_m = \left(m - \frac{1}{2}\right)\frac{\pi}{\theta_0}, \quad \tau = \ln\frac{r}{a}, \quad \beta_n = \frac{n\pi}{\tau_0}, \quad \tau_0 = \ln b/a.$$

The general solution Ψ contains six sets of arbitrary coefficients $A_m, B_m, C_m, D_m, G_n, H_n$ that permit to satisfy the six boundary conditions. Fulfilling all boundary conditions leads to the infinite system of linear algebraic equations like

$$X_n = \sum_{m=1}^{\infty} a_{nm} Y_m + b_n, \ n = 1, 2, ..., \quad Y_m = \sum_{n=1}^{\infty} c_{mn} X_n + d_m, \ m = 1, 2, ...,$$

with respect to the coefficients X_n, Y_m for the Fourier series on the complete trigonometric functions on the correspondinding intervals.

New principal result of our approach to solving of the infinite system is the asymptotic law

$$X_n = A + \frac{C}{n} + \mathrm{Re}\left(\frac{D}{n^{\lambda_1}}\right), \ n \to \infty, \quad Y_m = B + \frac{C}{m} - \mathrm{Re}\left(\frac{D}{m^{\lambda_1}}\right), \ m \to \infty,$$

with $\lambda_1 = 2.740 + \mathrm{i}1.119$ – the first root of equation $\sin\frac{\pi\lambda}{2} + \lambda = 0$ with $\mathrm{Re}\lambda > 0$, and A, B, C, D are some nonzero constants depending on the boundary velocity distibution (Krasnopolskaya et al. (1996)). That law permits both to considerably improve calculations and to describe analytically the local structure of Goodier (1934) –Taylor (1962) solution near the apex and the infinite sequence of Moffatt (1964) corner eddies.

3. Coherent Structures and Scales

The coherent structures of an initial circular dye blob deformation are revealed based on properties of fixed points which do not change in course of time, such as definite locations, types (elliptic or hyperbolic) and eigen values of periodic points. A method for finding all periodic points is developed and a detailed classification of periodic points is provided. Structures of unstable manifolds corresponding to hyperbolic periodic points of different orders are analyzed.

In Fig.1 four quater cavities in which $a = 0.5b$ and $\theta_0 = \pi/4$ are shown. Results of deformation of the initial circular dye blob inside every cavity are presented for a different location of the blob centre for the same protocol of motion. Centres of the blob correspond to hyperbolic point of first order (the upper cavity), to hyperbolic point of second order (the lower cavity) and to elliptic point of second order (left and right cavities). The smallest stretching corresponds to the best mixing (in the upper cavity) when the unstable manifold of hyperbolic point is more uniformly distributed inside the cavity than in other cases. The deformation process is controlled by unstable manifolds of one hyberbolic point of fourth order and three points of sixth order in the left and right cavities, so the stretching there is the biggest. The unstable manifolds serve as skeletons which form the main coherent structures of deformed four blobs in Fig.1. There exists, however, another structure connected with some small "rubber" region which is not destroyed at all. The origin of this coherent structure can be explained in terms of existence of the elliptic periodic points. There is one "rubber" in the middle of the upper cavity, created by three elliptic points of different periods located inside initial circular dyed blob.

Existence of the topological invariances of the blob deformation allows to use the "coarse grain density" conception of the Gibbs statistical mechanics and to introduce intensity and scales of segregation concepts for

Figure 1. Structures of flow in quater cavities after 10 periods (upper and lower devisions), after 9 periods (the left part) and after 9.5 periods (the right one).

quantitative estimates of the "goodness of (mechanical) mixing". Numerical data are presented showing the change in time of the statistical values of the square density, entropy, intensity and scales of segregation to their known values corresponding to the uniform ideal mixing.

References

Goodier, J.N. (1934) An analogy between the slow motion of a viscous fluid in two dimensions, and systems of plane stress, *Phil. Mag.* (ser.7), **17**, 554–576.

Krasnopolskaya, T.S., Melcshko, V.V., Peters, G.W.M., and Meijer, H.E.H. (1996) Steady Stokes flow in an annular cavity, *Q. J Mech. Appl. Math.*, **49**.

Moffatt, H.K. (1964) Viscous and resistive eddies near a sharp corner, *J. Fluid Mech.*, **18**, 1–18.

Taylor, G.I. (1962) On scraping viscous fluid from a plane surface, in M.Shäfer (ed.), *Miszellangen der Angewandten Mechanik*, Akademie - Verlag, Berlin, pp. 313–315.

VERTICAL MICROSTRUCTURE AND MIXING IN STRATIFIED FLOWS

J.M. REDONDO

Dept. Física Aplicada, U.P.C.
Barcelona 08034, Spain.

1. Introduction

A large part of the mixing processes that take place in geophysical situations have strong temporal and spatial variation of energy inputs, typical examples are sea and mountain brezes, tidal stirring and coastal mixing by waves as well as many industrial applications.

Several laboratory experiments have been made in order to study the vertical structure and mixing across sharp and linear density interfaces, see Fernando(1991) for further references. The laboratory experiments tend to fall into four categories: (i) Those where the mixed layer is driven over the non-turbulent layer thus creating a mean velocity shear. (ii) Those where the turbulence is generated by a horizontally or vertically oscillating grid or rod system and hence exhibit no mean velocity shear. (iii) Those produced by a sudden energy release, as in the dropping of a grid, where the turbulence is decaying and goes through a continuous change of scales. (iv) Those where the turbulence is generated by convective heating or cooling at a surface.

Phillips(1972) and Posmentier(1997) argued that a mechanism for producing sharp interfaces would take place if for high density gradients the vertical mass fluxes diminished, then large gradients would support smaller fluxes and the density interfaces would become sharper. The behaviour of the Flux Richardson number, R_f or mixing efficiency versus the gradient Richardson number, Ri was used by Linden (1979,1980) to describe a variety of experiments wich showed a maximum in $Rf(Ri)$. We will only consider here mixing in zero mean flows.

S. Gavrilakis et al. (eds.), Advances in Turbulence VI, 605-608.
© *1996 Kluwer Academic Publishers.*

2. Oscillating Grid and Rod Experiments

For oscillating grid experiments Turner (1968,1973) proposed that the entrainment velocity V_e, defined as $V_e = dD/dt$, where D is the depth of the turbulent layer, is given by a simple law of the form $E \propto Ri^{-n}$. where E, the entrainment rate is defined as $E = V_e/V$, V being some global or local reference velocity. The relevant Richardson number in terms of local parameters is:

$$Ri = \frac{g \frac{\partial \rho}{\partial z} \ell^2}{\rho \ u'^2} = \frac{g \ \Delta\rho \ \ell}{\rho \ u'^2} , \qquad (1)$$

for linearly or stepwise stratified flows, where $\Delta\rho$ is the buoyancy jump across a density interface, u' is the r.m.s turbulent velocity and ℓ is an integral length scale of the turbulence.

Turner (1968) found that the value of n was $3/2$ when the stratification was due to salt. When the density-stratification resulted from a temperature gradient, the value of n was found to be close to 1. Using $R_f = E Ri$ together with the experimental values of E gives $R_f \propto Ri^{1-n}$, and we can see that for salt experiments, there is a decrease in mixing efficiency with increasing Ri, while for heat, the mixing efficiency should be constant.

The thickness of the interfacial layer h both in heat and in salt-stratified experiments was determined by Crapper & Linden (1974), who used both a travelling conductivity probe and shadowgraph techniques. They found in their experimental range that for salt (high Pe) h/ℓ seemed to be independent of Ri and given by $\frac{h}{\ell} \approx 1.5$.

In experiments involving heat as a stratifying agent, Crapper & Linden (1974) found that molecular diffusion is important in the determination of the interfacial structure at $Pe \leq 200$. In this case they found that a diffusive core is formed at the centre of the interfacial layer, across which most transport occurs by molecular diffusion. Measurements made by Fernando & Long (1985) using a travelling conductivity probe, suggest that at low Ri an instantaneous density profile is not a good indicator of the average profile. In the literature there is some confusion regarding the definition of the thickness of the interfacial layer as well as on the shape and structure of the density profile. Some of these discrepancies might be partly due to the experimental methods employed. The use of a shadowgraph might be misleading in determing instantaneous values of h because the method averages over the entire width of the experimental facility, but it has the advantage of being a good global indicator of overall behavior and may be used to investigate in a simple fashion the pattern selection mechanisms in a lineraly stratified flow. Moreover, a travelling conductivity probe may produce some perturbations as it penetrates the density interface. Laser induced fluorescence (LIF) sheds new light on the instantaneous structure of the turbulence but it is not easy to mark linear stratified flows, and it is usefull only in the investigation of premixed sharp density interfaces, Hannoun & List (1988), Redondo (1995). LIF measurements show that the instantaneous interface is much sharper and depends on Ri.

The stratification, whether linear or produced by a sharp interface defines several overturning or buoyancy scales. When the integral turbulence scale ℓ and the buoyancy scale are of a similar size the stratification stops further vertical displacements of fluid and internal waves are usually generated. Ozmidov's lengthscale is $l_o = \left(\frac{\epsilon}{N^3}\right)^{1/2}$, where ϵ is the mean rate of viscous dissipation and N is the buoyancy frequency $N = (-\frac{g}{\rho}\frac{\partial\rho}{\partial z})^{1/2}$. Any motion larger than this scale would, in principle, not be able to overturn and would be restricted to wave-like motions.

There is some difference in the amount of internal wave generation between the vertically and the horizontally oscillated grids, these also produce lateral intrusions as in Browand et al.(1987) and Redondo & Cantalapiedra(1993). The use of horizontally oscillating rods has been used by Ruddick et al.(1989) showing little generation of internal waves.

3. Experimental Description and Results

A box made of 1 cm perspex plate, 0.255×0.255 m in base and 25 cm in height was used, after filling up to 18 cm of the box with a linear stratification using the two tank method. Heat, salt and sugar were used as stratifying agents, forming a central region of the tank with constant N. The experiments covered the range $N = 0.01 rad\ s^{-1}$ to $N = 1 rad\ s^{-1}$. A microconductivity probe described in Redondo(1987) with a DANTEC micropositioner were used to measure the evolution of the density profiles.

The turbulence in the midst of the tank was produced by oscillating laterally an array of vertical bars 0.003 m in diameter separated 0.03 m. The frecuency of oscillations ω was kept greater than N in order to avoid the radiation of internal waves. The amplitude of the oscillation was 0.015 m, producing localized mixing as in the the experiments of Browand et al. (1987).

For the mixing process we obtain an expression for the buoyancy flux as a function of n and ω as

$$B = \frac{g}{\rho}\overline{w'\rho'} \propto \omega^{(1/2-2n)} , \tag{2}$$

where n varies with the type of solute used, reflecting their respective dif fusivities κ_ρ.

The evolution of the microlayer formation as the turbulent decays is different for the different solutes, showing vertical scales between the rod diameters and the Ozmidov scale, which are diffused on a timescale of $l_o^2 \kappa_\rho^{-1}$. The layer thickness h and their velocity after a time N^{-1} were used to evaluate B, see table 1.

TABLE 1. Effect of solute

Solute	$\kappa_\rho(m^2 s^{-1})$	n	h/ℓ
Heat	1.4×10^{-7}	1	2.9
Salt	1.1×10^{-9}	3/2	1.4
Sugar	0.8×10^{-10}	5/3	0.8

References

Browand, F.K., Guyomar, D. and Yoon, S.C. (1987) The Behavior of a Turbulent Front in a Stratified Fluid: Experiments With an Oscillating Grid, *J. Geophysical Res.*, **92**, pp.5329-5341.

Crapper P.E. and Linden, P.F. (1974) The structure of density interfaces, *J. Fluid Mech.* **65**, pp. 45-64.

Fernando, H.J.S. and Long, R.R. (1985) On the nature of the entrainment interface of a two layer fluid subjected to zero-mean-shear turbulence, *J. Fluid Mech.* **151**, pp. 21-53.

Fernando, H.J.S. (1991) Turbulent mixing in stratified fluids, *Ann. Rev. Fluid Mech.* **23**, pp. 455-493.

Hannoun, I.A. and List, J.E. (1988) Turbulent mixing at a shear free density interface, *J. Fluid Mech.* **189**, pp. 209-227.

Linden, P.F. (1979) Mixing in stratified fluids, *Geophysical and Astrophysical Fluid Dynamics* **13**, pp. 3-23.

Linden, P.F. (1980) Mixing across a density interface produced by grid turbulence, *J. Fluid Mech.* **100**, pp. 691-703.

Phillips O.M. (1972) Turbulence in a strongly stratified fluid - is it unstable?, *Deep Sea Research* **19**, pp. 79-81.

Posmentier E.S. (1977) The generation of salinity finestructure by vertical diffusion *Journal of Physical Oceanography* **7**, pp. 292-300.

Redondo J.M. (1987) *Difusion turbulenta en fluidos estratificados*, Ph.D. Thesis. Univ. Barcelona.

Redondo J.M. and Cantalapiedra I.R. (1993) Mixing in horizontally heterogeneous flows, *Applied Scientific Research* **51**, pp. 217-222.

Redondo J.M. (1995) Turbulent mixing in the atmosphere and ocean, *Fluid Physics*, Ed. M.G. Velarde and C.I. Cristov, World Scientific, pp. 584-597.

Ruddick B.R., McDougall T.J. and Turner J.S. (1989) The formation of layers in a uniformly stirred density gradient, *Deep-Sea Research* **36** , pp. 597-609.

Turner J.S. (1968) The influence of molecular diffusivity on turbulent entrainment across a density interface, *J. Fluid Mech.* **33**, pp. 639-656.

Turner, J.S. (1973) *Buoyancy effects in fluids.* Cambridge University Press, Cambridge.

ANALYTICAL SOLUTION OF THE CONCENTRATION PROBABILITY DENSITY EQUATION IN DECAYING GRID–GENERATED TURBULENCE WITH A MEAN SCALAR GRADIENT

V.A. SABELNIKOV

Central Aero-Hydrodynamic Institute (TsAGI),
Zhukovsky, Moscow Region, 140160, Russia

Abstract

Self–similar closed–form solution of the equation for the one–point probability density function (pdf) of a passive scalar in decaying grid–generated turbulence with a uniform mean cross–stream scalar gradient is obtained. According to this solution the pdf is a function of the conditional expectations of the scalar dissipation and the transverse velocity. An exact expression that relates the conditional expectation of molecular diffusion to the conditional expectation of transverse velocity is derived.

Introduction

The impetus for this study stems from the recent experiment [1] and the subsequent theoretical work [2] which deal with the one–point probability density function (pdf) of a passive temperature fluctuations, with a uniform mean temperature gradient, in decaying grid–generated turbulence. As a solution for the one–point scalar pdf equation for a uniform mean scalar gradient case is still lacking, authors [1] compared their measured scalar pdf and conditional expectation of scalar dissipation with the asymptotic solution [3] in the absence of a mean scalar gradient. This asymptotic solution has a self–similar shape (time $t \to \infty$) for the scalar pdf equation and it is a unique function of the conditional expectation of scalar dissipation. In [1] the pdf was calculated from the asymptotic solution [3] by substituting in it the measured conditional expectation of scalar dissipation. The pdf thus calculated, as in [1] is stated, is in remarkably good correspondance with the measured pdf. Authors [1] claimed that observed fit may be somewhat fortuitous since, as they emphasized, the asymptotic solution [3] is for no mean gradient case. The above mentioned problem was further theoretically adressed in [2]. They remarked that the one–point scalar pdf equation contains, in the general case, two unknown statistical functions, both conditioned on the scalar value. The first is the conditional expectation of scalar dissipation, the second quantity of equal interest is the conditional expectation of velocity. The importance of modeling of the conditional expectation of velocity to describe the evolution of the scalar pdf in shear turbulent flows was also stressed in [4,5].

In the present work the self–similar solution of the equation for the pdf of a scalar, with a uniform mean scalar gradient, in decaying grid–generated turbulence is obtained. This solution is the generalization of the asymptotic solution [3] for the case with no mean gradient.

S. Gavrilakis et al. (eds.), Advances in Turbulence VI, 609-612.
© *1996 Kluwer Academic Publishers.*

Governing equations

The transport equation for the one–point pdf of a passive scalar, in decaying grid–generated turbulence with a mean cross–stream scalar gradient, neglecting molecular diffusion in physical space and the stream–wise turbulent flux term, is given by (see, e.g. [2,4,6])

$$u_0 \frac{\partial P}{\partial x} + \frac{\partial}{\partial y} \langle v|c \rangle P = -\frac{\partial^2}{\partial c^2} \langle N|c \rangle P, \tag{1}$$

where $P(c, x, y)$ is the pdf of a passive scalar c ; u_0 is the mean stream–wise velocity; x and y are stream–wise and cross–stream coordinates respectively; $\langle v|c \rangle$ and $\langle N|c \rangle$ are the conditional expectations of transverse velocity v and of scalar dissipation $N = D(\nabla c)^2$ respectively; D is the molecular diffusivity.

Eq. (1) can alternatively be written as (e.g. [4,6])

$$u_0 \frac{\partial P}{\partial x} + \frac{\partial}{\partial y} \langle v|c \rangle P = -\frac{\partial}{\partial c} \langle \omega|c \rangle P, \tag{2}$$

where $\langle \omega|c \rangle$ is the conditional expectation of molecular diffusion, $\omega = D \nabla^2 c$. The mean scalar profile for the uniform mean scalar gradient case is given by

$$\langle c \rangle = E(y - y_0), \tag{3}$$

where $E = \dfrac{d\langle c \rangle}{dy}$ and y_0 are constants; $\langle \ \rangle$ denotes the unconditional expectation.

Some insight into the nature of the scalar pdf may be obtained from analysis of the self–similar solution of equations (1) and (2). It has the following form

$$P(c, x, y) = \frac{1}{\sigma} g(s), \quad s = \frac{c - \langle c \rangle}{\sigma}, \tag{4}$$

where $\sigma^2 = \langle (c - \langle c \rangle)^2 \rangle$ is the scalar variance; experimental data [1] shows that σ^2 depends only on stream–wise coordinate x and increases with x. It is assumed that g is solely a function of the self–similar variable s. Physically, the self–similar solution is a limiting ($Re_\lambda \to \infty$, $x/M \to \infty$; Re_λ is Reynolds number based on Taylor microscale, M is the mesh length) solution. It can be observed when the influence of the initial conditions on the pdf is lost. The experimental verification of the self–similarity of the scalar pdf was done in [1]. They showed that (4) provides quite approximate collapse of the measured pdf's onto a single curve for the range $42.4 \le x/M \le 152.4$ of downstream locations (the grid is located at $x = 0$). The observed departures from self–similarity can be attributed to moderate value of Re_λ in [1] ($Re_\lambda \le 74.4$, and decayes with x).

For self–similar solution (4) to exist conditional expectations of the scalar dissipation $\langle N|c \rangle$, molecular diffusion $\langle \omega|c \rangle$ and velocity $\langle v|c \rangle$ must have the following form

$$\langle N|c \rangle = \langle N \rangle n(s), \quad \langle v|c \rangle = \frac{q}{\sigma} V(s), \quad \langle \omega|c \rangle = \frac{\langle N \rangle}{\sigma} \Omega(s), \tag{5}$$

where $q = \langle vc \rangle$ is the cross–stream turbulent flux , $\langle N \rangle$ is the unconditional scalar dissipation; q and $\langle N \rangle$ depend only on x; functions n, V and Ω, like function g in (4), are solely dependent on variable s. Space derivatives of the self–similar pdf (4) are (Eq. (3) is used)

$$\frac{\partial P}{\partial x} = -\frac{1}{\sigma^2} \frac{d\sigma}{dx} (sg)', \quad \frac{\partial}{\partial y} \langle v|c \rangle P = -\frac{q}{\sigma^3} (Vg)' E, \tag{6}$$

where $(f(s))'$ denotes the derivative of f with respect to s.

Inserting the Eqs. (4) -(6) into (1) and (2), one obtains:

$$(ng)'' + [(\alpha_1 s + \alpha_2 V)g]' = 0, \tag{7}$$

$$[(\alpha_1 s + \alpha_2 V)g]' = -(\Omega g)', \tag{8}$$

where α_1 and α_2 are constants for self–similar solution to exist. They are defined by

$$\alpha_1 = -\frac{u_0}{2\langle N \rangle}\frac{d\sigma^2}{dx}, \quad \alpha_2 = -\frac{Eq}{\langle N \rangle}. \tag{9}$$

It follows from the balance equation for the concentration variance

$$\frac{1}{2}u_0\frac{d\sigma^2}{dx} + \langle vc \rangle\frac{d\langle c \rangle}{dy} = -\langle N \rangle, \tag{10}$$

that the following equality is valid

$$\alpha_1 + \alpha_2 = 1. \tag{11}$$

Values of constants α_1 and α_2 experimentally measured in [7] are $\alpha_1 = -0.5$, $\alpha_2 = 1.5$.

Equation (8) lead to the relationship

$$\Omega = -(\alpha_1 s + \alpha_2 V), \tag{12}$$

for arbitrary self–similar pdf g. It means that the second form of writing of the equation for the scalar pdf (2) can not be used for the determination of self–similar solution. According to Eq. (12) one may conclude that for the self–similar solution (4) to exist the conditional expectation of molecular diffusion has to be closely linked with the conditional expectation of transverse velocity. Let us analyze now Eq. (7). First consider the case of no mean scalar gradient, i.e. $E = 0$, $\alpha_2 = 0$, $\alpha_1 = 1$ (due to Eq. (11)). Eq. (7) in this case can be integrated easily to yield (it is assumed further that $g(s)$ decreases sufficiently rapidly as $|s| \to \infty$)

$$(ng)' + sg = 0. \tag{13}$$

The solution of Eq. (13) is:

$$g(s) = \frac{A_1}{n(s)}exp\Big[-\int_0^s \frac{\zeta}{n(\zeta)}d\zeta\Big], \tag{14}$$

where the constant A_1 is determined by the normalization condition $\int g(s)ds = 1$. Eq. (14) is the well–known asymptotic solution found in [3] using a different method. The general solution of Eq. (7) is:

$$g(s) = \frac{A_2}{n(s)}exp\Big[-\int_0^s \frac{\alpha_1\zeta + \alpha_2 V(\zeta)}{n(\zeta)}d\zeta\Big]. \tag{15}$$

In order to extend the analysis, one needs information concerning function V. The conditional expectation of velocity $\langle v|c \rangle$ in the grid–generated turbulence with the uniform mean cross–stream scalar gradient was measured in [8]. It was shown that a linear dependence

$$\langle v|c \rangle = \frac{q}{\sigma^2}(c - \langle c \rangle) \tag{16}$$

holds reasonably well for the experimentally observed conditional expectation of velocity $\langle v|c \rangle$. In such a case, one obtains using Eqs. (4), (5) and (16) that V is the linear function

$$V(s) = s \tag{17}$$

Using Eq. (17) and identity (11) it is seen that the self–similar solution (15) for the uniform mean scalar gradient case looks exactly the same as the solution (14) for the case with no mean scalar gradient. This result explains why pdf's measured in [1] for a mean scalar gradient case are well described by asymptotic

612

solution [3] with no mean gradient (14). It is clear that the above agreement is a consequence of linearity of the conditional expectation of velocity, Eq. (17).

However, it should be emphasized, as the measurements [1] showed, that the physical meaning of the self–similar solutions for the two above cases is quite different. Indeed, for the case without a mean scalar gradient, the conditional expectation of scalar dissipation $\langle N|c \rangle$ is nearly const (so that $n \approx 1$) and the pdf's are close to Gaussian. For the mean gradient case the pdf's have exponential tails and the conditional expectation of scalar dissipation $\langle N|c \rangle$ becomes U–shaped and is quite far from being const, in fact $n(s) \sim |s|, |s| \gg 1$.

Let us return now to expression (12) for the conditional expectation of molecular diffusion. Using also of Eq. (17) with identity (11) leads to

$$\Omega = -s, \tag{18}$$

irrespective of the form of the pdf.

The success of the linear dependence (17) in describing the experimentally observed conditional expectation of velocity provides a possible explanation of why linear dependence (18) is quite a good approximation to the experimentally measured conditional expectation of molecular diffusion (see, e.g. measurements in turbulent wake [9]) even for strongly non–Gaussian pdf's. It follows also from expression (12) that deviations of the conditional expectation of molecular diffusion from linear dependence are linked with deviations of the conditional expectation of velocity from a linear function. Finally, it should be noted that in [10] another explanation is proposed as to why the experimental pdf's [1] are well described by formula (14). It is based on the use of Eq. (18). Both explanations are interrelated, since it was shown above that equations (17) and (18) are equivalent.

References

1. Jayesh and Z. Warhaft, "Probability distribution, conditional dissipation, and transport of passive temperature fluctuations in grid–generated turbulence", Phys. Fluids **A 4**, 2292 (1992).
2. A. Sahay and E.E. O'Brien, "Uniform mean scalar gradient in grid turbulence: Conditioned dissipation and production", Phys. Fluids **A 5**, 1076 (1993).
3. Y.G. Sinai and V. Yakhot, "Limiting probability distributions of a passive scalar in a random velocity field", Phys. Rev. Lett. **63**, 1962 (1989).
4. V. R. Kuznetsov and V. A. Sabelnikov, *Turbulence and Combustion* (Hemisphere, New York, 1980).
5. J.D. Li and R.W. Bilger, "A simple theory of conditional mean velocity in turbulent scalar - mixing layer", Phys. Fluids **A 6**, 605 (1994).
6. E. E. O'Brien, "The probability density function (PDF) approach to reacting turbulent flows", in *Turbulent Reacting Flows*, edited by P.A.Libby and F.A.Williams (Springer - Verlag, New York, 1980).
7. A. Sirivat and Z. Warhaft, "The effect of a passive cross–stream temperature gradient on the evolution of temperature variance and heat flux in grid turbulence", J. Fluid Mech. **128**, 323 (1983).
8. K. S. Venkataramany and R. Chevray, "Statistical features of heat transfer in grid generated turbulence: constant gradient case", J. Fluid Mech. **86**, 513 (1978).
9. P. Kailasnath, K. R. Sreenivasan and J. R. Saylor, "The conditional scalar dissipation rates in turbulent wakes, jets, and boundary layers", Phys. Fluids **A 5**, 3207 (1993).
10. S. B. Pope and E. S. C. Ching, "The stationary probability density functions: An exact result", Phys. Fluids **A 5**, 1529 (1993).

EXPERIMENTAL PROBABILITY DENSITY FUNCTIONS OF A PASSIVE SCALAR IN A TURBULENT GÖRTLER FLOW AND COMPARISON WITH THE POPE AND CHING MODEL

J. L. AIDER and J. E. WESFREID
Laboratoire de Physique et de Mécanique des Milieux Hétérogènes
URA CNRS 857
Ecole Supérieure de Physique et de Chimie Industrielles
10, rue Vauquelin 75005 Paris, France

In this communication we are interested in the experimental study of the mixing of a non-reacting passive scalar in an hydrodynamic turbulent flow in a curved channel (turbulent Görtler flow). We used Laser Induced Fluorescence technique [1] to visualize and analyze quantitatively the effect of the geometry and of the Reynolds number on the statistics of the concentration time series. The Fig. 1 is a visualization of the turbulent flow in a cross-section of the curved channel (60° from the leading edge) and we show the four regions corresponding to the different instabilities involved in the production of vortical structures:

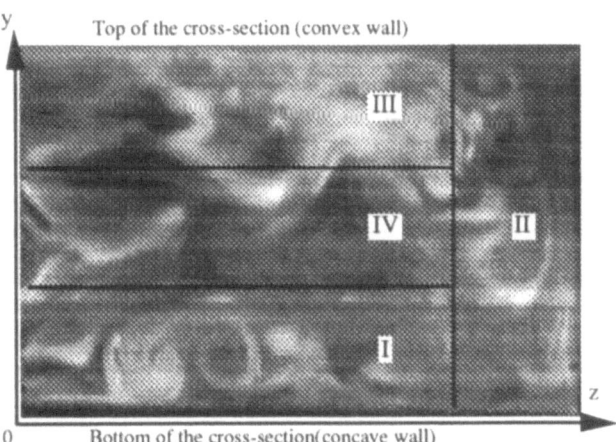

Fig. 1: Visualization of the flow in a cross-section.

- Görtler instability in the boundary layer over the concave wall (region I),
- Ekman cells in the corner of the duct (region II),
- detaching of the boundary layer under the convex wall (region III),
- homogenous mixing region in the middle of the cross-section (region IV).

613

S. Gavrilakis et al. (eds.), Advances in Turbulence VI, 613-614.
© *1996 Kluwer Academic Publishers.*

614

We represent on Fig. 2 the PDFs $P(x)$ [2] of the normalized concentration time series $x = \dfrac{c - \bar{c}}{\sigma}$, with $\bar{c}$ the mean and σ the standard deviation. The two PDFs correspond to the regions II (a) and IV (b) and are ploted on a semi-log graph. Stretched exponential tail for the PDF in high shear region (II), and nearly gaussian shape for the PDF in the homogenous region (IV) are obtained. They are compared to the theoretical proposition of Pope and Ching [3] who found the following form for $P(x)$:

$$ P(x) = \frac{C}{q(x)} \exp\left(\int_0^x \frac{r(x')dx'}{q(x')} \right) $$

with $r(x)$ and $q(x)$ functions calculated from conditional averaging of the time sries $x(t)$, which is found to fit very well the experimental data.

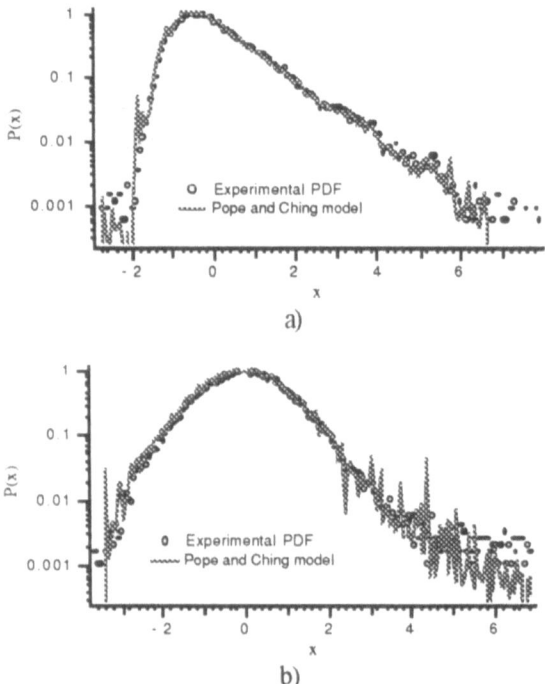

Fig. 2: Experimental and numerical PDFs for a point in region II (a) and in region IV (b).

References:

[1] J. L. Aider and J. E. Wesfreid, "Visualizations and PDFs of fluctuations of concentration", in *Experimental and Numerical Flow visualization*, A.S.M.E. FED-Vol 218, Book no. G00968, 123-130 (1995)

[2] S. Balachandar and L. Sirovich, *Probability Distribution Functions in turbulent Convection* ICASE Report No. 91-17 (1991).

[3] E. S. C. Ching and S. Pope, "Stationnary Probability Density Functions: an exact result", Phys. of Fluids A, **7**, 1529 (1993)

AN EQUATION FOR THE PROBABILITY DENSITY FUNCTION
OF TEMPERATURE INCREMENTS IN A BOUNDARY LAYER

L. DANAILA
I.R.P.H.E., 12 Av. Général Leclerc, 13003 Marseille

J. DUSEK
I.M.F., 2 rue Boussingault, 67000 Strasbourg

AND

P. LE GAL , F. ANSELMET
I.R.P.H.E., 12 Av. Général Leclerc, 13003 Marseille

Abstract.
We establish an equation for the probability density function of temperature increments in fully developed turbulence. This equation takes into account the properties of the mixing of a scalar in sheared turbulence more closely than that obtained in a previous study by the same group.

Various statistical approaches have been developed over the last years to account for the intermittent nature of turbulent energy transfer from the large energy producing scales to the small dissipative ones. In particular, (Vaienti *et al.*, 1994) have obtained an evolution equation for the probalility density functions (pdfs) of temperature increments $\Delta\theta(\vec{r}) = \theta(\vec{x}+\vec{r}) - \theta(\vec{x})$. However, the experimental analysis (Ould-Rouis *et al.*, 1995) of this equation showed that this theoretical work needs to be refined,since the assumptions of homogeneity and isotropy of the scalar field on which this theory is based are clearly not satisfied in usual shear flows such as boundary layers or jets.

The asymmetries in the pdfs are known to be induced by the mean gradients associated with the flow boundary conditions (Antonia *et al.*, 1978), and (Sreenivasan, 1991) . Therefore, we generalise the theoretical considerations to a sheared flow such as a heated boundary layer. In this case, homogeneity and isotropy approximations are rather well satisfied in any plane parallel to the heated wall, whereas more elaborate assumptions must

S. Gavrilakis et al. (eds.), Advances in Turbulence VI, 615-616.
© 1996 *Kluwer Academic Publishers.*

be used for the direction normal to the wall. We also introduce a mathematically more general strategy since we consider the problem in the space of distributions on differentiable rapidly decreasing functions $g(X, \vec{r})$, depending on one scalar $X(= \Delta\theta)$ and one vectorial argument $\vec{r}$ representing a translation parallel to the wall.

We express the action of the convection-diffusion operators on the g function. Instead of applying the statistical treatment in the whole R^3 space we limit ourselves to a (x, y) plane parallel to the wall. The resulting equation thus depends on the distance z from the wall and contains additional terms due to gradients normal to the wall:

$$-\frac{\partial}{\partial X}[q_3(r, z, X)P(r, z, x)] + \frac{1}{r^2}\frac{\partial}{\partial r}r^2[q_1(r, z, X)P(r, z, X)] +$$

$$2k_0\left\{\frac{\partial^2}{\partial X^2}[\tilde{q}_2(r, z, X)P(r, X)] - \frac{1}{r^2}\frac{\partial}{\partial r}r^2\frac{\partial}{\partial r}P(r, X)\right\}$$

$$+k_0\frac{\partial}{\partial X}[q_4(r, z, X)P(r, X)] + q_5(r, z, X)P(r, z, X) = 0. \qquad (1)$$

where the functions q_1, q_2, q_3, q_4, q_5 of r, z, X are conditional expectations of the following stochastic variables, respectively: $|\vec{Y}| = |(Y_x, Y_y)| = |(\Delta u_x(\vec{x}, \vec{r}, t), \Delta u_y(\vec{x}, \vec{r}, t))|$, $Z = [\nabla_{(x,y)}\theta(\vec{x}, t)]^2$, $W = \Delta[u_z\frac{\partial\theta}{\partial z}]$; $S = \partial_z^2\Delta\theta(\vec{x}, \vec{r}, t)$, $T = \partial_z u_z(\vec{x}, t)$.

The second refinement of the present analytical work is associated with the use of the velocity rms value u' and temperature Taylor microscale λ to make this equation dimensionless. This procedure clarifies the role of each term and allows more precise numerical investigations. The dimensionless equation writes, in the case of isotropy and homogeneity:

$$\left[\frac{2}{r} + \frac{\partial}{\partial r}\right](P \cdot q1) + 2.\frac{\overline{(\nabla\theta)^2}}{Pe}\nabla_X^2(P \cdot q2) - \frac{2}{Pe}\left[\frac{2}{r} + \frac{\partial}{\partial r}\right] \cdot \frac{\partial P}{\partial r} = 0. \qquad (2)$$

The main difficulties consist in too restrictive assumptions leading to 2 and in conditional expectations difficult to access by experiments in 1. DNS results are processed to obtain the terms in 1. The results will allow to test the local isotropy and homogeneity in various kinds of turbulence.

References

Vaienti S., Ould-Rouis M., Anselmet F. and Le Gal P. 1994 "Statistics of temperature increments in fully developped turbulence: Part 1: theory", *Physica D*, **73**, 99.

Ould-Rouis M., Anselmet F., Le Gal P. and Vaienti S., 1995, "Statistics of temperature increments in fully developed turbulence: Part 2: Experiments", *Physica D*, **85**, 405.

Antonia, R.A. and Van Atta, C.W. 1978, "Structure functions of temperature fluctuations in turbulent shear flows", *J.F.M.*, **84**, 561.

Sreenivasan, K.R. 1991, "On local isotropy of passive scalars in turbulents shear flows", *Proc.R.Soc.Lond. A*, **434**, 165.

KINEMATIC SIMULATION OF TWO-PARTICLE DISPERSION

J.C.H. Fung† & J.C. Vassilicos

DAMTP, University of Cambridge, United Kingdom

1. The Velocity Field

We generate on the computer a 2-D turbulent-like velocity field $\mathbf{u}(\mathbf{x}, t)$ that is identical to that of Vassilicos & Fung (1995, Phys. Fluids **7** (8), 1970) i.e.

$$\mathbf{u}(\mathbf{x}, t) = \sum_{n=1}^{N_k} [\mathbf{A}_n \cos(\mathbf{k}_n \cdot \mathbf{x} + \omega_n t) - \mathbf{A}_n \sin(\mathbf{k}_n \cdot \mathbf{x} + \omega_n t)], \tag{1}$$

where $\mathbf{A}_n = A_n(\cos \phi_n, -\sin \phi_n)$ and the wavevector $\mathbf{k}_n = k_n(\sin \phi_n, -\cos \phi_n)$. The angles ϕ_n are random and uncorrelated with each other and the velocity field (1) is incompressible because $\mathbf{A}_n \cdot \mathbf{k}_n = 0$ for all n. The positive amplitude A_n is chosen according to $A_n^2 = E(k_n)\Delta k_n$, where $E(k)$ is a prescribed energy spectrum of the form $E(k) = E_0 L(kL)^{-p}$ in the range $2\pi/L = k_1 \leq k \leq k_{N_k} = 2\pi/\eta$ and such that $E(k) = 0$ outside this range. $\Delta k_n = (k_{n+1}-k_{n-1})/2$ for $2 \leq N_k \leq N_k-1$, $\Delta k_1 = (k_2-k_1)/2$ and $\Delta k_{N_k} = (k_{N_k}-k_{N_k-1})/2$. The distribution of wavenumbers k_n is either algebraic ($k_n = k_1 n^\alpha$) or geometric ($k_n = k_1 a^{n-1}$), where α and a are dimensionless numbers which are functions of L/η and N_k because $k_{N_k} = 2\pi/\eta$. We experiment with two different models of unsteadiness (Fung & Vassilicos 1996, Phys. Fluids, submitted):

(a) a model where the unsteadiness frequency ω_n is proportional to the eddy-turnover time of wavemode n, i.e. $\omega_n = \lambda\sqrt{k_n^3 E(k_n)}$, where λ is a dimensionless constant;

(b) and a model where all the wavemodes are advected with a constant velocity velocity U, i.e. $\omega_n = U k_n$.

This kinematically simulated velocity field is 2-D only in the sense that it has two components. There are of course no dynamics in such simulations. In this paper the prescribed values of p are chosen between $p = 1$ and $p = 5$.

2. The Locality Assumption

The mean square distance $\overline{\Delta^2}(t)$ between two fluid elements that are advected by the turbulent-like velocity field (1) is a function of the following parameters ($\Delta_0 = \Delta(t = 0)$):

$$\overline{\Delta^2} = \overline{\Delta^2}(t, L, \eta, \Delta_0, E_0; p, N_k, \lambda) \tag{2a}$$

---

† Permanent address: Department of Mathematics, HKUST, Hong Kong

S. Gavrilakis et al. (eds.), Advances in Turbulence VI, 617-618.
© 1996 *Kluwer Academic Publishers.*

if the unsteadiness is simulated as in (i) and

$$\overline{\Delta^2} = \overline{\Delta^2}(t, L, \eta, \Delta_0, E_0, U; p, N_k) \tag{2b}$$

if the unsteadiness is simulated as in (ii). The locality assumption states that in the limit $Re \sim (L/\eta)^{4/3} \to \infty$ and in the range $\max(\eta, \Delta_0)/\sqrt{E_0} \ll t \ll L/\sqrt{E_0}$, the only dimensional parameters affecting $\overline{\Delta^2}$ are t and the energy density at $k = \sqrt{1/\overline{\Delta^2}}$, i.e. $E((1/\overline{\Delta^2})^{1/2}) = E_0 L(L^2/\overline{\Delta^2})^{-p/2}$. Hence, (2) may be replaced by $\overline{\Delta^2} = \overline{\Delta^2}(t, E((1/\overline{\Delta^2})^{1/2}))$ and dimensional requirements yield (Morel & Larcheveque 1974, J. Atmos. Sc. **31**, 2189):

$$\overline{\Delta^2} = G_\Delta (E_0 L^{1-p})^{2/(3-p)} t^{4/(3-p)}. \tag{3}$$

It is only for $p < 3$ that (3) can be deduced from the locality assumption. When $p = 5/3$, $\overline{\Delta^2} = G_\Delta (E_0^{3/2}/L) t^3$ and $E_0^{3/2}/L$ is equivalent to an average dissipation rate per unit mass of fluid. The constant G_Δ is a function of p, N_k and either λ or $U/\sqrt{E_0}$.

3. Summary of Results

Particle trajectories $\mathbf{x}(t)$ are obtained by integrating $d\mathbf{x}(t)/dt = \mathbf{u}(\mathbf{x}(t), t)$ numerically. The summary of our results is as follows (a more detailed account can be found in Fung & Vassilicos (1996, Phys. Fluids, submitted):

(a) $\overline{\Delta^2} = G_\Delta (E_0 L^{1-p})^{2/(3-p)} t^{4/(3-p)}$ is valid provided that $p < 3$ and that the unsteadiness is neither too strong nor too weak, i.e. $\lambda = O(1)$ or $U \sim \sqrt{E_0}$.

(b) If $p \geq 3$, $\ln(\overline{\Delta^2}(t)/\Delta_0^2) \sim (\sqrt{E_0}t/L)^q$; q decreases with p and $q = 1$ for $p = 3$.

(c) Individual realisations of turbulent-like flows are topologically different above and below $p = 3$. When $p < 3$, 2-D turbulent-like flows have a fractal-eddy structure that consists of cat's eyes within cat's eyes. When $p \geq 3$ no fractal-eddy structure exists, and no extra topological features appear by zooming into increasingly small scales inside eddying regions.

(d) When $p = 5/3$, $G_\Delta = O(10^{-2})$ as in experimental measurements (Tatarski 1960, Izv. Vyssh. Uchebn. Zaved. 3 Radiofizika **4**, 551). However, G_Δ can change by a factor of 2 simply by changing the distribution of modes in wavenumber space. G_Δ increases with unsteadiness, decreases with p and increases with the number of modes, N_k. The low value of G_Δ and the ways of these dependencies are consistent with the idea (proposed by Fung *et al.* 1992, J. Fluid Mech. **236**, 281) that two-particle dispersion is effectively happening in bursts when particle pairs meet straining regions. This idea is investigated quantitatively by measuring the skewness S and the flatness F of the second invariant II sampled along particle trajectories. S decreases to a constant value of -1.1 and F increases to a constant value between 8.5 and 9.0 with increasing N_k. Particles are therefore more often in eddying regions than in straining regions, but also more often in both eddying regions and straining regions than if the distribution of II sampled during their flight was Gaussian.

STRUCTURAL DIFFUSION IN 2D AND 3D RANDOM FLOWS

NADEEM A. MALIK

L.M.F.A., Ecole Centrale de Lyon, 69131 Ecully, France

1. The Flow Fields

We investigate *one*- and *two*-particle (relative) diffusion of fluid particles in incompressible 2D and 3D *steady* and *unsteady* random flows with prescribed self-similar power spectra of the form $E(k) = C_\alpha \varepsilon_*^{2/3} k^{-\alpha}$ for $k_{min} \leq k(= |\kappa|) \leq k_{max}$ and zero otherwise, for $1 < \alpha < 3$. C_α is a constant, and $\varepsilon_* = \varepsilon L_1^{(5-3\alpha)/2}$ where ε is the rate of energy dissipation per unit mass, and L_1 is a length scale – in our work we choose $L_1 = 2\pi/k_{min}$. The role played by the streamline topology is of special interest. The velocity fields are generated using Kinematic Simulation (KS), Fung, Hunt, Malik & Perkins 1992, JFM **236**, 281, viz

$$\mathbf{u}(\mathbf{x}, t) = \sum_{n=1}^{N_k} \left\{ \mathbf{a}_n \cos(\boldsymbol{\kappa}_n \cdot \mathbf{x} + \omega_n t + \phi_n^a) + \mathbf{b}_n \sin(\boldsymbol{\kappa}_n \cdot \mathbf{x} + \omega_n t + \phi_n^b) \right\}$$

and $\mathbf{a}_n \cdot \hat{\boldsymbol{\kappa}}_n = \mathbf{b}_n \cdot \hat{\boldsymbol{\kappa}}_n = 0$. The unsteadiness is proportional to the eddy-turnover frequency, $\omega(k) = \lambda \varepsilon_*^{1/3} k^{(3-\alpha)/2}$, and λ is a non-dimensional unsteadiness factor. The method of selecting the modes is described in the previous reference. $\langle u_i^2 \rangle = 1$, $i = 1, 2$ and also for $i = 3$ in 3D.

The 2D flows are characterised by a *closed streamline topology*. There are two sets of streamlines; (1) small streamlines, figure 1, with a scale $O(L_1)$, and (2) large streamlines with a scale $O(10^2 L_1)$. The large streamlines exist in much smaller density than small streamlines, and they posses a self-similar topology which for figure 2 gives a fractal (capacity) dimension of $D_k \approx 1.7$. Cases for $R_k = 5, 20$ are similar (Malik 1996, submitted).

2. Results and Discussion

In the *steady* 2D flows, the diffusion of particles on the two sets of streamlines are *statistically independent*, each characterised by disparate time scales with $T_2 \approx O(10^3 T_1)$. Our results show that the asymptotic limit $\langle x^2(t) \rangle \to Const$ takes a time $t/T_2 \gg 10$.

Unsteadiness, $\lambda > 0$, causes particles to migrate between the two sets of streamlines on a time scale T^m and for their motion to decorrelate rapidly on a Lagrangian time scale T^L. The trajectories are not closed and the diffusion asymptotes to the Taylor regime $\langle x^2(t) \rangle \sim Dt$ for $t \gg T^L$, $D = D(\lambda, \rho)$ is the diffusivity, and $\rho(t) = \rho_1(t)/\rho_2(t)$ is the ratio of the number density of particles on the respective streamlines. For very small unsteadiness, $0 < \lambda < 0.1$, the fractal toplogy of the large streamlines causes a change in $\rho(t) \neq \rho(0)$ for $t \gg T^m$. If $T^L < T^m$, we see an *anomalous double Taylor diffusion regime*, figure 3 sold line (1), characterised by the diffusivities $D(\lambda, \rho(0))$ and $D(\lambda, \rho(\infty))$. Large unsteadiness, $\lambda > 0.1$, smothers the anomalous diffusion and we see a single Taylor regime.

S. Gavrilakis et al. (eds.), Advances in Turbulence VI, 619-620.
© *1996 Kluwer Academic Publishers.*

2-particle diffusion scales on ε_* in both 2D and 3D, which gives $\langle \Delta^2(t) \rangle = G_\Delta(\alpha)\varepsilon_*^{\beta/3}t^\beta$ within a *large* inertial range ($R_k \gg 10^2$), with $\beta = 4/(3-\alpha)$. This is valid in 2D, provided there is *appreciable flow unsteadiness*, $\lambda \approx 1$; but in 3D it is true even for *steady* flow and the generalised Richardson constants G_Δ, figure 4, are *independent of the unsteadiness for $\lambda \leq 1$*. The 2D diffusion requires flow unsteadiness to overcome the correlating effect of closed streamlines, whereas the *open streamline topology* in 3D gives, inrinsically, a high degree of decorrelation to particle motion which is unaffected by even quite strong flow unsteadiness. This is important for physical models since eliminating unsteadiness is a significant simplifying assumption.

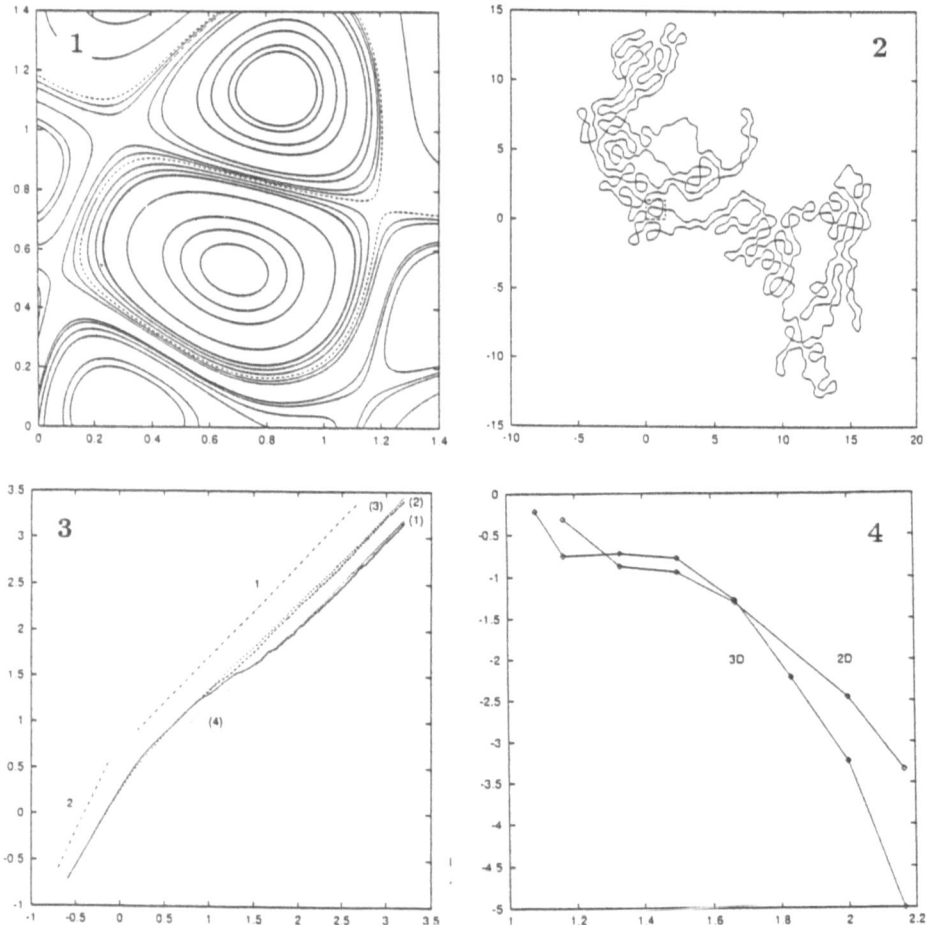

Figures 1 & 2, small and large streamlines, respectively, in 2D, $x/L_1 - y/L_1$. $R_k = 1$, $N_k = 50$ modes. The area in figure 1 is the dashed box in figure 2. **Figure 3**, log-log of $\langle x^2(t) \rangle /L_1^2$ against t/T_1. $R_k = 1$, $N_k = 50$, $\lambda =$ (1) 0.01, (2) 0.1, (3) 0.5, (4) 1. **Figure 4**, $\log(G_\Delta)$ against α for 2D ($\lambda = 1$) and 3D ($0 \leq \lambda \leq 1$).

Probability Distribution Function for the Gradient of a Passive Scalar Diffusing in Isotropic Turbulence: Mapping-Closure Model

Bhimsen K. Shivamoggi

University of Central Florida, Orlando, FL 32816

Abstract:

A mapping-closure model is used to derive the probability distribution function (PDF) for the gradient of a passive scalar diffusing in isotropic turbulence. This PDF describes the non-Gaussian features observed in experiments.

Obukhov-Corrsin theory of scalar turbulence (Obukhov, 1949 and Corrsin, 1951) assumes that the scalar-variance cascade is

- local in the spectral space, and
- involves a continuous loss of information.

The normalized statistics of band-limited scalar fields in the inertial-convective range would then be

- identical, and
- depend only on the mean dissipation rates x and ε

However, DNS of Reutsch and Massey (1991) and Pumir (1994) have shown that the most intense regions of the scalar gradient occur as large flat sheets.

This indicates a build-up of intermittency in the inertial-convective range of the scalar-variance cascade which has been confirmed in the laboratory experiments of Antonia et al. (1975 and 1984), Sreenivasan et al. (1988 and 1990), and Warhaft et al. (1992 and 1994).

Due to intermittency, the statistics of successively band-limited scalar fields become ever more non-Gaussian. This has been confirmed by the laboratory experiments of Antonia et al. (1984) and Warhaft et al. (192 and 1994). Here, we use the mapping-closure principle of Kraichnan (1990) to give the PDF for the gradient of a scalar diffusing in a random velocity field (Shivamoggi, 1995). This is based on the distortion of a Gaussian reference field $X_G(x)$ into a dynamically-evolving non-Gaussian field for the scalar gradient.

S. Gavrilakis et al. (eds.), Advances in Turbulence VI, 621-624.
© 1996 *Kluwer Academic Publishers.*

The model equation for the mapping function expresses the competition between the random convection and the diffusive relaxation processes:

$$\frac{\partial T}{\partial t} + v \cdot \nabla T = \kappa \nabla^2 T. \tag{1}$$

Let s be a transverse component of $\frac{\partial v_i}{\partial x_j}$ and X be a component of $\frac{\partial T}{\partial x_i}$, with the initial values

$$t = 0: \frac{\partial v_i}{\partial x_j} = s_0, \frac{\partial T}{\partial x_j} = X_0 \tag{2}$$

s_0 and X_0 being Gaussian distributed.

The initial states s_0 and X_0 are then assumed to evolve via local distortion through the action of non-stochastic effective stretching function $J(s_0, t)$ and $I(X_0, t)$ according to

$$\left. \begin{array}{l} s = J(s_0, t)s_0, \\ X = I(X_0, t)X_0 \end{array} \right\} \tag{3}a,b,$$

The stretching functions $J(s_0, t)$ and $I(X_0, t)$ are then postulated to evolve according to

$$\frac{\partial J}{\partial t} = |s_o| J^2 - v k_d^2 J^3, \text{ (Kraichnan, 1990)}$$

$$\frac{\partial I}{\partial t} = |s_0| JI - \kappa \hat{k}_d^2 I^3, \text{ (Shivamoggi, 1995)} \tag{4}a,b$$

The first terms on the right hand sides describe the growths due to inertial and convective stretching while the second terms describe the viscous and diffusive decays. The absolute values reflect the symmetry between positive and negative values of the transverse velocity gradient.

We have in the stationary state,

$$J = \frac{|s_0|}{v k_d^2}, I^2 = \frac{|s_0| J}{\kappa \hat{k}_d^2} \tag{5}a,b$$

Thus,

$$|s| = \frac{|s_0|^2}{v k_d^2}, |X| = \frac{|s_0| |X_0|}{\sqrt{v \kappa k_d^2 \hat{k}_d^2}}. \tag{6}a,b$$

The PDF's of s_0 and X_0 are assumed to be given by the Gaussian fields -

$$P(s_0) = \frac{1}{\sqrt{2\pi\sigma_1^2}} e^{-\frac{s_0^2}{2\sigma_1^2}}$$

and,

$$P(s_0, X_0) = \frac{1}{2\pi\sigma_1\sigma_2\sqrt{1-\rho^2}} e^{-\frac{1}{2(1-\rho^2)}\left[\frac{s_0^2}{\sigma_1^2} - \frac{2Ps_0X_0}{\sigma_1\sigma_2} + \frac{X_0^2}{\sigma_2^2}\right]}$$
(7)a,b

where,

$$\sigma_1^2 \equiv \langle s_0^2 \rangle, \sigma_2^2 \equiv \langle X_0^2 \rangle, P \equiv \frac{\langle s_0 X_0 \rangle}{\sigma_1\sigma_2}$$

The PDF's of s and X are then given by

$$\left. \begin{aligned} P(s) &= P(s_0)\frac{\partial s_0}{\partial s} \\ P(X) &= P(s_0, X_0)\frac{\partial |s_0 X_0|}{\partial X} \end{aligned} \right\}$$
(8)a,b

Then, following the usual procedure to calculate the PDF's of functions of random variables (Papoulis, 1965), we obtain

$$P(s) = \left(\frac{\upsilon k_d}{\pi\sigma_1^2 |s|}\right)^{1/2} e^{-\frac{|s|}{2\sigma_1^2/\upsilon k_d}}, \text{ (Kraichnan, 1990)}$$

$$P(X) = \frac{\sqrt{\upsilon\kappa k_d^2}}{\pi\sigma_1\sigma_2\sqrt{1-\rho^2}} e^{\frac{\rho|X|}{\sigma_1\sigma_2(1-\rho^2)/\sqrt{\upsilon\kappa k_d^2\hat{k}_d^2}}} K_0\left(\frac{|X|}{\sigma_1\sigma_2(1-\rho^2)/\sqrt{\upsilon\kappa k_d^2\hat{k}_d^2}}\right),$$
(Shivamoggi, 1995) (9)a,b

Note the asymptotic behavior of $P(X)$ -

$$P(X) \sim \frac{(\upsilon\kappa k_d^2\hat{k}_d^2)^{1/4}}{\pi\sqrt{\sigma_1\sigma_2}}\frac{1}{\sqrt{2\pi X}} e^{-\frac{|X|}{\sigma_1\sigma_2(1-\rho^2)/\sqrt{\upsilon\kappa k_d^2\hat{k}_d^2}}},$$
(10)

as $|X| \Rightarrow \infty$. Thus, $P(X)$, like $P(s)$, shows exponential tails!

The scalar gradient PDF $P(X)$ is compared in Figure 1 with the experimental data of

- Antonia et al. (1984) in a turbulent round jet
- Tong and Warhaft (1994) in a grid turbulence.

The agreement is seen to be very good.

624

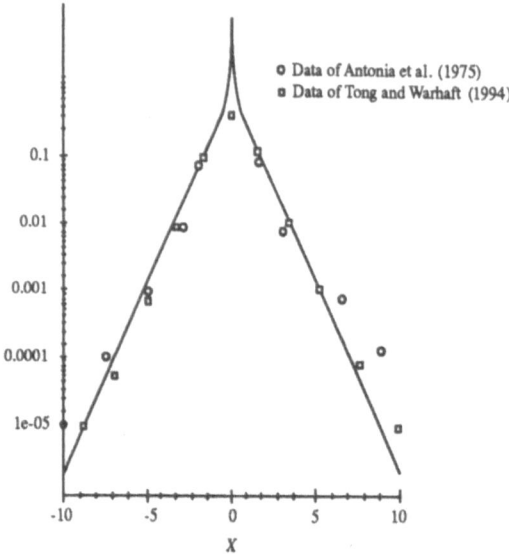

Figure 1. Comparison of the calculated PDF (13)b with the experimental data (the parameters ρ and $\dfrac{\sigma_1 \sigma_2}{\sqrt{\upsilon \kappa k_d^2 \hat{k}_d^2}}$ have been taken to be .6 and .59, respectively).

Acknowledgments

The author is thankful to Dr. R. H. Kraichnan for his valuable suggestions. This work was carried out at the Technische Universiteit Eindhoven under the auspices of the J. M. Burgerscentrum. The author is thankful to Professors P.P.J.M. Schram and G.J.F. van Heijst for their hospitality.

References

[1]   Antonia, R.A., Hopfinger, E.J., Gagne, Y., and Anselmet, F. *Phys. Rev.* **A30**, 2704 (1984).

[2]   Antonia, R.A., and van Atta, C.W., *J. Fluid Mech.* **67**, 273 (1975).

[3]   Kraichnan, R.H., *Phys. Rev. Lett.* **65**, 575 (1990).

[4]   Meneveau, C., Sreenivasan, K.R., Kailasnath, P. and Fan, M.S., *Phys. Rev.* **A41** 894 (1990).

[5]   Papoulis, A. *Probability, Random Variables and Stochastic Processes* (McGraw-Hill, New York, 1965).

[6]   Prasad, R.R., Meneveau, C., and Sreenivasan, K.R., *Phys. Rev. Lett.* **61**, 74 (1988).

[7]   Pumir, A., *Phys. Fluids* **6**, 2118 (1994).

[8]   Shivamoggi, B.K., *Phys. Rev.* **E51**, 4453, (1995).

[9]   Tong, C. and Warhaft, Z., *Phys. Fluids* **6**, 2165 (1994).

[10]   Warhaft, Z., and Jayesh, *Phys. Fluids* **A4**, 2292 (1992).

AUTHOR INDEX

Mechanics

FLUID MECHANICS AND ITS APPLICATIONS

Series Editor: R. Moreau

Aims and Scope of the Series

The purpose of this series is to focus on subjects in which fluid mechanics plays a fundamental role. As well as the more traditional applications of aeronautics, hydraulics, heat and mass transfer etc., books will be published dealing with topics which are currently in a state of rapid development, such as turbulence, suspensions and multiphase fluids, super and hypersonic flows and numerical modelling techniques. It is a widely held view that it is the interdisciplinary subjects that will receive intense scientific attention, bringing them to the forefront of technological advancement. Fluids have the ability to transport matter and its properties as well as transmit force, therefore fluid mechanics is a subject that is particularly open to cross fertilisation with other sciences and disciplines of engineering. The subject of fluid mechanics will be highly relevant in domains such as chemical, metallurgical, biological and ecological engineering. This series is particularly open to such new multidisciplinary domains.

Kluwer Academic Publishers – Dordrecht / Boston / London

Mechanics

FLUID MECHANICS AND ITS APPLICATIONS
Series Editor: R. Moreau

Kluwer Academic Publishers – Dordrecht / Boston / London

Mechanics

SOLID MECHANICS AND ITS APPLICATIONS

Series Editor: G.M.L. Gladwell

Aims and Scope of the Series

The fundamental questions arising in mechanics are: *Why?*, *How?*, and *How much?* The aim of this series is to provide lucid accounts written by authoritative researchers giving vision and insight in answering these questions on the subject of mechanics as it relates to solids. The scope of the series covers the entire spectrum of solid mechanics. Thus it includes the foundation of mechanics; variational formulations; computational mechanics; statics, kinematics and dynamics of rigid and elastic bodies; vibrations of solids and structures; dynamical systems and chaos; the theories of elasticity, plasticity and viscoelasticity; composite materials; rods, beams, shells and membranes; structural control and stability; soils, rocks and geomechanics; fracture; tribology; experimental mechanics; biomechanics and machine design.

1. R.T. Haftka, Z. Gürdal and M.P. Kamat: *Elements of Structural Optimization.* 2nd rev.ed., 1990   ISBN 0-7923-0608-2
2. J.J. Kalker: *Three-Dimensional Elastic Bodies in Rolling Contact.* 1990
                                                                               ISBN 0-7923-0712-7
3. P. Karasudhi: *Foundations of Solid Mechanics.* 1991   ISBN 0-7923-0772-0
4. *Not published*
5. *Not published.*
6. J.F. Doyle: *Static and Dynamic Analysis of Structures.* With an Emphasis on Mechanics and Computer Matrix Methods. 1991   ISBN 0-7923-1124-8; Pb 0-7923-1208-2
7. O.O. Ochoa and J.N. Reddy: *Finite Element Analysis of Composite Laminates.*
                                                                               ISBN 0-7923-1125-6
8. M.H. Aliabadi and D.P. Rooke: *Numerical Fracture Mechanics.*   ISBN 0-7923-1175-2
9. J. Angeles and C.S. López-Cajún: *Optimization of Cam Mechanisms.* 1991
                                                                               ISBN 0-7923-1355-0
10. D.E. Grierson, A. Franchi and P. Riva (eds.): *Progress in Structural Engineering.* 1991
                                                                               ISBN 0-7923-1396-8
11. R.T. Haftka and Z. Gürdal: *Elements of Structural Optimization.* 3rd rev. and exp. ed. 1992
                                                                ISBN 0-7923-1504-9; Pb 0-7923-1505-7
12. J.R. Barber: *Elasticity.* 1992   ISBN 0-7923-1609-6; Pb 0-7923-1610-X
13. H.S. Tzou and G.L. Anderson (eds.): *Intelligent Structural Systems.* 1992
                                                                               ISBN 0-7923-1920-6
14. E.E. Gdoutos: *Fracture Mechanics.* An Introduction. 1993   ISBN 0-7923-1932-X
15. J.P. Ward: *Solid Mechanics.* An Introduction. 1992   ISBN 0-7923-1949-4
16. M. Farshad: *Design and Analysis of Shell Structures.* 1992   ISBN 0-7923-1950-8
17. H.S. Tzou and T. Fukuda (eds.): *Precision Sensors, Actuators and Systems.* 1992
                                                                               ISBN 0-7923-2015-8
18. J.R. Vinson: *The Behavior of Shells Composed of Isotropic and Composite Materials.* 1993
                                                                               ISBN 0-7923-2113-8
19. H.S. Tzou: *Piezoelectric Shells.* Distributed Sensing and Control of Continua. 1993
                                                                               ISBN 0-7923-2186-3

Kluwer Academic Publishers – Dordrecht / Boston / London

Mechanics

SOLID MECHANICS AND ITS APPLICATIONS

Series Editor: G.M.L. Gladwell

20. W. Schiehlen (ed.): *Advanced Multibody System Dynamics*. Simulation and Software Tools. 1993   ISBN 0-7923-2192-8
21. C.-W. Lee: *Vibration Analysis of Rotors*. 1993   ISBN 0-7923-2300-9
22. D.R. Smith: *An Introduction to Continuum Mechanics*. 1993   ISBN 0-7923-2454-4
23. G.M.L. Gladwell: *Inverse Problems in Scattering*. An Introduction. 1993 ISBN 0-7923-2478-1
24. G. Prathap: *The Finite Element Method in Structural Mechanics*. 1993   ISBN 0-7923-2492-7
25. J. Herskovits (ed.): *Advances in Structural Optimization*. 1995   ISBN 0-7923-2510-9
26. M.A. González-Palacios and J. Angeles: *Cam Synthesis*. 1993   ISBN 0-7923-2536-2
27. W.S. Hall: *The Boundary Element Method*. 1993   ISBN 0-7923-2580-X
28. J. Angeles, G. Hommel and P. Kovács (eds.): *Computational Kinematics*. 1993   ISBN 0-7923-2585-0
29. A. Curnier: *Computational Methods in Solid Mechanics*. 1994   ISBN 0-7923-2761-6
30. D.A. Hills and D. Nowell: *Mechanics of Fretting Fatigue*. 1994   ISBN 0-7923-2866-3
31. B. Tabarrok and F.P.J. Rimrott: *Variational Methods and Complementary Formulations in Dynamics*. 1994   ISBN 0-7923-2923-6
32. E.H. Dowell (ed.), E.F. Crawley, H.C. Curtiss Jr., D.A. Peters, R. H. Scanlan and F. Sisto: *A Modern Course in Aeroelasticity*. Third Revised and Enlarged Edition. 1995   ISBN 0-7923-2788-8; Pb: 0-7923-2789-6
33. A. Preumont: *Random Vibration and Spectral Analysis*. 1994   ISBN 0-7923-3036-6
34. J.N. Reddy (ed.): *Mechanics of Composite Materials*. Selected works of Nicholas J. Pagano. 1994   ISBN 0-7923-3041-2
35. A.P.S. Selvadurai (ed.): *Mechanics of Poroelastic Media*. 1996   ISBN 0-7923-3329-2
36. Z. Mróz, D. Weichert, S. Dorosz (eds.): *Inelastic Behaviour of Structures under Variable Loads*. 1995   ISBN 0-7923-3397-7
37. R. Pyrz (ed.): *IUTAM Symposium on Microstructure-Property Interactions in Composite Materials*. Proceedings of the IUTAM Symposium held in Aalborg, Denmark. 1995   ISBN 0-7923-3427-2
38. M.I. Friswell and J.E. Mottershead: *Finite Element Model Updating in Structural Dynamics*. 1995   ISBN 0-7923-3431-0
39. D.F. Parker and A.H. England (eds.): *IUTAM Symposium on Anisotropy, Inhomogeneity and Nonlinearity in Solid Mechanics*. Proceedings of the IUTAM Symposium held in Nottingham, U.K. 1995   ISBN 0-7923-3594-5
40. J.-P. Merlet and B. Ravani (eds.): *Computational Kinematics '95*. 1995 ISBN 0-7923-3673-9
41. L.P. Lebedev, I.I. Vorovich and G.M.L. Gladwell: *Functional Analysis*. Applications in Mechanics and Inverse Problems. 1996   ISBN 0-7923-3849-9
42. J. Menčik: *Mechanics of Components with Treated or Coated Surfaces*. 1996   ISBN 0-7923-3700-X
43. D. Bestle and W. Schiehlen (eds.): *IUTAM Symposium on Optimization of Mechanical Systems*. Proceedings of the IUTAM Symposium held in Stuttgart, Germany. 1996   ISBN 0-7923-3830-8

Kluwer Academic Publishers – Dordrecht / Boston / London

Mechanics

Kluwer Academic Publishers – Dordrecht / Boston / London

Mechanics

From 1990, books on the subject of *mechanics* will be published under two series:

FLUID **MECHANICS AND ITS APPLICATIONS**
Series Editor: R.J. Moreau

SOLID **MECHANICS AND ITS APPLICATIONS**
Series Editor: G.M.L. Gladwell

Prior to 1990, the books listed below were published in the respective series indicated below.

MECHANICS: DYNAMICAL SYSTEMS
Editors: L. Meirovitch and G.Æ. Oravas

1. E.H. Dowell: *Aeroelasticity of Plates and Shells.* 1975   ISBN 90-286-0404-9
2. D.G.B. Edelen: *Lagrangian Mechanics of Nonconservative Nonholonomic Systems.*
   1977   ISBN 90-286-0077-9
3. J.L. Junkins: *An Introduction to Optimal Estimation of Dynamical Systems.* 1978
                                                                    ISBN 90-286-0067-1
4. E.H. Dowell (ed.), H.C. Curtiss Jr., R.H. Scanlan and F. Sisto: *A Modern Course in*
   Aeroelasticity.   *Revised and enlarged edition see under Volume 11*
5. L. Meirovitch: *Computational Methods in Structural Dynamics.* 1980
                                                                    ISBN 90-286-0580-0
6. B. Skalmierski and A. Tylikowski: *Stochastic Processes in Dynamics.* Revised and
   enlarged translation. 1982   ISBN 90-247-2686-7
7. P.C. Müller and W.O. Schiehlen: *Linear Vibrations.* A Theoretical Treatment of Multi-
   degree-of-freedom Vibrating Systems. 1985   ISBN 90-247-2983-1
8. Gh. Buzdugan, E. Mihăilescu and M. Radeş: *Vibration Measurement.* 1986
                                                                    ISBN 90-247-3111-9
9. G.M.L. Gladwell: *Inverse Problems in Vibration.* 1987   ISBN 90-247-3408-8
10. G.I. Schuëller and M. Shinozuka: *Stochastic Methods in Structural Dynamics.* 1987
                                                                    ISBN 90-247-3611-0
11. E.H. Dowell (ed.), H.C. Curtiss Jr., R.H. Scanlan and F. Sisto: *A Modern Course in*
    Aeroelasticity. Second revised and enlarged edition (of Volume 4). 1989
                                          ISBN Hb 0-7923-0062-9; Pb 0-7923-0185-4
12. W. Szemplińska-Stupnicka: *The Behavior of Nonlinear Vibrating Systems.* Volume I:
    Fundamental Concepts and Methods: Applications to Single-Degree-of-Freedom
    Systems. 1990   ISBN 0-7923-0368-7
13. W. Szemplińska-Stupnicka: *The Behavior of Nonlinear Vibrating Systems.* Volume II:
    Advanced Concepts and Applications to Multi-Degree-of-Freedom Systems. 1990
                                                                    ISBN 0-7923-0369-5
                                           Set ISBN (Vols. 12–13) 0-7923-0370-9

MECHANICS OF STRUCTURAL SYSTEMS
Editors: J.S. Przemieniecki and G.Æ. Oravas

1. L. Frýba: *Vibration of Solids and Structures under Moving Loads.* 1970
                                                                    ISBN 90-01-32420-2
2. K. Marguerre and K. Wölfel: *Mechanics of Vibration.* 1979   ISBN 90-286-0086-8